DEVELOPMENTS IN SEDIMENTOLOGY 12

CARBONATE SEDIMENTS AND THEIR DIAGENESIS

DEVELOPMENTS IN SEDIMENTOLOGY 12

CARBONATE SEDIMENTS AND THEIR DIAGENESIS

BY

ROBIN G. C. BATHURST

Jane Herdman Laboratories of Geology, University of Liverpool, Liverpool (Great Britain)

SECOND ENLARGED EDITION

ELSEVIER SCIENTIFIC PUBLISHING COMPANY
Amsterdam Oxford New York 1975

ELSEVIER SCIENTIFIC PUBLISHING COMPANY
335 JAN VAN GALENSTRAAT, P.O. BOX 211, AMSTERDAM

AMERICAN ELSEVIER PUBLISHING COMPANY, INC.
52 VANDERBILT AVENUE, NEW YORK, NEW YORK 10017

First Published 1971
Second Enlarged Edition 1975

LIBRARY OF CONGRESS CARD NUMBER: 77–135489

ISBN 0–444–41351–0 (CLOTHBOUND)
ISBN 0–444–41352–9 (PAPERBACK)

WITH 359 ILLUSTRATIONS AND 24 TABLES

COPYRIGHT © 1975 BY ELSEVIER PUBLISHING COMPANY, AMSTERDAM

PRINTED IN THE NETHERLANDS

To Bruno Sander of Innsbruck
for his example and inspiration

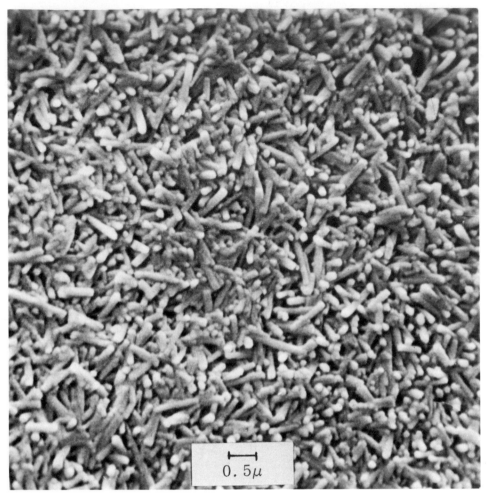

Surface of Recent ooid showing slightly blunted rods of aragonite with long axes statistically tangential to the surface. Scanning electron mic. Oolite shoal, Abu Dhabi, Persian Gulf. (Courtesy J.-P. Loreau.)

PREFACE

The prodigious expansion and diversification of carbonate sedimentology since the mid-1950's have made the writing of a critical synthesis a project of some urgency. In compiling this book I have confined my attention to those aspects of the subject wherein the most significant progress in understanding has been made, namely the Recent marine depositional environments, diagenesis, carbonate chemistry, and the microstructure and ultrastructure of calcareous skeletons. Reluctantly, I have not gone deeply into other matters of great interest, such as the interpretation of ancient limestones, because this would require a second book and several more years. As it happens, this is probably no great misfortune since the important progress in the study of the ancient has yet to gather momentum and, most likely, will come in the application of the newly developed areas I have just referred to. This book is not intended to be definitive or exhaustive: on the contrary, I have selected as examples to illustrate certain ideas only those topics which are supported by substantial and good published work. In this way I hoped to achieve my twin aims of introducing the subject to the graduate geologist and of helping him to see it whole. I emphasize the second aim. Carbonate sedimentology is not just a matter of thin-sections and point-counts, or of diving with mask and snorkel to count and conquer with statistics the living benthos, nor is it simply a question of ion activity products and Gibbs free energies, of isotopic equilibria or partition coefficients, nor again can it be just a business of nailed boots, a hammer and sound field mapping. It is all of these things, of course, but the whole is greater than the sum of the parts. Progress has already been made, for example, where the question of aragonite precipitation from sea water has been linked to codiacean productivity, where grain size distributions have been related both to fluid dynamics and to biological activities, where X-ray diffraction and isotopic analyses of Recent dolomites have been tied to painstaking field descriptions, and where the direct experience of the complexities of the Recent have induced humility and restraint in the interpretation of the ancient. It is, indeed, ground for satisfaction that the development of carbonate sedimentology as a distinct and self-conscious branch of geology, with bands of devotees holding specialized symposia in Liverpool, Heidelberg, Naples and Dallas, has not led to a narrow and parochial outlook, but,

instead, has inspired a broad dynamic interplay of stratigraphy, petrology, palaeontology and geochemistry in the best traditions of geology.

It is appropriate at this point briefly to draw attention to certain peculiarities
of carbonate sediments which distinguish them from the mechanically deposited
siliciclastic sediments. To begin with, the primary materials are almost all formed
near to the site of deposition (except for allodapic sediments and pelagic deposits)
as shells, faecal pellets, oöids and other grains, as coral-algal reef frames, or as
stromatolites. For the most part, these materials are dependent on biological
processes for their production and for their early modification on the sea floor.
The high solubility of carbonate minerals in water containing CO_2, combined
with variety of mineralogy and crystal size, renders most carbonate sediments
highly susceptible to diagenesis. Their diagenetic histories, commonly complex and
prolonged in comparison with the brief flickers of their exposures in the original
depositional environments, are among their outstanding characteristics. The
differences between siliciclastic sandstones and carbonate rocks are portrayed in
a revealing way in Table I, taken from the study of porosity by CHOQUETTE and
PRAY (1970).

In the description of carbonate rocks I have used the simpler terminology
of FOLK (1959, fig.4: biomicrite, pelmicrite, etc.), the scheme of DUNHAM (1962:
wackestone, packstone, etc.), and GRABAU's (1904) calcilutite and calcarenite
where data do not allow more precise description (ignoring Grabau's restriction
of these terms to hydroclastic, water-broken, grains). For sedimentary particle
size I have followed WENTWORTH (1922): for crystal size I have used
FRIEDMAN's (1965d) scale (micron-sized, etc.); for texture I have taken FRIEDMAN's
terms (1965d) (poikilotopic, etc.). Unconsolidated sediments are called lime or
carbonate muds, sands, etc. In referring to space-filling crystals which have grown
attached to a free surface, I have followed the sensible current trend (e.g., MURRAY
and PRAY, 1965) and have called everything cement on the grounds that to distinguish between intergranular and intragranular precipitated filling is to ignore the
identical nature of each. Cement is modified where necessary by such terms as
radial-fibrous or equigranular. The term drusy is not used. Other terms are discussed in the text and all are defined either in the text or in the Glossary.

In acknowledging the help of others, I must pay tribute to the influence of
two men on the evolution of my own understanding of carbonate sedimentation
and diagenesis. When, in 1953, I looked about for a guide to help me to disentangle
the varied and complex fabrics of Carboniferous limestones, I discovered the
English translation (1951) by E. B. Knopf of SANDER's (1936) paper on the depositional fabrics of Alpine Triassic carbonate sediments. Reading this from cover
to cover in a few days, I was given an unforgettable lesson in logic, discipline and
humility in the interpretation of fabrics, which has been my guide ever since. The
research worker pursues a lonely occupation and may falter without the encouragement of his peers. On many occasions over the last decade my spirits have been

TABLE I

COMPARISON OF THE POROSITY IN SANDSTONES AND CARBONATE ROCKS
(From CHOQUETTE and PRAY, 1970)

Aspect	Sandstone	Carbonate
Amount of primary porosity in sediments	Commonly 25–40%	Commonly 40–70%
Amount of ultimate porosity in rocks	Commonly half or more of the initial porosity; 15–30% common	Commonly none or only a small fraction of the initial porosity; 5–15% common in reservoir facies
Type(s) of primary porosity	Almost exclusively interparticle	Interparticle often predominates; but intraparticle and other types are important
Type(s) of ultimate porosity	Almost exclusively primary interparticle	Widely varied, due to post-depositional modifications
Sizes of pores	Diameter and throat sizes closely related to sedimentary particle size and sorting	Diameter and throat sizes commonly show little relationship to sedimentary particle size or sorting
Shape of pores	Strong dependence on particle shape—a "negative" of particles	Highly varied, ranges from strongly dependent "positive" or "negative" of particles to a form completely independent of shapes of depositional or diagenetic components
Uniformity of size, shape, and distribution	Often fairly uniform within homogeneous sandstone body	Variable, ranging from fairly uniform to extremely heterogeneous, even within a body made up of but a single rock type
Influence of diagenesis	Minor; usually minor reduction of primary porosity by compaction and cementation	Major; can create, obliterate, or completely modify porosity; cementation and solution important
Influence of fracturing	Generally not of major importance in reservoir properties	Of major importance in reservoir properties if present
Visual evaluation of porosity and permeability	Semiquantitative visual estimates often relatively easy	Variable; semiquantitative visual estimates range from easy to virtually impossible; instrument measurements of porosity, permeability often required Capillary pressure often needed
Adequacy of core analysis for reservoir evaluation	Core plugs of 1-inch diameter often adequate for "matrix" porosity	Core plugs often inadequate; even whole cores (~ 3 inches) may be inadequate for large pores
Permeability–porosity inter-relationships	Relatively consistent; often dependent on particle size and sorting	Highly varied; commonly independent of particle size and sorting

raised by the encouragement of my friend and colleague in carbonate sedimentology, L. C. Pray, whose wisdom and experience, generously offered, have sustained
me beyond measure.

The construction of this book has drawn me to the study of many fields in
which I had no detailed experience and, in order to reduce the consequent and
inevitable deficiencies and errors, I have had the good fortune to receive the aid of
a number of specialists. Without their help, the book, quite frankly, could never
have been written. I am, therefore, enormously indebted, for their prompt and
painstaking criticisms of the penultimate drafts of chapters or sections, to C. G.
Adams, M. P. Atherton, P. J. Brenchley, R. G. Bromley, P. R. Bush, J. E. Dalingwater, K. S. Deffeyes, H. V. Dunnington, G. F. Elliott, G. Evans, R. L. Folk,
G. M. Friedman, R. N. Ginsburg, S. Gólubić, L. V. Illing, E. Jamieson, W. J.
Kennedy, G. Newall, L. C. Pray, N. Rast, C. T. Scrutton, D. J. Shearman, G.
Skirrow, J. P. Swinchatt, J. C. M. Taylor, J. D. Taylor, I. M. West, P. K. Weyl,
and A. Williams. The entire book has been read by A. G. Fischer and D. H. Zenger
whose suggestions have been liberaly incorporated in the text. I acknowledge also
the acute comments of R. M. Pytkowicz, P. A. Scholle, and G. Bachmann, and
helpful discussions with many others.

In the preparation of my own photomicrographs I have depended heavily
on the craftsmanship of W. Lee, F.R.M.S., F.R.P.S., F.I.I.P., R.B.P., Head of the
Central Photographic Service, University of Liverpool, whose art and patience he
has placed at my service for the last twelve years. All photomicrographs have been
made by him unless otherwise acknowledged. I am indebted to the Trustees of the
British Museum of Natural History for allowing me to use the H. A. Nicholson
collection of thin sections of Stromatoporoidea, and to B. D'Argenio and R. L.
Folk for the loan of thin sections. All the line figures have been drawn by J. Lynch,
Senior Geological Cartographer in the University of Liverpool, and I am greatly
indebted to him for his understanding cooperation and wise advice.

Many photographs have been given to me by workers in the carbonate field,
for whose willing response I am exceedingly grateful. Illustrations are an essential
and substantial part of a book of this kind and my reliance on donated illustrations
has been considerable. My thanks go to the following, whose names appear in the
captions of the photographs they gave me: M. M. Ball, M. Black, R. G. Bromley,
H. Buchanan, J. Dalingwater, G. R. Davies, J. D. Donahue, G. F. Elliott, C. W.
Ellis, G. Evans, F. Fabricius, A. G. Fischer, M. Gatrall, C. D. Gebelein, T. F.
Goreau, Ch. Grégoire, J. M. Hancock, J. C. Hathaway, J. D. Hudson, W. J.
Kennedy, M. Kirchmayer, L. Land, C. MacClintock, R. K. Matthews, C. Monty,
N. D. Newell (photographs by R. Adlington), L. C. Pray, J. R. P. Ross, E. A. Shinn,
W. Schwarzacher, J. D. Taylor, G. E. Tebbutt, K. M. Towe, H.-E. Usdowski,
N. C. Wardlaw, N. Watabe, J. N. Weber, A. Williams, S. W. Wise, Jr., M. Wolfe,
J. L. Wray. My thanks go also to those donors whose photographs I have been
unable to use.

My thanks go to authors and editors for permission to reproduce material from the following publications: Bathurst, R. G. C., 1958, *Liverpool Manchester Geol. J.*, 2 (Fig. 336,337); Bathurst, R. G. C., 1959, *J. Geol.*, 67 (Fig.306); Bathurst, R. G. C., 1964, in: *Approaches to Paleoecology*, Wiley, New York, N.Y. (Fig.258, 302); Bathurst, R. G. C., 1966, *Geol. J.*, 5 (Fig.252, 285, 288, 289); Bathurst, R. G. C., 1967, *Marine Geol.*, 5 (Fig.156, 160, 161, 240); Bayer, F. M., 1956, in: *Treatise on Invertebrate Paleontology*, F, Geol. Soc. Am. and Univ. Kansas Press (Fig.42, 43); Black, M., 1965, *Endeavour*, 24 (Fig.105–108); Choquette, P. W. and Pray, L. C., 1970, *Bull. Am. Assoc. Petrol. Geologists*, 54 (Table I); Cullis, C. G., 1904, in: *The Atoll of Funafuti*, Roy. Soc. London (Fig.267–272); Davies, G. R., 1970b, *Am. Assoc. Petrol. Geologists, Mem.*, 13 (Fig.208–211); Elliott, G. F., 1955, *Micropalaeontology*, 1 (Fig.95, 96, 99); Fabricius, F., 1964, *Senckenbergiana Lethaea*, 45 (Fig.157); Fischer, A. G. and Garrison, R. E., 1967, *J. Geol.*, Univ. Chicago Press, 75 (Fig.277, 278); Goreau, T. F., 1959, *Ecology*, 40 (Fig.137); Goreau, T. F. and Hartman, W. D., 1963, *Publ. Am. Assoc. Advan. Sci.*, copyright 19, 75, (Fig.138, 139); Grégoire, C., 1957, *J. Biophys. Biochem. Cytol.*, 3 (Fig.1); Grégoire, C., 1962, *Inst. Roy. Sci. Nat. Belg., Bull.*, 38 (Fig.3, 25, 26); Gutschick, R. C., 1959, *J. Paleontol.*, 33 (Fig.85, 86); Hathaway, J. C., 1967, *Geotimes*, 12 (Fig.231); Hess, F. L., 1929, *Proc. U.S. Natl. Museum*, 76 (Fig.241); Konishi, K. and Wray, J. L., 1961, *J. Paleontol.*, 35 (Fig.103); Lindström, M., 1963, *Sedimentology*, 2 (Fig.291); MacClintock, C., 1967, *Peabody Museum Nat. Hist., Yale Univ., Bull.*, 22 (Fig.11, 13, 17); Matthews, R. K., 1966, *J. Sediment. Petrol.*, 36 (Fig.80); NEUMANN, A. C. et al., 1970, *J. Sediment. Petrol.*, 40 (Fig.149, 150); Ross, J. R. P., 1963, *Palaeontology*, 6 (Fig.112, 113); SCOFFIN, T. P., 1970, *J. Sediment. Petrol.*, 40 (Fig.144, 151); Storr, J. F., 1964, *Geol. Soc. Am., Spec. Paper*, 79 (Fig.132, 133); Taylor, J. D., Kennedy, W. J. and Hall, A., 1969, *Bull. Brit. Museum Zool., Suppl.*, 3 (Fig.7); Towe, K. M. and Cifelli, R., 1967, *J. Paleontol.*, 41 (Fig.68–74); Usdowski, H.-E., 1962, *Beitr. Mineral. Petrog.*, 8 (Fig.318); Wardlaw, N. C., 1962, *J. Sediment. Petrol.*, 32 (Fig.330); Watabe, N., 1965, *J. Ultrastruct. Res.*, 12 (Fig.5); Wise, Jr., S. W. and Hay, W. W., 1968b, *Trans. Am. Microscop. Soc.*, 87, (Fig.8, 9); Wolfe, M., 1968, *Sediment. Geol.*, 2 (Fig.321); Wood, A., 1957, *Palaeontology*, 1 (Fig.101); Wray, J. L., 1964, *Geol. Surv. Kansas, Bull.*, 170 (Fig.102).

A husband's creations are rarely his alone: my wife's collaboration has lent further delight to a labour inherently agreeable and my family's patience with my long-standing preoccupation has helped me in great measure. The text would have been notably less intelligible were it not for Diana's reading of all the drafts, and I recall thankfully her graceful compounding of acute criticism with the healing balm of encouragement. Long hours of proof reading, being shared, were made enjoyable, and the compilation of the References was, by her effort and skill, mercifully lightened.

So much depends on the day-to-day backcloth to a working life that I hasten finally to express my thanks and appreciation to my colleagues in the

Jane Herdman Laboratories of Geology for their tolerance and support, and to A. G. Fischer who, having drawn me to Princeton University as a National Science Foundation Visiting Scientist, then acted most effectively as midwife in the last stages of labour.

Department of Geological and Geophysical   The Jane Herdman Laboratories of Geology,
Sciences, Princeton University 1968–69   University of Liverpool 1964–68

FURTHER READING

Certain books of a general nature and a number of symposia are indispensable as background material for the study of carbonate sedimentology. These are the deeply thoughtful works by CAYEUX (1935) and SANDER (1936 or 1951), the symposia edited by LE BLANC and BREEDING (1957), HAM (1962), PRAY and MURRAY (1965), MÜLLER and FRIEDMAN (1968) and FRIEDMAN (1969), the section by FÜCHTBAUER (1970) on "Karbonatgesteine" in the textbook by Füchtbauer and Müller, and the collections of reviews edited by CHILINGAR et al. (1967a,b).

PREFACE TO THE SECOND EDITION

In writing the necessarily short Appendix for the second edition I have tried, within the space at my disposal, to strike a fair balance between commentary and bibliography. A complete account of relevant work published since the writing of the first edition would be impossibly long: the collection examined here is, therefore, selective. Moreover, even some outstandingly important papers are, I regret, given only brief mention. I have, however, taken the opportunity to emphasise four fields that were not treated in the first edition, but which clearly have great potential. These are: deep-sea sediments and their diagenesis, carbonate sediments in lakes, calcrete, and hydrogeology and karst.

I acknowledge warmly the help of my wife in the critical reading of an early draft and in the checking of proofs. To those many colleagues who have proferred advice and information I extend my profound thanks.

I add, as before, a note on certain publications of a general nature. The collection of papers on the Persian Gulf, edited by PURSER (1973) would alone serve as an excellent textbook on carbonate sedimentology. I suspect that no one seriously interested in this subject can afford to ignore the general sedimentological text by BLATT, MIDDLETON and MURRAY (1972). To these must be added the essays by FOLK (1973), the SNPA symposium edited by BALCON (1973), a short review of diagenesis by BATHURST (1973), vol. 18(3) of *Sciences de la Terre* (1973), the essay by GINSBURG (1974), and the revised and enlarged English translation of FÜCHTBAUER (1975).

The Jane Herdman Laboratories of Geology,
University of Liverpool 1974–75

CONTENTS

Chapter 1

PETROGRAPHY OF CARBONATE GRAINS 1:
SKELETAL STRUCTURES

INTRODUCTION

A very large part of our understanding of carbonate sediments and rocks is derived from studies made with the microscope. Field work lays bare the gross relationships, but is apt to be hampered by the failure of many limestones to reveal themselves clearly in the hand specimen—a serious handicap in so complex a group of rocks. Geochemical and X-ray studies, though profoundly influential, suffer not only from the length of time between question and answer, but, above all, from their inability to take cognizance of the complex fabrics which are of such critical importance in this group of multicomponent rocks. The immediacy of the microscopical approach has sustained this method as the major research tool throughout the rapid expansion of carbonate studies since World War II: the newer and more discerning classifications depend on it. The development of refined staining techniques, of replication, including the shadowing of acetate peels, combined with the use of the transmission and scanning electron microscopes, have shown, along with the subtle methods of cathodoluminescence, that the microscope has a rich future.

The basis of microscopical work is the determination of the shape, size and orientation of crystals and crystal groups and the spatial relationships between them—in a nutshell the study of mineralogy and fabric. While a relatively cursory knowledge of, for example, the wall structure of a shell will suffice for the recognition of whole or little damaged tests—where macroscopic forms such as ribs or septa are an adequate guide—a more detailed grasp of wall structure is essential for the determination of small fragmentary grains. These grains have shapes which offer no taxonomic clues and which are, indeed, commonly only a reflection of breakage and abrasion. Regrettably, though, even a painstaking study of the structure in small skeletal grains rarely allows taxonomic distinction below the level of family and is often valuable only in the separation of phyla. However, to the carbonate petrologist, grain identification is not the only goal of skeletal structure analysis. Carbonate rocks are as much the products of diagenesis as they are of primary deposition: consequently, an understanding of skeletal structure allied to mineralogy is indispensible as a basis for the varied investigations of diagenesis with which the carbonate petrologist nowadays concerns himself.

Unfortunately, the investigation of skeletal structure has been both uneven taxonomically and all too often lacking in rigour. The unevenness of investigation, the haphazard distribution of knowledge, are reflected in the following pages. Some phyla have not been included because useful information on skeletal structure, at the level of microstructure or ultrastructure, is lacking.

The treatment herein is restricted to wall fabrics, as they are seen with the light and electron microscopes in fresh or well preserved material. The gross, macroscopic form of skeletons—such as ribbing, organization of septa or arrangement of chambers—has been dealt with elsewhere in palaeontological texts and will not be repeated. The approach is primarily petrographic and the data here summarized are intended to help those concerned not only with taxonomy and the delineation of biofacies, but with such other matters as calcite/aragonite ratios, crystal growth processes, differential dissolution, selective dolomitization, recrystallization and a variety of problems requiring a knowledge of the crystalline and mechanical properties of sedimentary grains.

MOLLUSCA

The unusually varied and complicated structures in the shells of the molluscs became known largely through the fundamental studies of fine structure with the light miscroscope by W. J. SCHMIDT (1921–1929) and the combined structural and mineralogical studies of BØGGILD (1930). Only in recent years have important advances been made upon these great works. OBERLING (1964) has given us a detailed and thoughtful paper on the bivalve wall with emphasis on layering, tubulation, ribbing and the distribution of shell structures. MACCLINTOCK (1967) has produced a detailed and acute monograph on the patelloid and bellerophontoid gastropods and includes not only the distribution of structures in layers, but descriptions of the structures which are of use in the recognition of detrital grains. Regrettably, neither of these authors has supplied mineralogical data. The most recent, and most comprehensive, work on shell structure is a monograph on the Bivalvia by TAYLOR et al. (1969). This comprises a very full and detailed survey of earlier work and includes much new material related to both structure and layering. It is based on thin sections and peels observed with the light microscope, on photomicrographs taken with the transmission and the scanning electron microscopes, and on mineralogical determinations made by X-ray diffractometer.

Molluscan shells are constructed for the most part of organised aggregates of micron-sized crystals disposed in layers. These layers differ from one another in structure (see below), orientation of structure, and mineralogy. In any one unaltered species the layers are either all aragonite or interlayered aragonite and calcite: both high-magnesian and low-magnesian calcites occur (P. D. Blackmon in CLOUD, 1962a, p.61). In the Ammonoidea the shell is aragonite but the aptychus is calcite. The Ostreacea, long regarded as having purely calcitic walls, are now known to have aragonite in their myostraca, in the calcified part of the ligament and in the prodissoconch. This information follows the investigations of STENZEL (1962, 1963, 1964), whose results have been confirmed and extended by Taylor et al. (1969). Outside the Mollusca these two minerals are found together in the same skeleton only in the Hydrozoa, Octocorallia and Cirripedia (LOWENSTAM, 1964). Where they occur together in the Mollusca, however, they are never mixed but are laid down as separate monomineralic layers. The arrangement of carbonate material in layers is a basic characteristic of the molluscan shell. A shell may have from one to six layers depending on its systematic position.

The live shell is covered on the outside, at least at some period during growth, with a horny coat, the **periostracum**: this is a quinone-tanned protein of the keratin–fibrin group and is the site at which calcification is initiated. In some species, it bears hairs or bristles. The operculi (aptychi in the Ammonoidea) are horny or calcitic.

Structural nomenclature is derived from BØGGILD's (1930) classic study, with modifications by later workers. The variation of structure is considerable, both throughout the phylum and, not uncommonly, within the shell of a single species. This is no place for the presentation of detailed taxonomic criteria: these are available in the works quoted. The petrologist working with grains will be fortunate if he can identify molluscan debris at class, or occasionally family, level. The classification and description of structures given here express, therefore, a some-what simplified picture of shell structure consistent with the crude petrographic detail to be found in a detrital grain that normally represents only a part of the shell, has no known orientation, and has commonly been more or less neomorphos-ed. In describing a structure, no distinction will be made between its appearance in different taxa (indeed, little is known about this), but in the various publications referred to the reader will be able to find what information there is on this topic.

Molluscan structures

BØGGILD's (1930) terms for structures are used here with the addition of the term myostracal. Seven main structures are distinguished:

Nacreous	Crossed-lamellar
Prismatic	Complex-crossed lamellar
Homogeneous	Myostracal
Foliated	

It should be born in mind that the appearance of each structure varies considerably according to its orientation in the field of the microscope.

Nacreous structure. This structure is always aragonite (WADA, 1958, 1959) and is built of little tablets (TAYLOR et al., 1969) laid one on the other (Fig.1–7).

Fig.1. Oblique view of broken succession of sheets of nacre in the bivalve *Pinctada margaritifera* (Aviculidae). Interlaminar organic membranes seen as thin white lines. Shallow grooves on surfaces of laminae are boundaries between tablets. Replica. Electron mic. Recent. (From GRÉGOIRE, 1957.)

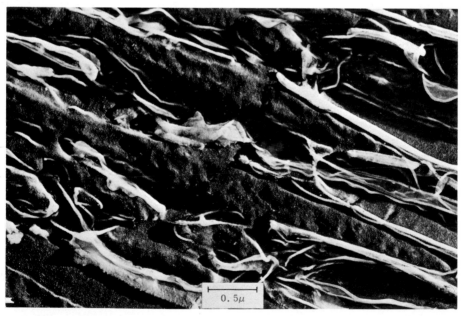

Fig.2. Vertical section through nacreous structure, polished and heavily etched, *Nautilus* sp. Interlaminar and intercrystalline membranes variously folded, torn or erected. Replica. Electron mic. Eocene. (From GRÉGOIRE, 1959a.)

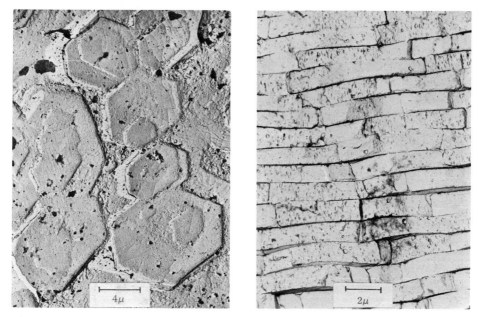

Fig.3. Slightly etched nacreous surface of a septum in *Nautilus pompilius*, showing oriented growth of tablets. The plane of the page contains (001), tablets are elongate parallel to *a* and join along well developed (010). Replica. Electron mic. Recent. (From GRÉGOIRE, 1962.)

Fig.4. Vertical section through lenticular nacre in a Jurassic mytilid, showing the vertical stacks. Replica. Electron mic. (Courtesy of J. D. Hudson.)

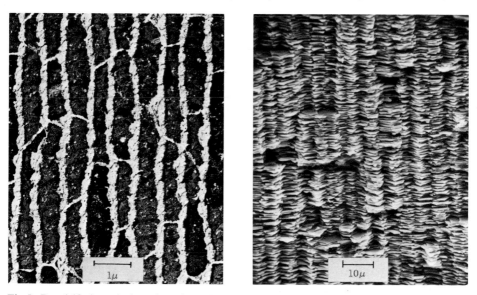

Fig.5. Decalcified vertical section through the nacre of the bivalve *Pinctada martensii*. White cords of interlaminar organic matrix run N–S: those of the intercrystalline matrix run E–W between the tablets. Electron mic. Recent. (From WATABE, 1965.)

Fig.6. Vertical section through lenticular nacre in the gastropod *Haliotis rufescens* showing vertical stacks. Scanning electron mic. Recent. (From WISE and HAY, 1968b.)

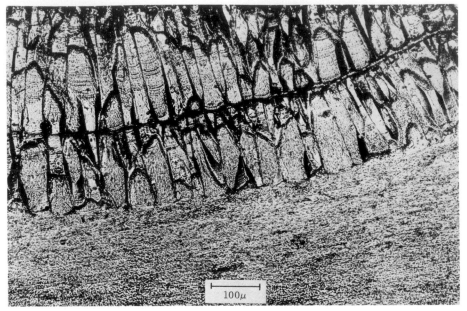

Fig.7. Radial section of outer layer of prismatic aragonite on middle layer of nacreous structure in the bivalve *Anodonta cygnaea*. Peel. Recent. (From TAYLOR et al., 1969.)

These are plates flattened parallel to (001); this basal pinacoid is in turn parallel or nearly parallel to the shell surface. Tablets may be rounded, or euhedral with (010) and (110) faces. Optical data are based on measurements with the light microscope and X-ray diffraction. The tablets are 0.4–3 μ thick (direction parallel to the c axis), and mainly 2–10 μ in breadth. They are arranged in continuous sheets (Fig.4, 5, 6, 8) to give either the brick wall pattern of Bøggild (sheet nacre) or vertical stacks (columns) of up to five or six tablets (lenticular nacre). The structures are clearly visible with the electron microscope (Fig.6, 8).

From the work of WISE and HAY (1968b) and WISE (1969a; 1970a,b) it seems likely that, in gastropods, the growth surface of the nacreous layer is always covered by tall conical stacks of aragonite crystals. In vertical cross-sections these tablets are arranged in vertical columns. In bivalves, lenticular nacre is known only in a few genera (notably *Pinna, Nucula*). Normally the tablets overlap at the growth surface and show brick-wall (sheet nacre) structure in vertical section. The Recent cephalopod, *Nautilus*, contains lenticular nacre (S. W. Wise, personal communication, 1969). W. J. Kennedy (personal communication, 1969) has also found lenticular nacre to occur widely in gastropods and cephalopods.

Under the light microscope care is needed to distinguish nacre from homogeneous fabric (below): it is, at best, seen in vertical sections as a cryptocrystalline material with vague lineation and a tendency to uniform extinction throughout the layer. With the naked eye the inner surface of the shell wall shows the familiar colours and lustre for which nacre is renowned.

Electron photomicrographs by GRÉGOIRE (1962, plate 2) and others reveal that in the Recent *Nautilus* the new tablets (Fig.3) tend toward hexagonal form: later they unite to form a mosaic of tablets with polygonal compromise boundaries (TOWE and HAMILTON, 1968a, fig.7, 8). The interlaminar and intercrystalline conchiolin (Fig.2, 5; GRÉGOIRE, 1957, plate 254–256) has the form of a reticulate lace-like sheet (Fig.25, 26; GRÉGOIRE, 1957, plate 253–257; 1960, plate 1–5). The organic trabeculae that compose the sheet "appear as knobby cords, resembling frequently the rhizomes of the garden iris" (GRÉGOIRE, 1957, p.801). WATABE (1965) suggested that in some bivalves each tablet is itself built of smaller blocks, each of which is enclosed in a conchiolin matrix and, furthermore, is impregnated with an intracrystalline matrix (this has been disputed, p.15). KOBAYASHI (1969) referred to broad tablets composed of aggregates of narrower tablets.

Prismatic structure. Taylor et al. (1969) distinguished in the bivalves between simple prisms which may be aragonite or calcite, and composite prisms which are only aragonite. Prismatic layers in the inner or middle shell layers of bivalves were regarded by Taylor et al. as myostracal.

The **simple prism** has much the same form whether it is aragonite or calcite (Fig.8–11, 24). Both prisms are striated: the calcite prism has transverse striations, is meniscus-shaped and convex downward, whereas the aragonite prism has diverging longitudinal striations in addition to the transverse ones. The shell surface is honeycombed, with the walls of intercrystalline organic matrix exhibiting a dominantly pentagonal arrangement. Grégoire has shown with the electron microscope that the aragonite prism is constructed of a stack of disc-shaped laminae each separated from the next by organic matrix. TAYLOR et al. (1969) have found aragonite prisms to be divided longitudinally into blocks. These probably represent the divergent striations.

The **composite prisms** usually lie with their long axes parallel to the shell surface and radiate from the umbo. They have square cross-sections and are built of second order needles which diverge toward the margins of the prism. These needles are, in turn, divided into smaller units.

MACCLINTOCK (1967) recognised bladed prisms up to 800 μ long and 5–100 μ wide, also fibrils up to 220 μ long, but 1–2 μ in cross-section. Both usually make an angle of about 50° to the surface. His fibrillar structure grades into the complex prismatic structure of BøGGILD (1930) which comes within the composite prismatic structure of TAYLOR et al. (1969). It is distinguished by the extreme fineness of its second order needles.

Homogeneous structure. This structure is very finely crystalline. In ordinary light little fabric can be seen, but with crossed nicols (Fig.12) large areas extinguish more or less uniformly. This structure is aragonite. The c axes have angles to the shell surface that vary from species to species, being generally normal, but other-

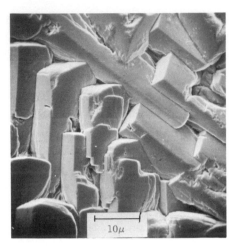

Fig.8. Vertical fracture surface through the wall of the gastropod *Cittarium pica:* a layer of prismatic structure overlies one of nacre vertically stacked. Scanning electron mic. Recent. (From WISE and HAY, 1968b.)

Fig.9. Aragonite prisms in the growth surface of the gastropod *Cittarium pica.* Scanning electron mic. Recent. (From WISE and HAY, 1968b.)

Fig.10. Oblique section of prismatic calcite in the bivalve *Pteria alacorri*, etched to show intercrystalline organic matrix. Slice. Recent. (Courtesy of W. J. Kennedy.)

Fig.11. Sagittal section through layer of composite prismatic structure (aragonite) in the gastropod *Lepeta concentrica*. Slice. Recent. (From MacCLINTOCK, 1967.)

wise oblique or parallel. Bøggild was not able to see individual crystals with the light microscope. TAYLOR et al. (1969) examined the bivalve *Arctica islandica* in detail with the electron microscope and saw irregular rounded granules, approximately $1 \times 3 \times 0.5 \mu$, each in an organic envelope. Strong grey-brown banding is present in places.

Foliated structure. This calcite structure is built of more or less regularly arranged tablets (Fig.13–16) joined side by side to make "fine sheets grouped in larger lenticular folia in turn forming larger units" (TAYLOR et al., 1969). The folia (Fig.14) are parallel or nearly so to the shell surface with an appearance in places resembling low-angle cross-strata (Fig.16). Some orientations are highly irregular, especially in the Pectinacea. The sheets are made up of tiny elongate tablets (also called laths) placed side by side (Fig.13, 15). Vertical sections across folia show a pseudopleochroism, straw-red brown, which is in part an effect of the different amounts of organic matrix exposed in different sections. Tablet dimensions quoted or measured by Taylor et al. are: length: 4–17μ; width: 2–3.5μ; thickness: 0.1–0.5μ. The c axis is normal to the main tablet surface which is (0001); b is parallel to the length. Each tablet seems to be an agglomeration of smaller laths.

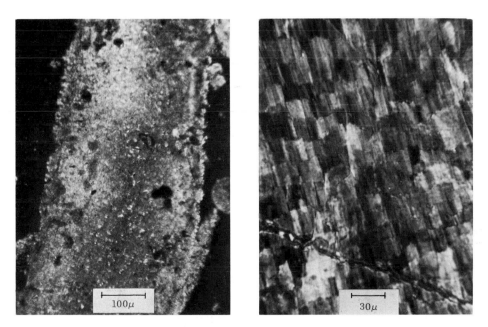

Fig.12. Patchy distribution of polarization colours in homogeneous structure within a fragment of molluscan shell. Slice. Crossed polars. Recent.

Fig.13. Section through foliated structure (calcite), parallel to the growth surfaces, showing bladed fabric, in the gastropod *Acmaea mitra*. Slice. Recent. (From MacCLINTOCK, 1967.)

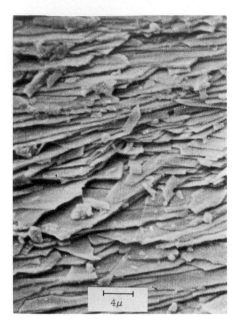

Fig.14. Oblique view of broken surface of foliated structure (calcite) in the bivalve *Ostrea edulis*. Scanning electron mic. Recent. (Courtesy of W. J. Kennedy.)

Fig.15. Inner surface of foliated structure (calcite) in the bivalve *Ostrea hyotis*. Dark outline around each tablet is organic matrix. Replica. Electron mic. Recent. (Courtesy of W. J. Kennedy and J. D. Taylor.)

Crossed-lamellar structure. This is always aragonite (Fig.17–24) and has a three-fold structural hierarchy consisting of first and second order lamels (BøGGILD, 1930) and, just distinguishable with the light microscope, the trace of third order lamels (KOBAYASHI, 1964; MacCLINTOCK, 1967; TAYLOR et al., 1969). The first order lamel is an elongate plate or lens with its longest axis parallel to the shell surface: these lamels are generally concentric (Fig.20, 21). In vertical radial section the first order lamel is usually rectangular, such that its longer axis is normal to the shell surface, but this axis may twist and turn and be locally sub-parallel to the shell surface. In this section also, the alternate lamels show a pseudopleochroism, straw-red brown, for reasons mentioned in the section on foliated structure. The lamels are packed together side by side so as to form an array of roughly parallel vertical sheets, having lengths up to several millimetres and widths about 0.5 mm. According to BøGGILD (1930) the plane containing the two longer axes is the optic axial plane and is normal to the shell surface or slightly oblique.

The first order lamels are constructed of second order lamels. These are plates, 1–1.5 μ thick (shortest axis), that lie with their tabular surfaces normal to the tabular surfaces of the first order lamels (Fig.20) but oblique to the shell surface

Fig.16. Vertical section of foliated structure (calcite) in a gryphaeid bivalve. Peel. Jurassic.

Fig.17. Median sagittal section of crossed-lamellar structure (aragonite) in the gastropod *Patella compressa*. Slice. Recent. (From MacClintock, 1967.)

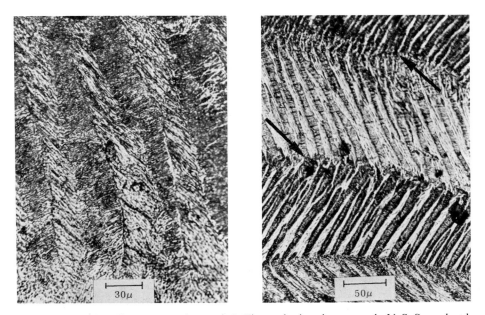

Fig.18. Crossed-lamellar structure (aragonite). First order lamels run nearly N–S. Second order lamels run NW–SE and NE–SW. Peel of a molluscan grain. Recent.

Fig.19. Layers of crossed-lamellar structure (aragonite). The margins of one layer are shown by arrows. Within this layer the first order lamels run NNW–SSE. In one set of first order lamels the second order lamels are clearly visible running ENE–WSW. In the other set the second order lamels lie nearly in the plane of the page and show an oblique lineation nearly parallel to the edges of the first order lamels. Peel of a molluscan grain. Recent.

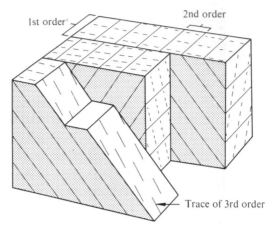

Fig.20. Diagram to illustrate the orders of lamels in crossed-lamellar structure.

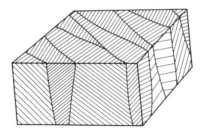

Fig 21. Diagrammatic illustration of crossed-lamellar structure to show its appearance in three dimensions.

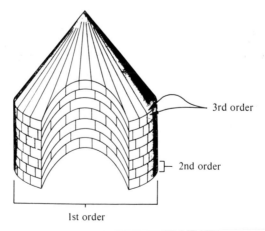

Fig.22. Diagrammatic illustration of the fundamental major prism of complex-crossed lamellar structure with its three orders of lamels. (After MacClintock, 1967.)

Fig.23. Section of crossed-lamellar structure (aragonite) in the bivalve *Trachycardium consors.* Three first order lamels each contain numerous second order lamels. Replica. Electron mic. (Courtesy of W. J. Kennedy and J. D. Taylor.)

Fig.24. Vertical section through the wall of the bivalve *Spondylus calcifer*. Layers, from left to right, are: fibrous calcite, crossed-lamellar structure, myostracum (arrow), crossed-lamellar structure, foliated structure (occupying most of the right-hand part of the picture). Slice. (Courtesy of W. J. Kennedy.)

with which they make an angle of between 15 and 45°. MacClintock (1967) gave an average intermediate axis of 15 μ in the patelloid wall. In adjacent first order lamels the dips of the second order lamels, measured from the shell surface, are opposed (Fig.20, 21).

The traces of the third order lamels are indicated in Fig.20. These lamels are hard to see and the reader is referred to Kobayashi (1964) and MacClintock (1967) for details.

Complex-crossed lamellar structure. Bøggild (1930) wrote of this aragonitic structure that "it is the most intricate of all and the details cannot, probably, be determined with full certainty". The reader is referred to Bøggild's acute descriptions and his many photomicrographs. Here, we may simply note that this structure is a complicated arrangement of units of opposed second order lamels of the type present in crossed-lamellar structure, commonly disposed so as to form a sort of prismatic structure. Taylor et al. (1969) described some variations in detail, and MacClintock (1967) interpreted this structure in the patelloid gastropods as consisting of roughly cylindrical major prisms each built of a series of cones stacked one upon another with their apices forming the axis of the prism (Fig.22). The prisms are equivalent to the first order lamels of crossed-lamellar structure, the cones (each about 1 μ thick) are equivalent to the second order lamels. Radial lamellae in each cone are third order.

Myostracal layers. Myostracum is the name given to the aragonite deposits secreted beneath the attachment areas of muscles (Fig.24). This structure varies between two extremes, from a thick layer of prisms to layers so thin that only a linear trace is visible in thin section (Taylor et al., 1969). The prisms are from 0.5–50 μ long, parallel to the c axis, and this is perpendicular to the shell surface. Prism contacts are very irregular. Compared with other structures, myostracal structure is light grey or colourless in transmitted light instead of the darker straw or red-brown of some other fabrics and is more transparent.

Tubules

Canals (or tubules) occur in the shells of some molluscan species (Oberling, 1955, 1964; Omori et al., 1962; Kobayashi, 1969; Taylor et al., 1969). They are circular in cross-section and generally 4–6 μ in diameter. They are open to the interior of the shell and penetrate through the layers including the periostracum.

The organic matrix

Intimately dispersed among the mineralogical units in the various shell structures is a complex and variable matrix of organic material (Fig.25, 26), commonly called

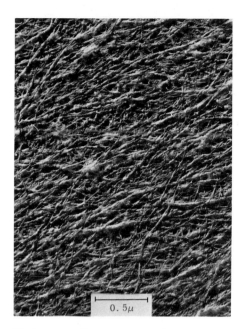

Fig.25. Posterior organic membrane coating the adapical convex side of a septum in the phragmocone of *Nautilus pompilius*, showing unoriented felt of fibrils and nodules. Replica. Electron mic. Recent. (From GRÉGOIRE, 1962.)

Fig.26. Fragment of organic membrane from the decalcified nacre of *Nautilus pompilius*. The lace-like reticulate sheet has the characteristic nautiloid pattern of sturdy, irregularly cylindrical trabeculae, studied with hemispherical protuberances separate by elongate openings of irregular form. Replica. Electron mic. Recent. (From GRÉGOIRE, 1962.)

conchiolin. This matrix is proteinaceous when fresh. Studies with the electron microscope have shown that conchiolin is interlaminar and intercrystalline (as mortar between bricks; GRÉGOIRE et al., 1955; WATABE and WADA, 1956; TSUJII et al., 1958; GRÉGOIRE, 1960, 1966a, b; TSUJII, 1960; HARE, 1963; DODD, 1963, 1964; LOWENSTAM, 1963, 1964; WILBUR, 1964; WATABE, 1965; HUDSON, 1968; KOBAYASHI, 1968; TAYLOR et al., 1969). TOWE and HAMILTON (1968a, b) have investigated certain artifacts which may arise in the preparation of very thin slices (600–800 Å thick) with an ultramicrotome. In particular, the subdivision of nacre tablets into blocks encased in intra-tablet organic sheets, discussed by Watabe, has been shown by Towe and Hamilton to be reproducible by varying the orientation of the diamond knife. Grégoire has found that the fine structure of the reticulate sheets and the trabeculae (see nacreous structure) have taxonomic value in the Gastropoda, Bivalvia and Cephalopoda. WISE (1970) has found, in gastropods, alternate disposition of nacre layers and organic membranes.

Amphineura

This class includes the chitons. Bøggild (1930) stated that no other group has such a complex structure, varying between species and in different parts of the same shell, yet always aragonite. Typically there are five layers:

(1) The uppermost layer: dark coloured and, unlike the underlying layers, found over the whole shell, crossed by rather large pores and with a very complicated fabric.

(2) Locally developed only, with homogeneous or faintly prismatic structure.

(3) Very complicated construction of vertical lamellae found mainly in the "outer parts of the shell, near the margin".

(4) Locally only, a "series" of subordinate layers of homogeneous or indistinctly prismatic structure.

(5) Lowermost layer: located especially in the middle (axial) part of the shell, makes up more than half the thickness of the wall. Uniform, very fine crossed-lamellar structure. The long axes of the 1st order lamels may attain 5 mm.

Gastropoda

The calcite layers, those most likely to be preserved in limestones, are all essentially alike throughout the Gastropoda. The aragonite layers are highly diverse. Calcite forms only the uppermost layer, usually with a structure that Bøggild described as "very irregularly prismatic". Most Gastropoda do not have a calcite layer: the distribution of the layer is sporadic and seems to follow no plan, turning up in a family here, in a few genera there. The calcite layer is usually, though not always, thicker than the aragonite layer.

In surveying briefly the distribution of structures, the taxonomic classification followed is that in the *Treatise on Invertebrate Paleontology* (Cox, 1960).

Bøggild found a close relationship in aragonite structure between the three subclasses of the Gastropoda (Cox, 1960), with certain exceptions. In general the shell walls of the Prosobranchia, Opisthobranchia and Pulmonata have three, less commonly four, layers of crossed-lamellar aragonite with alternating fabric orientation. Only in the Prosobranchia are occasional calcite upper layers to be found. Four taxa have peculiar fabrics of their own.

The Archaeogastropoda (Lower Cambrian–Recent), with the exception of the Patellina, are the most variable of the Gastropoda (Bøggild, 1930). The most characteristic feature is the presence of a nacreous layer, usually the lowest of the layers. The rest of the shell may consist of various mixtures of layers having any of the homogeneous, prismatic, crossed-lamellar or complex-crossed lamellar structures. Many of the Archaeogastropoda are extinct and aragonitic fabrics have not been preserved. In the superfamily Bellerophontoidea four structures are known, prismatic, foliated, crossed-lamellar and nacreous (MacClintock, 1967).

The superfamily Patelloidea has been divided by MacCLINTOCK (1967) into seventeen shell-structure groups. The number of layers in the shell varies from four to six. The dorsal layer is either prismatic (simple or composite), crossed-lamellar, or foliated (one species, *Acmaea scabra*). The ventral layer is either crossed-lamellar or complex-crossed lamellar, except for *A. scabra* in which it is foliated. The intermediary layers include a myostracum and composite prismatic structure, foliated structure, crossed-lamellar structure and complex-crossed lamellar structure. There is no nacreous structure, though the inner layer may be iridescent.

The walls in the Heteropoda and Pteropoda are aragonite with homogeneous structure: there is no calcite (BøGGILD, 1930).

Bivalvia

The results of TAYLOR et al. (1969) are summarized below. The classification is NEWELL's (1965). The shells have either two or three layers exclusive of the myostracum. Thus to each description must be added a myostracal layer of prismatic aragonite in the area of muscle attachment.

Palaeotaxodonta
> Nuculacea: outer layer prismatic aragonite, middle and inner nacreous.
> Nuculanacea: outer and inner layers homogeneous.

Cryptodonta
> Solemyacea: outer layer prismatic aragonite, inner homogeneous.

Pteriomorpha
> Arcacea: outer crossed-lamellar, inner complex-crossed lamellar.
> Limopsacea: as above.
> Mytilacea: either all nacreous, or outer layer of nacre with inner of complex-crossed lamellar; in some species outer layer is prismatic calcite, in some inner layer is prismatic calcite or aragonite.
> Pinnacea: outer layer prismatic calcite, middle and inner layers nacreous.
> Pteriacea: as above.
> Pectinacea: Pectinidae: foliated with or without a middle layer of crossed-lamellar.
>> Spondylidae and Plicatulidae: outer layer foliated, middle and inner crossed-lamellar.
> Anomiacea: outer layer foliated, inner complex-crossed lamellar.
> Limacea: outer layer foliated, inner crossed-lamellar or complex-crossed lamellar, with or without middle crossed-lamellar.
> Ostreacea: foliated, with outer layer of prismatic calcite on upper valves of some species.

Palaeoheterodonta
  Unionacea: outer layer of prismatic aragonite (lacking in cemented Ether-
      iidae), middle and inner nacreous.
  Trigoniacea: as above.

Dodd (1964) has given a description of the Recent *Mytilus californianus*
which is valuable for its detail. There are, one proteinaceous and four carbonate
layers:
  Periostracum: proteinaceous.
  Prismatic calcite: acicular crystals 1–3 μ wide, up to 35 μ long, with their
      outer ends reclined toward the beak at 0–45° to the surface commonly
      in cone-shaped groups.
  Nacreous aragonite: this forms a layer only within the pallial line and
      consists of tabular crystals, their longer axes parallel to the surface, in
      layers alternating with layers of proteinaceous matrix. Maximum
      crystal diameters are 5–6 μ with thicknesses about 1 μ.
  Prismatic calcite: an inner layer on the beak side of the nacreous layer,
      composed of acicular crystals 3–4 μ wide and 40–50 μ long, in cone-
      shaped aggregates.
  Blocky aragonite: locally as a myostracum under muscles with roughly
      equant crystals up to 60 μ in diameter.
Electron photomicrographs by GRÉGOIRE (1961a, b) of *Mytilus edulis* show
that the crystals in both the nacreous and the prismatic layers are clothed in sheets
of reticulate conchiolin.
    In the Hippuritacea (rudists) low-magnesian calcite is always a major
component of the shell wall, but in addition there may be direct or indirect evidence
of aragonitic structure. Prismatic calcite structure appears to be primary since it
occurs, for example, in rudists in the Chalk, where other calcite structures (e.g., the
outer layer of *Inoceramus*) are preserved but known aragonite structures (scaphopod,
cephalopod) have been destroyed. Crossed-lamellar structure and complex-crossed
lamellar structure (both aragonite) have been found by KENNEDY and TAYLOR
(1968) in *Hippurites* and *Plagioptychus* from the Upper Cretaceous of Austria.
The characteristic cellular structure of the calcitic radiolitid wall is shown in
Fig.27, 28.

Scaphopoda

Most scaphopods are built of crossed-lamellar structure with thin layers of
myostracal prisms (W. J. Kennedy and J. D. Taylor, personal communication,
1969).

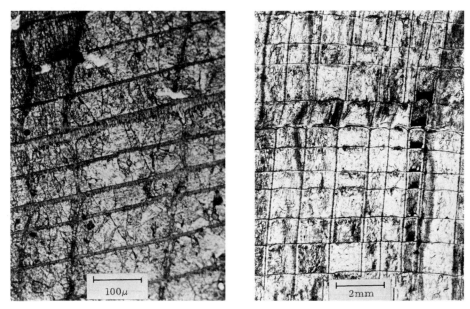

Fig.27. Section of wall of the calcitic rudist *Durania mortoni*. Cells filled with calcite cement. Slice. Cretaceous. (Courtesy of W. J. Kennedy.)

Fig.28. Section through wall of a radiolitid rudist.

Cephalopoda

These shells have a simpler structure than most other molluscs (TEICHERT, 1964) and are composed of aragonite except for aptychi which are calcite.

Nautilus. The external wall of the Recent *Nautilus* has three layers: an outer layer ("porcellaneous" of GRÉGOIRE, 1962; "spherulitic-prismatic" of MUTVEI, 1964), a rather thicker middle layer of nacre, and a very thin inner layer ("sutural infilling" of Grégoire; "semi-prismatic" of Mutvei). The outer layer is divisible into two parts, an outer granular part clearly separated from an inner part composed of composite prisms. In the nacre there is both sheet and lenticular structure. The inner layer, which is by far the thinnest of the three, varies in thickness and is not everywhere present. The septa are constructed of the same three layers but are dominated by the nacreous layer which is very much thicker than the outer and inner layers.

Ammonoidea. From their detailed study with the electron microscope of *Saghalinites* and *Scaphites*, BIRKELUND (1967) and BIRKELUND and HANSEN (1968) discovered that the shell wall has an outer "porcellaneous" (finely prismatic) layer with an inner nacreous layer, the arrangement in the protoconch being more complicated. The operculae (aptychi) are built of calcite.

MUTVEI (1967) in his work on *Promicroceras* stressed the dominance of the nacreous layer, with the spherulitic-prismatic and semi-prismatic layers absent or extremely thin.

The siphuncular tube in some cephalopods contains spicular aragonite.

Monoplacophora. ERBEN et al. (1968) detected an outer prismatic layer on a nacreous layer in *Neopilina* (Recent) and *Pilina* (Silurian). The former has an inner prismatic layer. Other species had structures difficult to determine. This study was based on the light microscope and transmission and scanning electron microscopes.

Coleoidea: Belemnitida. The rostrum of Jurassic and Cretaceous forms is made of prismatic calcite with varying degrees of orientation. The Triassic *Aulacoceras* has a rostrum of secondary calcite, presumably a replacement of aragonite. What little is known of the phragmacones shows the Cretaceous *Belemnitella mucronata* to have a wall of fine prismatic calcite, whereas the Triassic *Atractites* is now secondary calcite and must have been aragonite (BØGGILD, 1930). W. J. KENNEDY (personal communication, 1969) suspects that all phragmacones may have been aragonite initially.

Coleoidea: Sepiida. The shell of *Spirula* is all aragonite. The wall is prismatic with a thin outer homogeneous layer, but the septa are nacreous.

Cameral deposits. The cameral deposits in the chambers of the phragmocones of many fossil cephalopods are mainly of fibrous aragonite, locally altered to calcite with corresponding loss of the structure of growth laminae (GRÉGOIRE and TEICHERT, 1965; FISCHER and TEICHERT, 1969).

BRACHIOPODA

Sections through the brachiopod valve reveal a two-layered wall of low-magnesian calcite, clearly visible in ancient as in Recent shells (WILLIAMS, 1956, 1965, 1966, 1968a). In the Recent shell there is a third, organic layer, the **periostracum** which covers the upper surface of the calcite shell. Immediately under the periostracum is the **primary layer** made of fine fibres of calcite elongated perpendicularly to the shell surface. It is this layer that forms the shell edge. The optic axes are parallel to the elongation of the fibres. In life these crystals are not enclosed in organic sheaths, and so are extracellular in origin (WILLIAMS, 1965, p.11). The layer is thin and difficult to see except in well preserved shells. It is, however, generally clear (but is cryptocrystalline) in Spiriferida, Thecideacea, Pentameracea, the later Porambonitacea, Terebratulida and Rhynchonellida: it occurs sporadically in Orthida and Strophomenida. Under the primary layer lies the **secondary layer** (Fig.29–31). In this the long and slender fibres of calcite lie at low oblique angles to the inner surface of the primary layer: each fibre converges upon the primary layer posteriorly (shown unusually clearly with the scanning electron microscope by BRUNTON, 1969). The optic axis is not parallel to the long axis of the fibre but, as in the primary layer, is perpendicular to the outer shell surface. The fibres are deposited intracellularly in a greatly expanded cell membrane. This is a proteinaceous sheath and some residue of it commonly remains even after lithification of a limestone (JOPE, 1967a, b, 1969a, b). In cross section each fibre is bounded on its outer side by three arcuate surfaces, each concave into the fibre, the middle arc being referred to as a keel: on the inner side there is a single convex arcuate saddle (Fig.34, 35). All internal apophyses are constructed of the secondary layer: the dorsal median ridge, the teeth, crura and dental plate. A third layer is common in the Pentamerida and Spiriferida. It occurs on the inside of the secondary layer, as a distinctive modification of it, and is built of coarse calcite prisms in continuity with those in the secondary layer (Fig.36). The prisms are elongate parallel to the

Fig.29. Vertical section of endopunctate wall of a brachiopod. The secondary layer of calcite fibres is crossed by canals. Slice. Jurassic.

Fig.30. Vertical section of impunctate wall of a rhynchonelloid brachiopods howing the calcitic secondary layer. Slice. Jurassic.

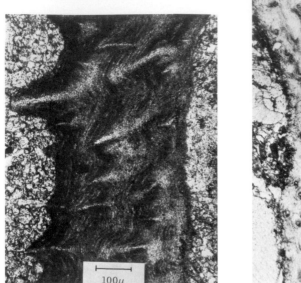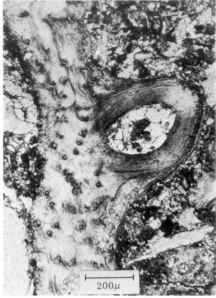

Fig.31. Vertical section of the pseudopunctate wall of a productoid brachiopod. Secondary layer of calcite fibres crossed by taleolae. Slice. Carboniferous Limestone.

Fig.32. Wall of pseudopunctate productoid brachiopod with attached spine filled with calcite cement. Slice. Carboniferous Limestone.

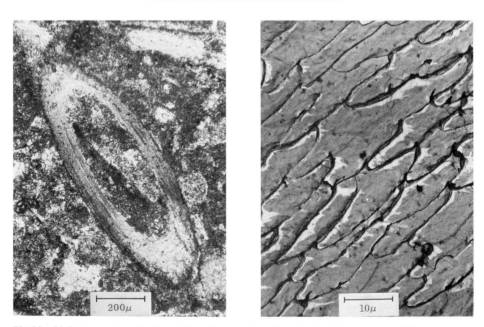

Fig.33. Oblique section of spine of brachiopod showing concentric structure of the secondary layer of fibrous calcite. Embedded in biomicrite. Slice. Carboniferous Limestone.

Fig.34. Vertical section through the secondary layer of a rhynchonelloid brachiopod, showing transverse sections of the calcite fibres with keels and saddles. Replica. Electron mic. Ordovician. (Courtesy of A. Williams.)

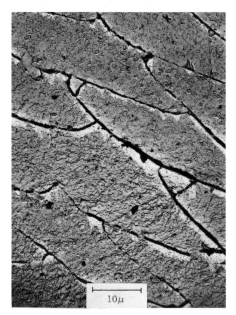

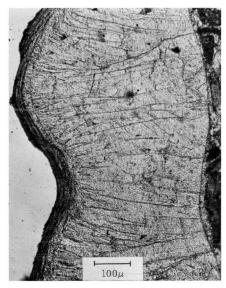

Fig.35. Vertical section through the secondary layer of an atrypidine brachiopod, showing transverse sections of calcite fibres with keels and saddles. Replica. Electron mic. Ordovician. (Courtesy of A. Williams.)

Fig.36. Vertical section through the impunctate wall of a pentamerid brachiopod. Thin secondary layer of calcite fibres underlain by thick tertiary layer of prismatic calcite. Slice. Silurian.

optic axis which is disposed perpendicularly to the surface of the shell. A fourth modification replaces the primary layer in the Plectambonitacea. Here the calcite crystals have the form of "narrow arcuate ribbons" (WILLIAMS, 1968a, p.38) which, instead of lying in alternate rows (*en echelon*) like the fibres, are packed one on the other and side by side. Williams has named this structure **laminae.**

ARMSTRONG (1969) has suggested that laminae could more significantly be described as bladed structure. A blade has a rectangular cross-section, is two to four times wider than high, and many times longer than wide. The blades lie parallel to each other forming continuous sheets, the longer axes of the blades lying parallel to the surfaces of the sheets. Sheets are superimposed, one on the other, in parallel array. Where the longest axes of the blades in one sheet are parallel to the longest axes of the blades in the overlying and underlying sheets, the structure could be called **parallel-bladed** (as in the plectambonitacean Strophomenida). Where the longest axes of the blades in a sheet are not parallel to the longest axes of the blades in the overlying and underlying sheets, Armstrong called this structure **cross-bladed** (as in non-plectambonitacean Strophomenida).

Three important transverse structures bring variation to this simple theme. Only one of them is ever present in a single species and they are correlated with taxonomy. Where a shell wall is **endopunctate** it contains canals up to 100 μ wide, once occupied by extensions of the mantle, which run through the wall in a direction perpendicular to the surface (Fig.29). In life these canals ended just below the periostracum, in the primary layer. In fossil shells the canals are generally filled with either micrite or calcite cement. A shell is described as **impunctate** if canals are not present (Fig.30). In a third structure, giving rise to a **pseudopunctate** wall, the fibres of the secondary layer are bent upward in a series of superimposed cones to give a linear structure which lies perpendicularly or obliquely to the base of the primary layer (Fig.31, 32). In some shells the cores of the pseudopunctae consist of rods of granular calcite called **taleolae** against which the fibres are deflected (Fig.31). The pseudopunctae commonly extend from an expanded circular end, just below the inner surface of the primary layer, to the inner surface of the secondary layer where they can be seen as tubercles. From the base of the primary layer they are inclined toward the anterior.

These three major structures should not be confused with superficial perforations which are associated with the pattern of ribs, or with the bases of spines, or surface pits, which are seen as folds **(exopunctata)** generally restricted to the primary layer. In some shells a number of lamellose calcitic skirts were deposited by a periodically retracted mantle. **Spines,** on the outside of the shell, are hollow with the long axes of the calcite fibres arranged concentrically (Fig.32, 33). Though they are circular in cross-section and parallel-walled in longitudinal section, oval oblique sections are generally seen. In some shells a layer of coarsely prismatic calcite grows over the base of the secondary layer in areas where muscles are attached.

The secondary layer, at first sight, can be confused in thin section with the foliated structure in the bivalve shell (also calcite and commonly preserved), but the opposed units of the latter, looking like cross-strata, are diagnostic. During diagenesis the secondary layer is least affected and commonly shows little alteration when viewed in thin section. The more porous primary layer, with crystals which are slender and unprotected by organic sheaths, is readily destroyed and its previous existence may only be revealed by trails of iron oxide, clay, etc. defining the lost upper surface.

Certain taxonomic generalizations can be made as follows (A. Williams, personal communication, 1968):

endopunctate	*impunctate*
Enteletacea	nearly all Rhynchonellida
Terebratulida	Pentamerida
some Spiriferida	most Spiriferida
some Thecideacea	all Orthida (except for the Enteletacea and Gonambonitacea)
a few Palaeozoic Rhynchonellida	pre-Devonian Davidsoniacea

pseudopunctate (with or without taleolae)
all Strophomenida (except for Pre-Devonian Davidsoniacea)
Gonambonitacea

ZOANTHARIA

Into this subclass fall the four calcareous orders of corals, the Scleractinia or stony hexacorals (Mesozoic – Recent), the Heterocorallia, a small order containing two genera (Devonian–Carboniferous), and the Rugosa and the Tabulata (Palaeozoic). It is convenient to begin with the Scleractinia because our knowledge of their fabrics is much more definite than it is for the older orders.

Scleractinia (calcareous Hexacorallia)

The fundamental skeletal elements are aragonite fibres, united as spherulites in linear series, into **trabeculae** (Fig.37), or into laminar sheets of needles with their long axes transverse to the sides of the sheet (VAUGHAN and WELLS, 1943; J. W. WELLS, 1956).

The spherulites, called **sclerodermites**, have centres that are dark in transmitted light (VAUGHAN and WELLS, 1943, p.32). From them the long axes of the

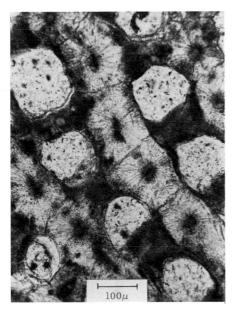
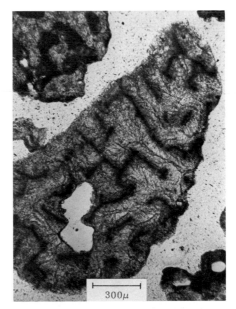

Fig.37. Section of trabeculae (running NNW–SSE) of the scleractinian *Siderastrea* (aragonite), with radiate sclerites, dark centres of calcification and empty pores. Slice. Pleistocene.

Fig.38. Section of scleractinian (aragonite) grain. The dark axes of the trabeculae are the centres of calcification. Radiate sclerites just visible in a few places. Slice. Recent.

aragonite fibres radiate in three dimensions (Fig.37). The centres (also called centres of crystallization or of calcification) were examined in a hermatypic and an ahermatypic species of modern Scleractinia by WAINWRIGHT (1964). In reflected light they are "white", but in transmitted light they are brown and, between crossed polars, dark. With the help of microradiodiagrams (X-ray absorption) he found indications that the centres have a content of aragonite from 5 to 8% less than the surrounding mass of fibres. Between crossed polars each fibre behaves as a single crystal except that when the axis is extinguished the edges are still bright. Transverse sections show sharp but irregular boundaries between fibres, not the polyhedral habit which would be expected of an aggregate of single crystals. X-ray microdiffraction analysis confirms that each fibre is a crystal aggregate with c axes having a preferred orientation within $10°$ of the axis of the fibre and with a and b axes randomly arranged. Finally, Wainwright found, with the electron microscope, that carbon replicas show that each fibre is, in fact, a bundle of acicular aragonite crystals.

Where the sclerodermites are arranged in a linear series making an almost continuous (commonly zig-zag, Fig.38) line of centres of calcification, the resulting trabecula has a circular cross-section equal to the equatorial section of a single sclerodermite. Fibres are directed upward in the direction of growth, radiating upward and outward from the axis formed by the centres. Simple trabeculae may be combined into a compound trabecula. A trabecula grows parallel to its axis and is youngest at one end and oldest at the other. In septa the trabeculae are combined into palisades or into fans. In so far as the trabeculae are tightly or openly packed the resultant septum has a laminar or fenestrate structure. Trabeculae may also grow separately as trabecular spines. The organic matrix was examined by WAINWRIGHT (1963).

The arrangement of fibres is rather different in the basal plate and the associated epitheca, dissepiments and stereome. There are no distinct centres of calcification, and the long axes of the fibres lie normal to the surface of the plate with a more or less distinct lamination parallel to the surface. There appears to be no fundamental difference between the fibres in the sclerodermites and those in the basal plates or the coenosteum (BRYAN and HILL, 1941; WAINWRIGHT, 1963, p.173). The orientation of the aragonite fibres seems to be determined by their invariable growth normal to the skeletogenic epithelium. In a sharply folded epithelium the fabric is trabecular, but in the flat epithelium which controls the plate, epitheca, dissepiments, stereome and coenosteal structures, the fibres are parallel.

Rugosa

The calcite fibres in the septa are oriented like those in the Scleractinia, as trabeculae, (Fig.39–41) but, in the tabulae and dissepiments, they lie parallel to each other and normal to the surface (HILL, 1936, 1956a). The trabeculae lie generally in the median

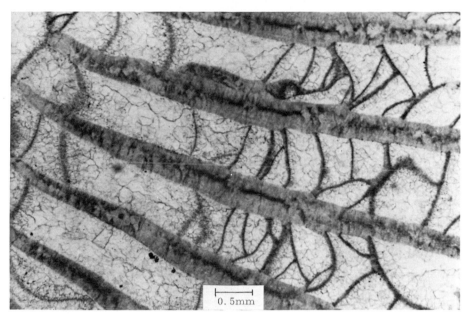

Fig.39. Transverse section through the rugose coral *Dibunophyllum*. The septa run WNW–ESE and have dark axial centres of calcification from which fibres radiate. The dissepiments run irregularly NNE–SSW and show a faint trace of fibres lying normal to their surfaces. The corallum is filled with calcite cement. Slice. Carboniferous Limestone.

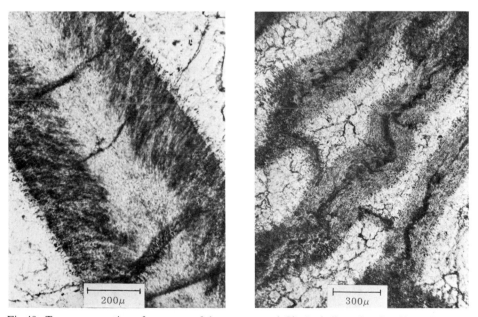

Fig.40. Transverse section of a septum of the rugose coral *Clissiophyllum*, showing fibres diverging from the axis. Slice. Carboniferous Limestone.

Fig.41. Transverse section of the rugose coral *Zaphrentis*. The septa run NE–SW and show dark centres of calcification. The dissepiments run NW–SE and also show centres of calcification. The corallum is filled with calcite cement. Slice. Carboniferous Limestone.

plane of the septum, projecting upward and inward toward the axis of the corallite where they are normal to the planar shape of the dissepiments. WANG (1950, p.182) found that the lengths of fibres in the sclerodermites vary from 50 μ to well over 500 μ. He also described the lamellar fabric of the basal and wall structures occurring in some Rugosa as consisting mostly, not of fibres, but of plates about 1 μ thick and 30 μ broad with subrounded shape (WANG, 1950, p.183) oriented with their two longer axes parallel to the lamellation, that is at right angles to the position of the fibres in the Scleractinia. In the caniniids and the plerophyllids, both in their septa and dissepiments, a lamellar structure is traversed by parallel fibres lying normal to the surface. This WANG (1950, p.185) called fibro-lamellar structure. KATO (1963), however, was convinced that the Rugosa have only two basic structures, the trabecular and the fibro-normal. Optic axes of the calcite fibres coincide with their long axes. Other structures, such as granular (ALLOITEAU, 1957) or zigzag, Kato believed to be secondary in origin and valueless for taxonomy.

Tabulata

Calcite fibres are the fundamental unit. In the tabulae, tabellae (as in *Michelinia*) and the dissepiments these are oriented with long axes perpendicular to the upper and lower surfaces of the plate (HILL and STUMM, 1956). In the rare septa, they are directed upward and outward from the axes of trabeculae. In general, the structures differ little from those in the Rugosa (KATO, 1968).

Heterocorallia

These Carboniferous corals, elongate and solitary, have fibres that are now calcite which lie perpendicularly to a weakly apparent growth lamellation (HILL, 1956b).

OCTOCORALLIA (ALCYONARIA)

Most species of alcyonarian bear discrete spicules of **sclerites**. Those in the rind are high-magnesian calcite (W. J. SCHMIDT, 1924b; CHAVE, 1954a; LOWENSTAM, 1964). LOWENSTAM (1964) has given X-ray data on the suborder Holaxonia (of the order Gorgonacea). In the Holaxonia, although the rind sclerites are calcite, the carbonate in the base (cemented to the substrate) and in the near-base axis consists of different proportions of aragonite and calcite depending on the temperature of the sea water. The percentage of aragonite varies from 0% to nearly 100% with increasing temperature. These mineralogical data are for bulk carbonate: the anatomical distribution of the two minerals is not known in detail. While many genera have characteristic sclerites, distinguishable by their size and form, the simpler forms of sclerite are common to a number of genera or orders. The basic type is a spindle-shaped rod, more or less pointed at both ends (Fig.42). With the microscope the sclerites show undulatory extinction unlike the single crystal sclerites of the Holothuroidea: Recent ones may be colourless or pink. Some idea of the variability of sclerital morphology and its taxonomic application can be gained from a consideration given below of the sclerites of the five spiculate orders (BAYER, 1956). The first four orders have sclerites more or less strongly sculptured, with spines, complicated warts and other protuberances.

Spiculate orders

   Stolonifera. Where present, sclerites are usually long slender, spinose rods. Cretaceous–Recent.
   Telestacea. Sclerites numerous, free, or partially or completely fused to form rigid tubes. Recent.
   Alcyonacea. (Known also as "soft corals"; the term "alcyonarian" has often been used in a restricted sense for this order.) Stout, spindle-shaped sclerites. Lower Jurassic–Recent.
   Gorgonacea. (Includes the families Plexauridae, the sea whips, and Gorgonii-

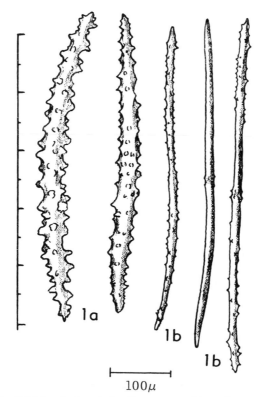

1a

1b

1b

1b

100μ

Fig.42. Calcitic sclerites of the Gorgonacea. (From BAYER, 1956.)

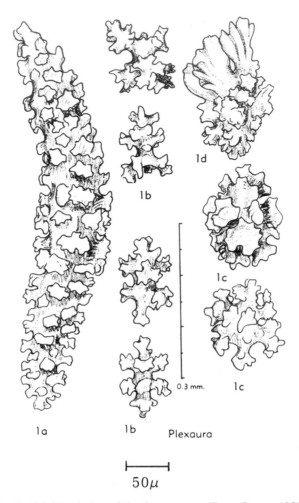

Fig.43. Calcitic sclerites of the Gorgonacea. (From BAYER, 1956.)

dae, the sea fans and sea feathers.) Sclerites immensely varied in form, either long, slender and spinose, or stout and coarsely warted, or capstans, or double clubs, crossed clubs, floral buds, flaming torch, or stellate thorns, etc. (Fig.42–45). Cretaceous–Recent.

Pennatulacea. Sclerites generally 3-flanged and smooth, rods or needles, rarely tuberculated, or small scales and plates. Cretaceous–Recent.

The order Coenothecalia (Cretaceous–Recent), without spicules, has "a massive skeleton of fibrocrystalline aragonite" (BAYER, 1956, p.193).

HYDROZOA

The calcareous orders in this class are the Milleporina and Stylasterina. They both construct skeletons composed of sclerites and trabeculae and were, for this reason,

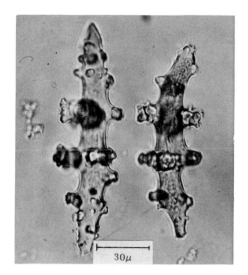
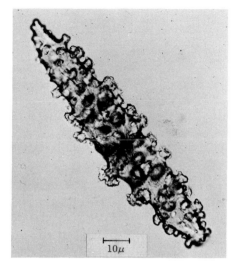

Fig.44. Calcitic sclerites of the Gorgonacea. Recent.

Fig.45. Calcitic sclerites of the Gorgonacea. Recent,

regarded earlier as belonging to the Anthozoa. Their soft parts indicate their affinities with the Hydrozoa. The calcareous skeletons contain either aragonite, calcite with aragonite or calcite. In the Milleporina all the species are warm water and are known only to contain aragonite. In the Stylasterina the mineralogy varies both taxonomically and with temperature, the ratio aragonite/ calcite increasing with temperature (LOWENSTAM, 1964).

STROMATOPOROIDEA

The Stromatoporoidea are a rather poorly understood group of carbonate-secreting organisms characterized by their habit of building layered coenostea. **Coenosteum** is the name given to the whole skeleton (Fig.46). The affinities of the group are not generally agreed. Some Palaeozoic and Mesozoic forms are comparable to the Recent order Milleporina (in the class Hydrozoa), especially *Stromatopora* (Ordovician–Permian) and *Stromatoporellina* (Jurassic–Cretaceous). Others may be close to the Recent family Hydractiniidae (also Hydrozoans), particularly *Actinostroma* (Cambrian–Carboniferous) and *Actinostromaria* (Jurassic–Cretaceous). Some stromatoporoids could be related to the algae, Corallinaceae. A live sponge, *Ceratoporella*, with a skeleton of fibrous aragonite and well developed astrorhizae has been described by HARTMAN and GOREAU (1966) from a Jamaican reef: this seems to have the closest affinity of any living organism to the Stromatoporoidea. The phylum is extinct and fossil coenostea are normally low-magnesian calcite.

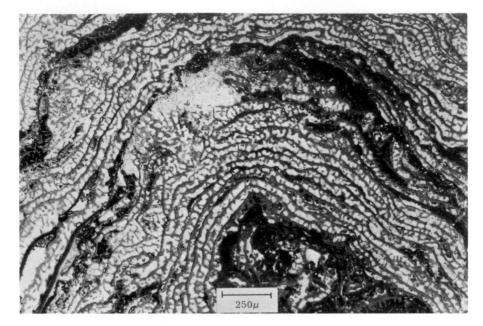

Fig.46. Vertical section through the stromatoporoid *Burgundia* showing concentric laminae and radial pillars. Dark areas are galleries filled with micrite. Other galleries are filled with calcite cement. Slice. Jurassic. (H. A. Nicholson coll.)

For the decade before the appearance of Stearn's paper in 1966, the established classifications of the Stromatoporoidea were those of LECOMPTE (1956) and GALLOWAY (1957). Galloway depended heavily on microscopic detail and his scheme would be appropriate in the study of grains in limestones were it not that skeletal and diagenetic fabrics were in places confused and the need for distinction between them not always appreciated. It is also not everywhere certain that Galloway distinguished between organic cellular tissue and crystal mosaics. Lecompte has not investigated the petrographic side in detail but has brought to bear his wide experience of sessile coelenterates and of reef ecology. He is cautious with regard to microfabrics and feels that they too often reflect merely conditions of preservation.

STEARN (1966) has been the first to attempt seriously a systematization of the intricate petrographic data apertaining to this hotchpotch of calcareous organisms. Regrettably he perpetuates the erroneous description of the dusty specks as "dark" and reviews possible origins such as iron sulphide, carbonaceous material or amino acid. In fact, for all that is known about them, they could be aggregates of finely crystalline carbonate or, as NICHOLSON (1886–1892) believed, pores. But Stearn is keenly aware of the need for more rigorous examination of the skeletal fabrics. He writes: "The small size and indifferent preservation of these microstructures seem to have discouraged objectivity and encouraged idealization so that they have been described in terms of what they might have been rather than what they are now." (STEARN, 1966, p.77). He gives a useful glossary of terms.

The skeletal walls

The skeleton is now calcite but may initially have been aragonite. Evidence bearing on this point is not conclusive. The material of which the skeleton is composed, commonly referred to as the **tissue**, consists either of overlapping dome-shaped **cyst-plates**, convex outward (Fig.47, 48), or of concentric **laminae** (Fig.52, 54) with various supports. Both of these structures lie with their flatter surfaces tangential to the surface of the coenosteum. They are supported and kept apart by more or less transverse, radial **pillars** (Fig.47, 48, 52, 53, 54). The spaces between these walls, now filled with cement or micrite, are called **galleries**. The skeletal material, seen in thin section by transmitted light, commonly has a brown colour, or has a dusty appearance which is due to a dense mass of translucent or opaque inclusions (specks) which are colourless by reflected light. It has been customary to speak of **primary** and **secondary** skeleton (tissue) in some cyst-plates and laminae, because some of them have a trilaminar internal structure (Fig.50). The axial or median part appears as a dark line by transmitted light and is called primary and forms a median lamina. It appears dark because it is finely crystalline. Enclosing the primary lamina there may be more coarsely crystalline, secondary tissue of a different character: this forms the two outer laminae.

Skeletal microstructure

Our understanding of microfabrics among the Stromatoporoidea is far from clear. The microfabrics themselves have mostly been quite inadequately described and little consideration has been given to the effects of diagenesis: the divorce of palaeontology from petrography is nowhere more complete than here. There is also an urgent need for an agreed terminology. The standard of description is very variable and commonly highly subjective. The dust of colourless, opaque inclusions has been described as dark or black by authors who failed to examine it in reflected light. Certain dark layers are referred to as pigmented, yet others as leached, where the difference appears to be simply a matter of crystal size and the darkening effect of a dense population of Becke lines. The whole subject is made exceedingly complex by the common occurrence of neomorphic fabrics.

Crystal fabric in the skeleton. The fabric of the skeleton, sharply distinct from the calcite in the galleries, may be built of fibrous crystals lying either transverse to the skeletal surfaces, or nearly parallel to the surfaces such that the fibres in pillars diverge upward from a median axis or plane giving rise to the so-called **water jet** structure seen in longitudinal section. The transverse structure, in pillars, is oriented radially at right angles to the axis of the pillar. Alignment of specks within the skeleton accentuates its fibrous structure. The fabric that fills the galleries, where it is sharply differentiated from the tissue, is commonly a typical calcite cement evolved by the inward growth of cement from the skeletal surfaces: alternatively it may be a detrital micrite. In some specimens the earlier cement crystals have grown in lattice continuity with transverse fibres in the adjacent tissue.

In places the fabrics of the skeleton and the gallery filling may not be clearly distinguishable. The entire specimen may be composed of a sparry calcite mosaic (Fig.51) that bears no discernable relation to the distribution of skeleton and gallery, because crystal boundaries cross impartially from the one to the other. Skeleton and gallery are separately recognizable only because of the presence of specks or brown colour in the skeleton.

Specks. The distribution of specks in the tissues is variable. Not only does it have, in some specimens, a linear pattern, but it may be irregularly patchy, and the blotchy appearance is described as **flocculent** (Fig.49, 50). Commonly the specks are concentrated into roughly circular (actually subspherical) masses surrounded by relatively clear tissue. Alternatively the patches of clear fabric are spherical and the surrounding fabric is dense with specks. The clear patches are regarded by many writers as one time voids.

Complications arise where fabrics that are normally characteristic of the skeleton appear in the galleries. Water jet structure has been described from galleries (STEARN, 1966, p.80). Some calcite mosaic in the galleries is rich in specks and STEARN (1966, p.79) has suggested, on no clear grounds, that the specks, or some organic precursor, "seem to have moved" into the calcite of the galleries. There is nothing unusual in a calcite cement dusted with inclusions: early cement mosaics are commonly so.

Compact fabric. Both LECOMPTE (1956, p.118) and GALLOWAY (1957, p.361) referred to a fabric they called **compact**, which is found in the Labechiidae and in the genera *Clathrodictyon* and *Actinostroma*, but which neither of them described. Galloway was the more expansive and referred to tissue which is "compact and homogeneous and flocculent", again without definition. Neither author gave any information on scale. STEARN (1966) described compact tissue as "composed of evenly distributed specks or evenly coloured calcite".

Maculate fabric. Another important fabric, recently surveyed by ST. JEAN (1967), was named **maculate** (Fig.49) by GALLOWAY (1957, p.363). NICHOLSON had called it "dotted or porous structure" also "minutely porous or tabulated" (1886–92, p.36, 90), LECOMPTE has "fibre squelettique cellulaire" (1952, p.263) or "cellular type" (1956, p.118), and ST. JEAN (1964, 1967) described maculae as spheres 12–25 μ in diameter. A careful reading of Galloway brings only partial clarification. He referred to tissue containing "spherical *light spots* which are surrounded by darker tissue" yielding a "spotted or maculate appearance". Alternatively "the dots do not show the white centres but appear only as *dark spots*" in grey tissue (italics are mine). The "maculae" he described as "fine" (200–300 μ), but the "dots" as "coarse" (30–60 μ) with, sometimes, a dark wall and light centre. The maculae, he wrote, are "not pores but dots" (GALLOWAY, 1957, p.363).

The above notes show how pressing is the need for accurate, unequivocal descriptions of fabric. The problem is made more difficult because, of the eighteen or so genera listed by Lecompte and Galloway together as having maculate or cellular fabric, nearly a half are accepted only by one author or the other. Lecompte's cellular tissue seems more widely interpreted than Galloway's macular.

Vacuolate tubulate fabric. GALLOWAY's (1957, p.361) third skeletal fabric is "compact and vacuolate, transversely fibrous or tubulate structure". A year earlier LECOMPTE (1956, p.118), in referring to the "cellular type", had written "the differentiation of subtypes of structure in this group . . . such as regular porous, spongy, irregular vacuolar, lacks value because these terms mostly reflect only conditions of preservation".

Latilaminae, astrorhizae. Both authors agreed on two other structures that may be discovered in thin sections. A **latilamina** is a group of laminae having lateral continuity but distinguished from the groups above and below by either an angular disconformity, a thin layer of mud, density of vertical stacking of laminae or other features indicating periodicity in growth. This layered appearance of the coenosteum is the most obvious feature by which a stromatoporoid is recognised in the field. On the surface of many stromatoporoids, or at certain levels among the laminae, there are **astrorhizae**, irregularly stellate or rootlike systems of canals. They have no proper walls but merge with the spaces between laminae. Some contain tabulae or curved cystose plates. The astrorhizal canals are probably more important than any other morphological feature in understanding the stromatoporoids and their relation to other organisms.

Palaeozoic stromatoporoids

Only for the Labechiidae were Lecompte and Galloway in agreement and this family is described below.

Labechiidae. Ordovician–Devonian. Coenosteum (Fig.47,48) laminar, massive, conical, columnar or fasciculate with or without an axial, cystose column. "Cystose" here is equivalent to Lecompte's vesicular and describes an arrangement of outwardly convex plates or domes stacked more or less regularly on one another

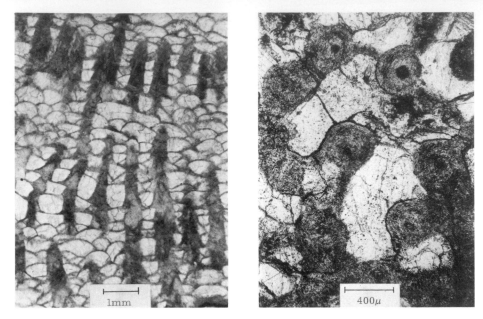

Fig.47. Vertical section of the stromatoporoid *Labechia* with cyst-plates convex outward. Pillars continue through several cyst-plates. Galleries filled with calcite cement. Slice. (H. A. Nicholson coll.)

Fig.48. Tangential section of the stromatoporoid *Labechia* showing cross-sections of thick pillars. Oblique sections through cyst-plates are visible as grey, dusty lines traversing the coarse calcite cement in the galleries. The heavy black lines are crystal interfaces. Slice. (H. A. Nicholson coll.)

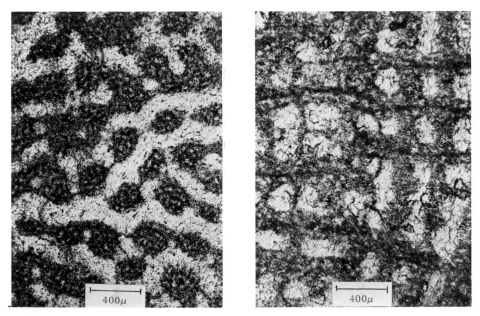

Fig.49. Transverse section of the stromatoporoid *Syringostroma*, showing cellular pillars in cross-section. Slice. (H. A Nicholson coll.)

Fig.50. Vertical section of the stromatoporoid *Syringostroma*. Laminae run N–S, pillars E–W trilaminar. Tissue cellular. Slice. (H. A. Nicholson coll.)

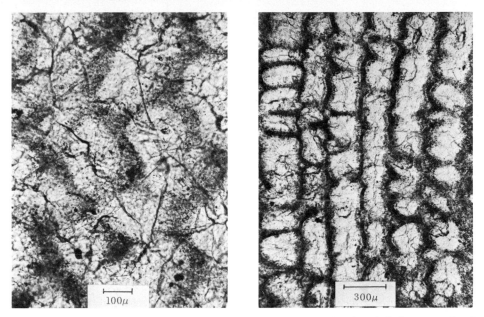

Fig.51. Oblique section of the stromatoporoid *Clathrodictyon* replaced entirely by a mosaic of sparry calcite. Trace of tissue remains as dusty bands. Slice. (H. A. Nicholson coll.)

Fig.52. Vertical section of the stromatoporoid *Clathrodictyon* with laminae running N–S and pillars E–W. Galleries filled with calcite cement. Slice. (H. A. Nicholson coll.)

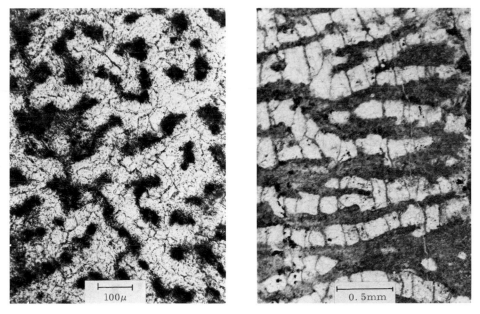

Fig.53. Tangential section through the stromatoporoid *Clathrodictyon* showing cross-sections of pillars. Galleries filled with calcite cement. Slice. (H. A. Nicholson coll.)

Fig.54. Vertical section of the stromatoporoid *Dehornella*. Thin laminae run N–S, thick pillars E–W. Galleries filled with calcite cement. Slice. Jurassic. (H. A. Nicholson coll.)

to give a cellular structure. The cyst-plates, forming a coarse vesicular structure, have a compact primary tissue, generally with inner and outer flocculent layers. They form latilaminae. Pillars may be absent or denticulate (as short spines on the upper surface of the plate, not extending to the plate above), or they may continue through several cyst-plates. STEARN (1966) recorded the cyst plates as having a thickness of 30–45 μ and being made of a single layer of compact tissue.

Other families. The classification of other families is not generally agreed and awaits clear distinction between diagenetic fabrics and those of taxonomic significance. Lecompte's classification, which takes into account both gross morphology and microfabrics, is probably superior to any other, but some workers, particularly those who support Galloway's work, have not wholly accepted it. With commendable restraint STEARN (1966) contented himself with redescribing a number of genera, after looking at the holotypes. Two descriptions are included here to illustrate some of the fabrics:

Syringostroma. Silurian-Devonian. Coenosteum (Fig.49, 50) an amalgamate structure within which may be distinguished thick, persistent pillars and continuous, generally thin, laminae. Pillars rod-like, traversing many laminae; cellular or reticulate. Laminae composed of dark or light microlaminae.

Clathrodictyon. Silurian. Coenosteum (Fig.51–53) composed of imperforate, continuous laminae, commonly undulant or crumpled, and short pillars that are confined to an interlaminar space. The pillars and laminae are a single uniform layer of compact, commonly speckled, tissue. Where the laminae are bent, the pillars extend from their downward inflexions and are cylindrical in form.

Mesozoic stromatoporoids

LECOMPTE (1956, p.128–138) included Jurassic and Cretaceous genera in the Stromatoporoidea. Two of these are given here as examples:

Burgundia. Jurassic. Coenosteum laminar, spheroid, domed or irregular (Fig.46). Well-defined continuous, thin laminae with pillars mostly short and straight, bifurcated at the top or expanded at one or both extremities, not crossing laminae. Other pillars are long and crooked, and cross laminae. The tissue is compact, microcrystalline or granular.

Dehornella. Upper Jurassic. Coenosteum small, encrusting (Fig.54). Continuous pillars, tortuous and very irregular in thickness, are joined by uneven cross-bars. Locally, pillars are restricted between laminae. The pillars have a dark core that "undergoes extinction in polarised light".

FORAMINIFERIDA

Following Wood's (1949) work on the structure of the test wall, the separation of the Foraminiferida into suborders, and of the Rotaliina into superfamilies, is now made primarily on the basis of the mineralogy and crystal fabric of the wall (as seen with the light microscope), the arrangement of lamellae, and the presence or absence of perforation (LOEBLICH and TAPPAN, 1964b). Of the five suborders, four are calcareous (all but two families being calcitic). They are the Textulariina (wall not always calcareous), Fusulinina, Miliolina and Rotaliina (Table II). The fifth, the Allogromiina, is membraneous and pseudochitinous and is not therefore considered here. Longest diameters of apparently discrete crystals vary generally from 0.5 to 5 μ. The calcareous test wall may have one of six fabrics: agglutinated, microgranular, porcellaneous (finer than microgranular, brown in transmitted light), radial-fibrous, spiculate, monocrystalline.

Attempts to correlate structures seen under the electron microscope with those seen with the light microscope have not yet been widely successful, though details are interesting (HAY et al., 1963). There are signs, however, that the existing classification based on wall structures suffers from some fundamental defects, born of too loose an interpretation of the optical data. A paper by TOWE and CIFELLI (1967) on the ultrastructure of the test wall contains, in addition to new and significant data from electron microscope studies, an essay on crystal fabrics and calcification which lucidly sets the stage for important new developments in the

TABLE II

CLASSIFICATION OF THE FORAMINIFERIDA BY FABRIC AND MINERALOGY, WITH THE TERMINOLOGIES OF LOEBLICH AND TAPPAN (1964a–c), AND WOOD (1949)

Loeblich and Tappan	Wood	Test wall
Textulariina	Agglutinating	Mineralogy varied, agglutinated
Fusulinina	Fusulinidea	Microgranular calcite
Miliolina	Porcellanea	Microgranular calcite, finer than Fusulinina, brown in transmitted light
Rotaliina	Hyalina	Radial fibres, perforate calcite *or* one to several large, perforate calcite crystals *or* granular, perforate calcite *or* concentric spiculate calcite *or* radial fibres, perforate aragonite: all walls are lamellar except in monocrystalline or spiculate forms

understanding of wall structure. This essay has important implications far beyond studies of the Foraminiferida, in the general area of the interpretation of skeletal petrography.

Towe and Cifelli emphasized the distinction between optical evidence for the crystallographic structure of the test walls, obtained directly with the polarizing optical microscope and X-ray diffractometer, and the morphological interpretation of that evidence in terms of crystal shape and its orientation. They proposed that the terms "radial" and "granular" be confined to their optical crystallographic senses and divested of all morphological implications. Morphological interpretation of the optical data for the Rotaliina has in the past led to the view that the multicrystalline wall is composed, either of crystals elongated perpendicularly to the surface, the elongation being parallel to the c axis, or of equant crystals randomly oriented, with the randomness including both shape and crystallography (Fig.66). This scheme is not supported by an analysis of the ultrastructure and, indeed, this shows few obvious relationships between optics and morphology. Electron photomicrographs of, for example, *Ammonia beccarii*, heretofore regarded as having a radial-fibrous perforate wall, show an array of plate-like structures arranged more or less parallel to the surface (Fig.68–70). Towe and Cifelli suggested that the plates are not separate crystals but are parts of larger crystals and mark the intersections of growth lines with the plane of the replica. The supposed fibrous structure turns out to be an artifact caused by the superimposition of numerous pores within the thickness of the section: it is not visible in peels of polished etched surfaces.

Careful examination of the granular (microgranular) wall with polarized light, for example in *Nonion labradoricum*, reveals that no crystals have c axes normal to the test surface. In a truly random array some crystals would show this orientation and its absence indicates that the apparently random orientation is an illusion. There must be, in fact, a preferred orientation such that c axes normal to the surface are excluded. Towe and Cifelli proposed an ingenious solution, whereby the steric configuration of the polypeptide chains in the organic substrate would act as a template for the nucleation of $(h0\bar{h}1)$ rhombohedral faces of calcite (see also WEBER, 1969, on echinoderms). In this way a preferred orientation would be maintained but the c axis could vary within a stereographic girdle, an interpretation which is supported by the etch patterns of the crystals. The resultant effect would be a granular mosaic with no apparent preferred orientation when viewed between crossed polars.

The porcellaneous structure, as seen with the electron microscope, differs totally from the radial and granular structures in being composed of a truly randomly oriented three-dimensional array of needles and laths. Towe and Cifelli gave details of this wall structure in *Spiroloculina depressa*, *Quinqueloculina seminulum* and *Q. subrotunda*.

Finally, all the tests examined by these authors had, on their outer surfaces, a veneer of specially oriented crystals differing in arrangement from the underlying wall structure (Fig.73, 74).

As a result of the work by Towe and Cifelli, the descriptions of wall fabrics given below must be treated with some caution and scepticism until such time as an improved system is devized which is more in line with current studies of ultra-structure.

The mineralogies of about 155 of the commoner Recent Rotaliina (Table III) have been examined by BLACKMON and TODD (1959). Each is either calcite, high-magnesian or low-magnesian, or aragonite, these minerals never being mixed in the same test. The content of $MgCO_3$ in the calcite was also measured (Table III) and falls mainly in two ranges, 0–5 mole % and > 10 mole %, the low-magnesian and high-magnesian calcites of Chave (p.235). Much more work in this field remains to be done and the authors stressed that their results can only show "trends and probabilities" when applied to classification.

Textulariina

The agglutinated wall consists of varying amounts of organic and mineral cement (both metabolic in origin) and of mineral grains derived from the sedimentary substrate. Extreme examples are almost pure microgranular calcite cement (crystal diameters 5–10 μ) and it may be difficult to distinguish these walls from those in some of the Fusulinina or recrystallized Miliolina, which both have microgranular walls. Walls composed largely of micrite are difficult to see if the test is embedded in similar micrite. The position of the wall can sometimes be established where coarser debris is absent from a narrow zone adjacent to the sparry cement that fills the chamber (Fig.55). Some agglutinated walls are pierced by fine tubules, 1–4 μ in diameter: these may be irregular or branching or with their long axes normal to the test wall (Fig.56). They end blindly at an inner organic membrane and are open at the outer surface of the wall. The mineral grains reflect always the composition of the substrate, but their selection is partly dependent on taxonomy. So also is their orientation in the wall. Selection of grains is influenced by their size, density and surface texture. Grains commonly employed are quartz, various heavy minerals, clay minerals, carbonates, and organic debris including such things as smaller foraminiferids, coccoliths, prisms of *Inoceramus*, radiolaria and sponge spicules. The cements may be carbonates or other salts or hydroxides of calcium, iron or silicon. Some walls possess a rough layering, having a fine outer layer, coarser middle layer and a smoothly finished inner surface.

Fusulinina

This suborder includes the three extinct Palaeozoic superfamilies Parathurammin-acea, Endothyracea and Fusulinacea.

TABLE III

WALL FABRIC IN THE SUPERFAMILIES[1] OF THE ROTALIINA, WITH $MgCO_3$ CONTENTS[2] OF SOME FAMILIES

Radial calcite or aragonite: perforate		bilamellar	Granular calcite: perforate monolamellar or bilamellar	Spiculate calcite	One calcite crystal (less usually two or three): perforate
monolamellar	rotaliid lamellar				

Nodosariacea
  Nodosariidae L
  Polymorphinidae L
Buliminacea
  Buliminidae HL
Discorbacea
  Discorbidae H
Some Spirillinacea
  (Rotaliellidae)
Robertinacea
(the only aragonite
superfamily;
Ceratobuliminidae,
Robertinidae)
	Rotaliacea
	  Rotaliidae HL
	  Calcarinidae H
	  Elphidiidae L
	  Nummulitidae H
		Globogerinacea
		  Heterohelicidae H
		  Globorotaliidae L
		  Globigerinidae L
		Orbitoidacea
		  Amphisteginidae L
		  Planorbulinidae H
		  Cymbaloporidae L
		  Homotrematidae H
			Cassidulinacea
			(monolamellar)
			are Nonionidae,
			Caucasinidae,
			Pleurostomellidae,
			bilamellar are
			Osangulariidae,
			Anomalinidae)
			  Cassidulinidae L
			  Nonionidae L
			  Anomalinidae HL
				Carterinacea	Some Spirillinacea
					(Spirillinidae)
					  Spirillinidae H

[1] LOEBLICH and TAPPAN (1964a–c). [2] BLACKMON and TODD (1959) with names changed to agree with the *Treatise*.
L = low-magnesian calcite, H = high-magnesian calcite.

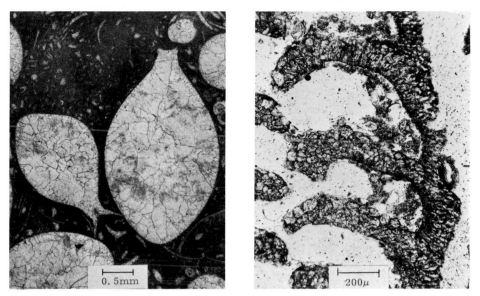

Fig.55. The agglutinating foraminiferid *Saccammina*. The wall can just be discerned along the right margin of the largest test where there is a dark micritic area in which skeletal debris is absent. The tests are filled with calcite cement and embedded in biomicrite. Slice. Carboniferous Limestone.

Fig.56. An agglutinating foraminiferid with a perforate wall. Slice. Recent.

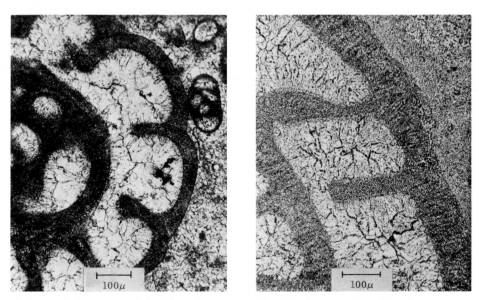

Fig.57. Endothyracea: radial section of test. The walls of finely crystalline calcite show traces of layered structure. Chambers filled with calcite cement. Slice. Carboniferous Limestone.

Fig.58. Endothyracea: section through test showing perforate walls with trace of layered structure. Slice. Carboniferous Limestone.

Parathuramminacea. These have a simple wall of finely granular calcite, only rarely perforate.

Endothyracea. These have either simple walls of finely granular, equant crystals of calcite, or two-layered walls, one layer being finely granular, the other having fine radial fibres of calcite. The granular layer may be the outer or inner layer. Commonly there are two layers, otherwise one or rarely three. Walls in some genera are perforate (Fig.57, 58). Some walls bear arenaceous material in a calcite cement.

Fusulinacea. The walls are perforate, of finely granular equant crystals of calcite and may be single-layered or multilayered (Fig.59–61). Secondary deposits of calcite occur as chomata, parachomata, tectoria or axial fillings. In primitive forms a thin spirotheca (tectum) is later covered, above and below, by tectoria (layers of secondary calcite). In more advanced forms tectum is supplemented by a variety of other layers, for example a transparent diaphanotheca or a thick honeycomb-like keriotheca.

 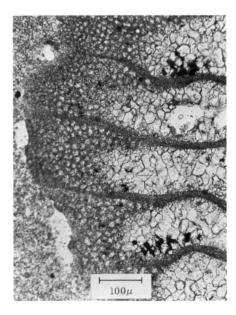

Fig.59. Perforate walls in one of the Fusulinacea. Chambers filled with calcite cement. Slice. Permian.

Fig.60. Perforate wall in one of the Fusulinacea. Chambers filled with calcite cement. Slice. Permian.

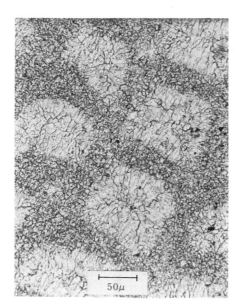
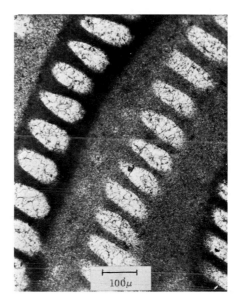

Fig.61. Section of walls in one of the Fusulinacea showing micron-sized equant calcite crystals. Chambers filled with calcite cement. Slice. Permian.

Fig.62. Miliolina: radial section through an Eocene test showing the imperforate, finely crystalline walls. Chambers filled with calcite cement. Slice.

Miliolina

Examined with the light microscope, the walls are of finely granular, apparently equant crystals of high-magnesian calcite and commonly have a pseudochitinous lining with or without some agglutinated material on the outer surface. They are imperforate, at least in the post-embryonic stages (Fig.62). The wall is brownish in transmitted light when fresh and shows first order grey polarization colours whatever its thickness, with no detectable preferred orientation, except for sporadic tiny patches. The positions of the patches vary in different genera and in parts of a single test (Fig.63). In *Marginopora vertabralis* there is an oriented fabric of slightly elongate crystals around the chambers parallel with the length of the walls (WOOD, 1949, p.235). Species of five families analysed by BLACKMON and TODD (1959) are all of high-magnesian calcite. The families are the Fischerinidae, Nubeculariidae, Miliolidae, Soritidae and the Alveolinidae.

Rotaliina

Examined with the light microscope (Fig.64–74), the wall fabric may consist of radially arranged calcite or aragonite fibres, or microgranular calcite, or calcite spicules commonly oriented parallel to the periphery, or a single crystal of calcite.

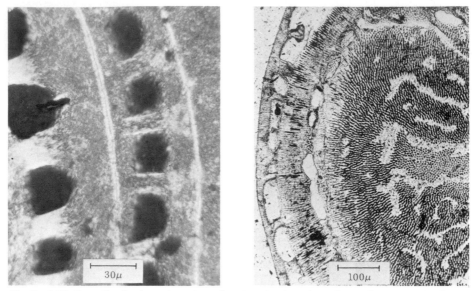

Fig.63. Miliolina: distribution of patches of preferred orientation of crystals in the wall of a test (light areas). Most of the wall shows random extinction. Slice with crossed polars. Recent

Fig.64. Rotaliina: radial section of a test. Perforate walls seen in transverse and longitudinal section. Slice. Recent.

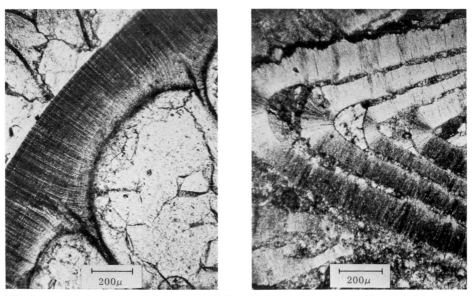

Fig.65. Radial section through a nummulitid showing perforate wall. Embedded in calcite cement. Slice. Eocene.

Fig.66 Radial section of a nummulitid. The distribution of polarization colours marks the preferred orientation of optic axes normal to the wall surfaces but the impression of a radial-fibrous structure is illusory. Embedded in biomicrite. Slice. Tertiary.

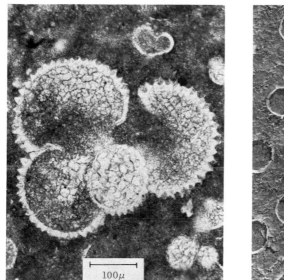

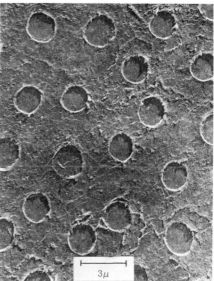

Fig.67. Globigerinidae: section of a spinose test showing perforate walls. Chambers filled with geopetal micrite and calcite cement. Embedded in micrite. Slice. Tertiary.

Fig.68. Polished and etched tangential section through the wall of *Ammonia beccarii* (Rotaliina) showing pores and plan view of plate-like units. Replica. Electron mic. Recent. (From TOWE and CIFELLI, 1967.)

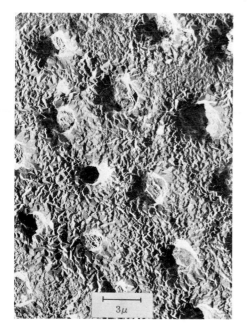

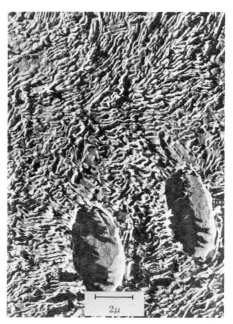

Fig.69. Ventral surface of the rotaliid *Ammonia beccarii* showing rhombic form of the calcite units. Replica. Electron mic. Recent. (From TOWE and CIFELLI, 1967.)

Fig.70. Polished and etched section oblique to the wall of *Ammonia beccarii* showing pore canals and plate-like calcite units. Replica. Electron mic. Recent. (From TOWE and CIFELLI, 1967.)

The calcite is high-magnesian or low-magnesian (Table III). All but the spiculate wall are perforate. The radial-fibrous structure, in some species at least, is an illusion as shown by studies with the electron microscope (p.40). All tests have lamellar walls except for the monocrystalline and spiculate forms. The structure of the outer veneer has been described (Fig.73, 74; TOWE and CIFELLI, 1967).

All the lamellar tests grow by the addition of a new layer that covers the whole of the previously existing test and is extended forwards to make the wall of a new chamber. Thus the wall of the latest chamber is continuous with an outer layer that encloses the whole of the older part of the test. As each new chamber grows the whole of the exposed part of the test is again secondarily thickened. The wall of the new chamber can grow in three ways that lead to three distinct types of lamellar arrangement.

A **monolamellar** wall is formed by the growth of a single new layer, which does not grow over the septum. A **rotaliid lamellar** wall arises where the new layer extends also over the septum, causing it to be two-layered. A **bilamellar** wall is made originally of two layers, where it forms the new chamber, but only its outer layer continues over the remainder of the test. It does not cover the septum, which keeps its primary bilamellar form.

In the Carterinacea the wall consists of fusiform spicules of calcite, elongate parallel to the c axis, with long axes commonly parallel to the periphery, embedded in a groundmass of calcite.

Details of wall fabric for the ten superfamilies in the Rotaliina, so far as these are known, are given in Table III. Not all the genera in these superfamilies have been checked for wall fabric and some redistribution may later be necessary.

Identification of Recent foraminiferal debris

MATTHEWS's (1966) experiments in the recognition of silt-grade debris gave the following details.

The porcellaneous test (Miliolina: of microgranular calcite) yields platy particles, yellow brown to greyish orange. The texture is micritic. Between crossed polars the wall is dark grey to dull brown. Pores, if present, are 5–20 μ in diameter.

The radial hyaline wall (Rotaliina) gave platy grains, colourless, light grey or pale yellowish grey. With polars crossed a pseudo-uniaxial figure may be visible in large particles: smaller fragments appear dark. The grains have no discernable microtexture ("smooth"). Where perforate the pores are 2–10 μ across.

Granular hyaline fragments are platy, colourless to pale yellow, with micritic texture. With crossed polars "the plates appear to be made of interlocking 5–10 micron crystals of irregular shape".

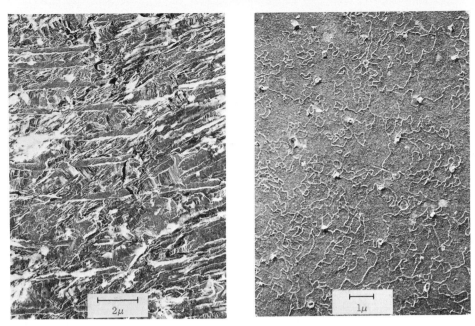

Fig.71. Broken section through wall of *Lenticulina calcar* (Rotaliina) showing crystal structure running obliquely to the pore canals, also residual organic matrix (white). Replica. Electron mic. Recent. (From TOWE and CIFELLI, 1967.)

Fig.72. Outer surface of *Lenticulina calcar* showing pores and sutures. Replica. Electron mic. Recent. (From TOWE and CIFELLI, 1967.)

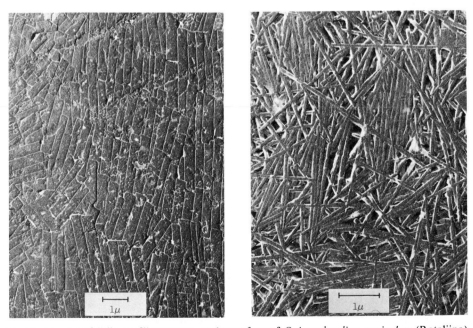

Fig.73. Veneer of "tile-roof" structure at the surface of *Quinqueloculina seminulum* (Rotaliina). Replica. Electron mic. Recent. (From TOWE and CIFELLI, 1967.)

Fig.74. Veneer of disorganized calcite needles at the surface of the rotaliid *Quinqueloculina seminulum*. Replica. Electron mic. Recent. (From TOWE and CIFELLI, 1967.)

ECHINODERMATA

Echinoidea, Crinoidea, Asteroidea, Ophiuroidea, Blastoidea

Our knowledge of the endoskeleton in these classes, summarized by RAUP (1966b), is uneven and comes largely from the Recent Echinoidea. The structure in Recent tests is mostly fenestrate, built of high-magnesian calcite, with a very high porosity that may exceed 50% (Fig.75). With few exceptions the separate skeletal elements (plates, spines, sclerites) behave in the light microscope as single crystals (Fig.76). Yet SCHMIDT (1929) argued that, theoretically, the crystals should be composite, and the results of X-ray analysis by GARRIDO and BLANCO (1947) and NISSEN (1963) certainly suggest that the optically uniform crystal is really a mass of submicroscopic crystals with their c axes almost perfectly aligned and the a axes aligned to within only a few degrees. Working with the transmission electron microscope, TOWE (1967) noted that the topography of natural and fractured surfaces of modern echinoid plates, tubercles and spines indicated that the interior of the stereom is a single crystal, but that it is covered by an outer layer of polycrystalline calcite having a preferred orientation of crystallites. In the arrangement of inter-connected pores there is a varying amount of order in pore size and shape, and there is always some deviation from geometrical perfection.

Ancient echinodermal elements are mostly low-magnesian calcite, yet they still behave as large single crystals. If the calcite was originally high-magnesian, then the Mg^{2+} must have been lost either by incongruent dissolution or by calcitization (p.337). The slow rate of lattice diffusion makes incongruent dissolution hard to justify. On the other hand, calcitization implies that there was a total exchange of cations and CO_3^{2-} groups with the pore water during diagenesis and that the associated pores in which the reactions took place were so small that structural detail was preserved. Under these conditions the means by which the original orientation of the c axis was preserved is not obvious. Is it possible that the new low-magnesian calcite nucleated epitaxially on pre-existing low-magnesian calcite cement which in turn preserved the original crystallographic orientation of the skeleton?

The c axis is nearly always related precisely to the morphology and symmetry of the animal (RAUP, 1959, 1960, 1962a, b, 1965, 1966a; RAUP and SWAN, 1967). Where the skeletal element has a strongly curved surface the orientation of the c axis may be related to the curvature. There is here a regular and continuous change of orientation, giving an orderly undulose extinction in the crystal. In a radial plate of a blastoid this may involve a total change in the disposition of the c axis amounting to 30°. This apparent curvature of a crystal lattice is understandable if the visible crystals are regarded as aggregates of much smaller crystals which vary progressively in their orientation. In the echinoid spine (Fig.78, 79) c is parallel to the long axis. Where spines are spatulate or fan-shaped, as in some

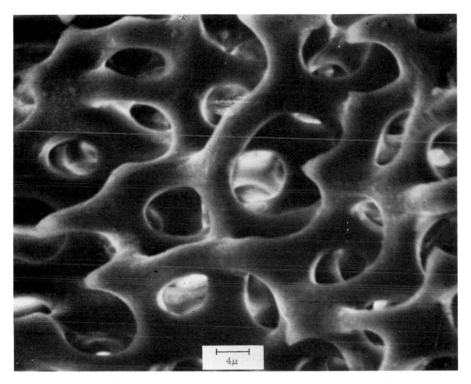

Fig.75. Structure of the spine (high-magnesian calcite) of the asteroid *Acanthaster planci*. Scanning electron mic. Recent. (Courtesy of J. N. Weber.)

cidaroid echinoids, the c axis varies regularly in orientation across the spine so that it is axially parallel to the central axis and also but superficially parallel to the diverging external surfaces. In the coronal echinoid plates c is generally either normal to the large surfaces, or nearly tangential to them and parallel to the plate columns (meridional). In both cases it lies in a plane that is radially disposed to the animal's symmetry. Some cidarids and arbaciids have c normal to the ambulacral plate but tangential to the interambulacral plate. In one species of echinoid the inclination of c to the plate changes progressively along any one column of coronal plates (Fig.81), from the oldest to the youngest, from about 80°, down to nearly 0° (tangential) and up again to 60°. In most echinoids the symmetry of the apical system is related to the bilateral symmetry of the pluteus larva and not to the adult morphological symmetry. The c axis tends to be transversely oriented (Fig.82) and is exceptional in that it does not lie in the radial plane of the main axis of the test.

In crinoids and blastoids both kinds of orientation are known, c normal to plate or tangential-meridional. There is a suggestion of a tendency in the Asteroidea and Ophiuroidea for c to have a dorso-ventral alignment. In the columnal of the crinoid stem the axis of canal is parallel to c.

The distribution of the Mg^{2+} in the calcite is similarly complex. WEBER

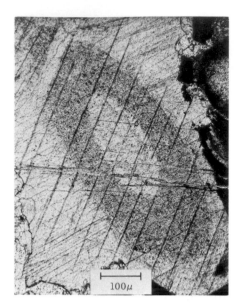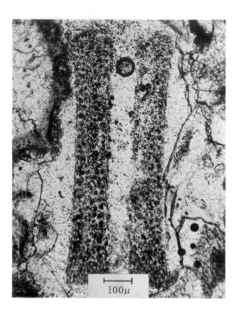

Fig.76. Oblique section of crinoid columnal, the wall being distinguished by the high content of dusty inclusions. The columnal together with the enclosing calcite cement forms a single crystal in so far as the area extinguishes uniformly and is traversed by continuous cleavages. Slice. Carboniferous Limestone.

Fig.77. Longitudinal section of a crinoid columnal, the wall marked by the content of dusty inclusions. The enclosing cement is syntaxial with the single crystal which forms the columnal. Slice. Carboniferous Limestone.

(1969) made about 1,800 analyses of parts of the tests of the living classes Echinoidea, Crinoidea, Asteroidea, Ophiuroidea and Holothuroidea. His data show that the content of Mg^{2+} varies (*1*) within a single skeletal component (e.g., an echinoid tooth, (*2*) within a single test (e.g., coronal plates versus spines), and (*3*) within homologous components in a single population. SCHROEDER et al. (1969) found a range of 3–43 mole % Mg^{2+} in echinoid teeth alone (p.235).

Identification of echinoderm particles. Recent fragmentary particles were found by MATTHEWS (1966) to have a characteristic porous structure with pores 25 μ in diameter. Silt-sized fragments commonly have concave edges, the remains of pores (Fig.80). Particles are colourless and extinguish uniformly. Fossil echinodermal grains, embedded in calcite cement or micrite (Fig.83, 84), are distinguished by their dusty appearance and uniform extinction, even where particle shape is no guide. The distribution of dusty material (partly micrite, or possibly inclusions, in the pores) may appear random or may reveal the fenestral pore pattern (Fig.78). Pore-filling cement is commonly syntaxial with the host skeleton, rendering the pores difficult to see or invisible. Grains are commonly enclosed in a syntaxial

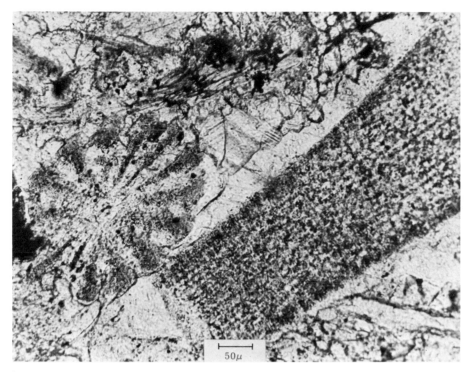

Fig.78. Transverse and longitudinal sections through echinoid spines in a matrix of biosparite. Slice. Jurassic.

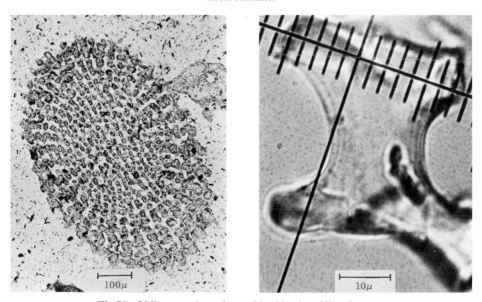

Fig.79. Oblique section of an echinoid spine. Slice. Recent.

Fig.80. Fragment of echinoderm with concave pore walls. Recent. (From MATTHEWS, 1966.)

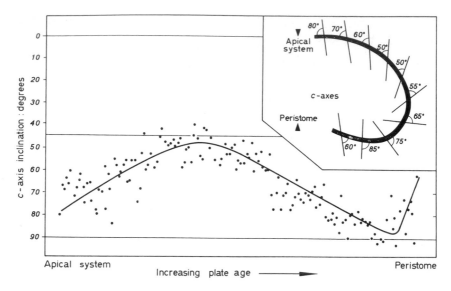

Fig.81. *C*-axis inclinations in the ambulacral plates of one specimen of the regular echinoid *Hemicentrotus pulcherrimus*. (After RAUP, 1960.)

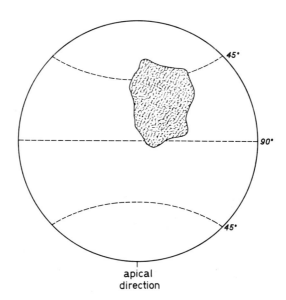

Fig.82. Stereographic projection of *c*-axis orientations in the ambulacral plates of *H. pulcherrimus*. (After RAUP, 1960.)

cement rim (Fig.76, 77) or in a syntaxial rim of neomorphic calcite which has replaced the embedding micrite (Fig.345).

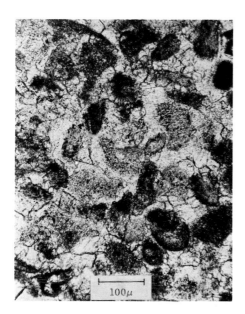 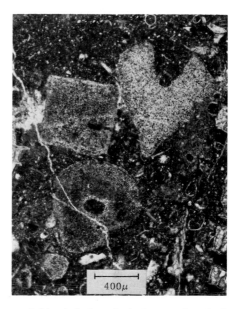

Fig.83. Grains of crinoidal columnals and plates, revealed by their dusty inclusions, each embedded in a syntaxial crystal of calcite cement, in a biopelsparite. Slice. Carboniferous Limestone.

Fig.84. Grains of crinoidal columnals embedded in a biomicrite. Slice. Silurian.

Holothuroidea

The skin, muscles and organs are, in most genera, supported by a variety of **sclerites**, bodies appearing under the light microscope as single uniformly extinguishing crystals of calcite, and shaped like microscopic anchors, hooks, wheels, tables, baskets, spectacles, rods, racquets, ladles, discs, plates or rosettes (Fig.85–88). In their exhaustive monograph, FRIZZEL and EXLINE (1955) remark: "Fossil sclerites have no definitive structures or characters, of which we are aware, that would serve to distinguish them with absolute certainty from remains of some other organisms. Their identification depends upon their similarity to the sclerites of Recent forms, their dissimilarity to parts of other fossils and, in some cases, their size range." Special care should be taken to avoid confusing them with the sclerites of certain other groups, such as some ophiuroid plates, some of the alcyonarian (Octocorallia) spicules, and the larval plates of pelmatozoans and, probably, of echinoids: in sponge spicules the axial canal can be seen. Holothurian sclerites range in greatest dimension from 0.05–1 mm. Their fabric is often perforate, occasionally laminar. The relation of the c axis to the mature sclerite is variable. A holothurian may have a variety of sclerites or only one type, or none.

  Owing to the fact that most sclerites in a sediment are not obviously related in origin to one animal, a dual classification is necessary to allow for those that

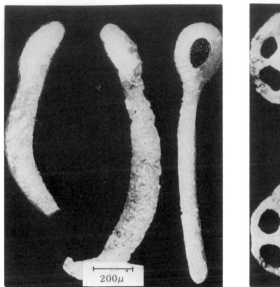 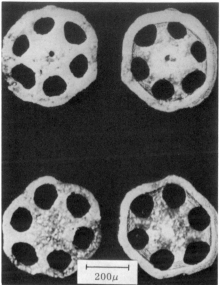

Fig.85. Holothurian sclerites, each a single crystal of calcite. Mississippian. (From GUTSCHICK, 1959.)

Fig.86. Holothurian sclerites, each a single crystal of calcite. (From GUTSCHICK, 1959.)

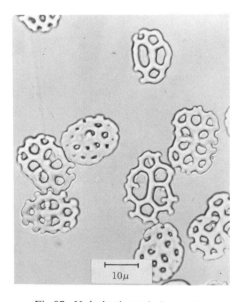 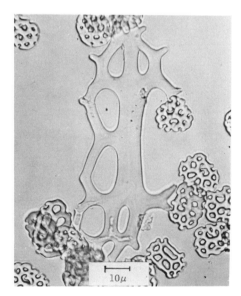

Fig.87. Holothurian sclerites, each a single crystal of high-magnesian calcite. Recent.

Fig.88. Holothurian sclerites, each a single crystal of high-magnesian calcite. Recent.

are known to belong to one species of animal (as in Recent species) and those that are found singly. The classification into nine families for disjunct sclerites (below) by FRIZZEL and EXLINE (1955, pp.59–155) puts them in a logical order based on morphology and includes some useful detailed description.

Families of disjunct holothurian sclerites

Stichoptidae. Simple or multiradiate rods, or modified rods; straight or curved simple rods (rods and C-rods); simple rods with eyes or discs at one or both ends (eye-bars or spectacles, spoons, racquets); and triradiate, tetraradiate, and multiradiate rods.

Calclamnidae. Perforate plates, rarely multilayered; relatively thin, flat or concavo-convex; rectangular, circular, hexagonal, or irregular in outline; without socket; perforations never denticulate.

Etheridgellidae. Imperforate discs and buttons; with pseudospire in type genus; surface sometimes with a pattern of polygonal depressions.

Achistridae. Hooks; with eye, shank and spear.

Calclyridae. Lyres; consisting of a central shaft, short neck, and two marginal arms; arms join shaft at base and below neck, with irregular perforations between arms and shaft.

Priscopedatidae. Tables; consisting of a perforate disc with spire and/or stirrup; disc circular to subcircular, irregular, cruciform, subquadrate, or sub-hexagonal; frequently with four central perforations; perforations elliptical to circular or polygonal; spire single or quadripartite, low to high, smooth or spinose; stirrup 2-footed, 3-footed or 4-footed.

Theeliidae. Concavo-convex wheels; each consisting of an outer rim, an inner central portion, and spokes connecting rim with central portion; interspoke spaces sometimes filled with secondary calcite that gives a false impression of an imperforate hollow hemisphere.

Synaptitidae. Perforate oval plates, flat or concavo-convex, typically with a socket at small end on concave side; socket single, double or absent; perforations usually denticulate, with fine teeth on margins.

Calcancoridae. Anchors; with shank, stock, and flukes; stock smooth or denticulate; flukes double or triple, smooth or with teeth on lower margins.

The thalli of some benthonic algae become calcified either wholly or in part. Calcite, generally high-magnesian, or aragonite is precipitated, either intracellularly as a calcification of the cell wall (Rhodophyta), or extracellularly (perhaps some intracellularly), in or around certain tissues, as a by-product of the plant's photo-synthesis (Chlorophyta). The different groups of calcareous algae can be recognised by the shapes and dimensions of their calcareous parts, such as cells, walls, cell layers, and tubes with various types of branching, and, in the Dasycladaceae, Charophyta and some Rhodophyta, by the mode of occurrence of reproductive bodies (sporangia of the palaeontologists). The preservation of detail is variable and depends primarily on the method and degree of calcification, and this varies both taxonomically and with the age of the plant at the time of death. Although some studies have been made of the mineralogy and chemistry of modern calcified thalli, scarcely any work appears to have been done on the crystal fabric of benthonic calcareous algae, Recent or ancient. It does seem, from a study of thin sections and photomicrographs, that both the calcite and aragonite are of micritic crystal size. It may well be that, in fossil algae, or indeed in grains of dead calcareous algae found today on the sea floor, the fabric has undergone alteration and is now coarser than it was in life. Some observations by the author on the codiacean *Halimeda* in the Bahamas show that in the fresh plant the aragonite is a compact mass of needles with no obvious orientation except in the immediate vicinity of the tubes where there is a lining about 2μ thick (Fig.93, 94). This consists of crystals with c normal to the wall of the tube and showing a faint suspicion of prismatic structure elongate parallel to c. After death, the fabric may be progressively replaced by coarser aragonite (p.384). Recent aragonite needles, as illustrated by CLOUD (1962a) and LOWENSTAM (1963), have dimensions mainly between $0.2 \times 3 \mu$ and $0.8 \times 7 \mu$. The fossil calcareous algae are now calcite and exhibit roughly equant crystals comparable in diameter with FOLK's (1959) micrite $(1-4 \mu)$.

Data on the primary mineralogy is derived almost entirely from living forms. There is at present no way of deciding whether the calcitic micrite in a fossil was originally calcite or aragonite.

The families of calcareous algae that are important as limestone builders are shown in Table IV. Mineralogy is taken from CHAVE (1954a) and JOHNSON (1961a), or inferred from the latter's data on $MgCO_3$ content, on the assumption that amounts of 3% or more in the total carbonate implies a calcite lattice.

Corallinaceae

Preservation of detail is particularly good because high-magnesian calcite (in Recent forms) is deposited both within and between cell walls. The thallus (Fig.89) is differentiated into two parts, the **hypothallus** that is built of larger cells than the

TABLE IV

GROUPS OF IMPORTANT LIMESTONE-BUILDING ALGAE
WITH THE MINERALOGY OF SOME RECENT GENERA

Phylum Rhodophyta (red algae)
    Family Corallinaceae (Carboniferous–Recent)
        Recent genera, composed of high-magnesian calcite: *Archaeolithothamnium, Melobesia,*
        Lithothamnium, Porolithon, Goniolithon, Lithophyllum, Amphiroa, Corallina, Jania
    Family Solenoporaceae (Cambrian–Miocene)
    Family Gymnocodiaceae (Permian–Cretaceous)

Phylum Chlorophyta (green algae)
    Family Codiaceae (Cambrian-Recent)
        Recent genera, composed of aragonite:
        Halimeda, Penicillus, Rhipocephalus, Udotea
    Family Dasycladaceae (Cambrian–Recent)
        Recent genera, composed of aragonite:
        Acetabularia, Neomaris, Cymopolia, Bornetella

Phylum Charophyta
    Family Characeae (Trias–Recent)
        Recent genera, composed of low-magnesian calcite:
        Chara

Phylloid algae (Upper Carboniferous–Permian)
        Archaeolithophyllym, Anchicodium, Eugonophyllum, Ivanovia

Phyllum Chrysophyta
    Family Coccolithaceae (Jurassic–Recent)
        Recent genera, composed of low-magnesian calcite:
        Podorhabdus, Coccolithus, Lithastrinus, Rhabdosphaera

perithallus, each part having also a characteristic cellular pattern. The spore matures in a case called a **sporangium**, and the sporangia in turn are usually combined in larger cases called **conceptacles**. The species of coralline algae are distinguished on the basis of the type and structure of the hypothallus and perithallus, the structure, arrangement and size of conceptacles and fine detail of the calcified tissue.

In crustose thalli the hypothallus may have one of three forms, in all of which the cells are arranged in layers. In the **simple** form there is a basal stack of layers that bend gently upward toward the perithallus (Fig.89). The **co-axial** form consists of a stack of curved layers, one before the other with a common axis parallel to the base, the whole structure overlain by the perithallus. In the **plumose** hypothallus the lower layers of cells curve downward and are surmounted by higher layers of cells that curve upward, the whole again covered by the perithallus. In a few species where these structures are not well developed there may be just one or several horizontal or irregular layers of cells. In many branching forms there is a thick central or **medullary** hypothallus enclosed in a thinner, outer perithallus. This is not always clearly developed, as in some species of *Lithothamnium*.

Fig.89. Vertical section of the crustose coralline alga *Lithothamnium*. Plumose hypothallus surmounted by perithallus having continuous horizontal walls. Slice. Recent.

Fig.90. Vertical section through the coralline alga *Archaeolithothamnium cyrenaicum* with layers of sporangia. Slice. Miocene. (Courtesy of G. F. Elliott.)

The cells of the perithallus are rectangular in sections normal to the growth surface and commonly rectangular or rounded in sections parallel to it.

The conceptacles (Fig.90) fall also into two groups depending on whether there is one or a number of apertures in the roof.

It is likely that the cell walls are built of calcite fibres, only a few tenths of a micron in breadth, with their c axes oriented normally to the surface of the cell wall (BAAS-BECKING and GALLIHER, 1931). Recent forms are high-magnesian calcite.

Solenoporaceae

The cells in this extinct family are generally larger than those in the Corallinaceae and, by contrast, in sections parallel to the growth, they mostly appear polygonal, rarely circular (Fig.91). The horizontal cell walls or cross-partitions are variously developed, and differentiation into hypothallus and perithallus is unusual and never sharply developed. The thalli tend to form rounded nodules rather than crusts (Fig.92).

Gymnocodiaceae

The thalli in this extinct family are segmented and, in different species, these

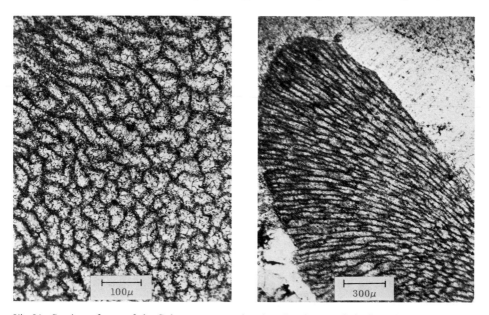

Fig.91. Section of one of the Solenoporaceae showing the characteristically polygonal shape of the cells. Slice. Ordovician.

Fig.92. Section through a rounded nodule of one of the Solenoporaceae showing cross partitions. Slice. Ordovician.

segments differ considerably in size, shape and the degree of calcification (Fig.99). The walls are perforate. Segments are hollow and may be cylindrical, oval, cone-shaped, spherical, barrel-shaped or finger-like in longitudinal section, and circular or oval in cross-section.

Codiaceae

The thallus is tubular and is built of a mass of interwoven, freely branching tubes, round in cross section (Fig.93, 94). Calcification (by aragonite in Recent species) begins either at the surface of the thallus or just below it and spreads inward. Consequently the youngest plants are uncalcified and the maximum calcification is found in the old plants. It is quite usual to find the outer part of a plant calcified but the inner part with no carbonate at all. Fossil crustose or nodular genera are distinguished on the basis of type of branching, the shape of the thallus and the diameter of the tubes. Fossil erect, branching genera are recognised by presence or absence of segmentation, and the sizes of the thallus, the segments and the filamentous tubes.

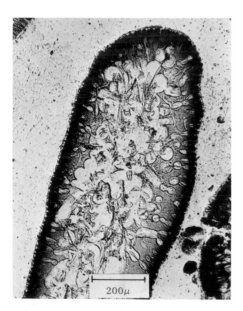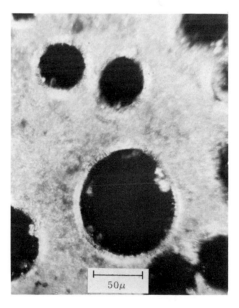

Fig.93. Longitudinal section of a segment of the codiacean *Halimeda* with utricles. Slice. Recent.

Fig.94. Section through *Halimeda* showing the preferred orientation of crystals around the circumference of the utricle. The light fringe is extinguished in the NW–SE and NE–SW positions. Slice with crossed polars. Recent.

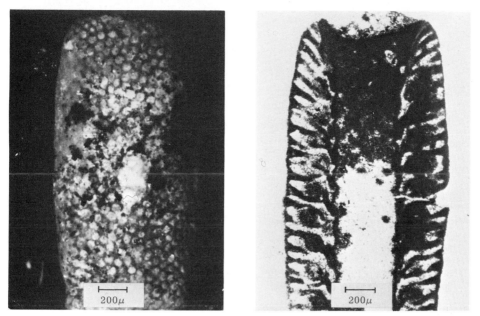

Fig.95. Exterior of a segment of the dasycladacean *Cymopolia kurdistanensis*. Reflected light. Palaeocene. (From Elliott, 1955.)

Fig.96. Longitudinal section of a segment of *C. kurdistanensis*. Slice. (From Elliott, 1955.)

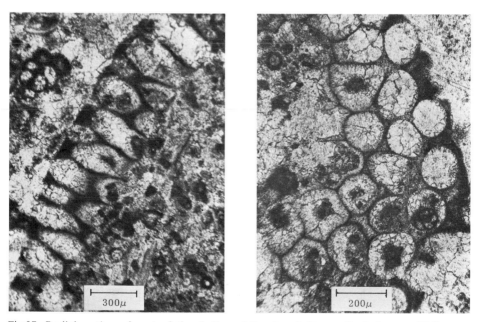

Fig.97. Radial section of part of the thallus of the dasycladacean *Koninckopora*. Walls now micritic. Cells filled with calcite cement. Slice. Carboniferous Limestone.

Fig.98. Tangential section of *Koninckopora*, cells being filled with calcite cement. Slice. Carboniferous Limestone.

Identification of Recent codiacean debris. MATTHEWS's (1966) work on silt-grade particles shows fragments of *Penicillus* and *Rhipocephalus* to be irregular to platy in shape, pale grey to pale yellow. They bear pores, 10 μ in diameter at intervals of 5–25 μ. With crossed polars they are dark grey to dark orange-brown. The particles are not unlike some porcellaneous foraminifera. Codiaceans break down finally to needles (Fig.231, p.277).

Dasycladaceae

The plant consists essentially of a relatively thick central stem from which grow slender branches (Fig.95, 96). In the more primitive genera the branches are not arranged on the stem in any regular way, but in most genera they are distributed in whorls that radiate from the stem, not unlike the spokes of a wheel. In a few genera they radiate in all possible directions from the stem, giving that part of the plant a spherical or egg-like shape. The primary branches commonly bear tufts of secondary branches which may bear tufts of tertiaries. Quaternary branches are known in a few genera. Most thalli are cylindrical in shape, others spherical, club-shaped, ovoid or segmented (e.g., Fig.95–98). Sporangia are spherical, ovoid or cylindrical and occur either in the stem, upon the stem, on the primary branches or among the secondary branches.

The plant may be more or less perfectly calcified, in Recent forms by arago-nite. The process of calcification may proceed in one of two ways. Either the carbonate is precipitated around the central stem and primary branches, with possible envelopment, in the older plant, of the secondaries, or even the tertiaries, or it forms a thin crust over the outer parts of the plant. In neither case is the plant ever completely calcified. It is important to realise that in the Dasycladaceae the cell walls and other tissues are not themselves calcified, as they are in the Rhodo-phyta or the Codiaceae or the Charophyta. They are merely enclosed in pre-cipitated carbonate, so that what remains when the tissues have decomposed is a partial mould of the original plant.

Charophyta

The large oögonia (female sex organs) are more or less calcified, showing walls made of radial-fibrous calcite. The spiralled gyrogonites in particular are apt to be well calcified (Fig.100), in fresh water by low-magnesian calcite.

Uncertain affinity

Though of somewhat uncertain affinity, *Girvanella* is probably a codiacean (Table IV). Its calcified remains consist of roundish to beanshaped masses of flexuous, filamentous tubes (Fig.101) commonly grown round a nucleus. The tubes

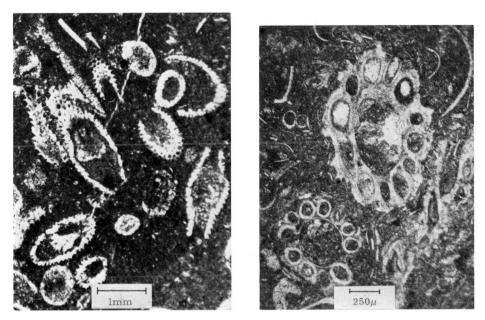

Fig.99. Biomicrite with *Gymnocodium bellerophontis*. Slice. Permian. (From ELLIOTT, 1955.)

Fig.100. Sections through the spiralled structures of the oögonia of *Chara*. Slice. Jurassic.

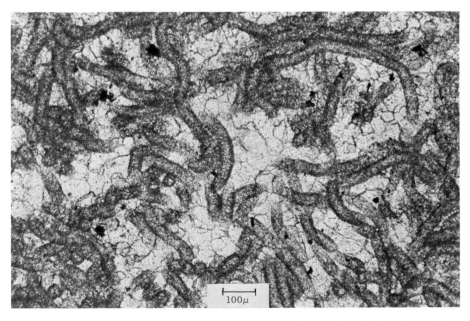

Fig.101. *Girvanella problematica*, most likely an extinct codiacean. Caradocian. (From WOOD, 1957.)

have circular cross-sections, a constant diameter, thick well-defined walls but no cross-partitions or perforations. Branching is rare in some species, common in others. The genus is subdivided on the basis of tube diameter and wall thickness (WOOD, 1957).

Phylloid algae

"Phylloid" is a convenient word to describe a group of very common, Late Palaeozoic, algal constituents (Fig.102, 103) which resemble each other in having leaflike shapes (PRAY and WRAY, 1963, p.209; WRAY, 1968). *Archaeolithophyllum* (Pennsylvanian) has a calcified undulating, encrusting thallus with thicknesses from about 250–800 μ. It has conceptacles and a cellular structure divisible into hypothallus and perithallus. It was probably a corallinacean (JOHNSON, 1956; WRAY, 1964). The sessile genera *Anchicodium*, (JOHNSON, 1946), *Eugonophyllum* (KONISHI and WRAY, 1961) and *Ivanovia* (KHVOROVA, 1946; MASLOV, 1956) were Pennsylvanian and Early Permian: they also have phylloid form, but their initial structures resemble those of codiaceans. The blades are generally more than 200 μ thick (Fig.103). The cortex bears regularly spaced utricles and the central medulla has weaving and branching filaments. The degree of calcification is variable.

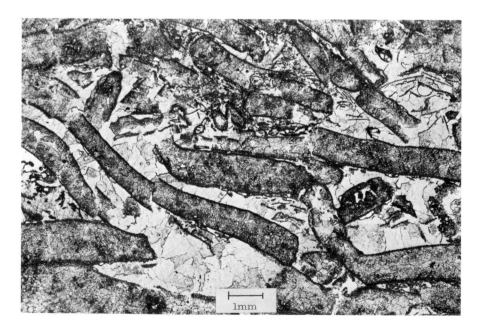

Fig.102. Thalli of *Archaeolithophyllum*, a phylloid alga, in a vertically oriented section of bio-sparite. Slice. Pennsylvanian. (From WRAY, 1964.)

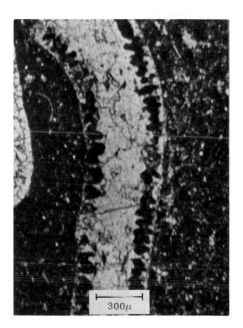 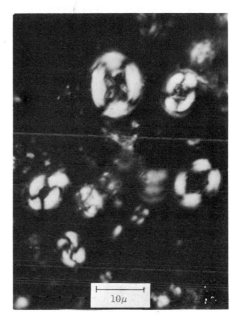

Fig.103. Transverse section of the phylloid alga *Eugonophyllum* showing utricles filled with micrite in the cortex. Interior of the blade now filled with calcite cement. Slice. Pennsylvanian. (From KONISHI and WRAY, 1961.)

Fig.104. Coccoliths seen with a high power, oil immersion objective, and crossed polars. Pseudo-uniaxial crosses indicate radial orientation of crystal axes. Slice. Recent.

PLANKTONIC CALCAREOUS ALGAE

Coccolithaceae

Fossil coccoliths are known in sedimentary rocks from the Jurassic onward (BLACK and BARNES, 1959, 1961; HAY and TOWE, 1962; HONJO and FISCHER, 1964; NOËL, 1965). Whole coccospheres, consisting of coccoliths packed around a single algal cell, are found in large quantities in the plankton of the open ocean (superb electron photomicrographs in MCINTYRE and BÉ, 1967). Little detail can be seen with a light microscope (Fig.104). Specimens are generally studied with the electron microscope as carbon replicas shadowed with a heavy metal such as gold, platinum or chromium. Recently, detailed photographs have been taken with the electron scanning microscope.

Each **coccolith** consists of an intricately organised structure composed of calcite crystals (Fig.105–109) which are commonly between 0.25 μ and 1 μ in diameter. Published data indicate a content of less than 1 mole % of $MgCO_3$ (THOMPSON and BOWEN, 1969). Each coccolith is usually about 2–20 μ broad in its

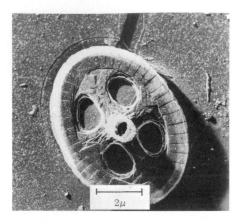

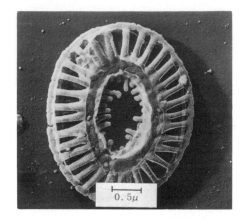

Fig.105. The coccolith *Podorhabdus albianus*. In the centre is a broken stalk. Replica. Electron mic. Gault. (From BLACK, 1965.)

Fig.106. The coccolith *Coccolithus huxleyi*, the most widely distributed species in oceanic oozes today. Replica. Electron mic. Recent. (From BLACK, 1965.)

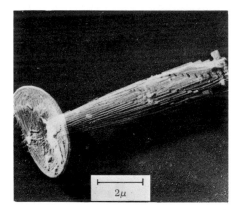

Fig.107. The coccolith *Lithastrinus* sp. simply constructed of overlapping calcite crystals. Replica. Electron mic. Gault. (From BLACK, 1965.)

Fig.108. The coccolith *Rhabdosphaera claviger* built of fibres that are elongate rhombohedra of calcite. Replica. Electron mic. Recent.(From BLACK, 1965.)

flattened plane. In many species the coccolith is a circular to oval (roughly elliptical) disc, the disc being not plane but gently curved about the minor axis of the ellipse. With this shape, like a bent penny, coccoliths can combine to form a coccosphere which encloses the living algal cell. The dominant crystal habit in the coccolith, not always distinguishable, is rhombohedral but crystal shape varies greatly with taxonomy. The rhombs are commonly stacked in imbricate fashion and this arrangement is associated with a rotary or spiral pattern in the construction of the coccolith. Thus circular or elliptical rosettes are usual. Other forms have hexagonal

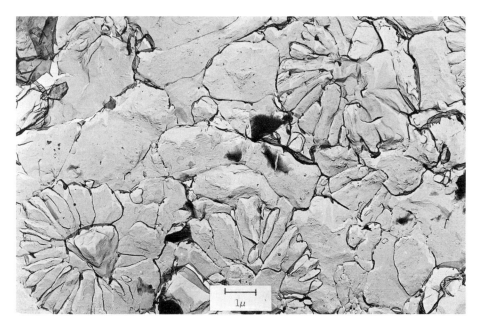

Fig.109. Coccoliths in micrite. Replica. Electron mic. Jurassic. (Courtesy of A. G. Fischer.)

or pentagonal symmetry, yet others are more or less ornamented rods.

Among the fossil nannoplankton (?), presumably assignable to the Coccolitha-ceae, are the hook-like, single crystal elements known as ceratoliths, and the star-shaped bodies called discoasters in which each ray of the star is, normally, a single crystal.

CALCISPHERES

First described by WILLIAMSON (1880), these are spheres constructed of a calcite wall enclosing a spherical chamber (Fig.110, 111). They are known from Devonian and Carboniferous limestones. The main wall lies between inner and outer con-centric surfaces, but in addition the outer surface may bear spines. The equatorial external diameter of the entire calcisphere varies in general from 60–225 μ. Thick-nesses of the main walls may be 3–30 μ. Spines which increase the thickness of the wall to a total of 50–170 μ, giving overall diameters of calcispheres about 500 μ, are figured by STANTON (1963). Looking, at first sight, like syntaxial diagenetic overgrowths, the spines nevertheless contain radial canals about 6 μ in diameter, visible both in longitudinal sections and cross-sections. BAXTER (1960) referred to radial partitions in the wall, 1–2 μ wide in a tangential direction and 4–6 μ long radially. The spines are elongate calcite crystals extinguishing with their vibration directions oriented radially and tangentially. The walls may be uniform in texture

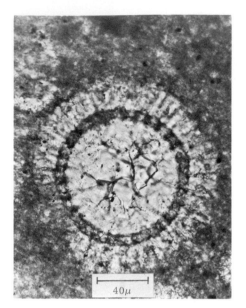 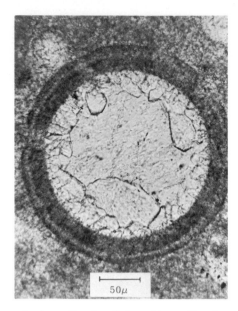

Fig.110. A calcisphere having two concentric layers in the wall and radial partitions. Chamber filled with calcite cement. Slice. Carboniferous Limestone.

Fig.111. A calcisphere with three layers in its wall. Chamber filled with calcite cement. Slice. Carboniferous Limestone.

or may have up to seven concentric layers distinguishable by alternations of uniform micritic texture and a fabric showing radial elements. The chamber may be filled with calcite cement, or more or less occupied by an internal sediment of micrite.

The inner surface of the wall, often marked by a dark line, is in some specimens separated from the micritic internal sediment by a concentric zone of sparry calcite, possibly cement. Stanton suggested that this zone represents an erstwhile innermost layer which has been removed. The inner surface of this postulated innermost layer is in places suggested by a line of inclusions.

The calcispheres have at one time or another been referred to a number of plant and animal taxa. STANTON (1963) noted their tendency to occur clumped together as aggregates, akin to those of plant spores or reproductive bodies. Recently RUPP (1967) has claimed that the non-spinose calcispheres resemble closely the reproductive cysts of the modern dasycladacian *Acetabularia* (Fig.143).

BRYOZOA

Russian work, especially that of Shulga-Nesterenko (references in ELIAS and CONDRA, 1957), has placed emphasis on the structure of the wall (Fig.112, 113) of

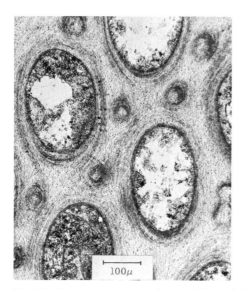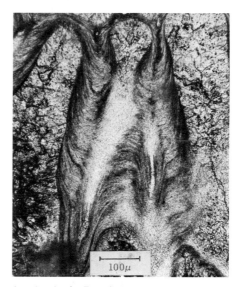

Fig.112. Transverse section of part of the peripheral region in the Permian trepostome bryozoan *Stenopora dickinsi*. (Courtesy of J. R. P. Ross.)

Fig.113. Transverse section of part of the peripheral region in the trepostome bryozoan *Stenopora dickinsi*. (From Ross, 1963.)

the **zoarium** (whole skeleton). Her careful and minute examination has been extended by ELIAS and CONDRA (1957) in a detailed analysis of the microstructure of the Permian *Fenestella* (of the order Cryptostomata, Ordovician–Permian). Ultrastructure in the Fenestellidae has been examined by TAVENER-SMITH (1969). Like the Foraminiferida, this group of animals must, for taxonomic classification, be studied in thin section. Nevertheless, despite the great lead by ULRICH (1890) in the last century, petrographic studies in thin section seem to be rare.

In *Fenestella*, regularly spaced, often bifurcating, **branches** gradually spread as a continuous **fenestrate** layer that may be plane or curved, depending on the shape of the zoarium and the extent to which it encrusts other organisms, pebbles, etc. The branches are joined, in the same plane, by regularly spaced cross-branches or **dissepiments**. In each branch there are two rows of **zooecia**, the now empty cells in which the single individual lived, complete with tentacles, mouth, stomach, intestine, anus, nerve ganglion and so on. These all open on the **obverse** side of the zoarium. The other side is the **reverse**.

The three fundamental components can be seen in Fig.114. Running throughout the entire zoarium, and forming the basis of the fenestrate structure, is the **colonial plexus** or **wall** of granular calcite. Enclosing the colonial plexus as a later, secondary deposit, is a thin inner, and a thick outer, **sclerenchyma** which is laminated calcite. Transverse **skeletal rods** (also called filaments, spicules or capillaries) radiate into the sclerenchyma, normally to its outer surface (Fig.114,

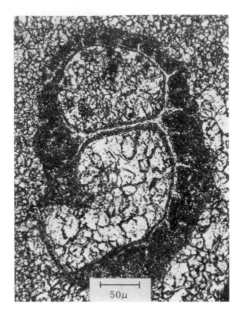 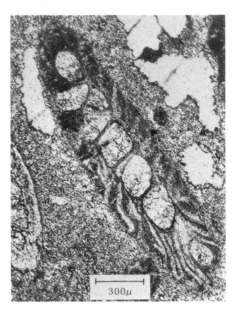

Fig.114. Paired zooecia in a fenestellid bryozoan. The colonial plexus appears as a white line separating the inner from the outer sclerenchyma: the plexus also extends radially as capillaries. Zooecia filled with calcite cement. Peel. Carboniferous Limestone.

Fig.115. Part of a fenestellid bryozoan showing zooecia and capillaries. Zooecia filled with calcite cement. Peel. Carboniferous Limestone.

115), probably formed by folds in the sclerenchyma itself, akin to the pseudo-punctate structure in brachiopods (p.24), the laminae curving outward against the axis of the filament. The sclerenchyma lines the zooecia, where it is relatively thin, and the dissepiments and external surfaces of the branches where it is thicker. ELIAS and CONDRA (1957, p.26) consider it to "correspond to the rhythmic growth lines of the brachiopods, mollusks, and other invertebrates". Modern bryozoans are built of either high-magnesian calcite or aragonite or of both.

ELIAS and CONDRA (1957, p.24) took thin sections of standard thickness (20–40 μ) and reduced the thickness further by smearing dilute hydrochloric acid with a finger tip over the section which was immediately washed. Examination with a microscope was followed by another acid smear until the desired thinness was attained. They found that the "lamination is made of numerous thin, continuous or interrupted, straight to wavy to plicated barlike laminae. The laminae are less transparent than the structureless substance in which they are embedded because of their apparent higher index of refraction. The intervals where present between the successive laminae range up to several times the laminal width. The width of the laminae is usually 0.8–0.9 μ. Under polarized light the structureless substance between the laminae tends to resolve into large irregular patches, each acting more or less as a crystalline unit and embracing several laminae. But the laminae do not act optically with the enclosing substance and remain light, usually pale orange colored, when the medium blacks out" (ELIAS and CONDRA 1957, p.37). They referred to similar fabrics in living Cyclostomata (to which they referred *Fenestella*) and in Palaeozoic Trepostomata. Ross (1960, p.1062), in her paper on cryptostome bryozoa, wrote of "laminate calcite which is also found in the Fenestellidae, Rhabdomesidae, many of the trepostomes, and the Recent Heteroporidae". In fragments the laminar structure is immediately obvious and is parallel to the surface (Fig.112, 113), unlike the secondary layer in the brachiopod. Commonly one or more zooecia are visible.

     In some cheilostome bryozoans (the most abundant order today, Cretaceous to Recent) the plane laminar structure of the frontal wall is progressively overgrown, during life, by a wavy, finely tuberculate set of laminae. The waves have broad rounded troughs and sharp crests that point outward. The structure is reminiscent of the outer sclerenchyma with filaments in *Fenestella*. In other species this overgrowth consists of mammilary structures built of radiate aragonite fibres (CHEETHAM et al., 1969).

     The mineralogy of bryozoans has been briefly studied by CHAVE (1954a) and LOWENSTAM (1954). Details of chemistry are given by SCHOPF and MANHEIM (1967). Detailed studies of mineralogy and fabric in some Recent cheilostome bryozoans have been made by RUCKER and CARVER (1969). They found that, in the older Anasca, the encrusting species are calcitic but the free-living species have aragonite in the outer basal lamina. In the order Ascophora, some species are all calcite, others all aragonite, but most have a mixed mineralogy with aragonite occurring as a late thickening of the frontal wall. The calcite is low-magnesian.

TRILOBITA

The mineralized part of the trilobite cuticle is constructed mainly of calcite crystals so small that individual crystals can rarely be distiguished with the light microscope. BØGGILD (1930) likened the fabric to his homogeneous structure (p.7). The whole cuticle shows a preferred orientation of the *c* axis more or less uniformly perpendicular to the cuticle surface (SORBY, 1879; CAYEUX, 1916; STØRMER, 1930). The report by VON ZITTEL (1887) of alternate layers of calcium carbonate and a calcium phosphate has not been substantiated by later workers. Neither CAYEUX (1916) nor BØGGILD (1930) could find such layers. RICHTER (1933) reported 30% of phosphate in the cuticle but CAYEUX (1933) in detailed analyses revealed only small amounts. The presence of much fine detail in the cuticle, combined with its small crystal size, suggests that there has been almost no neomorphic alteration in most specimens: RAW (1952) seems to have been alone in regarding the trilobite cuticle as a calcitic replacement of a primary aragonite. STEHLI (1956) analysed a single pygidium from a Permian asphalt, in which primary mineralogies seemed to

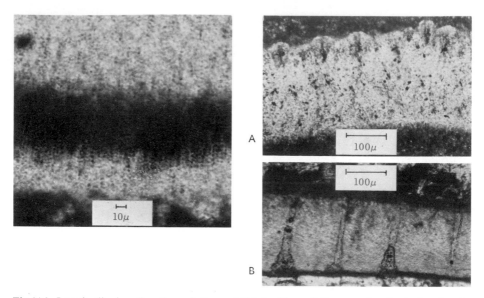

Fig.116. Longitudinal section through the pygidial doublure of *Bumastus barriensis* showing fine canals. Slice. (Courtesy of J. Dalingwater.)

Fig.117. A. Tubercles on the outer surface of the cuticle of the trilobite *Acaste*. Wenlock Limestone, Silurian. B. Canals in the cuticle of the trilobite *Brogniartella*. Caradocian, Ordovician.

have been preserved, and his X-ray results showed pure calcite.

Amino acids were reported by ABELSON (1954) from the cuticle but FUJIWARA (1963) found none in her material. DALINGWATER (1969) found traces of amino acids and obtained, after decalcification, a residue which may be a relic of original organic material.

Fine structures are preserved in many specimens. The thickness of the cuticle is commonly 300–400 μ, very thick compared with the cuticles of living arthropods. Fine vertical lines, radially disposed, were studied by STØRMER (1930) in the Trinucleidae. Viewed stereoscopically the lines (about 1 μ diameter) lie at various levels and might, he suggested, be canals. DALINGWATER (1969) compared these canals (Fig.116) with the pore canals of living arthropods which have similar dimensions. Wider canals (Fig.117B) have been described by many authors (e.g., by G. LINDSTRÖM, 1901 and CAYEUX, 1916), but it is difficult to decide if these are tegumental or setal ducts. Tubercles, seen in thin section (Fig.117A), were compared by DALINGWATER (1969) with those described from the arachnid cuticle by KENNAUGH (1968). Laminae have been noted by various authors (e.g., STØRMER, 1930; ROME, 1936; KIELAN, 1954). They are often picked out by the distribution of pyrite.

FURTHER READING

Certain works are outstandingly useful. They are the monographs by Bøggild (1930) and Taylor et al. (1969) on the Mollusca, the studies by Oberling (1964) of Bivalvia wall fabric, the superb electron-photomicrographs of Grégoire (1961a, b; 1962) and Wise (1969b), Mutvei's (1964) unsurpassed photographs with a light microscope and also a nice review by Wilbur and Simkiss (1968). The questions of environmental and biological controls of shell mineralogy are surveyed by Kennedy et al. (1969). Brachiopoda are dealt with in the recently published *Treatise* (Williams, 1965). For the Zoantharia and Octocorallia there are works in the *Treatise on Invertebrate Paleontology* (*F*), by Bayer (1956), Hill (1956a, b), Hill and Stumm (1956) and J. W. Wells (1956), also Hill (1936), Schindewolf (1942a, b), Alloiteau (1952), Schouppé and Stacul (1955) and Wells (1957b). Kato (1963) gives much fine detail and a useful historical sketch of work on the Rugosa. On the Stromatoporoidea Stearn (1966) is the obvious starting point. For the Foraminiferida Wood's (1949) paper is joined by the *Treatise* (Loeblich and Tappan, 1964a, b, c) and the work of Blackmon and Todd (1959) on test mineralogy. The thoughtful re-examination of the basis of classification by wall fabric, in the paper by Towe and Cifelli (1967), is of interest not only to the student of Foraminiferida, but to all who are interested in calcification in biological systems. Raup (1966b) and Weber (1969) have brought together the latest information on the echinoderm skeleton. On calcareous algae the ground is widely covered by Johnson (1961a) and, on the subject of coccoliths, Black (1965) gives a delightful introduction in Endeavour. Ancient coccoliths are particularly well illustrated in Fischer et al. (1967); Hay and Towe (1962) have useful references. More detailed introductions to coccoliths are by Black (1963) and Noël (1965). Calcispheres are covered by Stanton (1963). The papers by Elias and Condra (1957), Boardman and Cheetham (1969) and Tavener-Smith (1969) are useful introductions to the Bryozoa. On the Trilobita helpful papers are scarce. The work in Dalingwater's (1969) thesis has yet to be published. Useful details are to be found in Cayeux (1916), Størmer (1930), Rome (1936) and Kielan (1954).

Criteria for the recognition of skeletal debris in modern carbonate sediments are elaborated by Ginsburg (1956), Purdy (1963a) and Matthews (1966). A thorough background to the whole subject is available in Lowenstam's (1963) detailed and thoughtful review of microarchitecture, mineralogy and chemistry in modern and ancient biological systems. General data on mineralogy are to be found especially in Chave (1954a, b) and Clarke and Wheeler (1922), with chemistry in Vinogradov (1953). Feray et al. (1962) have written a revealing essay on the difficulties of identifying small skeletal fragments. Other references are given in the text.

Additional references not given in the preceding chapter

For general background there is the work by MAJEWSKE (1969) on fossil fragments in rocks, also CAYEUX (1916). Physiology of molluscs, WILBUR and YONGE (1964, 1966). On mollusc shell structure, WADA (1961), KOBAYASHI (1966) and KOBAYASHI and KAMIYA (1968). On organic materials in molluscs, GRÉGOIRE (1958a, b, 1959a, b, 1967), HUDSON (1967b) and GRÉGOIRE and VOSS-FOUCART (1970). On skeletal secretion in brachiopods, TOWE and HARPER (1966), HARPER and TOWE (1967) and WILLIAMS (1968b). On calcification in corals, GOREAU and GOREAU (1959, 1960a, b). On classification of alcyonarian sclerites, DEFLANDRE-RIGAUD (1957); on holothurian sclerites, GUTSCHIK (1954, 1959), LAGENHEIM and EPIS (1957), and SUMMERSON and CAMPBELL (1958); on skeletal secretion in the foraminiferids, BÉ and HEMLEBEN (1970), BÉ et al. (1966).

On calcareous algae, PIA (1926), WOOD (1941a, b, 1942, 1964), KONISHI (1958), JOHNSON (1961b, 1964), P. R. BROWN (1963a, b). On mineralogy of microfossils, SWITZER and BOUCOT (1955). On the use of the transmission and scanning electron microscopes, HAY and SANDBERG (1967), HONJO and BERGGREN (1967), KIMOTO and HONJO (1968) and WISE and HAY (1968a). On the mineralogy of prisms in some mollusc shells, LUTTS et al. (1960), and on the tests in some pelagic foraminiferids EMILIANI (1955).

Chapter 2

PETROGRAPHY OF CARBONATE GRAINS 2:
OÖIDS, PISOLITES, PELOIDS AND OTHER MICRITIC FABRICS

RECENT OÖIDS AND PISOLITES

Marine and lacustrine oöids

The Recent unaltered marine oöid is distinguished by the possession of an outer **coat** or **cortex** of concentrically lamellar, cryptocrystalline aragonite (Fig.118, 119). Within this coat a detrital **nucleus** may be recognized, commonly a grain of quartz, a peloid of micritic carbonate, a fragment of skeleton, or even a broken piece of oöid. In some Recent and in many ancient marine oöids, the concentric lamellar pattern is traversed by a more or less well-developed fabric of radially arranged calcite fibres. The unaltered aragonite oöids yield a positive pseudo-uniaxial figure with crossed polars (SORBY, 1879, p.74). The oöids of radial calcite give a negative figure. Lake oöids are less well known. The best known, from Great Salt Lake, Utah, appear to be aragonite more or less altered to calcite.

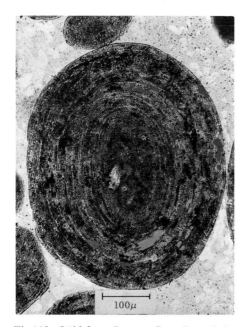 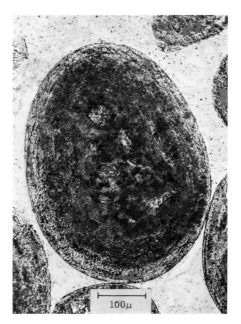

Fig.118. Oöid from Browns Cay, Great Bahama Bank. The dark patches are mainly endolithic algae. Slice. Recent.

Fig.119. Oöid with large nucleus. East Bimini, Great Bahama Bank. Slice. Recent.

Dimensions. Despite the large volume of published work on these tiny objects, so like the ova of fish ($\omega o \nu$ = egg), there are few data on the range of grain size among marine oölites. NEWELL et al. (1960, p.487) gave the range of diameters for the Recent Browns Cay oölite, Bahamas, as 0.3–1.0 mm. For an oölitic tidal delta in the Ragged Islands, Bahamas, an effective range of 0.2–1.0 mm was given by ILLING (1954, fig.4, sample B-39). For the oölite deltas of the Trucial coast embayment in the Persian Gulf, figures kindly supplied by D. J. Shearman of The Imperial College for a coarse and a fine oölite show that 90% of the oöids lie in the ranges 0.15–0.5 mm and 0.065–0.15 mm. RUSNAK (1960) recorded medians from 0.25–0.5 mm on the Laguna Madre beaches, Texas and sorting values for $\sigma\varphi$ of 0.20–1.36. EARDLEY (1938, p.1364), for the Recent Great Salt Lake oöids, gave 0.15–1.5 mm for the long diameters. The thickness of oölitic coats has been given by various authors. NEWELL et al. (1960, p.490) wrote of 20–90 more or less continuous lamellae, each having a thickness of 1–3 μ, with the thickness of the oölitic coat typically between 60–180 μ. RUSNAK (1960) found 8–10 lamellae in oöids of Laguna Madre. ILLING (1954, p.35) found that the coats in the Ragged Island area, Bahamas, fall mainly in the range of thickness 30–40 μ. Thin oölitic films have been recorded by BATHURST (1967a) from Bimini Lagoon and elsewhere in the Bahamas, and from Batabano Bay, Cuba and the Campeche Bank, Yucatan. In the Lagoon these range in thickness from 1 to 8 μ, but generally 1 to 3 μ, which is the thickness of a single lamella from Browns Cay (Fig.240).

Oriented lamellae with c tangential. Details of crystal fabric have been worked out by ILLING (1954, pp.36–37) for Bahamian Recent oöids. The mineral was shown to be aragonite by X-ray analysis. Optically the slower vibration direction is normal to the grain surface, the faster tangential. He was unable to detect individual crystals so that the optical properties are an average view of an aggregate of crystals. They are consistent with a distribution of aragonite needles with long axes and optic axes randomly oriented in the tangential surface of the grain. A single crystal could, of course, show the faster vibration direction radial and the slower tangential. This would happen if the crystal were viewed along the c axis and both c and b lay tangentially to the oöid surface. Illing found that the refractive index of the tangential vibration direction of the oöid was 1.605. This is a statistical value, the ray having passed through many cryptocrystalline aragonite crystals. The value agrees closely with an estimate based on the rough formula:

$$N_t = \tfrac{1}{2}N_p + \tfrac{1}{2}(N_m + N_g) = 1.606$$

where N_t is the refractive index of the affective tangential vibration direction in the oöid and the three refractive indices of aragonite are given as N_p: 1.530, N_m: 1.681, N_g: 1.685 (WINCHELL, 1946). Illing's value, being smaller, fits better with the flattening of the indicatrix normal to the c axis. His measurement of the refractive index of the radial vibration direction of the oöid yielded a value of 1.640 instead of 1.683 (the mean of N_m and N_g). The reason for this low figure is not obvious. Any tendency for the refractive index to be modified as a result of porosity or impurity in the cortex (ILLING, 1954) should apply to both vibration directions since they share the same ray.

I earlier described the petrography of the oölitic films around grains in Bimini Lagoon (BATHURST, 1967a). These films are identical petrographically with the lamellae of Illing and Newell et al. The film, or lamella, under crossed polars at a magnification of $900 \times$, is constructed of a mosaic of more or less equant crystals with diameters up to about 2 μ. The smallest detectable crystal (area that extinguishes uniformly) is about 0.5 μ, but smaller ones probably exist. Viewed with a gypsum first order red compensator, the mosaic as a whole shows the usual orientation of the slow vibration direction normal to the surface of the lamella. The separate crystals are best seen in the region of the black cross where the mosaic is mostly in extinction, and where slight differences of orientation between adjacent crystals are more clearly appreciated.

It is, of course, dangerous to interpret optical crystallographic data in terms of crystal shape, as Cifelli and Towe have shown (p.40). Direct evidence of shape must be obtained.

RUSNAK (1960, p.477) referred to crystals with a "size range" of 1–5 μ and negative elongation. On the other hand, SHOJI and FOLK (1964) studied the surfaces of Bahamian oöids with the electron microscope and saw aragonite needles, 0.5–2.0 μ long, lying tangential to the oöid surface. D. J. Shearman (personal communication, 1968) reported oöids from the Persian Gulf with aragonite needles 2 μ long. Tangential aragonite crystals are unusually clearly shown in electron photomicrographs of Bahamian oöids by FABRICIUS and KLINGELE (1970) and LOREAU (1970a).

Oriented lamellae with *c* radial. In the oöids of Laguna Madre, Texas, there are three kinds of lamellae (RUSNAK, 1960, p.477). Besides the two lamellae known elsewhere, with the *c* axes of the aragonite having either a preferred tangential orientation or none at all, Rusnak described a third fabric wherein the *c* axes are preferentially oriented radially to the surface of the oöid. The boundaries between lamellae are sharp. The commonest lamella is the unoriented one which also contains clay. Both the tangential and the radial lamellae give a black cross between crossed polars.

Unoriented lamellae. ILLING (1954), and NEWELL et al. (1960), found that some lamellae are made of aragonite with no discernible preferred orientation. FRIEDMAN (1964, p.790) and BATHURST (1966, p.21) have noted the similarity between these lamellae and the micrite envelopes formed by superficial alteration of skeletal and other carbonate grains. NEWELL et al. (1960) found that these lamellae generally contain algae. The lamellae may be continuous around the oöid, but are more usually discontinuous and lenticular. The crystals of which they are composed are larger than those in the oriented coat and can be delineated in thin section, especially at the edge of a wedge-shaped section. They appear to be a by-product of algal boring at the surface of the oöid (p.383).

Organic material. Recent oöids have a high organic content. NESTEROFF (1955b, 1956a, b), NEWELL et al. (1960, p.491) and SHEARMAN and SKIPWITH (1965) have dissolved oöids gently in very dilute acid until only a complex mass of mucilagenous material remains. The filamentous structure of this is characteristic of algal colonies, also fungi and actinomycetes. Algae have, in fact, been seen living in pits (up to 50 μ across) in the surfaces of oöids (NEWELL et al., 1960, p.490).

Asymmetrical oöids. An unusual kind of oöid has been recorded from the Laguna Madre (FREEMAN, 1962) in which the growth of the oölitic coat has been unevenly distributed over the surface of the nucleus. The coat forms one or more bosses, between which the nucleus may be exposed. Some degree of localized accretion is also apparent in the Browns Cay oöids.

Polished oöids. In reflected light the Recent oöids of the agitated surf zones have creamy, highly reflecting surfaces. Laboratory experiments have demonstrated that this intense polish can be produced by abrasion (NEWELL et al., 1960; DONAHUE, 1965). In cave oöids, which are agitated by the fall of water into the splash cup, topographically high areas on the oöid surface are more highly polished than other areas (DONAHUE, 1965, p.252).

Radial calcite fibres in oöids. Oöids of an unusual kind have been described from the floor of Great Salt Lake. Though not marine, they are important in view of the detailed and careful pioneer study of them made by EARDLEY (1938). Cryptocrystalline aragonite, mixed with clay, forms concentric lamellae which appear dark in transmitted light. Cutting radially across these are coarser clear crystals, large enough in places for their refractive indices to be measured and related to calcite (Fig.120). The radial crystals may cross all the concentric lamellae or only some, or they may be confined within a single lamella. The surfaces of the oöids vary from "pearly smooth" to mottled. It is noteworthy that those with the most highly developed radial fabric, with many calcite crystals extending to the surface, are the most mottled. The mixed aragonite–calcite composition is reflected in the content of Sr^{2+}. E. D. Glover (personal communication, 1969) recorded with an electron microprobe a mean value of 4,900 p.p.m. for a single oöid (Fig.225, p.262), point values varying in the oöid by a factor of 2. Similar mean

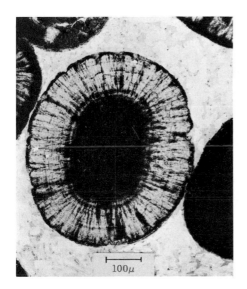 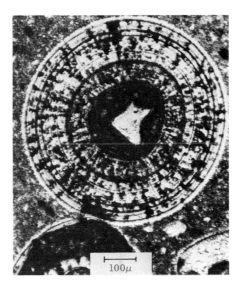

Fig.120. Oöid from Great Salt Lake, Utah, with radial fibres of calcite terminating in bumps at the surface which yield the mottled appearance in reflected light. Slice. Recent.

Fig.121. Oöid with both radial and concentric structure, now all calcite. Jurassic. Ain, France. (Slide lent by D. J. Shearman.)

values were obtained by ZELLER and WRAY (1956), ODUM (1957) and KINSMAN (1969).

Cave oöids and pisolites (cave pearls)

Cave oöids and pisolites seem to be generally constructed of radial-fibrous calcite (Fig.122). Pisolites with aragonite with c tangential have been described from a hot spring in Czechoslovakia (p.300). Kirchmayer has done pioneer work in this field (KIRCHMAYER, 1962, 1964; HAHNE et al., 1968). A single pisolite from a mine, 1 cm in diameter, measured by Kirchmayer, showed 230 rings, alternately light and dark, around the nucleus. These are probably summer and winter precipitates of calcite respectively (see also BAKER and FROSTICK, 1951). Comments by Donahue on pisolite fabric are given on p.306. The data of GRADZIŃSKI and RADOMSKI (1967) indicate that concentric lamination is a primary fabric which may be replaced by a radial-fibrous secondary fabric.

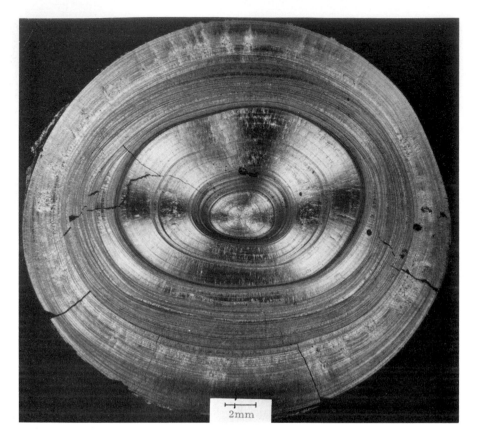

Fig.122. A cave pisolite of calcite showing pseudo-uniaxial cross between crossed polars. Slice. Carlsbad Caves, New Mexico, U.S.A. (Courtesy of J. D. Donahue.)

ANCIENT OÖIDS AND PISOLITES

Primary and secondary fabrics

Oöids are known in limestones as far back as the Precambrian, yet relatively little has been written about their petrography, doubtless owing to their relative uniformity. They are now calcite, although it is usual to suppose that they were once aragonite. Rather like the oöids of the Great Salt Lake, they can have one or both of two fabrics in equatorial section, concentric lamellae of dark-looking micrite (without preferred crystal orientation) traversed by coarser crystals that may show a radial elongation (Fig.121). The proportions of these two fabrics vary in different oöids, and between oölites. The descriptions and photomicrographs of CAYEUX (1935) have yet to be improved upon. The lamellae differ from one another in their degree of opacity, which is correlated with differences in crystal size and orientation:

see electron scan photographs by LOREAU (1969). There appears to be every gradation between pure concentric lamellar fabric and fully developed radial fabric, in which the crystals with their *c* axes radially oriented give, for the whole oöid, a negative pseudo-uniaxial figure between crossed polars.

Tangential sections show no lamellae. While many oöids are uniformly grey-looking in ordinary light, others are a golden-brown, probably owing to contained organic substances. This applies particularly to some that have an outer layer, perhaps tens of microns thick, made of golden-brown calcite fibres with a strong radiate orientation, overlying a more or less distinctly lamellar, grey-looking micritic interior. In contrast, the radiate, coarse fabric may traverse the whole thickness of the oölitic coat or it may be restricted to certain lamellae or to particular patches. Nearly always this fabric shows traces of the concentric lamellae, commonly as relic wisps of concentric micrite. Other oöids, less common, are built of a relatively coarse (sparry) calcite mosaic (Fig.123) with no obvious preferred orientation of crystal shape or optics. Various secondary, replacement fabrics are dealt with in another chapter (p.481). It is not unusual to see that one or more outer lamellae have become partly, or even wholly, detached from the oöid before the rock was cemented. Exfoliation also occurs as a result of compaction (Fig.319, p.463). Total dissolution of oöids leaves oolimoulds (Fig.124).

While it is unusual to see oöids less than about 0.25 mm in diameter, T. Freeman (personal communication, 1969) has discovered silt-grade oöids,

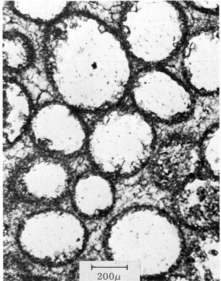

Fig.123. Oöid replaced by calcite cement. Slice. Cambrian. Pennsylvania, U.S.A.

Fig.124. Oömouldic porosity in dolomitized limestone. Slice. Jurassic. Arkansas, U.S.A.

about 0.06 mm diameter, as the finer fraction in a bimodal oölite in the Mississippian Fayetteville Limestone of Arkansas.

Vadose pisolites in caliche

Large and irregular pisolites, several millimetres in diameter, may be confused with caliche. THOMAS (1965) and DUNHAM (1969b) have produced evidence that the supposed lagoonal oncolites of the Permian Capitan reef complex (New Mexico and Texas) are, in fact, caliche. Dunham showed how adjacent pisolites interfered with each other's growth and so were immobile during growth. They show downward, geotropic elongation indicating that they were stationary; silt has perched in the upper parts of concentric growth layers.

PELOIDS

McKee has given us the useful term **peloid** to embrace all grains that are constructed of an aggregate of cryptocrystalline carbonate, irrespective of origin (McKEE and GUTSCHICK, 1969). It is necessary to have such an umbrella term because the origin of these aggregates is often in doubt, yet it must be possible to refer to them without implying any particular mode of formation.

Faecal pellets

There are those who, rather rashly, suspect that all peloids have been pressed and bonded into their present form during passage through the gut of a mud-feeding animal. Data on authenticated modern faecal pellets (Fig.158, 159 on p.131)

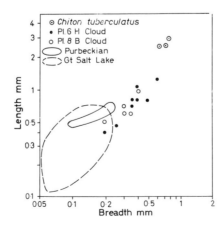

Fig.125. Dimensions of some faecal pellets. (From EARDLEY, 1938; P. R. BROWN, 1961; CLOUD, 1962a.)

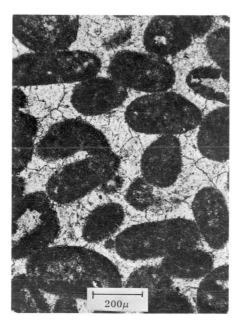

Fig.126 Peloids, probably some faecal pellets, in calcite cement. Carboniferous Limestone. Westmorland.

are scattered and observations somewhat fortuitous. On the photomicrographs in CLOUD (1962a) it is possible to make measurements which give some idea of the dimensions of faecal pellets on the Great Bahama Bank (Fig.125). ILLING (1954, p.24) gave longest diameters of 0.5–0.7 mm, within the range of examples by Cloud. KORNICKER and PURDY (1957) gave dimensions for *Batillaria minima*, a gastropod in the lagoon at Bimini, Bahamas, and EARDLEY (1938, p.1404) for the brine shrimp *Artemia gracilis* in Great Salt Lake, Utah. Recent faecal pellets are apt to contain a good deal of organic matter. ILLING (1954, p.25) found that pellets in the Ragged Island area, Bahamas, are greyish, owing to the presence of tiny flakes of a black, opaque substance. These pellets may have a vague concentric layering, but this is probably unusual. When the pellet is dissolved in acid the black specks remain enmeshed in a mucilagenous substance. Other pellets referred to by Illing are white. Fresh pellets are readily crushed, but older ones are hard and brittle (as in the Persian Gulf, p.199). In thin section faecal pellets contain an illsorted agglomeration of fine particles of various kinds (Illing's plate 3). This mixture changes gradually with time to a uniform micrite (PURDY, 1963a, p.343, plate 1).

It is widely felt that, in both Recent and ancient carbonate sediments, elongate peloids, ellipsoids of revolution, are faecal pellets (Fig.126). P. R. BROWN (1964, pp.257–264) has described presumed faecal pellets from the Upper Jurassic of Dorset, England. He found that in the Broken Beds they are ellipsoidal to spherical and composed of calcite crystals about 4 μ in diameter. The apparent minimum diameters (in thin section) of the pellets are from 40–300 μ. Most are pale grey in thin section with no internal structure. Others are mottled with little patches of extra fine-grained brown micrite. Acetate peels of this brown micrite show component crystals with diameters less than 2 μ. In the Hard Cockle Beds the pellets are spherical, also possibly rod-like, with apparent minimum diameters of 70–140 μ. The component crystals are less than 2 μ. In the Soft Cockle Beds the pellets, with apparent minimum diameters of 120–250 μ and longest diameters up to 700 μ, have a golden-brown colour, possibly an indication of organic matter. The component crystals (less than 2 μ) contrast with the coarser matrix of micrite that supports the pellets and which has crystals 3–4 μ.

Other peloids

Peloids that are roughly spherical or irregular in shape are more difficult to diagnose than elongate peloids. It is possible that some Recent ones are inorganic accretions (ILLING, 1954, p.27), but the experience of PURDY (1963a, p.347–348, 1968) and BATHURST (1966, p.21) in the Bahamas suggests that many of the irregular grains, at least, are skeletal particles that have been replaced by micrite as a result of processes associated with endolithic algae (p.389).

In ancient limestones peloids are common (BEALES, 1958), and have been described under the name *pseudoolithes* or *fausses oolithes* by CAYEUX (1935, p.267). Generally their internal structure is uniformly micritic and, unless they are obviously ellipsoidal, their origin remains doubtful. It seems likely that many are micritized skeletal particles of the type referred to in the previous paragraph. With the naked eye many are indistinguishable from oöids and it is necessary to section them in order to discover whether they have any structure. It has long been known that in some pelsparites the peloids tend to merge (BEALES, 1956, p.864) and to produce a flocculent structure. Such peloids *appear* to have coalesced leaving residual patches of spar amongst them. CAYEUX (1935, p.271, plate 18) has given us a description that is still the standard of comparison for this structure which he called *structure grumeleuse* (p.511). The apparent blurring of the outline of the peloid is important here. A more tightly packed flocculent clotted structure can be seen in many calcite-mudstones, if the diaphragm of the microscope is reduced to give heightened relief. SCHWARZACHER (1961, pp.1486–1487) has measured the clots in some Carboniferous mudstones of this kind: they have diameters of 20–150 μ. An analysis of the crystal sizes, in acetate peels, has yielded a crystal size frequency distribution that is bimodal (Fig.352, p.507). A peak around 2.25 μ represents the micrite in the clots and another around 5.85 μ the interstitial calcite cement.

Grapestone and lumps

In the Bahamas, in the Gulf of Batabano, and off the Trucial coast in the Persian Gulf, in the Khor al Bazam back-reef lagoon, carbonate sand grains of various kinds are clumped together to form compound grains (Fig.152, 246 on p.124, 312 resp.). These irregularly shaped grains, occasionally reminiscent of lithified bunches of microscopic grapes, have been called grapestone by ILLING (1954, p.30). Characteristically "*limited* numbers of grains become *firmly* cemented together" (J. C. M. TAYLOR and ILLING, 1969, p.80). The component sand grains are commonly rounded and intensely micritized (p.384). They are cemented together by micritic aragonite (ILLING, 1954, plate 1–4; PURDY, 1963a, plate 4).

CALCILUTITES

Muds (clay-grade): precipitated or algal

The true aragonite needle muds, or drewites (after G. H. Drew: see DREW, 1914), are too fine-grained to be studied with the light microscope, so that it is only recently that helpful petrographic data have begun to appear, based on photographs of replicas with the electron microscope (Fig.231, p.277). The main source of photographs is the paper by CLOUD (1962a) on the muds west of Andros Island, Bahamas. Here the aragonite appears as needles with, for the most part, a length/breadth ratio of 10/1. The dimensions of the figured needles vary from $0.1 \times 1.0 \mu$ to $0.5 \times 5.0 \mu$ (see also FOLK, 1965). CLOUD (1962a, p.97) wrote that the photographs "hint that the aragonite needles of known algal origin may be generally better formed and less likely to have ragged ends than those of the bulk sediments, the whiting filtrate, and the forced precipitate". But judgement is necessarily subjective at the present time. Indeed, the question of the chemical or algal origin of these muds is still a matter for dispute (p.276). It is possible that the study of replicas with the electron microscope may lead, as LOWENSTAM (1963, p.149) suggested, to the recognition of morphological differences between needles derived from different sources.

NEWELL and RIGBY (1957, p.61) have described mud-sized "amorphous aggregates" of calcite in fresh-water ponds of Andros Island, Bahamas, and SCHOLL (1963, p.1599) suggested that calcite mud may be precipitated in White-water Bay, Florida, in brackish-water.

Muds: clastic

Many muds must be products of mechanical breakdown. In time the organic matrix that binds together the crystals in a skeleton may decompose leaving an accumulation of crystals. Feeding predators, boring algae and breaking waves all help to reduce skeletal debris to clay or silt particles. Many skeletons are constructed of minute needles, prisms or plates of aragonite or calcite (Chapter 1). MAIKLEM (1968) has examined the hydraulic properties of these breakdown products.

CHAVE (1960, 1964) has shown that much fine material can be released from skeletons, in experiments with tumbling barrels. ILLING (1954, p.16), in describing muds from the Ragged Island area, Bahamas, noted that the grains in his muds ($< 5 \mu$) are all single crystals, most of them showing no trace of crystallographic shape. Since they are unlike the needles of precipitated aragonite he thought that they must be the products of abrasion. X-ray diffraction tests on these muds show them to be a mixture of aragonite and calcite. Unfortunately, despite attempts by many people, it has rarely been possible to relate any clay grade particles unambiguously to a unique source. MATTHEWS (1966) has given criteria for the recognition of silt-sized skeletal debris with a microscope. Some of his details are presented in the appropriate sections of Chapter 1.

Mudstone (clay-grade)

This is the calcite-mudstone of Dixon in DIXON and VAUGHAN (1911, p.516), the *diagenetisch verfestigte Kalkpelite* of NIGGLI (1952, p.412), the *Pelit* of SANDER (1936, p.32), *la gangue-sédiment* of CAYEUX (1935, p.151) and *la pâte* or *cryptite* of various French authors. It is the micrite (microcrystalline ooze) of FOLK (1959, p.8) with crystal diameters of 1–4 μ, the mudstone of DUNHAM (1962, p.117) and the lithographic limestone of many earlier writers. A type of lithology studied so widely, and for so long, might be expected to have given up some of its secrets. Yet, if anything, the business of interpreting this material is becoming more difficult. For example, HATHAWAY and ROBERTSON (1961, p.301) have shown, with electron-photomicrographs (Fig.354, p.510), that the plane intercrystalline boundaries, which might be thought to indicate quiet rim cementation of single crystals (p.429), can also be produced by subjecting a Recent aragonite mud at low pressure to 400 °C for as long as 63 days (p.509). Other calcite-mudstones show dominantly curved or irregular intercrystalline boundaries under the electron-microscope (Fig.348, 349, 351, 354 in Chapter 12). Studies of the frequency distribution of grain sizes are difficult, but SCHWARZACHER (1961, p.1487) has carried out an analysis of calcite silty mudstones in a Carboniferous knoll-reef in northern Ireland, using delicately prepared peels, viewed with a phase-contrast microscope. His histograms (Fig.352) show the degree of detail that can be realized in a study of this kind. BLACK (1953, p.lxxxi) found that the range of particle sizes in the finer, coccolith, fraction of disaggregated Chalk of northern Europe is 0.5–4 μ. BATHURST (1959a, p.365) gave the same range for crystal diameters in some calcite-mudstones of the Carboniferous Limestone of England and Wales. BANERJEE (1959, p.379) recorded the same range for calcite-mudstone in the Carboniferous Limestone of

North Wales. FOLK (1965) put the general upper limit for micrites at 3.5 μ. P. R. BROWN (1964, p.261, 263) gave a maximum diameter of 2 μ for some Upper Jurassic calcite-mudstones in Dorset, England.

One of the peculiarities of calcite-mudstones is a tendency to form clots. This is discussed in the section on peloids (p.86). Generally the clots contain some non-carbonate impurity, such as clay minerals, chert, pyrite and a variety of objects that float on the surface of the acid after the mudstone has been dissolved. These include spore-like objects (BANERJEE, 1959, p.385) and wing cases and other parts of insects.

A study of calcite-mudstone with the electron microscope was made by GRÉGOIRE and MONTY (1963). They described the rock (Fig.348) as a "calcaire à pâte fine" of Lower Viséan age in Belgium. With the light microscope they could see crystals with diameters 0.9–3 μ, but with the electron microscope the lower limit of visible diameter was down to about 0.3 μ. Most of the crystals with diameters of 0.5–0.9 μ are grouped together as patches of fine mosaic. They have simple outlines and are packed tightly together. The patches are separated from each other by a coarser mosaic of crystals with diameters up to 2 μ. The patches of finer mosaic have diameters of 10–30 μ and it is interesting to compare these with SCHWARZACHER's (1961, p.1487) mean diameter of 50 μ (range 20–150 μ) for peloids seen with the light microscope (p.86). The crystals within the patches are certainly finer than those in his peloids which have a modal diameter of about 2.25 μ.

Coarser calcilutites

The factors controlling crystal size in carbonates are different from those that impose textural limits on terrigenous material (Table I). We are not concerned with the products of weathering, the widespread maximum diameter of clay-mineral crystals at about 2 μ, or the limited size range of quartz crystals. Carbonate crystals are generally needles, rhombs, prisms or plates, formed either by break-down of skeletons or by precipitation. It is not surprising, therefore, that the break between clay and silt grade, at 4 μ in the Wentworth scale, which was designed for terrigenous sediments, has little relevance to carbonates. Recorded size ranges in the silty calcilutites include 5–35 μ from part of the Carboniferous Limestone in North Wales (BANERJEE, 1959, p.382), 2.5–14 μ from several knoll-reefs in England and Eire (BATHURST 1959b, p.509), 1–13.5 μ bimodal from a Carboniferous Limestone knoll-reef in northern Ireland (Fig.352 and SCHWARZACHER, 1961, p.1487), 4–15 μ in Silurian shales (FOLK, 1962c), about 4–10 μ in the Upper Jurassic of Dorset, England (BROWN, 1964, p.261, 263), and 0.5–20 μ in the Upper Jurassic of northwestern Germany (V. SCHMIDT, 1965, p.148). The crystals of many of these lutites are large enough for their intercrystalline boundaries to be examined. In my limited experience many of these boundaries are made of plane surfaces, a conclusion which could have important genetic implications (p.505). It is also certain that the crystal sizes of at least some of these coarser calcilutites (the microspars of Folk, p.513) are the result of diagenetic crystal enlargement.

MICRITE ENVELOPES

Many skeletal fragments, ancient and Recent, appear to be sheathed in an envelope of micrite (p.384). Careful examination, however, shows that, far from being an encrustation, this coat is a replacement. It is, in this sense, sharply distinct from the encrusting oölitic or algal coat. In ancient limestones the envelope is composed of micritic calcite. Recent envelopes in Bimini Lagoon, Bahamas, are made of aragonitic micrite (BATHURST, 1966, p.19), though elsewhere in the Bahamas and in south Florida WINLAND (1968) has found them to be high-magnesian calcite. The outer surface of the ancient coat has gentle, smooth contours. The inner surface, abutting against the unaltered skeleton or its cement cast or calcitized replacement, is irregular and commonly shows blebs and tubes of micrite. These correlate closely with the micrite-filled algal bores of the Recent envelopes. The envelope is believed to form largely as a result of the precipitation of carbonate in discarded algal bores (p.388). Ultrastructure is given by LOREAU (1970b).

ANCIENT STROMATOLITES

Although the deposition of ancient limestones is not dealt with in this book, the structures made by stromatolites are of such importance and yet so little known, that the addition of a brief note seemed a matter of some urgency. As can be seen in Chapter 5, the growth of a carbonate stromatolite leads to the development of a laminated rock, generally a micrite, highly porous with an irregular fabric consisting of laminae, partly broken, collapsed or destroyed, and a varied distribution of internal sediments and cement. Fenestral porosity is common. The large structures of domes, hemispheroids, etc. are described in Chapter 5.

    Mention must also be made of the discovery that filaments, probably algal, are preserved in limestones as old as the Cambrian. A demonstration of filaments extracted after very slow digestion in extremely dilute HCl was given at the Geologists' Association, London, by Esther Jamieson, G. Rees and D. J. Shearman in 1966. The limestones included the Jurassic Pea Grit and oölite from the Cotswold Hills, other pisolites from the Permian Magnesian Limestone of northeast England, and some Silurian oölites from the Malvern Hills (D. J. Shearman, personal communication, 1968). I have also prepared filaments from the stromatolitic micrite given to me by Dr. Jamieson from the Devonian of Canada. The filaments have cells and what appear to be reproductive bodies. Though I know that Jamieson is being cautious about reaching a definite conclusion as to whether the filaments are really so old, or just recent introductions, I join with her in finding it incomprehensible that recent filaments can have found their way into the heart of a piece of very dense micrite. At least this line of research is open to anyone prepared to immerse a chip of limestone in 1% HCl overnight and to stain the residue with methyl blue or malachite green. Recently these early claims for the preservation of non-calcareous algae in ancient limestones have been vindicated by the detailed studies made by DE MEIJER (1969) and HUDSON (1970) who digested limestones in 1% HCl and found algae in the insoluble residue. Groups identified so far are the Cyanophyta, Chlorophyta, Rhodophyta and Xanophyta. The ages of the limestones were Paleogene, Jurassic, Carboniferous and Cambrian.

FURTHER READING

The great volume by CAYEUX (1935) is still the obvious starting point for the study of oöids, peloids and calcilutites, supplemented by ILLING (1954), SORBY (1879) and FOLK (1965). On oöids, EARDLEY's (1938) work on the Great Salt Lake is a rigorous foundation, with the paper by NEWELL et al. (1960) on marine Bahamian oöids. The organic, mucilagenous material in oöids and on carbonate grains in general has been demonstrated by NESTEROFF (1955b; 1956a, b), by SHEARMAN and SKIPWITH (1965), CHAVE (1965), CHAVE and SUESS (1967) and CHAVE (in press). On cave oöids and pisolites basic petrographic work is by KIRCHMAYER (1962, 1964), GRADZIŃSKI and RADOMSKI (1967) and HAHNE et al. (1968). For a wide survey of modern carbonates prior to 1930 there is the comprehensive monograph by PIA (1933). Micrite envelopes are dealt with by BATHURST (1964b, 1966). MATTHEWS (1966) has an interesting study of skeletal lutite with petrographic detail, mineralogy and strontium distribution. Other references are given in the text.

Chapter 3

RECENT CARBONATE ENVIRONMENTS 1:
GENERAL INTRODUCTION AND THE GREAT BAHAMA BANK

> Environments are invisible. Their groundrules, pervasive structure, and overall patterns elude easy perception.
>
> *The Medium is the Massage*
> M. McLuhan and Q. Fiore (1967)

GENERAL INTRODUCTION:
RESEARCH ON RECENT CARBONATE SEDIMENTS

Stimulus of marine geology and stratigraphy

Studies of carbonate environments in contemporary shallow seas have drawn inspiration from two main incentives. For the stratigrapher there is the need to apply the new information to the interpretation of ancient sedimentary rocks: for the marine geologist, the clarity and warmth of tropical shallow waters make them ideal for the most direct study possible of sea-floor processes.

The most widespread of past carbonate deposits are those of ancient epeiric and marginal seas and geological interest has naturally been oriented rather strongly toward the task of relating these greater precursors to Recent carbonate floors. In his essay on time in stratigraphy, SHAW (1964) has defined as epeiric those vast seas that once lay across the central parts of continents, to be distinguished from marginal seas which lapped over the continental edges. Whereas the epeiric seas were beyond the range of oceanic tides, the marginal seas stretched from the inner limits of oceanic tides to the outer edge of the continental shelf. Marginal seas have been of more frequent occurrence than epeiric and they were, as they are now, tidal, rich in varied benthos, hydraulically energetic and displayed a fascinating budget of allochthonous (mainly terrigenous) and autochthonous (mainly intrabasinal carbonate) sediments.

To help in the interpretation of the carbonate deposits left by these ancient seas, we can now draw upon researches carried out in Recent shallow seas, for example, the Bahamas–Florida platform, the Gulf of Batabano off southern Cuba, the Campeche Bank off Yucatán, the continental shelf off British Honduras, the Trucial coast of the Persian Gulf, and the coasts of Western Australia and Queensland. Yet these areas are but small marginal seas, and are totally divorced in scale from the ancient epeiric seas which were a thousand, even several thousands of kilometres broad. Nevertheless, an attempt must still be made to understand epeiric sedimentation in the light of modern sea-floor studies. This can only be done, however, by extrapolating from our knowledge of the small marginal seas.

It is important in the discussions that follow, in this chapter and the next, to bear this limitation always in mind.

To swim with snorkel or aqualung over the carbonate sediments of a shallow sea is to watch at first hand the act of sedimentation. The swimmer is unhindered by the mud-clouded waters usual in regions of terrigenous (land derived) deposits, and no longer reliant on the indirect evidence of grab samples, echo-soundings and the small photographic field. He is able to examine with precision not only the sedimentary process, but the activities of numerous living things, and the intricate and immensely important details of topography on the scale of metres and centimetres. The general impression is, above all, one of vigorous life—of corals, algae, sea-grasses, sponges, alcyonarians, bryozoans, echinoderms, tunicates, foraminiferids, diatoms, molluscs, burrowing decapods and fish— and of the complex influence of these organisms on the formation and distribution of sediments.

Yet even here on the modern sea floor the link with stratigraphy is of continuing importance: as a constraint on our researches on Recent sediments, as a second body of complementary evidence, we must bring in the data of stratigraphy. Three-dimensional examination of the Recent sea-floor is inconveniently restricted to the small compass of cores with diameters of two or three centimetres, or to rather wider but much shallower box-cores. By a parallel consideration of well-exposed sections of ancient strata in three dimensions, the researches into the ancient and Recent join to give us as complete an understanding of the history of carbonate sedimentology as we are ever likely to contrive.

Summary of principal past researches

The intention in this and the next chapter is to examine in turn five selected shelf lagoons concerning which sedimentological knowledge is relatively advanced. The Bahamas–Florida platform is the best known of these, with a history of research going back to the work of Louis Agassiz and his son Alexander (AGASSIZ, 1894, 1896) on the reefs and sediments, in the late nineteenth century, and to the opening of the Tortugas Laboratory of Marine Biology where Vaughan began his far-reaching studies of the Recent and Tertiary history of Florida. VAUGHAN (1910) appreciated the variety of source for the grains in the carbonate sediments—cay rock (Pleistocene limestone), foraminiferids, corals and especially molluscs, and reef-rock debris resulting from surf action. He also noted the "white water" as a source of fine sediment, the sea near the reef becoming densely laden (white) with suspended material during storms. He referred to the high proportion of "amorphous carbonate of lime" (see G. C. Matson's table in the same paper) and emphasized the roles of agitation of the water and of photosynthesis in the deposition of lime mud. In a study of oöids VAUGHAN (1914a, b) succeeded in growing spherulites of aragonite in aragonite mud. He was influenced by the brilliant work of the young DREW (1914) who had demonstrated the precipitation of aragonite needles in Bahamian sea water through the action of denitrifying bacteria. Tragically Drew did not live to pursue his studies. Vaughan was joined by Field who later collab-orated in the International Expedition to the Bahamas (in 1930), introducing M. Black to the area. Both men added further to knowledge of the petrography of the sediments and their extent, drawing particular attention to the important aragonite muds. FIELD (1931) showed that the aragonite muds west of Andros Island are not planktonic in origin. Contrary to modern findings (p.276) he regarded them as products of erosion of the Andros coast. To BLACK (1933a) we owe the discovery and detailed analysis of the sediment-binding activities of filamentous algae (p.217). THORP (1936) made mechanical and compositional analyses of samples from the Bahamas and Florida and showed that the major part of the sediment was skeletal, with an order of decreasing frequency thus: coralline algae–molluscs– foraminifera–corals. He demonstrated, too, that on the reef flat and reef slope the sediment is coarser than in the less turbulent troughs. On the question of pellets he suggested that some pellets might be of faecal origin though he felt that other origins were not excluded. He measured the Mg/Ca ratio in the carbonates and related it to the dominance of the magnesian coralline algae.

    Since 1945, in the Bahamas, ILLING (1954) has provided the main basis for present-day petrological studies. The numerous researches of Newell, Imbrie and Purdy have greatly extended the areas mapped, and the amount of quantitative petrographic data, and have led to a closer understanding of the relationship between sedimentation and organisms. Smith and Cloud have greatly extended our knowledge of the chemical parameters. About the same time that these workers were opening up the Bahamas, Ginsburg, in Florida Bay and the reef belt, was measuring lateral changes in biofacies and sedimentary textures. From this period the work of Lowenstam was to be increasingly influential in the ecological field.

    Far to the south, off Cuba, DAETWYLER and KIDWELL (1959) were meanwhile recording sedimentary variation over the wide extent of Batabano Bay. Work outside the western Atlantic was slower to expand, but the investigations begun by Illing and Shearman in the Persian Gulf, on the reef–lagoon–sabkha environment, and by Houbolt on the adjacent off-shore sediments, have now provided an im-portant new perspective. Lastly, in this very cursory catalogue it would be wrong to ignore the inspiration of Eardley's most thorough work on the distribution, deposition and chemistry of Recent carbonate sediments. Though carried out in a lake, the influence of EARDLEY's (1938) studies in Great Salt Lake, Utah, has been far reaching.

The complex environment

In a region of shallow-water carbonate sedimentation, many processes, physical, biological and chemical, are in everchanging relationship. The partial pressure of CO_2 in the water fluctuates as a result of photosynthesis, respiration, precipitation of $CaCO_3$, evaporation, rainfall and fresh water run-off from the coast. Thus, as the solubility product for $CaCO_3$ fluctuates, there is a continuously varying balance between addition of carbonate ions to the water and their removal. Plants and animals not only secrete carbonate skeletons but break them and abrade them. Everywhere the water is moving under the influence of tides and winds, carrying with it dissolved carbonate and suspended calcilutite and, in places, dragging lime sands across the floor. In the surf zone on open shores sorting is most efficient and grains are abraded to a shiny polish. Skeletons, oölites, aragonite muds, and faecal pellets formed in one place may be moved elsewhere to be mixed with the sediments derived from other environments. Whereas some reef and back-reef detritus may remain within metres of where it originated, skeletal beach sands, oölite bars, or muds may move distances measured in kilometres. But most of the live participants in this restless scene will eventually disappear without trace and there will remain only a few mineralized and damaged skeletons as evidence of a once varied and numerous population and a dynamic environment.

THE GREAT BAHAMA BANK

Dimensions

The Bahamas Platform, of which the Bank is a part, is a barely submerged plateau off the coast of Florida (Fig.127) constructed of Pleistocene limestones on Tertiary and Cretaceous limestones and dolomites covered with a film of Recent carbonate sediments. It is some 700 km from north to south with its southern end, near the Ragged Islands and the Old Bahama Channel, lying about 22 °N, just south of the tropic of Cancer and due east of Cuba. From the Bimini Islands at its western edge to Eleuthera at the eastern edge is about 300 km. The platform is interrupted by three wide and deep channels—Providence Channel, Tongue of the Ocean and Exuma Sound. These have axial depths ranging from 600–3,500 m, 1,300–2,500 m and 1,600–2,000 m respectively, each deepening eastwards toward the Atlantic abyssal floor. Between this platform and the stratigraphically related Florida Platform, the Florida Straits descend only to 850 m.

The Great Bahama Bank lies between the Florida Straits and Tongue of the Ocean (a width of about 140 km). It is generally submerged to a depth of less than 10 m, the greater part of the floor being covered by less than 7 m of water. Only around the edges of the Bank does the floor slope steeply from 10 m to 200 m in a distance of 2 km or less. To the north the Bank is separated from Little Bahama Bank by Providence Channel, to the south from Cuba by the Old Bahama Channel, giving a length of about 450 km. The thickness of the unconsolidated Recent carbonate sediments, lying on the basement of calcite-cemented Pleistocene limestones, is probably nowhere greater than 3.5 m, though measurements have been

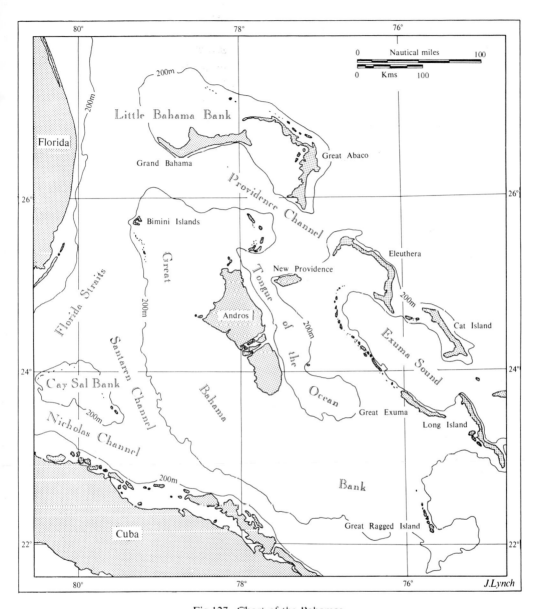

Fig.127. Chart of the Bahamas.

made in only a few widely scattered places. Radiocarbon dates indicate a probable late Würm (Wisconsin) age for the uppermost parts of the Pleistocene limestone. Recent carbonate sedimentation must have started, perhaps, a little earlier than 4,000 years ago after the beginning of the post-Würm glacial retreat and rise in sea level. (Discussion of radiocarbon data from Pleistocene and Recent sediments in CLOUD, 1962a and MARTIN and GINSBURG, 1965.)

The Great Bahama Bank is the part of the Bahamian plateau which has been studied in most detail and it is interesting to compare its size with some well-known regions of carbonate shallow-water deposition in the past. It is similar in dimensions to the part of Ireland covered by the shallow Viséan (Upper Mississippian) sea, between the southern fold belt and Donegal. The Bahama Bank also resembles in area the Jurassic Paris Basin and the Pennsylvanian Paradox Basin (Utah–Colorado), yet it is about a 1/20th part of the area of the Middle Mississippian (Osage) region of marine carbonate deposition in the United States.

Topography

The Bank is a broad but exceedingly shallow shelf sea (Fig.128), yet, for the most part, the depth of water is greater than the thickness of the underlying layer of uncemented Recent sediment. This sediment layer is pierced in many places by inliers of Pleistocene oösparite and biopelsparite. This is particularly so around the edges of the Bank, where the Pleistocene basement rises to form islands (cays) composed of cemented Pleistocene oölite dunes and beach ridges. These marginal rocky shoals, combined with the lateral extent of the Bank, greatly reduce the exchange of tidal water with the surrounding deeps and this isolation has led to the development of an extensive region of relatively sheltered, undisturbed water. Here even wind is powerless to raise a heavy sea, owing to the frictional resistance of the bottom. On the marginal Pleistocene shoals, particularly on the western edge of the Bank, Recent oölites have grown. East of Andros Island there is a line of coral–algal reefs. These Recent shoal accumulations have further restricted the mixing of ocean with Bank waters. Beyond the Bank edge, there is a narrow shelf the topography of which suggests drowned beach-ridges and erosional terraces. The true edge of the shelf lies at a depth of about 30–40 m and is succeeded by a "precipitous marginal escarpment" (NEWELL et al., 1959), interpreted by NEWELL (1955) as a drowned Tertiary reef. Two gently shoaling, broad rises cross the Bank (Fig.128), the Bimini axis running from the Bimini Islands to northern Andros Island and the Billy Island axis, from the island of that name near Williams Island in a direction almost due westward. Over these axes the sediment is apparently thinner and bare Pleistocene limestone is more commonly exposed than elsewhere.

East of Andros Island there is a discontinuous line of limestone cays and reefs, about 0.8–4 km in from the edge of the shelf, enclosing a lagoon 1–4 km wide with a depth of water from 1.5–5 m.

More than half the surface of the scattered islands is less than 3 m above mean sea level. The islands are concentrated along the edge of the Bank, particularly on the eastward side. Andros Island rises, in a few places along its eastern seaboard, to more than 30 m in a north-south trending ridge, overlooking the back-reef lagoon and the reef belt and, beyond them, the deep waters of Tongue

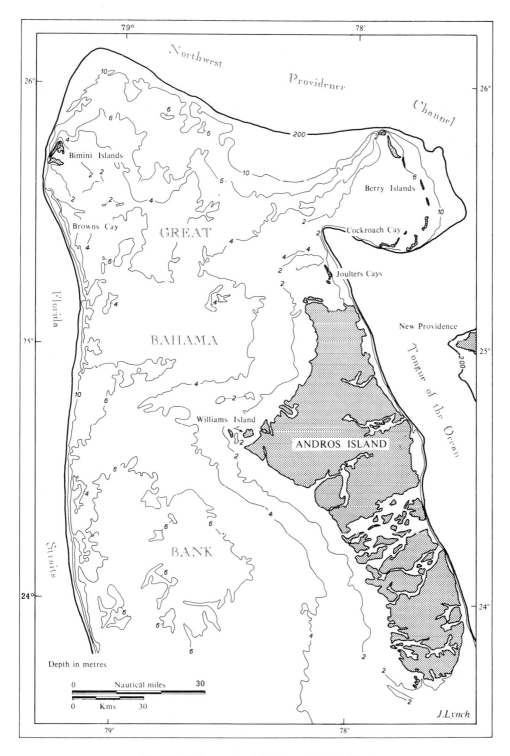

Fig.128. Bathymetry of the Great Bahama Bank.

of the Ocean. The main extent of this large island lies to the west of this ridge. It is a low and marshy region, up to 50 km wide, of mangrove swamp, ponds, tidal channels, sink holes, supratidal mud flats (p.534), and sporadic low mounds (locally named "hammocks") of Pleistocene limestone supporting a few pines (SHINN et al., 1969). Apart from Andros there are twenty principal islands in the Bahamas and thousands of cays and rocks. In general the coasts of these are either low limestone or lime mud cliffs, rising a metre or so above sea level, or calcarenite beaches.

Water movement

The movement of the water across the Bank is of prime importance, both in its effect on the biological communities, through the supply of nutrients, and in its influence on the distribution of carbonate in solution (SMITH, 1940; COSTIN, 1965; B. KATZ, 1965; TRAGANZA, 1967). Tidal water moves sluggishly on and off the Bank, in a more or less radial pattern, concentrated in tidal channels by the cays, oölite shoals, reefs and calcarenite bars: the flood current is stronger than the ebb. The tidal currents near the ocean are about 25 cm/sec (0.5 knot), except in the channels where they may exceed 100 cm/sec (2 knots). Though the mean tidal range in the open ocean is 0.78 m (Nassau), this falls off inward from the Bank edge so that the mid-Bank range is negligible. It is difficult to know how important these every-day tidal movements are in the net lateral transport of sand-sized grains. Recent work indicates that a far more effective, albeit very rare, influence is the occasional hurricane which can bring about a temporary rise of sea-level at the margins of the Bank as great as 3 m, allied to short term, high velocity currents (p.122; BALL et al., 1967).

The sluggish radial movement of tidal water is easily modified by winds. The Bahamas lie in the belt of trade winds and during the summer, from March or April to the end of August, the prevailing wind is from the east or southeast: in September it begins to veer toward the north. Throughout the winter months it blows from the northeast or east, with violent interruptions in November and December by northwesterly gales. As a result of this seasonal pattern of winds, there is, in the summer, a net westward drift of water across the Bank. In the winter the residual current is southward and, during the northwesterly gales, eastward.

Temperature and salinity

Throughout the year the temperature of the open surface water in the Bahamas has a range of mean monthly values (February–August) from 22 to 31°C. The rather high figure for this latitude is a result of the northward movement of the Gulf Stream, bringing warm water from the Caribbean (SMITH, 1940; CLOUD, 1962a; BROECKER and TAKAHASHI, 1968). However, it is as well to remember that locally, and for short periods, in a metre or less of water, the temperature range can be more extreme than this with important consequences for carbonate precipitation (p.288). The lateral variation in open water is slight: during May 1955 the surface temperature of the Florida Straits was 27.9°C (mean) and this increased across the Bank to reach a maximum of 28.4°C (mean) off Andros Island—a rise of only 0.5°C (CLOUD, 1962a, pp.12–13). The water over the Bank is everywhere turbulent enough, as a result of wind driven waves, to ensure uniform distribution of temperature vertically.

Over a wide area in the lee of Andros Island the various forces acting on the water throughout the year are to a large extent opposed to each other, so that the net drift of this water in any one direction is slight. The roughly concentric arrangement of isohalines, which develops in the summer months around Williams Island (Fig.235, p.285), is never quite destroyed by the winds of winter. A nucleus of hypersaline water persists in the neighbourhood of Williams Island from one year to another, escaping dilution by waters of lower salinity because of its remoteness from the ocean.

The reasons for the persistence of this salinity pattern (Fig.235) are threefold: (1) the small tidal exchange with oceanic water, (2) the shelter from the full vigour of the trades provided by Andros, and (3) the opposing directions of the winds in summer and winter. The negligible tidal range has already been mentioned. The sheltering effect of Andros Island is such that the strong currents moving westward across the Atlantic, driven by the trade winds, are deflected to the north and south of the island so that only a sluggish flow remains to follow the west Andros coast, southward from Joulters Cays and northward from the southern tip. The movements of these two flows along the coast, especially in the summer, are apparent from the parallel orientations of the isohalines (Fig.235). The flows converge on Williams Island and escape westward. Though they bring into the central area a supply of less saline water from Tongue of the Ocean, it is clear that this dilution is not enough to offset the effects of evaporation and to destroy the concentration of hypersaline water near Williams Island. In the summer the prevailing east and southeast winds draw the hypersaline waters gradually westward toward the Florida Straits (Fig.235). Progress in that direction is slow as the water is shallow and friction over the floor correspondingly effective: also the wind direction is variable.

During this period of sluggish westerly drift the intense evaporation caused by the tropical heat of summer regenerates the nucleus of hypersaline water, causing a steepening of the salinity gradient which rises from the peripheral oceanic deeps toward the region of Williams Island. In the summer when oceanic surface salinities are 36‰, values of 46‰ have been recorded just off the Andros coast (Fig.235).

As the northerly winds of winter break up the summer pattern of isohalines, the hypersaline mass begins to drift toward the southeast. This is particularly evident during the northwesterly gales. There is, in the winter, a movement of

oceanic water with salinity of about 36‰ on to the Bank and there is also some
dilution by rain water issuing from the Middle Bights of Andros Island. This fresh
water runs off Andros Island as a result of the annual rainfall of 100–150 cm,
mainly between May and September. The effect of this twofold dilution is to
restrict the salinity over the Bank to a maximum of 38‰, except for the immediate
lee shore of Andros Island, near Williams Island, where slightly higher values have
been recorded. It seems likely that excess Bank water escapes southeastwards.

Lithofacies, habitats, communities

Lithofacies here refers to a spread of sediment with lithological characteristics
that distinguish it from other neighbouring areas of sediments (Fig.129). These
characteristics are the nature of its constituents, their proportions, the grain size,
the sorting, degree of abrasion, attack by microscopic boring algae and so on.
Community refers to the sum of living organisms in a particular place (Fig.130).
Habitat means the environment in which an organism lives: it is characterised
not only by topography, temperature, salinity, turbulence and other physical and
chemical factors, but also by the other living things present and by the effect of the
organism itself upon its own surroundings. However, because of the difficulties of
describing such a complex affair, the word "habitat" is commonly used in current
Bahamian studies in the more restricted sense of topography and substrate.

Petrography and environment. Anyone who has compared the constituent
grain composition of a modern lime sand with the details of its associated biological
habitat and living community must surely have been struck by the poverty of
information preserved in the sand. Of the animated scene of green *Thalassia*
bending before the tide, its blades encrusted with foraminiferids, algae and
hydroids, interspersed with burrowing bivalves and annelids, echinoids, holothurians, calcareous and non-calcareous algae—or of the grandeur of a reef with
its several corals, and a variety of molluscs, echinoids, foraminiferids, fish, anemones and algae living in attractive niches, with alcyonarians and hydrocorallines
on the adjacent rock platform—of all this there remains in the final accumulation
of carbonate grains only a faint echo. It has been estimated that commonly less
than 5% of a local biota secretes hard parts that can survive as fossils in the ultimate
sediment (NEWELL et al., 1959, p.200). But this degree of preservation is high
compared with the scant evidence that remains of such factors as temperature,
salinity, turbidity and turbulence, or of predation, biocompetition and commensal-
ism.

At this point it is as well to bear in mind that, while the sediment is bathed in
sea water, no mineralogical changes take place as part of a move toward greater
mineral stability. TAFT and HARBAUGH (1964), BERNER (1966a) and TAFT (1968)
examined cores which penetrated 1–2 m below the sediment–water interface and

TABLE V

CORRELATION BETWEEN LITHOFACIES, HABITATS AND COMMUNITIES
ON THE GREAT BAHAMA BANK

Lithofacies	Habitat	Community
coralgal (with small areas of solid reef)	reef	*Acropora palmata*
	rock pavement	plexaurid[1]
	rocky shore	littorine
	rock ledges and prominences	*Millepora*
	unstable sand	*Strombus samba*
oölitic and grapestone	stable sand	*Strombus costatus*
oölite	mobile oölite	*Tivela abaconis*
mud and pellet mud	mud and muddy sand	*Didemnum candidum* (locally *Cerithidea costata*)

[1] Plexauridae (sea whips), a family in the order Gorgonacea.

found no signs of mineralogical alteration related to depth (p.361; see also SIEVER et al., 1965).

The data. Nevertheless there are distinctive relationships on the Bank between the distributions of lithofacies, habitats and communities. But before examining them certain concepts must be considered briefly. PURDY (1963b) has mapped the lithofacies north of latitude 23° 30′ N on the basis of point-counts of thin sections of 218 impregnated sediment samples. NEWELL et al. (1959) have described the communities and their habitats (Table V).

Relationships. The most obvious and direct relationship on the Bank has been found to be between communities and two physical factors—the **substrate** (rock, sand or mud in, or on, which the organism lives) and turbulence. This does not mean that, for example, temperature, salinity and light are unimportant, but over the Bank their lateral variation is gradual whereas the lateral changes in substrate and turbulence are relatively sharp. Depth, too, is important: locally it is related closely to both turbulence and current velocity, either by affording a measure of shelter from surf on the lower parts of the fore-reef slope or by restricting and increasing tidal velocities in narrow channels. The influence of depth on the vertical distribution of organisms is, however, in many areas, relatively unimportant because the maximum depths at which many, perhaps most, of the benthonic species are known to live elsewhere are not attained anywhere on the Bank.

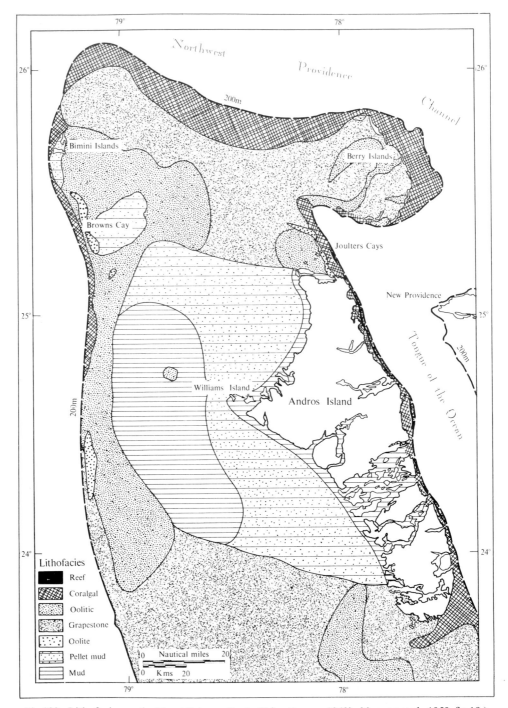

Fig.129. Lithofacies on the Great Bahama Bank. (After PURDY, 1963b; NEWELL et al., 1959, fig.10.)

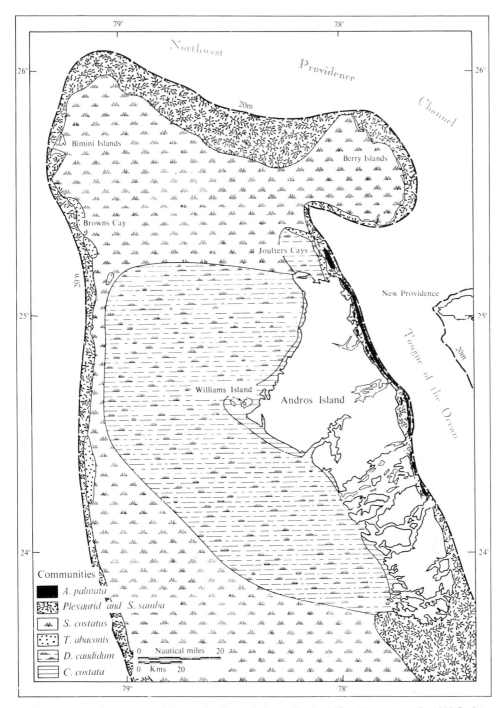

Fig.130. Organism communities on the Great Bahama Bank. (After NEWELL et al., 1959,fig.9.)

Boundaries. Communities on the Bank tend to have sharper boundaries between them than do habitats, and the least abrupt boundaries are those of lithofacies. It is apparent, therefore, that the sharpness of the boundary separating one community from the next cannot depend solely on a change in habitat, but must rely on other factors, such as biological competition. A number of communities do expand along a roughly circular growth front: clumps of *Thalassia* with their accompanying biota are an example. In parts of Bimini Lagoon these plants on the periphery of one community compete with patches of the red alga *Laurencia* or with thickets of the coral *Porites*, or the associated bathroom sponge *Ircinia*. The floor of Florida Bay seen from the air is remarkable for what appears to be a pattern of interfering circles and lobes of a great variety of sizes and colour, and it seems likely that these are competing communities. This pattern appears in air photographs of swamps west of Abaco Island (ILLING, 1954, p.86).

The more gradational boundaries between lithofacies arise not only from the fact that the sedimentary grains commonly relate to only a small fraction of the community, but also from the inevitable lateral transport of grains by tidal currents and waves. The resultant distribution of grains does not necessarily reflect at all closely the pattern of habitats or communities at the time that the sediment was accumulating—a situation that might be noted more often in the reconstruction of fossil lithofacies. In the working out of such reconstructions it is also too often assumed that the sediment was in equilibrium with its environment.

Looking down through the water at Bahamian sea beds it is by no means obvious whether or not the sediment on the floor is fully adjusted to the dynamic situation around it. FOLK and ROBLES (1964) in a remarkably interesting paper on part of the Alacran Reef complex, have shown that intertidal and shallow subtidal carbonate sediments, of varied skeletal composition, can possess textures that are closely adjusted to the local hydraulic regime. On the other hand, in more sheltered areas of sea floor, especially where the sediment is stabilized by the benthonic flora or by a subtidal algal mat, the degree of adjustment can be very small. Poorly adjusted sediments of this kind have been described by BATHURST (1967a, p.99) from Bimini Lagoon, Bahamas. The well adjusted sediments described by Folk and Robles have sorting values of about $\sigma = 0.5 \, \varphi$ (well-sorted; FOLK and WARD, 1957). The Bimini samples have sorting of about $\sigma = 1 \, \varphi$ (moderately to poorly sorted).

Unpublished data by Bathurst on sediments from the grapestone lithofacies near Cockroach Cay, Berry Islands (Fig.128), gave similarly poor sorting on an area of floor with a well developed subtidal algal mat. Folk and Robles showed clearly the influence on the sorting process of particle shape and effective density and, specially interestingly, the influence of the pore structure and permeability of the deposited loose sediment upon its stability. The degree of adjustment attained depends upon the duration of exposure to the various modifying processes, such as browsing, sorting, and surf abrasion. In all sediments, at the time that they are finally buried, there must be some failure in adjustment.

Relevance of lithofacies. Lithofacies do, of course, reflect both parent community and habitat, but in varying degrees. In Bimini Lagoon the *Halimeda* and peneroplid (*Archaias*) grains reflect the *Thalassia–Strombus costatus* community, and the poor sorting reflects the tranquil habitat. On an oölite shoal, life is scarce and the oöids recall more than anything the hydraulic and chemical aspects of the habitat. Particularly confusing is the accumulation of coral-coralline algal debris that is found as a sand but originates as a solid reef. Some habitats and communities are scarcely represented at all by sediment, as, for example, the sponge–alcyonarian–*Sargassum–Laurencia* community of some rock pavements.

The general relationships between Purdy's lithofacies and the habitats and communities of Newell and others are shown in Table V. The correlation is not everywhere as close as this, but the overall tendency is not in doubt (Fig.129, 130).

Fig.131. Aerial photograph of reef complex, off Walkers Cay, Little Bahama Bank. From top left to bottom right the environments are: back-reef lagoon with patch reefs (water depth 6–9 m), reef flat and crest (parts awash at low tide), fore-reef debris succeeded by groove (chute) and spur (mainly 12–23 m). (Courtesy of C. W. Ellis.)

The coralgal lithofacies and its environment

The reef habitat. The reefs are concentrated on the outer part of the marginal rocky platform (Fig.129), on the eastern, windward side of Andros Island. There they form, with cays of Pleistocene limestone, a line of surf-capped barriers, parallel to the coast, broken at many places by tidal channels (Fig.131). Smaller patch reefs are scattered about a lagoon between the reef and the mainland. Elsewhere on the Bank reefs are rare. The frame-builders are corals: they and the reef debris are strengthened and bound together by encrusting coralline algae. In niches on the reef and in its cavernous interior live a variety of molluscs, echinoids, foraminiferids, hydrocorallines, annelids, alcyonarians and fish.

All skeletal material is much bored by sponges (NEUMANN, 1966) and algae (RANSON, 1955c, d). Encrusting coralline algae have been found by RANSON (1955a) to dissolve the calcareous surfaces on which they grow.

On the reef-crest the golden, tree-like *Acropora palmata* (Fig.132), with stout, widespread branches, absorbs much of the wave energy. The tips of these corals are commonly just above water at low springs. Optimum conditions for growth are found, not at the surface of the water, but about a metre below the surface at low spring tide, owing to the damaging intensity of the ultra-violet light. *A. palmata* is a large colony of animals and some species are as much as 3 m high with trunks 30 cm thick. Seaward of the reef-crest, to a depth of about 3 m, *A. palmata* is

Fig.132. *Acropora palmata*. Relief about 1 m. Off Great Abaco. (From STORR, 1964.)

Fig.133. Sea fan, order Gorgonacea, family Gorgoniidae. Height of colony about 0.5 m. (From STORR, 1964.)

Fig.134. Sea whip, order Gorgonacea, family Gorgoniidae. Height of colony about 0.8 m. Water depth 1.5 m. Berry Islands, Bahamas. (Photo by R. Adlington, with the author.)

accompanied by *A. cervicornis* (Fig.172) and by waving purple and yellow fans of *Gorgonia flabellum* (Fig.133), also by small shrub-like alcyonarians (Fig.134), brown patches of the poisonous *Millepora* (Fig.135), and brain corals (*Diploria*; Fig.137) a metre or more across[1]. Bare rock and fragments of dead coral are encrusted by the coralline algae (Fig.136) *Goniolithon* and *Lithothamnium*. Everywhere there are fish. Farther down, in the dim light below 3 m or so, the large, massive, fungoid or mound-like *Montastrea* dominates the reef-front, with *A. palmata* having, in the less turbulent water, a more slender form. In more sheltered places the stubby, branching coral *Porites porites* (Fig.155, 156) helps to make niches for molluscs, annelids, foraminiferids and echinoids. Rather more exposed surfaces bear the massive corals *Porites astreoides*, *Diploria* and the small pebble-like *Siderastrea*. Vase-shaped sponges are numerous. Rock-ledges are encrusted by the knife-sharp, leafy coral, *Agaricia* (Fig.138). The ragged reef-flat leeward of the crest is at most 50 m wide and seems to consist largely of Pleistocene limestone encrusted with coralline algae. It lies a few centimetres below water level at low

[1] Photographs of Jamaican reefs are shown here in the absence of suitable pictures from the Bahamas.

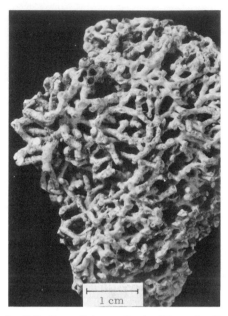

Fig.135. *Millepora* colonies with about 30 cm of relief. Great Bahama Bank. (Photo by R. Adlington, courtesy of N. D. Newell.)

Fig.136. A branching coralline alga, *Lithophyllum*. (Courtesy of G. F. Elliot.)

Fig.137. Hemispheroidal colonies of *Diploria strigosa*, with a low ratio of surface to mass in shallow water of the breaker zone (compare fig.138). Typical reef crest framework of *Acropora palmata* in background, with vertical corrugations of *Millepora alcicornis* in left foreground. Port Royal barrier reef, near South Cay, Jamaica. (From GOREAU, 1959.)

Fig.138. *Agaricia* on the fore-reef slope, at a depth of 50 m, Maria Buena Bay, Jamaica. High ratio of surface to mass, typical of deep water: compare fig.137. The outwardly solid appearance of the cabbage-like corals hides extensive boring by sponges in their bases. (From GOREAU and HART-MANN, 1963.)

springs. The reef as a whole has an open texture with a vigorous internal circulation of water.

The reef-front is broken up into alternate spurs and grooves running seaward perpendicularly to the line of reefs (as in the Florida reef belt, p.150). The grooves, easily recognised from the air (Fig.131) by the pale turquoise of their sandy floors, commonly ascend into the reef-crest. The spurs are vast buttresses, from 5–45 m across, faced with living corals and coralline algae. The grooves between them have vertical or overhanging sides and are up to 5 m wide and 3 m deep. They are floored at their upper ends by bare rock or reef with debris-filled

Fig.139. A submarine cliff, lately cleared of coral by a slide, has been recolonized by sponges, algae, and by the tree-like antipatharian on which sponges are growing. *Halimeda* is not visible. The Antipatharia, known also as black or thorny corals, are an order of the Zoantharia. Depth 35 m. Cardiff Hall, Jamaica. (From Goreau and Hartman, 1963.)

holes, while, lower down, ripple-marked lime sand is mixed with boulders of dead reef rock. The sandy-gravelly sediment in the grooves is rushed upslope during storms, but, in quiet weather, is gradually moved seaward. The fore-reef slope continues downward, richly populated with corals and giant bushes of *Halimeda*, possibly to a depth of 50 m or more. Fig.139 shows a fore-reef cliff in Jamaica.

On the lee side of the reef there is a pitted and corroded surface of Pleistocene limestone and cellular *Lithothamnium* rock, with erect blades of *Millepora* and a host of molluscs, foraminiferids, echinoids, annelids and alcyonarians. The surface slopes gently to the sandy lagoon floor.

The unstable sand–rock pavement habitat: back-reef lagoon. The water in the Andros back-reef lagoon has a depth generally between 2 m and 6 m (Fig.128–130). On the floor, patches of the turtlegrass *Thalassia* grow in a lime sand composed of skeletal debris, some faecal pellets and scarce lithoclasts of Pleistocene limestone. These *Thalassia* patches, increasing in size and number toward the mainland, are associated with a diverse biota of calcareous algae (such as *Halimeda*), burrowing bivalves, gastropods, holothurians, echinoids, annelids and crabs. The foraminiferids in the lagoon are dominated by peneroplids, rotaliids, miliolids and nonionids. The blushing *Rotalia rosea* is common. On the more stable sand there are conical mounds with apical exhalent holes and adjacent inhalent holes: their bases are as much as 20 cm across. They are probably made by a thalassinidean decapod such as *Callianassa* (p.127). In the expanses of bare, ripple-marked sand toward the reef, *Halimeda* and other calcareous algae are to be found, with sponges where the sand is thinly spread over the limestone: there are also the large herbivorous gastropod *Strombus samba* and the wafer-thin echinoid, *Mellita sexiesperforata* (sand dollar). Reef debris is swept into the lagoon, especially during storms. Calcilutite must be formed by the disintegration of limestone and sand grains by a variety of organisms, such as crabs, *Strombus*, the echinoid *Diadema*, the boring sponge *Cliona*, boring molluscs, filamentous and coccoid blue-green algae and parrot fish. This fine sediment is presumably carried out to sea since there is little in the lagoon sands.

Scattered irregularly over the lagoon floor are patch reefs, often tens of metres in diameter. Here a typical reef biota of the *A. palmata* association flourishes on exposed masses of Pleistocene limestone. The most vigorous growth is around the periphery of the reef, much of the middle consisting of dead coral encrusted with calcitic coralline algae. From the air it is apparent that most of the patch reefs are bounded by narrow girdles of loose calcarenite. Areas of rock pavement near the reef support a typical alcyonarian (plexaurid) community, like that on the rock pavement toward the edge of the shelf. The extensive algal stromatolites in the lagoon (MONTY, 1965, 1967) are described in Chapter 5.

The rock pavement habitat. Between the reefs and the edge of the shelf, out beyond the oölite shoals and the island beaches, the terraced floor of Pleistocene

Fig.140. Roundish masses of *Ircinia*, about 25 cm high, and other sponges in a rock and rubble channel. Depth 2.5 m. Bimini Lagoon, Bahamas.

Fig.141. *Halimeda*, without holdfast. (Courtesy of G. F. Elliot)

limestone slopes down from about 9 m to 50 m. This rocky substrate, covered in places by thin transient sheets of lime sand, supports a community adapted to a sessile life in a turbulent environment. There are patches of *Thalassia* in the sand, clearly visible from the air as deep purple areas against the paler surroundings. In the highly turbulent wave driven, shallow water only a few sponges (*Ircinia*, Fig.140) and massive corals such as *Montastrea* and *Diploria* can survive. Farther out, on the less disturbed bottom, these organisms are more plentiful and larger, and are joined by the aragonitic algae *Halimeda* (Fig.141), *Udotea*, *Penicillus*

(Fig.153, 156 on p.126, 130 resp.) and *Rhipocephalus*, all of which grow holdfasts into the sand. Living in this sand is the red, non-calcareous alga *Laurencia* (Fig.144) in little bushes about 10–20 cm high, the wide, flat-bladed *Thalassia* (Fig.153, 154 on p.126, 127 resp.) and the long green spikes of *Syringodium*. The rock-fixed community, above all the massive corals and the flexible alcyonarian, dominates the scene. The alga *Sargassum* also clings to the rock and its brown tendrils can often be seen floating on the water of the adjacent ocean: *Halimeda* bushes are profuse. Besides *Montastrea* and *Diploria* there are the little *Siderastrea* and the leafy *Agaricia*. In places the corals form knolls; generally they make appropriate niches for molluscs, echinoderms and bryozoans. Patches of this pavement with its typical community crop out in many parts of the Bank, wherever Pleistocene limestone is exposed, even for a few square metres.

The rocky shore habitat. Conditions in the rocky shore habitat (Fig.142) are more extreme than in any other. Here, according to the tide and the weather, the limestone surface may be subaerially exposed to intense sunlight, heat and drying, or it may be battered by the storm waves of deep water. At other times the rock surface is bathed in as much as a metre of calm water. The yellow zone (Fig.142) is

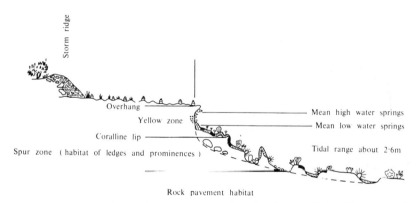

Fig.142. Inshore communities, Rabbit Cay Point, Bimini, Bahamas. (After NEWELL et al., 1959, fig.12.)

divided into upper and lower zones on the basis of a variety of chitons, algae, gastropods, bivalves and sponges (NEWELL et al., 1959, p.206). Typical of this kind of coast is the intricate and irregularly corroded and sharp-edged surface above the overhang, and the smooth vertical or concave surface of the yellow zone, with a lip of encrusting coralline alga (*Porolithon*) at its base that is just exposed at the lowest, spring tides.

The habitat of rock ledges and prominences. The top of this generally narrow habitat, up to 5 m wide and lying in 0.8–1.5 m of water, is just exposed at low spring tides. The surface of rocky ledges, prominences and pot-holes is subjected to intense wave turbulence. It is colonised by adhering brown blades of *Millepora*, and sponges, algae, alcyonarians, and many boring organisms, including algae and the sponge *Cliona*. There are shrubs of *Laurencia* and the little, branching, high-magnesian calcitic red alga, *Goniolithon*. Twenty species of gastropod and bivalve, a variety of corals, echinoids (especially *Diadema*), ophiuroids, anemones and crabs have been recorded.

The unstable sand habitat. Outside the narrow though widely distributed habitats of the reef, rock pavement, rocky shore and its ledges, there are extensive deposits of unstable carbonate sand making a continuous girdle round the Bank (Fig.129). This peripheral band of sand is for much of its length only about 5 km wide, but opposite Providence Channel, where it receives the full force of the more northerly winter winds and gales, and to the south of Andros Island, the width is between 20 and 40 km. Vigorous tidal action ensures that the sand floor is strongly ripple-marked. In many parts of the unstable sand habitat the sand has been built into cross-stratified submarine sand bodies (dunes) with relief in some places as great as 3 m. These bodies are commonly elongate, extending for 1 km or more, with surfaces decorated with ripples having wave lengths between 7 cm and 1 m. Some bodies form megaripples which are described on p.121. BALL (1967) has analysed

 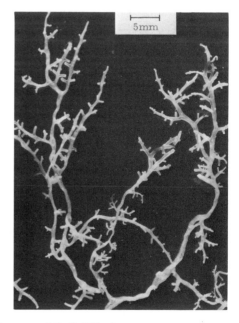

Fig.143. *Acetabularia*. (Courtesy of G. F. Elliot.)

Fig.144. *Laurencia*. (From SCOFFIN, 1970.)

the distribution and internal structure of Bahamian carbonate sand bodies and has shown how sensitive the direction of building (and thus the direction of cross-strata) is to the local bathymetry, to tidal and wind driven currents and wave refraction, and to the changing shapes of the sand bodies themselves.

On such an unstable substrate the vegetation is understandably sparse: there are occasional clumps of *Thalassia* (Fig.153), a scattering of aragonitic algae (*Halimeda, Udotea, Acetabularia*—Fig.143), bushes of *Laurencia* (Fig.144), some *Padina* (a brown alga) and the typical foraminiferid *Rotalia rosea*. At least nineteen species of bivalve have been recorded, also the sand dollar *Mellita sexiesperforata* and polychaete worms. Mounds left by burrowing shrimps are present, but rare compared with the number in the stable sand and mud habitats.

The petrography of the coralgal lithofacies. From the five sharply distinctive habitats one lithofacies has been described so far—the coralgal lithofacies of PURDY (1963b). It is interesting to examine the constituent composition of the coralgal sand and to ponder on the extent to which this information by itself succeeds in reflecting the nature of the communities and habitats from which it was derived. It is necessary to keep such a comparison mainly on the level of phyla or families because, outside the Foraminiferida, Algae and Bivalvia, it is commonly impossible to identify grains more accurately.

The characteristic organisms in these five habitats are those that live on rock bottoms or in unstable sand. They are mainly corals and coralline algae, alcyonarians, and a mixed collection of molluscs, echinoids, holothurians, crabs and red and brown algae and sporadic *Thalassia*.

The first thing that strikes one about the coralgal sand (Fig.145) is that more than half the grains coarser than 125 μ are non-skeletal. The origins of these are uncertain: they may be faecal pellets, inorganic accretions or skeletal grains micritized by boring algae (p.388). Of the remaining half of the grain assemblage it is worthwhile recalculating the recognisable skeletal grains to 100% (Fig.146). It can then be seen that they comprise four nearly equal groups, corals + coralline algae, *Halimeda*, foraminiferids, and molluscs.

The mean constituent composition of the sand was based by Purdy on 33 samples taken over an area of about 3,600 km² of lithofacies, so it would be unreasonable to expect a close correlation between skeletal composition and the composition of any single community. On the other hand, it is at this very level of scant information that the stratigrapher must often try to make an environmental interpretation of ancient carbonate rocks.

The significance of the grain counts. The dominance of the corals and coralline algae is clear, particularly in comparison with the constituent analyses of the other lithofacies (Fig.146). But the stratigrapher unversed in the composition of modern benthonic communities might be excused for supposing that the sea floor

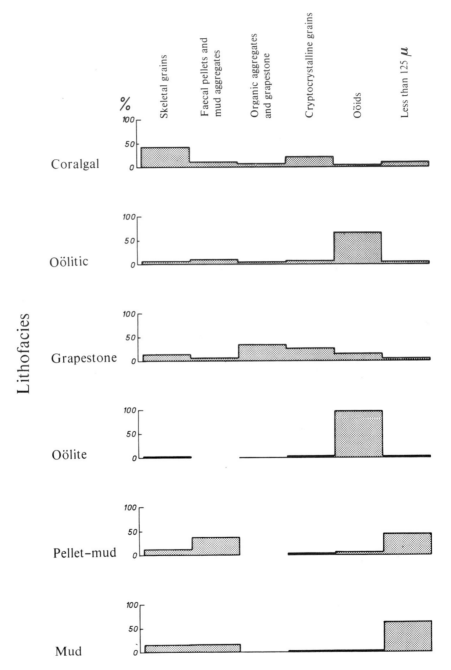

Fig.145. Percentages of *all* constituent particles, in the Bahamian lithofacies. (From data in PURDY, 1963b.)

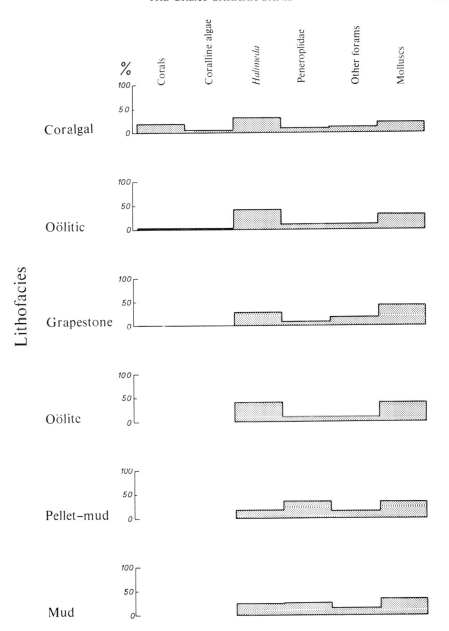

Fig.146. Percentages of skeletal particles only in the Bahamian lithofacies. (Data from PURDY, 1963b.)

was here inhabited by mixed thickets of corals and *Halimeda*. Corals are, of course, greatly under-represented compared with *Halimeda*. One reason for this may be that coral skeleton is not a good sand former, as T. F. Goreau (personal communication, 1964) has found, for it tends to disintegrate directly into the aragonite needles of the sclerodermites. Coralline algae, on the other hand, are enormously over-represented. In fact, they probably comprise as little as 1% of the reef (NEWELL and RIGBY, 1957, p.42), and they are not even recorded in the floral lists of the widespread communities of the rock pavement and unstable sand habitats (NEWELL et al., 1959, p.214, 218).

The constituent composition can be profoundly influenced by the rate of growth of a skeletal organism and its life span. The high percentage of *Halimeda* in the coralgal sand is doubtless a result of the rapid growth of these plants. Geologists are faced here with the same problem that confronts palynologists. Even if all skeletal material released by death in the community were preserved in the calcarenite, the proportions would still fail to represent those existing in the live community, because different organisms contribute skeletal debris to the lime sand at different rates.

Skeletal debris is subject not only to abrasion in turbulent water, but to a wide range of biological attack, by echinoderms, crustaceans, boring bivalves and sponges, fish, and boring coccoid and filamentous algae. Resistance of skeletons to attack varies between species (CHAVE, 1964) and thus selective breakdown leads to further distortion of the constituent composition.

Of the density of organisms in a community the constituent composition of the sand can tell us nothing. Yet, apart from the list of species, the density of their distribution in space is the living community's most signal characteristic. Nor is the association of organisms revealed. In the communities of the coralgal lithofacies there are certain intimate associations—a coral with its peculiar boring bivalves, *Halimeda* with its parasitic gastropods and fungi, and *Thalassia* with its encrusting biota—but these, by their very nature, cannot be reflected in a count of grains.

Certain organisms are not represented at all in the sand. Outstanding among these are the alcyonarians, and the non-calcareous algae such as *Laurencia*. Yet these comprise a large proportion of the biomass in several of the communities and exert an important modifying influence on the habitat. They are large tree-like or bushy structures, commonly 15–100 cm high, and reduce the velocity of water near the bottom, besides providing shelter and substrate for a variety of smaller plants and animals.

Attempts have been made to estimate the relative turbulence of various Recent (and ancient) carbonate habitats in terms of the wt.% of lime mud in the sediment. The value of such an index rests on the assumption that fine-grained sediment is produced in the habitat. It seems safe to assume that, in the environment of the coralgal lithofacies, lime mud is formed continually by mechanical abrasion, biological breakdown and the decay of sclerodermal, spiculate or needle-bearing tissues. It must not be forgotten, however, that the quantity of fines in the sediment reflects a balance between the rate of their production and the rate of removal. It cannot be simply a measure of the loss by winnowing, as PURDY and IMBRIE (1964, p.28) remind us. Indeed, in the coralgal lithofacies the mean wt.% of fines (10.8%) is twice that in the less turbulent oölitic and grapestone lithofacies of the stable sand habitat. This is the reverse of what would be expected from a comparison of turbulence. The discrepancy appears even more remarkable if we consider the higher rate of production of mud in the oölitic and grapestone facies by the disintegration of the green aragonitic algae *Penicillus*, *Udotea*, and *Rhipocephalus* and the probably higher rate of supply of material finer than 125 μ as a result of browsing in the rich and diverse *Strombus costatus* community. Against this it may be argued that the intense turbulence in the surf zone of the coralgal lithofacies may lead to a high production of fines by mechanical breakdown. Yet the correlation (Table VI) between the percentages of corals or of coralline algae (indices of turbulent habitats) with wt.% of fines throughout the Great Bahama Bank, though understandably negative, is very low ($r = -0.14$ and $r = -0.12$ respectively (PURDY, 1963a, p.352). What is clear from all of this is that the interpretation of the wt.% of fines is a difficult and, in our present state of knowledge, sometimes an impossible task.

The oölitic and grapestone lithofacies and their environment

Sand ripples. Within the stable sand habitat (Table V, Fig.128–130) the floor is shallower than 9 m and, on rare shoals, may be exposed at low water. Hydraulically it is quieter than the various habitats associated with the coralgal facies. Ripple-mark has been described by NEWELL et al. (1959, p.222) as "not so common here as in the more exposed seaward zones" and they pointed to its absence in the grapestone lithofacies. In sheltered places, such as Bimini Lagoon, there is ripple-mark only in the tidal channel and this limitation applies also to the Fish Cay–Cockroach Cay area of the Berry Islands. Nevertheless, at first sight, the bottom has, in a number of places, an appearance of mobility, particularly from the air, which is misleading: in tidal channels, and in other regions with strong tidal flow as on the floor just behind the oölite shoals of South Cat Cay and Browns Cay, megaripples are common (Fig.147). They have wave lengths of 50–100 m and amplitudes of about 45 cm. Yet, closer inspection shows that only the crests of some megaripples consist of unconsolidated ripple-marked lime sand (as in the mobile oölite habitat, p.135). Most of the bottom here, and in areas without megaripples, is covered by an "organic film" (NEWELL et al., 1959, p.220), a "thin brownish-colored organic film" (PURDY, 1963b, p.479), a "subtidal mat" (BATHURST, 1967c). This algal mat (below) takes a week or more to grow and is a decisive indication that the sand grains are not being moved in the normal day-to-day current regime. In one area in the Berry Islands it can be shown, by an inspection of air photo-

Fig.147. Aerial photograph of megaripples, wavelengths 50–100 m, off Walkers Cay, Little Bahama Bank. Dark areas are *Thalassia*. (Courtesy of C. W. Ellis.)

graphs, that the megaripples in two tidal channels of the grapestone lithofacies had not moved between 1941 and 1963. The hydraulic conditions that lead to the growth and movement of these megaripples are almost certainly the high tides and increased velocities associated with hurricanes (BALL, 1967; BALL et al., 1967). The effects of hurricane Betsey (in 1965) on the northeastern Bank are not yet recorded. Year-round observations are needed on sediment transport, ripples and the stability of the organic film. Owing to the structure of the university year most of the field work has unfortunately been carried out in summer vacations.

Subtidal algal mat. A study of the organic film, or subtidal mat as Bathurst, Ciriacks and Buchanan (unpublished data) have called it, is of critical importance in any analysis of bottom traction or of the feeding habits of the benthonic fauna (BATHURST, 1967c). Wherever I have examined the bottom, from Bimini to Browns Cay, or along the west coastline of Andros from Williams Island to the southern end, or in the southern Berry Islands the mat covers the stable bottom. It is absent

Fig.148. Subtidal algal mat, broken by foot-prints. Depth 1 m. Berry Islands, Bahamas. (Photo by R. Adlington, with the author.)

only in some, not all, tidal channels, and on the active mounds of burrowers (presumed callianassid) and on the mobile surfaces of oölite shoals. It occurs normally below water level, but has been seen on the tops of a few bars exposed at low tide.

The subtidal mat is pale brown generally, but can be pale green. The colour is as evident from the air as it is from a boat or when diving. The coherence of the mat can be demonstrated if a hand is slid obliquely under the sand and then gently lifted: the top 0.25 cm of the sediment tends to cohere (Fig.148). The coherence of this layer varies from scarcely discernible to such a degree of cohesion that a crust forms which can be eroded by currents and carried for 15 cm as fragments 2–3 cm in diameter. Where the mat is strong, small erosional channels, a few cm wide and 1–2 cm deep, cause it to overhang slightly. In the southern Berry Islands the mat is strongest on shoals that are only about 15 cm or less under water at low tide.

Under the microscope the mat can be seen to be a colourless, transparent, elastic, gel-like substance with the sand grains embedded in it. Besides skeletal debris, the mat is crowded with minute organisms, unlike the underlying sediment which is known only for its bacteria and microscopic boring algae. These organisms include micromolluscs, young and resting stages of foraminiferids, polychaete worms, arthropods, nematodes, platyhelminthids(?), ostracods, ciliated protozoa, hydroids, diatoms, tunicates, copepods and algal filaments (Fig.149–151). The

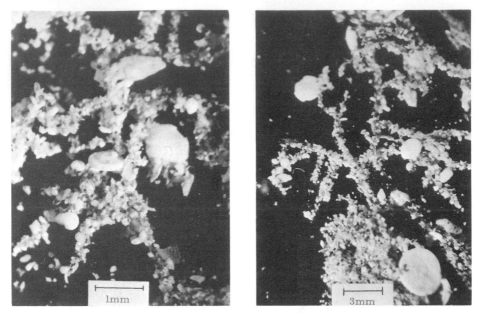

Fig.149. Lime sand grains in the subtidal algal mat, bound by harpacticoid copepod tubes. Abaco Sound, Bahamas. (From NEUMANN et al., 1970.)

Fig.150. Lime sand grains in the subtidal algal mat, bound by tunicate stolons. Abaco Sound, Bahamas. (From NEUMANN et al., 1970.)

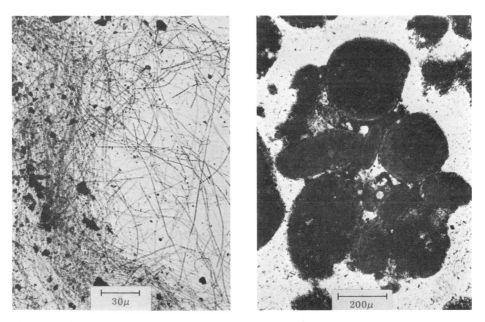

Fig.151. Lime particles in the subtidal algal mat bound by *Schizothrix* filaments and mucilage. (From SCOFFIN, 1970.)

Fig.152. Grapestone: lime sand grains cemented by aragonite micrite. Slice. Recent. Berry Islands Bahamas.

colour of the mat derives from the motile diatoms, brown or green, that lie on the upper surfaces of the sand grains, about ten on each grain. SCOFFIN (1970, p.269) has found the mat to be dominated by a filamentous alga, probably the blue-green *Schizothrix*.

A few small experiments were carried out one summer in the Berry Islands by Bathurst, Ciriacks and Buchanan and in Bimini Lagoon by Bathurst to discover the growth rate of the mat. Areas of 4 m², scraped free of mat, were still quite bare after a week. An unexpected discovery of great interest was the rapid appearance of ripple-mark on the scraped bottom after only a few hours. Once the mat had been removed from these static surfaces of calcarenite, the grains could be transported by tidal currents and so ripple-mark developed. Similar experiments by GEBELEIN (1969) gave the same results on the shallow sea floor off Bermuda.

Subtidal mats have recently been examined in Bimini Lagoon and on Little Bahama Bank by SCOFFIN (1970) and by NEUMANN et al. (1969, 1970). They have found that the mats are composed of various assemblages of green, red and blue-green algae, diatoms and animal-built grain tubes. Sedimentary particles are held and trapped by algal or diatom mucilages and by algal filaments (as described on p.224). SCOFFIN (1968) designed an underwater flume and, with the help of this, the erosion velocities of mats and sediments were measured. The value of the flume (basically an inverted flume, with top and two sides) is that water can be drawn through it at known velocities, passing over mats in their natural undisturbed states. Erosion velocities were determined for mat surfaces on the sea floor and then for the same sediment, treated with bleach to destroy organic matter, and redeposited in a tank in the laboratory. The same underwater flume was used for all measurements. It was found that the intact mat could withstand current velocities at least twice as high and, in some cases five times as high, as those that eroded the bleached unbound sediment. The intact mat surface could also withstand current velocities from three to nine times as high as the maximum tidal currents (13 cm/sec) recorded in the mat environment. The implications with regard to the stabilization of sediment surfaces are obvious.

On mat covered floors between North Fish Cay and Cockroach Cay, Berry Islands, the only lime sand that moves in the tidal current is that which forms the unconsolidated mounds, produced by callianassid decapods (p.127). The mounds comprise the only loose sand in the area and they tend to develop opposed slip slopes in the lee of the flood and ebb tidal streams. Some have truncated but rippled tops.

These observations, made in places of relatively quiet water away from tidal channels, point to the likelihood that many thousands of km² of mat-covered floor are stable, not because of feeble currents, but because the surface grains are embedded in the algal mat. The subtidal mat, rather than the current regime, may be responsible for the stability of the stable sand habitat. Purdy once commented that the grapestone bottoms have a characteristic brownish appearance. This may well indicate a widespread development of subtidal algal mat, in agreement with the known stability of these regions.

The subtidal mat is destroyed if buried. This is clear from study of the callianassid mounds. As a mound grows, its circular base advances over the mat. Dissection of mounds failed to show any preserved mat on the old, buried floor.

The mat is presumably the source of food for all the animals that browse on the sandy floor—echinoderms, holothurians, crustaceans, fish, and *Strombus*. It is difficult to see how these creatures could survive without the mat.

The importance of the mat has been stressed, because here is a substance, with a widespread, critical influence on the stability of the sand bed and on the nourishment of the fauna, which eventually disappears without trace. It seems sensible to suppose that equivalent algal mats have existed in past times with effects no less important, so that in stratigraphic interpretation allowance must be made for this. Above all, it is necessary to realise that, for example, an ancient biosparite, showing little sign of having been sorted, may well have been deposited on a floor where tidal currents were fast enough to sort the sand but were prevented from doing so by a mat.

Holdfasts. The stability of the lime sand is further enhanced by a number of plants. *Thalassia* covers the bottom (Fig.153) more or less completely in many areas: its blades, commonly about 15 cm long, reduce current velocity near the bottom,

Fig.153. *Thalassia* blades, *Penicillus* (Neptune's shaving brush, top centre) and hairy filaments of *Batophora*. Depth 1.5 m. Berry Islands, Bahamas. (Photo by R. Adlington.)

while its roots, which go down in places as far as 75 cm below the floor, give added stability to the sand. In places *Thalassia* is joined by *Syringodium*. Algae bind the sand grains together with their branching root-like holdfasts. Under *Batophora* (Fig.153) the grains are firmly united to a depth of 0.5–1 cm. The holdfasts of *Halimeda* may go down 10 cm or more. Other calcareous algae with holdfasts are *Rhipocephalus*, *Udotea* and *Penicillus*. Among animals, an orange colonial tunicate (Ascidian) is locally a sand-binder.

Mounds. Conical mounds made by a burrowing shrimp, *Callianassa*, are common in the stable sand habitat (Fig.154). The usual mound, resembling a miniature volcanic cinder cone, has a remarkably constant size, a diameter at base of about 20 cm, a height of about 6 cm, with flanks sloping at about 30°. At the apex of the cone there is a small exhalent vent about 3 mm wide. The inhalent hole lies in the adjacent sand floor, about 25 cm from the apex. Its diameter of about 2 cm is commonly enlarged by sliding. The active cone has a surface of loose sand, and grains can occasionally be seen leaving the exhalent vent. When freshly deposited the sand is grey, like that under the floor, presumably having emerged from a reducing environment. The sides of the cone are apt to be fluted by local slides of sand.

For many years the origin of these highly characteristic structures remained a frustrating mystery, until SHINN (1968c) investigated them, and identified the

Fig.154. *Callianassa* mound, showing central exhalent vent, also *Thalassia*. Coke can is 12.5 cm long. Bimini Lagoon, Bahamas.

mound maker as the crustacean *Callianassa* (of the order Decapoda). He did this partly by excavating the sediment with a device which withdrew water and sediment into a tube at high velocity, so exposing the buried structures while keeping the water free of suspended sediment, and partly by filling the burrow system with a catalyzing polyester resin. The resin, being heavier than water, was poured into the burrow: it then hardened and yielded a cast of the burrow system. The paired inhalent and exhalent holes meet at about 45 cm below the surface to make a central room from which four or five tunnels radiate horizontally for about 30 cm. One system may produce several mounds. The *Callianassa* burrow is easily recognized because it is lined with lime mud, generally finer-grained than the host sediment.

The callianassid mounds may be sparse (several metres apart) or so crowded that the cones interfere. The more densely arranged mounds seem generally to be interspersed with *Thalassia* and green calcareous algae. On old, presumably inactive, mounds the subtidal algal mat spreads up the flanks from the surrounding floor. Smaller mounds, possibly made by the shrimp *Alpheus* (SHINN, 1968c), occur where the thickness of unconsolidated sand on the Pleistocene limestone is reduced to 5–10 cm.

The mounds and burrows play an important part in the distribution of sediment. The process of burrowing causes the sand to be mixed vertically to a depth of more than a metre. As a result all bedding and all textures formed by bottom traction and deposition at the sand–water interface are destroyed. (GINS-BURG, 1957, p.87, introduced oligochaetes into a laminated sediment in an aquarium and within a month the laminae were obliterated.) This homogenization of the sand, throughout a metre or more of thickness, means that the history of deposition and with it the story of successive sea-floor environments is hopelessly jumbled. There remains only a blurred statement, a hotchpotch of data relating to grain type and size, reflecting in its structure and texture only burrowing activity and nothing of surface sedimentation. In the study of fossil limestones any interpretation of ancient environments must be tempered with reserve in the face of such wholesale destruction of evidence.

The mounds are also remarkable because, in some regions, they are the only source of loose grains. On a floor bound by the algal mat and a mixed vegetation, only these grains can be moved to and fro by the tidal currents. The distance travelled by grains in one direction or the other, according to evidence from the Berry Islands, is only, and very approximately, a matter of 3–4 cm at the most. The *net* distance travelled during one tidal cycle is a fraction of this. Nevertheless, as the flood tide is stronger than the ebb, the net movement over months or years may be significant. Mounds join with plants in reducing the smoothness of the bottom and slowing bottom currents.

Tidal channels. Both in the stable sand habitat and elsewhere on the Bank

Fig.155. Irregular surface of Pleistocene limestone on the floor of a rocky channel with the massive *Porites astreoides* and the branching *Porites porites* with a large, black sponge, *Sphecio-spongia*, in the background. Depth 2.5 m. Berry Islands, Bahamas. (Photo by R. Adlington.)

the tidal channels are of three kinds: (*1*) rock and pebbles, (*2*) rippled sand, (*3*) stable sand. The nature of the channel floor follows from two factors, the thickness of unconsolidated sand on the limestone and the current velocities. In type *1*, where sand is a centimetre or so thick, or absent, the bottom of the channel is rocky, with scattered pebbles, patches of ripple-marked sand, possibly some corals (*Porites astreoides* is typical; Fig.155), and sponges (*Ircinia* and *Spheciospongia*), and even some alcyonarians. In type *2*, the rippled sand channel, where the floor is a mobile sand thick enough for its surface to be unaffected by the underlying limestone, then the sand is ripple-marked and shows also megaripples. Vegetation is absent, but the suspension feeding bivalve *Tivela* lives in the sand. The last type *3* of channel bottom is a stable sand. It is colonized by patches of *Thalassia*, with green calcareous algae, and *Batophora* (Fig.153) and *Laurencia*. In between the plants the sand is bound by the algal mat. Megaripples, if present, are not mobilized by the daily tidal cycle. A vivid analysis of the flow regime and the deposition of carbonate sediments in a tidal channel has been made by JINDRICH (1969).

The *Strombus costatus* community. This community is relatively rich in variety of species and numbers of individuals. Especially where *Thalassia* covers much of the bottom (Fig.153) a habitat is provided for echinoids such as *Clypeaster*, *Mellita*, *Tripneustes* and *Lytechinus*, and for at least twenty-two species of bivalve

Fig.156. *Penicillus, Thalassia, Porites porites*, sponges, subtidal algal mat. Coke can 12.5 cm long. Depth 1 m. Bimini Lagoon, Bahamas. (From BATHURST, 1967a.)

Fig.157. *Manicena areolata*. (From FABRICIUS, 1964.)

and sixteen of gastropod, *S. costatus* being a large herbivorous gastropod. Green aragonitic codiacean algae are common, especially the sand-forming *Halimeda* and the mud-forming *Penicillus, Rhipocephalus, Udotea*, also the dasycladacian aragonitic *Acetabularia* (Fig.143). Of the red algae, *Laurencia* is important and the high-magnesian calcitic *Goniolithon*. *Thalassia* is joined, in shallower waters, by the other common phanerogam, *Syringodium*. On the upper surfaces of the *Thalassia* blades there is a dense population of minute encrusting foraminiferids, encrusting coralline algae, filamentous algae and hydroids. The polychaete *Arenicola* is reported, though the maker of the mounds is *Callianassa*. Holothurians are plentiful in sheltered lagoons. Sponges are numerous and various: the large ones are *Ircinia* (bathroom sponge, Fig.140), *Spheciospongia* (manjack, Fig.155, p.129), and *Tedania ignis* (fire sponge). The large asteroid, *Oreaster*, is ubiquitous. Crabs are busy burrowers and disturbers of sedimentary structure (Fig.195, 196 on p.196, 197 resp.). Where the limestone base is barely covered, *Porites* takes hold (Fig.156): in general, corals need a rocky substrate (as in the Persian Gulf, p.183). The flower like coral *Manicena* (Fig.157), on the other hand, grows in the loose sand. Oncolites are abundant on the sheltered intertidal floor west of Cockroach Cay, Berry Islands.

The petrography of the oölitic and grapestone lithofacies. These two lithofacies are remarkable for their high content of non-skeletal sand grains, 89% in the oölitic (distinct from the oölite facies, Table V) and 83% in the grapestone facies (Fig.145). These totals are compounded of faecal pellets (ellipsoidal peloids, Fig.158, 159), mud aggregates (irregularly shaped peloids), organic aggregates (sand grains bound together by a brown cellular material), grapestone (sand grains bound together by micrite, Fig.152, 246), flakes (p.317), cryptocrystalline grains (irregularly shaped, brown peloids), and oöids. The two lithofacies differ largely owing to the high content of oöids (67%) in the oölitic facies, and of grapestone, with the related organic aggregates, and cryptocrystalline grains (61%) in the

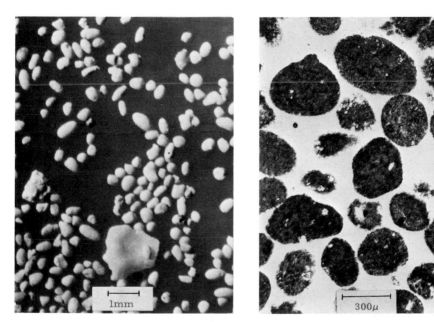

Fig.158. Faecal pellets. Reflected light. Recent. Berry Islands, Bahamas.

Fig.159. The faecal pellets of Fig.158 in thin section.

grapestone facies. The oölitic lithofacies lies in regions where a shoal bottom or proximity to the Bank edge leads to stronger tidal currents. It has been found in a band running southward from South Cat Cay parallel to the Florida Straits, also on the shallow Bimini axis, behind Joulters Cays and on the Berry Islands plateau (Fig.128). Elsewhere, in water less turbulent, because deeper or far removed from the ocean, grapestone has accumulated. It is important to note that, in the water most sheltered from the trade winds and isolated from the ocean, in mid-Bank for 80 km west of Andros Island, neither oölitic nor grapestone lithofacies is known, and aragonite mud is the dominant deposit.

So far as is known, oöids are not forming in the oölitic lithofacies. The water here is not as strongly agitated as it is over the oölite shoals of the mobile oölite habitat, and the oöids have mat surfaces unlike the polished grains of the shoals. The presence of oöids in the oölitic facies (*contra* oölite facies, Table V) is presumably a consequence of their carriage across the Bank from the oölite lithofacies, by flood tides to regions whence the slower ebb tides cannot return them. Oöids are less common or absent in the coralgal, pellet-mud and mud lithofacies. The correlation of grapestone lithofacies with less turbulent water may follow from the need for grains to lie in undisturbed contact for long enough for them to be cemented together (p.317).

In addition to the grapestone grains there are flakes of the same material, from 1–2 mm thick and up to 3 cm across, possibly a cemented algal mat. The bedding, in sand bars in this facies where plants and animals have not destroyed it, is typical of the kind formed during rare storms or hurricanes in the Bahamas (IMBRIE and BUCHANAN, 1965; BALL, 1967). The stability of megaripples for tens of years in this lithofacies has been referred to (p.122).

Significance of the grain counts. In thinking about the proportions of different skeletons in the sands (Fig.160), the limitations have already been discussed (p.120). The low content of coral and coralline algae in both oölitic and grapestone facies is consistent with their remoteness from reefs or rock pavement. The rather higher value in the oölitic facies fits with an influx of at least some of these grains from the Bank margin. Otherwise the skeletal content of these two lithofacies differs little from the coralgal lithofacies (Fig.145) despite the very different nature of the bottom. Indeed, owing to the impossibility of registering density of population in a community, the skeletal composition is also similar to that of the oölite lithofacies which has a vastly lower population density. Again, it is impossible to know to what extent the proportions of the skeletal types relate to proportions of live benthos or even to the rates of contribution from different species. This difficulty is aggravated by the high content of cryptocrystalline grains (Fig.161) in the grapestone lithofacies: the content is, in fact, much higher than is shown in Fig.145 because the component grains of the grapestone aggregates are commonly cryptocrystalline themselves.

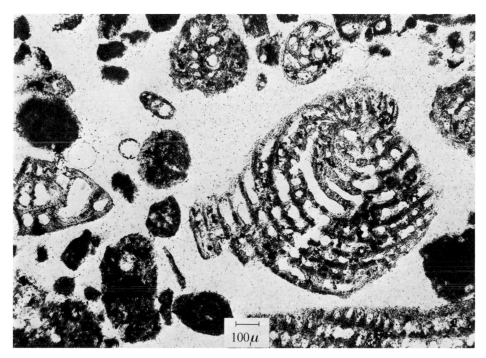

Fig.160. Miliolid particles and micritized skeletal debris. Slice. Recent. Bimini Lagoon, Bahamas.
(From BATHURST, 1967a.)

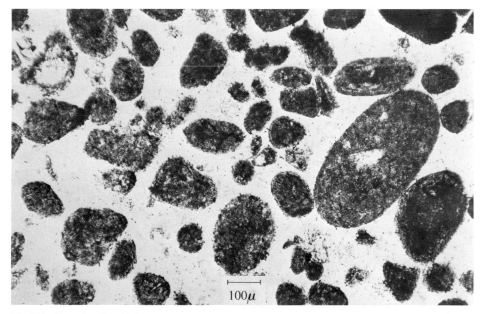

Fig.161. Faecal pellet (right) and micritized skeletal debris. Slice. Recent. Bimini Lagoon, Bahamas. (From BATHURST, 1967a.)

In a petrographic study of grains in the North Fish–Cockroach Cay area, Berry Islands, I have found that, although dry mounts of sand looked at with a microscope in reflected light contain plenty of skeletal-shaped debris, in peels and thin sections of the impregnated sand identification of the skeletons is rarely possible: the grains, whether single or aggregated into grapestone, are pale brown and cryptocrystalline. It seems here as if micritization of skeletons by algae (p.384) or other micro-organisms is important. In these sands from the Berry Islands it has so far been impossible to compare the skeletal compositions of the various habitats because an insufficient number of skeletons have survived micritization; this despite the relatively rich living benthos. Micritized and unaltered skeletal debris from Bimini is shown in Fig.160, 161. The micritization of skeletons on such an extensive scale poses the question whether some of the peloid sands (the "bahamites" of BEALES, 1958) among ancient limestones may not be composed of altered skeletons. It is worthwhile remembering that grains bored by algae are weakened peripherally and so are easily abraded and rounded (BATHURST, 1967a). In this way all evidence of skeletal shape disappears.

NEWELL et al. (1959) included all of what was later to be Purdy's oölitic and grapestone lithofacies in the one stable sand habitat (Table V). It is clear from PURDY's (1963b) analysis that a gradation exists between the more turbulent oölitic and the quieter grapestone kind of bottom. It does seem that, as an indicator of the level of turbulence, the oöid can now be joined by grapestone, the one indicating high turbulence, the other low. In stratigraphy, the application of such a measure as, say, the percentage of oöids or grapestone in a calcarenite, would thus have obvious significance. It is interesting that the amount of sediment finer than 125 μ is slightly higher in the oölitic facies than in the grapestone, 5% against 4.5%, and so, again, gives no information about relative turbulence.

Several of the essential aspects of the stable sand habitat are not reflected at all in the petrographic analysis. The *Thalassia* and *Syringodium*, the mat, the holdfasts—these elements which so profoundly influence the environment cannot appear in a grain count. It might be thought that mounds could be detected by examining the fabric of cores, or of ancient limestones. Certainly cores studied by IMBRIE and BUCHANAN (1965) from mound areas are conspicuously lacking in bedding, but we have yet to learn to recognize the presence of mounds, or their accompanying burrows, in ancient limestones, in the large quantities in which they should be expected to occur. In the grain-counts echinoderms do not figure, nor is there any record of the needles or spicules from codiaceans or sponges. In cores, and in subsequent limestones, megaripples can be preserved as cross-strata.

The oölite lithofacies and its environment

The mobile oölite habitat. At various places along the margins of the Bank, at a distance of 1–6 km in from the 180 m (100 fm) line, there are shoals of almost pure oölite. An example of their distribution can be seen in Fig.247 (p.315). The tidal range of 0.78 m in the adjacent ocean, combined with the shallowness of the water over the oölites, causes strong tidal currents across the shoals with flood velocities up to about 150 cm/sec. The changes of some variables across a typical oölite shoal are indicated by PURDY (1964a, p.256). The grains are in nearly constant movement, unless exposed at low water, and the surface of the oölite is modified by ripple-mark and megaripple. In this mobile substrate no macroscopic vegetation can grow and only the rapidly burrowing bivalve, *Tivela abaconis*, can live in the sand. The content of lutite and organic matter within the oölite is negligible, so this animal is necessarily a suspension feeder.

The oölites are clearly products of the present environment. A series of radiocarbon dates of the outer 10% of the volume of oöids in the modal size fraction (0.35–0.59 mm) from nine scattered localities gave a general age of less than 160 years (MARTIN and GINSBURG, 1965).

The growth of the oöids is clearly related to the turbulence and to the passage of a thin layer of hypersaline water, to and fro over the shallow margin of the Bank (p.100). The highest contents of oöids in the oölite, approaching 100%, were found by NEWELL et al. (1960) at about 1.8 m below low water, suggesting that here may be optimum conditions for oöid growth.

The almost barren floor of the oölite shoal drops seaward to the unstable sand habitat, where the suspension feeding bivalve, *Glycimeris*, is abundant, with some *Syringodium*, but little *Thalassia*. There are the aragonitic algae, *Halimeda*, *Penicillus* and *Rhipocephalus* and the calcitic *Goniolithon*. In places where the sand is thin on limestone, the alcyonarians of the rock pavement habitat are to be found. The sand is generally ripple-marked.

On the Bank side of the oölite shoal the floor falls away more gradually, through a zone of lightly populated megaripples, to the *Thalassia* carpet of the stable sand habitat. These megaripples, away from the turbulence of the oölite shoals, with wave-lengths up to 100 m and amplitudes of about 0.5 m, have active ripple-mark only on their crests (see p.121). The rest of the floor, between the crests, is bound by the subtidal mat and bears scattered codiaceans, *Syringodium*, the asteroid *Oreaster* and the sand dollar, *Mellita*. There is also a species of sea anemone and one of a worm. In the stable sand habitat *Thalassia* is commonly dense, with *Halimeda*, *Penicillus*, *Rhipocephalus* and some *Acetabularia* (Fig.143), also sponges and the herbivorous *Strombus*. There is a Bankward increase in the quantity of grains finer than 125 μ.

All three types of tidal channel (p.129) occur in the mobile oölite habitat. On the floors of some *Thalassia* channels oncolites attain diameters of about 5 mm. ILLING (1954, p.62) has shown how tidal deltas (sand-tongues) are built on the Bankward sides of tidal channels between cays.

The petrography of the oölite lithofacies. The constituent composition is unique in that it reflects, with nearly 100% efficiency, the nature both of the habitat and the community. The high polish on the grains has been shown to be the result of abrasion (NEWELL et al., 1960, p.490; DONAHUE, 1965; GRADZIŃSKI and RADOMSKI, 1967), but no one has yet found a way of detecting this polish in ancient oölites. PURDY (1963b) gave 93% of oöids for the mean constituent analysis. Nevertheless, it is revealing to find that the proportions of *skeletal* types differ little from those of the oölitic and grapestone lithofacies and, ignoring corals and coralline algae, from the coralgal lithofacies as well (Fig.146).

The pellet-mud and mud lithofacies and their environment

The muddy-sand and mud habitats. The region most sheltered from the trade winds and farthest from the strong tidal activity of the Bank edge occupies about 10,000 km^2 west of Andros Island (Fig.129, 130). Here the movement of the tenuous layer of Bank water is at its most sluggish with a net flow of about 0.8 km a day even in gusty December. The floor is an aragonite mud with a variable proportion of peloids (mainly faecal pellets) made of the same material. The mud particles are aragonite needles and the proportion of this needle mud in the peloid-mud mixture increases towards the Andros coast. Ripple-mark is virtually absent, but callianassid mounds are abundant. The vegetation cover is sparse compared with the stable sand habitat though *Thalassia* is locally dense. The most conspicuous members of this, the *Didemnum* community, are the creeping chlorophyte *Caulerpa*, the candle sponge *Verongia* and the pale-grey colonial tunicate (Ascidian) *Didemnum candidum*. Other algae are *Halimeda*, *Rhipocephalus*, *Penicillus*, *Udotea*, *Avrainvillea*, *Acetabularia*, *Dictyosphaeria*, *Batophora* and *Laurencia*. Besides *Thalassia* there is *Syringodium*. The fauna is also sparse, in numbers and in variety. Only two species of mollusc have been recorded, two echinoids, the coral *Manicena* that grows in loose sediment, a bryozoan and three sponges, fish, crabs and shrimps. One of the sponges, the soot-black to brown *Hircinia*, is inhabited by a mixed population including ostracods, copepods, isopods, malacostracans, amphipods and decapods—in all about thirty species. This is a fitting reminder of the influence of vast numbers of tiny organisms which, invisible to the diver and the under-water photographer, yet play a role in the ecology and in the food chain.

The subtidal algal mat was evident on the floor where the author dived near Williams Island and elsewhere along the coast of Andros Island southward from there. The mud felt soft and feathery around the feet as they pierced the weak mat. Polyethylene core tubes were easily forced down to the limestone 3 m below. CLOUD (1962a) records this as the greatest thickness attained anywhere by the mud. Along the west coast of Andros, for about 2.5 km offshore, the very high salinity (43–46‰) is at times drastically reduced by an influx of fresh water from the mangrove swamps and channels of the island. Apparently in these conditions only an exceptionally sparse population survives, in which a gastropod *Cerithidea costata* and a bivalve *Pseudocyrena* are conspicuous. *Batophora* (Fig.153), the green non-calcareous alga, is particularly abundant.

Whitings. The origin of the aragonite mud is arguable (Chapter 6). It may be an accumulation of needles left by decomposition of codiaceans and dasyclada-ceans (LOWENSTAM and EPSTEIN, 1957; STOCKMAN et al., 1967, NEUMANN and LAND, 1969) or it could be a precipitate (CLOUD, 1962a). Phenomena of special interest in this connection are the whitings. These are areas of milky-white water which are best seen from the air as elongate patches some tens to hundreds of metres long. They are dense suspensions of aragonite needles (CLOUD, 1962a, records 11.5 mg/l). Their origin is probably varied: they are known to be caused by fish stirring the bottom and similarly by the screws of vessels. It is possible that others are formed by precipitation of aragonite as a suspension. In the Persian Gulf (WELLS and ILLING, 1964) it is possible that precipitated whitings appear suddenly as a result of CO_2 up-take during diatom bloom. Cloud (1962a, p.19) has described a diver's view of a Bahamian whiting: "Suspension with free diving apparatus beneath its [the whiting's] center gave the sensation of weightless fixity in the middle of a sunlit cloud bank. It was impossible, without resting motionless, to detect buoyancy and drift, or to tell up from down or sideways. The brilliant lighting was so dispersed that a hand, invisible at armslength, had to be kept extended to grope for bottom and avoid collision on surfacing."

The petrography of the pellet-mud and mud lithofacies. The sediment is 88–94% aragonite. The remaining 6–12% is calcite with rare hydrous mica and traces of kaolinite, chlorites and quartz. Most of the calcite is high-magnesian ($MgCO_3$ 11–16 mole%) of skeletal origin. Toward Andros Island there is more low-magnesian calcite ($MgCO_3$ 0–5 mole%), probably derived from fresh-water precipitates in the pools on the island. There is much less skeletal debris than in the muds of Florida Bay. Most of the mud and silt particles are aragonite needles, single or weakly aggregated. Other fine particles include the calcitic spicules of *Didemnum*.

Among the sand grains, the total amount of cryptocrystalline and grapestone grains, organic aggregates and peloids, is only about 4% (mean) near Andros, increasing westwards to about 8%. Pellets of presumed faecal origin comprise 17–38% of the sediment, the quantity rising from east to west. The skeletal content is around 14% (mean) and differs from all the other lithofacies in the high proportion of Peneroplidae (mainly *Archaias*) relative to *Halimeda*, molluscs and other foraminiferids (Fig.146). The origin of the aragonite mud is discussed in Chapter 6.

The pellets. Faecal pellets are the main constituent of the carbonate sand fraction. Relating these to their organisms is not easy as different animals may produce rather similar pellets. It seems likely, though, that an important contributor in this habitat is a ubiquitous polychaete, *Armandia maculata* (CLOUD, 1962a), which has been seen to produce pellets of the appropriate shape and size. The gastropod *Batillaria minima* produces characteristic pellets (KORNICKER and PURDY, 1957), but these are restricted to mangrove swamps or their vicinity.

PURDY (1963b) gives the mean content of all sand grains in the pellet-mud and mud lithofacies, respectively, as approximately 57% (including 6% of oöids presumably carried in from the west) and 38%. Cryptocrystalline grains amount to no more than 2%. The pellet-mud has 11% skeletal and 38% faecal grains, and the mud has 17% of each. Mud aggregates are included with faecal pellets as they probably have the same origin.

It is particularly necessary to realise that in this sediment of the pellet-mud lithofacies, as in no other Bahamian sediment except the oölite, we have an apparently undiluted autochthonous carbonate accumulation. Obviously, the presence of the pellet sand grains does not indicate bottom traction, any more than the roundness and good sorting of the pellets suggest turbulence. On the contrary, the angularity of the skeletal debris points to local breakage by predators. The smooth ellipsoidal shape and good sorting among faecal pellets merely draws attention to the passage of mud through guts and to the fact that, among the local worms and molluscs, there is but slight variation in anal cross-section.

The lack of bottom traction draws attention to a difficult question. How, in the absence of current depositional and erosional surfaces, can bedding planes be formed? Calcite-mudstones in limestone sequences are, after all, commonly bedded. Or are we dealing here, west of Andros Island, with an unbedded, massive sediment, a sort of giant bioherm? There is no evidence at all bearing on the possible development of bedding as this pellet-mud continues to accumulate.

Retrospect: the significance of lithofacies

In the preceding pages an attempt has been made to examine, among other things, the relation of Purdy's lithofacies to the habitats and communities of Newell et al. Certain aspects now deserve a momentary re-emphasis. First of all, it is clear that a petrographic analysis scarcely, if ever, reveals such things as the proportion of organisms, population density, or even the very existence of some plants or animals. In addition, it will be seen that four lithofacies or pairs of lithofacies relate to nine habitats and nine communities (Table V). This is a measure of the low sensitivity of even a painstaking, lengthy and detailed petrographic analysis of a group of Recent carbonate sediments. How much less informative are the data culled from ancient limestones that have been diagenetically changed. As for the interpretation of such data, the immense difficulties, many perhaps insurmountable, must now be apparent. On the other hand, the lithofacies do have a geographic pattern which quite closely follows those of the habitats and communities. So, in the interpretation of ancient habitats and communities, the *pattern* may be discoverable, but assessment of the more intimate details must depend on rule-of-thumb comparisons with modern environments. The older the limestone formation, the more hazardous must the comparisons be.

A number of influential materials disappear as a sediment is buried, e.g., the subtidal algal mat, *Thalassia*, alcyonarians and mounds. We can, however, expect to glean from many ancient limestones some information about the substrate— sand, mud, a mixture—little enough in all conscience, but profoundly important from the ecological point of view.

There is, nevertheless, a serious problem concerning the preservation of such a wealth of petrographic detail in ancient limestones. Judging by their petrography the Bahamian sediments contain from 60% to over 90% aragonite. If sediments of this kind, with porosities probably around 60%, are cemented, and if the necessary carbonate is supplied by dissolution of aragonite grains, then a very large part of the petrographic evidence must disappear (p.327; BATHURST, 1966).

The relation of the lime mud content to turbulence is certainly not a simple one, and it is not certain that mud has been trapped by baffles of *Thalassia*. Unlike the *Thalassia* on the mud banks in Florida Bay (p.161), the denser carpets of *Thalassia* in the Bahamas are to be found mainly in areas of high water energy, especially on the bottoms of broad tidal channels, so that the baffle system is not effective in trapping lime mud, even though this be locally derived from the epibiota on the *Thalassia* blades. Possibly the currents outside the pellet-mud and mud lithofacies are too strong. The remarkably low mud content in the oölitic and grapestone habitats, where the production of needles by codiaceans is high, is surprising. The *Penicillus* plant lives only two to three months and thus quickly adds its heap of needles to the sediment. The failure of aragonite mud to accumulate (it is very scarce in the quiet Bimini Lagoon) has been attributed to winnowing by ebb currents that carry the needles off the Bank: evidence in support of this fate is needed. There are good reasons for concluding that the concentration of mud west of Andros is a consequence of the low turbulence and shelter from the trade winds and not necessarily a result of high salinity. Purdy (1964a, p.28) writes: "It seems clear from the relationships evident in the Bahamas and elsewhere in the Caribbean region that the only proper inference to be drawn from the occurrence of an ancient calcilutite is that the original mud was deposited faster than it was removed. Analogies based on a Bahamian model as to the salinity of this environment or as to the origin of the mud-sized constituents are totally unwarranted without evidence additional to the mere occurrence of the calcilutite."

The large number of cryptocrystalline grains, probably reflecting a high intensity of attack by boring algae, is a disappointment, since so much biological evidence is thereby lost.

Lateral mixing between lithofacies has not too seriously disguised the pattern of habitats, but on a smaller scale of metres it may do so considerably.

Little attention has been paid to the confusion of the lithological record

TABLE VI

MATRIX OF CORRELATION COEFFICIENTS AMONG MAJOR BAHAMIAN SEDIMENTARY CONSTITUENTS
(n = 216, after PURDY, 1963a)

	Halimeda	Peneroplids	Other foraminiferids	Corals	Molluscs	Faecal pellets	Mud aggregates	Grapestone	Oölite	Cryptocrystalline grains	Fraction < 1/8 mm
Coralline algae	0.330	0.089	0.224	0.701	0.143	-0.187	0.029	-0.098	-0.205	0.094	-0.119
Halimeda		0.147	0.295	0.369	0.420	0.012	0.216	-0.211	-0.360	-0.055	0.166
Peneroplids			0.600	-0.010	0.453	0.303	0.321	-0.255	-0.404	-0.168	0.668
Other foraminiferids				0.101	0.454	0.117	0.323	-0.129	-0.508	0.113	0.549
Corals					0.292	-0.165	0.058	-0.118	-0.210	0.022	-0.141
Molluscs						0.127	0.232	-0.141	-0.546	0.013	0.454
Faecal pellets							0.517	-0.379	-0.368	-0.428	0.663
Mud aggregates								-0.325	-0.400	-0.253	0.546
Grapestone									-0.314	0.642	-0.391
Oölite										-0.436	-0.425
Cryptocrystalline grains											-0.357

caused by the reworking of a sediment if, as a result of changes of sea level, it comes to lie at various water depths which differ from that at which it originated. At such times, if the sediment has not been covered by younger sediment, it is not only reworked but also restocked with benthonic and pelagic biota. LOGAN et al. (1969) have illumined this fundamental facet of sedimentation, in their study of carbonate sediments on the Yucatán shelf which have been subjected to post-Wisconsin changes of sea level. The question that must be asked of any sediment is: to what extent is it in equilibrium with its original sea floor environment as distinct from one or more subsequent sea floor environments? Controlling factors are the rates of deposition, the rate of sea level change and the rate of transgression. The problem is well summarized in LOGAN et al. (1969, p.128).

PURDY (1963a, p.352) elegantly summarized the relationships between constituents (Table VI), by the use of product moment correlation coefficients. Using his 216 (out of 218) samples from stations all over the Bank, he took the percentage of a constituent in a sample and paired it with the percentage of another constituent in the same sample. Doing this, sample by sample, he ended with 216 pairs of, say, *Halimeda* and mollusc percentages, from which a correlation coefficient was calculated. This operation was carried out for all the 66 possible pairs of grain types, using a high speed digital computer, with the results shown in Table VI. It can be seen that the percentages of corals and coralline algae have a high positive correlation, also that, whereas faecal pellets are positively correlated with mud aggregates, they show a negative correlation with cryptocrystalline grains. Grapestone and oölite each show negative correlation with everything. Peneroplidae are strongly and positively correlated with the content of fines. Purdy then went on to erect a hierarchy of reaction groups (Fig.162) following the method of SOKAL and MICHENER (1958). This scheme shows neatly the correlations among constituents and groups of constituents.

FURTHER READING

There are excellent introductions to the Great Bahama Bank by NEWELL and RIGBY (1957), PURDY and IMBRIE (1964) and COOGAN (1969). The classic basis for all subsequent petrographic studies is by ILLING (1954). Habitats and communities are dealt with in detail in NEWELL et al., (1959) and descriptions of the lithofacies are in PURDY (1963a, b). Much additional information is contained in PURDY (1961), CLOUD (1962a), IMBRIE and BUCHANAN (1965), MONTY (1965, 1967), BALL (1967), BATHURST (1967a), BROECKER and TAKAHASHI (1968) and SCOFFIN (1970). The penetrating analysis of the Alacran reef complex by KORNICKER and BOYD (1962) is valuable complimentary reading; also the Bermuda symposium on organism–sediment relations (GINSBURG and GARRETT, 1969). As this book went to press there appeared a major study of the Andros Island carbonate tidal-flats by SHINN et al. (1969). Other references are given in the text.

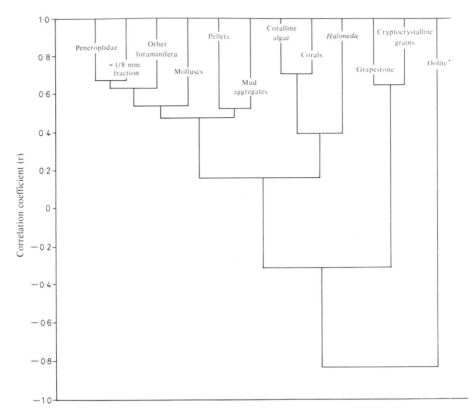

Fig.162. Reaction-group hierarchy of major sedimentary constituents of Bahamian lithofacies, based on data in Table VI. For example, foraminiferids and molluscs tend to be abundant in the same samples ($r = + 0.53$), but the abundancies of skeletal grains and grapestone with crypto-crystalline grains tend to be inversely related ($r = - 0.32$). (From PURDY, 1963a.)

Additional references not given in the preceding chapter

Papers of general interest are by NEWELL (1951) on reefs and oölites, NEWELL et al., (1951) on eastern Andros Island, MILLIMAN (1967a, b) on Hogsty reef, CLOUD (1955, 1962b, 1965) on chemical origin of aragonite muds, BAARS (1963) on petrology, FIELD (1928) on Bahamian sediments, IMBRIE and PURDY (1962) on their classification, PILKEY (1964) on the carbonate fractions of southwestern Atlantic sediments, BEALES (1966) and TAFT (1967) on carbonate sediments in general, STORR (1964) on some very remarkable patch reefs off Abaco Island on Little Bahama Bank, MULTER (1969) on the sedimentology of the Bahamas–Florida platform and NEWELL (1955) on Bahamian platforms, with an up-to-date discussion of structural history by GOODELL and GARMAN (1969).

Papers on the Bimini area are by NEWELL and IMBRIE (1955) and BATHURST (1969), both general, KORNICKER (1958b) on the ostracods, G. Y. CRAIG (1967) on size distributions of molluscs, SQUIRES (1958) on corals, FABRICIUS (1964) on *Manicena areolata*, SEIBOLD (1962a) on water chemistry and TUREKIAN (1957) on salinity.

There is a detailed and thorough study of the growth and construction of sea grass carpets and their modification by MOLINIER and PICARD (1952).

There is an important symposium on penetration of $CaCO_3$ substrates by lower plants and invertebrates edited by CARRIKER et al. (1969).

Erosion and boring by molluscs, echinoids and the sponge *Cliona* are described in NORTH (1954), GOREAU and HARTMAN (1963), and R. F. MCLEAN (1967a, b). BROMLEY (1970) gave a general review of borings.

Two bore holes are covered by FIELD and HESS (1933) and GOODELL and GARMAN (1969).

Comparisons of Bahamian environments with ancient limestones have been made by BEALES (1958, 1961, 1963), LAPORTE and IMBRIE (1964), HOFFMEISTER et al., (1967) and LAPORTE (1968).

Works about sedimentation on neighbouring deep ocean floors are by BUSBY (1962), PILKEY (1966) and GIBSON and SCHLEE (1967) on Tongue of the Ocean, RUCKER (1968) on Exuma Sound, HARRISS and PILKEY (1966) on the Bermuda Apron, ANDREWS (1970) and ANDREWS et al. (1970) on the Great Bahama Canyon, SCHNEIDER and HEEZEN (1966) on the Caicos outer ridge, Bahamas.

Anyone who wishes to understand the transport and deposition of shallow water carbonate sediments should read FOLK (1962a, b, 1967) and FOLK and ROBLES (1964), WILSON (1963, 1967), ALLEN (1968, 1969, 1970), MAIKLEM (1968), FORCE (1969) and JINDRICH (1969).

Papers on Recent reef complexes

For the reader anxious to pursue further the subject of Recent reef complexes the following references will be found helpful.

A good general introduction to coral reefs can be had from LADD (1950) and LADD et al., (1950) on Recent reefs, and LADD (1961) on reef building, also the chapter on coral reefs in KUENEN (1950): a more biologically biassed treatment is given by H. B. MOORE (1964). More specialist papers of great interest are the delightful study by ODUM and ODUM (1955) of a small coral reef community on Eniwetok, the important study of some Jamaican reefs by GOREAU (1959), the broadly based analysis of the Alacran reef-complex by KORNICKER and BOYD (1962), the detailed survey of reef structure and its development on the Yucatán shelf by Logan (in LOGAN et al., 1969, p.129), and GOREAU on calcification in corals (1961, 1963). An important new work on ecology is by STODDART (1969).

Other interesting papers are the reports on the Kapingamarangi lagoon by MCKEE (1958) and MCKEE et al. (1959), the Alacran by HOSKIN (1963, 1966), those on the activities of the boring sponge *Cliona* by GOREAU and HARTMAN (1963) and

NEUMANN (1966), STORR's (1964) report on the Abaco reef track and the hollowed out patch reefs, and MILLIMAN's (1967a, b) two studies of the Hogsty Atoll. The Bikini Atoll is dealt with in an interesting general way by LADD et al. (1950) and in greater detail by EMERY et al. (1954).

There are three extensive and very interesting studies of the Mahé reef by LEWIS (1968, 1969) and J. D. TAYLOR (1968). The investigations by ORR (1933) and ORR and MOORHOUSE (1933) are two particularly relevant papers on the physical and chemical conditions on the Great Barrier Reef, recently augmented by the work of MAIKLEM (1970) and MAXWELL and SWINCHATT (1970).

Other papers of interest

CHESHER (1969): Destruction of Pacific corals by a starfish.
DALY (1917): Origin of reefs.
DARWIN (1962) on coral islands (reprinted).
FAIRBRIDGE (1950, 1967): Australian reefs.
FAIRBRIDGE and TEICHERT (1948): Great Barrier Reef.
FOLK (1967) on sandy cays, Alacran.
GINSBURG (1964) on Florida carbonate sediments.
GOREAU (1964) on mass expulsion of zooxanthellae after a hurricane.
GOREAU and YONGE (1968) on mobile corals on muddy sand.
GUILCHER (1952, 1956) on reef morphology.
HOFFMEISTER et al. (1964) on living and fossil reefs, Florida.
HOSKIN (1968) on magnesium and strontium in Alacran reef complex.
J. A. JONES (1963) on ecology of Florida patch reefs.
JORDAN (1952) on Gulf coast reefs off Florida.
LADD (1956) on coral reef problems in the Pacific.
LADD et al. (1953, 1967) on drilling in Pacific atolls.
LAUBIER (1965) on coral reef in the Mediterranean.
LEWIS and TAYLOR (1966) on Seychelles sediments and communities.
E. R. LLOYD (1933) on coral reefs and atolls.
MACNEIL (1954) on reefs, banks and sediments.
MAXWELL, (1968) on Great Barrier Reef.
MAXWELL et al. (1961, 1963, 1964) on Heron Island reef.
MUNK and SARGENT (1948, 1954) on adjustment of Bikini atoll to waves.
MULTER and MILLIMAN (1967), KIRTLEY and TANNER (1968) on sabellariid reefs.
NESTEROFF (1955a) on Red Sea reefs.
NORRIS (1953), PUFFER and EMERSON (1953) on Texas oyster reefs.
NUGENT (1946) on some Pacific reefs.
RIGBY and MCINTIRE (1966) on Isla de Lobos reefs, Veracruz.
SARGENT and AUSTIN (1954) on biological economy, Bikini and nearby reefs.
SEIBOLD (1962b) on coral reef problems.

SQUIRES (1964) on coral thickets in New Zealand.

STODDART (1962a, b, c; 1965a) on Hurricane damage to Honduras reefs.

STODDART (1965b) on the shape of atolls, and atoll sediments (1964).

TEICHERT (1958), MOORE and BULLIS (1960), and STETSON et al. (1962) and SQUIRES (1965) on cold water and deep water coral banks.

TRACEY et al (1948) on Bikini reefs.

TURNER et al. (1969) on colonization of man-made reefs.

UMBGROVE (1947) on reefs of East Indies.

VON ARX (1948): Bikini and Rongelap circulations.

VAUGHAN (1919) on corals and reef formation.

WELLS (1957a) on coral reefs.

WIENS (1962) on atoll environment and ecology.

YONGE (1930): A year on the Great Barrier Reef.

YONGE (1951) on coral reef form.

YONGE (1968): an essay on living corals.

MILLIMAN (1965): bibliography of recent papers on corals and coral reefs.

PUGH (1950): bibliography of organic reefs, bioherms and biostromes.

RANSON (1958): bibliography of corals and coral reefs.

Chapter 4

RECENT CARBONATE ENVIRONMENTS 2:
FLORIDA, GULF OF BATABANO, PERSIAN GULF, BRITISH HONDURAS

SOUTHERN FLORIDA

Recent work in Florida has led to a greater understanding of reefs, of sedimentation in back-reef lagoons and of the remarkable interplay of carbonate mud mounds and tidal channels (GINSBURG, 1956, 1964; SHINN, 1963; SWINCHATT, 1965). In this richly varied region many geologists have had their first wide-eyed and enchanting experience of the modern carbonate sea floor under the patient tutorship of Herbert Alley of Tavernier Key to whom grateful acknowledgement is here made.

Topography

The region (Fig.163) is dominated by an arcuate string of green islands, the Florida Keys, which make a discontinuous barrier 280 km long, and up to 3.5 m above sea level, separating the reef tract and the Straits of Florida from Florida Bay. From the bottom of the Straits, at 800–1000 m (about 450–550 fm), the floor rises increasingly steeply to the edge of the Portales Plateau which lies at about 460 m (250 fm). Thence it slopes gently upward to about 45 m (25 fm), where it again steepens as it approaches the reefs and the plateau of the back-reef lagoon. Where there are no reefs, the slope flattens at about 8 m (4.5 fm) and then dips slightly landward, as the back-reef lagoon deepens, before approaching the shoals off the Florida Keys. At other places, just within the 9 m (5 fm) line, the spurs and grooves of coral reefs rise abruptly to a terrace at about 5.5 m (3 fm) and thence to the reef-flat (Fig.164). In a few places the flat is awash at low tide. The back-reef lagoon varies in width from 5 to 15 km.

The longer axes of the islands of the Florida Keys are mostly parallel to the 9 m line. The islands are rarely more than a kilometre wide, except near Key West where there are about a dozen larger islands variously oriented: the biggest of these, Big Pine Key, has an area of 20 km². A number of the islands are skirted by mangrove swamp. The elongate islands are made of Pleistocene Key Largo Limestone. The larger ones have narrow bands of Key Largo Limestone against the back-reef lagoon, but have westerly extensions built of the other Pleistocene limestone, the Miami Limestone (HOFFMEISTER and MULTER, 1968).

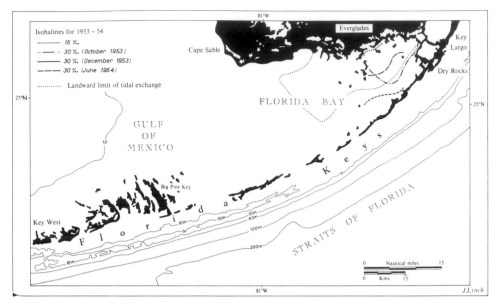

Fig.163. Southern Florida. The isohalines in Florida Bay show the invasion of fresh water, from the swamps of the Everglades, in the autumn as a delayed result of the rainy season August–October. (After GINSBURG, 1956.)

To the west and north of the Keys lies Florida Bay, 3,500 km² of shoal lime mud banks, stretching away westward to the 3.6 m (2 fm) line where the floor dips more steeply into the Gulf of Mexico. The distance from Cape Sable to the nearest Key is just over 40 km. Bathymetry and the distribution of the *Thalassia*-covered, carbonate mud banks are shown in Fig.171 (p.155).

Water movement

In mid-Straits the Florida Current flows northward to become, off Cape Hatteras, the Gulf Stream. Toward the Florida shore an intermittent counter current flows southward along the edge of the platform at the 9 m line. The tidal range at the 9 m line is about 0.7 m. In the easterly half of Florida Bay it is very small, a mere 15 cm: changes in depth much greater than this are caused by winds which either pile the water up against the Keys, raising the water level in the Bay, or blow it away into the Gulf of Mexico. Tidal exchange between the Bay and the back-reef lagoon is limited to the southeastern margin of the Bay (Fig.163). Over a large part of the Bay, owing to its extreme shallowness and the relatively great influence of friction between water and bottom, and because the outer banks act as barriers, there is no tidal exchange at all, either with the Straits or the Gulf (Fig.163; see MCCALLUM and STOCKMAN, 1964, for detail). Run-off from the Florida peninsula, though increasingly modified by the demands for water for agriculture and building, flows southward through the Everglades.

Hurricanes have an influence on erosion and deposition out of all proportion to their short duration. (BALL et al., 1967, on hurricane Donna of 1960; PRAY, 1968, on Betsy of 1965.) Winds of more than 120 km/h (75 mph) may blow for hours at a time, winds of 220 km/h (140 mph) for several minutes and sharp gusts attain 290 km/h (180 mph). On the reefs the wave amplitude reaches 3–5 m and blocks of reef rock up to 1 m across are thrown about, and live corals are commonly buried by rubble. The massive corals are the most resistant: the *Acropora palmata* with *uniformly* oriented branches (p.150) suffers little damage, but *Acropora palmata* with *variously* oriented branches is much broken. *Goniolithon* scrub, with its very open network, is scarcely affected. *Thalassia* carpets are apt to be invaded by lobes of cross-bedded lime sand, but are otherwise little damaged (THOMAS et al., 1961). Mangroves are defoliated and uprooted. Vast quantities of lime mud are put into suspension, exposing the roots of *Thalassia*, yet Ball and coworkers and Pray found that the mud mounds suffered no more damage than the removal of a few centimetres of mud: they were neither breached nor washed away. They certainly seemed to be a good deal more resistant to violent water movement than coral–algal reefs or the sand islands. One of these islands, Sandy Key in Florida Bay, was cut into two as water drove across it and is now a pair of islands separated by a tidal channel with well marked spillover lobe (delta) of lime sand. Vertical successions of quiet water lutites are interrupted by thin beds of sand containing an exotic biota. For example, a core taken on Cross Bank, in the Bay, consisted of silt-grade lutite interrupted by a 5 cm layer of sand with much *Halimeda*, an alga that does not grow in the Cross Bank area (MÜLLER and MÜLLER, 1967, also in English in MULTER, 1969; see also STODDART, 1962c, 1965a, for damage to Honduras reefs).

Temperature and salinity

Between the 9 m line and the Keys the annual temperature range of the water is about 15–33 °C. In the Bay, in shallower water less than 15 cm deep on the banks, the range is greater, about 15–40 °C. In the adjacent deeper channels it is, perhaps, 20–30 °C.

Salinities between the 9 m line and the Keys range approximately from 32 to 38‰, while in Florida Bay the salinity pattern is complex with the considerable range of 10–55‰. The low value is caused by the influx of fresh water where this flows from the Everglades (Fig.163) as a result of the high rainfall of summer. The high figure was found by TAFT and HARBAUGH (1964, p.23) in a channel $1\frac{1}{2}$ h after high tide, when the volume of water in the channel was greatly reduced and the lime mud banks were exposed. Peak values of this order (or even up to 70‰) make an exceedingly inhospitable habitat for the less tolerant organisms.

The reefs

Toward the outer edge of the plateau of Pleistocene limestone, which extends from about the 9 m line to the Florida Keys, there arises from its surface a discontinuous line of coral reefs and ridges of the limestone, no reef being more than a mile long. Between these elongate barriers, and occupying the greater part of the length of the outer reef-arc of GINSBURG (1956), loose carbonate sand covers the basement of Pleistocene limestone.

SHINN (1963) has described the reef called Key Largo Dry Rocks in some detail. Its major axis, parallel to the edge of the plateau, is 730 m long. Its width is about 152 m. The reef-flat, which occupies about a third of the plan area of the

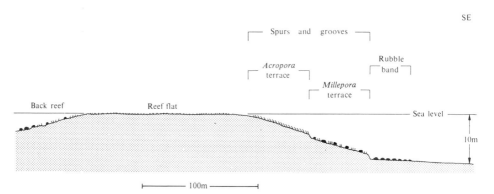

Fig.164. Reef topography, Key Largo Dry Rocks, Florida reef arc, summarized diagrammatically. (Data from SHINN, 1963.)

reef (Fig.164, 165), is composed of dead *Acropora palmata* and *Acropora* shingle. Water depth at mean low water is 0.6 m. Seaward of the reef-flat there is a 60 m wide terrace of live *Acropora palmata* (elkhorn or moose horn coral, Fig.166). This is the most actively growing part of the reef and the corals are oriented with their branches pointing toward the Keys. It seems that, with this growth habit, the polyps at the tips of the branches find maximum protection from scour by the incoming waves. At the outer edge of the *Acropora* terrace the floor drops almost vertically for about 2.5 m to the *Millepora* terrace. This is about 30 m wide and is covered mainly by *Millepora* blades (Fig.167) and heads of *Montastrea*. Beyond this second terrace the bottom again drops 2.5 m to the rock terrace of Pleistocene limestone. Here, at the foot of the second cliff, there is an 18 m band of rubble (Fig.168), mainly cobble-sized debris of *Acropora* and *Millepora*, bound by the coralline alga *Lithothamnium* and holdfasts of the sea fan *Gorgonia*. Beyond the rubble, scattered coral growth extends to a depth of 25 m or more.

Cutting across both terraces, in a direction perpendicular to the reef arc, is a system of grooves and spurs (Fig.169). The grooves are 3–6 m wide and about

Fig.165. Reef flat. Surface controlled by level of low spring tides. Mostly dead *Acropora palmata in situ*. Five sergeant major fishes. Key Largo Dry Rocks, Florida. (Courtesy of E. A. Shinn.)

Fig.166. Spreading branches of *Acropora palmata*, large sea fans and bush-like gorgonians. Florida reef. (Courtesy of E. A. Shinn.)

Fig.167. Tree-like gorgonians left; *Millepora* right, especially below the swimmer's fins and along overhanging ledge. Florida reef. (Courtesy of E. A. Shinn.)

Fig.168. Coral rubble near outer margin of Key Largo Dry Rocks, product of attack by storms and hurricanes. (Courtesy of E. A. Shinn.)

Fig.169. Aerial view of Sand Key reef off Key West, Florida, showing groove (chute) and spur passing, toward the right, into the reef flat. (Courtesy of E. A. Shinn.)

3 m deep and are floored by gravel and ripple-marked sand. Between them are wide spurs, 30–60 m across, with surfaces of dead or living *Acropora*.

The *A. palmata* on the first (*Acropora*) terrace is peculiar in that the branches point *away* from the deep sea toward the reef flat. Here, where the wave thrust measurements indicate a maximum, the coral has adapted its form as the polyps grow preferentially on the sheltered leeward surfaces. The spurs probably arise by random attachment and growth of corals followed by rapid growth of any new colony that finds itself sheltered in the lee of an earlier one. SHINN (1963): "An oriented colony protected in the lee of another colony is able to grow seaward and connect; this increases the linearity of growth centers. The unpopulated areas between the oriented colonies fill with sand, which is unfavorable to colonization by corals." This process leads to a preferential development of linear spurs oriented at right angles to the front of the reef and the accumulation of detritus between the spurs encourages their persistence by providing a mobile substrate unsuitable for the attachment of coral planulae. The spur, as a result of its growth from branching coral and its overgrowth after death, has an interior (revealed when dynamited by Shinn) that is labyrinthine and honeycombed, constructed of dead corals overgrown by *Millepora* and *Lithothamnium* and enclosing many voids and quantities of trapped debris. This combination of framework, binder and detritus, so characteristic of reefs, present and past, is a rare, though distinctive occurrence, in ancient limestone successions. Its rarity is understandable if we note that the area occupied by reefs, in the combined area of the reef tract and Florida Bay, amounts to about 0.01%.

Back-reef lagoon

Environments. The curved ribbon of shoal floor sheltered between the plateau edge (9 m line) and the Keys is open to the diurnal cycle of tides. Immediately behind the reef-flat, in the more sheltered water, there is a community of corals including *Porites*, (Fig.155, 156 on p.129, 130 resp.), *Montastrea*, *Siderastrea* and the brain coral *Diploria* (Fig.137, p.110), some having heads a metre or more across. There are also clumps of *A. prolifera* and the staghorn coral *A. cervicornis* (Fig.172). These large colonial animals provide niches for smaller organisms such as the green aragonitic codiacean, *Halimeda*, (Fig.141, p.114), and alcyonarians (Fig.166, 167), smaller corals, and molluscs, foraminiferids and bryozoans. Here and elsewhere in the back-reef lagoon, *Acropora palmata* is absent.

The major part of the back-reef lagoon (Fig.171) is divisible into an inner (landward) zone, dominated by *Thalassia*, and an outer (seaward) zone (Fig.172). The outer zone comprises a limestone basement, bare or covered by a thin ripple-marked lime sand (Fig.170), and patch-reefs (Fig.173) rooted to the limestone. Growing on the bare limestone is a community of sessile organisms, the corals *Porites* and *Siderastrea*, alcyonarians, sponges, chlorophytes and rhodophytes,

Fig.170. Rippled lime sand in the outer back-reef lagoon at a depth of 8 m. Florida. (Courtesy of E. A. Shinn.)

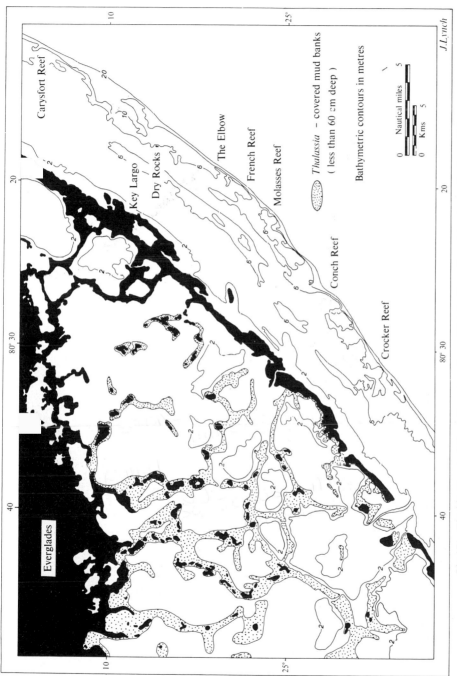

Fig.171. Florida Bay and the adjacent reef-arc. Distribution of mud banks in the Bay. (After GINSBURG, 1956.)

Fig.172. *Acropora cervicornis:* measuring branches for a study of growth rates. *A. palmata* in background, *Thalassia* and lime sand nearer the camera. Rear of Key Largo Dry Rocks. (Courtesy of E. A. Shinn.)

Fig.173. Edge of a patch reef to left, with gorgonians. *Thalassia* carpet to right and in foreground with patches of bare lime sand. Back-reef lagoon, Florida. (Courtesy of E. A. Shinn.)

molluscs and smaller organisms. The ripple-marked sand is a habitat for annelids, crustaceans, burrowing molluscs and echinoids. The patch-reefs vary in diameter from a few tens of metres to closely spaced concentrations of several hundreds of metres, and they tend to be equant, even circular, in plan. Large, elongate concentrations, off Key Largo, are about 1.5 km long and about 400 m wide. Usually a peripheral rim of live coral surrounds a central surface of dead coral overgrown by alcyonarians (sea whips and sea fans) and encrusted by *Millepora* and coralline algae. A characteristic of the patch-reef is the intense biological attack around the base, by boring and burrowing organisms, causing coral heads to split and topple over. The growing part of the patch reef is constructed of large heads, commonly more than a metre across, of such corals as *Montastrea*, the massive *Porites astreoides*, *Siderastrea* and *Diploria*. Sheltered between them are smaller corals, the branching *Porites porites*, the leafy *Agaricia* (Fig.138, p.111) and others.

In the inner zone of the lagoon the *Thalassia* varies in density from plants half a metre apart to an almost continuous cover. As in the Bahamas, *Thalassia* is denser in deeper water, forming a nearly continuous carpet below 2 m. There is a great variety of algae, molluscs, echinoids and smaller corals: several species of *Halimeda*, with *Penicillus* and *Udotea* (all aragonitic codiaceans), the coralline algae *Goniolithon* and *Amphiroa* (high-magnesian calcite), and nine abundant molluscan genera besides others more difficult to find. Echinoids include *Clypeaster*, *Lytechinus* and *Tripnuestes*. In depressions in the *Thalassia*-carpeted floor bare sand may be populated by the long spined, blue-black echinoid *Diadema*, by worms, burrowing molluscs and crustaceans. On the *Thalassia* leaves, which are up to 30 cm long, there is a rich encrusting microbiota of coralline algae, foraminiferids, bryozoans and other organisms. W. E. MOORE (1957) noted that the foraminiferal population of the back-reef lagoon is largely Soritidae (*Peneroplis*, *Archaias*), Amphisteginidae, Anomalinidae, toward the reef, and Miliolidae and Nonionidae toward the Keys; the population is richer in families, species and genera than that of Florida Bay.

Geologists have been much attracted by the large carbonate mud banks in this inner zone, such as Tavernier and Rodriguez Banks. Rodriguez Bank (TURMEL and SWANSON, 1964) lies about 1.5 km east of the axis of the Keys and west of French Reef: it appears as an island in Fig.171. It is elongate, about 1.5 × 0.5 km, low and flat topped. The surface of the top varies around mean sea level, from 0.3 m above to 1 m below. The Bank is clothed axially with the black mangrove, *Avicennia nitida* and marginally with the red mangrove, *Rhizophora mangle*. On the adjacent sea floor three biological zones are apparent, concentric with the Bank, being specially well developed on the windward side. An inner zone of *Thalassia* and codiaceans is succeeded outward, at a distance of about 250 m, by narrow zones of branching *Porites* and sharp, spiky thickets of *Goniolithon*. These organisms are about 5–10 cm high. The Bank is situated over a topographic low in the underlying Pleistocene limestone, and it seems probable that mud, in which

Thalassia thrived, first accumulated in this hollow, and that this primary colonization was succeeded by the growth of mangrove as the level of the mud approached
sea level. The Bank is probably nearly self-supporting as regards the supply of
sediment of skeletal origin (see p.164).

Petrography and its implications. Petrographic variation among the grains
of the back-reef lagoon, in a direction perpendicular to the line of Keys from the
fore-reef shoreward, reveals an interesting pattern: moreover, it is one that could
be discerned in ancient limestones. GINSBURG (1956) made point-counts of grains
retained on the 125 μ sieve (Table VII). These were artificially impregnated and
thin-sectioned. It is apparent that, outside the reefs, grains of coralline algae
and foraminiferids make up the major part of the sand. On the reef coral debris is
dominant but, in the lagoon, there is a preponderance of grains of *Halimeda* and
molluscs. The relations between skeletal debris and the communities from which
they are derived are as unprecise here as they are in the Bahamas (p.120). There
are, unfortunately, no data on the relative abundances of live organisms, so that
it is not possible to say to what extent the proportions of grains in the sands
represent proportions in living communities. It could be that here, as off the
Andros reefs, coralline algae are seriously over-represented compared with the
living biomass (p.120). Again, as in the Bahamas, *Thalassia* and the alcyonarians
are not represented in the counts, though some alcyonarian spicules were recorded
(GINSBURG, 1956, p.2425). It cannot be assumed that point-counts give information

TABLE VII

CONSTITUENT COMPOSITION OF SEDIMENTS, WITH CALCILUTITE CONTENTS: FLORIDA BAY AND REEF
TRACT
(After GINSBURG, 1956)

Location / Constituent	Florida Bay (17 samples)		Reef and back-reef (25 samples)	
	mean (%)	range (%)	mean (%)	range (%)
Point-counts of grains > 125 μ				
Algae	$^1/_2$	0–1	42	7–61
Molluscs	76	58–95	14	4–33
Corals	–	–	12	2–26
Foraminiferids	11	1–32	9	3–32
Non-skeletal	3	0–3	12	3–24
Miscellaneous	$^1/_2$	0–4	9	2–23
Unknown	1	0–3	8	4–15
Ostracods	2	1–6	—	—
Quartz	6	0–20	—	—
Weight calcilutite (< 125 μ)	49	10–85	17	0–68

on the detailed composition of the communities, their population densities or the rates of contribution by different organisms (p.102). The nature of the problem is such that no form of petrographic analysis of sediments, Recent or ancient, can reveal these important parameters. What Ginsburg has shown is the existence of a compositional gradient of a kind which *reflects* the distribution of the communities and one that could equally well be determined in an ancient limestone. The central problem remains: how would one interpret such a fossil gradient in terms of environment. Hence the need to correlate, on modern sea floors, the petrographic data with the environmental.

SWINCHATT (1965) has taken further the analysis of the petrographic gradient. His point-counts of grains coarser than 62 μ, in sliced impregnated samples, illumine in an interesting fashion the distribution of communities and the transport of sand in the back-reef lagoon. In Fig.174 and 175 it can be seen that, from the 9 m line to a point about half way across the lagoon, the frequencies of *Halimeda*, corals and molluscs are either fluctuating, falling or rising. From this half way point to the Keys the frequencies are relatively steady. The same kind of variation can be detected for mean grain size and standard deviation (sorting) of the sand fraction (> 62 μ). The coarser sands of the reef are succeeded shoreward by finer

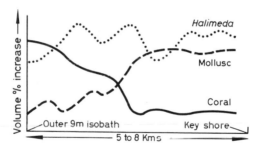

Fig.174. Variation in composition of grains coarser than 62 μ, from 9 m line to Florida Keys, by point-count. (After SWINCHATT, 1965.)

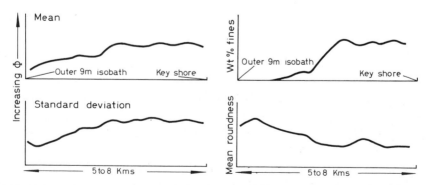

Fig.175. Variation in mean grain size, sorting, content of fines (< 62 μ) and roundness, from the 9 m line to Florida Keys. (After SWINCHATT, 1965.)

sands, but across the inner part of the lagoon the mean grain size remains nearly constant. The moderate to good sorting in the outer part of the lagoon is replaced, in the inner part, by persistently poor sorting. Roundness is high in the seaward part but poorer in the shoreward part. Taking whole samples, Swinchatt measured the wt. % of grains less than 62 μ and this value rises shoreward from the 9 m line until, approximately half way across the lagoon, it becomes steady.

The gradients found by Swinchatt for skeletal composition, texture and roundness are based on eight traverses. They indicate a division of the back-reef lagoon into two "sub-environments". The boundary between these lies at the outer edge of the *Thalassia* carpet. The petrographic transition is abrupt, being complete in a few metres at the most (SWINCHATT, 1965, p.85). That the sudden change is due to the presence of *Thalassia* is indicated by its abruptness. If it were the fines ($< 62 \mu$) that controlled the growth of the *Thalassia* then the relatively sudden change from "little or no" fines to a content from 10 to 55 wt.% would not be expected. SWINCHATT (1965, p.85) offered the following reconstruction of the development of the two sub-environments as the sea level rose after the last glaciation: "Possibly, at a lower stage of sea level, waves were damped out in shallow, back-reef waters resulting in the accumulation of fine sediment throughout the area. Dense growth of grass was developed on the fine substrate. As sea level rose, wave turbulence and wind-driven currents spread coarse, reef-derived material over the fine sediment. As sea level continued to rise, the coarse sediment progressively encroached over the finer substrate, possibly only during storms, to its present extent. In this speculative model, the initial cause of distribution of grass would be the distribution of a suitable fine bottom sediment. Once established, however, grass growth tends to control sedimentation." It would be interesting to test this hypothesis by making traverses, like those of Swinchatt, but with cores to the Pleistocene surface. These would reveal, also, to what extent the thickness of the unconsolidated sediment influences the spread of *Thalassia*.

From his collected data Swinchatt depicted an outer sub-environment where the water is turbulent, causing rounding of grains and loss of carbonate mud. The sand contains much reef (or patch-reef) debris, and its mobility makes it unsuitable as a substrate for the rich biota of the inner sub-environment. By contrast, in the inner sub-environment the slowing of water near the bottom by the baffle effect of the *Thalassia* blades (GINSBURG and LOWENSTAM, 1958) leads to the accumulation of carbonate mud, partly winnowed from the outer sub-environment and partly locally derived by skeletal decay, predation and boring. The finer, more loosely packed sediment, with a high content of carbonate mud (and surely an accompanying high content of non-skeletal organic detritus?) provides a habitat suited to a variety of surface and burrowing organisms.

Swinchatt has a useful comment on the distribution of skeletal particles larger than 2 mm. These shells, he found, were unabraded (commonly unbroken) and must therefore be of local origin. This certainly makes sense. In Bahamian sediments, although perhaps the large particles are rather commonly broken, they too are certainly not abraded. Sometimes all the fragments of a shell may be found within an area of a few cm², suggesting breakage by a predator (a star fish or a ray?), which ate the contents but abandoned the skeletal debris.

Another valuable outcome of this study of the back-reef lagoon is the unequi-vocal demonstration that mixing of sand-sized grains between the two sub-en-vironments is negligible, being restricted to a few metres of overlap.

It is necessary to note here that subsurface samples of sediment, taken from below the sediment–water interface as far down as 1–2 m, showed no sign of mineralogical diagenesis (TAFT and HARBAUGH, 1964; BERNER, 1966a). The mineral assemblage seems stable in this upper subsurface environment (a matter discussed more fully on p.361), yet FLEECE and GOODELL (1963) reported significant downward reduction in high-magnesian calcite in the upper 2.4 m of Florida Bay.

Florida Bay

Environments. Florida Bay (Fig.171) is not entirely isolated from the back-reef lagoon. In the eastern part of the Bay the tidal fluctuation is directly caused by the flow of water through deep channels between the Keys. As a result of this diurnal exchange with ocean water (Fig.163) some of the organisms of the back-reef lagoon have been able to spread for a few kilometres into the Bay. These include *Halimeda* and *Penicillus*, and *Porites porites* var. *divaricata* (the slenderest of the branching varieties of this coral) and *Siderastrea*. Both corals depend here on a rock substrate. Otherwise the Bay communities are different from those in the back-reef lagoon or on the reef, being poorer in species and genera. The dominant organisms are *Thalassia*, *Halodule* (one of the marine Zosteraceae), molluscs and foraminiferids (especially *Peneroplis* and *Archaias*). A very thorough study of the sedimentary dynamics of one of the tidal channels through the Keys has been made by JINDRICH (1969).

Of the distinctive topography (Fig.171) GINSBURG (1956, p.2400) wrote: "In the southeast part of the Bay the mud banks are very narrow, they have numerous channels, and they surround relatively large areas of deeper water. In the central and west parts of the Bay the banks are generally wider, more irregular in plan, and the areas of deeper water smaller than in the southeastern Bay. Along the mainland, and in the northeast corner of the Bay the banks are replaced by an almost continuous rim of mangrove islands [SCHOLL, 1963, 1964; TAFT and HARBAUGH, 1964], which may represent the final results of the island-building activities of these plants."

The banks are dominantly of aragonite calcilutite (but see p.263) with a content of carbonate mud (silt and clay grades) that can exceed 80% by weight. On many banks the water is only a few centimetres deep at low tide and on some the floor is actually exposed. Some bank surfaces are populated by *Thalassia* and *Halodule*, which encourage the accumulation of lime mud. In the sediment there is a rich and varied molluscan fauna (Table VII).

The origin of lime muds, both Recent and ancient, has long been a matter of contention, between those who favour a dominantly inorganic origin, by direct precipitation from sea water, and those who regard skeletal breakdown as the main producing process. The question of the source of the lime muds of southern Florida has lately been investigated quantitatively in the field by STOCKMAN et al. (1967), who reached the conclusion that the rate of production of lime mud by breakdown of a variety of indigenous skeletons has been more than enough to account for the total accumulation of mud in Florida Bay and the inner, landward stretch of the back-reef lagoon. The investigation consisted primarily of a quantitative estimation of the production of aragonite mud ($< 15\ \mu$) by the ubiquitous codiacean *Penicillus*. Over a period of one year square plots on the sea floor, of 0.5 m side, were continually observed and the appearance of new *Penicillus* plants and the death of plants were recorded. The mean life span for *Penicillus* at different stations varied from 30 to 63 days. From the combined data on life span, abundance, and wt.% of aragonite/plant, an estimate of the rate of lime mud production in unit area was made. By making allowance also for the breakdown of other codiaceans, of calcareous algae in general, and of molluscs, corals and foraminiferids (particles up to 62 μ), the total production of lime mud was calculated on the assumption that the production over the last few thousand years has not varied significantly. Work by NEUMANN and LAND (1969) off Great Abaco in the Bahamas (Fig.127, p.97) not only confirms the adequacy of the supply of algal mud but indicates that surplus mud probably contributes to neighbouring sea sedimentation.

Nevertheless, the Sr^{2+} concentrations of lime muds from Florida analysed by STEHLI and HOWER (1961) have values 2,700–4,600 p.p.m. These are distinctly closer to molluscan levels of Sr^{2+} at 1,100–4000 p.p.m. (Fig.225) than to codiacean at 8,740 p.p.m. Unfortunately, lack of data on the particle sizes of the samples analysed makes it impossible to know to what extent the low Sr^{2+} reflects a high content of molluscs in the coarser fractions, so disguising the codiacean contribution in the fraction less than 15 μ.

The channel floors may be bare Pleistocene limestone or limestone covered by only a few cm of loose lime sand: though rarely more than 1.75 m deep, the channels reach 4.5 m locally. They support a less varied molluscan fauna than the banks (GINSBURG, 1956, p.2403).

The foraminiferids in the Bay are dominated by Soritidae (*Peneroplis, Archaias*) and Miliolidae. W. E. MOORE (1957) reported that he found no live specimens of Soritidae and discussed the possibility that the Soritidae are not indigenous but have been transported, from the back-reef lagoon or some part of the Bay which was not sampled. He noted that the distribution of genera is related to the grain size of the sediment, the smaller tests lying in the finer sediments: thus the larger tests, including the peneroplids, are separated from the smaller tests, the miliolids, rotaliids and nonionids. The absence of live specimens may stem from the fact that the Soritidae live attached to *Thalassia* or *Halodule* blades and that grabs and dredges (W. E. MOORE, 1957, p.729) tend to slide over the *Thalassia* carpet and to retrieve sediment preferentially from the bare floor.

A. J. Gancarz (personal communication, 1969), working at Princeton University, on the relation between foraminiferids and substrates has found that live tests are common on codiacean algae. *Penicillus* and *Halimeda* carry a considerably larger foram population than *Udotea*. On *Penicillus* the tests are concentrated on the basal parts of the stalk and especially on the filaments of the hold fast at the sediment–water interface. On *Udotea* the foraminiferids are attached mostly to the blades. *Halimeda* supports as large a crop as *Penicillus*, but the tests are distributed all over the plant. Gancarz found live tests of five genera of the Soritidae (*Archaias*, *Peneroplis*, *Monalysidium*, *Sorites*, and *Spirolina*), including juveniles and adults, showing that they not only live in the Bay but also breed there.

Petrography and its implications. The petrographic differences between the sediments of the Bay and those of the reef and back-reef lagoon are summarised in Table VII. Most remarkable are the low contents of calcareous algae and corals in the Bay and the high contents of molluscs and calcilutite. TAFT and HARBAUGH (1964) described the sediment as having 90% or more calcium carbonate, with some quartz, radiolaria, sponge spicules and organic material. GINSBURG (1956, p.2425) added to the composition in Table VII the further components of echinoids, bryozoans, worm tubes, *Millepora*, crustaceans and alcyonarian spicules. TAFT and HARBAUGH (1964) found the grain size frequency distributions are bimodal with peaks at 350 μ and 6 μ. Presumably the higher peak is skeletal debris and the lower one aragonite needles. These needles are an important component in the 49% calcilutite of Table VII. A certain amount of low-magnesian calcite was found by TAFT and HARBAUGH (1964) as euhedral, rather etched crystals with intermediate diameters about 20 μ. These come from the sawgrass marshes in the Everglades where this mineral is possibly precipitated. Up to 5% dolomite occurs in western sediments of the Bay as rhombs (diameters 62–16 μ), somewhat etched. The rhombs are dead to ^{14}C and are therefore older than 40,000 years B.P.

Florida Bay is a highly distinctive environment, with its peculiar marine biota, its shallowness and small exchange with waters of the open ocean. There seems little doubt that reworking by currents is slight and that the sediment is largely autochthonous. It is obvious that the petrographic analyses distinguish the Bay lithofacies sharply from those of the back-reef and reef. It is, of course, impossible to gauge from such analyses the roles of *Thalassia*, or of the bank and channel topography. Information on structures seen in cores is urgently needed if we are to recognise similar banks in ancient limestones. It is hard to imagine what a section through lithified Bay sediments would look like in a quarry face. There is no obvious process for the making of bedding planes in the muds. Lithification would presumably give a 4 m layer of bioturbated micrite cut by irregular cal-carenite-filled channels and overlying a thin basal calcarenite.

Finally, it should be remembered that here, as in the back-reef lagoon (p.161), there is no evidence of mineralogical diagenesis in the first 1 m of sediment below the sediment–water interface (TAFT and HARBAUGH, 1964; also dealt with on p.361).

A note on carbonate mud banks

The study of Recent submarine **banks** of calcilutite touches the heart of an outstandingly obdurate problem, namely the origin of the many calcilutitic bioherms that are known in the geological column. Familiar examples are the Silurian Niagaran "reefs" of the Great Lakes area (LOWENSTAM, 1950), the Mississippian bioherms of the southwestern United States (PRAY, 1958), the Waulsortian type "reefs" of the Devonian in Belgium (LECOMPTE, 1936, 1937) and "knoll-reefs" of the Carboniferous in England (PARKINSON, 1957; BATHURST, 1959b) and of the Republic of Ireland (LEES, 1961, 1964; SCHWARZACHER, 1961). Despite uncertainty about the actual amount of relief possessed by these ancient bioherms, there seems to be no doubt that they were true topographic highs. A major difficulty in understanding their origins has been the question of the maintenance of the relief in a presumably waterlogged lime mud. There is, so far, no convincing evidence of how permanent mud slopes were supported: the fossil remains of frame-builders or sediment-binders are disconcertingly absent.

From the quantitative estimates of the production of aragonite mud by codiacean algae (STOCKMAN et al., 1967), allied to the earlier study of the trapping and stabilization of mud by *Thalassia* (GINSBURG and LOWENSTAM, 1958), it is now plain that lime mud banks can develop as largely self-supporting systems. They supply their own sediment from the calcareous benthos and this is trapped among the sea-grass blades and stabilized by their roots and rhizomes. The tightly packed roots are able to support slopes which are steep and nearly vertical along the sides of tidal channels. Additional resistance to erosion is provided by the presence of a subtidal algal mat and by the low threshold velocity of mud-sized particles and the low bed roughness.

A recent study of lime muddy sand banks, in remarkable detail, has been carried out by DAVIES (1967, 1970a). In eastern Shark Bay, Western Australia, a continuous bank of lime muddy sand extends for 130 km, along a north-south axis and parallel to the coast. It has a mean width of about 8 km and is crossed by more than 50 tidal channels, the longest being 13 km long. In three dimensions, the bank is wedge-shaped, thinning shoreward, with its maximum thickness of 7.6 m at its steeper seaward margin. The sediment is mostly composed of skeletal car-bonate with a little terrigenous quartz: it is sand grade but has up to 30 wt. % of its grains in the silt–clay size ranges. Mineralogically it is largely high-magnesian calcite and the grains are heavily micritized (p.383). The surfaces of the mound and channels are richly populated by the sea-grasses *Cymodocea* and *Posidonia* and a number of codiacean algae. Growing on this flora, especially on the sea-grass blades, is an epibiota of foraminiferids, coralline algae and bryozoans. Faecal pellets are produced by a grazing benthos. Davies has described how *Posidonia* thrives in a relatively high energy substrate and, with its roots and rhizomes, stabilizes the walls of tidal channels. On the adjacent levées a *Posidonia*-algal association is responsible for much trapping of sediment and the algae are themselves a source of new sediment. In the broad basins on the bank between levées, *Cymodocea* meadows harbour a rich epibiota and benthos. Quantitative estimates show that the on-bank supply of skeletal material, whole and broken, combined with the baffle effect of the leaves and the stabilization by roots and rhizomes, enables the bank to grow as a closed sedimentary system, requiring only water, sunlight, and nutrients from outside. Given the necessary geological developments, it seems that this great bank could be preserved as a bioherm.

One of the most interesting aspects of the work on Recent lime mud banks has been the discovery of their self-sufficiency. They produce their own sediment and their own sediment-binders. To what extent this kind of almost closed system was possible in earlier times is a matter of fascinating debate, particularly with regard to those lime mud bioherms which grew before the extensive spreading of angiosperms in the Lower Cretaceous. It would be interesting to know if the algae alone, calcareous and non-calcareous, were able to support the development of mud banks unaided.

GULF OF BATABANO, CUBA

In the Gulf of Batabano an intensive study, based altogether on more than a hundred core, grab and dredge samples, has revealed an unmistakable pattern of lithofacies (sediments: DAETWYLER and KIDWELL, 1959; molluscs: HOSKINS, 1964; foraminiferids: BANDY, 1964). It has shown, also, interesting relations between biological communities, substrate and turbulence.

Topography, water movement, salinity

Off the southern coast of Cuba, near the western tip of the island, a large lagoon, the Gulf of Batabano (Fig.176), stretches in an east–west direction for 290 km with a maximum breadth from north to south of 130 km. It has an area of about half that part of the Great Bahama Bank studied by Purdy (p.96), and three times that of the Recent carbonate region of southern Florida. East of the Isla de Pinos a line of nearly continuous linear reefs and cays stretches for about 120 km, sep-

Fig.176. Bathymetry of the Gulf of Batabano, Cuba. (After HOSKINS, 1964.)

arating the lagoon from the ocean. West of the Isla two small groups of patch reefs and cays lie at the southwestern edge of the Gulf. Along the seaward edge of the Gulf the floor is at about 2 m and, beyond the edge, it slopes to about 1,800 m in a distance of 5 km. Over the lagoon itself the mean depth is about 7 m, though locally the floor may descend to 14 m. The area is scattered with cays, and with banks that are commonly less than 2 m below water level. Most of the mainland coast is mangrove swamp. Into the large bay north of the Peninsula de Zapata flows the fresh water of the Rio Hatiguanico.

The mainland of Cuba skirting the bay is made up of Quaternary alluvial and marine sediments on a Tertiary succession consisting mainly of limestones. West of Havana, the mountains of the Sierra de los Organos are made of phyllites and quartzites overlain by limestone, all probably of Jurassic-Cretaceous age. In the interior of the Isla de Pinos, ridges of schist and marble of the same age make an inlier in the younger subaerially eroded rocks of a plain which occupies most of the island. These younger rocks are either Oligocene or Pleistocene. Beneath the unconsolidated sediments of the Gulf is a level platform of Tertiary limestones, eroded and leached at a time of lowered sea level. There is thus, around the Gulf, no source of siliceous sediment to contaminate the carbonate sediments forming in the lagoon.

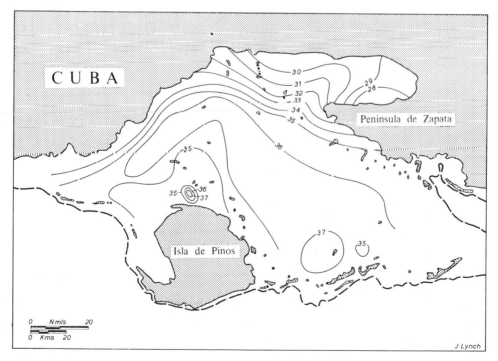

Fig.177. Isohalines (‰) for bottom water, Gulf of Batabano, Cuba. August, 1958. (After HOSKINS, 1964.)

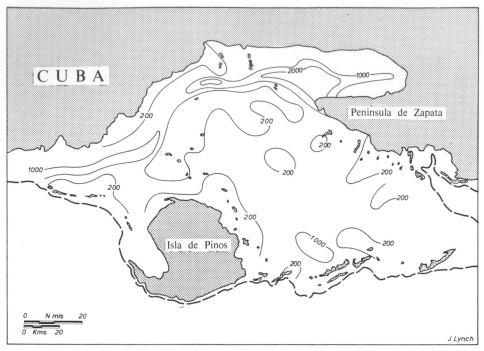

Fig.178. Abundance of molluscs (dead specimens/300 g of dry sediment), Gulf of Batabano. (After HOSKINS, 1964.)

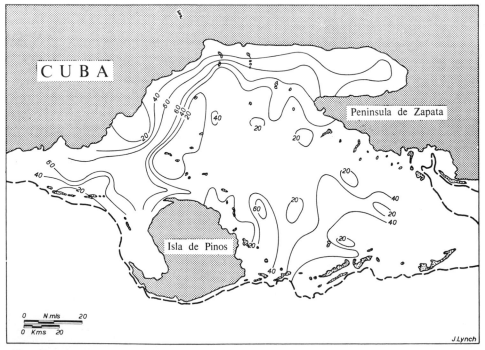

Fig.179. Diversity of molluscs (dead species/300 g dry sediment), Gulf of Batabano. (After HOSKINS, 1964.)

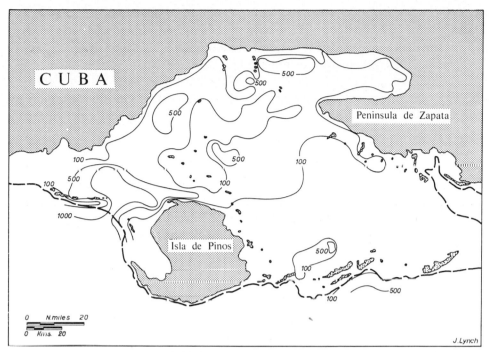

Fig.180. Abundance of foraminiferids (dead and live specimens/g wet sediment), Gulf of Batabano. (After BANDY, 1964.)

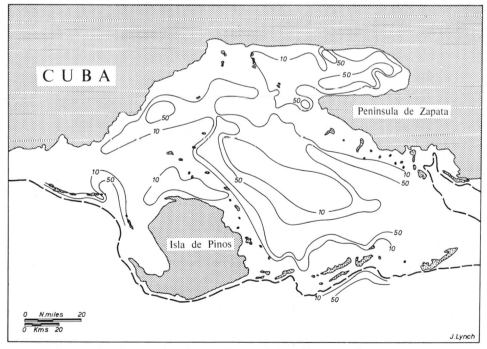

Fig.181. Ratio of abundances foraminiferids/ostracods (dead and live specimens), Gulf of Batabano. (After BANDY, 1964.)

The tidal range along the seaward margin of the Gulf lagoon is from 0.6 to 0.75 m, but it falls off northward until, in the northern lagoon, it is only about 9 cm. The tidal currents near the seaward edge of the lagoon are faster than 50 cm/sec (more than 1 knot) and set perpendicularly to the line of cays. Over the lagoon they are slower than 25 cm/sec (0.5 knot) and as a result of these small tidal currents (see p.100), and of the shallowness of the lagoon water, the wind exerts the major control over water movement. The prevailing wind is from the east and causes an anticlockwise movement of water in the lagoon.

Salinity at the surface of the open ocean is 35–36‰. Values for bottom water over the lagoon are given in Fig.177, but are largely true for the whole water mass as winds ensure good mixing and aeration. The 36‰ contour illustrates the anticlockwise movement of oceanic water. The effect of the fresh water from the Rio Hatiguanico is also apparent. A maximum salinity of 39.2‰ was observed in February 1959 just north of the Isla de Pinos.

Organisms, substrates

The organisms that inhabit the floor of the Batabano lagoon are corals and coralline algae, the calcareous algae *Halimeda* (Fig.141, p.114), *Penicillus* (Fig.153, p.126) and *Acetabularia* (Fig.143, p.116), at least 87 species of bivalve and 104 of gastropod. There are numerous foraminiferids, benthonic in the lagoon, mainly planktonic outside the reefs, also echinoids, ostracods, bryozoans, annelids and sponges. *Thalassia* is abundant over much of the northern and central lagoon. HOSKINS (1964) has produced some interesting maps (Fig.178, 179) showing the abundance of molluscs (expressed as specimens of dead mollusc per 300 g of dry sediment), their diversity (species of dead mollusc per 300 g) and distribution of molluscan biofacies (Table IX). BANDY (1964) has added equally valuable maps for the number of foraminiferids per gram (Fig.180), the ratio of foraminiferids to ostracods (Fig.181) and biofacies (Table IX). Both authors give also distribution maps for selected species of mollusc and foraminiferid.

The lime sands in the southeastern part of the lagoon and along the southern edge are rarely more than 0.6–0.9 m thick. The lime muds of the central and northern parts are much thicker and may locally attain 6 m. The sediments near the mainland and the Isla de Pinos contain some quartz. DAETWYLER and KIDWELL (1959) have given a map of the sediment types which was slightly modified (Fig.182) by HOSKINS (1964). The constituents are briefly considered here (Table VIII).

Skeletal grains. These are the calcareous skeletons, whole or broken, of a variety of organisms. Quantitatively important are the aragonitic alga *Halimeda*, also foraminiferids and coralline algae (high-magnesian calcite), corals (aragonite) and molluscs (aragonite and low-magnesian calcite). Quantitatively insignificant are echinoids, ostracods, bryozoans, annelid tubes, sponge spicules.

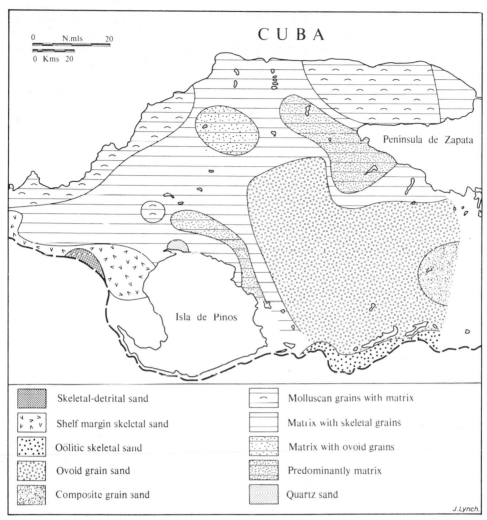

Fig.182. Lithofacies in Gulf of Batabano. (After DAETWYLER and KIDWELL, 1959; HOSKINS, 1964.)

Ovoid grains. These are well-rounded ellipsoids of revolution composed of micritic carbonate. Some contain black or brown flakes and even some recognisable skeletal debris. Length/breadth ratios are about 2/1. The grains fall in the sieve range 700–30 μ, and are particularly abundant in the range 500–250 μ. They appear to be faecal pellets.

Irregular non-skeletal grains. These are micritic grains which lack the ellipsoidal shape of ovoid grains and are generally without black or brown flakes. They are irregular in shape and rounded to subangular. They occur largely in the sieve fractions below 500 μ, especially in the range 125–62 μ. They seem to be akin to the cryptocrystalline grains of the Bahamas (p.117).

TABLE VIII

PETROGRAPHY OF SEDIMENT TYPES IN BATABANO BAY
(After DAETWYLER and KIDWELL, 1959)

Sediment type	Grains > 62 μ (point-counted %)	Matrix < 62 μ (wt. %)
Skeletal-detrital sand	*Halimeda* > coral + coralline algae > foraminiferids > molluscs; detrital limestone up to 50	0–25
Shelf margin skeletal sand	*Halimeda* > coral + coralline algae > foraminiferids > molluscs; detrital limestone a trace	0–25 (locally to 50)
Oölitic skeletal sand	(After HOSKINS, 1964: not described)	
Ovoid grain sand	Dominated by ovoid and irregular non-skeletal grains; composite non-skeletal grains locally up to 5; more skeletons than composite grain sand	5–50
Composite grain sand	Composite non-skeletal grains > 50; ovoids + irregular non-skeletal grains subordinate; superficial oöids < 5; skeletal grains minor	0–5
Molluscan grains with matrix	Molluscs > foraminiferids > *Halimeda* (locally absent); quartz locally	25–50
Matrix with skeletal grains	Skeletal grains dominant; ovoid + irregular non-skeletal grains minor; detrital quartz locally	50–75
Matrix with ovoid grains	Ovoid and irregular non-skeletal grains rather > skeletal grains; molluscs are dominant skeletal grains; foraminiferids and *Halimeda* locally	50–75
Predominantly matrix	Skeletal grains usually dominant; ovoid and irregular non-skeletal grains present	> 75

Composite grains. These are aggregates of ovoid grains, with some skeletal grains, cemented with micritic carbonate. They are like the grapestone of the Bahamas (p.87).

Oöids. Like the superficial oöids of Illing (p.78) these in the Gulf of Batabano have a thin oölitic coat on a skeletal or non-skeletal nucleus. They fall in the sieve range 500–125 μ.

Fig.183 and 184 show the distributions of carbonate mud and non-skeletal grains.

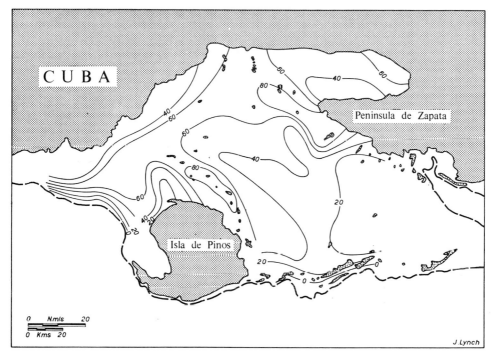

Fig.183. Distribution of carbonate mud (matrix: wt. % < 62 μ), Gulf of Batabano. (After HOSKINS, 1964.)

Environments

As a basis for a discussion of the biofacies, lithofacies and habitats, it is convenient here to divide the Gulf of Batabano into six environments (Fig.184), four large ones, and two small ones lying immediately behind the reefs. The outstanding characteristics of these six environments, based on information from DAETWYLER and KIDWELL (1959), BANDY (1964) and HOSKINS (1964), are given in Table IX. The data necessarily present a simplified version of a complicated pattern.

To begin with, the large quantities of lime mud (Fig.183) in the northwestern, northern and central environments point to the absence of lateral transport of sand grains by bottom traction. Thus it can be assumed that the sediments of these environments at least are autochthonous and that the point-counts of dead mollusc shells indicate the distribution of living communities over the past thousand years or so.

The **northwestern** and **northern environments** (Fig.184) are the most hospitable for molluscs which are both plentiful and varied. The mud, put into suspension by wind-driven currents, contains much nutritious, fine organic material. The low to moderate turbulence is clearly not a function of depth of water since this varies widely from beach to 12 m. Rather it must be a result of distance from the open

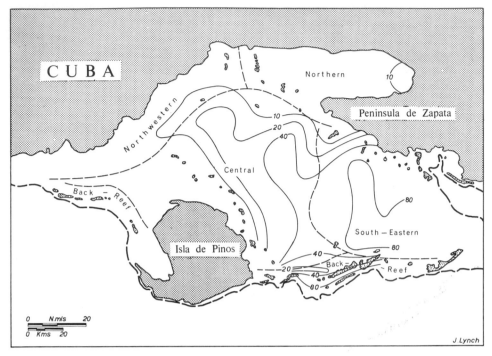

Fig.184. Distribution of non-skeletal grains (wt. %) and approximate boundaries of environments, Gulf of Batabano. (After HOSKINS, 1964.)

ocean and shelter from the prevailing east wind, the restriction of the stronger anticlockwise currents to a region nearer to the 36‰ salinity contour, and the abundance of *Thalassia* which reduces velocities near the bottom. The northwest environment, with salinity both normal and stable, is a habitat suited to the *Tegula–Arcopsis* molluscan subassemblage (Table IX) and to a mixture of this with the *Anomalocardia–Corbula* subassemblage; also to *Halimeda* and *Penicillus*, and to a mixed foraminiferal assemblage of *Elphidium* with its symbiotic algae and *Discorbis* and some miliolids. The northern environment, with low and fluctuating salinity, supports an *Anomalocardia–Corbula* subassemblage of a composition known to occur generally in places of below-average salinity. The euryhaline foraminiferid *Ammonia beccarii tepida* is concentrated between the Peninsula de Zapata and the mainland. Some of the species are poikilohaline (*Batillaria minima* and *Anomalocardia brasiliana*). The molluscan assemblage is composed of fragile species adapted to life on or in a muddy substrate or attached to vegetation. Miliolids are at least as plentiful among the foraminiferids as *Ammonia*, *Elphidium* and *Discorbis*.

    The **central environment** (Fig.184) shows a gradation from a western peripheral zone of shallower water to a central and eastern zone of deeper water. This gradation is accompanied by a change from a *Tegula–Arcopsis* molluscan sub-

TABLE IX

CHARACTERISTICS OF SIX ENVIRONMENTS DESCRIBED FOR BATABANO BAY, CUBA

Characteristic	Northwestern	Northern	Central	Southeastern	SE back reef	SW back reef
Sediment type	molluscan grains with matrix	matrix with skeletal grains and molluscan grains with matrix	matrix with skeletal grains: some matrix with ovoid grains: some predominantly matrix	ovoid grain sand with some composite grain sand	ovoid grain sand: passing seaward into oolitic skeletal sand	shelf marginal skeletal sand: some skeletal detrital sand
Substrate	usually firm	firm	firm to shifting	firm to shifting	shifting	shifting
Median grain	sand	sand and silt	silt: locally clay	sand	sand	sand
size ϕ	2–4	3–8	4–>8	<1–4	<3	1–4 falls seaward
Sorting ϕ	poor	poor	fair	good to fair	excellent	excellent to fair falls seaward
Standard deviation	3–4	>3	2–3	1–2.5	<1	<1–3 falls seaward
Wt. % matrix	20–60	40–60	40–80	<40	<20 falls seaward	0–20 falls seaward
Salinity (‰)	33–36	28–32 fluctuates	35–37 locally 39	35–37	35	35
Mollusc abundance specimens per 300 g	high 200–2000	high 200–>2000	medium 200–1000	medium 200–1000	low <200	low <200
Mollusc diversity species per 300 g	high 20–60	high 20–60	medium 20–40	low 20–40	low <40	low <40
Mollusc subassemblages	TA + AC	AC	TA + TE	TE	Cerithium	Cerithium
Foraminifer assemblages	miscellaneous, some Miliolid	Miliolid, miscellaneous	Miliolid, some miscellaneous	Archaias; Miliolid marginally	Archaias and Amphistegina	Archaias and Amphystegina
Depth (m)	<12	<2	2–9	<9	<20	0–20
Vegetation	abundant Thalassia	abundant Thalassia	medium Thalassia	sparse Thalassia	sparse calcareous algae	sparse calcareous algae
Turbulence	low to moderate	low to moderate	low to high	moderate to high	high	high

assemblage (Table IX) with miliolids, in the north, west and south (locally mixed with the *Anomalocardia–Corbula* subassemblage), to a *Tellina–Ervilia* molluscan subassemblage with *Elphidium* (BANDY, 1964, fig.4B, C) in the east. This lateral variation within the area is correlated with a fall in the content of lime mud and a slight rise in salinity. There is also an increase in bottom turbulence and mobility of the substrate as the density of *Thalassia* declines eastward. Neither the abundance of molluscs nor their diversity is as high as in the northwestern and northern environments. Salinities are higher and there is a marked concentration of lime mud compared with other environments.

The **southeastern environment** (Fig.184) is distinguished by the *Tellina–Ervilia* molluscan subassemblage and three genera of the family Soritidae, namely *Archaias*, *Peneroplis* and *Sorites*. The sediments here are coarser and better sorted than those to the west, the water is more turbulent and the substrate less stable. The region is exposed to the prevailing east wind. BANDY (1964) has demonstrated a strong positive correlation between the frequencies of species of his *Archaias* assemblage and the proportion of lime sand in the sediment.

The turbulent **back-reef environments** (Fig.184) are colonized by the *Cerithium* molluscan subassemblage (Table IX), by *Halimeda*, by the *Archaias* assemblage of foraminiferids, and another comprising *Amphistegina*, *Asterigerina* and *Rotorbinella* (BANDY, 1964, fig.6). Mollusc abundance and diversity are both low, and grains of corals and coralline algae are abundant. The turbulence of the water and mobility of the substrate are higher than in any other part of the Gulf. *Thalassia* is unimportant. The two back-reef environments differ in their sediments: that in the southeast contains some oölite and the ovoid grain sand derived from the lagoonal hinterland, while the southwestern environment has the highest concentration of foraminiferids in the Gulf (BANDY, 1964, fig.2B).

Discussion

Unlike the studies made on the Great Bahama Bank and in Florida, the biological data for the Gulf of Batabano are not based on observations of living communities. Except for a few general details they stem from grain counts of dead skeletons and are, in fact, raw petrographic data from which conclusions about the distribution of living communities may or may not properly be drawn. The reliability of such conclusions must now be considered.

The stability of the sediments of the northwestern, northern and central environments is denoted on the grounds of the high mud content. It is further supported by the large size range of the fauna, mainly from about 15 mm down to 1 mm and less, from one of the larger bivalves to a miliolid: this is not a typically current-sorted size distribution. Again, the broad expanses of the environments, coupled with the relative constancy of faunal content within them for thousands of square kilometres, make it necessary for us to conclude that the animals lived in the environments in which they are found. The distances are great and we cannot therefore estimate, from the available data, the effectiveness of bottom traction over metres or kilometres. Here it would be as well to bear in mind Swinchatt's opinion regarding the Florida sediments (p.160), that skeletons bigger than 2 mm, except in the surf zone or reef, are autochthonous—particularly in a silt or clay.

We have to face the possibility that the ovoid grain sand of the southeastern environment, with its "moderate to high" mobility, is transgressing westward into the central environment. Such a process would doubtless be helped by strong flood tides backed by the prevailing east wind. It would be interesting to know if the fauna of the *Tellina–Ervilia* subassemblage really lives in the deeper water toward the west or is carried there when dead. Two factors make this degree of transport highly unlikely. One is the still high amount of mud in the substrate, especially in the west where it exceeds 20% by weight and the other is the considerable size range of the shells of the *Tellina–Ervilia* subassemblage. These skeletons range in diameter from about 70 mm to 5 mm and, if we add the tiny *Archaias* and the miliolids, this is certainly not a size distribution showing good current sorting. As in the Bahamas, the presence of ovoid sand grains is related not to bottom traction but to the activity of animals capable of transforming the muddy substrate into pellets.

It is important to remember that the percentages of different components in the sediments are relative. A rise of molluscan abundance in a certain direction may reflect an increasing density of population or simply a reduction in the contribution from other organisms. The maps (Fig.178–181) do not show changes of absolute densities of the various subassemblages or phyla. Nor, of course, do they show the proportions of live organisms. In addition, grain counts can take no account of the many organisms that leave no skeleton.

The petrography of the sands in the back-reef environments, especially the southeast, suggests some mixing at their northern boundaries with the sediment of the lagoon to the north (Table IX). Despite this the sands in these environments remain distinctive. Their effective isolation from the adjacent lagoonal environment is indicated by the steep gradients shown by certain parameters along their northern boundaries, and this is, indeed, why the boundaries were drawn there. Both back-reef environments show a rapid fall northward in the percentages of *Cerithium litteratum*, of coral and coralline algae (HOSKINS, 1964) and an abrupt rise in the amount of mud (Fig.183). Behind the southwestern back-reef environment there is a considerable drop in the total quantity of foraminiferids (BANDY, 1964, fig.2B), and behind the southeastern back-reef environment a sudden drop in the percentage of *Ervilia nitens* (HOSKINS, 1964, fig.16).

In conclusion, it can be said that the grain counts and measurements of lime mud content show a pattern of lithofacies that is probably closely correlated with a similar pattern of biological communities. It must be emphasized, though, that the assumption that lithofacies and living communities correspond is based here

on the known highly muddy nature of the sediments and their generally poor
sorting, leading to the assumption of minimal transport of the sand grains. Thus,
similar correspondence in ancient limestone successions can only be expected in
biomicrites or biopelmicrites. It should also be noted that the lithofacies have each
a very big areal extent and that, for many studies of palaeogeography in smaller
areas, such a pattern would be exceedingly crude. In this area of about 25,000 km^2
only four environments can be picked up with a distribution of about 120 samples.
Finally, it is plain that a study of the lithofacies alone, through examination of
sliced impregnated sediments, equivalent to a thin section survey of ancient
limestones, would yield an even cruder picture than the one presented here.

THE TRUCIAL COAST EMBAYMENT, PERSIAN GULF

During the last few years exploration along the Trucial coast of Arabia has
brought to light an extensive region of Recent carbonate sediments which, though
having much in common with carbonate areas of the Caribbean, shows interesting
peculiarities of its own. HOUBOLT (1957) has given a vivid picture of the relation
between petrography and bathymetry off Qatar. Geologists of Koninklijke/Shell
Exploratie en Produktie Laboratorium in The Netherlands and V.C. Illing and
Partners of London have investigated the shallower deposits around Qatar,
including the lagoons and tidal flats (A. J. WELLS, 1962; ILLING et al., 1965;
SHINN, 1969; TAYLOR and ILLING, 1969). A group from the Imperial College of
Science and Technology, London, led by D. J. Shearman, has concentrated on the
Abu Dhabi area. On Abu Dhabi the main background is to be had from the theses
of KINSMAN (1964b), BUTLER (1965), KENDALL (1966) and SKIPWITH (1966), with
papers by KINSMAN (1966b), KENDALL and SKIPWITH (1968a, b) and BUTLER (1969).
From this collective Dutch and British effort has come new understanding of
sedimentation on oölite shoals and in lagoons, of the relation between sediment
type and bathymetry and, above all, of the growth of dolomite and gypsum on
supratidal salt flats.

Topography

The Persian Gulf (Fig.185), an arm of the Indian Ocean, is 970 km long and varies
in width from 340 km to 55 km in the Strait of Hormuz. It is rarely deeper than
90 m. The axis of the trough lies toward the mountainous coast of Iran, while, on
the southern side, the floor shelves gently upward to the Arabian coast. Here,
between the Qatar and Oman Peninsulas, the Arabian coast bends southward to
embrace a broad region of shallow water, the Trucial coast embayment. The
Arabian hinterland is low lying desert, except for the mountains of the Oman
Peninsula which rise to 1,800 m, and no rivers drain into the Trucial coast em-

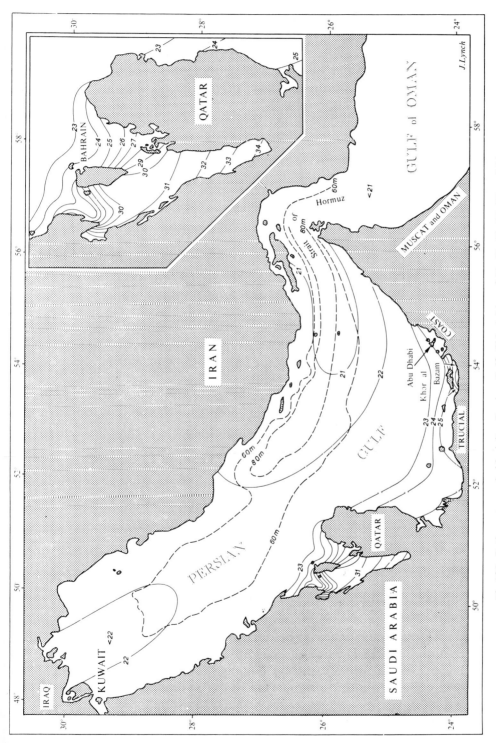

Fig.185. Bathymetry and chlorinity, Persian Gulf. Chlorinities of 21, 22, 23, 24, 25 and 31 are equivalent to salinities of 37.8, 39.6, 41.5, 43.2, 45.0 and 55.8. (After SUGDEN, 1963b.)

bayment. The Gulf is bounded on the west by the Arabian Precambrian shield covered by Palaeozoic, Mesozoic and Cainozoic sediments. East of the Trucial embayment lie the Cretaceous and Jurassic strata of the Oman Mountains. Over-looking the northern shore of the Gulf, the Zagros Mountains of Iran are a fold belt aligned NW–SE and involving Mesozoic and Cainozoic strata with Palaeozoic appearing in fault slices. Superimposed on the areal pattern of structure are a number of salt plugs, particularly in southwest Iran and the southern Gulf: these are primarily of Upper Tertiary age. The main structural pattern of the mountains is a result of Mio-Pliocene folding. The area is still tectonically active as shown by Quaternary raised beaches and uplifted planed surfaces.

Carbonate deposition was general and widespread from the Permian to the Miocene, varying lithologically from deep water marls to shallow water calcarenites and oölites, with occasional evaporites. Tertiary folding was followed by deposition of a thick mass of continental sediments.

Off the eastern Trucial coast embayment in the Abu Dhabi region there is a multitude of small shoals and islands which give shelter to extensive lagoons and inlets (Fig.192, 193). Reefs and oölite deltas occur sporadically off the seaward coasts of the islands. Beaches facing the open sea tend to accumulate lime sands but, in the lagoons, the dominant sediment is peloidal lime sand becoming muddy with aragonite in channels in the inner lagoon. Along the shore of the mainland there are broad areas of supratidal salt flat known locally as "sabkha". Lagoons and sabkha are also developed around the Qatar peninsula. The sabkha lies above the level of high tide, but during high springs, especially when helped by a following wind, it is liable to extensive flooding. West of the Khor al Bazam, as far as the northern point of Qatar, the coast is irregular: generally low cliffs backed by sand dunes.

The uncemented carbonate sediments of the embayment are, so far as has been determined, underlain by Quaternary limestones, and these are locally exposed as bare rock. These limestones commonly cap or lie against Miocene rocks landward of the sabkha.

The islands in the embayment have a variety of origins: some are Recent accumulations of mud, others are sand bars or reef flats. Yet others have followed upon colonization by mangroves with subsequent entrapment of sediment.

Water movement, temperature, salinity

Owing to the intense evaporation in this unusually hot climate, the surface waters of the open Gulf are highly saline with values mainly around 39–40‰. The heavy saline water moves downward so that bottom salinities are more than 41‰, especially around the southern margin of the Gulf. This dense water then flows out into the Gulf of Oman along the southern side of the Strait of Hormuz, and is replaced by water of normal oceanic salinity (35‰) flowing in on the north side of the Strait. This incoming, less dense water spreads over the surface of the Gulf, where it makes up for the loss by evaporation and replaces the dense outflow (EMERY, 1956).

Off Qatar the tidal range is 0.4–0.7 m (HOUBOLT, 1957). Off Abu Dhabi on the open shelf the range at *springs* is 2.1 m: this is about 20% above normal. In mid-lagoon it is 0.8 m and, in the inner lagoon 0.5–0.6 m (KINSMAN, 1964b). At night land "breezes" regularly drive the shallow lagoon water seaward, exposing large areas of mud floor (D. J. Shearman, personal communication, 1967). Both Houbolt and Kinsman found that, in the shallower water, wind has a great influence in raising or lowering the water level. The dominant wind is from the northwest. Blowing at gale force in the winter, it is called the "shamal".

The Gulf is probably the warmest sea in the world, the temperature at its surface, in the open sea, varying from about 20 °C in February to 34 °C in August (EMERY, 1956). In the Abu Dhabi area the range of surface water temperature on the open shelf is 23–34 °C, and in the inner lagoon is 22–36 °C. Shallow water masses tend to have a high diurnal range— as much as 10 °C. The sabkha ground water may reach 40 °C: these exposed salt flats have nearly the same extreme variation as the neighbouring desert.

Surface salinities of Gulf water are shown in Fig.185. Values as high as 45‰ were found outside the lagoons of the Abu Dhabi area (SUGDEN, 1963b), but in the inner lagoon the range is high, 54–67‰ (EVANS and BUSH, 1969). Rain is infrequent but torrential, especially near the Oman Mountains, helping to flood the sabkha and dissolving the salt crust. It also dilutes the lagoonal water. Annual rainfall is highly variable: in the period 1958–1964 the heaviest was 6.73 cm, the lightest 0.33 cm. A high humidity aided by nightly dew makes possible the persistence of a flora on the sand dunes.

The open sea floor off Qatar

HOUBOLT (1957) examined and discussed samples of sediment from about four hundred stations near the Qatar peninsula (Fig.186). The samples had been collected, with van Veen grab and dredge, by a party under C. Kruit in the marine concession of the Shell Company of Qatar Limited. Bathymetry had also been surveyed. Houbolt had no information on the biota other than the Foraminiferida. The interest of Houbolt's results lies in the relations between bathymetry, sediment type and dead foraminiferal assemblage. PILKEY and NOBLE (1966) analysed 50 bottom samples collected mainly to the north of the southernmost 60 m bathymetric line (Fig.185). These were grab samples and cores collected by the U.S. Naval Oceanographic Office.

Petrography. On the open Gulf floor PILKEY and NOBLE (1966) found that there is a striking dichotomy between, on the one hand, fluviatile, wind-blown

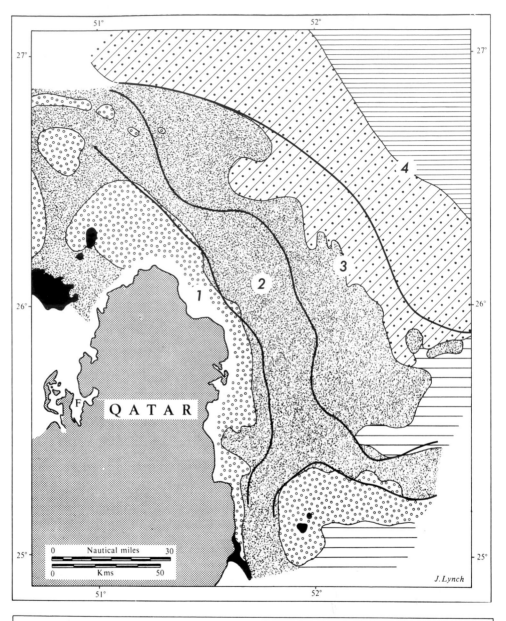

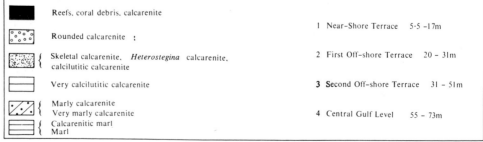

	Reefs, coral debris, calcarenite	
	Rounded calcarenite	1 Near-Shore Terrace 5·5 –17m
	Skeletal calcarenite, *Heterostegina* calcarenite, calcilutitic calcarenite	2 First Off-shore Terrace 20 – 31m
	Very calcilutitic calcarenite	3 Second Off-shore Terrace 31 – 51m
	Marly calcarenite / Very marly calcarenite / Calcarenitic marl / Marl	4 Central Gulf Level 55 – 73m

Map labels: 51°, 52° (top and bottom); 27°, 26°, 25° (left and right)

QATAR

F

1, 2, 3, 4

0 Nautical miles 30
0 Kms 50

J.Lynch

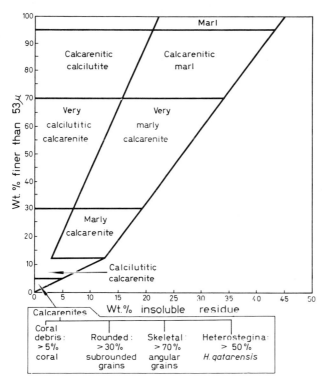

Fig.187. Classification of sediments off the Qatar Peninsula, Persian Gulf. (After HOUBOLT, 1957.)

lutite consisting of low-magnesian calcite and dolomite and, on the other hand, sand grade or coarser material composed of dominantly fresh, skeletal, high-magnesian calcite and aragonite. The lutite content of the sediments is greater in the deeper water toward the Iranian shore, and so, naturally, is the content of clay minerals (mainly illite and a mixed layer illite–montmorillonite), which is locally as high as 50 wt.%. The sediment is progressively richer in lime sand toward the Trucial coast where the shallow floor supports a relatively rich calcareous benthos. Corals and coralline algae flourish where hardgrounds (p.371) are exposed or covered by only a few centimetres of sediment (SHINN, 1969, p.143). The sediments exhibit a very wide range of grain size, and about half of the floor is composed of sediments with more than 30 wt.% of silt and mud (clay) grade material. In these fine sediments the content of matter insoluble in dilute hydrochloric acid varies over a range of about 10–50% by weight. Houbolt therefore constructed a twofold classification of sediments based on grain size and wt.% of insolubles. The lime sands he divided further, on the basis of grain type and degree of abrasion. These relations are summarized in Fig.187. For sediments with more than 12% by weight

Fig.186. Bathymetric levels and lithofacies off the Qatar Peninsula, Persian Gulf. *F* is Dohat Faishakh. (After HOUBOLT, 1957.)

of silt and mud (material finer than 53 μ or 4.3φ) there is a dual scheme. This separates mixtures of lime sand and lime mud from marls. In the marls there are the same carbonate materials with the addition of terrigenous, siliceous grains, mainly of silt and mud size. The separation at 53 μ (4.3 φ) was made by wet sieving.

In the whole fraction finer than 1 μ the dominant mineral is calcite, with lesser quantities of clay mineral dominated by illite. The absence of aragonite is in striking contrast with the shallow water lagoonal muds behind Abu Dhabi Island which, in places, contain more than 90 wt.% of aragonite. Quantitative estimates of the composition of the coarse fraction (coarser than 53 μ) were made on loose material with a stereoscopic polarizing microscope.

Coral debris (Fig.186) is associated with coral reefs, which border the peninsula and a number of islands, and with patch reefs on the near-shore terrace.

Rounded calcarenite (HOUBOLT, 1957, p.44) contains "over 30 per cent of more or less rounded carbonate grains, most of which have an etched granular surface texture" (Fig.186, 188, 190). There are also some quartz grains derived from the coastal dunes. Houbolt suggested that the mat surfaced grains are derived from the surf zone, where they were abraded. The surfaces of the grains are certainly much abraded, so much so that the superficial microarchitecture of the skeletons is mostly destroyed. Thus in reflected light there is little to see and skeletons were recognized in thin section. The turbulence responsible for the abrasion would also prevent deposition of lutite. The rounded calcarenite lies on the near-shore terrace (Fig.186) and passes seaward into skeletal calcarenite.

The **skeletal calcarenite** has no more than 5% of lime mud or marl (finer than 53 μ or 4.3 φ), and much of this is trapped inside skeletal chambers, which leads to the conclusion that, like the rounded calcarenite, the skeletal calcarenite lies on a floor from which silt and clay-sized grains are winnowed. The skeletons which are dominantly molluscan, are mostly broken but the edges of the fractures are sharp (Fig.189). Thus the skeletons cannot have been broken by surf action, which would have rounded them, but must owe their broken condition to the activities of feeding animals. The lower turbulence of the water over the floor of the first and second offshore terraces (Fig.186, 190) is also shown by the poorer sorting of the skeletal calcarenite compared with the rounded calcarenite (Fig.188).

Heterostegina calcarenite occurs sporadically at the transition from rounded calcarenite to skeletal calcarenite. This sediment with a high concentration of the foraminiferid *H. qatarensis* is, therefore, itself transitional in type (Fig.186).

Lithofacies, bathymetry, Foraminiferida off Qatar. Referring to bathymetric data based on echo soundings or, in places, on the length of the grab line, HOUBOLT (1957, p.34) divided the topography of the bottom into four levels. In Fig.186 the relation between these levels and the distribution of sediment types (lithofacies) is shown. Their relations to the dead foraminiferal assemblages are given in Fig.190. Starting from the bathymetry, as the variable showing the most marked disconti-

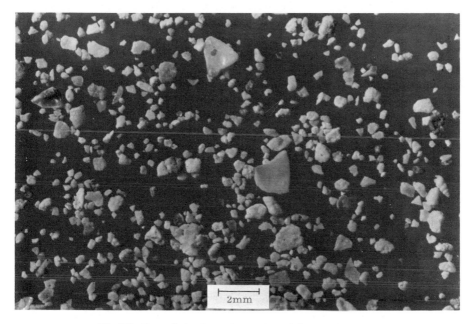

Fig.188. Rounded calcarenite. Reflected light. Off Qatar.

Fig.189. Skeletal calcarenite. Reflected light. Off Qatar.

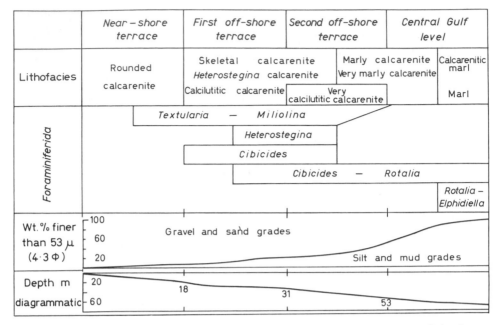

Fig.190. Relation between bathymetry, lithofacies and foraminiferal assemblages off the Qatar Peninsula. (Data from HOUBOLT, 1957.)

nuities, the lithofacies and Foraminiferida can then be expressed in terms of the four levels.

It is plain that there is an overall relation between lithofacies, foraminiferal assemblage and depth. There is a marked break in lithofacies at the outer edge of the near-shore terrace and this is correlated with the inner limits of *Cibicides*. Such a break could reasonably be attributed to the outer limit of a surf zone (zone of breaking waves) where an unstable substrate gives way to a stable, possibly at a time of lower sea level. The next transition, involving both lithofacies and Foraminiferida, occurs within the second offshore terrace where the content of fine sediment is rapidly increasing. This may represent a significant change from a sandy to a dominantly muddy substrate. On the other hand, the apparent correlation of Foraminiferida with substrate (Fig.190) may reflect here the temperature gradient rather than the change in substrate. The more obvious trends in modern foraminiferal ecology are those that relate to changes in depth and temperature (BANDY, 1964).

The Miliolina in Fig.190 include *Quinqueloculina* and *Triloculina*, genera typical of bays and inshore waters. *Textularia* is a common inner shelf genus. Peneroplids, so abundant on the shallowest sea floors and often living attached to vegetation, are also recorded off Qatar from the near-shore terrace. *Cibicides* is better known elsewhere from depths greater than 50 m, and *Elphidiella* from depths less than 50 m (BANDY, 1964).

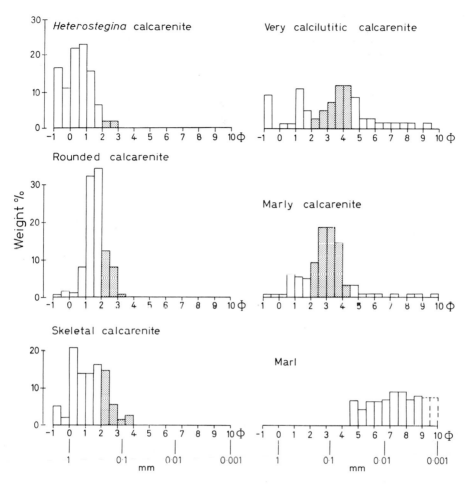

Fig.191. Typical grain size frequency distributions of some of the sediments off the Qatar Peninsula. The stippled classes represent sediment probably winnowed from the two inner terraces and deposited on the two outer terraces (Fig.186). (Size data from HOUBOLT, 1957.)

The unusually simple trend of lithofacies away from the shore is probably, in part, related to a systematic winnowing of the finer sands from the inner terraces to the outer (HOUBOLT, 1957, p.55, 94). The topography is, of course, relatively uncomplicated. In Fig.191 all the calcarenites are markedly poor in grains finer than 2–3 φ, whereas the very calcilutic calcarenite shows a subsidiary peak, and the marly calcarenite a peak in the range 2–4 φ. It can be assumed that fine debris is provided on the inner terraces as a result of abrasion and grain breakage: thus its absence is an indication of its removal. An interesting point is that rounded quartz grains between 2–3 φ are the least stable hydraulically, having the lowest threshold velocities. Thus, even allowing for differences of shape and specific gravity, it is likely that one of the controlling factors here is the ease with which carbonate grains of this size are eroded by bottom traction and are moved from the inner terraces to the outer. There is, unfortunately, no information on the sizes of the Foraminiferida, so we are not able to consider the possibility of their seaward transport after death. A decrease in turbulence of bottom waters with depth is the likely controlling factor in the deposition of the pelagic marls, rich in land-derived, siliceous silt and clay mineral mud. The high calcite composition of the fraction less than 1 μ is a direct result of the contribution of wind-born dust that reaches the Gulf from Saudi Arabia and Iran. EMERY (1956, p.2367) analysed a sample that had accumulated on the deck of a research vessel during three days and found it to contain 83% calcite. The remainder was, dominantly, felspar with a lesser amount of quartz.

Briefly, the sea floor around Qatar reveals an unusually direct and simple relation between lithofacies, turbulence, depth, temperature and benthonic Foraminiferida. It is important to realise, at the same time, that this simple seaward relation does not persist all along the shelf off the Trucial coast. East of Qatar, numerous banks and depressions are floored, respectively, with relatively well sorted calcarenite and muddy calcarenite.

Note on aragonite whitings. The blue surface of the sea in the Persian Gulf is locally broken by milk-white patches, up to 10 km across. These have been investigated by WELLS and ILLING (1964). From the air the milky patches, or **whitings**, can be seen to appear and grow within a few minutes and to persist thereafter for several hours. They would seem to be akin to the whitings over the muddy-sand and mud habitats west of Andros Island, Bahamas (p.137). For the swimmer visibility is reduced to less than 0.5 m or even to a few centimetres. The water bears a suspension of aragonite needles and some pelagic organisms and it has a density of about 1 g of solid/100 l. Wells and Illing made seven points of special interest:

(*1*) The milkiness appears at the same moment over an area of many square kilometres. The region is normally devoid of fish which, even where they disturb the bottom, only muddy the water in patches a few hundred metres across.

(*2*) The whitings form at the surface and overlie clear water.

(*3*) They occur over any depth of water in the Gulf but are very frequent, though smaller, over depths of less than 10 m.

(*4*) The aragonite crystals flocculate, and sink and disappear from sight, in less than a day, whereas stirred bottom mud takes several days to settle.

(*5*) The underlying bottom mud contains up to 50% calcite, but in the whiting there is "very little calcite".

(*6*) Analyses of water immediately before and after the appearance of a whiting suggest that there is a very slight drop in Ca^{2+} and Ca^{2+}/Mg^{2+}. This could mean that calcium had been removed from the solution.

(*7*) Water stratification in terms of temperature, salinity and pH remains stable and this points to the absence of turbulence and bottom disturbance.

Wells and Illing advance the hypothesis that, in sea water known to be saturated for aragonite (WEYL, 1961), the sudden, rapid growth of a population of diatoms (bloom) could remove CO_2 from the water on such a large scale and so rapidly that equilibrium could only be restored by widespread precipitation of $CaCO_3$. Stirring of bottom sediment, as oil tankers expel large quantities of light northern latitude sea water, which hits the bottom and floats upward bearing suspended mud, cannot yet be wholly discounted. Traverses of pH measurements are needed (p.286). Laboratory experiments by DE GROOT (1965) indicate that the maximum possible rate of precipitation of aragonite from sea water is far too slow to account for the known *sudden* appearance of whitings in the Persian Gulf. The formation of a precipitate, hastened in the laboratory, still took at least two weeks. De Groot also found that the mineralogy of the whitings and the local bottom sediment were "very similar".

Undoubted precipitates of aragonite needle mud seem to have been monitored successfully from surface waters in the Dead Sea. Whitings appear at intervals of a few years and NEEV and EMERY (1967, p.93) investigated one of them. When the suspended precipitate appeared the concentration of HCO_3^- in the water decreased from 0.25 to 0.22 g/l, the visibility was reduced from 3 to 0.5 m and the concentration of suspended mud (mostly aragonite) increased from 0.2 to 0.4 g/l. Whitening of the near-surface water seems to occur at times of highest temperature in August. The changes in concentration of HCO_3^- began at the surface and progressed gradually downward to a maximum depth of 40 m. Isotopic data from the laminated sediments of the Dead Sea floor yield values for $\delta^{18}O$ and $\delta^{13}C$ which indicate that the white laminae, presumed deposits of the whitings, are preferentially enriched in the heavier isotopes. These data are consistent with a hypothesis of precipitation. This reaction would be brought about by near-surface concentration of HCO_3^- as a result of evaporation, with selective loss of the lighter isotopes of oxygen and carbon.

A mystery to be solved is the fate of the aragonite muds from the Trucial coast whitings. KINSMAN (1964b) has proposed that some of the suspended mud is carried shoreward and deposited in the lagoons. On the other hand, if the flocculated needles in whitings settle so quickly, they should be an important constituent in the deeper marls and lime muds described by Houbolt and by Pilkey and Noble. But aragonite was not detected in the finer fractions of these deposits.

The Abu Dhabi complex

Much of the character of the Abu Dhabi region stems from the array of off-shore islands that give shelter on their landward side to extensive lagoons (Fig.192, 193; SUGDEN, 1963a; KINSMAN, 1964b, c; EVANS et al, 1964a, b; G. EVANS, 1966a, b; KENDALL and SKIPWITH, 1969b; EVANS and BUSH, 1969).

Fig.192. Depositional environments of the Abu Dhabi complex. *Ch* = channel, *D* = delta,
R = reef, *C* = creeks, *A* = algal stromatolites. (After KINSMAN, 1964b.)

The islands are composed largely of unconsolidated, cross-stratified lime sand and it is possible that they are relics of old seif dunes similar to those in the adjacent desert. The only consolidated deposits are some lightly cemented dune calcarenites, some thin crusts under the lagoonal and intertidal sediments and beach rock. The island structure is complicated by an association of mud flats, mangrove colonies and their entrapped sediment and sand bars. The seaward beaches of these islands are aligned along a front of about 70 km, with a few scattered fringing and barrier coral reefs. Between the islands, tidal channels allow water to move in and out of the lagoons, and encourage the growth of oölite deltas at their seaward mouths. The coast of the mainland and the landward coasts of the islands are, for the most part, mangrove swamp and algal flats broken by creeks.

The petrography of the carbonate sediments is dealt with in detail by KENDALL and SKIPWITH (1969a) with special emphasis on diagenesis and alteration caused by endolithic algae (p.383).

The inner shelf. From the 6 m contour a limestone platform slopes gently seaward (Fig.193). It is more or less covered by lime sand and sparse coral. In the sand small foraminiferid grains are persistent but do not exceed 10%. Most of the grains are molluscan and it seems that bivalve grains are more abundant than gastropod, unlike the lagoonal shell debris. Other contributors are echinoderms, calcareous algae, ostracods and bryozoans. The algae so characteristic of the Caribbean (*Halimeda, Penicillus, Rhipocephalus* and *Udotea*) are unknown in the Gulf, but coralline algae are derived from the reefs. Approximate wt.% of carbonate minerals (mean of 10 samples) are aragonite: 81%, low-magnesian calcite: 8%, high-magnesian calcite: 3%, and dolomite: 1.6%. Insoluble material averages 13.4% (EVANS and BUSH, 1969). Many shells are thin-walled yet unbroken and many bivalve shells remain articulated after death, abrasion being slight and broken shells angular, testifying to attack by feeding animals rather than by waves. The distribution of grain sizes is critically affected by changes in bottom topography as slight as 1 m: the coarser sands lie on topographic highs and muds collect in hollows. In the finer grades, non-carbonate grains average 28% with a modal diameter of about 130 μ, and include rounded quartz grains of dune origin. Oöids are rare except near the deltas, and there is a trace of derived limestone grains. Most sand and gravel grains have been bored by algae and this has led to their further rounding.

Reefs. At several places on the shallower slopes of the inner shelf, coral reefs rise to within a metre or less of the water surface (Fig.192). A large part of one reef, off the Ras (headland) al Khaf, is exposed at low tide. The distribution of reefs is irregular, there being no well-oriented linear pattern such as is seen off Andros Island, Bahamas, or in the Florida Keys. The base of the reef is generally at 6–7 m. Its construction may be dominantly of massive or branching *Acropora*, with

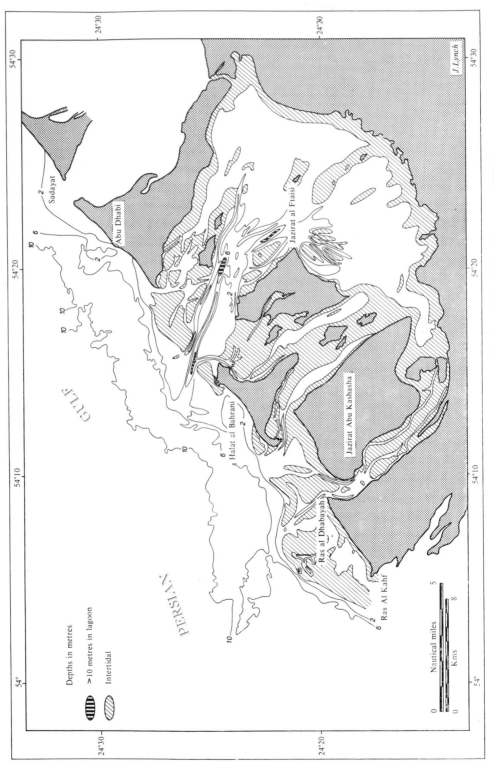

Fig.193. Bathymetry in the Abu Dhabi complex, Persian Gulf. (After KINSMAN, 1964b.)

subsidiary *Siderastrea* and *Porites*, all growing best between 10–15 m. Scattered corals occur both seaward of the reefs and in the tidal channels and back-reef lagoons (Fig.192). Other organisms include a small, branching coralline alga, *Lithophyllum*, foraminiferids, an abundance of *Echinometra* with other echinoids, and sponges belonging to the Haploscleridae (spongin, not calcareous). The corals and the loose sediment are bound together by a liberal encrustation of *Lithophyllum*. *Archaeolithothamnium* is present locally and *Jania* is ubiquitous. Mineralogically these sediments are like those of the inner shelf, but are slightly enriched in high magnesian calcite.

The reef top is irregular and is locally cut by channels. In hollows, rimmed by live coral, the finer sediments accumulate, including muds, and on the larger reefs there are extensive bars of coarse detritus.

A peculiarity of the corals in this region is their unusual tolerance of high salinity and temperature (KINSMAN, 1964a). They are not the species familiar in the Caribbean, but have been likened by KINSMAN (1964a) to *Acropora pharaonis*, *Porites lutea* and *Siderastrea liliacea* and others. In all, eleven coral genera are known in the Abu Dhabi area. The conditions of growth of hermatypic corals outside the Persian Gulf, particularly in the West Indies, indicate that they flourish in a temperature range of 25°–29°C and salinities of 34–36‰: for short periods only they can survive between about 16–36°C, and they can tolerate salinities of 27–40‰. Here, off Abu Dhabi and the other islands, the upper limits are considerably higher, being at least 40°C and 45‰. The massive form of *Porites*, known in the Caribbean for its high tolerance, here withstands salinities as great as 48‰

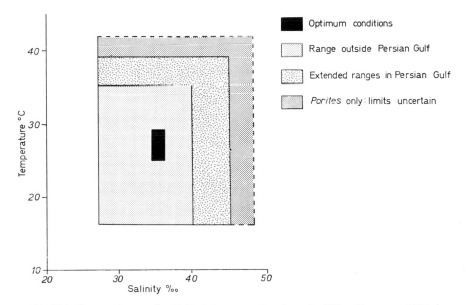

Fig.194. Temperature and salinity tolerances of reef corals. (After KINSMAN, 1964a.)

(Fig.194). The inhospitable nature of the Persian Gulf environment generally is apparent from the small number of genera known, 15 compared with 80 in the Indo-Pacific fauna (KINSMAN, 1964a).

Landward of the reefs the sandy floor slopes down to the weed-covered muds of the back-reef lagoon lying at depths varying from 1.8 to 6 m. The carpet of *Halodule* (one of the Zosteraceae, the Persian Gulf equivalent of the Caribbean *Thalassia*) is broken locally by scattered coral heads. Patches of coral growth of various sizes are scattered behind all the tidal deltas, generally in areas sheltered from waves by islands, but coral hardly penetrates into the lagoon. Large reefs lie off Jazirat al Ftaisi, northeast of Halat (Fig.192) where, along the sides of the deep tidal channels, branching corals are prolific. The reef top is below low-water except for a few coral heads. Growth is patchy and much of the reef bears a luxuriant cover of brown algae.

Seaward beaches of islands. The beaches on the northwesterly shores of the islands (Fig.193) are exposed to the full tidal range of the Gulf (1.2 m neap, 2 m spring) and to the northwesterly winds, especially the shamal (p.181). Below low-water mark fine sands, with variable amounts of carbonate mud, slope gently seaward (SUGDEN, 1963a). Extensive patches of floor are stabilized by the *Halodule* while, between these patches, loose sand forms a ripple-marked surface. This pattern of vegetation and rippled sand is similar to that in the Bahamian rock pavement habitat (plexaurid community, p.113). Above low-water mark, the surface of the beach is steeper and is composed of coarse sand. This contains mainly eolian quartz, molluscan grains, oöids and derived limestone grains: the carbonate is almost entirely aragonite, with a very little calcite. The steep sandy surface is parallel to its internal bedding planes and is constructed by waves. Toward low-water mark the surface is cuspate. Above high-water mark there may be one or more berms or the surface may fall gently into a lagoon. Inland, dunes are succeeded by areas of exposed limestone, coarse lag deposits or exposed surfaces of marine sediment.

At about high-water level there is a considerable accumulation of dead algal and other plant material, especially *Halodule*, and on an accreting beach this can be buried to form peat. (In Florida and the Bahamas *Thalassia* accumulates in the same way.) Above high-water, burrowing crabs are active. The large crab *Ocypoda* digs holes as wide as 10 cm and 1 m deep and these animals work on such a scale that bedding is largely destroyed (Fig.195).

Tidal channels. Water movement in and out of the lagoons is concentrated in tidal channels of which there are three groups (Fig.192, 193). These are directed between the three islands of Sadayat, Abu Dhabi, Halat al Bahrani and the peninsula, Ras al Dhabayah. Between Sadayat and Abu Dhabi a single, wide channel reaches a depth of 7 m: between Abu Dhabi and Halat, where two channels join, the bottom is at 13 m, and northeast of Jasirat al Ftaisi it is in places even deeper. Between Halat and Ras al Dhabayah the channel floor nowhere exceeds 4.6 m. The cross-section of a channel is related to the area of lagoon that it serves: depth/width ratios vary from 1/40 to 1/> 80. The sediments vary according to the flow and form of the channel: the largest channels, between Abu Dhabi and Halat, are floored by a coarse lag deposit of coral and shell debris thickly encrusted by *Lithophyllum* and bryozoans, all bored and abraded, and there are numerous haplosclerid sponges. In other channels, *Halodule* grows with leaves up to 20 cm long, and the sediment is finer and the course of the channel less well defined. Some channels in the inner lagoon appear to be filling with lime sand mixed with up to 50% lime mud. All the channel sediments are dominantly aragonite: insolubles average 7 wt.%. Where sediment is thin or absent the limestone is encrusted with live coral.

The pattern of flood tide circulation is such that the water moves in at right angles to the shelf, as in the Bahamas, but is then influenced by topography in the shallower water. The pattern is reversed during ebb flow.

Tidal oölite deltas. Enormous spreads of mobile, ripple-marked oölite lie seaward of the islands, about the mouths of the tidal channels (Fig.192). Their outer limit is the 1.8 m contour where there is generally a conspicuous line of breakers: beyond this the floor slopes steeply down to the inner shelf at 6 m. Still farther out, the 7.2 m line follows the form of the delta front, the orientation of which is controlled in part by the prevailing and dominant NNW wind. Toward the outer boundary of the oölite there are numerous megaripples with wave-lengths of 200–300 m and amplitudes of 0.6–1.2 m. The longer crests are straight and parallel, but shorter crests tend to be either straight or curved and to lie *en echelon*: locally there are petalloid lobes. (For a discussion of similar bed forms in the Bahamas, see BALL, 1967.) KINSMAN (1964h, p.68) wrote of the oölite deltas: "Water in the channels, particularly on the ebb tide, exhibits extreme turbulence. Ebbing waters from adjacent shoal banks are much less turbulent and usually flow into the channels almost at right angles to the channel current direction. Ripple marking on the bank top and channel side are usually found to be at right angles". As in the Bahamas some of the troughs of the megaripples are immobile and support a flora, in this case brown seaweed. In places bare limestone is exposed, showing that the oölite accumulation, though extensive, is very thin. Large areas of the deltas are exposed at low tide (Fig.193).

The oölite consists of at least 60% oöids with subordinate grains of thick shelled molluscs which are intensely abraded. At the seaward edge of the delta the grains are highly polished. Farther back they lack polish (except on surf beaches) and are bored by algae and in this they resemble the oölites in and around Bimini Lagoon, Bahamas. There are few grains of reworked limestone, but from 2–15 wt.% of non-carbonate grains which are specially abundant in the finer sands.

Most of the nuclei are skeletal grains or peloids. The mean of 6 samples gave aragonite: 89.4%, low-magnesian calcite: 4.1%, high-magnesian calcite: 2.8%, dolomite: 1.4%; insolubles average 3.3% (EVANS and BUSH, 1969).

Inner shores of islands. These are gently shelving regions except where tidal channels run close to the shore. Bars and spits are extensively developed: pointing into the lagoon, they are influenced mainly by flood tides. The shores are swampy with tidal creeks, and are widely colonized by a bushy plant, *Arthrocnemum glaucum* (Fig.198) and a mangrove *Avicennia marina* (Fig.199). *A. glaucum* does much to stabilize sediments in the upper parts of the intertidal zone. The topography, sedimentation and biota are similar to the creek-algal facies of the mainland shore and are described in that section (p.199).

Lagoon. A large part of the floor of the lagoon is exposed at low tide, particularly in mid-lagoon (Fig.193), and on the terraces which border the main islands and the mainland. The intertidal flats are generally bare lime sand which, when traced below low-water, is commonly mixed with aragonite mud and has a carpet of *Halodule*. Mud is abundant only in settling basins such as the landward ends of channels or discarded channels. It is important to appreciate, though, that the lime sand in the lagoon is largely faecal pellets and is, therefore, composed of dominantly aragonite mud redistributed in pellet form as a lime sand. The *Halodule uninervis*,

Fig.195. Crab burrow with pellets on a seaward beach. Trucial coast, Persian Gulf. (Courtesy of G. Evans.)

with its grass-like leaves 20 cm long, forms a tangled mat of roots in the top 10–15 cm of the sediment and does much to stabilize it. Filamentous, non-calcareous algae crowd the upper surfaces of the leaves so that these are better able to trap fine particles and to reduce the velocity of water near the bottom. Both the shamal wind and the strong afternoon onshore wind cause considerable turbulence in the lagoon. On the other hand, bottom traction of sand is slight, since only a few of the shallower sand banks having ripple-marked surfaces.

Crabs play an important role in the development of fabric in the intertidal sands, though the debris of their carapaces constitutes only an insignificant fraction of the sediment. The burrowing crab *Scopimera* completely covers the sand with its pellets while the tide is out (KINSMAN, 1964b), but the next flood tide destroys them: crab pellets in general have a loose texture and are associated with a radial pattern of linear tracks around the burrow (Fig.195, 196). The pellets are not preserved. The effect of the burrowing on the sediment is to destroy the bedding and to make the surface uneven: many crabs burrow to 0.5 m below the surface. Twenty-three

Fig.196. Small crab *(Scopimera)* pellets on a tidal flat, outer lagoon. Abu Dhabi. (Courtesy of G. Evans.)

species of crab were recorded in the Abu Dhabi area and numerous smaller ones remain unidentified. Apart from a few species of mollusc and worm, crabs are by far the most abundant organism in the inner lagoon and creeks.

Coralline algae are ubiquitous (e.g., *Jania*, which grows on *Halodule*), and so are green algae, but neither contribute much to sediment volume. Foraminiferids, especially peneroplids, make up less than 10% of the sediment.

Molluscs are very active in the lagoon and most of the sand is composed of their debris: gastropods outnumber bivalves (unlike the innershelf). Common gastropod genera are *Cerithium*, *Mitrella* and *Nassarius*. These three are found all over the Abu Dhabi area, but in the innermost lagoon, where conditions are extreme, *Cerithium* occurs almost to the exclusion of other forms as *Cerithidea* does off the west coast of Andros Island, Bahamas (p.136). The gastropod population is particularly dense along the mid-lagoon coasts: here, between tides, the surface can be seen littered with live and dead shells, trails and faecal pellets (Fig.197). These conditions are similar to those in the northern part of Bimini Lagoon where *Batillaria* is common. These gastropod flats are in the mid intertidal zone, below the crab flats.

The intertidal lime sands on the shallower banks and along coasts exposed to northwesterly winds are largely of gastropod debris and faecal pellets. On some sheltered beaches and below low-water mark the sands are muddy with aragonite needles, with the other constituents, gastropods and some foraminiferids, playing

Fig. 197. Cerithiids and their tracks in pellet lime sand, tidal flat, inner lagoon. Abu Dhabi. (Courtesy of G. Evans.)

a minor role. Most of the muds contain pellets, and the shallower the water the greater is the proportion of hardened pellets, so that on intertidal banks they are quite hard and mud is virtually absent. All stages are known from weakly agglutin-ated mud to pellets fully cemented by micritic aragonite (KINSMAN, 1964b). The muddy sediments are extensively exposed during low tides and, if these low tides should coincide with the heat of the day, the muds are desiccated to a degree that puts a sharp limit on the types of organism that are able to survive in this habitat. Corals, for example, cannot live here.

The Foraminiferida. J. W. MURRAY (1965a, b, 1966a, b, c) has shown the need to distinguish between the distributions of dead and live foraminiferids in the Persian Gulf embayment (see also p.162). There is a pattern in the distri-bution of dead foraminiferids which might well be detectable in a fossil limestone with similarly diverse origins. The frequencies of *Triloculina*, *Rosalina* and *Ammonia* appear to increase with turbulence, whereas *Peneroplis* tends to accumulate on broad stretches of quiet, shallow floor (as in Batabano Bay, Florida and the Baha-mas). The live animals, however, do not live on the surface of the sediment but are attached to algal bushes, *Halodule*, *Zostera*, etc. Suitable habitats appear to be epiphytic growths of a *Jania*-like alga, various "digitate weed growths" on dead coral and the leaves of *Halodule* or *Zostera* with their attached filamentous algae (J. W. Murray, personal communication, 1966). Species that live attached belong to the genera *Peneroplis*, *Quinqueloculina*, *Triloculina*, *Clavulina*, *Elphidium*, *Buliminella*, *Miliolinella*, *Glabrotella*, *Rosalina*. The areas where the foraminiferids live are, therefore, sharply localized: only after death are the tests redistributed by currents. Their final resting places reflect the current pattern and the sizes and shapes of the tests. The dead tests in the lagoon tend to move landward, those on the oölite deltas are washed seaward, while transport in the channels may be either way.

Creeks, swamps, algal flats. Along much of the mainland and inner island shorelines the intertidal mud flats or sandy mud flats are broken by tidal channels. The dominant vegetation on the flats is a low bushy halophyte, *Arthrocnemum glaucum*, which is 30–40 cm high (Fig.198). In the seaward, lower mud flats the mangrove *Avicennia nitida* (Fig.199) occurs sporadically: around parts of Jazirat al Ftaisi and Jazirat Abu Kashasha it grows densely enough to yield mangrove swamps. Extensive areas of the upper flats are covered by a tough, rubbery mat (algal stromatolite) of blue-green, non-calcareous algae (Fig.200). Lime sands are concentrated along beach ridges, strand lines and on the floors of the tidal creeks, and are constituted either of entire and little abraded cerithiids or of faecal pellets.

The black to brown algal stromatolite is generally broken into polygons as a result of shrinkage during desiccation (Fig.200, 202). These are commonly upturned at their edges and have been locally overturned by flood water. Elsewhere the clefts between the polygons have been the sites of vigorous algal growth so that

Fig.198. Bushes of *Arthrocnemum*, with algal stromatolites in the foreground. Abu Dhabi. (Courtesy of G. Evans.)

Fig.199. The black mangrove, *Avicennia*, with pneumatophores in aragonite mud, inner lagoon, Abu Dhabi. (Courtesy of G. Evans.)

Fig.200. Algal flat, inner lagoon, Abu Dhabi, with polygonal zone (centre) and cinder zone (left and middle right). (Courtesy of G. Evans.)

they are surmounted by rounded ridges. The algal mat may be washed by many tides a month and water lies in shallow pools, in runnels and between polygons. The algal filaments protrude upward from the surface of the mat and trap sedimentary grains. Variation in the supply of sediment or in algal growth gives rise to a laminated sediment quite unlike the bioturbated muds and sands of the rest of the lagoon floor (see Chapter 5). Whole gastropods may be trapped between the laminae. Under the mat, in some places, aragonite and gypsum are precipitated from the pore water. The upper 1–2 cm of the mat may, indeed, be cemented by a mixture of these minerals.

Although the algae play an important part in the reduction of bottom water velocities and the trapping of mud grains, these shallow intertidal flats are also ideal situations for the action of the settling-lag effect of Postma, elaborated by VAN STRAATEN and KUENEN (1958). As a flood tide wanes the water carrying a particular aragonite needle in suspension will move increasingly slowly. A velocity is reached, at a place A, where the needle can no longer be carried. It starts to fall but, as the water is still moving, it eventually reaches the bottom some distance inland of A at a point B. At B the flood current velocities are below those that permit erosion of needle muds. So also are the ebb current velocities, particularly as on the flats they are slower than the flood velocities. So, even if a needle had not quite settled after the turn of tide, it would continue to settle in the ebbing water. The fact that the critical erosion (threshold) velocity of mud grains is higher than their settling velocity further reinforces the tendency of mud to be deposited, by ensuring that most mud grains, once deposited around *B*, will remain on the floor.

Algal flats of the Khor al Bazam. The best developed algal flats (see also p.227) are on the shores of the Khor al Bazam lagoon, immediately west of the Abu Dhabi lagoon. KENDALL and SKIPWITH (1968a, b) have given a particularly detailed description of these stromatolites. There are four algal mat zones running parallel to the water's edge, in a belt up to a kilometre or more wide, and differing from one another in the growth forms of the algal mat. They are situated thus between the bare, seaward lime sand and the landward salt flats of the sabkha.

bare lime sand
cinder zone
polygonal zone
crinkle zone
flat zone
sabkha

algal mats

Each zone passes gradually into the next. Aerial photographs show the seaward margin of the algal flat to be formed of a number of lobes each of which coincides with an area of active seaward growth of the mat.

The **cinder zone** (Fig.201) occupies the seaward fringe of the algal flat, and at high water it is the most deeply submerged of all the zones. Seaward of its lobate edge there are skeletal and pelletal sands. It is noticeable that the grains in these sands are more densely infested by chlorophyte and cyanophyte algal cells than are those in the deep lagoon or in the intertidal areas or on the shelf. This zone resembles a warty, black layer of volcanic cinder. Under the rubbery surface layer there is generally an irregular mixture of aragonite mud with gypsum crystals and mat-derived peat. The cinder zone is narrow compared with the other zones, but dispersed developments of this cindery algal growth appear as patches on the adjacent sands, and on the crests of bars especially on their sheltered inner slopes.

The **polygonal zone** (Fig.202) is characterized by algal mats broken into polygons which are separated from each other by sediment-filled cracks. Polygons increase in diameter seaward from 3 cm to 3 m. The cracks are the posthumous reflection of a system of shrinkage cracks, and the edges of the polygons are commonly upturned as a result of drying. The result is a pile of vertically stacked polygonal saucers (concave upward), inverted stacked hemispheroids, the reverse of the stacked hemispheroids (SH) of LOGAN et al. (1964), which are convex upward (p.220). A new code (SH-I) is recommended for this inverted type by Kendall and Skipwith (1968b). Under the live surface layer there is generally 15 cm or more of laminated peat with scattered gypsum crystals, a relic of earlier growth layers caused by alternate sedimentation and algal growth.

Fig.201. Algal stromatolites of the cinder zone, with scattered cerithiids and other molluscs. Abu Dhabi. Matchbox is 7.5 cm long. (Courtesy of G. Evans.)

Fig.202. Algal stromatolites of the polygonal zone. Lagoon. Abu Dhabi. The felt pen is 8 cm long. (Courtesy of G. Evans.)

The **crinkle zone** has a leathery-looking algal skin disposed as crinkles or blisters. Typically the skin is only loosely attached to an underlying mush of gypsum crystals. Superimposed on the blistered surface it is possible to make out a polygonal pattern in the topography which becomes more pronounced toward the polygonal zone.

The algal surface of the **flat zone** is smooth with a skin about 3 mm thick. Under the skin there is either a brown quartz-rich lime sand or a grey to white calcilutite consisting of a subhydrate of calcium sulphate. Both sediments contain gypsum crystals. The flat is divisible into a lower wet sub-zone and an upper dry sub-zone. The wet one is pink and the dry one grey with more often than not a covering of halite. Under the wet pink surface the material is bright pale green. Landward of the flat zone lies the sabkha.

From the distribution of the algal mats Kendall and Skipwith were able to draw certain conclusions about the parameters that control growth. The mats are always intertidal but are restricted to those beaches that are protected from severe wave action and strong tidal currents. Such areas are typically found in lagoons and in sheltered bays. They concluded that mats will only survive where the turbulence is below a critical value. When this value is exceeded, as during storms, the cinder zone in particular is dislodged and in places "the leathery skin becomes rolled up like a carpet" (SKIPWITH, 1966). The growth rate of an algal mat is greatest where it can receive maximum sunlight yet remain moist. The supply of sunlight to the cinder zone is unduly restricted by the water: the crinkle and flats zones are too dry and growth there is also, it seems, hindered by deposition of wind-blown sediment. It is likely that optimum conditions for growth exist in the polygonal zone. It is plain, from the distribution of evaporites under and on the mats, that the algae can withstand very high salinities.

Inorganically or physiologically precipitated aragonite? The chlorinity of the water in the Abu Dhabi region rises progressively toward the Arabian shore (KINSMAN, 1966b; BUTLER, 1969). The ratios of K^+, Na^+, Sr^{2+}, Mg^{2+} and SO_4^{2-} to Cl^- are constant. Sr^{2+} and Mg^{2+} are, presumably, being withdrawn from the water to take part in new aragonite and calcite lattices, but the amounts lost are masked by the errors involved in analyzing such small quantities as exist in sea water. On the other hand, the Ca^{2+}/Sr^{2+} ratio and the titration alkalinity (p.234) fall markedly from the open shelf, through the lagoon, to the main shore.

This trend in Ca^{2+} removal is the outstanding factor bearing upon the origin of the aragonite of the lagoonal muds. These accumulations of needles (1–2 μ long) and of irregular grains have Sr^{2+} concentrations of around 9,390 p.p.m. (Fig.225, p.262; KINSMAN and HOLLAND, 1969). They cannot have been derived from calcareous algae since these are far too scarce in the area. Nor are the muds products of shell disintegration, because Kinsman has found that the local molluscs are composed mainly of an aragonite with Sr^{2+} 1,500–2,000 p.p.m. Corals yield values similar to those of the muds but are too limited in geographic distribution to be the source of so much fine material. Finally, the Sr^{2+} value of 9,390 p.p.m. is close to the expected value, if aragonite containing some $SrCO_3$ were to be precipitated in equilibrium with the sea water at the known temperatures (p.261). As a result of these considerations Kinsman concluded that at least 80–90% of the mid and inner lagoon muds and pellets are inorganic precipitates. It is by no means clear at what period of the year the needles are precipitated, nor at what time of day. The lagoon waters early on winter mornings can be as cold as 18 °C, hardly a condition con-ducive to extreme evaporation and precipitation. The oöids have an Sr^{2+} content (9,590 p.p.m.) as for inorganic precipitation (p.261). In weighing the likelihood of inorganic chemical precipitation it is necessary to consider the widespread and probably profound influence of a biologically formed mucilaginous substance (p.252) that penetrates virtually all grains, particularly those that have been exten-sively bored by algae: the effect of this substance on ion transport of any kind could be important. Its origin is uncertain but it may be a product of blue-green algae.

There is no doubt that algae, particularly endolithic algae, are exceedingly active in the region, and the opinion of KENDALL and SKIPWITH (1969a), that the precipitation of aragonite needles is probably physiologically controlled, may very well be justified. The possible influence of algal proteins is considered on p.253.

The sabkha. One of the most far reaching advances in recent sedimentological studies has been the discovery of sabkha diagenesis, in particular the subsurface growth of gypsum, anhydrite, and dolomite. In the Persian Gulf this new work has been carried out by CURTIS et al. (1963), SHEARMAN (1963) and KINSMAN (1964b, 1965, 1966b) in the neighbourhood of Abu Dhabi, and by ILLING et al. (1965) on Qatar. Inland from the algal flats and above the level of normal high tides, a halite-encrusted surface rises very gradually toward the exposed Tertiary limestone farther inland. A similar surface comprises the greater part of the offshore islands. This surface forms a band of salt flats, called locally sabkha (or sebkha), running parallel to the coast and in places as wide as 16 km. When high offshore winds (the shamal, p.181) combine with high spring tides, very broad areas of the sabkha are flooded. In these circumstances, especially when accompanying rain dilutes the sea water, the halite is dissolved, and water may lie in pools for two or three weeks at a time. The intense evaporation brings about not only a concentration of salts in surface capillaries but also an upward movement of brines from the permanent water table. The very high salinities attained lead to the precipitation of gypsum and aragonite with a consequent rise in the Mg^{2+}/Ca^{2+} ratio that in turn encourages subsurface dolomitization of the dominantly aragonite sediment (Chapter 13).

The accumulation of aragonite pellets and mud on the swamps and algal flats, and of lime sand on exposed beaches, has caused a general rise and seaward advance of the intertidal surface (EVANS et al., 1969). This progradation of the shore is assisted by spring tides and storms that permit deposition of sediment above the normal high tide level. The process is similar to that which leads to the extension and coalescence of islands around Andros Island, Bahamas. A prograding shore of this kind leaves behind it a widening belt of unconsolidated sediment, either sandy or muddy, which is characterized by high porosity and permeability and an internal laminated structure that reflects the influence of algal mats.

The whole region is subject to wind erosion and deposition. Even the muddy sabkha is mostly covered by several centimetres of windblown carbonate sand. The depth of wind erosion depends on the amount of the capillary pore water so that the muddy sabkha, having a higher water content, is more resistant than the sandy sabkha. Erosion is also effected by wind driven currents in the shallow flood waters. Most of the time the sabkha surface is dry: the water table is a little above sea level and lies mainly at depths from 0.3 to 1.2 m below ground level and falls gently seawards. Inland its depth is fairly constant but toward the algal flats it is apt to respond to the rise and fall of the tide.

Sabkha diagenesis. The subsurface environment of the sabkha has some very special characteristics as a result of which eight, possibly nine, authigenic minerals are now forming, either directly from the pore water or by metasomatic replacement of the original sedimentary grains. The crystal growth pressure exerted by the new minerals is so great that in places it is "fairly certain" that the surface of the sabkha is lifted by anything from 0.3 to 1 m (KINSMAN, 1964b). The authigenic minerals are:

halite, aragonite, calcite (possibly), gypsum, anhydrite, dolomite, celestite, magnesite, huntite [$Mg_3Ca(CO_3)_4$].

The temperature at the sabkha surface, particularly under the porous, dark and moist algal mat, may in the summer reach $60°C$ or more. Thus evaporation is everywhere intense so that the chlorinity of the brines in the sabkha is very high, mainly between 135 and $165‰$. The range for the subsurface water of the adjacent, down-slope algal flats is about $35–110‰$ and, for the lagoon waters, it is only $24–37‰$. Ground water values were determined from samples taken in pits dug to just below the water table.

If the chlorinity of sea water is raised to $65‰$ (salinity $117‰$) gypsum starts to precipitate. Of more interest in this situation is the molar ratio Mg^{2+}/Ca^{2+}. This ratio in the lagoon waters, from which aragonite is being precipitated, is around 5.5. In the brines from the algal flats, the ratio ranges from 6 to 11 and so gypsum is precipitated. On the other hand, values for sabkha ground waters range from 8 to 2 in a landward direction showing that either Mg^{2+} is being lost or Ca^{2+} gained, or both (KINSMAN, 1966b; BUTLER, 1969).

The distribution of the authigenic minerals can be profitably examined. A more detailed consideration of the dolomitization processes is given in Chapter 13. Here it will be convenient to examine the occurrence of the five dominant minerals, halite, aragonite, gypsum, anhydrite, and dolomite.

On the surface, where brines are readily evaporated to dryness, **halite** is precipitated as the dominant salt: straight evaporation of normal sea water would produce a mixture of salts of which about 78% by volume would be halite. Pools on the sabkha evaporate to leave halite crusts 7–8 cm thick. Provided that there is an unbroken connection between surface pore water and that of the water table, capillary action will ensure an upward movement of brine to replace the loss by evaporation, yielding at the same time further halite. Where the upper part of the sabkha is dry, precipitation of halite takes place at some lower level in the sediment which is governed by the upper limits of capillary movement. The halite precipitate is, nevertheless, ephemeral as it is promptly dissolved when the surface is flooded or washed by rain.

Aragonite forms as a cement in the upper centimetre or so of the algal flats. Its habit is needle-like with long axes less than 5 μ or equant. The amount of this new, subsurface, aragonite falls off inland where, as a result of its replacement by dolomite, it disappears in some places completely. Small needles less than 1 μ long are especially liable to be lost in this way.

Gypsum is common in the upper parts of the sediment in most of the algal flats: some of the sediments have almost pure "gypsum crystal 'mushes' extending 2–3 feet (up to 1 m.) below the surface" (KINSMAN, 1964b). In the sabkha, gypsum can make up more than 50% of the top metre of sediment. As far down as the sediments were cored, to 1.2 m, layers of nearly pure gypsum alternate with gypsum-free or gypsum-poor layers. The crystals are mostly discoidal and flattened, perpendicular, or nearly so, to the c axis. Some are deeply corroded following partial dissolution. Some are clear, others contain zonally arranged carbonate inclusions, even whole gastropods, or traces of sedimentary lamination typical of the stromatolitic deposits of the algal flats. Crystal diameters up to 1 cm are common: other habits can attain 10–25 cm. A late development of gypsum takes the form of discoidal crystals with cores of anhydrite nodules. Near-surface gypsum is liable to wind erosion and dissolution, and mechanical concentrates are known, even dunes composed of more than 70% gypsum. Poikilotopic gypsum sand crystals locally litter the surface as a lag deposit.

The earliest record of Recent **anhydrite** is by CURTIS et al. (1963) who described a core collected by Kinsman from the mainland sabkha of the Abu Dhabi area in 1962. So far, all the anhydrite discovered in the sabkha has been separated from the intertidal flats by a zone of gypsiferous sediment: none has been found in the offshore islands (KINSMAN, 1966b). The earliest formed anhydrite in the diagenetic succession occurs within a few centimetres of the surface as small, brilliantly white nodules 1–2 mm across, commonly associated with gypsum mush. Diagenetic anhydrite is known as far down as cores penetrated (1.2 m), but is mostly developed above the permanent water table. In some places the upper metre of sediment is more than 50% anhydrite, but the distribution overall is very variable.

Generally in the sabkha the nodules vary in diameter from less than 1 mm to more than 15 cm and in shape from spherical to strongly flattened. They occur singly or as complex masses containing hundreds of nodules with a variable degree of packing: in this way layers are formed which may be horizontal or oblique. It is probable that much of the layered anhydrite did not originate as nodules. Nodules are generally free of included grains, though where these are present the anhydrite is brown. In the nodules the lathe-like colourless crystals are elongate, $2\ \mu$–4 mm long, parallel to c and flattened in (100): some are as thin as 2–3 μ: none are thicker than 60 μ. Nodules are to be found in the eroded debris at the sabkha surface.

Though some anhydrite appears to be a simple, direct precipitate, the brine analyses suggest that most of its growth is dependent upon prior dissolution of gypsum. There is, however, no evidence that the anhydrite is an *in situ* replacement of gypsum: partially replaced gypsum crystals are lacking, as are anhydrite pseudo-morphs of gypsum (KINSMAN, 1966b). Its precipitation is helped by the supply of Ca^{2+} released during dolomitization.

Dolomite has been found throughout the sabkha sediments, on mainland and islands. It is rare, however, in the sands and is mainly concentrated in the originally muddy sabkha. Fine grained, wind-blown dolomite has been determined in all the sediments of the marine area. The earliest authigenic crystals are in the top 2–3 cm of sediment, at the algal flat-sabkha junction, in the capillary zone above the water table. Inland the depth at which the mineral occurs increases to at least 0.7 m and below the water table. In any vertical section the quantities of dolomite and aragonite are inversely related (as dolomite and calcite in the Coorong, p.521). The aragonite is mainly in the form of needle-rich mud (much of it aggregated into pellets) but locally there are whole skeletons. Both needles and skeletons are dolomitized. The dolomite takes the form of little rhombs and equant grains less than 3 μ in diameter. Its composition shows a slight excess of calcium, ranging from $Ca_{50}\ Mg_{50}$ to $Ca_{55}\ Mg_{45}$.

The detailed process of dolomitization is examined in Chapter 13. Briefly, the precipitation of aragonite and gypsum raises the Mg^{2+}/Ca^{2+} ratio of the pore solution to a condition where aragonite immersed in the solution is no longer stable (Fig.357, p.526). The large surface area of the tiny needles and the high temperature enhance the reaction rate as exchange takes place between the Mg^{2+} of the brine and the Ca^{2+} of the needles. It is apparent that, as water is lost from the sediment by evaporation, some sea water moves in below the water table to replace it. Replacement of evaporated water is also effected by flooding. The depth of dolomitization in the sabkha may well depend on the extent to which the heavier brines near the surface move downward as density currents in the pore system. This might lead to the often discussed reflux process of dolomitization (p.532).

Commentary on Abu Dhabi. A study of the Abu Dhabi carbonates is more than usually profitable because it includes within its compass a number of clear cut cases of inorganic precipitation. Not only are oöids and aragonite muds forming, but a number of minerals are growing in the subsurface sediments of the sabkha, modifying the composition of the brine as they grow, thus bringing about further changes leading to the growth of yet other minerals. It is, of course, true that this is an over-simplified statement. For example, the oöids are extensively bored and, over much of the delta, encrusted by tiny live blue-green algae, and the effect of these on the immediate chemical environment can scarcely be negligible.

An important structural distinction is apparent between lagoonal muds and tidal flat muds, the first being stirred thoroughly by crabs, the others retaining the lamination derived from its deposition as part of an algal mat. The bioturbating crabs seem to exert much the same destructive effect on the sedimentary structures as *Callianassa* in the Bahamas and Florida (p.128). *Halodule* destroys bedding, too, and is important as a stabilizer and a baffle. Yet crab debris is scarcely detectable in the sediment and *Halodule* leaves no skeleton at all. The deposition of early aragonite cement in the algal flats is yet another example of intertidal cementation (p.367): the fate of this cement in rain water is worth looking into. The association of carbonates with evaporites is of immense importance on account of the widespread occurrence of this association in ancient successions. It may well be, as SHEARMAN (1966) has suggested, that many of the evaporites in fossil carbonate sequences are also of subsurface, sabkha origin. Finally, the little understood but ubiquitous mucilaginous coating on grains, not unlike the subtidal mat of the Bahamas in some of its properties, may have effects as important as its occurrence is widespread (p.252).

The general sedimentary association of the sabkha certainly has a lithological individuality which has made fossil examples readily distinguishable (e.g., SHEARMAN, 1966; LAPORTE, 1967; MATTER, 1967; SCHENK, 1967; WEST et al., 1968).

Sabkha of the Qatar Peninsula

Another region of sabkha has been studied by ILLING et al. (1965) on Qatar (Fig.186). Along the indented coastline of the northwestern part of the Qatar peninsula many of the lagoon floors pass landward into intertidal algal flats and supratidal sabkha. A detailed examination was made of the sabkha of one of these lagoons, the Dohat Faishakh (Fig.203).

Between Qatar and Bahrein to the west, the water is seldom deeper than 5.5 m. Offshore of Qatar there are lines of low reefs constructed of a coralline alga, *Neogoniolithon*, and these protect the intertidal areas from wave action. The lagoon of Dohat Faishakh is almost landlocked and is joined to the open sea only by a narrow, winding channel. As the salinity of the surface water between Qatar and Bahrein (Fig.185) reaches the high mean annual value of 56‰ (SUGDEN, 1963b) it

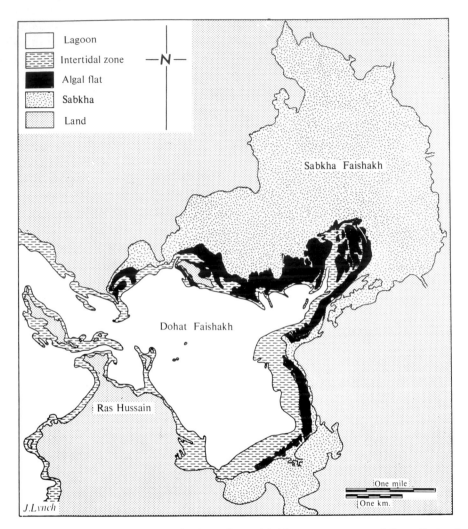

Fig.203. Sedimentary environments in the Dohat Faishakh area, west coast of Qatar Peninsula. (After ILLING et al., 1965.)

is not surprising that even higher salinities characterize the lagoonal water.

In the Dohat Faishakh the water in mid lagoon is about 1.25 m deep and has a tidal range of about 0.6 m. Here, as in so many shallow lagoons, the water depth frequently responds more to wind than to tidal forces. The carbonate sediment is a muddy silt with small bivalves, gastropods and peneroplids. The lower intertidal floor, especially to the north, is muddy lime sand in which the sand grains are made of the faecal pellets of cerithiid gastropods.

To the north, in the upper parts of the intertidal zone, there are widespread algal flats with the same rubbery surface as the flats near Abu Dhabi. These algal stromatolites have similar polygonal growth forms, arising originally from a system of desiccation cracks, and also have irregular bulbous surfaces. In places, the rubbery mat forms low, delicate baffles that trap the flood water in pools. The sediment is laminated and, when dry, curled fragments of algal mat or sedimentary laminae may be eroded and blown about by the wind. From the algal mat the following species of blue-green algae were identified:

Microcoleus chthonoplustes (Mcrt.) Zanard
Schizothrix cresswelli Harv.
Lyngbya aestuarii (Mert.) Lyngb.
Entophysalis deusta (Menegh.) Dr. and Daily.

The sabkha. The change from algal flat to sabkha (Fig.203) may be sudden, or transitional over as much as 100 m. The brown surface of the sabkha lies just above normal high water level and rises almost imperceptibly inland. Slight differences in topography cause some areas to be wet and boggy, others dry and firm. The surface is dark brown, a halitic-gypsiferous-carbonate sediment that is a mixture of wind blown material and water borne sediment from occasional tide or wind driven floods. Below the surface there is a change within 0.5 m to grey sediment. Five minerals in particular grow in or on the sabkha. There is a **halitic crust** which is deliquescent and remains damp. It forms afresh each day and is nightly dissolved in dew. **Gypsum** is found both in the sabkha and in the nearby algal flats. Small crystals of gypsum are common in the upper part of the sediment in the algal flats, a region that dries between tides. These little flattened crystals tend to be concentrated at particular levels down to about 0.3 m. As they grow they push aside the unconsolidated sediment round about them. On the sabkha surface shallow depressions, flooded more frequently than surrounding areas, are floored with a wet mush of tiny gypsums with selenite habit. In dolomitized sediment gypsums flattened normal to c may have diameters as great as 12 cm. These crystals increase in size landward: many have carbonate inclusions. Locally gypsum is poikilotopic and acts as a cement. The mineral accumulates in some parts of the sabkha sediment as concretions several centimetres across that join, in places, to form layers. In the landward sabkha, a white, fine, sugary gypsum, which is porous, is sometimes associated with halite near the surface.

Traces of **anhydrite** have been found as inclusions in gypsum at the sabkha surface. Sheaves of **celestite** crystals less than 0.2 mm long are known in the upper metre of the landward sabkha.

Dolomite occurs in the stiff mud that bears the large gypsum crystals. The mud is tan coloured where oxidized near the surface, but grey lower down in the reducing environment and is a variable mixture of aragonite, dolomite and calcite. As at Abu Dhabi the dolomite has a slight excess of calcite, from $Ca_{53}Mg_{47}$ to $Ca_{55}Mg_{45}$. Apart from the top 5 cm or so, dolomite is most abundant in the top 0.6–1 m of the landward sabkha. The content decreases rapidly, below 1 m, though

5–10% remains, partly of detrital origin, either locally reworked or windblown out of the neighbouring carbonate deserts.

X-ray data suggest that the dolomite, in a general way, grows at the expense of aragonite. With a microscope it is possible to see that, first of all, the aragonite mud between the pellets is replaced by dolomite and that later the pellets are attacked. The final texture is one of shells and wind derived quartz grains suspended in a matrix of dolomitic mud. Microtextures indicate that gypsum and dolomite grow more or less at the same time, both replacing some of the original carbonate. Calcite shells are leached in places, but are not dolomitized, though algal bores in them, once filled with micritic aragonite (p.383), are selectively dolomitized.

BRITISH HONDURAS

Skeletal calcilutites

A recent study by MATTHEWS (1966) of lime muds in the lagoon between the offshore reef and the mainland, south of Belize, is of special interest because it directs attention to carbonate muds of mixed skeletal rather than of inorganic or codiacean origin. For too long have students of Recent carbonate sediments and ancient limestones thought in terms of inorganic precipitation as the origin of clay grade carbonate deposits to the virtual exclusion of skeletal destruction. The Honduras calcilutites occur on a reef complex and in a lagoon between this north–south trending reef complex and the mainland 30 km to the west (Fig.204). They are dominantly carbonate lutites with 5–30 wt.% of non-carbonate terrigenous mud. The lagoon is deepest in the lee of the reef complex where depths vary from 18 m to 64 m. Matthews paid special attention to the carbonate grains in the size range 62–20 μ. (Smaller grains rarely have features that enable their parentage to be determined.) These silt-sized grains vary greatly in mineralogy (aragonite, or calcites with a range of $MgCO_3$), in crystal size and fabric, in the size of contained pores or chambers and in other microarchitectural details. The origins of many grains cannot be identified under the microscope: in particular, grains of mollusc and coral are apt to be indistinguishable. Much skeletal debris, even when unaltered, reveals a frustrating degree of cryptocrystalline anonymity, but after it has been bored intensively by algae and replaced by micritic aragonite (p.383) it retains no trace of its origin at all. Taxonomic identification is generally limited to the level of phyla, though in the Foraminiferida it may attain to the suborder. FERAY et al. (1962, p.26) have made a study of the lower size limits of recognizable grains.

From direct visual observation of the Honduras muds MATTHEWS (1966) could see that they have been formed by the physical breakdown of skeletons. He listed four causes of breakdown: (*1*) decomposition of binding organic matter (e.g.,

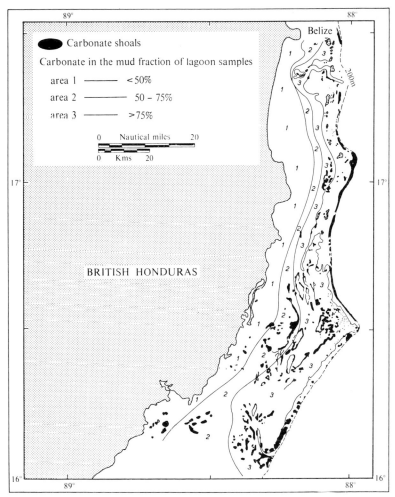

Fig.204. Distribution of calcilutite in lagoonal sediments south of Belize, British Honduras. (After MATTHEWS, 1966.)

conchiolin) from the crystal fabrics, (2) the weakening of fabrics by boring algae, (3) feeding by the predatory benthos, and (4) breakage and abrasion in agitated water. A detailed examination of skeletal breakdown, and its control by abrasion, organic decomposition and skeletal microstructure, has been made by FORCE (1969). NEUMANN (1965, p.1019) has drawn attention to the large quantity of mud produced by the boring sponge *Cliona* off Bermuda. But it is necessary to know not only the nature of the source but its geographic position. Matthews set out to discover to what extent the lagoonal muds are autochthonous or have been carried in from the reef complex. At the level of recognition possible with a microscope many of the skeletal grains of the lagoonal benthos and reef benthos cannot be distinguished. Corals and the codiacean alga, *Halimeda*, live almost exclusively on

the reef complex and ought to be useful indices of reef-derived mud but, unfortunately, a considerable volume of coral debris may have been missed in the point-
counts because it cannot be distinguished from molluscan debris, and much
Halimeda may well be lost among the cryptocrystalline grains which occupy about
25% of the sediment by point-count.

    These difficulties were to a large measure resolved by bringing in mineralogical
data and analyses of strontium content. The lagoonal lutite contains 30–70% of
high-magnesian calcite, a useful indication that inorganic precipitation of aragonite,
if it is active, is only of minor significance. The only important skeletons, volumetrically, with high strontium contents are corals and *Halimeda* (aragonite with
Sr^{2+} range 6,300–8,500 p.p.m.). Molluscs in the sediments have less than 2000
p.p.m. strontium. By demonstrating that strontium in the mud falls off gradually
from the reef complex through the lagoon toward the mainland (Fig.205), Matthews
showed the strong likelihood that high-strontium aragonite is being transported
from the reef westward into the lagoon. In Fig.205 the Sr^{2+} content in the coarse
carbonate fraction (coarser than 62 μ) falls off rapidly westward to a concentration
of 2,000 p.p.m. and persists at this level across the lagoon. Thus, sand particles

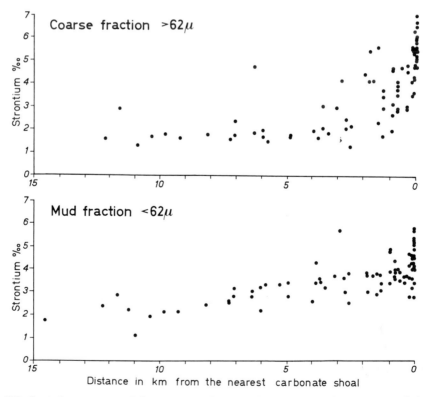

Fig.205. Strontium content of the carbonate fraction plotted against distance west of the reef
complex, for sediments of the lagoon south of Belize, British Honduras. (After MATTHEWS, 1966).

throughout most of the lagoon cannot be reef debris of *Halimeda* and coral, but must be of local origin. On the other hand, the content of Sr^{2+} in the total mud ($< 62 \mu$) fractions falls gradually westward and so indicates transport of mud particles in that direction. The content of terrigenous lutite increases toward the mainland indicating a westerly source. The limited conclusions that Matthews was able to draw on the basis of unaided point-counts, and the fuller interpretation possible with the help of mineralogical and strontium data, are indicated briefly in Table X. From his complete data MATTHEWS (1966) was able to show that autochthonous lutite is forming in the lagoon as a result mainly of the breakdown of molluscs and radial hyaline foraminiferids, but that there is a continuous movement into the lagoon of allochthonous mud from the reefs, consisting mainly of coral and *Halimeda*.

In a later paper BILLINGS and RAGLAND (1968) confirmed and extended Matthews's results, demonstrating a reefward increase of Sr^{2+} and Mg^{2+} in the acid-soluble fraction of the lagoonal sediments and a landward increase of Fe^{2+}, Mn^{2+}, Zn^{2+} and K^+ associated with the clay-mineral fraction.

This use of strontium content in the identification of skeletal sands has, more recently, been adopted by FRIEDMAN (1968a, p.913), who, in addition, distinguished between high-magnesian aragonite (corals) and low-magnesian aragonite (molluscs) in Red Sea reef sediments (p.235).

TABLE X

COMPARISON OF INTERPRETIVE PRECISION POSSIBLE ON THE BASIS OF POINT-COUNTS ALONE AND AIDED BY MINERALOGICAL AND STRONTIUM ANALYSES: FOR CALCILUTITES ON THE BRITISH HONDURAS SHELF[1]

Component grains (a selection only)	Mean particles (%)			
	Results by point-count unaided		Results by point-count with help of mineralogy and strontium (%)	
	lagoon	reef complex	lagoon	reef complex
Cryptocrystalline	25.0	24.0		
Coral or mollusc	29.8	35.9		
Chlorophyte algae	2.5	4.6		
Coral	3.6	6.9	14.0	33.1
Mollusc	4.4	2.0	23.6	9.4
Halimeda			7.9	11.3
Other codiaceans			1.4	2.7

[1] For technical reasons the point-count data are restricted to the silt grade whereas the threefold data are for the silt + clay grades. Selected from tables in MATTHEWS (1966, p.441, 450).

FURTHER READING

General introductions to southern Florida are by BAARS (1963) and GORSLINE (1963), GINSBURG (1964), SCHOLL (1966) and MULTER (1969). The basic papers for detail on the area are those by GINSBURG (1956), SHINN (1963), SWINCHATT (1965). The Gulf of Batabano is covered by DAETWYLER and KIDWELL (1959), BANDY (1964) and HOSKINS (1964). The Persian Gulf is introduced by EMERY (1956), G. EVANS (1966b) and PILKEY and NOBLE (1966). For the Qatar Peninsula there are two basic papers, by HOUBOLT (1957) and ILLING et al. (1965). The Abu Dhabi region is introduced by EVANS et al. (1964a, b), G. EVANS (1966a) and EVANS and BUSH (1969). Details of Abu Dhabi are in KINSMAN (1966b), KENDAL and SKIPWITH (1968b) and BUTLER (1969), with foraminiferids covered by J. W. MURRAY (1965–1970b). After this chapter was completed two detailed and extremely interesting reports were published on the Khor al Bazam lagoon, a westerly extension of the Abu Dhabi complex, by KENDALL and SKIPWITH (1969a, b). On the British Honduras lagoon there is the paper by MATTHEWS (1966). Other references are given in the text.

Additional references not given in the preceding chapter

Carbonate deposition in Florida Bay by SCHOLL (1966), marine ecology of southern Florida by TABB et al. (1962) and J. A. JONES (1963).

On sabellariid worms as reef-builders MULTER and MILLIMAN (1967) and KIRTLEY and TANNER (1968); on the distribution of *Thalassia* in the United States D. R. MOORE (1963); on burrowing in the lime sediments SHINN (1968c); LLOYD (1964) on oxygen and carbon isotopes in molluscs; bathymetry of Florida Straits by HURLEY et al. (1962), and a study of the paleoecological implications by KLEMENT (1966) and LAPORTE (1968).

The evaporite–dolomite facies in the Persian Gulf is mentioned briefly by ILLING (1963), BUTLER et al. (1964) and ILLING and WELLS (1964).

Chapter 5

RECENT CARBONATE ALGAL STROMATOLITES

INTRODUCTION

Carbonate algal stromatolites are laminar deposits of an earthy, leathery-looking or jelly-like material, commonly green, brown or black on the upper surface, which tend to grow either subaerially in moist places, or under water, in warm climates (Fig.206, 207). Their growth is controlled dominantly by a dynamic relation between living blue-green algae, forming an algal mat, and the entrapment and precipitation of $CaCO_3$. On the Bahamas–Florida platform their widespread occurrence is striking. Mats, with or without significant sediment content, are usually to be found on the floors of fresh-water or brackish lakes or at their wet margins, in brackish pools in the sawgrass swamps of the Everglades, in rocky intertidal pools along the coast, on supratidal aragonite muds or on the quieter submarine, subtidal sediment floors (p.122). It is characteristic of some of these growths that they can withstand

Fig.206. Supratidal algal stromatolite on Crane Key, an island in Florida Bay. The mat is blackish on a pale tan sediment. To the left of the box (6 cm long) a flap of mat has been turned back.

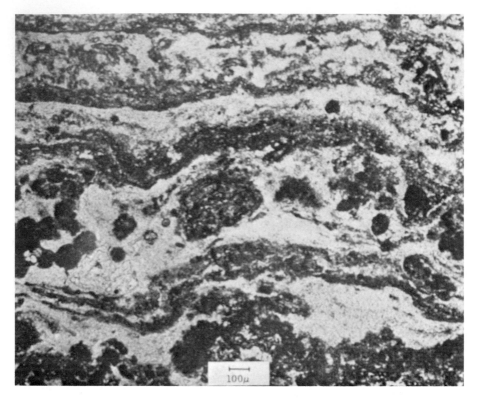

Fig.207. Vertical section through a *Schizothrix* algal stromatolite. Calcified layers are grey, separated by voids or masses of uncalcified algal filaments. Andros Island, windward lagoon. Slice. (Courtesy of C. Monty.)

prolonged drought or subaerial exposure. The word *Stromatolith* was introduced by KALKOWSKI (1908).

Although stromatolites may be composed of whatever sediment is available in the environment, those built of carbonate sediments tend particularly to form more or less permanent structures, stabilized by early precipitation of $CaCO_3$. These typically undulating laminar structures in limestones and dolomites have been well known to geologists since the latter part of the last century. An appreciation of their origin was surprisingly delayed, considering their widespread Recent occurrence and the simple tools needed for their investigation.

THE WORK OF BLACK

A true understanding of the nature of stromatolites was made possible through the publication by BLACK (1933a) of his outstanding, pioneering studies on the binding of calcilutites by blue-green algae into characteristically laminated sediments, on the west coast of Andros Island, Bahamas. He showed how colonies of coccoid and filamentous algae, living in and on the sediment, can trap sediment on their filaments by virtue of the sticky mucilaginous sheath within which the algal trichome is enclosed. The stromatolites examined by Black are lacustrine, supratidal and intertidal (subtidal stromatolites were recorded later, by Monty in 1965). He found that more than fifteen cyanophyte species are involved in the discontinuous entrapment of carbonate mud. Grain size in the laminae varies according to tide, storm and rainfall: sedimentation may be interrupted by erosion or modified by the growth rate and form of the algal species in residence. Exposed muds may crack and then curl, so imposing on the stromatolite both discontinuities and topographic irregularity. Influenced by the cracks, the piling of laminae is apt to evolve into separate dome-shaped "algal heads" or polygons, further modified by the curled edges into a variety of growth forms (Fig.200–202, Chapter 4). For these are only growth forms, reflecting a twofold laying down of sediment and algae, whose several species, each in its peculiar fashion, control through their modes of growth the form of the developing structure. Forms known for many years as *Cryptozoon* (HALL, 1883), as *Collenia* (WALCOTT, 1906, 1914), as PIA's (1926) Spongiostroma with subdivisions Oncolithi and Stromatolithi, and numerous other "genera", can now be matched with Recent forms built by different algal species, and by associations of species, as they adapt themselves to, and are affected by, the fluctuating exigencies of their environments (see p.199 on the stromatolites of the Abu Dhabi sabkha).

MORPHOLOGY AND GROWTH

Classification

Black's results have been extended by a number of workers, particularly CLARKE and TEICHERT (1946), GINSBURG and LOWENSTAM (1958), LOGAN (1961), D. J. Shearman and co-workers (in: KINSMAN, 1964b), MONTY (1965, 1967) and GEBELEIN (1969). It has for some time been obvious that a Linnean binomial nomenclature is inappropriate for algal stromatolites, and attempts have been made to classify growth forms according to their geometry (e.g., ANDERSON, 1950). The most satisfactory scheme so far, by LOGAN et al. (1964), draws upon their experience of both ancient and Recent stromatolites.

The algae concerned are generally a mixture of coccoid and filamentous Cyanophyta. The carbonate laminae, up to a millimetre or so thick (Fig.208), accumulate to give undulose, spheroidal, columnar or club-shaped forms composed of grains from clay to sand grade. One form may pass into another as growth proceeds, *Collenia*-like forms giving rise, by separation of mats into isolated patches, to columnar or club-shaped *Cryptozoon*-like forms. LOGAN et al. (1964) based their geometrical classification on the hemispheroid (domeshaped) and spheroid (oncolite). They described three dominant types: laterally linked hemispheroids

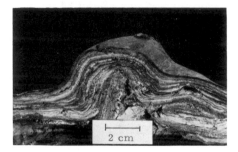

Fig.208. Section through stromatolitic carbonate sediments of eastern Shark Bay, 26°S 114°E, Western Australia. The coarse layer is a storm deposit. (From DAVIES, 1970b.)

Fig.209. Domed structure formed by growth of a mat-sediment sequence over a tilted block of soft stromatolite. Shark Bay, Western Australia. (From DAVIES, 1970b.)

(LLH), separate vertically stacked hemispheroids (SH) and spheroidal structures (SS). Other categories are compounded of these basic forms. In modern stromatolites the form is intimately related to the local physical situation and so makes the detailed study of form a potentially valuable tool in the interpretation of ancient stromatolites in terms of their past environments. Doming may be caused by such factors as the isolated growth of colonies (commonly monospecific), greater accretion on crests, evolution of gas under the mat or growth over pre-existing irregularities, such as the curled-up edges of mats (Fig.209–213). Where lateral linkage is inhibited by widening of cracks, stacked, domed hemispheroids develop. Continuous flat sheets are generally constructed by multispecific colonies. Where the original substrate is free and mobile, as a mud flake or shell, stacked hemispheroids grow all round the nucleus to give the familiar oncolite (SS). Other important physical controls, in addition to those given by Black, are: sporadic wetting by tides or during storms, desiccation, erosion during storms, burial as a result of sediment influx, and bioturbation by animals. The rate of induration is important, also the inhibiting effect on the growth of some algal species of prolonged drying, wetting or changes in salinity.

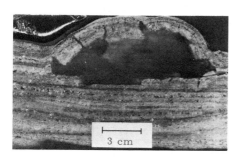

Fig.210. Gas dome in algal stromatolite. The roof is made of sediment and gypsum crystals bound by a thin algal film, and supported by gas pressure in the cavity. Shark Bay, Western Australia. (From DAVIES, 1970b.)

Fig.211. Gas domes on a supratidal flat (see Fig.210). Shark Bay, Western Australia. (From DAVIES, 1970b.)

Fig.212. Head-shaped stromatolites along the shore of an intertidal pond. Cape Sable, Florida (Fig.163). Inch scale. (Courtesy of C. D. Gebelein.)

Fig.213. Section through head-shaped stromatolite showing growth of youngest lamina over upturned edges of older desiccated laminae. Cape Sable, Florida. (Courtesy of C. D. Gebelein.)

The spheroid–hemispheroid system of nomenclature, valuable though it is, is a somewhat crude way of identifying forms which are complex in origin and arise in such varied circumstances. There is good reason to suppose that similar growth forms, as defined by LLH, SH and SS and their combinations, may arise by the growth of different algal associations and in a variety of environments. The system necessarily ignores, moreover, internal structures, for example the thickness of laminae, their lateral continuity and the separation between them, in what is, as sediments go, a remarkably porous and permeable structure liberally provided with internal sediments of varied nature. Yet it is clear that any refinement of the descriptive system must await a more thorough knowledge of present day stromatolites than we have at this moment.

Finer depositional fabrics

Examination of the depositional fabrics of Recent stromatolites shows them to consist of sedimentary grains mixed or interlayered with algae and mucilaginous polysaccharides. This binary construction, described by BLACK (1933a) and further illumined through the researches of MONTY (1965, 1967) and later by GEBELEIN (1969), gives the deposits not only a unique range of fabrics but a post-depositional and diagenetic history of unusual complexity.

  Organic components. The organic components are dominated by blue-green algae, both filamentous and coccoid. The taxonomy of the algae that form mats is at present in a highly tangled and controversial condition (GOLUBIĆ, 1969) largely because so little is known about them, and it is impossible to refer to species with certainty. Of outstanding importance in the Andros area, Bahamas (MONTY, 1965, 1967) and Bermuda (GEBELEIN, 1969), is one of the Oscillatoriaceae, *Schizothrix* sp. (*S. calcicola*?). Unpublished work by C. D. Gebelein and P. Hoffman has shown this species (Fig.151, p.124) to be the major cyanophyte in the stromatolites on intertidal mud banks of Lake Ingraham, Florida, just inshore of Cape Sable (Fig.163, p.148). *Schizothrix* has a world wide association with carbonate accumulation (MONTY, 1965) and is probably a universal component of Recent stromatolites (GEBELEIN, 1969).

  Schizothrix is filamentous (Fig.151) with cells about 1–2 μ in diameter. Morphologically it varies from small, isolated filaments to colonies of filaments densely arranged in mucilaginous films. The filament consists of the cellular trichome in a thick mucilaginous sheath, to which sedimentary grains adhere. Filaments commonly grow together in a manner resembling woven strands in rope. There is also a common mucilaginous material which fills much of the space between the filaments. In the construction of stromatolites the role of the filaments is twofold. The sheaths first trap particles which then become entangled among the filaments. The filaments of this genus, which grow quickly during the daytime, are also able to move upward out of their sheaths in excess of growth. *Schizothrix* is one of the most slender of the filamentous algae and it does not calcify its sheath. Thus direct evidence of its occupation cannot be preserved in ancient stromatolites.

  Another very important filamentous genus is *Scytonema*. This is both fresh water and marine and is commonly found in places subject to periodic wetting by rain or floods, such as intertidal marsh or marine splash zones. It forms cushions, discs or heads made of erect or prostrate filaments according to the amount of moisture present. It has a thick laminated sheath up to 30 μ across, which protects the trichome from desiccation, and which may be calcified.

Among the coccoid cyanophytes, a commonly occurring genus is *Entophysalis* which generally forms mucilage-bound clusters of unicells or sheets of unicells, each cell 1–2 μ in diameter. The cells have sheaths that may be active sites of carbonate precipitation (MONTY, 1967, p.67). An important characteristic of the sheets is their custom of coating sedimentary particles, so making them sticky and, at the same time, isolating them to a large extent chemically from the sea water (p.252).

The organic component in the stromatolites is an *association* of microorganisms, as BLACK (1933a, p.168) pointed out, dominated by blue-green algae, but containing also diatoms which deposit their own mucilages, and cysts of flagellates and green algae, bacteria, etc.

Lamination and serial events. In so far as the fabric of the stromatolite is laminar (Fig.208), it depends for this discontinuity on serial events during its growth, such as fluctuations in moisture content, in salinity of the water, in light (day-night), and in current velocity, sedimentation rate and abrasion. The causality of stromatolites is complex and, though certain influences may be dominant in one situation, others will be important also. A few examples follow.

On Andros Island, BLACK (1933a, p.178) noted that an alternation between higher and lower salinities produced a corresponding alternation in the dominance of *Schizothrix*, the sediment trapper and binder, and *Scytonema* which commonly calcifies its sheaths. In the brackish intertidal setting on Andros Island, MONTY (1967, p.81) found that algal-rich (hyaline) laminae, built of long filaments of *Scytonema* through which run bundles of *Schizothrix*, represent normal tidal conditions. These normal laminae alternate with carbonate-rich laminae, possibly formed by deposition of sediment during storms, or of faecal pellets left by browsing fish during high tide, or by precipitation of $CaCO_3$ following evaporation during neaps.

From the Andros windward lagoon MONTY (1965; 1967, p.89) has described subtidal, paired laminae forming at 2 m below low water. By scattering boiled carborundum as a marker over the surface of the algal mat he discovered that alternate hyaline laminae (200–900 μ thick) and sediment-rich laminae (up to 100 μ) grow in a single day-night cycle. After sunrise new bundles of *Schizothrix* filaments grow by cell division and vertical extension of the trichome inside the ever-elongating sheath. They vary in attitude from erect to variously inclined, and form a reticulate framework containing a few particles of carbonate. This is the hyaline lamina. Later in the day, and possibly during the night, the habit of the *Schizothrix* changes to a closely crowded layer of horizontal filaments in which carbonate particles are concentrated.

Different results from those of Monty, on the growth of paired laminae, were obtained by GEBELEIN (1969) who measured the growth of paired laminae by sprinkling insoluble Venetian red pigment (ferric oxide) on the growing surfaces, also on a subtidal floor, off Bermuda. Unlike Monty, he found that the algal-rich laminae formed at night and the sediment-rich laminae during the day, the particles being trapped by the rapidly growing filaments of *Schizothrix* in sunlight. At night the filaments of *Schizothrix*, combining with those of *Oscillatoria*, grew horizontally and slowly, giving a thin, strong hyaline lamina. Again, unlike the paired laminae in the Andros lagoon, the sediment-rich laminae off Bermuda were the thicker (about 1 mm) and the algal-rich laminae were only about 100 μ. In a recent study of intertidal stromatolites near Cape Sable, Florida, GEBELEIN and HOFFMAN (1968) found that the paired laminae formed in a single tidal cycle (p.229).

Mechanical trapping. The extent to which carbonate sediment is either trapped or precipitated in the stromatolites seems variable. Black emphasized the dominance of mechanical trapping, although he noticed the occurrence of unidentified carbonate crystals, possibly precipitated, in the sheaths of *Scytonema* (BLACK, 1933a, p.174). GEBELEIN (1969), too, has found evidence only for adherence and binding by mucilage and filaments. On the other hand, LOGAN (1961) and MONTY (1965, 1967) discovered, respectively, in Shark Bay, Western Australia, and in the windward lagoon of Andros Island, that carbonate accumulates not only by entrapment but by precipitation, a matter that will be considered below. With regard to mechanical trapping, Logan described stromatolites composed of an outer algal lamina or mat, enclosing an interior built of laminated, fine calcarenite or calcilutite, porous and semi-lithified. The green-brown superficial mat is soft and moist (or friable and blackened when dry) and the degree of induration increases progressively inward. Mats and their resultant carbonate laminae vary in thickness from 1 to 10 mm. The trapping of particles depends on an irregular supply of sediment carried into the tidal area and spread as thin layers over the mat. GEBELEIN (1969), working on the shallow subtidal floor off Bermuda, noticed that where sedimentation is too slow, only an algal mat (hyaline lamina) forms on the floor: no carbonate-rich laminae accumulate. If the particles are too large, as coarse sand or gravel, they cannot be trapped by the filaments so no stromatolite develops. Both Logan and Gebelein emphasized that the rate of bottom traction is important. It is obvious that a highly mobile sediment escapes entrapment: thus stromatolites do not grow on beach sands or on actively ripple-marked subtidal sands. Moreover, if bottom traction of grains is too fast, then the delicate algal meshes are abraded by grain impact. Abrasion of this kind leads to asymmetrical building of some stromatolites and so leaves a permanent record of the direction of fastest or most persistent bottom traction during the period of growth. DAVIES (1970a) has found an asymmetry in the overfolding of curled up algal mats which correlates with a strong on-shore flood tide, in eastern Shark Bay, Western Australia. Trapping of sediment is selective: the algal mucilages and meshes seem to be able to hold only the finer particles (BLACK, 1933a, p.176; GEBELEIN, 1969).

Precipitation of carbonate. Logan's stromatolites, in Shark Bay, when viewed in thin section, reveal sedimentary particles cemented together by coatings of acicular aragonite which also line body chambers of shells and foraminiferids. The more indurated parts of the stromatolites are "well cemented and hard" and, in this intertidal situation, they show an interesting resemblance to beach rock (Fig.276, p.367). Logan made this analogy and seemed to regard the cementation as entirely inorganic. This early cementation is critically important for the growth of some of the stromatolites in Shark Bay, particularly the club-shaped ones (looking so like a collection of discarded capstans), because without it such forms with overhanging walls would collapse.

Monty's precipitated $CaCO_3$ differs profoundly from Logan's (Fig.214, 215). It appears to be a biologically controlled but extracellular process, presumably related, at least in part, to the removal of CO_2, during algal photosynthesis, from water saturated for $CaCO_3$. MONTY (1967, pp.73–76) discussed the origin of certain particles of high-magnesian cryptocrystalline calcite, which are intimately associated with the algae. The particles, aggregates of clay grade crystals, range in diameter from a few tens of microns to 2 mm. They are common in the supratidal stromat-

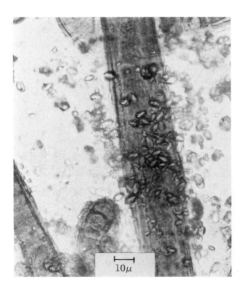 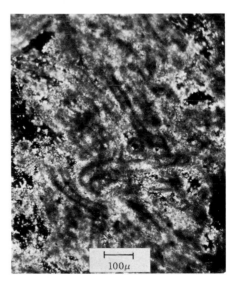

Fig.214. Rhombs of high-magnesian micritic calcite on filaments of blue-green algae. Supratidal stromatolite in east lagoon, Andros Island, Bahamas. (Courtesy of C. Monty.)

Fig.215. A more advanced stage of the early development in Fig.214. Algal filaments are embedded in a matrix of micrite. Supratidal stromatolite in east lagoon, Andros Island, Bahamas. (Courtesy of C. Monty.)

olites, which are built by alternating or mixed communities of *Schizothrix* and *Scytonema*. This calcite contains 5–10 mole% of $MgCO_3$ according to X-ray diffraction data. Its mineralogy distinguishes it from the Recent carbonate sediment which, in this area, is a mixture of aragonite and magnesian calcites with a distinctly higher content of $MgCO_3$, derived from the skeletons of echinoids, coralline algae and peneroplids. The nearby Pleistocene limestones are made of low-magnesian calcite. The magnesian calcite in the algal mats occurs only in algal sheaths or mucilages (at least when new) and occupies three kinds of site: (*1*) in the sheaths of unicells and of filaments, where it forms a crust, (*2*) outside the sheath but in the mucilaginous material laid down by the algae, including the felts of *Schizothrix*, and (*3*) as a replacement of dead coccoid cells, especially *Enterophysalis*. Calcified colonies of *Enterophysalis* show commonly a "core of dead cells loaded with cryptocrystalline calcite surrounded by a thin peripheral layer of living cells". Monty concluded that this magnesian calcite cannot be detrital in origin but was precipitated in the mat, his main grounds being the special mineralogy and the siting of the crystals.

Strong evidence in favour of precipitation of aragonitic micrite in stromatolitic sediments of Baffin Bay, Texas, is given by DALRYMPLE (1966). The micrite occurs in the algal mats as grains composed of cryptocrystalline aragonite mixed with local skeletal debris. It is possible that precipitation is not a direct result of algal activity but a result of the action of anaerobic bacteria in the lower reducing zones of the algal mats.

Clotted micrite. Clots reminiscent of *structure grumeleuse* (p.511) were described by Monty as evolving variously: as replaced unicells, as broken *Scytonema* sheaths, and as a generally erratic precipitate in the mucilage. The high frequency of this peloidal material in the stromatolites of eastern Andros (and probably elsewhere) is important because peloids are a major component in the composition of many ancient stromatolites.

Non-serial deposition. A large part of the accumulation of carbonate sediment in the stromatolites is not related to the serial, successive building of lamina on lamina, but takes place within the considerable porosity between the laminae at various times after they have been built. It is emphatically not possible to assume that the relative times of deposition of all the carbonate components can be determined by applying the law of superposition. Mention has already been made of the clots (peloids) resulting from the breakdown of *Scytonema* sheaths. Monty has seen how these fragments form in the deeper laminae as a result of collapse under load of the delicate algal framework. Ageing and compaction of stromatolites is brought about by desiccation: the total thickness is reduced by a selective process whereby the hyaline laminae are more compacted than sediment-rich laminae. Compaction is, of course, influenced by the degree of cohesion resulting from calcification and, possibly, from some early inorganic precipitation of cement.

Complication also occurs where the positively phototropic algal colonies, which remain sheltered below the surface when the outside environment is inhospitable, move up to the surface. MONTY (1967, p.72) wrote: "The calcified layers are supposed to form during 'drier' periods when *Scytonema* grows on a reasonably moist substrate while *Schizothrix* fills the humid innerspaces of the mat. During wetter periods, *Schizothrix* rises to the surface where it forms a gelatinous film by active cell division. As the wet period comes to an end, the *Schizothrix* bloom stops and is progressively by-passed by the slower but continuously growing *Scytonema*." Remnant algal cells are not found within the stromatolites (GEBELEIN, 1969). Thus, as living cells move upward phototropically, the structure they vacate develops its own peculiar collapse structures. Stromatolites eventually develop very high porosities and permeabilities, as would be expected in a sedimentary structure which has some rigidity owing to calcification but from which the algal material must, in the long run, be removed mostly (but not altogether) as a result of decay and oxidation. This means that ground water, gas or oil, can move easily through the structure, eroding or depositing material as it does so. More or less flat-topped internal sediments (so well illustrated by SANDER, 1936, 1951) are usual in fossil stromatolites and are probably the result of such processes as roof and wall erosion, deposition of trains of faecal pellets by infauna and general redistribution of sediment. Superimposed on the irregular accumulative processes there is the intense boring of detrital grains, especially by *Entophysalis* (MONTY, 1967, p.78), and their micritization into cryptocrystalline, unrecognizable peloids. Even apparently tidy lamination can give an overly simple appearance of serial deposition. Growth of laminae is irregular during the year, being more vigorous in the high summer, so that counting of laminae can yield only a minimum duration for the formation of the stromatolite. Using his carborundum tracers, MONTY (1967, p.91) found that growth would stop for a few days in one place while the adjacent colonies continued to grow. In this way topographic highs and lows evolve: a dome may grow rapidly and then pause while growth shifts to the adjacent depressions.

Morphology and distribution of stromatolites

It is plain that the distribution of the mat–sediment complexes, their forms and their internal structures, depend on a great variety of factors. Probably any moist exposed piece of ground, or lake or sea floor, in the tropics or subtropics, is potentially mat-bearing. Whether a mat can incorporate sediment in its structure, and so take on a more permanent form, depends on the supply of carbonate sediment and on the balance of algal growth, abrasion and induration. On some of the shallow water floors off Bermuda, increased thicknesses of purely organic mat are found where sedimentation is reduced, either in deeper waters where currents are slower, or near *Thalassia* beds where the baffling effect of the blades reduces water velocity (GEBELEIN, 1969). The same author noted a relation among form, current velocity and sediment supply. Where little sediment moves across the mat, only simple mats occur, following closely the pre-existing contours of the floor. Where sediment supply is greater, ellipsoidal nodules of stromatolite form, 1–10 cm long. They are rooted to the floor by algal filaments. Gebelein found that some kind of irregularity on the floor was a necessary prerequisite for the growth of these nodules, for example a small ripple-mark, a stromatolitic flake or an exposed *Thalassia* rhizome. Where sediment movement and the supply of carbonate-building material is even greater, elliptical domes form with long axes from about 30 cm to 300 cm aligned parallel to the main flow. When the rate of delivery of sediment to the area is too rapid, the mat is buried and destroyed.

In some environments flakes of stromatolitic material are common; they are about 1–3 mm thick and 1–5 cm across. MONTY (1967, p.77) has described these as primary structures, *incipient* mats, which he discovered in brackish intertidal settings, on the upper parts of beaches, among the roots of mangroves and in tidal pools. GEBELEIN (1969), on the other hand, encountered flakes of similar dimensions which he could demonstrate to be the *eroded* fragments of mat. This twofold mode of formation of flakes is probably true of those described by PURDY (1963a, p.348) from the grapestone lithofacies on the Great Bahama Bank (p.317). I have seen soft, algal-rich flakes and hardened, brittle flakes in the grapestone area of the Berry Islands, but have rarely been able to find any intermediate stages. In this connection it is interesting to speculate on the origin of grapestone. Does it, perhaps, evolve as algal-bound clusters of grains (p.317) which never coalesce to form a continuous lamina?

Scour influences the form of stromatolites, especially the club-shaped masses in parts of Shark Bay, where their narrow bases are partly a consequence of under-cutting during tidal flow and ebb. Once topographic diversity has developed, the differences between highs and lows will be accentuated as growth proceeds rapidly on the tops, which are moist and not too steep, but continues slowly in the depressions where the floor is scoured and algal growth of some species is inhibited by the longer periods of immersion.

In Shark Bay, there is a distinct zoning of the various forms of stromatolite, from flat mats on protected intertidal mud flats, to sinuous domed structures interspersed with the flat forms around high water mark, to discrete, club-shaped growths with overhanging sides situated on headlands. LOGAN (1961) wrote of an interplay of four factors: (*1*) upward growth of mat, (*2*) doming over irregularities, such as mud cracks, breccia fragments, older domes, (*3*) preferential growth on highs, and (*4*) inhibition of growth in depressions. GEBELEIN (1969) looked at the matter from a rather different viewpoint and produced five critical factors: (*a*) quantity of sediment brought to site (the inorganic building bricks), (*b*) the rate of bottom traction and consequent erosion, (*c*) grain size, only the finer grains being suitable for incorporation, (*d*) current velocity, influencing both sediment supply and erosion, and (*e*) wave turbulence which inhibits growth. The distribution of stromatolites in the Bermudan shallows shows a close relation to bathymetry, tidal movement and fetch. Finally, it must not be forgotten that the algal mat profoundly modifies its own environment and, as regions of the floor are colonized by mats, so it becomes easier for adjacent regions to be colonized as well.

Behind Cape Sable, Florida (Fig.163, p.148) there is an extensive complex of intertidal carbonate mud flats colonized in part by stromatolite-forming algae. This region, called Lake Ingraham, is connected to Florida Bay by man-made channels: it has a tidal range of 1 m and a salinity variation of 31–35‰. GEBELEIN and HOFFMAN (1968), in a detailed study of the stromatolites, have found a marked lateral zonation of algal morphology and organisms. At the edges of tidal channels levées have been built by the accumulation of hemispherical, laminated stromatolites which drape over deep prismatic cracks. Farther from the channels, on slightly lower ground, there is a broad zone of flat laminated stromatolites. Farther still into the interior there are unlaminated muds with tufts of the green alga *Caulerpa* and, beyond these, an extensive area of unlaminated, mottled muds with ridge and channel morphology much burrowed by polychaete worms. The history of the laminae throws useful light on the process of stromatolite accumulation. The laminae are paired, each pair consisting of a sediment-rich lamina, about 1 mm thick, and a layer rich in the filamentous alga *Schizothrix*, about 0.4 mm thick. Particles in the sediment-rich lamina are enmeshed by filaments of *Schizothrix*. The rate of accumulation was determined by marking the laminae with a red powder and the results showed that each pair is deposited during one tidal cycle, that is to say at the rate of two pairs a day. No net accumulation of sediment occurred in the absence of stromatolites and, when the algae in stromatolites were killed with formalin, accumulation ceased. The algae are, therefore, responsible for trapping the sediment. Thicker laminae were found where submergence under water was longer. Stromatolites were eroded during rainstorms, by mechanical washing of the surface and partial dissolution of the algal sheaths. Thus, net deposition occurs during the dry winter and net erosion during the rainy summer. Laminae are destroyed or prevented from growing in the lower tidal flats by burrowing polychaetes.

In this discussion of stromatolites the reader will have detected an obvious kinship between stromatolites and the Bahamian subtidal algal mats described in Chapter 3 (p.122). They have many genera of algae in common, both filamentous and coccoid, and the processes of trapping and binding seem to be the same in each case. The subtidal mat does not, however, progress through the accumulation of sediment, to the status of stromatolite. It appears, nevertheless, to be indistinguishable from the simple, non-accumulative mat, described by Gebelein from the neighbourhood of Bermuda. It lies at one end of a range of deposits at the other end of which are the laminated domes and algal heads.

SUMMARY

The carbonate algal stromatolites, and their related non-accumulative algal mats, develop great variation in morphology, both external and internal, in response to complex and ever changing growth environments. The primary growth fabric is a plexus of filamentous and coccoid trichomes in mucilagenous sheaths and in a matrix of more generalized mucilage. Under appropriate conditions, sedimentary particles adhere to the mucilage and are entangled in the filaments. In certain circumstances high-magnesian calcite is precipitated in close association with the algal cells. The dynamic balance between organic growth and sediment accumulation is influenced by numerous environmental factors to yield stromatolites, or mats, with a very varied range of morphologies.

These influencing factors include light, wetness, salinity, bottom traction, sedimentation rate, abrasion by particles, rate of induration—and the changing morphology of the stromatolite itself as, piling lamina on lamina, it modifies its own environment. Although the stromatolite is laminated, the law of superposition cannot be applied in a simple way to the interpretation of the history of its internal structures. In its porous interior, vacated by the upward moving algae, a variety of internal sediments accumulates in a framework characteristically given to irregular collapse, as algal tissues or mucilages decay, as overburden increases and desiccation proceeds.

The results so far published are but the early steps in the evolution of a comparatively new subject and represent only a few studied aspects of the broad field of algal stromatolites. The enormous geological influence of these, especially in Proterozoic sedimentation, can be gauged from the widespread and thick stromatolites on the Canadian shield, as in Bathurst Inlet near Hudson Bay. The very complexity of the carbonate stromatolite is a measure of its sensitive reaction to its environment—an indicator to us of the potential value of ancient stromatolites as a guide in interpreting ancient environments.

FURTHER READING

The fundamental papers are those by BLACK (1933a) on the stromatolites of western Andros Island, LOGAN (1961) and DAVIES (1970b) on Shark Bay, MONTY (1967) on eastern Andros Island, KENDALL and SKIPWITH (1968b) on Persian Gulf stromatolites, GEBELEIN (1969) on offshore Bermuda and, for classification of structures, LOGAN et al. (1964). The paper by NEUMANN et al. (1970) on subtidal algal mats is highly relevant, also SCOFFIN (1970). Other references are given in the text.

Chapter 6

SOME CHEMICAL CONSIDERATIONS

> The farther and more deeply we penetrate into
> matter, by means of increasingly powerful
> methods, the more we are confounded by the
> interdependence of its parts... It is impossible to
> cut into this network, to isolate a portion without
> it becoming frayed and unravelled at all its edges.
> *The Phenomenon of Man*
> P. T. DE CHARDIN (1959)

If we are to understand the many processes that culminate in the production
of a lithified carbonate rock, we must know the conditions in which crystals of
calcium carbonate grow or are dissolved in aqueous solution. This knowledge is as
necessary when considering the precipitation of an oölitic lamella or an aragonite
needle as when trying to unravel the progress of cementation or dolomitization.
If the system CO_2–H_2O–metal carbonate were a simple one, containing only a
single mineral, with a solubility product that varied only with P–T, then the problems
would be few. Instead, the various natural environments are highly complex. The
matter will be simplified to some extent by restricting the discussion to three of the
numerous anhydrous carbonates, calcite with a range of Mg^{2+} content, aragonite
and dolomite, and, in the first place, to the simplest possible system, CO_2–H_2O–
$CaCO_3$.

THE SIMPLE SYSTEM CO_2–H_2O–$CaCO_3$

Equilibria

The solubility of $CaCO_3$ in pure water without CO_2 is slight, about 14.3 mg/l
for calcite and 15.3 mg/l for aragonite (in POBEGUIN, 1954, p.95) in the P–T range
of the surface of the sea or in near-surface ground waters. With the addition of
CO_2 to the water, solubilities can reach hundreds of mg/l. The solubility product
$[Ca^{2+}]\cdot[CO_3^{2-}]$ increases with rising partial pressure of CO_2, decreasing temperature, and is greater in NaCl solutions and in sea water (DEER et al., 1962).
Six equilibria relate the dissolution or precipitation of $CaCO_3$ in water to the
invasion or evasion of CO_2:

$$CO_2 + H_2O \rightleftharpoons H_2CO_3 \tag{1}$$

$$H_2CO_3 \rightleftharpoons H^+ + HCO_3^- \tag{2}$$

If the solution contains free CO_3^{2-} then the proton released in reaction 2 reacts with the carbonate ion to give more bicarbonate:

$$H^+ + CO_3^{2-} \rightleftharpoons HCO_3^- \tag{3}$$

At the interface between the solution and solid $CaCO_3$ the equilibrium is:

$$CaCO_3 \rightleftharpoons Ca^{2+} + CO_3^{2-} \tag{4}$$

If these equilibria move to the right, both CO_2 and $CaCO_3$ are dissolved. If they move to the left, as a result perhaps of evaporation or photosynthesis of aquatic plants, CO_2 is removed and $CaCO_3$ is precipitated. The net result can be summarized as:

$$CO_2 + H_2O + CaCO_3 \rightleftharpoons Ca^{2+} + 2HCO_3^- \tag{5}$$

The restricted reversibility of this last reaction in sea water or any solution with a high Mg^{2+}/Ca^{2+} ratio is a matter of prime importance to carbonate equilibria (see p.255; also CHAVE and SCHMALZ, 1966, p.1038 and WEYL, 1967, p.200).

A major part of the CO_2 in sea water (pH near to 8.2) is held in HCO_3^- with a lesser amount in CO_3^{2-} (Fig.216, also p.255). Reactions 2 and 3 are very rapid. Reaction 1, the hydration reaction, is relatively slow but is followed by the instantaneous reaction:

$$H_2CO_3 + OH^- \rightleftharpoons HCO_3^- + H_2O \tag{6}$$

A rapid and simple graphical method of following the related changes in the critical variables has been designed by DEFFEYES (1965). The carbonate system is completely defined by the alkalinity and total carbonate carbon and, using these two variables as ordinate and abscissa respectively, the Deffeyes diagrams show the related variations among pH, HCO_3^-, CO_3^{2-} and P_{CO_2}.

If the pH were to exceed 9, two other reactions would become important:

$$CO_2 + OH^- \rightleftharpoons HCO_3^- \tag{7}$$

$$HCO_3^- + OH^- \rightleftharpoons CO_3^{2-} + H_2O \tag{8}$$

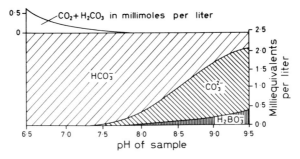

Fig.216. Variation of the alkalinity components of sea water with pH. (After CLOUD, 1962a.)

Reaction 7 is slow, whereas 8 is instantaneous and raises the CO_3^{2-} content of the water (Fig.216).

In pure water CO_2 remains mostly as dissolved gas. Up to about 30 milli-moles per litre may be dissolved before CO_2 appears as a separate gas phase (CLOUD, 1962a, p.100). Only where dissolved $CaCO_3$ causes the pH to rise above 6 does the HCO_3^- form in quantity. The CO_3^{2-} content is significant only when pH exceeds about 8 (Fig.216). Thus as pH rises so does the product $aCa^{2+} \cdot aCO_3^{2-}$.

The system is described theoretically by five constants which vary with temperature and pressure. These are the Henry's Law constant for CO_2:

$$H_2CO_3 = P_{CO_2} \cdot K_{CO_2} \tag{9}$$

the first and second ionization constants for carbonic acid:

$$K_1 = \frac{[H^+] \cdot [HCO_3^-]}{[H_2CO_3]} \tag{10}$$

$$K_2 = \frac{[H^+] \cdot [CO_3^{2-}]}{[HCO_3^-]} \tag{11}$$

the solubility product for $CaCO_3$:

$$K_c = [Ca^{2+}] \cdot [CO_3^{2-}] \tag{12}$$

and the ionization constant for water:

$$K_w = [H^+] \cdot [OH^-] \tag{13}$$

The real system is commonly studied in the field or in the laboratory by monitoring P_{CO_2}, pH, and finding $CaCO_3$ or Ca^{2+} or total inorganic CO_2. The value of direct measurements in the environment with the Weyl saturometer (WEYL, 1961) is increasingly appreciated. Though the constants can be read from tables, they are best derived internally (see, for example, PYTKOWICZ, 1965b).

pH and buffering

The pH of the solution depends on reactions 2, 3, 6 and 8 which control the quantity of H^+, HCO_3^- and CO_3^{2-} (p.231). For each mole of water and newly dissolved CO_2 that react, some yield H^+ and HCO_3^-, but a considerable proportion forms undissociated H_2CO_3. In addition some of the CO_2 is held in CO_3^{2-}. The addition of n moles of CO_2, instead of releasing 2n moles of H^+ in reaction 2, actually frees a much smaller quantity. In this way the system is **buffered**. As a result the pH of sea water varies little, despite changes in CO_2 tension (see PYT-KOWICZ, 1967). Values of pH for sea water rarely fall outside the range 7.8–8.3. By contrast, land waters having low alkalinity are poorly buffered.

Alkalinity

The alkalinity of sea water depends on the presence of HCO_3^-, CO_3^{2-} and $H_2BO_3^-$ (Fig.216). Following Brønsted–Lowry theory, a base is any compound that can accept a proton. So, in equation 2, HCO_3^- is a base, but in equation 3 it behaves as an acid while CO_3^{2-} is the base. The alkalinity relies on the two equilibria 6 and 8. Here water behaves as an acid, donating protons to the two bases to make the conjugate acid H_2CO_3 and conjugate base OH^-. **Titration alkalinity** is distinguished from **carbonate alkalinity**, the former being total alkalinity determined by titration with, for example, H_2SO_4 or HCl, the latter being a corrected value found by removing the alkalinity caused by $H_2BO_3^-$ (SMITH, 1941, p.236; H. W. HARVEY, 1955, p.165).

The simple system CO_2–H_2O–$CaCO_3$ is complicated by such factors as the existence of $CaCO_3$ polymorphs, the solid solution of $MgCO_3$ in the calcite lattice, surface chemistry, complexing and the effects of pressure in the deep ocean. Some aspects of these and other matters will now be touched upon.

THE COMMONER CARBONATE MINERALS

Polymorphs of CaCO₃

Solid $CaCO_3$ exists as three different ionic structures: calcite (trigonal system, hexagonal scalenohedral class), aragonite (orthorhombic system) and vaterite (hexagonal; also called μ-calcite). Details of unit cells are given by DEER et al. (1962). At 25 °C aragonite is stable only above a hydrostatic pressure of about 4 kb: at 600 °C the lower limit of the stability field is at about 14 kb (P. S. CLARK, 1957). At lower pressures, in the stability field of calcite, aragonite is metastable and inverts to calcite when heated (JOHNSTON and WILLIAMSON, 1916; LANDER, 1949; JAMIESON, 1953, 1957; MACDONALD, 1956; P. S. CLARK, 1957; BOETTCHER and WYLLIE, 1968). During inversion, which is a dry solid state reaction, the Ca^{2+} ions change their positions and the CO_3^{2-} ions rotate. There is an accompanying volume increase of a little over 8%. BURNS and BREDIG (1956), DACHILLE and ROY (1960) and JAMIESON and GOLDSMITH (1960) have noticed that when calcite is ground dry at room temperature in a mechanical mortar it changes to aragonite. The same process has been followed by M. Atherton (personal communication, 1967) at Liverpool University. It seems that shear stresses cause a reorganization of the lattice. Aragonite is more soluble than pure calcite and this is to be expected since it is the polymorph with lower symmetry, but it is less soluble in shallow sea water and distilled water than some high-magnesian calcites (Fig.218). In sea water the thermodynamic solubility product constants for the equation:

$$[aCa^{2+}] \cdot [aCO_3^{2-}] = K$$

are estimated as $K_{aragonite} = 6.9 \times 10^{-9}$ and $K_{calcite} = 4.7 \times 10^{-9}$ (both: LATIMER, 1952). The standard free energy of formation for aragonite is thus less than for calcite, since it is related to the equilibrium constant K by:

$$\Delta F°_f = RTlnK.$$

GARRELS et al. (1960) estimated that $\Delta F°_{f\ aragonite} = -269.6$ kcal./mole, and $\Delta F°_{f\ calcite} = -269.8$ kcal./mole.

FRIEDMAN (1968a) has distinguished between low-magnesian aragonite (less than 0.1‰ Mg^{2+}) which he finds in molluscs in carbonate sediments in the Gulf of Aqaba, and high-magnesian aragonite (more than 1‰ Mg^{2+}) which he records from the corals. Likewise MATTHEWS (1966) referred to molluscan aragonite as low-strontium and coral and codiacean aragonite as high-strontium (as in Fig.225).

Magnesian calcites

In 1922 Clarke and Wheeler published an extensive series of analyses of invertebrate skeletons which showed that calcite can hold more than 20 mole % $MgCO_3$, whereas in aragonite amounts greater than 2 mole % are rare. Bøggild in 1930, combining his study of molluscan shell fabric with the data of Clarke and Wheeler, concluded that three distinct kinds of carbonate were to be found, aragonite, low-magnesian calcite (2–3 mole % $MgCO_3$) and high-magnesian calcite (12–17 mole % $MgCO_3$). This apparent natural division received support from CHAVE (1954a) who published analyses of skeletal material from a variety of marine skeletons gathered live (also WEBER, 1969, on Echinodermata). Chave's results show that aragonite rarely contains more than 1.5 mole % of $MgCO_3$. More recent support for this restricted distribution has come from P. D. Blackmon (in CLOUD, 1962a, p.56) who found that in sixteen samples selected to represent the sediments west of Andros Island, Bahamas, the $MgCO_3$ content of the calcite lies characteristically either in the range 0–5 mole % or 11–19 mole %. Samples in the range 6–10 mole % are rare. SCHROEDER et al. (1969), using an electron microprobe and X-ray diffraction, found calcite with up to 43 mole % of $MgCO_3$ in echinoid teeth in the Aristotle's lantern. There was no sign of dolomite ordering. Peaks in the X-ray diffractograms indicated at least two phases containing Mg^{2+}, Ca^{2+} and CO_3^{2-}. Useful analytical information is in NEUMANN (1965) and WEBER (1968b).

From Fig.217 it is apparent that the $MgCO_3$ mole % is, in a general way, proportional to the temperature of the environment. Thus it is related in all to three factors, mineralogy, taxonomy (physiology) and temperature. It has also been shown that, at least for an echinoid and some brachiopods, there is a positive correlation between $MgCO_3$ mole % in calcite and the salinity of the surrounding water (PILKY and HOWER, 1960; LOWENSTAM, 1961).

The relation between $MgCO_3$ content and taxonomy has been illumined by

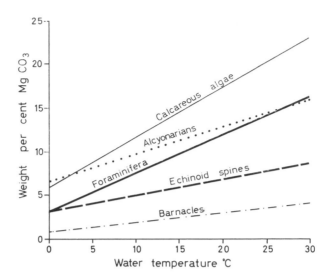

Fig.217. Relation between temperature and wt.% MgCO₃ in CaCO₃ skeletons of several taxa. (After CHAVE, 1954a.)

KITANO and KANAMORI (1966) who synthesized magnesian calcite at room temperature and pressure. They found that certain organic substances in the solution from which CaCO₃ was precipitated influenced the identity of the polymorph. Small quantities of sodium citrate and sodium malate favour the precipitation of calcite even in the presence of Mg^{2+}. Probably they form complexes with Ca^{2+}, so reducing the growth rate of the solid carbonate and allowing Mg^{2+} ions to be fitted into the lattice (a reversal of the Mg^{2+} inhibiting process described on p.243). The amount of $MgCO_3$ in the calcite lattice increased, with increasing Mg^{2+}, sodium citrate and sodium malate in the solution, up to 12.5 mole %. In these same solutions the amount of $MgCO_3$ in the lattice also increased with a rise in temperature. TOWE and MALONE (1970) used $(NH_4)_2CO_3$ to make high-magnesian calcite.

In order to discover the relative importance of mineralogy, as distinct from physiology or temperature, in controlling the $MgCO_3$ content of skeletal $CaCO_3$, it is necessary to compare coexisting calcite and aragonite in the same skeletons. Detailed work by LOWENSTAM (1964) on skeletons of Hydrozoa, Octocorallia, Cirrepedia and Bivalvia indicates the dominant role of mineralogy in determining the $MgCO_3$ content of the skeleton.

A further factor has been introduced by MOBERLY (1968) who found a marked correlation between $MgCO_3$ in the calcite and the rate of growth in the high-magnesian skeletons of certain coralline algae and bivalves. Growth rates in the algae were assessed through the determination of layer thickness, the degree of calcification of the cell walls, and the apparent seasonal production of sporangia.

CHAVE (1952), SPOTTS (1952) and GOLDSMITH et al. (1955) have given X-ray

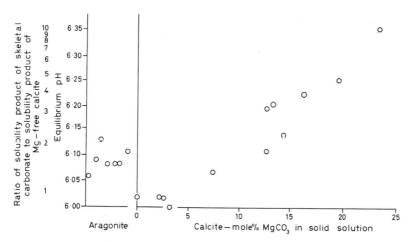

Fig.218. Equilibrium pH (solubility) of various skeletal carbonates in distilled water at 25°C, saturated with CO_2 at 1 atm. (From CHAVE et al., 1962.)

diffraction evidence that the $MgCO_3$ and the host calcite are in solid solution. This accords with the known tendency for divalent cations with ionic radii less than calcium to occur in the calcite lattice rather than in the aragonite (e.g., Mg, Fe^{2+}, Mn, in calcite, Sr, Pb, Ba in aragonite).

Values for the increasing differences in spacing of the (01$\bar{1}$4) reflection, from pure calcite through a range of precipitated calcites with $MgCO_3$ 5–14 mole %, are given by KITANO and KANAMORI (1966).

Solubilities of magnesian calcites are related by CHAVE et al. (1962) to the solubility of aragonite in Fig.218. JANSEN and KITANO (1963) reached similar results for natural high-magnesian calcites.

The dissolution of high-magnesian calcites is discussed on p.335. In the laboratory, GLOVER and SIPPEL (1967) have precipitated a range of magnesian calcites, $Ca_{80}Mg_{20}$–$Ca_{39}Mg_{61}$, at room temperature.

In ancient limestones the distribution of $MgCO_3$ in calcite is variable and is not necessarily related to taxonomy, as if there had been some post-depositional redistribution of Mg^{2+}. In carbonate sediments generally there is a change with time from unconsolidated mixtures of aragonite and high-magnesian calcite on the sea floor to cemented limestones of low-magnesian calcite having local concentrations of dolomite (STEHLI and HOWER, 1961). The cement appears to be normally a low-magnesian calcite. It seems that calcite skeletons which originally had some $MgCO_3$ must have lost variable amounts during diagenesis, while other constituents have gained some. Generally limestones are "virtually free of Mg" (GOLDSMITH et al. 1955, p.225). Some experiments by Friedman, Land and Schroeder on the dissolution of magnesian calcites are described on p.335.

Dolomites and protodolomites

Ideal dolomite is a 1/1 cation ordered carbonate wherein layers of cations and triangular CO_3^{2-} groups alternate and the cation layers themselves are composed alternately of pure Ca^{2+} and pure Mg^{2+}. It is this last peculiarity, the alternation of the two cation layers, which gives dolomite its high degree of order. In X-ray analysis the reflections caused by these alternating cation planes are known as "ordering reflections". Much of the research into the nature of Recent dolomites is concerned with the degree of ordering in the lattice, since it is the ordering which determines whether the calcium-magnesium carbonate is a dolomite or not. Carbonates with composition $Ca_{50}Mg_{50}$ are known which, lacking order reflections, are not dolomites (trigonal system, rhombohedral class) despite their chemical composition (GLOVER and SIPPEL, 1967).

    The term protodolomite was first used by GRAF and GOLDSMITH (1956) to describe the poorly ordered transitional stage through which their dolomites went during hydrothermal synthesis in the laboratory. Above about 200°C perfectly ordered dolomites formed, but at lower temperatures the carbonate, though bearing a resemblance to dolomite, was non-ideal. Graf and Goldsmith examined a number of naturally occurring transitional dolomites which, having structures which are, in principle, the same as those in their experimental protodolomites, they have called by the same name. These natural **protodolomites**, examined by X-ray diffraction, have three characteristics: (1) an expanded unit cell, (2) reflections with a strong c axis component (i.e., from planes normal or nearly normal to the c axis) are more diffuse than those with strong a axis component, (3) compared with the reflections common to both calcite and dolomite the principal order reflections of dolomite "appear in many cases to be somewhat weakened" (GOLDSMITH and GRAF, 1958a, p.681). These three characteristics are interdependent and follow from another characteristic of protodolomites, the possession of an excess of the larger, Ca^{2+} ion (radius 1.06 Å) over the smaller Mg^{2+} ion (radius 0.78 Å). It is, therefore, necessary in identifying protodolomites to ensure that these effects are not due to substitution of Fe^{2+} (0.83 Å) for Mg^{2+}. The stability of protodolomites has been ascribed to the similarity, in terms of crystal energy, of the Ca^{2+} and Mg^{2+} positions. It is to be expected that a lattice will be difficult to crystallize if it contains two or more cations occupying non-equivalent positions, especially if the cations have similar sizes, valencies and chemical properties. Indeed, this complex structure may be the reason for the poor success of experimental attempts to precipitate dolomite at room temperature and pressure (GOLDSMITH, 1953; GRAF and GOLDSMITH, 1956). Put differently, the small difference in energy between a protodolomite and a fully ordered dolomite means that the final stages of cation diffusion are unlikely to be achieved, in geologically reasonable time, unless the temperature is elevated to, say, 200°C. Up to 500°C the Ca/Mg ratio of dolomite in equilibrium with $CaCO_3$ or $MgCO_3$ does not differ significantly from unity, so that protodolomite is metastable at room temperature and pressure. The high degree of stability of protodolomites was indicated by the need to use a flux to remove the excess Ca^{2+} at between 520 and 832 °C from a protodolomite from Eniwetok and by the great abundance of natural protodolomites containing about 5 mole % excess Ca^{2+} (GOLDSMITH and GRAF, 1958a).

Atomic coordinates and X-ray structures for dolomite are given in DEER et al. (1962). A detailed account of the technique of quantitative powder X-ray analysis of mixtures of calcite, dolomite and aragonite is presented by D. L. Graf and J. R. Goldsmith in SCHLANGER (1963) and by NEUMANN (1965); the relation between lattice constants and composition by GOLDSMITH and GRAF (1958b).

MINERAL TRANSFORMATIONS

Dry transformation (inversion) of aragonite to calcite

It is apparent from the work of LINCK (1909), JOHNSTON and WILLIAMSON, (1916) and others that a temperature of about 400 °C is needed for aragonite to invert rapidly to calcite by a dry, solid state reaction. This dry process is correctly called "polymorphic inversion" (SPRY, 1969, p.87).

The inversion temperature depends on the history of the sample (e.g., the storage of energy in a deformed lattice) and on its crystal size. Below 400 °C the inversion still goes on but at rates that can be slow even judged against the geological time scale. FYFE and BISCHOFF (1965, p.9) summarized data which show, for example, that below 100 °C the time required for completion of the dry reaction is of the order of tens of millions of years. The experimental results of DAVIS and ADAMS (1965) suggest that this may be a low estimate. Fyfe and Bischoff watched the dry transformation optically and noted that the first calcite to nucleate was always in the prism zone of the aragonite. A band of calcite then grew in the direction of the aragonite c axis, but this growth was commonly stopped at cracks or other imperfections in the aragonite. Experiments by CHAUDRON (1954), BROWN et al. (1962) and LAND (1966) have shown that transformation rates vary according to the aragonite used.

Wet transformation of aragonite to calcite

Wet transformation of aragonite to calcite, in water lacking free Mg^{2+}, is much more rapid than dry transformation. The catalytic influence of the water is great and reduces the activation energy which must be exceeded for the transformation to proceed. In pure water it has a rate curve of the general form given by FYFE and BISCHOFF (1965, fig.8). This is shown here with more recent curves determined experimentally by TAFT (1967) in Fig.219. At first the reaction accelerates. This is

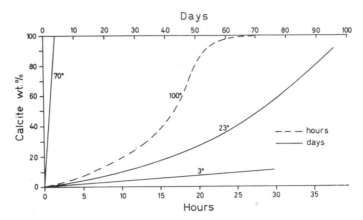

Fig.219. Rate curves for wet transformation of aragonite to calcite. (100°C after FYFE and BISCHOFF, 1965; others after W. H. Taft in CHILLINGAR et al., 1967a.)

consistent with a twofold process involving (*1*) the steady production of nuclei, and (*2*) continuous growth of existing nuclei. In the wet transformation the surface area available for the growth of nuclei increases, whereas in the dry reaction it remains essentially constant. The tail at the completion end of the curve is "poetic licence" (FYFE and BISCHOFF, 1965, p.10), since it cannot be accurately fixed, yet it follows from the necessary reduction and final disappearance of the aragonite. The same two authors noted that the dissolution of aragonite cannot be the rate controlling process because, if it were, the transformation would start by decelerating. The calcite product consists of "rhombohedra of amazingly uniform size" (see also NANCOLLAS and PURDY, 1964, p.18), and this suggests that small crystals grow quickly but older, larger ones grow slowly albeit steadily. "Growth on line or point defects is suggested." (FYFE and BISCHOFF, 1965, p.11.) LAND (1966) has shown how variable the transformation rate can be, even among replicate samples. Variation of the conditions seems to have little effect.

Later results by BISCHOFF and FYFE (1968) indicate that, during wet transformation of micron-sized aragonite crystals at 108°C, the calcite probably nucleated heterogeneously, either on the surface of the aragonite or within the adjacent double layer. The nucleation rate was proportional to the surface area of the aragonite. As the reaction rate increases with time, the rate controlling processes must be the nucleation and growth of calcite (BISCHOFF, 1969).

Some experiments on the wet conversion of aragonite to calcite were conducted by PRUNA et al. (1948), working with finely powdered aragonite in a cylinder at 350–400°C. The progress of the transformation was followed by recording the expansion of the powder specimen.

Pruna and his co-workers noted that the time spent by the specimen at elevated temperature was divisible into two parts, a period of induction or germination during which no expansion took place, followed by a period of transformation in which all the expansion occurred (see also CHAUDRON, 1954). The addition of calcite (intimately mixed) was marked by a reduction of the germination period and an increase in the rate of change in the transformation period. The effect was greatest when the calcite was most finely ground. The substitution for calcite of $NaNO_3$ (which is isomorphous with it) had a similar effect (noted also by CHAUDRON, 1954).

The authors interpreted their data as indicating growth of calcite either on fresh nuclei formed during the germination period or on crystals of calcite or $NaNO_3$ added to the mixture.

It appears that, when calcite crystals were emplanted in the mixture, growth of new calcite took place on these by heterogeneous nucleation. It is well known that inoculation of supersaturated solutions with seed crystals causes further growth to take place on the new crystals in preference to the formation of new nuclei. Ions must have been transported from the aragonite to the calcite and this implies a process of dissolution and precipitation.

The presence of Sr^{2+} in aragonite has been thought, by a number of writers, to give the mineral additional stability (STROMEYER, 1813, p.34, or in MELLOR, 1923, p.815; JOHNSTON and WILLIAMSON, 1916; SIEGEL, 1960). FYFE and BISCHOFF (1965, p.6) neatly stated the case for the stabilization of a particular polymorph as a result of the absorption of foreign ions. They considered the two polymorphs, aragonite and calcite, each of which can hold some Sr^{2+}. They went on to consider a P–T state in which the two pure phases are in equilibrium. If either one or both should be impure while in this state, then the more impure polymorph will be stabilized. In this particular example aragonite would normally be the more impure if Sr^{2+} were the foreign ion, because this large ion fits more readily into the aragonite lattice. The shift of equilibrium may be regarded as a reflection of the conditions of growth of aragonite and calcite crystals in a solution containing Sr^{2+}. Owing to the greater ease with which Sr^{2+} enters the orthorhombic lattice, the growth of aragonite rather than calcite will be encouraged by the combined activities of Ca^{2+} and Sr^{2+}. The free energy of mixing of the Sr^{2+} in the aragonite will be greater than in calcite, thus favouring the aragonite. However, this only works if the activity of Sr^{2+} is high enough. The quantity of $SrCO_3$ in the aragonite must exceed a critical value of at least 15 mole % if the mineral is to be stabilized relative to calcite, in P–T conditions where the pure minerals are in equilibrium (MACDONALD, 1956; GREEN, 1967). This value is far in excess of the maximum known content, which is only 3.87 mole % (DEER et al., 1962). Thus it is unlikely that aragonite precipitated in a sedimentary environment could be made stable relative to calcite in this way. Indeed, KITANO and KAWASAKI (1958) precipitated $CaCO_3$ from solutions with a range of molar ratios Sr^{2+}/Ca^{2+} from 0.01 to 0.6 (open sea water is about 0.009) and found that, as the concentration of Sr^{2+} was increased, the amount of aragonite in the combined aragonite-calcite precipitate decreased. Data for one series of experiments (their table 2) are summarized in Fig.220.

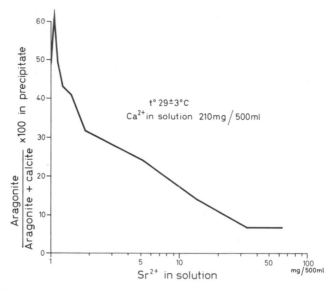

Fig.220. Precipitation of CaCO₃ from solutions with different contents of Sr²⁺, showing that the amount of aragonite in the precipitate of aragonite and calcite decreases as the concentration of Sr²⁺ increases. (From KITANO and KAWASAKI, 1958, table 2.)

Fyfe and Bischoff also drew attention to the effects that dissolved salts or CO_2 may have in accelerating the transformation of aragonite to calcite. Added ions increase the ionic strength of the solution and so reduce the activity coefficients of Ca^{2+} and CO_3^{2-}. Thus to maintain the activity product $aCa^{2+} \cdot aCO_3^{2-}$, the concentrations of these ions in the solution must be increased. The addition of CO_2 drives equations 1, 2 and 3 on p.231 to the right and increases the capacity of the solution for Ca^{2+} and CO_3^{2-}. At equilibrium with the aragonite, the ion activity product $aCa^{2+} \cdot aCO_3^{2-}$ is constant for all solutions, but $aHCO_3^-$ plays no direct part in this equilibrium. It can, therefore, vary independently. All accelerating additives increase the product $aCa^{2+} \cdot aHCO_3^-$. The rate controlling process here may be nucleation or growth, or both.

FACTORS AFFECTING CARBONATE EQUILIBRIA

Surface layer reactions

The seemingly anomalous equilibria in sea water between three or more solid carbonate phases which, in terms of the thermodynamic parameters of their main crystal lattices, are metastable, has been remarked upon by many workers. There is clearly a steady state equilibrium among these carbonate phases which has persisted for thousands of years, since there is almost no petrographic evidence that one phase has been transformed into another. (Electron photomicrographs referred to on p.247 suggest a very slow dissolution of magnesian calcite over possibly thousands of years yielding, by corrosion, a topographic relief on grains of the order of 10 mμ.) If the surfaces of the solid phases had the same thermodynamic properties as the relatively vast masses of their interiors, then the multiphase assemblage on the sea floor should change to low-magnesian calcite and dolomite (p.250). Since this has not happened, despite the small crystal size and large surface area of, for example, the Florida Bay muds, it is necessary to look for factors that make for equilibration among these mineral phases.

That the surface layers of a crystal, with their unneutralized bonds, have different free energies of formation from the main body of the crystalline interior has long been apparent. LENNARD-JONES and DENT (1928) established that, in an ionic crystal, the lattice near the surface experiences a contraction amounting to 5% of the normal inter-atomic distances. Anything that causes a distortion of the lattice must make it more soluble, whether the distortion be a simple contraction or is brought about by the absorbtion of a foreign ion or is the result of an accumulation of dislocations during rapid growth. Growth of a face is also affected by the surface adsorption of ions foreign to the lattice. BROOKS et al. (1951) were able to inhibit the growth of calcite crystals by introducing Calgon into the solution. As more Calgon was added the adsorption of phosphate eventually stopped calcite growth and the ensuing supersaturation was accompanied by the precipitation of the metastable hexahydrate, $CaCO_3.6H_2O$. WEYL (1967) has recently emphasized the crucial need to bear in mind that the solution can only "see" the surface layer of the crystal and that this surface may have a stability significantly different from the main volume of the crystal (of which it forms an infinitesimally small part). Furthermore, as Weyl showed by his experiments on oöid growth (p.303), the behaviour of a surface in a solution depends on the past history of the surface. Much that has previously seemed inexplicable in the solution kinetics of carbonates has become comprehensible when the behaviour of crystal surfaces has been taken into account.

Inhibiting effect of Mg^{2+} on calcite precipitation

In attempting to understand why aragonite is precipitated in some conditions and calcite in others CURL (1962) summarized the situation thus: "Calcium carbonate nuclei promote the same polymorph by epitaxy or oriented overgrowth; super-saturation with respect to *both* calcite and aragonite is a necessary condition for aragonite precipitation; the inhibition of calcite growth is consequently a necessary condition for aragonite precipitation." During inorganic precipitation the chief inhibitor appears to be the Mg^{2+} ion.

An important approach to the problem of the stability of the $CaCO_3$ polymorphs in sea water has come from those who have experimented with solutions containing selected ions in addition to the Ca^{2+} and CO_3^{2-}. These have included Li^+, K^+, Na^+, Mg^{2+}, Sr^{2+}, Pb^{2+}, Ba^{2+}, NH_4^+, Cl^-, NO_3^- and SO_4^{2-}. The presence of these ions in the solution affects the mineral equilibria (e.g., SAYLOR, 1928; ZELLER and WRAY, 1956; WRAY and DANIELS, 1957; LIPPMANN, 1960; KITANO, 1962a; KITANO and HOOD, 1962; SIMKISS, 1964; FYFE and BISCHOFF, 1965). The ion that has the greatest ability to inhibit the growth of calcite is Mg^{2+} and its influence was known to Leitmeier as long ago as 1910. Kitano and Simkiss found that Mg^{2+} had a controlling effect on the form of the precipitate. Kitano evaporated solutions of $Ca(HCO_3)_2$ with and without various salts in the proportions in which these are found in sea water. In the presence of $MgCl_2$ only aragonite was formed, in its absence only calcite. USDOWSKI (1963) found that increased Mg^{2+}/Ca^{2+} in the solution furthered the precipitation of aragonite rather than calcite.

LIPPMANN (1960) had earlier noted the inhibiting effect of Mg^{2+} on calcite growth. He precipitated $CaCO_3$ at atmospheric pressure and various temperatures including room temperature using the reaction:

$$KOCN + 2H_2O \rightarrow K^+ + NH_4^+ + CO_3^{2-} \tag{14}$$

and introducing Ca^{2+} and Mg^{2+} as either the chloride or sulphate. He supposed that the well known reluctance of calcite to grow in aqueous solutions containing Mg^{2+} arises from the difference in the standard free energies of formation of the hydrated Ca^{2+} and Mg^{2+}. He noted that $\Delta F°$ of hydration is for Ca^{2+} -428 kcal./mole and for Mg^{2+} -501 kcal./mole. As the hydrated Ca^{2+} and Mg^{2+} gather at the surface of a crystal, Ca^{2+} is more readily released from its water dipoles and joins with CO_3^{2-} to construct the least soluble lattice, which is calcite. If the activity of Mg^{2+} in the solution is high enough, the lattice so built is necessarily of magnesian-calcite. The Mg^{2+} ions, still hydrated, are adsorbed on the surface of the nucleus. However, the work that must be done to dehydrate them is such that the $\Delta F°_f$ of aragonite (which holds very little Mg^{2+}) is, in these circumstances, greater than for magnesian-calcite. So aragonite crystals grow while the growth of calcite nuclei is prevented by the obstructive adsorption of hydrated Mg^{2+}. DAVIES and JONES (1955) reached a similar conclusion as a result of observations on the growth rate of AgCl seed crystals. The reaction appeared to be interface-controlled and they postulated an adsorbed monolayer of hydrated ions at the crystal surface. They suggested that growth took place as a result of simultaneous dehydration of Ag^+ and Cl^- ions.

It is possible to quantify the hypothesis of Lippmann, in a crude fashion, and the resultant calculations (BATHURST, 1968) suggest that, in sea water wherein the choice of precipitate is between aragonite and a high-magnesian calcite, the conditions favour the precipitation of aragonite. In this argument we are concerned with two successive reactions, the endothermic dehydration of cations and the exothermic formation of a crystal lattice.

Lippmann's central point was the greater free energy of hydration of Mg^{2+} over Ca^{2+}. It seems reasonable, also, to assume that the free energy of formation of any high-magnesian calcite is less than that of pure calcite, because the high-magnesian calcites are more soluble, at atmospheric P–T, and because the free energy of formation of magnesite (the extreme high-magnesian calcite?) is less than that of calcite.

As a preliminary check, Lippmann's values for free energies of hydration, -501 kcal./mole for Mg^{2+} and -428 kcal./mole for Ca^{2+}, can be compared with values for the standard enthalpies of hydration given by the U.S. Bureau of Standards. These are -456 kcal./mole for Mg^{2+} and -377 kcal./mole for Ca^{2+}. The difference in each case is closely similar and, since Lippmann's values have a smaller difference and so give less support to his hypothesis, they will be used.

It will be assumed purely for convenience that the high-magnesian calcite, if it were precipitated, would have 10% of Mg^{2+} in its cations. If all the cations adsorbed on the crystal face were Mg^{2+} then the free energy of hydration would increase above that for calcite by $-(501 - 428) = -73$ kcal./mole. As the Mg^{2+} is only 10% of the cations, it will be assumed that the increase would be -7.3 kcal./mole. Therefore, the estimated free energy of hydration of the adsorbed cations $(Ca_{90}^{2+} + Mg_{10}^{2+})$, which will take part in the construction of the magnesian calcite lattice, is $-(428 + 7.3) = -435.3$ kcal./mole. This compares with -428 kcal./mole for the unadulterated Ca^{2+} involved in the growth of pure aragonite.

Although the standard free energies of formation of magnesian calcites are not known, the value for magnesite can be used as a rough guide. This is -246 kcal./mole, lower by $+23.8$ than the value for pure calcite which is -269.8 kcal./mole. Thus we can guess that the standard free energy of formation of our magnesian calcite is smaller than the value for pure calcite by an amount equal to one tenth of the difference between the two, so $-269.8 + 2.38 = -267.4$ kcal./mole (a figure confirmed by WINLAND, 1969; see also p.258).

Treating the crystal growth as a two stage process involving endothermic dehydration $(+ \Delta G_h)$ and exothermic lattice building $(- \Delta G_f)$, the growth of our aragonite and high-magnesian calcite can now be expressed, in terms of ΔG, the change in Gibb's free energy, as follows:

$$\Delta G_{aragonite} = + \Delta G_h - \Delta G_f \qquad \Delta G_{magnesian\ calcite} = + \Delta G_h - \Delta G_f$$

$$= + 428 - 269.6 \qquad\qquad\qquad = + 435.3 - 267.4$$

$$= + 158.4\ kcal./mole \qquad\qquad = + 167.9\ kcal./mole$$

These estimates favour the precipitation of aragonite by a balance of about 10 kcal./mole. The preceeding calculation is admittedly crude and inelegant, but it serves to illustrate the principles involved. The standard free energies of formation of pure calcite and pure aragonite are close together, differing only in the

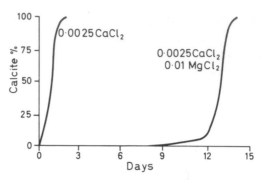

Fig.221. Rate of wet transformation aragonite → calcite in a solution of CaCl₂ and a solution of CaCl₂–MgCl₂, at a temperature near to 100°C. (From FYFE and BISCHOFF, 1965, fig.10.)

first decimal place, so that relatively small changes in the environment, such as the introduction of Mg^{2+} into the solution, would be expected to have important consequences.

Of particular interest are the data by FYFE and BISCHOFF (1965, p.11) on the inhibition of the wet transformation of aragonite to calcite by dissolved Mg^{2+}. They studied the wet transformation of aragonite to calcite at 100°C in solutions of 0.0025 molar $CaCl_2$ (to give some acceleration) with and without 0.01 molar magnesium. The transformation without Mg^{2+} is shown on the left in Fig.221. When Mg^{2+} was added the transformation at first went slowly, for nearly three hours, before continuing as if no Mg^{2+} had been present. Clearly Mg^{2+} is withdrawn from the solution during the early stages of the reaction, and this is born out by the discovery that the early formed calcite contains $MgCO_3$ in solid solution. They explained this on the grounds that, where the embryonic calcite nucleus has a high Mg^{2+} concentration, the probability is high that it will form dolomite or a magnesian calcite. It is also apparent from the work of BISCHOFF (1968b) that, where small quantities of Mg^{2+} are used, the Mg^{2+} is removed from the solution during the induction period and must therefore be absorbed into the calcite phase. (All aragonite is removed by the end of the transformation.) It seems that, as the new calcite nuclei form, the small quantity of Mg^{2+} is continuously removed from the solution until its concentration is too dilute for further interference in calcite growth. Thence calcite growth proceeds at the rate normal in the absence of Mg^{2+}.

WEYL (1967, p.180) measured the rates of equilibration of calcite with respect to solutions of various compositions and noted that the presence of the divalent cations Ca^{2+} and Mg^{2+} in the solution caused the rate of equilibration to be slower. The effect of NaCl (the major constituent of sea water) was relatively slight. He went on to show that the delayed equilibration only took place in the presence of solid carbonate, being absent in reactions between solutions. He concluded that the rate-inhibiting mechanism must be at the solid-solution interface and must involve the two cations, Ca^{2+} and Mg^{2+}, which take part in the construction of the magnesian calcite lattice.

BERNER (1967), in searching for a clearer understanding of the inhibiting effect of dissolved Mg^{2+} on the equilibration of carbonate grains with their enclosing solutions, bubbled CO_2 through a suspension of carbonate mud in solutions of various compositions at 30 °C and atmospheric pressure. Equilibration was taken as reached when the pH attained a steady state. The mud, from Florida Bay (p.161), consisted of aragonite, a high-magnesian calcite with 14 mole % $MgCO_3$, and a little low-magnesian calcite. No solid phase was ground and the surfaces were, therefore, free of strain of the kind described on p.234, 257. Many of the equilibrated solutions were analyzed chemically for Ca^{2+} and Mg^{2+}. The reactions were carried out at P_{CO_2} = 1 atm and P_{CO_2} = $10^{-2.5}$ atm: at the lower P_{CO_2} the reactions were much slower, as would be expected from equations 1 to 6. It should be noted that equilibration was approached from *undersaturation*. The solution properties of the Florida Bay mud were compared with those of carefully standardized calcite and dolomite. Electron photomicrography had already shown that the surfaces of many calcite mud grains were apparently corroded and this suggested that simple reversible solubility was not to be anticipated. Comparisons of the theoretical equilibrium pH for aragonite, calculated from standard free energy data, and the equilibrium pH actually obtained in the experiments, showed large and variable differences. On the basis of an examination of these Berner reached three conclusions: (*1*) the sediment does not exhibit simple reversible equilibria, (*2*) the steady state pH in distilled water rises as the ratio (A_c) of the solid surface area to dissolved carbonate is increased, and (*3*) in solutions bearing Mg^{2+}, the steady state pH is depressed always below the calculated pH. Relevant data are given in Fig.222 where IAP (ion activity product = $aCa^{2+} \cdot aCO_3^{2-}$) is a function of pH.

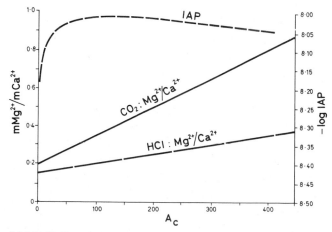

Fig.222. Plot of Mg^{2+}/Ca^{2+} and Ion Activity Product against A_c (explained in text) for Florida Bay sediment in distilled water. CO_2 plot for long runs, 5–25 h, in CO_2. HCl plot for rapid leaches, about 3 min, in dilute HCl. (After BERNER, 1967.)

The non-reversibility of the equilibrium is also shown by X-ray diffraction analyses of the muds before and after partial dissolution in sea water: these reveal a preferential loss of high-magnesian calcite.

Interpreting the increase of Mg^{2+}/Ca^{2+}, pH and IAP with increasing A_c in distilled water, Berner remarked that rapid leaches of the mud with dilute HCl for 2–3 min yielded some increase of Mg^{2+}/Ca^{2+} with A_c (Fig.222), but distinctly less than that shown by the longer runs with CO_2 (5–25 h). If the HCl plot can be assumed to represent rapid, nearly congruent dissolution, then the CO_2 plot must record definitely incongruent dissolution of the high-magnesian calcite (p.336). It seems that the partial dissolution mainly involves the magnesian calcite and not aragonite. The dissolution of Mg^{2+} responsible in part for the levelling of the IAP must be a rapid, short-term effect. Cations of Mg^{2+} can be rapidly removed from a restricted thin zone near to the solid surface, but diffusion of Mg^{2+} from deeper in the solid lattice is a very much slower process. Thus the release of Mg^{2+} by superficial, incongruous dissolution is strongly dependent on the surface area of the solid in contact with the solution. The greater the area, the more likely it is that a steady state pH will be reached before subsequent congruent dissolution can yield much Ca^{2+} to the solution. Two IAP's are involved, $aMg^{2+} \cdot aCO_3^{2-}$ and $aCa^{2+} \cdot aCO_3^{2-}$. Relatively rapid increase of the first, by fast dissolution of $MgCO_3$, will further restrict the slower dissolution of $CaCO_3$ because of the common ion effect of the CO_3^{2-} in the solution.

The depression of the steady state pH by Mg^{2+} in 3 (above) Berner attributes to a reduced dissolution rate of $MgCO_3$ from the calcite resulting from the common ion effect of Mg^{2+} in the sea water. There must also be a slowing of dissolution owing to the inhibition by Mg^{2+} of the precipitation of low-magnesian calcite.

BERNER (1966b) has obtained direct evidence of a tendency for the appropriate divalent ions of sea water, Mg^{2+} and Ca^{2+}, to be adsorbed on the surfaces of carbonate grains. He took clay-grade sediment from Florida Bay (p.161), consisting of aragonite with 15-mode mole % of magnesian calcite, and washed it free from soluble sea salt in distilled water. He then bathed the sediment in a solution, either of $MgCl_2$ or $CaCl_2$, strong enough to yield, after the grains had reached equilibrium, ion ratios in the solution such that mMg^{2+}/mCa^{2+} or mCa^{2+}/mMg^{2+} $\approx 50/1$. He then treated these reacted samples with very small volumes of solutions of, respectively, $CaCl_2$ or $MgCl_2$, small enough for the subsequent uptake of Mg^{2+} or Ca^{2+} from the grain surface to be readily measurable by titration with EDTA. He found that, when the grains were immersed in the large volume of solution, the solute cation was taken up by the grains, presumably on their surfaces, to be released again when they were treated with small volumes of solution of the other cation. During treatment of grains with the large volumes of solution, equilibrium between solution and grain surfaces was reached between 10 min and 24 h. The adsorption of cations was exactly reversible when the solutions were exchanged, adsorbed Mg^{2+} being replaced by adsorbed Ca^{2+} and vice-versa. The ratio mMg^{2+}/mCa^{2+} on the grain surfaces after equilibrium was very close to the ratio in the solution, indicating a high capacity for ion exchange. This capacity must be related to the fine grain size of the sediment ($< 2\ \mu$) and the rough and porous surface of the grains (as seen with an electron microscope). The distribution of ions over the surfaces is not known. Certain sites may be favoured and Berner suggested a preferential adsorption of Mg^{2+} on magnesian calcite. He concluded on the basis of his experimental data that, since the molar ratio in sea is $mMg^{2+}/mCa^{2+} \approx 5/1$, the ratio on the surfaces of the Florida Bay muds must *average* about 3/1.

FYFE and BISCHOFF (1965) found evidence for the blocking of active sites on a calcite face by Mg^{2+}. It is helpful to compare these ideas on cation adsorbtion with the order of decreasing solubility at low P–T: high-magnesian calcite–aragonite–low-magnesian calcite (Fig.218). The Mg^{2+}, by virtue of its greater $\Delta F°$ of hydration, must escape more readily than Ca^{2+} from the lattice into solution and so enhance the solubility of high-magnesian calcite.

Studies of nucleation of $CaCO_3$ at room temperature and pressure by PYTKOWICZ (1965a) showed that the presence of Mg^{2+} in solution caused a very large increase in the induction period (time between addition of Na_2CO_3 to natural and artificial sea waters and onset of precipitation). For example, after the addition of a certain quantity of Na_2CO_3, precipitation in the magnesium-free artificial sea water began after 1 min, but in natural sea water it was delayed for 6–7 h. Pytkowicz, thinking in terms of homogeneous nucleation, suggested that "many more collisions of carbonate ions are necessary to form nuclei in the presence of magnesium . . . The concentration of magnesium ions in sea water is five times the concentration of calcium ions, and collisions that form magnesium–calcium-carbonate aggregates are more probable than collisions that form pure calcium carbonate nuclei." There was some evidence that complexes of $MgCO_3°$, dissolved ion-pairs, obstruct the growth of pure $CaCO_3$ (aragonite) nuclei. While these results illustrate the effectiveness of Mg^{2+} in inhibiting the growth of $CaCO_3$, Pytkowicz also showed, by extrapolation of his data to natural sea water, that homogeneous nucleation of $CaCO_3$ would have an induction period of more than 70,000 years. He concluded, therefore, that inorganic precipitation by homogeneous nucleation must be insignificant in sea water. Shorter induction periods, of the order of hundreds of hours, were likely if nuclei (such as oöids) were available for heterogeneous nucleation of overgrowths, more especially if vigorous photosynthesis of marine plants caused rapid abstraction of CO_2 from the water.

Interesting experiments by DE GROOT and DUYVIS (1966) showed that Mg^{2+} and Ca^{2+} are equally subject to adsorption on calcite, but Mg^{2+} is adsorbed on aragonite to a much lesser extent than Ca^{2+}.

AKIN and LAGERWERFF (1965b) found that the addition of Mg^{2+} or SO_4^{2-} to synthetic and natural solutions of $CaCO_3$ enhance its solubility. They devised a model based on Langmuir adsorption of all the ions on a crystal surface. The theoretical surface consists of adjacent patches of calcite and a more soluble magnesian calcite. Ions can diffuse freely over the surface from one patch to another. The theoretical equation relates the product $aCa^{2+} \cdot aCO_3^{2-}$ to the concentration ratios $[Mg^{2+}]/[Ca^{2+}]$ and $[SO_4^{2-}]/[CO_3^{2-}]$ at equilibrium. It yields the ratio of solubilities for the calcite and magnesian-calcite surfaces. The theoretical data are in good agreement with solubility data for synthetic and natural solutions, despite the fact that the Langmuir surface is a simplification of the system of mobile screw dislocations at a crystal face. BERNER (1967, p.55) remarked that part of the enhanced solubility attributed by these authors to surface reaction with Mg^{2+} and/or SO_4^{2-} during precipitation can be explained by the formation of the dissolved ion-pairs $MgCO_3°$ and $CaSO_4°$. He emphasized, also, that excess strain caused by grinding, or initial supersaturation, could give the energy for the rapid development of supersoluble, Mg-enriched surface overgrowths.

WOLLAST (in press) emphasized that the free energies of formation of calcite and aragonite, and therefore their dissolution and precipitation, depend on the amount of Sr^{2+} and Mg^{2+} in solid solution. Basing his calculations on the premise that Sr^{2+} and Mg^{2+} ions in solution respectively promote the growth of aragonite and calcite, he went on to show by calculation that the stability fields are determined by the ratio Sr^{2+}/Ca^{2+} in aqueous solution.

The simplest explanation of the inhibition of calcite growth in the presence of Mg^{2+} is by P. K. Weyl (personal communication, 1969) and follows from the results in his 1947 paper. The Mg^{2+} ions enter the new calcite lattice so that the overgrowth is a high-magnesian calcite which, being more soluble than aragonite, is unstable.

There is a useful discussion of the possible effects of absorbed Mg^{2+} on carbonate solubility in BERNER (1966a, p.30).

Inhibiting effect of Mg^{2+} adsorption on dolomite growth

From the known composition of sea water (Table XII, p.272), if the water evaporates then dolomite should, theoretically, be the first mineral to precipitate (Fig.223). What happens is that the first mineral to come down is neither dolomite, nor even calcite which, after all, is the least soluble of the calcium carbonates. It is, instead, aragonite. The precipitated needle muds and the cement in beach rock, for example, are a mineral which, at 25°C, is theoretically only stable at a pressure greater than 4 kb. According to the standard free energies of formation of aragonite and dolomite, any aragonite in sea water should react with Mg^{2+} to give dolomite, thus:

$$2CaCO_3 \quad + Mg^{2+} \rightarrow CaMg(CO_3)_2 + Ca^{2+}$$

$$- 2 \times 269.6 - 108.99 \quad - 520.0 \quad - 132.18 \tag{15}$$

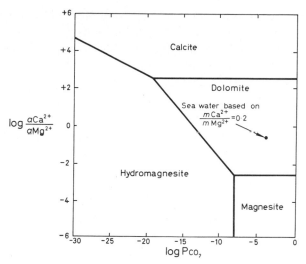

Fig.223. Stability of some calcium-magnesium carbonates in aqueous solution, at 25 °C and 1 atm, related to Ca^{2+}/Mg^{2+} ratio and partial pressure of CO_2,. (From GARRELS et al., 1960, fig.5.)

The net reduction of free energy is only -3.99 kcal./mole, however, so that the driving force behind the reaction is small. So it is not, perhaps, surprising that the behaviour of this simple system is readily upset. In fact, aragonite is not dolomitized until log aCa^{2+}/aMg^{2+} falls to about -2.95 (Fig.223). This happens in the sabkha along the Trucial coast of the Persian Gulf (i.e., when $mMg^{2+}/mCa^{2+} = 11$; KINSMAN, 1964b). Assuming that the partial pressure of CO_2 in the sabkha solutions is not greatly different from that of the adjacent lagoon, the critical log aCa^{2+}/aMg^{2+} lies about the boundary between the stability fields of dolomite and magnesite in Fig.223. This may be no coincidence. It is necessary to bear in mind that the growth of dolomite involves the construction of basal cation planes containing alternately Ca^{2+} and Mg^{2+}. It may be that the tightly hydrated Mg^{2+} ions cluster about the growing Ca^{2+} planes of dolomite just as Lippmann suggested they do on calcite (p.244). In so doing they would hinder the arrival of Ca^{2+} and CO_3^{2-} from the solution. When the activity of Mg^{2+} is high enough for magnesite to grow, presumably the hydration of Mg^{2+} is no longer an obstacle. GARRELS et al. (1960, p.418) wrote that "our whole experience with magnesium ion in solutions has impressed us with its reluctance to form solids from saturated solution". They, too, made the point that Mg^{2+} may be abnormally firmly hydrated. They also suggested that it may form an ion-pair in solution ($MgCO_3°$) and in a later paper (GARRELS et al., 1961) they gave experimental evidence in support of this (p.255).

    In discussing the wet transformation of aragonite to calcite in the presence of Mg^{2+} (p.246), FYFE and BISCHOFF (1965, p.11) remarked that, where the calcite nucleus has a high concentration of Mg^{2+}, it will tend to form dolomite or a magnesian calcite. But dolomite grows extremely slowly as a result of the severe statistical requirements which must be satisfied if ions are to take up positions in such a highly ordered structure (p.238). The ordering must, therefore, be accompanied by a large negative entropy of formation compared with that for a magnesian-calcite. "In effect, each nucleus converted to a dolomite or protodolomite nucleus is removed from the battle" (FYFE and BISCHOFF, 1965, p.12). The magnesian calcite lattice grows so much more quickly that alone it will contribute to the early stages of the transformation, during which the Mg^{2+} is withdrawn from the solution. Nevertheless, as the extent of disordered solid solution in magnesian calcite is restricted at such low temperatures (GOLDSMITH, 1959, p.338), the rate of growth will still be slower than for pure calcite.

Effect of history of crystal surfaces on dolomite–calcite equilibria

WEYL (1967, p.194) carried out some experiments similar to the one with oölites (p.303), involving the pH-monitored flow of sea water as it passed through packs of calcite and dolomite at 30 °C and 1 atm with P_{CO_2} at 1 atm. The results showed that the past history of the mineral surfaces is of critical importance. Acid-cleaned surfaces showed calcite to be more soluble than dolomite. Overgrowths on the clean surfaces were then precipitated by adding NaOH to the sea water to increase the CO_3^{2-} as in equation 8 and the sea water was finally analysed for pH, total CO_2, Mg^{2+} and Ca^{2+}. These data showed that the situation was reversed, and that the overgrowth of dolomite was more soluble than the overgrowth of calcite. Moreover, as in the oölite experiment, the successive layers of overgrowth were progressively more soluble, until the final accumulations were so soluble that any addition would immediately have gone into solution.

This relation is important with regard to the question of dolomitization by sea water. If clean crystals are put into sea water, the dolomite will begin to grow at the expense of the more soluble calcite, but soon its overgrowth will be equally soluble with the calcite surface. At this stage the surfaces of the two minerals can coexist in a pseudoequilibrium. Dolomite cannot dissolve because pure dolomite is less soluble than calcite. Dolomite cannot grow because its overgrowth would be more soluble than both clean calcite and overgrowth of calcite.

Influence of organic films on particle–water reactions

It is becoming increasingly apparent that chemical reactions between particles and their enclosing solutions, more especially in sea water, are either prevented or modified by the existence of a film of organic matter which covers, more or less completely, the surfaces (external and internal) of sedimentary particles (CHAVE, 1965; SHEARMAN and SKIPWITH, 1965). The surfaces of particles also bear colonies of a variety of micro-organisms, such as diatoms, blue-green algae, yeasts, bacteria and the early stages of brown seaweeds (Phaeophyta). These all tend to lodge in hollows and cracks in the surface (MEADOWS and ANDERSON, 1966, 1968). The organic substance is commonly pale brown: it may be rendered visible by dyeing with malachite green or methyl blue. Chave showed that natural carbonate particles in sea water did not react with the water when it was acidified with dilute HCl, although clean calcite did react. Attempts to get particles to react with sea water, by heating cold waters or cooling warm waters (stirring the while), failed to cause precipitation or dissolution. Chave remarked on the anomalous stability of aragonite and a range of magnesian calcites in sea water, and upon the supersaturation of near-shore Bermudan waters for all but the most soluble phases in summer, and undersaturation in winter. In a later paper, CHAVE and SUESS (1967) showed that marine carbonate particles failed to react with sea water *unless* they were first treated with H_2O_2, when reaction took place. Again, in a recent paper, CHAVE (in press) found, in attempting to increase the level of supersaturation in sea water and so to cause precipitation, by adding $CaCl_2$ or Na_2CO_3, that $CaCO_3$ does not precipitate until *after* some of the dissolved organic compounds have been precipitated. These latest experiments are believed to indicate that $CaCO_3$ cannot precipitate from sea water containing the normal amount of dissolved organic compounds. In algal mats (p.202 and 217) the barrier between carbonate particle and sea water must be especially restrictive. SUESS (1968) has determined in the laboratory that organo-carbonate associations form between stearic acid in a hexane solution and calcite and dolomite, also between albumin and aragonite, calcite and magnesian calcite. More or less complete organo-carbonate monolayers form on carbonate particles (see also SUESS, 1970). It is apparent that sea waters with different concentrations of dissolved polar organic compounds show different degrees of inorganic chemical reaction with carbonate: the ratio of the surface area of carbonate grains to the concentration of dissolved organic compounds is obviously a critical factor controlling the extent to which particles are wholly or only partly coated. The coating process seems, moreover, to be mineral-selective. Possible organic compounds involved in the coating of particles in sea water are fatty acids, fatty esters and fatty alcohols.

These ideas lately received impetus from work by MITTERER (1969) in which he drew attention to the close similarity of the protein compositions of mollusc shells, the skeletons of calcareous algae and oöids as revealed in chromatograms. If organisms use proteins either as templates for heterogeneous nucleation of mineral carbonate or as a means of attracting the appropriate inorganic ions electrostatically, then the organic matter in oöids (MITTERER, 1968; TRICHET, 1968) could serve a similar purpose. They are known to be stuffed with algae. There seems no doubt that in organisms a protein made up largely of acidic amino acids is important in calcification. Could similar proteins be influential outside the organism? It seems that we shall have to pay greater attention in the future than heretofore to the extensive researches being carried out into the processes of intracellular calcification. Valuable clues to the understanding of so-called inorganic precipitation may there be found, as in work by STOLKOWSKI (1951), KITANO (1964) and KITANO and HOOD (1965).

The occurrence of organic matter in ancient limestones (p.275) is well known (DANGEARD, 1936; NESTEROFF, 1955b, 1956b; GUNN and COOPER, 1964; CHAVE, 1965; SHEARMAN and SKIPWITH, 1965). Its effects on subsurface diagenesis are likely to have been as critical as those of its counterparts on reactions on the present sea floor or among suspended particles. In this scarcely developed field of the influence of organic films on particle–water inorganic reactions, it is probable that future work will show that organo–carbonate reactions are at least as important as the inorganic surface reactions discussed earlier in this chapter.

Enhanced solubility of very small grains

The question of the enhanced solubility of very small crystals, compared with the solubility of large masses of stable lattice, can be approached quantitatively. The various factors responsible for the extra free energy of small spheres have been combined into a single term by FYFE and BISCHOFF (1965, p.6) to give the total free energy ΔG_r:

$$\Delta G_r = \Delta G + \frac{2\sigma}{r} V \qquad (16)$$

where σ is the surface tension of the solid, V its volume per mole. The value of σ has to be determined experimentally. SCHMALZ (1963) has found empirical values for calcite such that:

$$- G_{r \; calcite} = \Delta G + \frac{1.2 \times 10^{-3}}{r} \qquad (17)$$

Significant increase in solubility can only arise where the radius is less than 10^{-3} to 10^{-4} cm, entering the colloidal range.

CHAVE and SCHMALZ (1966, p.1045) concluded from their own experiments, that particles of calcite with diameters less than 0.1 μ are quite unstable in water and that an enormous increase in solubility is to be expected for particles between 0.1 μ and 0.01 μ. The possible effect of enhanced solubility of small particles on the level of saturation is discussed on p.258.

Complexing in sea water

That the activity of CO_3^{2-} in sea water is a small fraction of its molality has long been established. The reason for this lies in the formation of dissolved complexes (GREENWALD, 1941) such as $MgCO_3^{\circ}$ within which large quantities of CO_3^{2-} are incarcerated and can therefore play no part, as free ions, in the solution. (The superscript ° indicates a dissolved ion-pair.) Ideas about the role of complexes in reducing the activities of CO_3^{2-} and HCO_3^- in sea water have been summarized and considerably developed through experimental work by GARRELS et al. (1961).

In order to study the activity coefficients of HCO_3^- and CO_3^{2-}, the authors made up solutions of $NaHCO_3$ and $Na_2CO_3 + NaHCO_3$, containing HCO_3^- and $HCO_3^- + CO_3^{2-}$ respectively. They then added various quantities of NaCl and $MgCl_2$, representing the dominant cations in sea water, either singly or as mixtures.

Within the range of ionic strengths to be expected in sea water (up to an ionic strength of 1), they found that the activity coefficient γHCO_3^- falls to around 0.4 when NaCl or $MgCl_2$ are added to the solution: this compares with an expected minimum value, from Debye-Hückel theory, of 0.65 for solutions without complexes. The reduction of γCO_3^{2-} is very much greater, and it moves from about 0.1 (without Mg^{2+}) to the low value of about 0.01 for an ionic strength of 1.0. Mixtures of $NaCl-MgCl_2$ resembling sea water yield values of γCO_3^{2-}/ionic strength close to those found for sea water (SVERDRUP et al., 1942, p.200). It must be assumed, therefore, that since these two cations are the dominant ones in sea water, it is they that have the main influence on the behaviour of the two activity coefficients: it is apparent that Mg^{2+} has a stronger effect than Na^1.

Investigating possible complexes, the authors found that $MgCO_3^°$ and $NaCO_3^-$ best fitted their data. They went on to calculate the equilibria:

$$K_{MgCO_3^°} = \frac{aMg^{2+} \cdot aCO_3^{2-}}{aMgCO_3^°} \quad \text{and} \quad K_{NaCO_3^-} = \frac{aNa^+ \cdot aCO_3^{2-}}{aNaCO_3^-}$$

Where these constants are plotted against ionic strength as in Fig.224, it is seen that they differ by a factor of 100, giving best fit dissociation constants of $K_{MgCO_3^°} = 4 \times 10^{-4}$ and $K_{NaCO_3^-} = 5.6 \times 10^{-2}$. GARRELS et al. (1961) estimated that aCO_3^{2-} is reduced to about 2% of its molality as a result of the addition of the appropriate amounts of NaCl and $MgCl_2$. The action of electrostatically charged particles of finite size (as considered in Debye-Hückel theory) reduces aCO_3^{2-} only to about 20% of its molality. The effect of the complex $NaCO_3^-$ is small compared with that of $MgCO_3^°$. Their calculations indicate that the total CO_3^{2-} determined by titration is roughly distributed as follows:

10 mole % is free CO_3^{2-}
15 mole % is in $NaCO_3^-$
75 mole % is in $MgCO_3^°$

These figures are remarkable indeed, remembering that the molality of Mg^{2+} in sea water is a mere 0.054 compared with Na^+ at 0.48.

The formation of the $MgCO_3^°$ complex greatly alters the apparent second ionization constant of H_2CO_3 (GREENWALD, 1941; also see p.250).

Complexing of calcium with organic compounds is also known to take place.

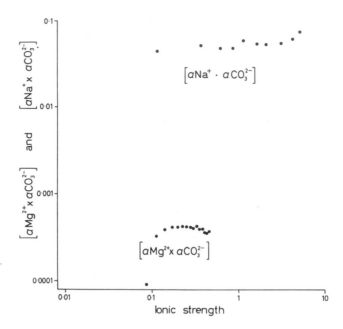

Fig.224. $[a\text{Mg}^{2+}\cdot a\text{CO}_3{}^{2-}]$ and $[a\text{Na}^+\cdot a\text{CO}_3{}^{2-}]$ plotted against ionic strength. (From GARRELS et al., 1961, table 2, 3.)

SMITH et al. (1960, p.430) referred to complexing of CO_2 with proteinaceous substances in sea water. KITANO and KANAMORI (1966) have used small quantities of organic compounds in Mg^{2+}-bearing solutions to permit precipitation of magnesian calcite (p.236).

Relation between saturation of sea water for $CaCO_3$ and mineralogy of the sediment

The level of saturation in sea water can depend, in a very sensitive manner, on the presence of particular carbonates in suspension or on the floor. SCHMALZ and CHAVE (1963) took samples of sea water, and of the adjacent bottom carbonate sediments, off Bermuda. Using the invaluable saturometer method of WEYL (1961), they measured the pH of the water and then added an excess of the sediment with which the water had been in contact. If the pH rose ($+ \Delta$pH) they took this to mean that the water was initially undersaturated for the added solid. If it fell ($- \Delta$pH), this indicated initial oversaturation. The results (Table XI) show that all the water samples, except the deepest from Harrington Sound, are manifestly supersaturated for their associated sediments. They may, however, be in equilibrium with a more soluble carbonate that is absent from the bottom sediment. It is plain that when other solids, prepared from pure skeletons common in the area, were added instead of bottom sediment to the sea water samples, the magnitude of $- \Delta$pH fell as the solubility of the added carbonate increased. The relative solu-

TABLE XI

Locality and depth	Initial pH	Adjacent sediment	Pure and low-magnesian calcites	Aragonite, medium-magnesian calcite	Coarse high-magnesian calcite	Fine high-magnesian calcite
Tobacco Bay (open coast after rain)	8.22	−0.28				
Ferry Reach (shallow, semi-restricted)	8.14		−0.52	−0.22	−0.46*	+0.17
Harrington Sound (enclosed lagoon)						
1.0 m	8.23	−0.40	−0.50	−0.25	−0.22	−0.07
4.5 m	8.22	−0.46				−0.03
24.0 m	7.44	−0.01				+0.21

* Apparently anomalous reading unexplained.

bilities of aragonite and various magnesian calcites are shown in Fig.218, on p.237, after CHAVE et al. (1962). JANSEN and KITANO (1963) have also found that natural high-magnesian calcites are more soluble than aragonite and that this solubility increases with the $MgCO_3$ content. The Bermudan waters were most nearly in equilibrium with the fine high-magnesian calcite which had been ground (Table XI). The deepest water from Harrington Sound is distinctly undersaturated for this material but nearly in equilibrium with its local sediment. It is significant that it is the shallow waters of the surf zone that are most nearly in equilibrium with the finely ground high-magnesian calcite. This seemed to the authors to imply that the shallow waters are in equilibrium with *suspended* carbonate, in a form even finer than the finely ground high-magnesian calcite—a suspended carbonate formed by grain-to-grain abrasion in the surf zone. (Planktonic skeletal calcite is low-magnesian.) Owing to its derivation by grinding action this suspended carbonate would be sufficiently disordered to have an enhanced solubility. They appealed to the observation by GARRELS et al. (1960, p.416) that prolonged grinding (12 h or more) raised the solubility of dolomite, possibly as a result of disordering at the grain surface. The same process could have an extreme effect on the solubility of colloidal sized debris formed by grain collisions. Indeed, these particles would have their solubility enhanced not only by grinding but also by their high surface energies (p.254). NEUMANN (1965, p.1025) noted that fine-grained, high-magnesian calcite is conspicuously lacking from the deeper sediments in Harrington Sound, although it must be produced by lithothamnioid algae, holothurians, serpulids, benthonic foraminiferids, bryozoans and echinoids. Doubtless its dissolution has contributed to the high concentration of $CaCO_3$ in the water.

The effect of grinding on the solubility of calcite has been investigated by CHAVE and SCHMALZ (1966, p.1045) who concluded: "It is evident . . . that considerable energy can be added to carbonate crystals by simple grinding, and that in just 10 min the activity of aragonite, 1.50, may be exceeded." They further demonstrated the enhanced solubility, caused by a combination of grain size reduction and mechanical strain, on reagent calcite, and the aragonite of *Macoma* and *Millepora*.

This enhanced solubility may be offset by the action of organic films on the particles. Reactions between suspended carbonate lutite and sea water have been followed by CHAVE (1965), who remarked on the lack of reactivity, despite attempts to hasten reaction in the laboratory by varying temperature and pH. From other experiments it seems possible that the finer lutite grains are enveloped in a coat of organic material, which prevents reaction: samples treated with H_2O_2 acquired a greatly enhanced reactivity (CHAVE and SUESS, 1967). On the other hand, observations on the mineralogy of different size fractions revealed smaller quantities of the more soluble carbonates (aragonite and high-magnesian calcite) in the finer fractions, as if these fine grains had reacted and dissolved more rapidly than the coarser grains.

Stability of $CaCO_3$ polymorphs in sea water in the light of partition coefficients for Sr^{2+} and Mg^{2+}

Equilibrium between aragonite and calcite particles in sea water is theoretically attained when the chemical potentials of the ions common to the two solid phases are equal (WINLAND, 1969). To determine the concentrations of the different ions at the theoretical equilibrium, it is necessary to know the partition coefficients of ions present in significant quantities, Mg^{2+}, Sr^{2+} and Ca^{2+}. Values for $(K_{Sr})_{aragonite}$ are known (p.261) and WINLAND (1969) has made some experimental first approximations of $(K_{Mg})_{calcite}$, in an attempt to define the equilibrium in sea water, at surface temperatures and pressures, between aragonite and calcite bearing, respectively, Sr^{2+} and Ca^{2+} in solid solution. Using the two partition coefficients, Winland found that aragonite in equilibrium with sea water (0.9 mole % $SrCO_3$) will have a free energy of formation near to that of pure aragonite, -269.5 kcal./mole; calcite in equilibrium with sea water (10 mole % $MgCO_3$) will have a free energy of formation of -267 kcal./mole (close to the value estimated by Bathurst, p.245). Thus, at equilibrium, aragonite has the greater free energy of formation and should be thermodynamically the more stable phase, in shallow seas, as is demonstrated in a different manner on p.244. Winland goes on to remark that this relation would encourage wet transformation ("recrystallization") of high-magnesian calcite to aragonite in sea water. This does, of course, beg an important question. So far as is known, high-magnesian calcite is stable in sea water. Early impressions of "recrystallization" to aragonitic micrite are now believed to be the result of passive precipitation in vacated algal bores (p.388; BATHURST, 1966; PURDY, 1968). Taylor and Illing reported the replacement of aragonite by high-magnesian calcite (p.373). Yet WINLAND (1969) showed a partly micritized peneroplid with no signs of boring although the high-magnesian calcite had changed to aragonite. He has observed coralline algae "recrystallized" at interior sites with well preserved margins.

Control of calcite–aragonite precipitation in pure water solutions in contact with air

POBEGUIN (1954) studied the precipitation of aragonite and calcite in the simplest possible circumstances using pure water solutions. She precipitated $CaCO_3$ by making use of the double decomposition of solutions of ammonium carbonate and calcium acetate at temperatures of 16–18 °C and atmospheric pressure. By varying the design of the containing vessels she was able to arrange for the two solutions to mix by diffusion at different rates. Mixing times were related to mineralogy of the precipitate as follows:

    10–15 days : calcite only
     6– 8 days : calcite, with a little vaterite and a little aragonite
    24–26 hours: calcite, vaterite and much aragonite

Pobeguin noticed that calcite began to precipitate as soon as the solution attained supersaturation, that aragonite only precipitated in a solution supersaturated for calcite and nearing supersaturation for aragonite, and that an apparently amorphous form precipitated only when the solution was supersaturated for all three minerals.

POBEGUIN (1954, p.95) concluded: "De ces expériences, nous pouvons, semble-t-il, tirer la conclusion suivante: dans les conditions envisagées, le facteur essentiel déterminant la précipitation de l'une ou l'autre forme de calcaire, en particulier de calcite ou d'aragonite, est *la vitesse de diffusion des ions* CO_3 *et Ca*, c'est-à-dire, en quelque sorte, la quantité d'ions en présence simultanément, en un point donné et à un moment donné." ("These experiments enable us, it seems, to draw the following conclusion: under the given conditions, the main factor which determines the precipitation of one or other of the forms of $CaCO_3$, in particular the precipitation of aragonite or calcite, is *the rate of diffusion of the ions* CO_3^{2-} *and* Ca^{2+}, that is to say in other words, the number of ions which are simultaneously present at a given point and at a given moment.") It is plain that the higher supersaturation favours the more soluble, higher energy, less ordered polymorph. The same view was expressed by VATAN (1947) with regard to the silica minerals. He suggested that the usual order of crystallization is opal (low-temperature cristobalite)–chalcedony–quartz. This sequence corresponds to an increase in lattice order and in the purity of the solution, and a slowing down both of the rate of nucleation and of crystal growth. FYFE and BISCHOFF (1965, p.8) refer to the crystallization of amorphous silica at 300°C, when the polymorphs form in the order cristobalite–tridymite (or silica-K, a monotropic modification)–quartz. The phases appear in order of increasing stability. An analogous sequence is seen in many calcite cements where, in the first generation, the crystals are numerous, commonly rich in inclusions and bounded by the higher index faces of the scaleno-hedron. In the second generation they are few, clear and rhombohedral in habit. USDOWSKI (1963) has found experimentally that increased concentration and a rise in the ratio Ca^{2+}/Mg^{2+} encourage the growth of radial-fibrous calcite rather than equant rhombohedral. CLOUD has pointed out (1962a,) that the order of precipitate vaterite–aragonite–calcite as supersaturation decreases (p.374) is an expression of Ostwald's rule, whereby precipitation takes place in steps from the least to the most stable solid phase (OSTWALD, 1897). Modifications of this rule to fit carbonate equilibria are discussed by BROOKS et al. (1951) and GOTO (1961). The possibility that the monohydrate, $CaCO_3.H_2O$, is an intermediate step in the precipitation of aragonite in some circumstances has been examined by MARSCHNER (1969).

Strontium in aragonite and calcite

The partition coefficient. A crystal may contain a variety of information, related to its growth environment, which was buried in the atomic lattice as the crystal grew. Isotopic ratios for oxygen and carbon (p.280 and 339), or trace element concentrations of magnesium (p.235) and strontium, can offer valuable clues to the history not only of the growing crystal, but of the solution from which it was precipitated. Some recently developed concepts of strontium behaviour in car-bonate sediments are considered in this section.

The strontium concentrations in different types of sedimentary particle, or in various generations of cement, in neomorphic minerals or in pore fluids past and present, become more meaningful when they are considered in terms of the partition (or distribution) coefficient for Sr^{2+} and Ca^{2+} (McINTIRE, 1963). Our understanding of this concept as applied to carbonate sediments has been greatly enhanced by the work of Holland, Kinsman and others, culminating in two im-portant papers (KINSMAN, 1969; KINSMAN and HOLLAND, 1969).

If crystals of aragonite and calcite are grown in solutions having the same ratio of molar concentrations, mSr^{2+}/mCa^{2+}, then, other factors being equal, the aragonite will take into its lattice a higher proportion of Sr^{2+} than the calcite, following the well known preference of the Sr^{2+} ion for the orthorhombic rather than the trigonal lattice. We can define the partition coefficient for aragonite as the ratio of the molalities of Sr^{2+} and Ca^{2+} in the crystal and in the solution from which it grows thus:

$$(K_{Sr})_{aragonite} = \frac{(mSr^{2+}/mCa^{2+})_{aragonite}}{(mSr^{2+}/mCa^{2+})_{solution}}$$

and we can go on to conclude, from our experience of the greater Sr^{2+} uptake in aragonite, that the partition coefficient for aragonite is higher than the coefficient for calcite. The most recent work (summarized and extended in KINSMAN and HOLLAND, 1969) gives experimental partition coefficients, at 25°C, as:

$$(K_{Sr})_{aragonite} = 1.12 \pm 0.04 \quad \text{and} \quad (K_{Sr})_{calcite} = 0.14 \pm 0.02$$

where the coefficients are followed by the standard deviations of replicate analyses. Taking the ratio $(mSr^{2+}/mCa^{2+})_{sea\ water}$ as equal to $(0.86 \pm 0.04) \times 10^{-2}$ (away from near-shore contamination), Kinsman has calculated a *predicted* ratio $(mSr^{2+}/mCa^{2+})_{aragonite}$ for *inorganically precipitated* aragonite in sea water. Ideally, this predicted ratio can be compared with ratios that have been determined for various naturally occurring aragonites. Where these ratios are similar to the predicted ratio, we might then conclude that precipitation was inorganic, but where they differ we could, perhaps, assume biological modification of the process of precipitation—the "vital effect" noted in connection with the selective uptake of oxygen isotopes (p.280).

In predicting mSr^{2+}/mCa^{2+} for inorganically precipitated aragonite, allowance must be made for the known decrease of $(K_{Sr})_{aragonite}$ with rising temperature (KINSMAN and HOLLAND, 1969, fig.1; also in HALLAM and PRICE, 1968b). Variation of the rate of precipitation during the experiments of Kinsman and Holland produced no detectable change in the partition coefficient for aragonite, though this rate is known to have a strong modifying effect on the coefficient for Mg^{2+} (p.236). Kinsman stated that the coefficient is fairly insensitive to a wide range of solution compositions and concentrations. Other than temperature variation, the only other known strong influence on the partition coefficient is biological fractionation.

Conventionally it has become usual to express the ratio mSr^{2+}/mCa^{2+} as the Sr^+ concentration (p.p.m.) in the aragonite or calcite. Predicted concentrations for inorganically precipitated aragonite, with concentrations for a variety of natural carbonates, are given in Fig.225.

The significance of Sr^{2+} data in Recent sediments. The figures in Fig.225 suggest that, for corals, the biochemical fractionation may be very slight. The apparently slight biochemical fractionation in codiacean algae throws interesting light on the question of the origin of aragonite muds and their fractionation of oxygen isotopes discussed on p.279. By far the greatest biological fractionation is shown by molluscan aragonite. The data of outstanding interest are those for oöids. Their precipitation is, it would seem, subject to some fractionation. These figures do not necessarily mean that oöids are intracellular physiological precipitates,

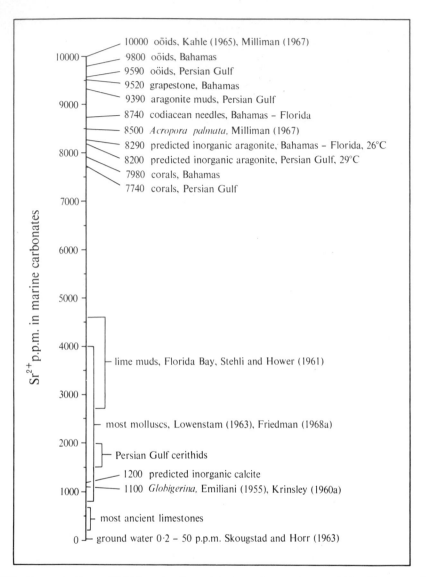

Fig.225. Concentrations of Sr²⁺ in natural precipitates and ground water. (After KINSMAN, 1969, and other authors where acknowledged.)

but they do suggest that the inorganic transfer of Sr^{2+} and Ca^{2+} from the sea water to the lattice encounters some interference. The most likely modifying influence would appear to be the mucilaginous film with which oöids are clothed (p.298), in common with other carbonate grains (p.252). Bahamian grapestone appears to show considerable biological fractionation, a fact of particular significance bearing in mind the intensive algal micritization of these agglomerations of grains. The Sr^{2+} data indicate that the precipitation of micritic aragonite in deserted algal bores could be a biochemical process, a conclusion which must extend to this same process in the algal micritization of carbonate grains generally (p.381). The restriction of submarine cementation to small cavities and other sheltered sites, where biological influences may be dominant, is discussed elsewhere (p.364).

Kinsman has examined the origin of the aragonite muds of the Abu Dhabi Lagoon, Persian Gulf (p.204), in the light of the data in Fig.225. That they cannot be break-down debris of molluscs is quite clear. Codiaceans are absent from the Persian Gulf: corals are too limited in their distribution. An origin related to the biologically influenced (?) precipitation of oöids must obviously be considered, possibly an inorganic process modified by a local microbiological situation.

The "lime muds" in Florida Bay (particle size not specified) appear, on the data available in Fig.225, to consist very largely of molluscan debris. Presumably this reflects the composition of the silt-grade fraction, if the clay-grade fraction is dominated by algal aragonite (p.161). The volumetric importance of molluscan shells in the sedimentation of the Florida "lime muds" seems, on the basis of strontium analyses, to outweigh by far the codiacean contribution emphasized by Stockman et al.

In the back-reef lagoon, British Honduras, the Sr^{2+} levels in the sediment point to a westward movement into the lagoon of coral and *Halimeda* debris, high in Sr^{2+}, derived from the reefs (p.214)."

The use of Sr^{2+} data in the study of diagenesis. The molar ratio mSr^{2+}/mCa^{2+} at the face of the growing crystal is directly related to the ratio in the solution. Changes in the composition of the solution in one particular pore depend directly on: (*1*) the ratios of ions being removed at the faces of adjacent growing crystals, (*2*) the ratios of ions being released by dissolving crystals, and (*3*) the rate of flow of the solution out of the pore and its replacement by solutions of different compositions. The magnitude of the change of molar ratio in the solution, in which crystals are growing, is related to the departure of the partition coefficient from unity, in other words to the mineralogy and temperature. The effect of dissolving crystals on a solution, to which they are not maternally related, depends on the different ratios of mineral and solution. It is clear, therefore, that, during crystal growth or dissolution, the ratio mSr^{2+}/mCa^{2+} in the solution is normally changing. Armed with these simple concepts, an attempt can be made to reconstruct something of the diagenetic history of limestones in so far as this has involved dissolution-precipitation.

A pioneer, classic study of the migration of strontium, during a complex sequence of diagenetic changes, was carried out by SHIRMOHAMMADI and SHEARMAN (1966 abstract), in the early 1960's, though only recently published in full (SHEARMAN and SHIRMOHAMMADI, 1969). In the southern part of the French Jura there is an association of limestones, dolomites and calcitized dolomites ("dedolomites"). By the study of thin sections the authors traced quantitatively the progressive change from calcarenite to dolomite and thence from dolomite back to a limestone composed of secondary calcite (p.543). These changes are reflected in the values for Sr^{2+} p.p.m. measured by X-ray fluorescence spectrometry.

Involved in the diagenetic interchange were three components, given below with their Sr^{2+} contents (p.p.m.):

original undolomitized limestone	160–380
pure dolomite	110–200
calcitized dolomite	< 5

The limestones have the normal values for limestones generally: the range 160–380 p.p.m. includes the majority, though a few lie between 70–630 p.p.m. The Sr^{2+} analyses of the mixture of limestones, partly dolomitized limestones, dolomites and partly calcitized dolomites make sense when related to the petrography. Sr^{2+} was first lost from the rock as the calcite of the limestone was replaced by dolomite which could hold, approximately, only a little over half the amount of Sr^{2+}. The thin section data show a close relation between increasing dolomite content and falling Sr^{2+}, roughly from 250 p.p.m. down to 150 p.p.m. Calcitization of dolomite took place in rocks in which the extent of the previous dolomitization varied. Consequently, these rocks, with more or less calcitized dolomite, contain all three major components—some unaltered limestone, some dolomite and some calcitized dolomite. Again, Sr^{2+} content closely reflects the proportions of these three components. Calcitization of a fully dolomitized limestone yielded a rock composed purely of metasomatic calcite with Sr^{2+} at less than 5 p.p.m. On the other hand, calcitization in a half dolomitized limestone (Sr^{2+} 200 p.p.m.) gave a rock containing half residual limestone (mean Sr^{2+} 250 p.p.m.) and half calcitized dolomite (say Sr^{2+} 5 p.p.m.), so that total Sr^{2+} in the rock is about 130 p.p.m. Shearman and Shirmohammadi showed, therefore, that the Sr^{2+} behaved in a logical fashion and that dolomitization and calcitization depended on dissolution-precipitation, rather than solid state diffusion. Incidently, this work is a reminder of the benefits of close association between chemical and petrographic studies.

Application of Sr^{2+} analysis to the early diagenesis of carbonate sediments, to the determination of the dissolution of aragonite, the release of Sr^{2+} to ground water and the precipitation of calcite cement, is nicely illustrated by the work of HARRIS and MATTHEWS (1968) described on p.329. The difference in Sr^{2+} content among some Recent carbonate sediments and ancient limestones is shown in Fig.225, indicating a fall from a dominant 3,000–6,000 p.p.m. in the sediments to 50–150 p.p.m. in lithified limestones.

Two points of special importance must be noted here briefly. In a provocative discussion of the behaviour of Sr^{2+} in carbonate diagenesis, KINSMAN (1969) concluded that something like 100,000 pore volumes of solution must have passed through a carbonate sediment during its change to limestone—a figure close to those estimated in Chapter 10 (p.440). Secondly, in considering the change in Sr^{2+} content of aragonite crystals as they grow, KINSMAN and HOLLAND (1969) have found that this change (reflecting a change in the solution) will be preserved in the crystal. Adjustment of the outer part of the crystal to the changing Sr^{2+} content in the solution by solid state diffusion is likely to be too slow, so that changes of the ratio mSr^{2+}/mCa^{2+} in the solution will be reflected in a zoning of the crystal.

Caution: uncertainties in the Sr^{2+} partition concept. Despite the theoretical interest of the Sr^{2+} partition approach, the method at present lacks the desirable sensitivity that would permit its successful application to a number of problems. A glance at the range of values for $(K_{Sr})_{aragonite}$ shown by KINSMAN and HOLLAND (1969, fig.1) reveals that, at 30°C for example, the experimental values for $(K_{Sr})_{aragonite}$ varied from 1.08 to 1.25, despite a small analytical error of ± 0.025. If a similar range of $(K_{Sr})_{aragonite}$ be assumed for 25°C, then the predicted concentration of Sr^{2+} for inorganically precipitated aragonite in the Bahamas (8,290 p.p.m.) can in fact lie anywhere within a range of about 8,200–9,600 p.p.m. Adjusting for an analytical error of about 1,000 p.p.m., the total range within which the true concentration of inorganically precipitated Sr^{2+} lies is 7,200–10,600 p.p.m. This embraces all oöids, aragonite muds and grapestone within a range of possible inorganic origin. The range is further extended because Bahamian water temperatures have a range of at least 22–31 °C. It would seem that the Sr^{2+} data at present available are suggestive but certainly not conclusive. An interesting discussion of factors that may influence the partitioning of Sr^{2+} is given by KINSMAN and HOLLAND (1969). Further complications are introduced by SCHROEDER's (1969) experimental results which suggest that Sr^{2+} is incorporated in the aragonite in several ways, i.e., lattice positions, absorbed in lattice interstices, or as inclusions (p.338).

Solubility and pressure

The solubility of $CaCO_3$ in water with a constant wt.% of dissolved CO_2 and at constant temperature has been shown to increase in two ways as the pressure rises: (1) following an increase of the hydrostatic pressure on the system, and (2) following an increase of linear pressure in the crystal lattice with accompanying elastic strain. The case of enhanced hydrostatic pressure is applicable to the study of $CaCO_3$ solutions in deep ocean waters and in deeply buried limestones. The case of linear pressure relates to pressure-solution along stylolites and grain-to-grain contacts. Both these cases are complicated in nature by variations in the content of dissolved CO_2 which, in the sea at least, undoubtedly varies with depth (Fig.226). The distribution of the CO_2 in the sea depends partly on exchange between water and atmosphere, partly on its liberation during bacterial decay of organic matter and on respiration and photosynthesis, and partly on its reaction with pelagic

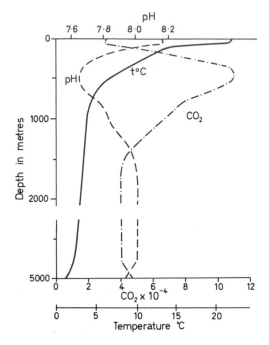

Fig.226. Vertical distribution of P_{CO_2} (10^{-4} atm), pH and temperature in middle latitudes of the Atlantic according to Wattenberg. (After DEFANT, 1961, fig.41.)

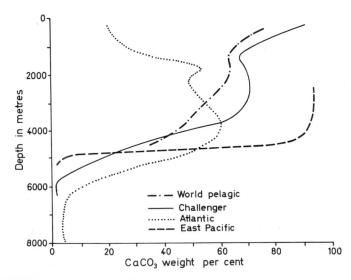

Fig.227. CaCO₃ content of sea floor sediments related to depth. (After KUENEN, 1950, fig.153—data from Pia, Trask, etc.—and BRAMLETTE, 1961, fig.4.)

$CaCO_3$. Further complexities arise through the large scale turbulence of ocean waters and the mixing of water masses. These matters and the related problems of residence time, examined with the help of radiocarbon data, are usefully discussed by SKIRROW (1965).

The process of pressure-solution is dealt with in Chapter 11 and illustrates the importance of differential solution where strain is unevenly distributed over the surface of a grain.

Dissolution of $CaCO_3$ in the deep ocean. It has long been known that the concentration of $CaCO_3$ in sediments on the floor of the open ocean decreases with depth of water, but the reasons for this change have not been securely established. This decrease in concentration is particularly rapid between 4,000–5,000 m (Fig.227). About this level there must be a **compensation depth** at which supply of solid $CaCO_3$ equals its loss by dissolution. Below 6,000 m less than 4% of $CaCO_3$ generally remains in the sediment. BRAMLETTE (1961) claimed that the dissolution of carbonate particles while settling is negligible: coccoliths (about 5 μ diameter) settle to 5,000 m in about 10 years, yet they accumulate on the bottom with foraminiferids which settle to the same depth in a few days. This is at variance with the researches of Peterson (p.268) and it seems likely that most coccoliths are eaten and travel rapidly to the bottom either in faecal pellets or in animal carcasses. Certainly the good correspondence between the biogeographic provinces of living Coccolithaceae and the distribution of dead coccoliths on the sea floors (MCINTYRE and BÉ, 1967) compels the conclusion that the fall time is short. The downward loss of $CaCO_3$ in sediments is not a simple reduction in all kinds of carbonate. It is clear that the progressive loss of aragonite proceeds in shallower depths than the loss of calcite. The implications of this have been studied (p.404) by HUDSON (1967a). So far no systematic difference has been noted between the frequencies of high-magnesian and low-magnesian calcites, but scarcity of evidence here reflects lack of investigation. MURRAY and HJORT (1912, p.173) found that the tests of Pteropoda, which are aragonite, begin to disappear from ocean floor sediments at about 2,700 m and are practically absent below 3,600 m. CHEN (1964) discovered that, in samples of sediment having less than 70% $CaCO_3$, pteropods are virtually unknown on the Bermuda Pedestal below about 4,200 m. FRIEDMAN's (1965e) work in the Red Sea has shown that, despite an existing supply of dead planktonic pteropods, their tests are absent from sediments below 956 m. What he found on the floor below this depth, and above 2,012 m, was mixed skeletal debris of calcite, some of it high-magnesian and some, especially *Globigerina*, low-magnesian. The one-time existence of pteropod tests, presumably on the bottom, is indicated by the presence in the sediment of calcite casts of the body chamber (FRIEDMAN, 1965e, plate 1) indicating post-depositional diagenesis.

It has recently been shown that the correlation between $CaCO_3$ content of sediments and water depth is not, after all, particularly strong, when examined on a world-wide basis (SMITH et al., 1968). There is no world-wide compensation depth and distance from land is an important influence. The authors suggested that a complicating factor is the coating of particles with organic compounds (p.252).

The survival of high-magnesian calcite, in deep water in the Red Sea (above) after aragonite has dissolved, is surprising since, in shallow marine waters, the order of decreasing solubility is high-magnesian calcite–aragonite–low-magnesian calcite (Fig.218). Work on deep marine carbonates of other areas reveals the same tendency for aragonite to dissolve at lesser depths than calcite. The unusual shallowness of the sediments in which aragonite is depleted in the Red Sea may be a result of the fine grain size (silt and clay grades) of the sediments (FRIEDMAN, 1965e).

Some notes by Friedman on the composition of calcarenites and calcilutites in cores taken from around 4,500 m on the Bermuda Apron hint at the greater solubility of high-magnesian calcite over low-magnesian. The calcarenites, presumed turbidites, contain a mixture of all three carbonates, akin to the composition of the shallow Bermudan near-shore sands (Fig.254, p.331). It must be assumed that rapid burial removed them to the safe haven of a saturated and stagnant pore solution. By contrast, the interbedded pelagic calcilutites which accumulated slowly are pure low-magnesian calcite: this polymorph, then, must be the least soluble. The trend of recent work, therefore, is to suggest an order of decreasing solubility in deep sea water aragonite–high-magnesian calcite–low-magnesian calcite.

The reason for the greater relative solubility of aragonite in deep water, compared with high-magnesian calcite, is not obvious. It is perhaps significant that WEYL (1959a, p.223) found in laboratory experiments that, as P_{CO_2} rises from 10^{-3} molal initial concentration to 10^{-2}, the solubility of aragonite diverges from and exceeds that of calcite by 11%. It may be, therefore, that the known increase of CO_2 content with depth in the upper few thousand metres is affecting solubility more than the increase of pressure.

A factor of unknown consequence is the stability of the various carbonate minerals in the pore water below the deep sediment surface. Diagenetic changes of considerable significance may go on in this environment.

A radically new outlook on the reasons behind the mineralogy of deep sea floor sediments is likely to develop in the next few years following publication of the descriptions of cores taken by the Glomar Challenger in the JOIDES Project (Joint Oceanographic Institutions Deep Earth Sampling). Among other things, the cores have yielded extensive information about rapidly translated carbonate turbidites of shallow water origin.

Sea water is undersaturated for $CaCO_3$ at a depth of only a few hundred metres, as shown by the field studies of PETERSON (1966) and BERGER (1967, 1970), who suspended calcite spheres and foraminiferids at various depths in the central Pacific and measured their loss in weight. Peterson concluded that large foraminiferids will fall to the bottom at 5,000 m in a few days, but small foraminiferids and other tiny particles of $CaCO_3$ will dissolve totally on the way down. Peterson and Berger recorded sharp increases in the rate of dissolution at about 3,700 and 3,000 m.

The question of the depth at which sea water is undersaturated for $CaCO_3$ has been examined in the laboratory by PYTKOWICZ and FOWLER (1967). Previously a major obstacle to progress had been the inadequacy of data on the increase of the activity of CO_3^{2-} as hydrostatic pressure increases. That the activity should be greater at depth was shown by DISTECHE and DISTECHE (1967), and by PYTKOWICZ et al. (1967) who estimated that, at about 10,500 m, 57% of the $Na_2CO_3^\circ$, $CaCO_3^\circ$ and $MgCO_3^\circ$ ion-pairs, present at the surface of the sea, should be dissociated. Recent new determinations of the pressure coefficients of the apparent dissociation constants of carbonic acid in sea water have been made by CULBERSON et al. (1967) and DISTECHE and DISTECHE (1967). Armed with these new data PYTKOWICZ and FOWLER (1967) have examined the solubility of foraminiferids in artificial sea water in a high-pressure cell with *in situ* electrodes. Their results confirm that sea water should be undersaturated at depths much shallower than the compensation depth, also that the compensation depth represents, not a sudden change from saturation to undersaturation, but the results of rate effects. What the rates of dissolution are for different types of sedimentary carbonate particle, and what factors control the rates, are still to be discovered. Turbulence is obviously one such factor, during fall or on the bottom, since it affects the rate of renewal of water at the particle surface. The authors showed that different species of foraminiferid have different solubilities. Rates of decay of organic films (p.252) may be important or a water layer saturated for $CaCO_3$ on the sediment (PYTKOWICZ, 1970).

The depth at which the sea is 100% saturated for $CaCO_3$ certainly varies from place to place. Estimates of this saturation depth in the Pacific indicate 3,000 m at about 50°S and 1,000 m at about 50°N (HAWLEY and PYTKOWICZ, 1969). The pH depends partly on the age of the water mass, because organic oxidation gradually lowers the pH. Thus a local reduction of pH can be caused by inflow of old water from a neighbouring ocean.

The degree of saturation of sea water for $CaCO_3$ at depth was investigated by BERNER (1965a, p.960). He drew some interesting conclusions from his work on the activity coefficients of Ca^{2+}, CO_3^{2-} and HCO_3^- at near-surface conditions. For extrapolation down to 5,000 m only minor corrections to the activity coefficients are necessary to allow for the increased pressure: the effects of reduced temperature and salinity, on ionic strength and the formation of ion-pairs, are insignificant (BERNER, 1965a). He plotted (Fig.228) the ion activity product, $aCa^{2+} \cdot aCO_3^{2-}$, against depth for three sets of pH data, for the equatorial Atlantic (H. W. HARVEY, 1955, p.37), the north Pacific (MOORE et al., 1962) and his own estimation of a maximum pH limit. He wrote: "The distribution of pH with depth is not well known, but results of most studies suggest a consistently lower pH at depth, than at the surface, due to increased partial pressures of CO_2 resulting from bacterial

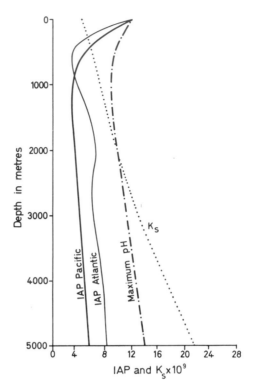

Fig.228. The ion activity product (IAP = aCa^{2+}·aCO$_3{}^{2-}$), maximum pH and the solubility product constant for calcite (K$_s$) in sea water related to depth. (After BERNER, 1965a, fig.5.)

activity, and to the effect of increased total pressure on the dissociation constants of carbonic acid." The higher values of aCa^{2+}·aCO$_3{}^{2-}$ for the maximum limiting pH are a result of the increase of the ratio aCO$_3{}^{2-}$/aHCO$_3{}^{-}$ with rising pH (p.232). In the same figure Berner showed the variation with depth of the thermo-dynamic solubility product constant [aCa^{2+}]·[aCO$_3{}^{2-}$] for pure calcite. He suggested that these various data indicate that the oceans may be able to dissolve CaCO$_3$ at depths up to only a few hundred m. Though his value for the solubility product may be substantially too high at the compensation depth, since it takes no account of the increase in the activity of CO$_3{}^{2-}$ with depth following the break-down of the ion-pairs MgCO$_3{}°$ and CaCO$_3{}°$ (below), it needs little modification for a few hundred metres. Berner's conclusions are born out by the experimental work of Pytkowicz and Fowler (above) and they agree with the results of Peterson and Berger (p.268). There is no sign of a sudden change around 4,000–5,000 m and Berner suggested that it is more likely that the rapid change in the CaCO$_3$ content of sediments at those depths is due to a change in the rate of dissolution. It certainly does not appear to be the result of a change from supersaturated to undersaturated conditions. On the other hand, the divergence between the ion activity product and the solubility product constant does become markedly exaggerated in Fig.228 from 3,000 m downward and this may be important as well.

The role of pressure in the deep ocean. The effect of increased hydrostatic pressure on the solubility of $CaCO_3$ has been investigated both theoretically and experimentally (PYTKOWICZ, 1968). Theoretically it has been estimated that, at the approximate compensation depth of 4,700 m (about 500 bars and 0 °C), the stoichiometric solubility product $[mCa^{2+}] \cdot [mCO_3^{2-}]$ should be between 2 and 3 times its value at the surface. Such an increase in solubility product depends on the increased dissociation constants of H_2CO_3 under pressure. The equilibria 2 and 3 on p.231, 232 will shift to the right as pressure increases. The estimated increase follows from the thermodynamic calculations by OWEN and BRINKLEY (1941) using partial molar volumes and compressibilities of ionic solutes for pure water and NaCl solution (no CO_2), and calculations by ZEN (1957) based on a comparison of molar and partial molar volumes. However, Dr. Pytkowicz has kindly pointed out to me that the thermodynamic estimates are too large by a factor of about 2 since they neglect the increased activity of CO_3^{2-} caused by the dissociation under pressure of the dissolved ion-pairs $CaCO_3°$ and $MgCO_3°$.

The CO_3^{2-} in these complexes is, of course, included in the stoichiometric ion product. As pressure increases, the released CO_3^{2-} necessarily shifts the equilibrium $CaCO_3 \rightleftharpoons Ca^{2+} + CO_3^{2-}$ toward the left and so reduces the solubility (PYTKOWICZ, 1963). The stoichiometric product should, instead, increase by a factor of about 1.5, and this figure is supported by the experiments of PYTKOWICZ and CONNERS (1964) which yielded an interpolated factor of about 1.75.

At this point it is necessary to distinguish carefully between the stoichiometric solubility product and the actual solubility, particularly in view of certain apparent contradictions in the literature (BATHURST, 1967b). From straight forward arithmetic it is obvious that, if the product $mCa^{2+} \cdot mCO_3^{2-}$ is to be doubled, and if the two ionic molalities are equal, then each must be multiplied by 1.41 (i.e., $\sqrt{2}$) and the value $mCaCO_3$ will also increase by the same amount. Thus the solubility will increase by 41%. Where, on the other hand, the two ionic molalities differ greatly, as in sea water, then the increase in solubility will be extremely small. The stoichiometric solubility product for average sea water at 1 atm, approximately saturated, can be regarded (from Table XII) as:

$$mCa^{2+} \cdot mCO_3^{2-} = 0.010 \times 0.00027 = 2.7 \times 10^{-6}$$

When, under pressure, the product is increased to twice this value, the ionic molalities must each increase by the same number of moles, since $CaCO_3$ dissolves congruently. This number is 0.00265 moles, and the new product is:

$$mCa^{2+} \cdot mCO_3^{2-} = 0.010256 \times 0.000526 = 5.4 \times 10^{-6}$$

Thus, despite a 100% increase in the stoichiometric solubility product, the

TABLE XII

MOLALITIES OF MAJOR DISSOLVED ELEMENTS IN SEA WATER
(From GARRELS and CHRIST, 1965)

Ion	Molality (total)	Free ion (%)
Na^+	0.48	99
K^+	0 010	99
Mg^{2+}	0.054	87
Ca^{2+}	0.010	91
SO_4^{2-}	0.028	54
HCO_3^-	0.0024	69
CO_3^{2-}	0.00027	9
Cl^-	0.56	100

solubility rises less than 5%. Actually, the solubility will increase rather more than this because some of the additional CO_3^{2-} will be converted to HCO_3^- as in equation 3 and Fig.216. The result is that, while the term mCO_3^{2-} increases almost two fold, the mCa^{2+} changes only very slightly. The figures in PYTKOWICZ (1965b, table 1) show how, in spite of the considerable increase of undersaturation at depth (by a factor of 7), the increase in alkalinity (caused by addition of HCO_3^-, CO_3^{2-} and other Brønsted bases) is detectable only in the second decimal place.

The distinction between the increase of the stoichiometric solubility product and of the solubility has a bearing on the results and conclusions that have been derived from recent experiments on the effect of pressure on the solubility of $CaCO_3$, and I am indebted to Dr. R. M. Pytkowicz (personal communication 1967) for clarifying this matter for me. It is now possible to reconcile the claim made by SHARP and KENNEDY (1965, p.402), that the solubility at 4,700 m (0°C) is rather more than twice that at 1 atm, with their experimental results and those of SIPPEL and GLOVER (1964), all of which indicate only a very slight increase in solubility with pressure. Sharp and Kennedy, in applying the data of OWEN and BRINKLEY (1941) and ZEN (1957), are dealing with the stoichiometric solubility product—not, as they state, the solubility. The relatively tiny influence of pressure on the solubility at 0°C is clear from Fig.229 and 230.

It is apparent, therefore, from the results of experiments at high pressures and with various CO_2 contents, both from Sippel and Glover (Fig.229) and from Sharp and Kennedy extrapolated (Fig.230), that the effect of pressure at 0°C on the solubility of $CaCO_3$ is not great. Sippel and Glover found an increase in solubility of about 20% at 500 bars, a pressure equivalent to a depth of 4,700 m. The extrapolation of the curves by Sharp and Kennedy toward 0°C is uncertain, but there is no doubt that the isobars converge.

The role of CO_2 is interesting, because both pairs of authors have shown that the CO_2 content has a relatively large effect on solubility. (The values of P_{CO_2} used, especially by Sharp and Kennedy, are larger than those occurring naturally.)

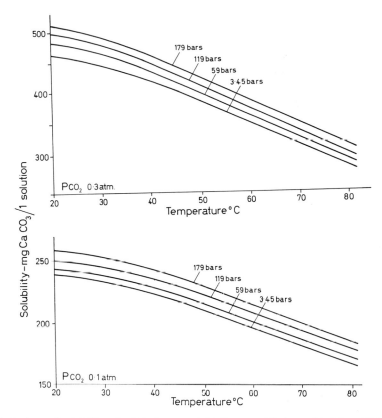

Fig.229. Isobars of solubility of calcite in water initially in equilibrium with 0.3 and 0.1 atm CO_2. (After Sippel and Glover, 1964, fig.4,5.)

Solubility in the deep crust. The increase of hydrostatic pressure with depth of burial in the crust, unlike the change in the sea, is accompanied by a big increase in temperature. The data of Sippel and Glover (1964) in Fig.229 show that for shallow depths, assuming the hydrostatic pore pressure to be a direct function of depth, the effect of pressure on the solubility of $CaCO_3$ is considerably less than that of temperature. For example, at a depth of 1,000 m (say 119 bars) and 220 mg/l of $CaCO_3$, a 10% increase of pressure at constant temperature would cause a negligible increase in concentration, whereas a 10% increase of temperature, at constant pressure, would produce a decrease of about 30 mg/l in the concentration (Fig.229, lower). As depth increases the effect of a change in pressure becomes relatively stronger. The thick line in Fig.230, indicating the approximate pressure-temperature gradient in the crust, is extrapolated from the gradient used by Sippel and Glover, simply to give an idea of possible relationships. The true nature of the gradient does vary, of course, both geographically and with depth.

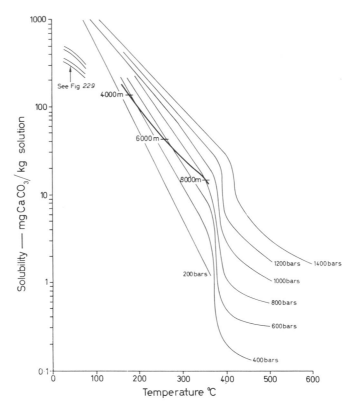

Fig.230. Isobars of solubility of calcite in water. (After R. F. Sippel and J. E. Glover in Fig.229, and SHARP and KENNEDY, 1965, fig.5, 0.4 wt.% CO_2.)

Preservation of skeletal aragonite in ancient limestones

It is well known that the abundance of skeletal aragonite in limestones decreases generally with the age of the host rock. Clearly there is a tendency for aragonite to be removed and for some of this lost aragonite to be replaced by sparry calcite. BØGGILD (1930) demonstrated that, in molluscan shell walls replaced by calcite, the new fabric is coarse, sparry and utterly unlike the original microcrystalline shell fabric (p.329). Thus, the recognition of secondary calcite, and therefore of the previous existence of aragonite, is straightforward. It is necessary, in addition, to distinguish calcite cement, which has filled a dissolution mould of the initial shell (p.417), from calcite neomorphic spar which is the product of calcitization of an aragonite fabric *in situ* (p.486).

The preservation of aragonite is selective, and is related to the permeability of the host sediment. Where permeability is low, as in finely crystalline limestones, argillaceous and bituminous limestones, and, above all, in bituminous clays and shales, the preservation of aragonite is the most complete (FÜCHTBAUER and GOLDSCHMIDT, 1964). In biosparites the aragonite content is, with rare exceptions, nil. The rapid change of aragonite to calcite in the presence of fresh water is described on p.327. The generally accepted reason for this reaction is that reduced permeability results in reduced movement of pore water, and thus in lower rates of dissolution. It may be asked, nevertheless, whether dissolution, restricted thus by low permeability, can ever be so utterly slow as to be ineffective over tens of millions of years. Skeletal aragonite is known to have survived for at least 300 million years in Lower Carboniferous shales (HALLAM and O'HARA, 1962) and for at least 350 million years in Upper Devonian sediments (GRANDJEAN et al. 1964). By comparison, metamorphic rocks undergo great mineralogical changes in relatively short periods, despite permeabilities which are a fraction of a millidarcy.

KENNEDY and HALL (1967) emphasized the role of the primary organic matrix of the skeleton in the subsequent preservation of the aragonite. They had found aragonite in Tertiary sediments with a considerable range of lithologies, and they felt that more attention should be paid to the conservational role of the proteinaceous films in which the aragonite crystals are embedded (e.g., the organic matrix in molluscan shell, p.14). The proteins are believed to break down in time to peptides and thence to amino-acids. The authors remarked that the distribution of fossil aragonite is closely related to the distribution of reducing substances, which would inhibit the oxidation and destruction of the organic sheets. They cited the particularly clear relationship in the Wealden of northwest Germany, where aragonite molluscan grains coated with oil are preserved, but those in water-saturated sediments have changed to calcite (FÜCHTBAUER and GOLDSCHMIDT, 1964).

Kennedy and Hall suggested a mechanism by which an amino-acid, adsorbed on the aragonite surface, could prevent its dissolution. Amino-acid molecules have the general form

$$\oplus \; R-\underset{\underset{H}{|}}{\overset{\overset{NH_2}{|}}{C}}-COOH \; \ominus$$

The side-chain, R, can be one of a number of groups, such as CH_3 (in alanine) or C_3H_7 (in valine). The amino and carboxyl groups are polar, with positive and negative charges respectively. In water they are hydrophilic and attract water dipoles, whereas the other two side-chains are hydrophobic. At the surface of the aragonite the charged groups in the amino-acid molecules would be attached to the unsatisfied CO_3^{2-} and Ca^{2+} ions, and "the resulting surface layer of amino-acid molecules would then present an outer surface of hydrophobic groups, which would prevent the access of water to the crystals" (KENNEDY and HALL, 1967, p.254). (This hypothesis is relevant to the stability of micrite envelopes, p.334.)

Dr. P. Weyl (personal communication, 1969) has suggested another process which may play an important role in stabilizing the aragonite in ancient sediments. In a sediment from which the magnesium-bearing sea water has not been flushed, the Mg^{2+} may react with the surfaces of existing carbonate grains to produce a high-magnesian carbonate of solubility equal to aragonite, so no transformation takes place (p.252). Other views of the role of Mg^{2+} in conserving aragonite are presented earlier (p.243).

ORIGIN OF THE BAHAMIAN ARAGONITE MUD:
INORGANIC OR PHYSIOLOGICAL PRECIPITATE?

Textbooks have for many years taught that, in the geological past no less than today, calcium carbonate muds have been inorganically precipitated from sea water. It is not easy to judge how much of this happy certainty is the inspired intuition of genius, how much a blind repetition of outmoded shibboleth. One thing is plain. Among specialists in this important but intricate field, opinions differ on the extent to which carbonate muds are being precipitated inorganically at the present time. Only the aragonite muds in the Abu Dhabi lagoon appear to be demonstrably inorganic in origin (p.204).

Over most of the Great Bahama Bank calcium carbonate is certainly precipitated in many ways—as algal needles (Fig.231), faunal skeletons, oölitic coats, the matrix of grapestone, and the filling of microscopic bores vacated by algae (p.388), depending on the local conditions. Aragonite mud is present in vast quantity, but its origin is not obvious. Whereas the growth of physiological (algal) needles can be observed easily, the inorganic precipitation of needles cannot and its detection depends on circumstantial chemical evidence. In this section the question of the inorganic or physiological precipitation of the aragonite mud will be considered, but much of the argument is equally relevant to the growth of oöids and grapestone, and possibly to the precipitation of micritic aragonite in algal bores.

The aragonite muds west of Andros Island at present occupy a central position in the argument over the origin of $CaCO_3$ muds in general, ancient and modern. They came to hold this position in the debate not, regrettably, because of any ideal simplicity in their environment, but because early investigators were first drawn, in the 19th century, to the Florida–Bahamas platform. The region is, in fact, complex and the muds could have a multiple origin. The case for an algal origin rests partly on estimates of needle-forming crops and partly on a comparison of oxygen isotope ratios between mud and algal needles. Water chemistry is consistent with inorganic precipitation.

Carbonate muds of undoubted skeletal origin have been described by Matthews from British Honduras (p.212). "Whitings", possibly spontaneous

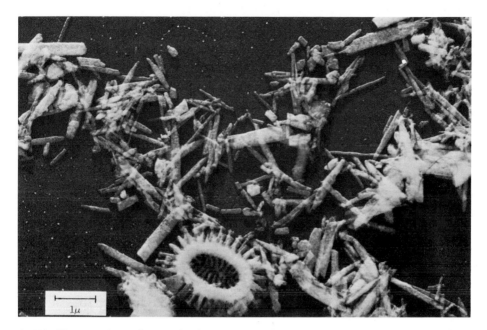

Fig.231. Electron photomicrograph of aragonite needle mud with coccolith. Bahamas. (From HATHAWAY, 1967.)

precipitates of aragonite as a cloudy suspension in sea water, are discussed on p.137 and p.188.

The sediment

Out of 58 mechanical analyses of surface sediments, from the pellet-mud and mud lithofacies (by P. D. Blackmon in CLOUD, 1962a, p.37) 39 have distributions that are clearly bimodal with peaks approximately at 250 μ (fine sand) and in the fraction finer than 2 μ (clay). The sand fraction consists of peloids which are themselves mostly aggregates of clay and silt (p.137). The composition of the fundamental needle mud is about 65 wt.% of clay grade carbonate and 15 wt.% of silt. The clay fraction has > 95 wt.% aragonite and, in the silts, there is generally 82–88 wt.% aragonite (P. D. Blackmon in CLOUD, 1962a, p.50). Total aragonite in the sediment is 88–94 wt.%, mainly in the form of needles and laths a few microns long. Both low-magnesian calcite (0–5 mole % of $MgCO_3$) and high-magnesian calcite (11–16 mole % of $MgCO_3$) are present, but the high-magnesian variety is strongly dominant and is derived largely from foraminiferids. Molluscs contribute to both calcites. There are also traces of quartz, hydrous mica, chlorite and kaolinite. Some further details are given on p.137.

The calcarenite fraction is autochthonous, native to the muddy-sand and mud habitats. This is apparent from the low content of grapestone, cryptocrystalline grains or oöids (under 4 wt.%) so common in the adjacent areas, and from the absence of ripple-mark and megaripple. As for the mud fraction, the nearest possible source outside the area is Andros Island, but the dark colour of its tidal channels seen from the air is against the likelihood of westward movement of suspended white mud in sufficient quantity. On the contrary, it seems more probable that mud is carried in the reverse direction, on to Andros, during extra-high tides and accumulates there on the supratidal flats. Mud in adjacent areas, such as the stable sand habitat, must be in part winnowed oceanward by the ebb tidal currents (p.100) and once off the bank it may sink beyond reach of the returning flood. On the other hand, there is the possibility that mud is moved centripetally into the region west of Andros Island by flood tide currents and is deposited there according to the settling-lag effect of Postma (p.201), elaborated by VAN STRAATEN and KUENEN (1958) and re-examined critically and quantitatively by McCAVE (1969).

Skeletal contribution to the sediment

The contribution of *recognisable* skeletal debris to the total sediment is readily determined as 11 wt.% (CLOUD, 1962a, p.93). The total mass of material derived from skeletons must be greater than this, probably 15–20 wt.%. Cloud had an ingenious scheme for determining the amount of aragonite in the mud bequeathed by dead calcareous algae. This calculation depended rather heavily on a simple experiment in which *Halimeda* fragments were crushed and their contribution to the silt and clay fractions noted. Unfortunately there is so little evidence relating to the natural contribution of *Halimeda* to the different size fractions, that any estimate based on such a laboratory experiment may be wildly out. Although the plant disintegrates in a solution of sodium hypochlorite (LOWENSTAM, 1955), the extent of its disintegration on the sea floor is quite unknown. The rate of disintegration must be related to grazing and, at least in Bimini Lagoon, it is retarded by the onset of micritization (p.384). Micritized grains of *Halimeda*, which form many of the ubiquitous cryptocrystalline grains in all Bahamian habitats, are well indurated and are unlikely to disintegrate, except superficially by abrasion, as in the surf zone or in the guts of grazers. Moreover, we still know next to nothing about the rate at which this plant, or any other, donates free needles to the environment. Other algae are present, *Penicillus*, *Rhipocephalus*, *Udotea*, *Acetabularia*, though in lesser quantity. But their needles are only weakly aggregated and so the plants would be expected to yield needles in greater amounts per cm^3 of plant than *Halimeda*. CLOUD (1962a, p.96) appeared to assume that all the aragonite of *Halimeda* is destined to form clay and silt debris. Considering that *Halimeda* is renowned as a sand grain former in many parts of the world today, this may be something of an exaggeration. Indeed, as *Halimeda* plants more than equal in number all other calcareous algae counted, any process such as micritization, tending to conserve the skeletons as resistant sand grains, must seriously reduce their contribution to the aragonite mud. Thus Cloud's estimate that 4–5 wt.% of the total sediment has originated as *Halimeda* needles seems, if anything, an over estimate.

The remarkably short life of *Penicillus*, 1–3 months, (STOCKMAN et al., 1967), may indicate important contributions from algae other than *Halimeda*. During this short life *Penicillus* may attain a height of 20 cm and could yield a considerable quantity of mud. We can use some of CLOUD's field measurements (1962a, table 30), and have a shot at an estimate of algal contribution based on *Penicillus*, assuming that all the aragonite from the plant eventually disintegrates to mud. Cloud recorded, in an area of 25 m² of floor, 17 *Penicillus* plants. These gave 1.8 g of aragonite excluding material held in or on the holdfasts. Very crudely, if the plants were replaced every 2 months, the annual contribution of mud would be 6 × 1.8,= 10.8 g. Correcting for the specific gravity of aragonite (2.94) and allowing a porosity in the mud of 50%, this yields an annual contribution of 7.4 cm³ of mud. Spread over an area of 25 m², this would provide, in a thousand years, a thickness of 0.3 mm. Even if this figure were too small by a factor of ten, the *Penicillus* plants would still give only a tiny fraction of the known accumulation of about 3 m of mud and pellets in 3,800 or so years. Finally, by adding the aragonite derived from all the other calcareous algae (mainly *Halimeda*) collected by Cloud in the sample area of 25 m², and by making similar assumptions, probably unjustified, about rapid growth and death, the estimated rate of accumulation can only be increased by a factor of 130 to about 4 cm in 1,000 years, or 15 cm in 3,800 years, say 4% of the required thickness. Only by coincidence is this close to Cloud's estimate of 4–5%.

A much more detailed and better controlled estimate of the rate of supply of aragonite mud by algae, based on *Penicillus*, has been made for Florida by STOCKMAN et al. (1967), who found the algal contribution to be an adequate source of all the Recent aragonite mud (p.162). The mean density of the Florida *Penicillus* communities over a year in different sampling stations ranged from 0–30 plants/m². A few communities achieved 76–108 plants/m², but over most of the area the density was probably about 2 plants/m². West of Andros Island Cloud's estimated density is much smaller, being 0.7 plants/m² (17 plants/25 m²).

Recognition of algal needles by $^{18}O/^{16}O$ ratios

In 1955 Lowenstam reported some experiments on the production of aragonite needles by digestion of codiacean algae in sodium hypochlorite and suggested that breakdown in natural conditions could be a significant source of needle mud. Subsequently LOWENSTAM and EPSTEIN (1957) measured the $^{18}O/^{16}O$ isotope ratios of aragonite needles from codiacean algae in order to discover whether these have unique ratios that might allow recognition of their remains in aragonite muds. This resourceful approach to the problem follows from the method they established earlier with Urey (UREY et al., 1951). The results, though not conclusive, illumine in an interesting way some of the difficulties and some of the applications of this use of isotopic analysis with the mass spectroscope.

The ratio $^{18}O/^{16}O$ in a marine calcium carbonate depends on three factors: (1) it follows the $^{18}O/^{16}O$ ratio in the water, (2) it decreases with rising temperature at the site of precipitation, provided that the carbonate is in equilibrium with the water, and (3) it varies with the taxonomy of the organism; some animals and plants may precipitate carbonate from a solution which, as a result of their metabolism, is preferentially enriched in one isotope or the other, generally ^{16}O.

Lowenstam and Epstein used codiacean algae collected from two areas, the muddy-sand and mud habitats west of Andros Island (Fig.129, p.104) and the oölite shoals of Schooner Cays, just off Eleuthera (Fig.127, p.97). They determined the $^{18}O/^{16}O$ ratios for bulk samples of needles and oöids, calculated as:

$$\delta^{18}O_c = \frac{^{18}O/^{16}O_{sample} - {}^{18}O/^{16}O_{PDBI\ standard}}{^{18}O/^{16}O_{PDBI\ standard}} \cdot 1,000 \qquad (18)$$

(see UREY et al., 1951; CRAIG, 1953). They had in mind the interesting possibility that algal aragonite might stand out from others either because it grew only in a restricted range of temperature and/or because of the "vital effect" (3) referred to above. It is also useful to compare the calculated temperatures of precipitation, but before temperatures can be calculated which are comparable between samples, allowance must be made for the influence on the carbonate ratio $\delta^{18}O_c$ of the sea water ratio $\delta^{18}O_w$. This is the factor (1) and it naturally varies from place to place. The value $\delta^{18}O_c$ is corrected so as to yield a standardized value $\delta^{18}O_s$ such that $\delta^{18}O_s = \delta^{18}O_c - \delta^{18}O_w$. The standardized values are expressed according to the amount by which they exceed or fall short of the $\delta^{18}O_w$—as if all the carbonates were precipitated in water with $\delta^{18}O_s = 0$. The operations of standardization and calculation of temperature are combined in the empirical equation of EPSTEIN and MAYEDA (1953):

$$t°C = 16.5 - 4.3\,(\delta^{18}O_c - \delta^{18}O_w) + 0.14\,(\delta^{18}O_c - \delta^{18}O_w)^2 \qquad (19)$$

Values found for $\delta^{18}O_s$ and calculated temperatures for the precipitating solution are shown in Fig.232. All the values must be presumed to be average values, representing the total variation of $\delta^{18}O_s$, caused by changes in $\delta^{18}O_w$ and in temperature, throughout the period during which the carbonate in the sample was precipitated.

A difficulty arises here. In this particular research programme sea water samples were not available for the measurement of $\delta^{18}O_w$ for each alga or for each sediment sample. Instead values of $\delta^{18}O_w$ were estimated from the known relation between salinity and $\delta^{18}O_w$ in the Bahamas (Fig.233). Yet, without knowing at what time of year the algae, mud needles, or oöids grew it is not possible to use salinity data very accurately for determining $\delta^{18}O_w$. The annual salinity variation in the waters west of Andros Island may well be 4‰, equivalent to a range

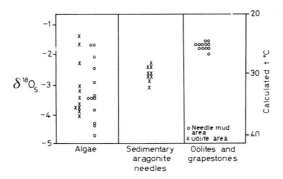

Fig.232. $\delta^{18}O_s$ values and calculated temperatures for different carbonates in various areas, after correction for summer sea-water temperatures. (After LOWENSTAM and EPSTEIN, 1957.)

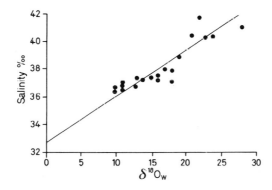

Fig.233. Relation between $\delta^{18}O_w$ and salinity of water samples on the Great Bahama Bank. (From LOWENSTAM and EPSTEIN, 1957, fig.2.)

in $\delta^{18}O_w$ of about 1.25. This in turn would alter $\delta^{18}O_s$ and yield a range of temperature of about 6°C. This is a very large penumbra of error in a region where the annual range of water temperature is only about 8°C (22–30°C). If, on the other hand, it be assumed (on the scantiest of evidence) that the algal growth and mud-oöid precipitation took place in the summer months of April–September, during which time the pattern of isohalines on the Bank is relatively static, then this error could be reduced from ±6°C to, say, ±3°C. To expect greater accuracy than this at present seems unreasonable in view of the known variability of salinity, diurnal or otherwise, during even the summer months.

The calculated temperatures in °C of the precipitation solutions, from eq. 19 were determined by Lowenstam and Epstein as:

aragonite muds   27.6–31.7
oölites   24.0–25.7
algae   22.8–39.8

To these figures must be added the error of ±6°C. Looking at these it is conceivable that the oöids were precipitated inorganically in equilibrium with the sea water. CLOUD (1962a, p.14) recorded late spring evening temperatures of 28 °C in the Florida Straits. The water on the Bank is generally only about 0.5 °C warmer, but short term maxima in excess of 28.5 °C may be important on oölite shoals at low tides on summer afternoons. Winter temperatures, on the other hand, are down to 22 °C. Thus the calculated temperatures of oöid growth (24.0–25.7 °C) could represent an average growth for many oölites throughout the whole year, but hardly growth limited to the summer months. Some confirmation for the calculated temperatures of oöid growth comes from those calculated for deposits of serpulid worms in the same area, Schooner Cays. Serpulids are known to precipitate in equilibrium with sea water and their calculated temperature is 25.4 °C (LOWENSTAM and EPSTEIN, 1957, p.371).

The algae used by Lowenstam and Epstein were collected in May, so it can perhaps be assumed that they grew during February–May in water temperatures between 24–25 °C (SMITH, 1940, p.157). Their calculated temperatures fall between 22.8 °C and 39.8 °C so that it is immediately clear that some, possibly all, did not grow in isotopic equilibrium with the enclosing sea water at any time of the year.

The interpretation of the $\delta^{18}O_s$ values for the algae, in terms of their possible contribution to the needle mud values, runs into other difficulties. The collected specimens (ten west of Andros, twelve from Schooner Cays) give some idea of the amount of variation of $\delta^{18}O_s$ (Fig.234) to be expected for May. (Individual oöid lamellae might have given an equally great range.) There is, however, no indication at present of the weight to be given to these values in terms of the different rates of contribution to the muds (volume/unit time) by different plants. Judging from the figures given by LOWENSTAM and EPSTEIN (1957, fig.5), relatively large contributions by *Halimeda* or *Udotea* would give high negative values for $\delta^{18}O_s$ whereas a heavy subvention by *Penicillus* or *Rhipocephalus* would give mainly low negative to low positive values. The $\delta^{18}O_s$ figures for the muds could mean they are a liberal mixture of several algal species. Nevertheless we have no means of knowing in what proportions the different plants might subscribe to the mud. Knowing neither the density of different species of codiacean on the sea floor, nor their growth rates, nor expectation of life, we cannot say what range of $\delta^{18}O_s$, or mean value, we should expect if the muds were purely algal in origin. Therefore we cannot judge if the known values of $\delta^{18}O_s$ for the muds are reasonable in terms of algal accumulation. Thus although the figures do not eliminate the possibility that the needle muds are algal, they do not help us to reach a firm conclusion. The question of origin remains open. In the circumstances he would be bold who followed LOWENSTAM and EPSTEIN (1957, p.374) in concluding that "the most reasonable conclusion would be that . . . the sedimentary aragonite needles are algal in origin".

We should not turn from the consideration of algal precipitated muds without at least remarking the strange fact that the temperature range for oöids is

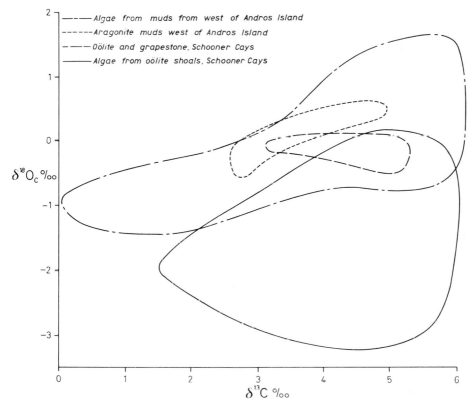

Fig.234. Relation between $\delta^{18}O_c$ (uncorrected for water) and $\delta^{13}C$ for different carbonates of the muddy floor west of Andros Island and of the oölites of Schooner Cays. (After LOWENSTAM and EPSTEIN, 1957, fig.3,4.)

entirely outside and below the range for aragonite muds (Fig.232). The difference in temperature between the waters over the oölite shoals and those over the muds is generally less than 0.5 °C. Can this large anomaly be a hint that the muds are not inorganic but algal, and that their higher temperatures reflect the vital effect (3) mentioned on p.280?

It should also be remembered, however, that in areas of sheltered water, with prolific algal growth (as in Bimini Lagoon) there are, nevertheless, no muds.

CLOUD (1962a, p.98) has commented on the large water corrections made by Lowenstam and Epstein, in excess of those indicated by the $\delta^{18}O$—salinity relationship given in their paper. The water corrections can be recalculated according to the salinities given by the authors (1957, table 1) for the area west of Andros Island (mean salinity = 39.7‰) and for Schooner Cays (mean salinity = 36.9‰). When this is done the calculated temperatures are reduced and become (°C):

aragonite muds   24.2–28.0
oölites   22.8–23.7

Equilibrium precipitation of needle muds can now be regarded as an all year round process, but equilibrium oöid growth seems, oddly, restricted to the winter months. Could this reflect the turbulence of the winter gales? The recalculated temperature ranges for muds and oölites are also closer together.

Inorganic precipitation

After considering the possible contributions to the mud from sources outside the area and from skeletal debris, and taking the oxygen isotope data into account, CLOUD (1962a) reckoned, that some 5 wt.% of the mud is detrital and 17–20 wt.% is skeletal debris: the skeletal sediment includes an estimated 4–5 wt.% of algal aragonite. There remains a further 75 wt.% of the sediment to be accounted for in some other way, and it is necessary to examine the possibility that this major portion has been inorganically precipitated directly from the water.

Although AGASSIZ (1894, p.50) first called serious attention to the lime muds west of Andros Island ("a very fine chalky ooze, resembling plaster of Paris which has just been mixed for setting") it was DREW (1914) who put forward the earliest considered argument for their origin, claiming that they were precipitated by the action of denitrifying bacteria. He was supported by KELLERMAN and SMITH (1914). LIPMAN (1924,1929) was able to show that, although suitable bacteria are present in the water over the Bank, their quantity is too slight for this purpose. BAVENDAMM (1932, p.205) thought that in mangrove swamps the high concentration of bacteria can be presumed to bring about substantial precipitation of calcium carbonate. Yet water samples from two lakes in the mangrove swamp three miles inland from the Andros coast near Williams Island, collected by SMITH (1940, p.184), showed high chlorinity (21.75‰ and 23.4‰) indicating intense evaporation, also high P_{CO_2} (4.3 and 2.5 × 10^{-4} atm), but the product $mCa^{2+} \cdot mCO_3^{2-}$ is high as well (2.81 and 4.32 × 10^{-6}). So precipitation in the lakes adjacent to mangrove swamp seems to be unimportant.

The work of Black. BLACK (1933b), who was a member of the International Expedition to the Bahamas in 1930 (with Bavendamm), made three traverses across the shoal area west of Andros Island, taking water samples. From these he measured the chlorinity and, deriving the salinity by calculation, produced the southernmost isohalines in Fig.235. His results led him to postulate that the hypersaline water in mid Bank is brought to this condition by evaporation in the hot shallows, and that the volume of water lost is continuously replaced as normal sea water moves in from the surrounding deep ocean. The high concentration of salts and the con-comitant loss of CO_2 combined, he suggested, to bring about precipitation of tiny aragonite crystals. The precipitation was not continual but depended on the sus-pension in the water of suitable nuclei at times when the bottom muds were stirred by wind-induced turbulence. Ingeniously joining rates of evaporation from reser-

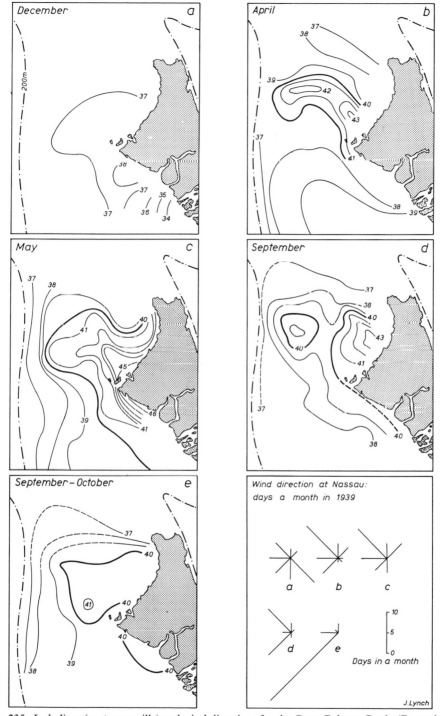

Fig.235. Isohalines (parts per mille) and wind directions for the Great Bahama Bank. (December, September and October from SMITH, 1940; April from SMITH, 1940, also BLACK, 1933b; May from CLOUD, 1962a; wind from Nassau Meteorological Station in SMITH, 1940.)

voir studies to the known solubility of $CaCO_3$ in sea water, he estimated the rate of precipitation. His figure of 0.0338 mm/year, even when combined with his large estimate of 80 wt.% skeletal and detrital grains, only gives 0.166 mm of sediment per year, or a little over 0.5 m in 3,800 years (probable age of deepest mud by radiocarbon, in CLOUD, 1962a, p.36). This would have to be increased by about 6 to yield the 3 m or so of mud on the Pleistocene basement. Considering the poverty of data available to Black, the order of magnitude was closely approached.

The work of Smith: chemical gradients. A major advance in quantification and physico-chemical theory was made by SMITH (1940) who, in the course of conservation work for the Bahamas Sponge Fishery Investigation Department, took the opportunity "to make such observations as were possible, with the time and equipment available" (SMITH, 1940, p.147). His observations, taken at various times of the year over a wide area, have ever since provided the main basis for discussions of the inorganic precipitation in the sea of aragonite mud—discussions much stimulated by Smith's acute chemical perception.

He was the first to show that the salinity gradients converge, in the period April–September, on a region near Williams Island (Fig.235, 236). He also found that the partial pressure of CO_2 reaches there a maximum level which is twice that in the Florida Straits, the pH falling correspondingly. His data revealed for the first time that, from the Straits toward Andros Island, the concentration of Ca^{2+} rises with the chlorinity yet the ion product $Ca^{2+} \cdot CO_3^{2-}$ and the concentration of CO_3^{2-} both fall. The four groups of samples, on which his chemical conclusions were based, were each collected miraculously within short periods of a few days. The short intervals of collection mean that the samples closely represent the true distribution of chemical and physical parameters at one moment over a very extensive area of sea. The direct measurements were of temperature, chlorinity, titration alkalinity (p.234), mCa^{2+} and pH. Other values were calculated from tables.

Smith was then faced with the intriguing problem of how to arrange his data. He was impressed by the shifting of the winds throughout the year (p.100; tracks of drift buoys in TRAGANZA, 1967) and wrote that "it is therefore apparent that once a body of water drifts onto the bank it may be shifted to and fro by conflicting wind influences for a long period without making much mileage in any one direction" (SMITH, 1940, p.156). His isohaline charts (Fig.235) support this idea. The lengthy residence time of water over the mid Bank, combined with the intense evaporation and the peripheral dilution by oceanic tidal water, causes the development of a steep salinity gradient and the pattern of summer isohalines (p.101).

Smith's model. The data for early October are the most complete and Smith selected these for detailed discussion. He began by emphasizing that the regional pattern of chemical parameters in early October was the outcome of processes

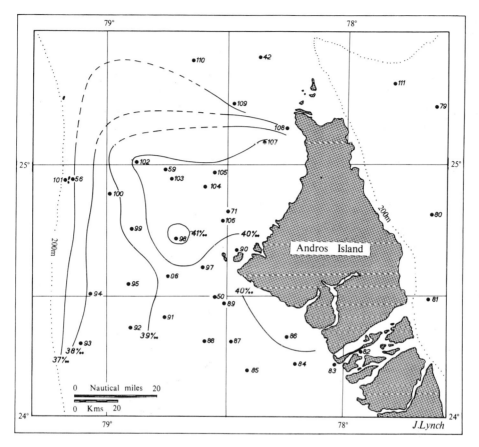

Fig.236. Isohalines for September–October, 1939, for the Great Bahama Bank with stations occupied by Smith. (After SMITH, 1940, fig.48.)

that had been acting throughout the preceding summer months on a nearly stagnant body of water. These processes include a persistent trade wind strengthening and backing gradually from just south of easterly to northeasterly (Fig.235), evaporation over the whole area, and mixing of Bank and ocean water along the Bank margins. While evaporation tends to raise the concentration of all ions, peripheral mixing imposes on the region a salinity gradient between Bank and ocean (Fig.235). Mixing with ocean water is least around Williams Island where distances from the Florida Straits and Providence Channel are greatest. Thus it is here that the salinity is highest. Superimposed on this pattern is the effect of the trades, tending to move the water sluggishly to the west and southwest throughout the summer. At times the northeast winds drive a wedge of less saline water southward from Providence Channel. This is shown by the break in the 40‰ salinity isopleth (Fig.235).

Of particular importance are Smith's figures for the relationship of chlorinity with mCa^{2+}, $mCa^{2+} \cdot mCO_3^{2-}$ and P_{CO_2}. He considered a series of stations (Fig.236) of increasing distance from the ocean and increasing chlorinity, in other words a traverse up the salinity gradient. The stations show a gradual fall in $mCa^{2+} \cdot mCO_3^{2-}$ toward Williams Island and a corresponding rise in P_{CO_2}, chlorinity and mCa^{2+}. These gradients he interpreted as the result of evaporation, the effects of which are greatest away from the ocean. Evaporation causes a rise in the partial pressure of CO_2 to above saturation toward Williams Island, also a rise in mCa^{2+} (SMITH, 1940, table 1) along with the general rise in concentration of all ions expressed as chlorinity. Yet at station 101 the degree of supersaturation for $CaCO_3$ in open ocean water is so far exceeded that aragonite is precipitated, and the rate of precipitation rises in the direction of Williams Island. This is indicated by the fall in $mCa^{2+} \cdot mCO_3^{2-}$ as chlorinity increases. It is important to remember that these considerations apply to a body of water which, except for tidal movements and short term changes of wind driven currents, had been drifting sluggishly *westward* all summer.

What of the influence of the trade winds on the pattern of chemical parameters which would, after all, have developed in the absence of a westerly drift? Here Smith drew attention to stations 98, 103 and 102 (Fig.236). They show a relation between chlorinity and P_{CO_2}, also chlorinity and $mCa^{2+} \cdot mCO_3^{2-}$, which is markedly discordant (Fig.237, 238). Within only a small range of chlorinity there is a relatively large fall in P_{CO_2} and a rise in $mCa^{2+} \cdot mCO_3^{2-}$. Noting that these stations lie toward the western edge of the 40‰ isohaline (Fig.236) and that the region of highest salinity (41‰) has already drifted westward, he suggested that these three stations illustrate the effect of the westerly drift. Mixing with ocean water is slight because the chlorinity remains at the high level of the innermost stations 97, 90 and 86. But CO_2 is being lost as the lobe of saline water moves towards the Straits and, as a result, the $mCa^{2+} \cdot mCO_3^{2-}$ is rising and,

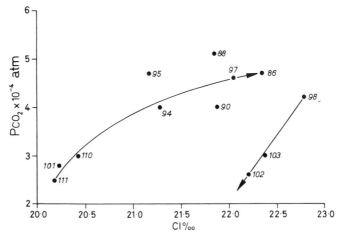

Fig.237. Relation between P_{CO_2} and chlorinity for twelve stations on the Great Bahama Bank, September 29–October 9, 1939. (From data in SMITH, 1940, p.174.)

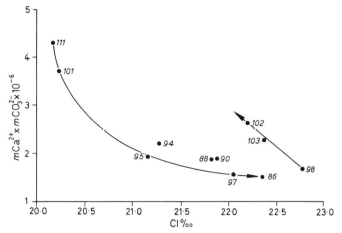

Fig.238. Relation between $m \, Ca^{2+} \cdot m CO_3^{2-}$ and chlorinity for eleven stations (not 110) on the Great Bahama Bank, September 29–October 9, 1939. (From data in SMITH, 1940, p.174.)

consequently, the precipitation of aragonite is on the wane. Why the fall in P_{CO_2} is so rapid is not altogether clear. Possibly this depends on slower evaporation in the lower temperatures of October so that the CO_2 equilibrium across the air–water interface has more time to achieve a balance. Increased turbulence caused by the stronger autumnal winds may also be a factor. Certainly the October winds begin to break up the summer pattern of isohalines.

A similar pattern is apparent for April, but the figures for December and January not unexpectedly show a rather confused distribution with a hint that water is moving out of the area, southward along the Andros coast.

Smith calculated the amount of $CaCO_3$ precipitated at different times of the year, basing his results on observations at four stations in the area where the product $mCa^{2+} \cdot mCO_3^{2-}$ is minimal near Williams Island. Though, in the absence of adequate information on the volume of water that moves through this area, it is impossible to estimate rates of sedimentation, his results do show clearly that there is a pronounced seasonal variation in the rate of abstraction of $CaCO_3$ (Table XIII). TRAGANZA (1967) reached similar conclusions and, making assumptions for

TABLE XIII

CALCIUM CARBONATE PRECIPITATED NEAR WILLIAMS ISLAND, BAHAMAS
(From SMITH, 1940, table IV)

Date	Temperature (°C)	$CaCO_3(mg/l)$
22 April 1939	25.7	49.6
4 October 1939	30.4	59.1
13 December 1939	21.7	22.1
19 January 1940	20.2	21.6

mean water depth and rate of replacement of Bank water, calculated a precipitation rate of 9 mole/m^2 year. The high rates for the summer months agree with the
corrected $\delta^{18}O_s$ values on p.283 which, if the needles are inorganic, imply precipitation mainly in the summer.

The work of Cloud. Cloud in his comprehensive 1962a paper (with Blackmon,
Sisler, Kramer, Carpenter, Robertson, Sykes and Marcia Newell) directed his
major incursion into the realm of calcium carbonate withdrawal from the sea water.
"To confront the problem of aragonite muds", he wrote, "is to confront the problems of the solubility relations of calcium carbonate and of its polymorphism in
the solid state." (CLOUD, 1962a, p.111.) Standing on the work of Smith he added a
further set of physico-chemical data for stations occupied in May 1955 and used
these in subsequent computations. His view was that if the behaviour of calcium
carbonate in sea water is to be closely followed and its relation to the precipitation
of particular solid polymorphs properly understood, it is necessary to know how
much of the Ca^{2+} and CO_3^{2-} is involved as activities in the attainment of supersaturation, and how much is preoccupied in other reactions, for instance the
formation of complexes (p.254). Cloud assumed that the minimum value for $Ca^{2+} \cdot$
CO_3^{2-} in the sea water near Williams Island represented a close approach to
the solubility product for aragonite. By somewhat unorthodox use of Debye-
Hückel values he arrived at a value for the solubility product constant and went
on to plot the variation of $aCa^{2+} \cdot aCO_3^{2-}$ across the Bank for May 1955. As
chlorinity increases eastward, from Florida Straits toward Williams Island,
this ion activity product first increases and then falls steadily. Everywhere the
activity product is greater than the solubility product constant. Across the Bank
the effects of evaporation are found to be increasingly severe as mixing with
ocean water is less efficient. So the chlorinity rises steadily toward Andros Island.
Along this same gradient the withdrawal of $CaCO_3$ from the supersaturated water
is progressively less compensated for by replacement from oceanic water, and so
the activity product falls steadily eastward.

The work of Broecker and Takahashi. The generalized positive relationship
between increase of residence time and salinity and the accompanying fall in the
product $aCa^{2+} \cdot aCO_3^{2-}$, as aragonite is precipitated, which was revealed by
SMITH (1940), has been investigated and elaborated in a quite different way by
BROECKER and TAKAHASHI (1966). In the years immediately following 1960 there
had been, as a result of nuclear tests, a continuous increase in the amount of
bomb-produced ^{14}C in the atmosphere, causing a corresponding rise in the ratio
$^{14}C/^{12}C$ in atmospheric CO_2. The geological usefulness of this rise in ^{14}C, for the
estimation of residence time ("mean time elapsed since the water molecules in any
given sample entered the banks from the adjacent ocean"), arises from the slow
adjustment of the ocean surface waters to the new atmospheric ratio—slow on

account of the slow transfer of CO_2 to and fro across the air–water interface and because of mixing with deeper water. Thus, in June 1963, when Broecker and Takahashi completed their field work, the increase in $^{14}C/^{12}C$ for the surface of the deep ocean was only about 12% of that in the atmosphere. Ocean water flowing onto the Bank was characterized, therefore, by a low ratio $^{14}C/^{12}C$, but, during its residence on the Bank as part of a very thin sheet of water, only about 4.5 m deep with a large ratio surface/volume, the water changed rapidly in the direction of equilibrium with the atmosphere. Broecker and Takahashi devized a model which allowed the ratio $^{14}C/^{12}C$ in a water sample to be translated into residence time. They then plotted residence time against salinity for June 1963 and made the encouraging and pleasing discovery that the plots fitted closely to a straight line. Differences in the plots for 1963 and 1962 indicate that this linear relation shifts slightly, parallel to the salinity axis, according to the amount of rainfall in the previous months. Estimated correspondence of residence times with salinities for June 1963 and 1962 are given in Table XIV. The residence times, like the salinities, increase progressively toward a maximum in the neighbourhood of Williams Island.

Other data collected by Broecker and Takahashi were P_{CO_2}, total dissolved inorganic CO_2, and the $CaCO_3$ saturation (by saturometer, WEYL, 1961). From these they calculated the rate of $CaCO_3$ precipitation: this falls approximately exponentially with increases in residence time and salinity. Estimated precipitation rates vary from 80 mg/cm^2 year toward the Bank edge to 30 mg/cm^2 year near Williams Island. The relation between $CaCO_3$ loss (expressed as mole/l) and salinity agrees well with the earlier data of SMITH (1940) and CLOUD (1962a). Estimation of the levels of supersaturation for aragonite in the Bank waters, and $aCa^{2+} \cdot aCO_3^{2-}$ plotted against $CaCO_3$ loss, fit well with SMITH (1940, 1941) and SEIBOLD (1962a) and show that the water is everywhere supersaturated. The activity product, at its lowest value near Williams Island, indicates a 15% supersaturation. The much

TABLE XIV

RELATION BETWEEN RESIDENCE TIME (BY ^{14}C METHOD) AND SALINITY ON GREAT BAHAMA BANK IN JUNE[1]

Days	Salinity (‰)	
	1963	*1962*
211	44	41
156	42	39.5
101	40	38
46	38	36.5

[1] Taken from BROECKER and TAKAHASHI (1966, fig.4); 1962 very approximate, reflecting unusually heavy summer rainfall.

lower activity products found by Cloud, indicating saturation near Williams Island, have considerable scatter and almost certainly reflect serious errors in pH measurements (BROECKER and TAKAHASHI, 1966, p.1591).

In discussing the mechanism of precipitation, whether it is inorganic or physiological, Broecker and Takahashi noted that the rate of $CaCO_3$ precipitation is proportional to the degree of supersaturation. This is to be expected for inorganic precipitation whereas, if it were physiological, such a relationship need not hold. Certain lacunae remain, however. In the first place, 25% of the sediment is demonstrably organic in origin (p.278). Also, at present-day rates of $CaCO_3$ loss from the water, the calculated mean rate of precipitation of 50 mg/cm^2 year could account for only 1.5 m of sediment in the past 5,000 years, whereas the actual accumulation has been about 3 m in 3,800 years. It is interesting that Broecker and Takahashi are unable to balance the CO_2 budget: about 50% more CO_2 is lost from the Bank water than can be accounted for by the calculated precipitation. The CO_2 is not taken up by photosynthesizing plants since the organic content of the muds is far too low: nor can it be assumed that this organic matter has been oxidized because by this process it would simply return to the water. It cannot be supposed that the CO_2 is used in the building of calcareous skeletons because this loss of $CaCO_3$ is allowed for in the calculated rate of precipitation. It does seem likely that large quantities of CO_2 are removed mechanically from the area in the form of suspended organic matter in the water, which is swept off the Bank and dispersed in the deep ocean where its concentration is much less.

The problem of the budget. Despite the theoretical models of various authors, it is still impossible to budget accurately for the effects of three processes, as WEYL (1963) pointed out. These are: (*1*) a net movement of water onto the Bank to replace that lost by evaporation, (*2*) a drift of ocean water across the Bank as a result, mainly, of wind-driven currents, and (*3*) the diurnal flow of tidal water on and off the Bank. No one has yet collected sufficient data, including current movements, for this budget to be fairly estimated. No satisfactory model has been designed and tested which would yield the roughly concentric pattern of isohalines and residence times combined with a centripetal decrease in $aCa^{2+} \cdot aCO_3^{2-}$ toward Williams Island. It is worth noting TRAGANZA's (1967) estimate that a wind of hurricane force could sweep all the resident water off the Bank in one day.

FURTHER READING

INGERSON's (1962) broad review of carbonate geochemistry is a good starting point. Outstanding as a guide to thinking in carbonate chemistry are the simple introduction by JOHNSTON and WILLIAMSON (1916) and the fuller treatment of fundamentals by GARRELS and CHRIST (1965) and KRAUSKOPF (1967) and the rapid graphical methods of DEFFEYES (1965). Essential background is provided by SKIRROW (1965) in his essay on CO_2 in aqueous solution, by CLOUD (1962a) in his study of inorganic precipitation, and by FYFE and BISCHOFF (1965) in their admirable example of the application of physical chemistry to problems of carbonate diagenesis; also by LOWENSTAM's (1963) sweeping biologically oriented study and the thoughtful work of GOTO (1961). WEYL's (1967) approach to the reactions between carbonate minerals and solutions is particularly rewarding. Much detailed information is contained in GRAF's (1960, parts I–IV B) series of papers on the geochemistry of carbonate sediments and sedimentary rocks. There is an exceedingly helpful article on the kinetics of crystal growth by NANCOLLAS and PURDIE (1964), and BERNER's two papers (1966a, 1967) on reversible adsorption of Mg^{2+} and Ca^{2+} on calcite surfaces, and its consequences, are highly relevant, also PYTKOWICZ (1965a) on inorganic nucleation. Problems of $CaCO_3$ dissolution in the deep ocean are examined at length in an essay by PYTKOWICZ (1968). On the origin of aragonite muds, SMITH's (1940) paper is basic, CLOUD's two works (1962a, b) are important, with LOWENSTAM and EPSTEIN (1957) on the relevance of oxygen and carbon isotopes, and BROECKER and TAKAHASHI (1966) on bomb ^{14}C and residence times and, with TRAGANZA (1967), further treatment of the level of supersaturation. Useful papers on the transfer of CO_2 across the sea water–air interface are by WYMAN et al. (1952), BOLIN (1960), KANWISHER (1963), SUGIURA et al. (1963) and SKIRROW (1965). Other references are given in the text.

Additional references not given in the preceding chapter

Reference books by SVERDRUP et al. (1942), DEGENS (1965) and RILEY and SKIRROW (1965).

Solubilities, transformations, syntheses of $CaCO_3$ minerals by REVELLE (1934), REVELLE and FLEMING (1934), SUNAGAWA (1953), REVELLE and FAIRBRIDGE (1957), ELLIS (1959), LIPPMANN (1959), KITANO et al. (1962), AKIN and LAGERWERFF (1965a), GOTO (1966), GREEN (1967) and BISCHOFF (1968a).

On molalities and ion-pairs in sea water, GARRELS and THOMSON (1962); on organic pore-chemistry, Bermuda, THORSTENSON (1969).

On the effects of pressure, and oxygen use by organisms, on the pH in deep sea water, PARK (1966, 1968).

On strontium and other trace elements in carbonate minerals, skeletons and sediments, KULP et al. (1952), STERNBERG et al. (1959), KRINSLEY (1960a, b), PILKEY and GOODELL (1963), CURTIS and KRINSLEY (1965), KAHLE (1965a), LERMAN (1965a, b), SIEGEL (1965), CULKIN and COX (1966), HALLAM and PRICE (1966, 1968a), KINSMAN (1966a). See also the next section.

On strontium in ground waters, SKOUGSTAD and HORR (1963); on magnesium, KITANO (1959), KITANO and FURUTSU (1959), PILKEY and GOODELL (1963), LERMAN (1965a, b), CULKIN and COX (1966) and LIPPMANN (1968a–c).

On trace elements in Recent shells, THOMSON and CHOW (1955), KRINSLEY and BIERI (1959), KRINSLEY (1960a, b) and WANGERSKY and JOENSUU (1964); on hot springs, KITANO (1959, 1962b, 1963).

A selection of references on the relation of trace-elements and isotopes to salinity and temperature

CHENEY and JENSEN (1965)

CHOQUETTE (1968)

H. CRAIG (1953)

DANSGAARD (1954)

DEGENS and EPSTEIN (1962, 1964)

DODD (1963, 1964, 1965, 1966a, c, 1967)

EPSTEIN et al. (1951, 1953)

FAURE et al. (1967)

GROSS and TRACEY (1966)

HODGSON (1966)

KEITH et al. (1964)

KEITH and PARKER (1965)

KEITH and WEBER (1964)

LERMAN (1965a, b)

LLOYD (1964)

LOWENSTAM (1954, 1961, 1963)

LOWENSTAM and EPSTEIN (1957)

MALONE and DODD (1967)

MOOK and VOGEL (1968)

PILKEY and GOODELL (1964)

ROSS and OANA (1961)

RUBINSON and CLAYTON (1969)

SIEGEL (1961)

TARUTANI et al. (1969)

TUREKIAN (1957)

TUREKIAN and ARMSTRONG (1960, 1961)

TUREKIAN and KULP (1956)

UREY et al. (1951)

WEBER (1967, 1968a)

WEBER and RAUP (1966a, b, 1968)

WEBER and SCHMALZ (1968)

WEBER and WOODHEAD (1969)

Chapter 7

GROWTH OF OÖIDS, PISOLITES AND GRAPESTONE

> Il est impossible d'embrasser d'un coup d'oeil le
> vaste sujet des oolithes calcaires, sans être frappé
> de la rareté des données positives définitivement
> acquises, en dépit des nombreuses études dont il a
> été l'objet.
>
> *Les Roches Sédimentaires*:
> *Roches Carbonatées*
> L. CAYEUX (1935)

THE HISTORY OF OÖLITE THEORY

Specialists on the subject of oöid formation have been few and the literature is made up, for the most part, of contributions by those who had only a passing interest in the origin of the strange little spheroids and who were moved, as some particular aspect caught their attention, to place their intuitions on paper. The subject has been, thus, an ideal topic for single papers. (Here I stand equally guilty.) There is no doubt that progress has suffered owing to a lack of students prepared to involve themselves deeply in what is, after all, an extraordinarily interesting problem. It is remarkable, as Cayeux wrote thirty years ago, that so large a body of writing should contain so small a measure of definite achievement. In a matter so intimately related to questions of nucleation (in the molecular sense) and crystallization it is sad, and a reflection upon earlier interdisciplinary barriers in science, that the pioneer studies of oölites (and their larger equivalents, pisolites) were apparently quite unaffected by the brilliant thinking in physical chemistry in the first years of the century.

Despite the incisive lead given by LINCK (1903) in his paper on the inorganic precipitation of oöids, geological thinking is still dogged by the suspicion either that they are the direct products of organic activity or that their growth is necessarily dependent on this. Parallel with this uncertainty of outlook, there was for many years a confusion between the oöid (p.77) and what we would now call an oncolite (p.220). There remains at the present time a tendency for experimental studies to ignore the very special peculiarities of the crystalline microfabric within the marine oölitic cortex. This fabric is described on p.77.

The role of algae

The marine oöid contains within its lamellar cortex (p.78), and in its detrital nucleus, a variety of live and dead tissues which have been identified definitely as algae (Fig.118, p.77). Bacteria and possibly fungi are also present. This association encouraged the hypothesis, in the 1890's, that the marine oöid is of algal origin. ROTHPLETZ (1892), gazing at the snowwhite oöids on the beaches of the Great Salt Lake, noticed that farther out, below water level, they were entangled in masses of blue-green algae, colonies of Schizophyceae (Cyanophyta). He took dry oöids from the beach and gently dissolved them in acid, and found that the insoluble residue contained dead and shrivelled remains of the same algae. On this basis he wrote that the oöids "may, therefore, be understood as dead algal-bodies" and "the oölites of the Great Salt Lake are, therefore, indubitably the product of lime-secreting fission-algae" (ROTHPLETZ, 1892, p.280). It is interesting that in extending his studies to the oöids of the Red Sea (earlier described by WALTHER, 1888, p.481; 1891, p.527) Rothpletz distinguished between the "minute granules" that remained after digestion with acid, which he regarded as the builders of the oöids, and certain thread-like boring algae which he thought were foreigners that used the oöid as a convenient home.

WETHERED (1890, 1895) on the other hand regaided the vermiform, branching calcareous algal threads as the essential building material of oöids. He was attracted by the then recently described *Girvanella* and in his papers he figured numerous Jurassic and Dinantian calcareous bodies of sand size which, in thin section, were interlaced with very varied arrangements of tubular stiuctures. In his 1890 paper he simply pioposed that at least some oölitic structures are of organic origin, but by 1895 his views had hardened: "Minute fragments of remains of calcareous organisms, such as corals, polyzoa, foraminifera, crinoids, etc., collected on the floor of the sea. These became nuclei to which the oölite-forming organisms attached themselves, gradually building up a crust ... At the same time calcareous material was secreted, and the interstitial spaces between the tubules were filled." (WETHERED, 1895, p.205.) This, suiely, is a vivid description of the growth, not of oöids, but of oncolites.

These early convictions on the algal origins of oöids have not stood up well to subsequent criticism. When EARDLEY (1938) cairied out perhaps the most penetrating and thorough of all studies of modern oölites, in the Great Salt Lake (Fig.120, p.81), he wrote: "If Rothpletz found an algal colony in which even a single oölite grain was embedded, it must be rated as a very rare occurrence." (EARDLEY, 1938, p.1385.) Of Wethered's vermiform grains—whose only similarity to real oöids is their size—CAYEUX (1935, p.255) iemarked that encrusting algal structures of *Girvanella* type (Fig.101, p.65) are, in fact, extremely rare in oölites. This opinion must be echoed by anyone with experience of oölites of various geological ages. Cayeux also went to the heart of the matter when he insisted that it is geometrically impossible to construct with tubes a body that would show in all sections a concentric lamellar structure. Any section would be bound to show a mixture of longitudinal and transverse sections of tubes.

Despite the tenuous nature of the early theories of algal encrustation, an interest in the alga-oöid association has very properly persisted. DANGEARD (1936) made imaginative use of the delicate technique, whereby a thin section is dissolved in very dilute acid to produce a residue of organic material, which is then revealed in fine detail by staining. It is a pity that this valuable method, already known in the 19th century, has been so rarely used. The examination of thin sections of limestones has too often followed the habitual methods of the igneous petrographer with heavy emphasis on viewing with crossed nicols. ELIAS and CONDRA (1957) applied dilute hydrochloric acid to their sections of bryozoans, wiping the liquid across the slide with finger-tips, to reduce the thickness nicely to show the fine structure. Dangeard distinguished sharply between boring, endolithic algae and encrusting calcareous algae. He was at pains to ensure freedom from the accusation that the algae were modern plants, by sectioning limestone cores from bore holes. His results led him to propose, as a working hypothesis to be investigated further, that any completely physico-chemical theory for the origin of oöids is too absolute and that room should be left for biological influence. NESTEROFF (1955b, 1956a, b) took up the theme, using similar methods, and urged, from a wide experience of Recent and ancient oölites, that they are the constructions of algae, probably Cyanophyta.

NEWELL et al. (1960) leached oöids from the Browns Cay area, Great Bahama Bank (p.135), and found algal colonies in the residues. Encapsulated colonies in the oöids were identified as a blue-green alga, *Entophysalis deusta*, and a green alga, *Gomontia polyrhiza*. Both species were in various states of vegetative growth and deterioration. The algal filaments were infested by what were probably bacteria and or fungi, particularly those filaments deeply buried in the oöids. Newell and his colleagues saw that these algae cut across the lamellae but, at the same time, many of them are buried under unbroken lamellae. Though they are all boring species, their occurrence so deep in the oölitic cortex would seem to be a result, not of their deep penetration, but of their burial by oölitic aragonite deposited after the colony had been emplaced. "There is no suggestion that they play a direct part in the precipitation of calcium carbonate." (NEWELL et al., 1960, p.492.) SHEARMAN and SKIPWITH (1965) have found the oöids of the Trucial coast tidal deltas, in the Persian Gulf (p.195), to be so impregnated by algal bores, both full and empty, that many oöids can be crushed readily with the point of a needle (personal communication). Bahamian oöids, on the other hand, are dense and uncompactable, and fail largely by fracture (FRUTH et al., 1966).

As a rule oöids, Recent and ancient, when digested with weak hydrochloric or other suitable acid (and provided the organic matter has not been destroyed by severe diagenetic or metamorphic processes), do show an organic residue which mimics closely the concentric fabric of the oöid. There is no reason to suppose, however, that this pattern is other than a direct and simple result of the time relation between algal colonization and oölitic growth. NEWELL et al. (1960) have drawn attention to the concentrations of organic material and algal colonies in the unoriented layers (p.79) of Browns Cay oöids. They have proposed an alternation of conditions between oölitic growth in turbulent water and the adherence of aragonitic detritus rich in organic material during times of quiescence, either low turbulence or temporary burial. They also observed the similarity between the unoriented aragonite and "recrystallized" skeletons which have been altered to micritic aragonite. BATHURST (1966, p.21) has recorded that the lamellae of unoriented aragonite in the cortex of Recent oöids from Browns Cay are identical in appearance with the micrite envelopes of Bahamian grains in general (p.384). These lamellae have smooth, concentric outer surfaces where they are covered by oriented aragonite. Their inside surfaces are irregularly digitate against the inner lamellae of oriented aragonite. As ILLING (1954, p.49) put it: "The initial orientation of the crystallites was similar in each part, but the outer layer appears to have been added after the inner zone had suffered disorientation. In other oöliths the reverse situation is found, and the destruction of the concentric pattern has proceeded inward from the surface, leaving a turbid border on an oölitically coated nucleus." This relation would be expected if the history of an oöid were divided into alternate periods of oölitic growth and algal boring.

     An interesting outcome of the work by SHEARMAN and SKIPWITH (1965) in the Trucial coast embayment has been the discovery that the oöids there are not only penetrated by a network of algal mucilaginous material but are probably partly coated by the same material. Such coating may be a widespread phenomenon (p.252). The influence of this mucilaginous stuff on the growth of oöids has yet to be discovered. It could be critical if the coating were found to act, say, as an adhesive in the accumulation of aragonite detritus, or as a membrane through which ions must diffuse between the water and the oöid surface. The presence of this mucilage does require that we look cautiously at any hypothesis of simple chemical precipitation on a free surface—particularly in view of Mitterer's discovery (p.253) that the proteins in oöids are closely akin to those in carbonate secreting organisms.

     Against the exclusively organic control of oöid growth must be set the undeniable truth that oöids can grow in dark caves, in industrial water systems, under controlled inorganic conditions in the laboratory—even in kettles (Fig.239). In all these situations the action of photosynthesis is excluded.

The role of bacteria

The role of bacteria in the precipitation of calcium carbonate has had a fascination for geologists ever since the publication of DREW's paper in 1914. Drew had grown denitrifying bacteria in sea water and produced a precipitate of aragonite needles among which were a few finely laminate concretions with nuclei. These, he suggested were like some of the Bahamian oöids. VAUGHAN (1914b) took up this point and related it to the known fact that, under certain laboratory conditions, $CaCO_3$ precipitated from solution appears first as a gel within which aragonite spherulites grow. Vaughan collected samples of aragonite mud from the sea floor west of

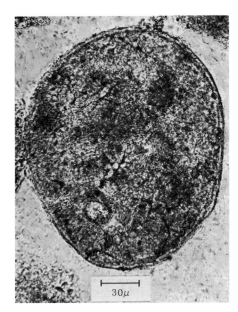

Fig.239. Oöid of calcite that grew attached to a wire mesh in a kettle used for boiling water in a hard water district. Slice. (Material given by Dr. J. A. E. B. Hubbard.)

Fig.240. Peloid coated with an oölitic film about 3µ thick. Bimini Lagoon, Bahamas. (From BATHURST, 1967a.)

Andros Island, Bahamas, where Drew had been working. After keeping these for at least three months he found in the mud little spherulites of aragonite with diameters up to 0.17 mm. They had nuclei of carbonate sand grains including foraminifera but were soft and easily crushable. Vaughan took these observations as an indication that marine oöids were all formed in aragonite muds as a direct result of the vital activities of denitrifying bacteria. In fact, all known Recent marine oöids have a tangential arrangement of optic axes and concentric lamination: they reveal no trace of spherulitic (radial-fibrous) structure, except where they have been diagenetically altered to calcite.

In recent years MONAGHAN and LYTLE (1956) have grown aragonite spherulites in Bahamian sea water inoculated with aragonite mud containing sulphate reducing bacteria. After six weeks of anaerobic incubation, spherulites and needles of aragonite were found in a white scum on the surface of the solution. LALOU (1957a, b) took black euxinic mud from the bay of Villefranche-sur-Mer, off the French Mediterranean coast, and enriched it with glucose. A film grew on the surface bearing within it spherulites of aragonite, with some calcite and dolomite mud.

These various observations are valuable in that they demonstrate the kind of processes by which marine oöids clearly do not grow. All the different processes begat spherulites in which the aragonite needles were radially oriented. There is no evidence to indicate that the same processes could yield oöids with a preferred tangential orientation of aragonite c axes.

Inorganic precipitation

From colloidal suspension. BUCHER (1918) addressed himself to the matter of oöid growth after he had been impressed by the work of Schade on urinary calculi. Schade had found that when crystals grew in an emulsion of the same mineral they formed concretionary bodies and that the orientation of the crystals was dependent on the purity of the emulsion. Pure colloidal suspensions gave radial arrangement of crystals, impure suspensions a tangential. Bucher recalled that various colloidal preparations of minerals were known to give rise to concretions, with crystal orientations ranging from radial, through imperfectly radial and tangential, to perfectly tangential. Drew's spherulites were among the examples he cited, also hailstones which showed both orientations. Bucher suspected that oöids grow in a colloidal mud. He added the pertinent comment that algae and bacteria are important only in so far as they help to create the necessary chemical conditions for growth. He also proffered criteria for the recognition of secondarily deposited oölites, composed of oöids that had been moved away from the mud in which they had grown. His three criteria were: absence of mud matrix, good sorting and cross-bedding.

Nevertheless, the association of oölites with aragonite muds on the present sea floor is not a close one, the two sediments accumulating in quite distinct depositional regimes. NEWELL and RIGBY (1957, p.54) drew attention to the great distance between the vast area of oölite shoals at the southern end of Tongue of the Ocean, Bahamas, and the nearest mud area. They also remarked that oöids were not found by them in any aragonite muds.

From supersaturated solution. Bucher emphasized what should be a central consideration—the search for a process which, despite considerable variety of algal and bacterial association, will nevertheless produce a concretion with a specific type of crystal fabric. Oöids grow in a variety of environments, not only in the sea but in caves and mines where photosynthesis is impossible. Yet, in some deposits of hot springs, as in the sea, oöids are found with tangentially oriented aragonite, as in the famous Karlsbad springs in Karlovy Vary, Czechoslovakia (HATCH et al., 1938).

CAYEUX (1935) was impressed by the great regularity and fineness of detail in the oriented aragonite layer (Fig.118, p.77). This is a fabric of extraordinary perfection, more typical, he averred, of inorganic than of organic growth. Certainly such perfection (Fig.122, p.82) is inconsistent with the spiralling of algal tubes, or the haphazard adherence of fine skeletal debris as SORBY (1879) proposed. Nor does it appear likely that any process of recrystallization could transform a casual accumulation of adherent detritus into a strongly oriented fabric of this type.

The necessary conditions for inorganic growth have been enumerated by CAYEUX (1935, p.261) as elevated temperature, abundant supply of $CaCO_3$, saturation of the solution, agitation and a source of nuclei. DONAHUE (1965), on the basis of his knowledge of cave oölites and his own laboratory experiments, put the requirements as four:

(1) supersaturation of a $CaCO_3$ solution;

(2) available detrital nuclei;

(3) agitation of grains;

(4) a splash cup.

These are in agreement with the opinions of recent authors who have worked on modern oölite shoals, for example ILLING (1954), NEWELL et al. (1960) and D. J. Shearman and his co-workers at the Imperial College, London.

The supersaturation is necessary for crystal growth. The detrital nucleus provides a surface, which may or may not contain aragonite crystals, on which aragonite can grow. Agitation prevents the cave pearl from becoming a stalagmite, and ensures generally that the whole surface is presented, on suitably frequent occasions, to the supersaturated solution. The splash cup keeps the oöid from straying out of the growth-promoting environment: its marine equivalent is the system of topographic and hydraulic controls that keep the oöid on the oölite shoal.

Two of the four requirements, the supersaturation and the agitation, may be looked at a little more closely. It is known, from the work of Smith (p.286), Cloud (p.290) and Broecker and Takahashi (p.290), that water near the Browns Cay oölite shoals, on the western edge of the Great Bahama Bank, shows an eastward, bankward, increase in salinity accompanied by a fall in the product $aCa^{2+} \cdot aCO_3^{2-}$. This trend, indicating precipitation of $CaCO_3$, is thought to result from evaporation and loss of CO_2 from the warm, shallow Bank water (p.287). A similar gradient has been found by Kinsman for the Abu Dhabi region in the Trucial coast embayment, from open shelf through oölite shoals to lagoon (p.204). Yet the Bahamian oölite shoals are only about 5 km from the deep water of the Florida Straits and it is by no means obvious how a body of tidal water could achieve a significant increase in supersaturation in about 5 h (travelling 5 km at 25 cm/sec). It seems probable that the critical factor in marine oöid growth, given the normal level of supersaturation for $CaCO_3$ in waters of low latitudes, is the existence of a physical system that permits the play of Weyl's alternating process of precipitation and reorganization, during bottom traction and burial, as described on p.303.

The role of agitation. The amount of agitation needed is far from clear. In parts of the Laguna Madre oölitic deposition on nuclei is uneven (FREEMAN, 1962) so that parts of a detrital nucleus can be exposed while the remainder is covered by irregular bosses of oölitic cortex. Presumably this unequal growth reflects unequal exposure of the oöid surface to the supersaturated solution: the mechanics of such

a partial shielding are not obvious. NEWELL et al. (1960, p.490) also recorded localized, rounded protuberances on some Bahamian oöids. Where the grains have a
surface polish, as at Browns Cay, this is caused by abrasion (as various workers
have shown experimentally, e.g., NEWELL et al., 1960, p.490; DONAHUE, 1965;
GRADZIŃSKI and RADOMSKI, 1967) and an origin in the zone of breaking waves is
indicated. On the other hand, I have found (BATHURST, 1967a) that carbonate sand
grains in southeast Bimini Lagoon, and elsewhere (p.309), are coated with oölitic
films about 3 μ thick (Fig.240). These films are petrographically indistinguishable
from the oölitic lamellae in the thick coats of the Browns Cay oöids, but have mat
instead of polished surfaces. Yet their host grains are immobilized by a subtidal
mat in an environment of low turbulence in a lagoon, where the only agitation is
caused by the burrowing of *Callianassa*, the browsing of parrot fish or the rare
hurricane once a decade or so.

   Thus we must look again at the conclusions of Cayeux (1935, p.251) and
NEWELL et al. (1960, p.495): that growth necessarily takes place while the oöid is
suspended freely in the solution. Nevertheless there seems no doubt that the more
turbulent water is, in the sea, associated with the development of thicker oölite
coats. NEWELL et al. (1960, p.487) have found that, around a depth of 2 m on the
oölite shoal, the proportion of coated nuclei is highest. They claim that this depth
gives optimum growth conditions for the area. TAFT et al. (1968) found the most
thickly coated oöids on the New Providence platform just below low-water level.
This matter is discussed further on p.314.

   Growth in the laboratory. The preceding considerations leave untouched the
problem of how oöids grow. What are the steps, in terms of nucleation of new
lattices, of the lodgement of ions on the surface of the crystal lattice, of possible
readjustments of the lattice with age, of the cementation of needles to give a hard
surface that will take a polish?

   Attempts to grow oöids in the laboratory have not yielded lamellae of
tangentially oriented aragonite needles. MONAGHAN and LYTLE (1956) precipitated
aragonite from sea water by adding CO_3^{2-} ions. Above a critical concentration of
CO_3^{2-}, spherulites, instead of single needles, were produced. Dr. Monaghan has
kindly confirmed in a letter that the similarity between the laboratory spherulites
and Bahamian oöids referred to on p.113 of their paper does not apply to the
crystal fabric inside the spherulites. The laboratory products were spherulites in the
strict sense and differ from the Bahamian oöids not only in size but in fabric, as
they are constructed of radiating needles. USDOWSKI (1963) also grew optically
negative spherulites. DONAHUE (1965) reproduced in a general way the environment
of cave oölite growth, in the laboratory. He dripped a supersaturated solution of
$CaCO_3$ into a porcelain evaporating dish in which there were some calcite pisolites
with diameters about 4 mm to act as detrital nuclei. A calcite cortex 0.17 mm thick
composed of five lamellae was deposited in six months. The new fabric showed no

preferred crystal orientation, possibly, Donahue thought, owing to the rapid growth rate. This was 1.5–2.5 times greater than rates recorded for natural cave pearls.

The coherence of the aragonite fabric. Electron photomicrographs of Persian Gulf oöids were examined by Twyman at The Imperial College, London (unpublished data, 1963). These photographs, and those by Fabricius and Klingele, and Loreau, reveal an agglomeration of tangentially arranged needles. At a fracture surface these are seen to have been separated and pushed around rather like the hairs in a carpet that has been brushed against the pile: clearly they were not so tightly cemented together that they could resist being disoriented. The apparent lightness of their coherence has also been demonstrated by D. J. Shearman (unpublished) at The Imperial College. He bathed Trucial oöids in a solution of hydrogen peroxide in an ultrasonic vibrator. They disintegrated to a needle mud.

Kinetics of oöid growth. A profoundly revealing series of tests was carried out by WEYL (1967; experiments C and D). He added dry oöids to sea water in a saturometer, the modified pH meter he had designed some years earlier (WEYL, 1961). The pH dropped immediately, indicating that carbonate had been precipitated. A variety of other grains was added, including fragments of aragonite and calcite skeletal debris, even grains of carborundum, with the same result (P. K. Weyl, personal communication, 1964). So long as visible nuclei (above colloidal size) were provided, precipitation took place. Two important implications follow from these observations. One is that nucleation is here heterogeneous, aragonite (since this is sea water) on calcite, or on carborundum or other presented substrate. The second is that, despite the presence in the sea water of a variety of suitable nuclei of colloidal dimensions, the factor that brought about a major withdrawal of $CaCO_3$ was the arrival of a *large* (sand grade) surface area.

Weyl arranged a system wherein sea water could be passed through a sample of oölite in a U-tube. The pH of the water was measured before (at A) and after (at B) its passage through the oölite, using the same pH meter with a by-pass. When flow began the pH at B fell at once and then rose gradually until it was only slightly less than the pH of the fresh sea water entering the oölite at A. There it remained. Flow was then stopped for at least 16 h. When it was started again the pH at B promptly fell as before and then rose slowly until it reached the same steady state, about 1 mv below that at A. If flow was stopped for a shorter period, the initial fall in pH at B was either less or did not take place.

The experiment was then carried a stage further. Taking oöids from an oölite shoal off Eleuthera Island, Bahamas (Fig.127, p.97), Weyl discovered a fundamental dichotomy in the response when these were added to the local sea water in a saturometer. Oöids that had been moving in the tidal current, on top of a bar, when collected, caused only a slight change in pH, equivalent to the precipitation of 2 p.p.m./3 min. Yet oöids taken from the bottom of a channel, where they had been stationary for an unknown period, showed a precipitation of 19 p.p.m./3 min. Finally, dry oöids caused a precipitation of 23 p.p.m./3 min. The recently active oöids, after standing in sea water, were reanalyzed in fresh sea water after 4 h and 9 h: their precipitation rates had increased to 6 p.p.m./3 min and 12 p.p.m./3 min respectively. Tests on calcarenites off Bonaire Island (p.530) yielded similar results, in the local sea water: for grains from the surf zone a precipitation rate of 3 p.p.m./3 min, but, for dry grains higher up the same beach, 13 p.p.m./3 min.

In the experiment where sea water was passed through oölite, the thickness of the aragonite precipitated on the oöids could be calculated from the surface area of the grains, the volume of sea water passed and the fall in pH. Such calculations indicate the precipitation of about 200 Å of aragonite before the steady state is reached, after which there is an approximate rate of precipitation of 0.03 Å/min, assuming that growth took place uniformly over the whole surface area. The crystal orientation of the overgrowth has yet to be determined.

Weyl has suggested a hypothesis to explain the sudden fall in the pH of water newly passed through the oölite, the subsequent rise in pH and the need for a resting period of about 16 h. The idea is that during the resting period the overgrowth recrystallizes. It is conceivable that the aragonite is precipitated so rapidly that the new crystal lattice contains many impurities and imperfections (say water, Mg^{2+}, dislocations). The distortion of succeeding lattice overgrowths, and thus their solubilities, must increase until a stage is reached when the rising solubility product constant of the overgrowth equals the ion activity product $aCa^{2+} \cdot aCO_3^{2-}$ of the sea water and further precipitation is impossible. During the resting stage the new aragonite recrystallizes, as movements of ions within the lattice enable the whole structure to adjust itself toward a more ordered lattice. When this change has gone far enough, rapid crystal growth is again possible. A period of burial, when the oöid is in a saturated (not a supersaturated) pore solution, so that precipitation is nil, would allow the entire metastable overgrowth to recrystallize. Such a condition must characterize the bottom of the channel off Eleuthera, where the oölite sample collected for the experiment presumably consisted mainly of sub-surface oöids from a stationary sediment. Weyl concluded that oöid growth follows a cycle, passing from a stage of rapid precipitation as oöids move in a traction carpet, through a stage when the overgrowth is increasingly soluble and precipitation becomes progressively slower, to a resting stage when the oöid is buried—and finally erosion followed by renewed rapid growth. The lengths of the stages would depend on the local conditions of oöid transport and the building of sand bodies (as in BALL, 1967). The critical requirement seems to be an opportunity for the oöid to be buried.

Finally, in reviewing the question of inorganic precipitation, we must remember that the $^{18}O/^{16}O$ ratios of Bahamian oöids (p.280) found by LOWEN-

STAM and EPSTEIN (1957) and the Sr^{2+} content in Trucial coast oöids (p.205) measured by KINSMAN (1964b) are consistent with the hypothesis that aragonite is precipitated close to equilibrium with sea water. KINSMAN's (1969) data indicate that the Sr^{2+} content of oöids from Cat Cay (near Browns Cay), Bahamas, is consistent also with growth in equilibrium with the water. Both sets of data fit with an inorganic process possibly modified slightly by biological activity (p.261). Though there can be no doubt as to the importance of organic processes in modifying locally the chemistry of sea water, especially in microenvironments a few tens or hundreds of microns across, there seems no reason for regarding organic activity as other than a modifying factor in an inorganic, non-metabolic chemical system.

Growth in caves and mines

Oöids and pisolites grow in limestone caves (Fig.122, p.82) and mines as mobile forms of stalagmite, having much the same relation to stalagmite as oncolite does to sheet algal stromatolite. The development of a fixed stalagmitic form is prevented where the "cave pearls"—so named by HESS (1929), also called in German *Höhlenperlen*—move frequently enough to avoid being cemented to the floor. Claims by EMMONS (1928) and by DAVIDSON and McKINSTRY (1931) that cave pearls need not roll, that calcite is precipitated all round a nucleus as it rests on the

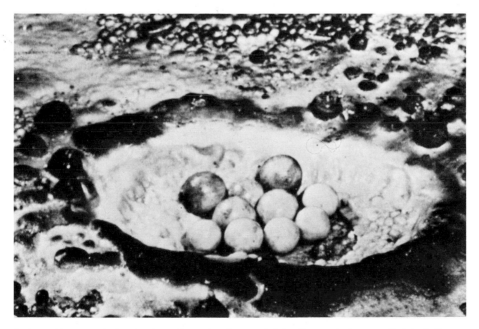

Fig.241. Cave pearls (each about 2 cm in diameter) lying in a splash cup. Carlsbad Cavern, New Mexico, U.S.A. (From HESS, 1929.)

cave floor and that crystal growth pressure (p.467) lifts the nucleus, probably represent one extreme of a range of agitation of the nucleus, the other extreme being the polished and agitated pisolite of HESS (1929; Fig.241) lying in a splash cup. In a study of the fabrics of cave oöids and pisolites DONAHUE (1969) found that the more agitated accretions are characterized in thin section by distinct concentric lamination and compact structure and show a pseudo-uniaxial cross with crossed polars. Surfaces tend to be smooth, even polished. The less agitated accretions have indistinct, porous laminae and lack preferred crystal orientation. Similar results were obtained by GRADZIŃSKI and RADOMSKI (1967). The essential requirements for well polished, concentric pearls are supersaturation, nuclei, agitation, splash cup (p.301).

Calcite oöids and pisolites have a fabric of radial-fibrous crystals with a concentric pattern marking time-dependent discontinuities of layer thickness, fabric or colour (as in KELLER, 1937; KIRCHMAYER, 1964; Fig.242, 243). These concentric discontinuities, in mines, have been related to annual variations in hydrology and cave meteorology (KIRCHMAYER, 1962). Stages in development of neomorphic spar in layers of different ages, up to more than 100 years old, are given in HAHNE et al. (1968). Conditions governing the precipitation of calcite and aragonite in stalactites and stalagmites—and so presumably in oöids and pisolites— were investigated in terms of pH, temperature and the ionic composition of the ground water by J. W. MURRAY (1954), SIEGEL (1965) and SIEGEL and REAMS (1966).

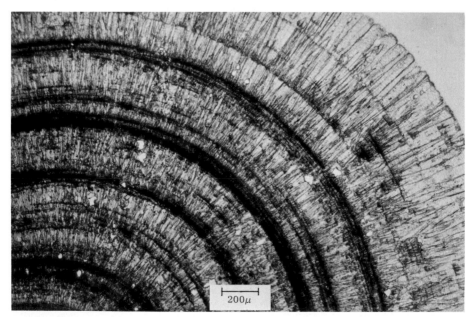

Fig.242. Mine oöid. Layers of radial-fibrous calcite alternating with thinner dark layers of equant calcite, each pair of layers marking the precipitate of one year. Slice. Styria, Austria. (Courtesy of M. Kirchmayer.)

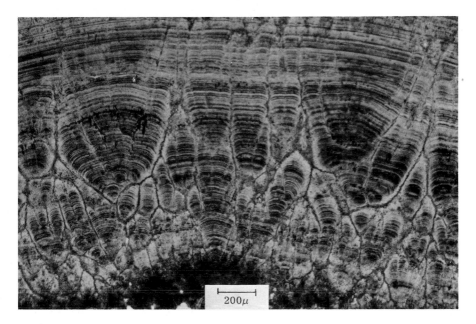

Fig.243. Mine oöid. The more even radial-fibrous growth of aragonite in the upper part of the picture was preceded, in the lower part, by growth of discrete bundles of radial-fibrous aragonite, giving rise to a mamillary surface. The larger bundles grew over adjacent smaller bundles and buried them: a fabric characteristic of speleothems and linings of hot water pipes in hard water districts. Slice. Ruhr Coal Basin, Germany. (Courtesy of M. Kirchmayer.)

FURTHER DEVELOPMENT OF OÖLITE THEORY

On the basis of past work on oölites, discussed in the preceding pages, the matter of oöid growth can be further developed. Further thoughts on the questions of growth kinetics, the control of crystal fabric, nucleation, the growth situation and limitation of grain size, will occupy most of the remaining pages of this chapter.

The building of oölitic aragonite in the sea

The nature of the marine oölitic fabric poses three main questions:

    (1) Why is it aragonite and not calcite?

    (2) What is responsible for the preferred tangential orientation of c axes?

    (3) How do the Ca^{2+} and CO_3^{2-} ions in sea water become organized into a rigid, collision-resistant framework?

    The question of the restriction of precipitation of $CaCO_3$ to aragonite has been dealt with in Chapter 6, where the role of Mg^{2+} is emphasized.

Tangential growth of aragonite

Sorby's snowball. The conditions for growth of tangential aragonite are of special interest since this fabric is an attribute of all optically studied Recent marine oöids. Various authors have related this orientation of needles to the agitation of oöids which is so common a factor in their growth environment. Either we must follow SORBY (1879, p.74) and regard the growth process as a mechanical contact-adhesion of loose detrital needles of aragonite mud (as oncolites) when they encounter the surface of the oöid, or we must accept that the needles grow rooted to a nucleus already embedded in the surface. A close look at the adhesion or snowball process prompts the suspicion that adequate cohesive forces are not available. Crystals in a liquid will not adhere unless, either there is a cementing material, or the crystals are colloidal and are attracted to one another as a result of surface charges, in which case the process is not mechanical. Though a mucilaginous pellicle (p.252) may act as an adhesive it is not invariably present. There is one serious obstacle to the acceptance of such a mechanical snowball process. In many oölite shoals the oöids are highly polished and various workers have shown that the polish is a result of abrasion. Bearing in mind the hard, rigid framework of the oölitic coat, intuition encourages the conclusion that any mechanical adhesion of needles must be prevented by the abrasive process. Thus growth by pure mechanical adhesion seems profoundly unlikely.

While on the question of the snowball mechanism it is necessary to remember that, in one important group of carbonate grains, this process undoubtedly plays some part. In the oncolites, mobile and more or less spherical stromatolites (p.220), growth progresses by the trapping of local detritus among algal filaments as the oncolite rolls over the bottom. It is doubtful whether such a process could often produce the high degree of preferred crystal orientation found in oöids.

Just as this book was going to press, the comfortable feeling of certainty on this question was disturbed by the publication of a paper by JENKYNS (in press) on the fabrics of the cortex in some Jurassic oöids in Sicily. The oöids are micritic in texture and show the usual well developed concentric lamination. Yet—within the cortex there are coccoliths and their debris. The question must be asked again—does the oölitic growth process involve some mechanical trapping or are these Sicilian oöids simply oncolites of unusual geometrical perfection?

Modification of radial-fibrous growth. Suppose, instead, that needles grow by the addition of ions, first to suitable crystalline nuclei embedded in the oöid surface, and thence by normal accretion on crystal faces. (A crystalline or lattice nucleus in this kinetic sense is the initial ordered aggregation of molecules from which a crystal grows. It is to be distinguished from the detrital nucleus of the oöid which may be a shell, pellet, etc., and which contains many thousands of potential crystal nuclei in its surface. The lattice nucleus may not be aragonite.)

This process would lead to the evolution of a radial-fibrous fabric (p.422) with a preferred orientation of c perpendicular to the oöid surface. Such an orientation arises naturally as a result of competition between crystals that grow most rapidly in a direction parallel to one crystallographic axis. Yet this is not the crystal orientation that we find dominating the oöids. We know, moreover, that aragonite can grow in the sea with c axis and longest axis perpendicular to a wall: on the Great Bahama Bank this preferred orientation is common (p.364), inside the chambers of Foraminiferida and the utricles of *Halimeda*. It is also typical of the aragonite cement in beach rock (p.369). Nevertheless, when aragonite accretes on the outer, exposed surfaces of *free* calcarenite grains, the preferred orientation is c parallel to the wall. This type of growth fabric occurs not only in the conventional oöid (as at Browns Cay), but in 3 μ oölitic films on a variety of calcarenite grains outside the environment of the oölite shoal—in the coralgal, stable sand, grapestone and mud habitats on the Great Bahama Bank, in eastern Batabano Bay and on the Campeche Bank off Yucatán (BATHURST, 1967a, p.91).

DONAHUE's (1965) work on cave oöids and on oöids grown in the laboratory led him to specify four necessary conditions for oöid growth (p.301). The only obvious condition among these which distinguishes the oöid surface from the foraminiferal chamber or beach rock pore is agitation. The walls of the foraminiferal chamber and the beach rock pore are not subject to bombardment by other grains, whereas the oöid is involved in a succession of collisions. Donahue regarded agitation as essential because in no other way could the growing oöid avoid being cemented to the surface of the splash cup by the precipitation of carbonate. Agitation alone, though, does not yield oölitic coats with c parallel to the surface. Many cave oöids (and pisolites) have c radial. The scant data so far published indicate that the calcite cortex is always radial-fibrous, but that the aragonite cortex can be either radial-fibrous or tangential. Examination of this generalization is urgently needed. There is no known correlation between the degree of agitation and the degree of preferred orientation of *aragonite* needles. Indeed, comparison of the much agitated oöids of Browns Cay, Bahamas, with the grains with oölitic films in Bimini Lagoon twenty miles to the north, shows no detectable petrographic difference. Even the appearance of the lamellae of oriented aragonite is identical. This despite the intense abrasion at Browns Cay, yielding polished oöids, and the low agitation in Bimini Lagoon where the oöids have mat surfaces.

However, it would be unwise at the present stage of investigation to ignore the possibility that, as a result of grain collisions, the radially disposed needles are preferentially broken or abraded so that the surviving fabric is dominated by tangential c axes. The tangential fabric of marine oöids may persist simply because it is the fabric that is most resistant to abrasion, as suggested by RUSNAK (1960) and USDOWSKI (1963).

Implantation of new nuclei. It would be wrong, though, to assume that the aragonite crystals grow rooted to the oöid surface from the very first. It is conceivable that nuclei are formed to begin with in the water, separately as colloidal aragonite mud, and later become attached to the oöid as a result of an inorganic surface effect. After all, new nuclei must be continually supplied to the system, otherwise growth would lead to a single large crystal. In the sea water there must be numerous colloidal aragonite particles, the product of abrasion, with dimensions measurable in tens or hundreds of Ångstroms. These potential nuclei of colloidal size may become adsorbed on the oöid surface (BATHURST, 1967b, p.452).

BATHURST (1968, p.1) suggested that the nuclei of the new needles might be pieces of ionic monolayer, with aragonite structure, elongate (owing to the (010) and (110) cleavages) parallel to the c axis. These colloidal nuclei, once suspended in the sea water, might become adsorbed on the oöid with their longest axes parallel to the oöid surface, so giving a tangential arrangement of c axes.

Effect of ion activities. F. Lippmann has considered the influence on crystal orientation of the movements of ions as they approach the crystal face (unpublished paper given at the Fourth Meeting of Carbonate Sedimentologists, University of Liverpool, 1967). The Ca^{2+} ion has a nearly spherical symmetry, so that its orientation as it arrives at the growing face cannot be critical. The CO_3^{2-} group, on the other hand, has approximately the form of a triangle. In the crystal of aragonite, this is flattened in a plane perpendicular to the c axis: consequently any constraint upon its orientation as it approaches the growing crystal could influence the orientation of the growing lattice. In sea water the activity of Ca^{2+} greatly exceeds that of CO_3^{2-}: thus surfaces of aragonite crystals will have an excess of Ca^{2+} adsorbed on them. CO_3^{2-} groups arriving at a face will be guided by the numerous, positively charged cations especially where these happen to have the 5.74 Å spacing of the c dimension of the unit cell. Lippmann has estimated that, in such circumstances, the CO_3^{2-} groups will settle into positions giving, statistically, a highly preferred tangential orientation of c. Aragonite crystals in solutions that have an excess of CO_3^{2-} over Ca^{2+} will be coated with CO_3^{2-} groups. Without guidance from Ca^{2+} ions, these will lie with an infinite variety of orientations and will provide an equal variety of substrates for epitaxial growth. By the law of competitive growth, a radial-fibrous fabric will develop.

Lippmann referred to some experimental evidence in support of this hypothesis. When he was precipitating aragonite by hydrolysis of KOCN in the presence of $MgSO_4$, peculiar aggregates appeared resembling double-ended sheaves. The straight middle part may be supposed to have grown while the crystal surfaces were covered with Ca^{2+}: the radial-fibrous excrescences should have begun to grow after the neutral point was passed, owing to consumption of Ca^{2+} and greater concentration of CO_3^{2-} following continued hydrolysis of the KOCN.

Origin of radial-fibrous calcite in certain Jurassic limestones

SHEARMAN et al. (1970) observed the anomalous fact that in many ancient oöids the detrital nucleus, presumably once aragonite, has been replaced by calcite cement, yet the oölitic coat, which was also aragonite, retains its original concentric structure upon which has been superimposed a pattern of radial-fibrous calcite crystals. Why should the nucleus have undergone dissolution and precipitation involving a visible cavity stage (p.327), whereas the oölitic coat appears to have escaped the cavity stage? How did the oölitic coat pass from the aragonitic stage shown in Fig.244a to the radial-fibrous calcite stage shown in Fig.244b, yet preserving its concentric structure?

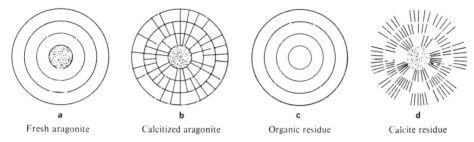

a	b	c	d
Fresh aragonite	Calcitized aragonite	Organic residue	Calcite residue

Fig.244. Diagram to illustrate the experiments of SHEARMAN et al. (1970). a. Aragonite oöid with tangential arrangement of optic axes—black lines are organic-rich layers. b. Concentric pattern partly replaced by radial-fibrous calcite. c. Organic residue after leaching with HCl. d. Calcitic residue after dissolution of organic material with H_2O_2.

In the Jurassic Ambléon Limestone in the Département of Ain, southern French Jura, the oöids bear traces of the original concentric structure as well as a late radial-fibrous structure (Fig.121). The concentric structure is revealed by thin dark lines that, on closer inspection, are seen to be made of micron-sized crystals, the lines commonly being only one crystal wide. Peels show the concentric structure more clearly than thin sections, as very closely spaced lines, separating concentric zones. The radial structure is shown partly by radially distributed dusty inclusions, which rarely cross from one zone to the next, and partly by very closely spaced radially arranged Becke lines which never cross from one zone into another. With crossed polars the fast direction is radial although the extinction is somewhat speckled by a scattering of randomly oriented crystals. When the limestone was stained with malachite green, in aqueous solution, the stain was accepted by the organic matter in the oölitic coats but not by the cement. The concentric lines showed clearly (Fig.244c), intensities of green varying between zones, and it was obvious from the distribution of the stain that organic matter was intimately mixed throughout the oölitic coat. Determinations indicated that organic matter in the total limestone is certainly more than 1 wt.%, so that in the oölitic coats it must be considerably higher. Limestone surfaces were polished, gently etched in acid, washed in water, immersed in hot H_2O_2 to dissolve organic matter, washed again and finally dried. When peels were made of the prepared surface, tiny acicular needles came away with the peel. It became apparent that needles were arranged in concentric layers, as in Fig.244d. Needles were up to $10\ \mu$ long, showed first order grey polarization colours (thus less than $1\ \mu$ thick) and were optically fast. One oöid gave 39 zones of acicular calcite, with a mean zone thickness of $10\ \mu$. The clear spaces between zones (Fig.244d) are the sites of the organic matter seen in Fig.244c, which was revealed by the green stain.

SHEARMAN et al. (1970) suggested that, during diagenesis, the original aragonite was dissolved but the organic matter remained as a template on which tiny crystals of calcite cement grew with the typical preferred orientation that accompanies competitive growth (p.422). Owing to the concentric arrangement of the main substrate surfaces the resultant fabric was radial-fibrous.

It is common in oölites to see that the outermost layer of radial-fibrous calcite has spalled off during compaction. What Shearman et al. noticed was that the pieces of spalled outer layers had broken cleanly—parallel to the radial-fibrous structure (Fig.245). Thus it was clear to them that the dissolution-precipitation process which had produced the radial-fibrous calcite had acted before compaction. It was,

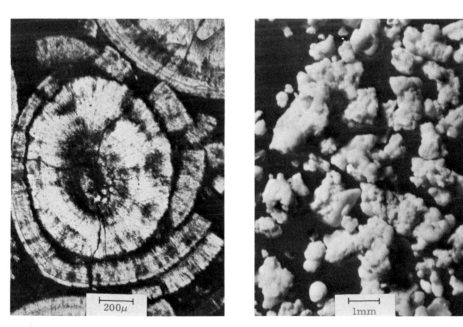

Fig.245. The outer layers of this oöid were disrupted during compaction. The radial fractures parallel to the secondary radial-fibrous fabric indicate that this grew before compaction and before cementation of the micritic matrix. Slice. Upper Mississippian. Montana, U.S.A. (Slice lent by R. L. Folk.)

Fig.246. Grapestone. Berry Islands, Bahamas.

therefore, an early, near-surface diagenetic process, the eogenetic stage of CHOQUET-
TE and PRAY (1970).

Grain size limits

A problem that has fascinated numerous geologists is the reason for the high
degree of sorting in oölites. Why are there no marine oöids a centimetre in diameter? Why an upper limit commonly in the region of 1 mm? CAROZZI (1960,
pp.241–244) has presumed that after an oöid has reached a critical size, it can no
longer be moved by the local currents and so it ceases to grow. Though there is some
truth in this, in that generally the smaller the grain in any oölite the more likely it is
to be coated, there is no doubt that, off Browns Cay, the entire oöid bed is in
motion at some stage of the tide. The surface is extensively ripple-marked, with
wavelengths from a few centimetres to a meter or so. To think in terms of a critical
amount of agitation that must be exceeded before oölitic growth can proceed, is to
ignore the fact that the largest oöids off Browns Cay (supposedly too large to
continue growing) are yet vastly more agitated than the carbonate grains with
oölitic films in Bimini Lagoon or off the west coast of Andros Island, Bahamas
(BATHURST, 1967a). The Bimini and west Andros oölitic coats are much thinner
than those of Browns Cay (3μ against more than 100μ) and it may be that there is
a correlation between agitation and growth rate: the more a grain moves the faster
it will grow. It seems likely, on the scant available evidence, that the rate of growth
is proportional to the amount of time spent by the oöid in motion. All marine oöids
divide their time between the traction carpet and the subsurface (p.316), but the
oöids in Bimini Lagoon, for example, will spend a greater proportion of their
existences in the subsurface environment than those off Browns Cay: in this
stagnant environment, saturation of the pore water must be achieved in a matter
of seconds or minutes after burial, when no further precipitation is possible.

The good sorting of oölites may, therefore, be attributed partly to the
following (BATHURST, 1968). Small oöids rapidly become larger because, being the
most mobile, they spend more time in motion than the larger grains. The growth of
larger oöids is slower because, as they grow, they spend a progressively smaller
proportion of time in motion. The largest in the Browns Cay oölite shoal nevertheless still move with the bed load. One may argue here that the shoal is a young
one, only about two thousand years old, and that the largest oöids have not yet
reached their maximum size. However, the present surface is underlain by other
cross-stratified Quaternary oölites of similar grain size. This obvious relationship
is a reminder that one factor determining the upper size limit must be the time at
which the oöids are finally buried. The more rapid their accumulation, the quicker
the burial and the smaller the grain size. The coarser the oölite the longer it has
been exposed in a growth environment, other things being equal.

A reason for the high lower-size limit on the crest of the shoal becomes obvious as one watches the tidal current pouring over the Browns Cay shoals, and the waves rippling and breaking in the sunlight. The smaller grains cannot be deposited there. Their cessation velocities are too low and if such grains were introduced onto the bank they would immediately be winnowed away.

This prompts an awkward question. If tiny grains cannot reside on the crest of the shoal, where do the young, embryonic oöids grow? To claim that they all begin as relatively large grains because the nuclei are large is to forget that some have nuclei so small that they escape detection under the microscope. Certainly many oölitic coats occupy as much as 80% of the radius of the oöid. It is plain, therefore, that oöids begin to grow in the deeper water and move toward the crest of the shoal as they become larger. NEWELL et al. (1960, pp.487–489) found that, on the Browns Cay oölite shoals, the proportion of coated grains increases toward the crests, also the number of lamellae on the nuclei. This means that the older oöids reside nearer the crest of the shoal. This region acts as a receptacle where the older oöids accumulate. The evidence does not indicate that the most agitated water of the crest is necessarily the place of most rapid growth. The newer, smaller oöids must be growing in the deeper, less turbulent water, below the 2 m depth at which NEWELL et al. (1960) found a sudden decrease in the proportion of coated grains.

The mechanics of the process of accumulation follow from the known movement of flood and ebb waters. In a situation like the Browns Cay shoals, oöids will be carried up the westward slope as flood water moves Bankward from the Florida Straits. The older and larger the oöid the less the probability that it will move over the crest of the shoal or make the complete return journey with the ebb tide. So, a lag deposit of mature oöids is gathered in the shallowest water. A similar process must be acting on the Bankward slopes, though the eastward facing lobes of oölite seen in air photographs (Fig.247) indicate a net eastward migration of oölite.

The relatively sharp upper size limit for oöids in an oölite is unlikely to be due solely to the high inertia of the largest oöids or to the cessation of growth following burial. The most important change that takes place, as medium sized oöids continue to grow, is the increase in mass. This increase, equivalent to the growth in volume, rises as the third power of the radius. At the same time the resistance of the water to the motion of the grain increases only as the square of the radius. Oöids of this size will obey closely the impact law (RUBEY, 1933). The force exerted by one grain on another, which is a function of mass and acceleration, will itself go up as the third power of the radius, but the fluid resistance to grain movement will rise only as the power of two. Thus the amount of physical damage in the way of abrasion that oöids can inflict on one another must mount rapidly as they grow (BATHURST, 1969, p.7). As the surface area affected by collision depends on the tangential meeting of two spheres, the *proportion* of total surface of an oöid damaged will remain the same as the oöids grow. The damage will simply become

Fig.247. Aerial photograph of spillover lobes of oölite, migrating eastward (toward the left) over skeletal sands with dark patches of *Thalassia:* wave-lengths of megaripples 50–100 m. Westerly edge of the Great Bahama Bank. (Courtesy of M. M. Ball.)

more severe. It seems logical to conclude that at some stage the rate of loss of mass by abrasion will equal the rate of gain by oölitic growth. Then the net growth would be zero. Though a maximum size would be fixed, largely according to the level of turbulence on the shoal, it should be noted that paradoxically, turbulence assists both growth (by encouraging motion) and abrasion.

An observation that would appear to emphasize the importance of abrasion as a controlling factor springs from the study of oöid shape. Where oölitic growth continues over a non-spherical nucleus, such as a rod or plate, the outer surface of the oöid becomes increasingly spherical, partly for geometrical reasons, but partly because, although growth takes place all over the surface, net accumulation is greater on the flatter surfaces away from domes and coigns. (Differences in radii of curvature for such large objects as oöids can have no significant effect on relative rates of crystallization and dissolution.) The greater susceptibility to abrasion of curved ends and coigns over flat surfaces is well known. DONAHUE (1965, p.252) noted that some cave oöids are preferentially polished over their "high points". This gives rise to "localized zones of polish along raised areas". We may, therefore, conclude that the balance between precipitation and abrasion over different parts of the oöid surface favours more rapid net growth on the less sharply convex and on the concave surfaces. Experimental verification of this conclusion would be welcome.

It is interesting to estimate the proportion of its residence time on the oölite shoal that the oöid will spend actually growing (BATHURST, 1967b, p.452). Field observations on the Browns Cay oölite shoal offer some clues. The oölites are distributed as lobate areas of megarippled sand. The shapes of these spillover lobes, visible from the air, indicate net Bankward drift (Fig.247). The megaripples have wavelengths of 50–100 m and amplitudes of 0.5–1 m: depth of water is about 2 m. The shallowest oölites are at the oceanward margin and are extensively rippled-marked (wave lengths of a few cm to several metres) and are moved almost continually by tidal currents and waves. Away from the Straits edge, however, only the crests of the megaripples are made of loose grains. I have swum over these megaripples. Their crests are covered with small scale ripple-mark, the grains being moved by the ebb and flow of the tide.

In the troughs between the crests the grains are lightly bound together by the subtidal algal mat which prevents movement of a bed load. There are scattered codiacean algae and blades of *Thalassia*. The surface appears to be immobile. The megaripples do not move in the tidal currents and I can only suppose that they are formed in some different current regime, perhaps during a hurricane. I have found, from a comparison of aerial photographs, that elsewhere in the Bahamas megaripples can remain without detectable migration, their individual crests separately recognizable, for more than twenty years (p.121). It must be supposed therefore that, on the Browns Cay oölite shoals, oöids can remain buried for decades before once more finding themselves dancing in the bed load on a mega-ripple crest. These crests of loose grains comprise only a small part of the total area of floor, say 5% at the most, so that more than 95% of an oöid's existence in that area may be spent in the subsurface environment. This must be an important consideration in the estimation of rates of oölitic growth. An oölitic cortex which has been growing, off and on, for 2,000 years may have spent a tiny fraction of that time actually growing. As oöids may be temporarily buried 30 cm or more down, this fraction of time may well be much less than 1%.

THE ORIGIN OF GRAPESTONE

The origin of grapestone (Fig.152, 246), the clusters of carbonate sand grains cemented by micritic aragonite (p.87), has so far aroused little interest, no doubt because grapestone has not been widely recognised in limestones and because few geologists have had the opportunity to see Recent examples in the field. If the present day distribution of this sediment in shallow carbonate areas can be taken as a guide, it has not only been one of the major limestone-forming materials, but is a valuable index of depositional environment.

Detailed petrographic studies of grapestone have been carried out by ILLING (1954, p.30), PURDY (1963a, b), KENDALL and SKIPWITH (1969a); also by R. G. C. Bathurst (unpublished) on material from the area between Cockroach Cay and

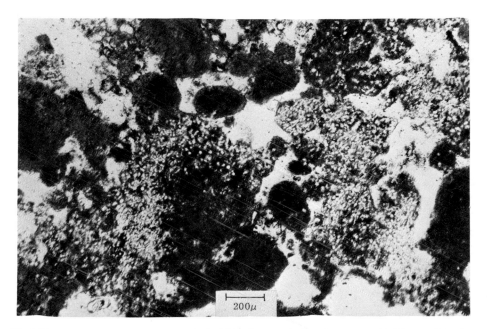

Fig.248. Grapestone, much bored by endolithic algae, and extensively micritized by precipitation
of micrite in the emptied bores. Slice. Berry Islands, Bahamas.

North Fish Cay, Berry Islands, Bahamas. Petrographic details are given on p.87.
The floor here is part of Purdy's grapestone lithofacies (p.121). The grains, examined in thin section and in acetate peel, have been intensely micritized as a result of
the filling of vacated algal bores with micritic aragonite (p.383). So widespread are
the results of this process (Fig.248) that unaltered skeletal grains are very rare. Live
boring algae are ubiquitous and numerous and must profoundly influence the
chemical microenvironment.

It is of interest that associated with the grapestone are flakes of the same
material, up to 2 cm or 3 cm across and 1–2 mm thick (also recorded from the
Campeche Calcilutite of Wisconsin age on the Yucatán shelf by LOGAN et al.,
1969, p.64). It is tempting to associate these hard, brittle flakes with the soft,
delicate flakes of subtidal mat (p.121). The soft flakes form by erosion of
a subtidal mat that is unusually coherent where it lies on shallow shoals in
the Berry Islands, Bahamas. The serious difficulty here is that no intermediate
stages of flake—between the soft and the hard—have been found, with the exception of a possible intermediate form which I saw in a brief visit with a field
party and was unable to examine closely. As a working hypothesis it can be proposed
that carbonate grains immobilized for long periods in a relatively tough subtidal
mat, and in pore water depleted of CO_2 by the metabolism of numerous boring
algae—are welded together in bunches or sheets by the precipitation of micritic
aragonite. Support for this hypothesis has lately come from the work of MON-

(1967, p.82) who found similar calcarenite aggregates (algal lumps) in the back-reef lagoon, east Andros Island, Bahamas. The sand grains are bound together by *Schizothrix*-type filaments (p.222). The aggregates are coated either with a dense brown layer of coccoid algae, or a gelatinous material, or with *Scytonema*-type filaments. "The skeletal grains within the lumps become completely unrecognizable because of algal boring and appear as cryptocrystalline pellets." (MONTY, 1967, p.83.)

MILLIMAN (1967a) and KINSMAN (1969) have noted the high strontium content of grapestone in the Bahamas which Milliman attributed to inorganic (extra-cellular) precipitation of a micritic matrix of aragonite and Kinsman to inorganic precipitation biologically modified. Their data are consistent with micritization associated with endolithic algae (p.383) and precipitation encouraged by intense algal photosynthesis.

Another aspect of grapestone has been emphasized with the discovery by H. D. Winland (personal communication, 1968) of encrusting ophthalmid forami-niferids in the matrix. Earlier recognized by Illing and Purdy, these foraminiferids have been revealed by Winland in great quantity, by a method of staining high-magnesian calcite (WINLAND, 1968). It is not yet clear to what extent the ophthal-mids contribute to the binding of the grains in the first place or, simply taking advantage of a convenient home, are incidental to grapestone genesis.

NEEDLE, OÖID OR GRAPESTONE?

Having examined the three modes of inorganic precipitation of aragonite in shal-low carbonate seas, it is proper to ask what the conditions are that favour one form rather than another—supposing some of the needles are inorganic.

The growth of aragonite needle muds and of oölites is, it would seem, en-couraged by turbulence sufficient to keep the particles moving in sea water super-saturated for sand-sized $CaCO_3$ nuclei. The growth of grapestone, by contrast, looks as though it may be dominated by algal photosynthesis on a sand floor immobilized by a subtidal mat (p.121).

Tentatively, therefore, we can guess that the series oöid–needle–grapestone represents a gradient of decreasing grain agitation, but *not* of water turbulence (BATHURST, 1968). On the oölite shoal, owing to the high degree of turbulence, nuclei of clay grade are scarce. The level of supersaturation of the water is main-tained at a high level by the evaporation and by the continual introduction of colloidal supersoluble debris from grain abrasion. The only nuclei available in quantity are sands such as skeletal grains, pellets and allochthonous quartz grains. Precipitation must, perforce, take place on these. In the quieter areas of needle deposition, turbulence is minimal but adequate to put the needles occasionally to suspension. Nuclei of clay grade are plentiful and precipitation on these is easier than on the sand grains which, in this environment of weak currents, scarcely move at all. In the Bahamas grapestone region, turbulence is greater than in the mud region though less than on the oölite shoals. However, the grains are not only bound by the subtidal algal mat but are isolated by it from the overlying sea water. They are stationary for long periods and are enclosed in a microenvironment from which CO_2 is abstracted rapidly by boring algae. There is no abrasion to cause a preferred tangential orientation of aragonite on the grains and, possibly owing to interference by the algae, radial-fibrous orientation is rare. On warm shallow banks, in water that is only a few centimetres deep at low tide and is locally exposed, this metabolically dominated process can proceed rapidly. Here our guesswork must end and await more substantial support from field and laboratory studies.

FURTHER READING

The fascinating section on marine oölites by CAYEUX (1935) is still the most acute, readable and complete survey of the subject. To bring it up to date the reader can turn to EARDLEY's (1938) masterpiece and the careful researches of ILLING (1954) and NEWELL et al. (1960), PURDY (1961, 1963a, b) and WEYL's (1967) fundamental experiments on the chemistry of growth. The growth and neomorphism of cave oöids and pisolites are well surveyed by KIRCHMAYER (1964), GRADZIŃSKI and RADOMSKY (1967) and growth conditions are examined by DONAHUE (1962, 1965, 1969). Other references are given in the text.

Additional references not given in the preceding chapter

On ancient oöids, USDOWSKI (1962), LABECKI and RADWANSKI (1967); on cave oöids, ERDMANN (1902), MACKIN and COOMBS (1945) and OTTEMANN and KIRCH-MAYER (1967); KAHLE (1965a, b) on aspects of oölite diagenesis.

Chapter 8

DIAGENESIS IN THE SUBAERIAL, FRESH WATER ENVIRONMENT

> Our reasonings grasp at straws for premises and
> float on gossamers for deductions.
>
> *Adventures of Ideas*
> A. N. WHITEHEAD (1942)

INTRODUCTION

Rooted in past habits of geological thought there still lingers the legend that lithification can only be effected by the action of heat and pressure—that sediments can only be changed into rocks when they are deeply buried. However true this may be for terrigenoclastic sediments, one of the refreshing trends of the last twenty years has been a growing awareness that large scale cementation of carbonate sediments can proceed at the familiar temperatures and pressures of the subaerial environment. The early onset of carbonate lithification (some might call it penecontemporaneous) is demonstrated not only by the very existence of Pleistocene and Recent limestones but also by the evidence of hardground formation (p.394) at innumerable levels in the stratigraphic column, and by the characteristic absence of compaction phenomena in limestones generally. Once it had been established that a hardened limestone need not have been deeply buried, logic demanded that we ask of every limestone the place of its conversion, whatever its subsequent tectonic history may have been. In each rock we must search for evidence of the environmental controls that obtained while the process of lithification was going on. This is a quest as yet scarcely begun, but in future years it should be possible to identify, for each limestone member, the diagenetic environments in which dissolution and precipitation took place. Moreover, it should not be surprising to find, within one limestone, evidence for lithification in two or even three diagenetic environments, namely the near-surface submarine, the near-surface subaerial, and the deeper crustal environment where pressure-solution plays a significant role. In the pages that follow, the near-surface fresh water, subaerial diagenesis of emergent carbonate sediments will be examined.

A word of warning is necessary here. The sequence of diagenetic changes that has emerged from studies of Tertiary to Recent limestones makes a gratifyingly pretty picture, but this result should not be allowed to hide the fact that the evidence comes from a small number of limestones of peculiar, post-Glacial history. In other words, it has yet to be shown that Bermuda and other islands are, diagenetically speaking, the microcosm of the vast tracts of limestone formed after deposition in the great epeiric seas of the past.

EVIDENCE FOR DIAGENESIS DURING SUBAERIAL EXPOSURE

In 1957, Ginsburg made the following significant comment: . . . "the late Pleistocene Miami Oolite is thoroughly cemented by calcite at the exposed surface and below the ground-water level. But where it is still in the marine environment or above the ground-water table it is so friable that it can easily be broken down into individual ooliths." (GINSBURG, 1957, p.96.) This clear expression of the role of

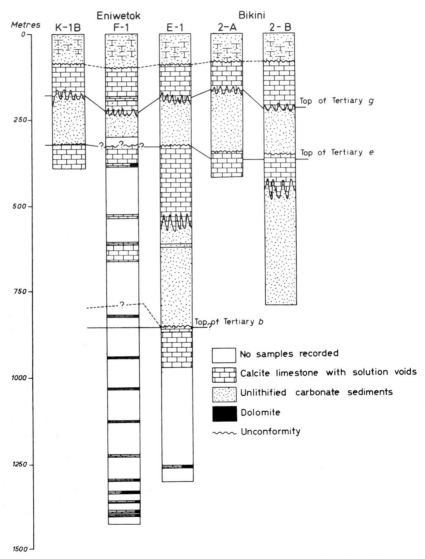

Fig.249. Lithological and stratigraphical logs from bore holes at Eniwetok and Bikini Atolls, Marshall Islands, Pacific. (After SCHLANGER, 1963.)

fresh water in the lithification process aroused widespread endorsement and, in the absence of strong advocacy of other diagenetic processes, the transformation of unconsolidated carbonate sediments to limestone by exposure to fresh water has achieved the popular status of chief limestone-forming process. Whether the supremacy of this role is justified remains to be seen, particularly in the context of research on submarine cementation. Yet there is no doubt that vast masses of Pleistocene limestone have been cemented in this way.

SCHLANGER's (1963) survey of two bore-holes through the Recent to Eocene carbonates of Eniwetok Atoll, in the Marshall Islands of the tropical Pacific, provides an unusually full documentation of the stratigraphic evidence as well as petrographic detail. In these holes (Fig.249) there is evidence of alternating zones of unconsolidated or very lightly cemented, friable carbonate sediment and of hardened, well cemented limestone (Fig.250). The friable sediment consists of a mixture of unaltered marine calcarenite and calcilutite, including primary aragonite, whereas the limestone is characterized by moulds and solution channels and an absence of aragonite: only calcite remains. The tops of the better developed hardened zones are sharp and coincide with discontinuities in the stratigraphic succession as revealed by the larger Foraminiferida (COLE, 1957). The cause of the unconformities was not peculiar to Eniwetok because unconformities of the same type and age have been recorded in bore-holes in Bikini Atoll, 350 km away to the

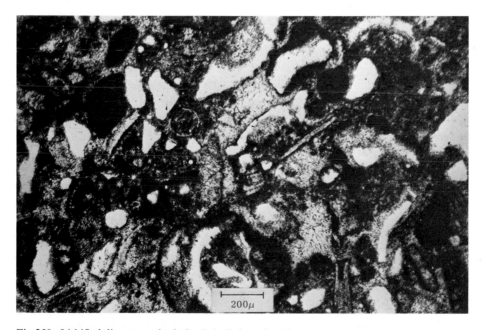

Fig.250. Lithified limestone in hole *F-1*, Eniwetok (Fig.249). $CaCO_3$ has been redistributed yielding a cemented micrite traversed by numerous vugs, some clearly formed by dissolution of molluscan shell. Slice.

east (EMERY et al., 1954). Such widespread contemporaneous erosion can only have been the result of a general fall in sea level. Extensive and prolonged subaerial exposure at these times is suggested by LADD's (1958) discovery of shells of land snails in the Bikini subsurface sediments: similar shells are known today on high forested islands (see also LADD and TRACEY, 1957). Ladd recorded dense concentrations of pollen and spores from land plants. Thus, in view of the emergence of the Cainozoic carbonate sediments at various times, it is highly probable that the hardening of the limestone horizons coincided with their concomitant subaerial exposure to rain water.

LOGAN et al. (1969, p.65, 68), have described how, on the Yucatán shelf, Mexico, the carbonate sediments of Wisconsin age that lie beyond the 90 m bathymetric contour are cemented with aragonite, whereas those in shallower outcrops are cemented with calcite. L. C. Pray (personal communication, 1969) has drawn attention to the significance of this distribution of cement, pointing out that the deeper Wisconsin sediments have always lain below the presumed 90 m deepest Wisconsin sea level and have, therefore, escaped the fresh water meteoric diagenesis suffered by sediments nearer the shore.

The Pleistocene limestones of Bermuda have attracted much attention, and studies by FRIEDMAN (1964), GROSS (1964), LAND (1966, 1967) and LAND et al. (1967) further support the contention that lithification of Bermudan carbonates coincided with subaerial exposure of the sediment surface. Barbados carbonate sediments have been similarly affected (HARRIS and MATTHEWS, 1968; MATTHEWS, 1968; PINGITORE, 1970).

In addition to these examples of lithification of emergent marine carbonates, it has been shown in many areas that the hardening of Pleistocene carbonate sediment is not only directly attributable to subaerial exposure but is superficial. On South Bimini Island in the Bahamas it is possible to dig downward through a crust about a metre thick to friable sediment underneath. LAND (1966, p.102) recorded that the early hardening, the change from his grade II to grade III (p.326), commonly takes place at the carbonate surface which is freshly exposed in man-made road cuts. This general type of alteration has been known as "case-hardening" and its progress in the Bahamas was referred to by ILLING (1954, p.48) who wrote that outcrops which are, "sufficiently cemented on the surface to have a distinct ring when struck or walked on (hence the local name of 'clinkstone'), are commonly soft enough at a depth of a few feet to be dug out with a spade. On New Providence Island during the war, roadways were bulldozed through these dunes after the hard crust had been removed. A year or two later, the freshly exposed surface was as hard as ever." Presumably this early, superficial lithification is dependent both on the direct action of incident rain water and on older meteoric water drawn to the surface of the sediment by a combination of evaporation and capillary action. KORNICKER (1958a) and MULTER and HOFFMEISTER (1968) have described limestone crusts related to soils.

It is obvious, therefore, that despite the rather slight and patchy information at present available on the progress of near-surface lithification, there is already a strong case for the view that the process goes on wherever carbonate sediments have contact with fresh water. This case receives support from petrographic studies made with the microscope, also from X-ray data on mineralogy, from laboratory experiments, and from measurements of the contained oxygen and carbon isotopes. These various approaches to the problem will be considered in the next few sections.

FABRIC AND MINERALOGICAL EVOLUTION OF RECENT AND PLEISTOCENE LIMESTONES:
BERMUDA AND FLORIDA

In the preceding discussion use has been made continually of the words "lithification" and "hardening" rather than the word "cementation" which, at first sight, might appear more appropriate. This is because the process whereby a loose accumulation of carbonate grains becomes a limestone is a complex mineralogical and fabric adjustment involving dissolution and precipitation, the migration of Mg^{2+}, Fe^{2+} and other ions and also gravity collapse, compaction and general spatial rearrangement of material. The detailed fabric changes in Pleistocene carbonates have been studied in Bermudan limestones by FRIEDMAN (1964), BATHURST (1966), LAND (1966) and LAND et al. (1967); also in Eniwetok and Guam Atolls by SCHLANGER (1963, 1964) and in British Honduras by TEBBUTT (1967, 1969).

Fabric evolution of Bermudan limestones

It is convenient to follow here the idealized scheme for diagenesis in these rocks (Fig.251), put forward by LAND (1966) and shown in LAND et al. (1967), with additions after FRIEDMAN (1964) and GROSS (1964), since this scheme summarizes and amplifies the sequence of changes seen in thin sections and described earlier by Friedman (1964, p.783) and Bathurst (1966, p.23). LAND's (1966) petrographic study is also the most complete (see also LAND et al., 1967). He divided the development into stages I to V.

Stage I: the initial sediment. This stage represents the unconsolidated lime sand which was composed (by analogy with the Recent) mainly of a mixture of molluscan, coral and *Halimeda* debris (aragonite), foraminiferids, coralline algae, and echinodermal fragments (high-magnesian calcite), and some lithoclasts of Pleistocene limestone (low-magnesian calcite). This assorted mineral assemblage is metastable in sea water, probably because of a combination of five factors: (*1*) the tropical shallow sea water is supersaturated for all three forms of $CaCO_3$, (*2*) the concentration of Mg^{2+} in water inhibits precipitation of calcite (p.243), (*3*) the high concentration of Mg^{2+} in the water presumably protects the high-

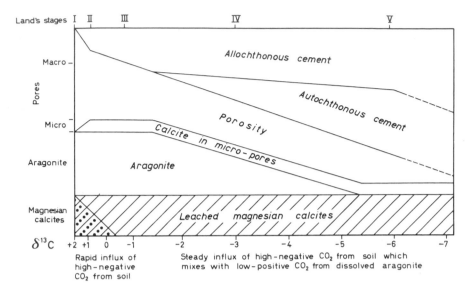

Fig.251. Idealized scheme of diagenetic change in Bermudan carbonates. (After LAND, 1966,
with δ^{13}C/low-magnesian calcite fitted approximately from data in FRIEDMAN, 1964.)

magnesian calcite to some extent from dissolution, (4) actively growing syntaxial
overgrowths may be considerably more soluble than their substrates (Weyl's
experiment, p.252), and (5) probably all grains outside the surf zone are more or
less completely encased in organic mucilaginous envelopes which isolate them
from the pore solution (p.252).

Stage II: the first cement. This is marked by the appearance of low-magnesian calcite cement on the surfaces of the grains near their points of contact.
DUNHAM (in press) has pointed out that, where the interface between the cement
and the pore has the form of a meniscus (as in Fig.252), this implies precipitation in
the vadose zone from water localized around the points of grain contact. The curved
outline of the water meniscus would control the outline of the precipitated cement.
The pores of echinodermal structures are filled with optically continuous calcite
cement and these grains may be encrusted with a thin layer of optically continuous
rim cement. Visible intragranular pores, such as tubes in *Halimeda*, cells in *Amphiroa* or chambers in foraminiferids, are partly or entirely filled with calcite cement.
The primary grains of aragonite and high-magnesian calcite are unchanged and the
rock is friable, the grains being easily separable. The restriction of the early cement
to the regions of the point contacts does not apply to all carbonate sediments of
other places and ages, many of which are first cemented by a uniformly distributed
fringe of aragonite (as in beach rock, Fig.276, p.367).

Stage III: the loss of Mg^{2+}. This stage is commonly hard to distinguish visibly from stage II, but it is clearly identifiable with the help of a magnesium spot test and by X-ray diffraction (LAND, 1966). The only minerals present are aragonite and low-magnesian calcite, and the grains are commonly coated with a thin fringe of calcite cement which constitutes up to 20% of the rock volume. The Mg^{2+} in the high-magnesian calcites has gone. These one-time grains of high-magnesian calcite are now scarcely distinguishable in thin section from their primary magnesian-rich precursors, but may show a slight increase in translucency, possibly owing to the decay of organic matter. The loss of Mg^{2+} from the high-magnesian calcite grains of coralline algae, without noticeable fabric change as viewed in thin section, has been remarked by FRIEDMAN (1964), GROSS (1964), LAND (1966), LAND et al. (1967), PURDY (1968), and GAVISH and FRIEDMAN (1969), so it must be assumed that any porosity that may have been developed in the course of the alteration was submicroscopic in dimension. Schlanger, on the other hand, noted the susceptibility of foraminiferids and coralline algae, in the Pleistocene limestones of Guam, to replacement by a more coarsely crystalline calcite spar (p.348).

It is important to realise that in stages II and III there is no visible sign of dissolution of $CaCO_3$ and, thus, the cement must have been derived from outside the rock, presumably from above. Possibly the appearance of cement is related to the dissolution of some overlying limestones, and the development of red soils, some of which are preserved in Bermuda. The allochthonous source of the $CaCO_3$ of the cement, here and in other limestone successions, poses questions of fundamental importance about the amount of carbonate sediment that must be dissolved in order to release $CaCO_3$ for cementation. Must all limestone stratigraphic successions be incomplete simply because they are cemented? Just how fragmentary is the history that is presented by an apparently continuous succession of limestones?

Stage IV: dissolution-precipitation. During this stage the aragonite dissolves. The porosity remains reasonably constant as if development of secondary porosity is balanced by the deposition of calcite cement (LAND, 1966; LAND et al., 1967). The dissolution-precipitation process may, therefore, involve ion transport over small distances of only a few millimetres: there is no reason to regard the $CaCO_3$ of the cement laid down during stage IV as anything but autochthonous in derivation. Land described these rocks as "tenaciously indurated." He found, too, that strontium (KAHLE, 1965a) and organic matter are lost at this stage (LAND, 1966), as on Guam (p.347). Presumably it is at this stage that uranium in the aragonite is lost (HAGLUND et al., 1969).

Secondary porosity is maintained, until filled with calcite cement, so long as a mould of the initial aragonite grain is formed. Where the aragonite grain is coated with a layer of calcite cement, this layer persists after the aragonite is dissolved and forms a mould which is later filled with a cast of precipitated calcite. The micrite envelope (p.384) performs a similar function, staying as a stable residue (p.333) after the aragonite has dissolved (Fig.252A–D). In both cases the trend is toward

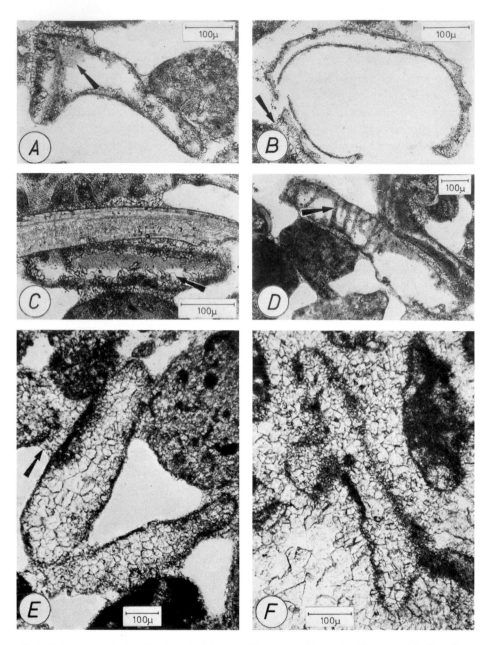

Fig.252. Stages in the lithification of Pleistocene limestones, Bermuda. The cement in the primary pores, around the grain contacts, is the meniscus cement (p.326) of R. J. Dunham and is indicative of precipitation in the vadose zone. All slices. A. Micrite envelope enclosing partly dissolved residue of molluscan shell (arrow). Some early calcite cement around grain contacts. B. Micrite envelope nearly emptied of molluscan shell. Some early calcite cement (arrow) around grain contacts. C. Micrite envelope enclosing partly dissolved residue of molluscan shell. Some early calcite cement around grain contacts and inside the envelope (arrow). D. Micrite envelope enclosing partly dissolved molluscan shell. Large micrite-filled bores remain (arrow). Some early calcite cement around grain contacts. E. Micrite envelopes filled with calcite cement. Primary pores still lightly cemented. (arrow). F. Fully cemented limestone with micrite envelopes and primary porosity filled with calcite cement. (From BATHURST, 1966.)

preservation of the form of the aragonite grain (Fig.252E) though its internal structure is lost. Some aragonite is replaced by calcite *in situ* via a system of solution films (p.347), with consequent conservation of the original lamellar structure of the shell wall. This process has been of frequent occurrence in the Pleistocene limestones of Florida and British Honduras where wet transformation of aragonite to calcite has been recorded by LAND (1966) and TEBBUTT (1967), especially in massive corals and bivalves.

The relation between dissolution of skeletal aragonite and precipitation of low-magnesian sparry calcite in the Pleistocene limestones of Barbados has been studied quantitatively by PINGITORE (1970). He confined his examination to the coral *Acropora palmata* and recorded the change from skeletal wall constructed of aragonitic sclerites to walls altered to a sparry calcite mosaic with associated pore-filling sparry calcite cement. The many terraces on Barbados are built of successively higher and older Pleistocene reefs, yet the amount of sparry calcite (replaced wall and cement) in the *A. palmata* does not increase progressively as the terraces are ascended. Pingitore found that the volume of sparry calcite, in the fully altered coral, closely approximates to that expected from dissolution of the aragonite skeleton followed by precipitation of the same mass of $CaCO_3$ as calcite spar. He found, too, that this process was a once-and-for-all event—the volume of sparry calcite had not continued to increase, after the aragonite had disappeared, in the older terraces. In fact 80% of the volume of the sparry calcite could be attributed to dissolution of the coral skeleton and reprecipitation. The 8% volume increase fitted the amount of space-filling calcite cement quite closely. This surplus volume, which follows the change from the denser aragonite to the less dense calcite, may well account for the first generation of calcite cement encountered in many limestones (p.434; as predicted by STEHLI and HOWER, 1961). The precipitation of second generation cement must wait upon some other, allochthonous source of $CaCO_3$ (p.441).

The results of HARRIS and MATTHEWS (1968) indicate that, in the vadose zone of the Pleistocene reef limestones in Barbados, most of the dissolved skeletal aragonite is reprecipitated locally as sparry calcite. These authors traced the movement of strontium during the dissolution-precipitation process. The unaltered reef aragonite in Barbados contains 6–8 p.p.m., but the calcite only 2–3 p.p.m. Thus, surplus strontium must accumulate in the ground water. So by measuring the ratio Sr^{2+}/Ca^{2+} in the ground water, the efficiency of the process can be assessed. After correcting for soil leaching and dilution by sea water, Harris and Matthews estimated that the dissolution-precipitation process is more than 90% efficient.

DODD (1966b) has made a careful study of Cretaceous limestones of Texas, in order to relate the period of dissolution-precipitation (as in Land's stage IV) to the degree of lithification obtaining at that time. A low degree of lithification is implied where micrite envelopes have collapsed or where moulds of shells have been invaded by sediment. An advanced degree of lithification is indicated where calcite cement in primary pore and adjacent mould can be shown to have grown outward from what is now a common surface.

Stage V: culmination in low-magnesian calcite. This stage marks a limestone consisting wholly of calcite and the remaining porosity (Fig.252F). In the Bermudan limestones this is a low-magnesian calcite (FRIEDMAN, 1964, p.784), with 0.01–0.03 mole % of $MgCO_3$ (LAND, 1966). The limestone now has a porosity of about 20%, with 40% of grains and 40% of cement. This residual porosity raises some interesting problems. To begin with, if the rock were ever to become more fully cemented with a porosity of 1–2%—such as is found in many limestones—a further allochthonous source of $CaCO_3$ would have to be provided since the existing mineral fabric of stage V is apparently stable under these particular conditions. Where would this source be? After all, a 20% porosity would absorb a great quantity of $CaCO_3$. In other words, we should not assume, when looking at a thin section of any fully cemented biosparite, that the cement is simply the redistributed $CaCO_3$ of the original aragonite. It may be so, in some cases, but, judging by the Bermudan and Barbados limestones, a considerable supply of allochthonous $CaCO_3$ is required for the complete cementation of some carbonate sediments.

The source of allochthonous $CaCO_3$. Some indication of the source of the allochthonous dissolved carbonate which provided the early cement in stage II was presented by LAND et al. (1967). They referred to certain limestones in the Bermudan succession which have altered to stage IV, in that the aragonite has been lost, but have almost no cement. Carbonate released by dissolution of the aragonite was largely carried away elsewhere. As these limestones (donor limestones, p.451) are most commonly to be found immediately under red soils, both they and the soils, the authors pointed out, are probably the remains of the carbonate sediments that supplied, by downward leaching, the allochthonous stage II cement of lower horizons.

Mineralogical evolution of some Bermudan and some Floridan limestones

STEHLI and HOWER (1961) pointed out that the chief mineralogical differences between the Recent and the Pleistocene marine carbonate sediments, from Florida and the Bahamas, are the dominance of aragonite and high-magnesian calcite (mostly 13–16 mole % $MgCO_3$) in the Recent material and of low-magnesian calcite in the Pleistocene (Fig.253). FRIEDMAN (1964) found a similar relationship (Fig.254, 255) in Recent and Pleistocene samples from Bermuda (his high-magnesian calcites had mostly 12–14 mole % $MgCO_3$) and this was confirmed by the studies of LAND (1966) and LAND et al. (1967). All the data show that the first mineralogical change in the fresh water regime is the loss of magnesium from the

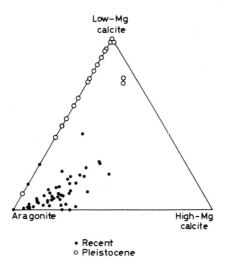

• Recent
○ Pleistocene

Fig.253. Mineralogy of Recent carbonate sediments from Florida (41) and the Great Bahama Bank (8) and of Pleistocene limestones from Florida (20). (After STEHLI and HOWER, 1961.)

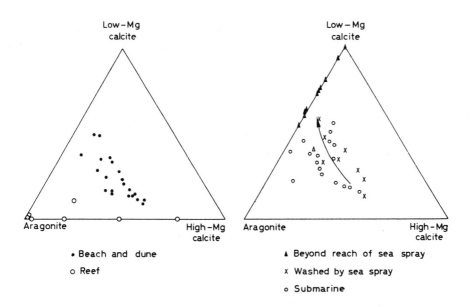

• Beach and dune

○ Reef

▲ Beyond reach of sea spray

x Washed by sea spray

○ Submarine

Fig.254. Mineralogy of Recent carbonate sediments from Bermuda. The beach sands are contaminated with lithoclasts of Pleistocene limestone containing low-magnesian calcite. (After FRIEDMAN, 1964.)

Fig.255. Mineralogy of Pleistocene skeletal calcarenites from Bermuda. Arrow indicates direction of diagenetic change as a result of exposure to fresh water. (After FRIEDMAN, 1964.)

magnesian-calcite and that only later does aragonite disappear gradually. The Pleistocene limestones of Florida and Bermuda are still diagenetically immature as a group, having a range of aragonite roughly from 0–50% (Fig.251). As the aragonite dissolves so the strontium content falls (STEHLI and HOWER, 1961; LAND, 1966) because this ion, occurring mainly in the aragonite, is less readily accommodated in the calcite lattice. The course of strontium removal during wet transformation of aragonite in corals (Pleistocene *Montastrea, Diploria,* and

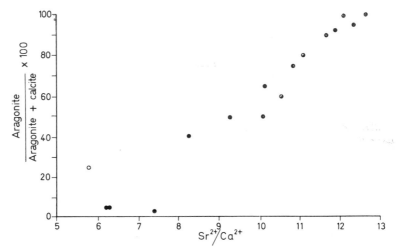

Fig.256. Relation of strontium/calcium ratio to the percentage of aragonite in Pleistocene corals. (After SIEGEL, 1960.)

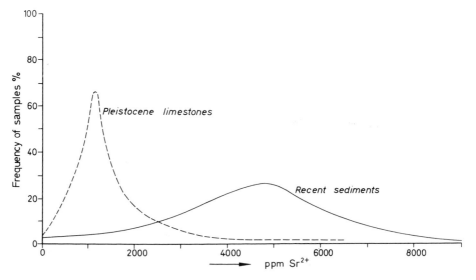

Fig.257. Percentage of samples with different values of Sr^{2+}, from 41 samples of Recent carbonate sediments and 20 samples of Pleistocene limestones. (After STEHLI and HOWER, 1961.)

Porites of Florida) is shown by the work of Harris and Matthews (p.329) and by the data of SIEGEL (1960) in Fig.256. Siegel interpreted this relation as showing inhibition of the inversion of aragonite to calcite by the Sr^{2+} ion, but it seems that he had in mind the dry, solid state transformation, particularly as he appealed to kinetic considerations. As these Pleistocene limestones have been subjected not to high temperature but to a bath of fresh water, it is more probable that the decrease of Sr^{2+}, as the proportion of calcite increases, simply records the loss by dissolution of the hospitable aragonite lattice. Various workers have shown that stabilization of aragonite by strontium is both theoretically and pragmatically improbable (p.241). A comparison of Recent with Pleistocene carbonates, in terms of Sr^{2+} content, by STEHLI and HOWER (1961) is shown in Fig.257.

One of the implications that flow from these various studies of mineralogical change is that, during near-surface lithification, large quantities of Mg^{2+} and Sr^{2+} are released. The subsequent fate of these ions is a matter of considerable importance but little understood.

The role of the micrite envelope in diagenesis

The micrite envelope, apparently formed centripetally in carbonate grains by precipitation of micrite in vacated algal bores (p.383), is initially composed of micritic aragonite or high-magnesian calcite, with some impurity depending on the amount of residual primary carbonate. In ancient limestones it is a low-magnesian calcite micrite (Fig.284, p.384) and, if it should have formed around an aragonite grain while on the sea floor, then that aragonite core has normally been replaced by calcite spar (Fig.252). The fabric of the spar seems generally to be typical of cement and, not infrequently, the envelope shows signs of fracture and collapse (Fig.258; BATHURST, 1964b). Both these data show that at one stage, during diagenesis, the skeletal core had been dissolved away leaving the micrite envelope empty of mineral material.

It is interesting that the micrite envelope should so often have remained as a rough mould of the aragonitic grain, subsequently to be filled with a cast of calcite cement (Fig.252, 259). This is one of the more important of the processes in the post-depositional history of many biosparites, enabling the original form of aragonitic grains to be preserved, albeit with loss of skeletal wall structure (BATHURST, 1966). It also poses two questions. Why should the micrite envelope survive, particularly if it is initially made of aragonite? What gives the envelope its peculiar stability?

KENDALL et al. (1966) drew attention to the high content of organic matter in the micrite envelope, in addition to that coating the grain (p.252). In the laboratory they had been impressed by the mechanical strength of the organic residues in carbonate grains: they suggested that, when during diagenesis the grain was dissolved, it was the organic matter of the envelope that remained as a mould.

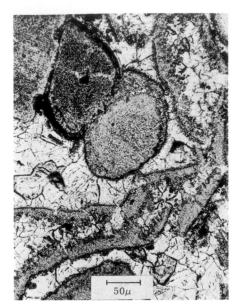 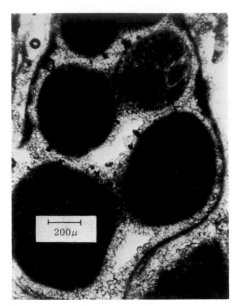

Fig.258. Collapsed micrite envelope enclosing what remains of a cement-filled cavity once occupied by a fragment of molluscan shell. Two upper grains are pressure-welded. Peel. Jurassic. Yorkshire. (From BATHURST, 1964b.)

Fig.259. Gastropod shell, enclosed in a micrite envelope, with chamber filled with micrite. The shell has been totally dissolved leaving a cavity which has been partly filled with calcite cement. Slice.

In subsequent discussions with Shearman it seemed likely that the micritic crystal size of the final calcite envelope is the result of a replication process. The micro-cellular structure of the organic matter, from which the micritic aragonite crystals had been dissolved, could perhaps have acted as a template. Thus, when calcite was eventually precipitated in the voids left by the aragonite, its growth was restricted to the same micritic crystal size as the original aragonite. A careful investigation of this possibility is greatly needed.

     The discovery by WINLAND (1968) that some Recent micrite envelopes are high-magnesian calcite has led him to suggest that the stability of the micrite envelope follows from its mineralogy. The high-magnesian calcite would not dis-solve, but would simply loose its Mg^{2+}, as in Lands' stage III. It is tempting to sympathise with this view, particularly as collapsed micrite envelopes, broken during compaction, look in thin section as though they were brittle at the time of fracture, not simply elastic organic matter. On the other hand, Winland's work, and my own, have shown that some Recent micrite envelopes are aragonitic, and these could not be transformed to calcite in the diagenetic environment without first being dissolved. It may well be that the aragonite is stabilized by being transformed to high-magnesian calcite, as it is in the hardgrounds of the Persian Gulf (p.373): TAYLOR and ILLING (in press) offered this solution of the difficulty.

Experiments on dissolution of magnesian calcites

Experiments of Friedman. In view of the changes undergone by high-magnesian calcites during diagenesis, between Land's stages II and III, the behaviour of these minerals during dissolution in the laboratory is of particular interest. FRIEDMAN (1964, p.810) investigated the matter in two series of experiments. In the first he kept some Recent sediments, mixtures of aragonite and high-magnesian calcite, in two continuously stirred solutions, initially distilled water and sea water, at 35 °C. Samples of the immersed grains were analysed by X-ray diffraction at monthly intervals for a year and a half. At the end of this period no mineralogical changes could be detected in any of the sediments. Two new series of experiments were then started, using sediments in distilled water and in sea water as before, but this time CO_2 was bubbled continuously through the water. In the solution which began as distilled water, low-magnesian calcite appeared and, over the three months of the experiment, the sediments were progressively enriched in this mineral. In the sea water no low-magnesian calcite was detected with certainty.

As high-magnesian calcites are more soluble than low-magnesian calcites (Fig.218, p.237) the expected result, if the processes were simple, would be for high-magnesian calcite (and presumably some aragonite) to dissolve and for low-magnesian calcite to precipitate. The solution would be undersaturated for one calcite, supersaturated for the other. That this twofold process only took place where CO_2 was present in the solution, but the cations of sea water were absent, is a reflection of the importance of HCO_3^- in the dissolution of carbonates and, in all probability, of the role of Mg^{2+} in inhibiting the growth of calcite (p.243).

Experiments of Land. LAND (1966; 1967, p.922) has taken the investigation further and has studied the dissolution of a range of high-magnesian calcites. He prepared finely ground samples and annealed them to remove the strain caused by grinding (such as had been observed by Garrels et al., p.257). This finely divided calcite was then put either into distilled water or into sea water, through which CO_2 was bubbled, and left there at 25 °C for at least 24 h. Dissolution was considered to have reached a steady state if, after 24 h, no change of pH could be detected over a period of 2 h. Analysis of the ground calcite by X-ray diffraction at the end of each run showed that the molar ratios Mg^{2+}/Ca^{2+} in the solid and in the steady state solution were essentially the same and that, in this sense, the dissolution of the magnesian calcites had been congruent. In fact, the solution over *Goniolithon* powder showed a slight enrichment in Mg^{2+} suggestive of the beginning of incongruent dissolution. On the other hand, since brucite is present in the skeleton of *Goniolithon* (SCHMALZ, 1965; WEBER and KAUFMAN, 1965), the anomalous result may record the preferential dissolution of $Mg(OH)_2$. In the same experiments the greater solubility of the more magnesian calcites was also confirmed, both in distilled water and sea water. In these short term studies by Land (24 h against

Friedman's 3 months) no low-magnesian calcite was precipitated: instead an equilibrium was reached between the solution and the metastable carbonate.

The peculiarity of this result should not go unremarked. The solution should ideally become progressively enriched in Mg^{2+} as the thermodynamically stable low-magnesian calcite is precipitated, so that eventually the ratio Mg^{2+}/Ca^{2+} in the solution is high enough for dolomite to be precipitated as well. For reasons referred to above and discussed elsewhere (p.250) this did not happen.

It is also important to remember that the reaction:

$$Ca_xMg_{1-x}CO_3^{2-} \rightarrow xCa^{2+} + (1-x)Mg^{2+} + CO_3^{2-}$$

is not reversible, whereas the reaction with pure calcite, in the absence of dissolved Mg^{2+}, is reversible.

It is, however, Land's work on longer term dissolution of magnesian calcites that has the more direct bearing on geological questions. While congruent dissolution seems to attain a steady state in about 24 h, the much slower course of seemingly incongruent dissolution displays a different behaviour. LAND (1967, p.923) placed sand-sized grains of high-magnesian calcites in distilled water sealed with "Parafilm" (to prevent evaporation but allow CO_2 exchange), and he left the mixture on a shaker for 4 months. After 1, 2 and 4 months the solutions were analysed for Ca^{2+} and Mg^{2+} by EDTA titration. The final solid was analysed in an X-ray diffractometer. Three of the types of material used had been involved also in the 24 h tests: these were three species of Rhodophyta, *Goniolithon* (29 mole % $MgCO_3$), *Amphiroa* (22 mole % $MgCO_3$) and *Lithophyllum* (15 mole % $MgCO_3$). A fourth, new, species employed was the foraminiferid *Homotrema* (11 mole % $MgCO_3$). For all four species the ratio Mg^{2+}/Ca^{2+} in the solutions became

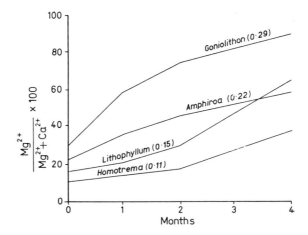

Fig.260. Changes in the compositions of the solutions bathing various magnesian calcites over a period of four months in the laboratory. Zero months marks the $MgCO_3$ mole % after 24 h. The initial compositions of the solid carbonates are given in brackets. (After LAND, 1966.)

progressively greater over the months: the final X-ray data showed that all but *Homotrema* remained as single magnesian calcite phases (Fig.260). *Homotrema* behaved differently and the final solid showed calcite as an additional phase: since this could not be found as free crystals Land assumed that calcite had replaced the primary $MgCO_3$ in the *Homotrema* skeleton. No changes could be seen in any of the materials in thin section with a microscope. The rate of transfer of Mg^{2+} to the solution was related to the initial $MgCO_3$ content of the skeleton, the higher magnesian calcites losing their Mg^{2+} more rapidly (Fig.260). All the final solutions were supersaturated for dolomite though none was precipitated.

Land regarded this process of selective Mg^{2+} removal from skeletons as incongruent dissolution, pairs of Mg^{2+} and CO_3^{2-} ions leaving the lattice while the primary Ca^{2+} and associated CO_3^{2-} ions remain as a permanent solid framework. The details of this process are not clear. To preserve the balance of charges it must be assumed that the Mg^{2+} ions released to the solution were each accompanied by CO_3^{2-} ions. The loss of all the $MgCO_3$ from the *Goniolithon* (Fig.260) in four months represents a removal of about 25% of the volume of the skeleton, yet this change was undetectable visually. Microscope examinations of naturally altered high-magnesian calcites (p.327) and of grains used in these experiments suggest that the high-magnesian calcite may not dissolve as a whole— where the solution is saturated for calcite and congruent dissolution cannot take place—but loses only $MgCO_3$ which is replaced by calcite. The absence of calcite replacement in the Rhodophyta has not been adequately explained, but was believed by Land to be exceptional. It would be interesting to have some indication of the mechanism of this incongruent dissolution. Do Mg^{2+} ions diffuse to the surface of the lattice to be replaced through a counter diffusion of Ca^{2+}? Even if the freed Mg^{2+} ions originate deep in the lattice, presumably the large CO_3^{2-} anion groups that accompany them must surely be derived from the crystal surface. The importance of surface area in incongruent dissolution has been noted by Berner (p.247). The geologically slow rate of lattice diffusion may be offset by the extremely small crystal size in the skeletons and the large resultant surface area exposed to ground water after organic matrices have decayed. It is necessary to bear in mind also the possible composite nature of the "crystal" unit seen with a petrographic microscope, noting the remarks by Towe and Cifelli (p.39) and Schroeder and Siegel (below).

There is another possibility. The transformation of high-magnesian calcite to low-magnesian calcite may be a process of calcitization. If the reaction undergone by *Homotrema* is the norm, as Land suggested, then, at least in the natural environment, there is enough $CaCO_3$ available in the pore water for the primary skeleton to be replaced entirely by a low-magnesian calcite through a process of congruent dissolution followed by precipitation. Such a process would explain the discovery by LAND and EPSTEIN (1970) that high-magnesian calcites that have been altered to low-magnesian calcite during fresh water diagenesis have been completely repopulated with oxygen and carbon isotopes in equilibrium with the pore water. If aragonite is replaced in this way (p.347), why not a high-magnesian calcite? The fact that this change happens early in diagenesis may reflect only the greater solubility of the higher magnesian calcites. The true nature of the change of high-magnesian calcites during diagenesis may be decipherable through examination of their isotopic history. A skeleton altered by incongruent dissolution will keep the primary CO_3^{2-} anion groups that were associated with Ca^{2+} ions, whereas a skeleton altered by total congruent dissolution and precipitation will acquire a totally new complement of CO_3^{2-} groups with new values of $\delta^{18}O$ and $\delta^{13}C$.

The problems raised by the transformation of echinoderm skeletons to low-magnesian calcite are touched upon briefly on p.50.

Comparison of Land's data with Friedman's cannot be direct and is unfortunately inconclusive. Friedman worked with unaltered Recent sediments, mineralogically heterogeneous: Land used homogenized powders. Land also points out that in all his experiments some Mg^{2+} was released, yet a few individual grains actually become enriched in Mg^{2+}.

Relating Friedman's and Land's work to the fabric and mineralogical evolution of the Bermudan limestones, it appears probable that the magnesian calcites lose $MgCO_3$ relatively rapidly between stages II and III (Fig.251, p.326). The virtually complete loss of Mg^{2+} represents an exchange of Ca^{2+} for Mg^{2+} which affects something like one cation in five, where the original composition was about 20 mole % of $MgCO_3$. The exchange of CO_3^{2-} between skeleton and pore water is important because it involves a change in the values of $\delta^{13}C$ and $\delta^{18}O$ in the solid phase.

Experiments of Schroeder and Siegel. Working at room temperature and pressure on Recent carbonate sediments for periods of up to 240 days, SCHROEDER (1969) and SCHROEDER and SIEGEL (1969) found that carbonate particles of aragonite and high-magnesian calcite dissolve not only in fresh water but in sea water as well. Moreover, the dissolution is incongruent.

The rates of dissolution of Ca^{2+}, Mg^{2+} and Sr^{2+} indicate that these ions are accommodated in the aragonite and calcite lattices in three ways—in lattice sites, in lattice interstices and as inclusions. The implication is that there is more than one mineral phase in the skeletal material studied. The incongruent dissolution of cations established an order of decreasing solubility Mg^{2+}—Ca^{2+}—Sr^{2+}. Factors found to influence the direction and degree of incongruency include mineralogy, the organism's biochemical environment during life, and the chemical composition of the water and the volume of water involved in the dissolution. The authors cast doubt on reliance on the contents of Mg^{2+} and Sr^{2+} as indicators of palaeoenvironment.

Isotopic evolution of Bermudan limestones

The isotopic composition of the $CO_3{}^{2-}$ anion group has proved to be an instructive index of the progress of lithification in the limestones of the Bermuda Islands (also Bikini and Eniwetok Atolls; GROSS and TRACEY, 1966). The ratios $^{18}O/^{16}O$ and $^{13}C/^{12}C$ are expressed as parts per thousand in the form $\delta^{18}O$ and $\delta^{13}C$ according to eq.18 on p.280. Skeletons all have characteristic isotopic ratios and so do calcite cement, sea water and rain water. When they react together the changes in isotopic composition can be traced. In the Bermudan limestones HAGLUND et al. (1969) have found an approximately linear relation between aragonite content and uranium content, each being reduced as lithification proceeds.

     The work of Gross: isotopic progression from skeleton to cement. GROSS (1964) applied isotopic analysis to the study of the diagenetic history of the Bermudan carbonates and showed that the mineralogical and fabric changes were accompanied by a progressive enrichment of the lighter isotopes of carbon and oxygen. His data are summarized in Fig.261. The range of $\delta^{13}C$ among the skeletons of the various common species of marine organism off the Bermuda Islands

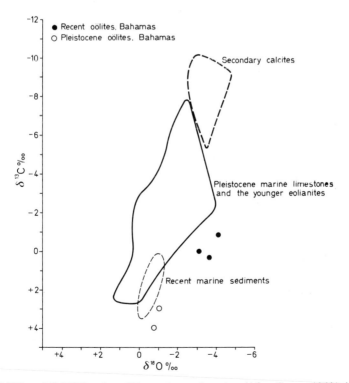

Fig.261. $^{18}O/^{16}O$ and $^{13}C/^{12}C$ ratios of Bermudan carbonates. (After GROSS, 1964.) Also data for some Recent and Pleistocene oölites in the Bahamas. (After FRIEDMAN, 1964.)

today is large, from $+4$ to -8; for $\delta^{18}O$ the range is large also, 0 to -4 (GROSS, 1964, fig.2). The ranges in the ground, homogenized sediments are much smaller since these powders are composed of mixtures of species and give mean isotopic values. Those for the Recent beach and near-shore homogenized sediments are apparent in Fig.261. The Pleistocene limestones have a wider range, both of $\delta^{13}C$ and $\delta^{18}O$, except for the oldest formation, the Walsingham eolianite, which has a limited range of relatively high negative values. The secondary calcites in Fig.261 are cement casts of once aragonite shells and veins: they show even more extreme negative values.

The studies of Friedman and Land (p.330, Fig.254) have revealed that the Recent sediments of Bermuda are mainly mixtures of skeletal debris composed of aragonite and high-magnesian calcite with some low-magnesian calcite as Pleistocene lithoclasts. The mean isotopic ratio $\delta^{13}C$ is around $+2$, and that for $\delta^{18}O$ is around -0.5 (Fig.261). On the other hand, secondary calcite, precipitated from rain-derived ground water, should have values of $\delta^{13}C$ which reflect contamination with soil CO_2, which has $\delta^{13}C$ between -20 and -30. It should also have values of $\delta^{18}O$ which relate to the $\delta^{18}O$ of Bermudan fresh water which has a mean of about -3.5 (see discussion and sources in Gross, 1964, p.187). Thus, it is to be anticipated that the Pleistocene limestones, which exhibit a range of fabric and mineralogical adjustment, will also have a range of $\delta^{13}C$ and $\delta^{18}O$ extending from scarcely altered sediment to the pure calcite cement. Such a variation is shown in Fig.261. The fact that the values for $\delta^{13}C$ and $\delta^{18}O$ in the most altered limestones and in the secondary calcites do not attain the extreme levels, either of the soil CO_2 or the fresh water oxygen, is in accord with the assumption that the CO_3^{2-} of the cement is derived from two sources. These are: (1) the original sedimentary

TABLE XV

RANGES OF ^{13}C AND ^{18}O FOR RECENT AND PLEISTOCENE LITHOTHAMNIOID ALGAE AND FORAMINI-
FERIDS, BERMUDA
(After GROSS, 1964)

	Lithothamnioid algae			
	Recent (4 specimens)		Pleistocene (4 specimens)	
^{13}C	$+1.9$	$+1.0$	$+1.6$	-5.8
^{18}O	$+0.3$	-2.1	$+0.5$	-2.3
	Foraminiferids			
	Recent (4 specimens)		Pleistocene (3 specimens)	
^{13}C	$+3.6$	$+0.2$	$+0.5$	-2.7
^{18}O	$+0.0$	-0.6	-0.8	-2.0

carbonate, and (2) the rain and soil waters. Gross also recorded isotope ratios for two groups of high-magnesian calcite skeletons, from the Recent sediments and from the Pleistocene limestones. These ratios, for lithothamnioid algae and for foraminiferids, are shown as ranges in Table XV. It is not at present possible to say how much of this change in isotopic content in the skeletons during diagenesis is the result of precipitation of space filling cement, and how much the outcome of an exchange of CO_3^{2-} at the surface of the skeletal calcite lattice as Mg^{2+} ions go into solution and Ca^{2+} ions are accommodated. Certainly the isotopic changes have

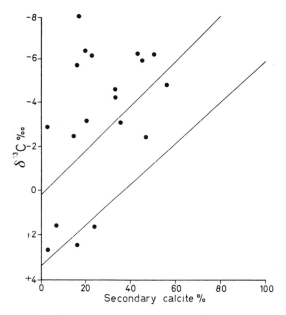

Fig.262. Relation between $\delta^{13}C$ and the content of secondarily precipitated calcite in Pleistocene limestones of Bermuda. (After GROSS, 1964.) The region enclosed by the two lines indicates the relation expected if lithification were solely a matter of precipitation of cement in pores.

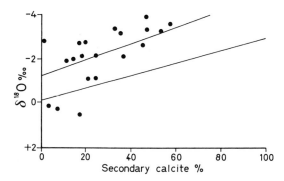

Fig.263. Relation between $\delta^{18}O$ and the content of secondarily precipitated calcite in Pleistocene limestones of Bermuda. (After GROSS, 1964.) The region enclosed by the two lines indicates the relation expected if lithification were solely a matter of precipitation of cement in pores.

been greater than would be expected if calcite cement had alone been the bearer of the more negative values of $\delta^{13}C$ and $\delta^{18}O$. Fig.262 and 263 show the relation between $\delta^{13}C$ and $\delta^{18}O$ and the content of secondary precipitated calcite in the Pleistocene limestones. Even allowing for failure to recognise all of the calcite cement with the microscope, Gross found that there is a definite discrepancy between the isotope values expected from cementation alone (limits indicated by two lines in Fig.262, 263) and the observed higher negative values. It does look rather as though some CO_3^{2-} ions with high negative isotopic ratios have been introduced into the skeletal calcite in place of some original CO_3^{2-} ions with more positive values. This trend has been confirmed by Land and Epstein (p.337). Complications arising from dissolution of aragonite are unimportant here, since the molluscan aragonite has an isotopic composition close to that of the bulk sediment: *Halimeda* has low negative values so that its loss would cause a slight positive shift in the bulk composition (GROSS, 1964, p.184).

The work of Friedman: influences of sea and rain waters. FRIEDMAN (1964, p.786) published mineralogical and isotopic data which indicate that samples of Pleistocene limestone taken from rock that is submarine or intertidal and wetted by sea spray show less diagenetic alteration than samples taken farther inland

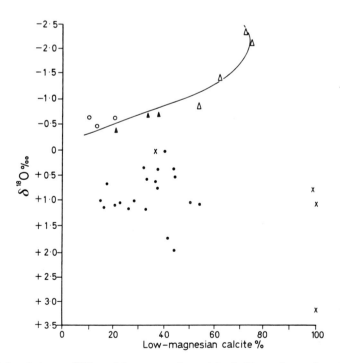

Fig.264. Relation between $\delta^{18}O$ and low-magnesian calcite in Bermudan carbonate sediments. For legend see Fig.265. (After FRIEDMAN, 1964.)

(e.g., Fig.255, 264, 265). This hypothesis is born out by the fabrics which he allowed me to inspect in his thin sections. LAND (1966) disputed this as a valid generalization, pointing out that exposure of the Bermudan carbonates to sea spray or fresh water has occurred on a number of occasions, as sea level has fluctuated, so that it is essential, also, to consider not only the present disposition of the limestones but their history. Some Bermudan limestones are lightly cemented and friable, though located inland, others are at stage III although in the spray zone, but generally speaking, Land suggested, there are two main trends: (*1*) the older limestones are the more altered, and (*2*) cementation shows a spatial relation with the development of soils. FRIEDMAN (1964, p.788) discussed this difficulty, noting that Pleistocene limestones which were presumably flushed thoroughly with fresh water at times of lowered sea level yet contained high-magnesian calcite and aragonite. Friedman's data are very convincing, embracing as they do the degree of dissolution of aragonite, the degree of loss of Mg^{2+}, the content of calcite cement and isotopic data for $\delta^{13}C$ and $\delta^{18}O$. He showed clearly that the precipitation of low-magnesian calcite is closely related to the general negative shift of isotopic values: there is a correlation between $\delta^{13}C$ and the content of low-magnesian calcite both in Recent sediments (containing Pleistocene lithoclasts) and in the Pleistocene limestones (Fig.264, 265). As Friedman concluded, there remains here

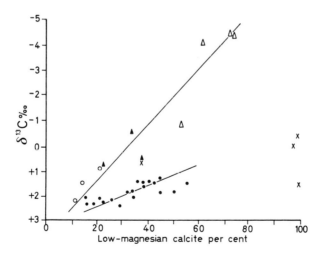

Δ Skeletal sands (Pleistocene) beyond reach of marine spray

▲ Skeletal sands(Pleistocene) marine spray zone

○ Skelelal sands (Recent) marine near shore

× Carbonate lutite (Pleistocene) deep sea

• Skeletal sands (Pleistocene) deep sea

Fig.265. Relation between $\delta^{13}C$ and low-magnesian calcite in Bermudan carbonate sediments. (After FRIEDMAN, 1964.)

a disturbing loose end: why have Pleistocene carbonates in the modern shoreline survived in the metastable phase if they were not covered by sea water during Illinoisian and Wisconsin times?

The isotopic data of Gross and Friedman give general support to the fabric and mineralogical changes supposed to have accompanied the lithification of the Bermudan limestones, though anomalies remain. One of these is the unique course of isotopic change indicated by Friedman's analyses of some Recent and Pleistocene oölites from the Bahamas (Fig.261). The $\delta^{13}C$ in the lithified oölites scarcely exceeds zero though the $\delta^{18}O$ attains the not unusual value of about $-4\%_0$. Could this be connected with the high organic content of oöids?

Summary of Bermudan limestone diagenesis

The probable diagenetic history of the Bermudan limestones can now be summarized (Fig.251). Assuming that the Pleistocene carbonate sediments had a mineralogical composition similar to the Recent ones, then Land's stage I, the unaltered sediment, was characterized by a mixture of high-magnesian skeletal calcite, aragonite skeletons and some lithoclasts of low-magnesian calcite derived from older limestones. The mean $\delta^{13}C$ was about $+2$ and $\delta^{18}O$ about -0.5. Between stages I and II the carbonate sediment was exposed to a downward moving body of fresh water, containing rain-derived O_2, also CO_2 from the atmosphere. This water was low in Mg^{2+}. In the sediment there was a relatively rapid growth of allochthonous low-magnesian calcite cement: the CO_3^{2-} in this was derived from two sources, younger overlying limestones which had been dissolved in the soil-forming process, and plant-CO_2 in the soil. The soil CO_2 had a $\delta^{13}C$ perhaps between -20 and -30, and a $\delta^{18}O$ around -3.5, with the result that there was a rapid increase in the negative values of both isotopes as allochthonous cement accumulated around the grain contacts.

Between stages II and III the Mg^{2+} ions of the high-magnesian calcite skeletons dissolved in the pore water, each accompanied by a CO_3^{2-} anion group. The pore water thereafter contained CO_3^{2-} derived both from the dissolved grains and from the overlying soil. To replace the Mg^{2+} there was a substitution of Ca^{2+} ions in the high-magnesian calcite lattice, each cation being accompanied by a CO_3^{2-} anion from either of the two sources, thus bringing to the new crystal fabric a mixture of low-positive and high-negative ratios of $\delta^{13}C$ and $\delta^{18}O$.

In stage IV aragonite was dissolving and the Ca^{2+} ions so released were being reprecipitated in autochthonous low-magnesian calcite cement. In this process the CO_3^{2-} released from the skeletal aragonite was mixed with the soil-derived CO_2 in the ground water, so that the CO_3^{2-} mixture from which the cement was precipitated had medium-negative values of $\delta^{13}C$ and $\delta^{18}O$, reflecting these two sources. The success of this dissolution-precipitation process depended to a large extent on the preservation of open moulds of the original aragonite grains, either by previous deposition of a rind of allochthonous cement or formation of a micrite envelope.

By stage V the limestone had a fabric constructed only of low-magnesian calcite: this consisted of degraded high-magnesian calcite skeletons, calcite cement and the original lithoclasts. The negative values for $\delta^{13}C$ and $\delta^{18}O$ mark a balance between the conservation of original CO_3^{2-} and the influx of CO_3^{2-} derived from the atmosphere and soil during the period of subaerial exposure of the limestone surface.

This simple story of diagenetic change does not, of course, express the full complexity of the natural process. Some hints of the true variety of diagenetic environments and their petrographic, chemical and isotopic products have been given in a provocative paper by LAND (1970), in which he examines the possible courses of change in the vadose, phreatic and beach rock zones of the Pleistocene Belmont Formation of Bermuda. His data suggest that phreatic diagenesis is much more rapid than vadose or beach rock diagenesis.

Comparison with Recent–Pleistocene carbonate sediments of Israeli Mediterranean coast

In a study of diagenesis in Recent and Pleistocene carbonate sediments along the coast of Israel between Haifa and Tel Aviv, GAVISH and FRIEDMAN (1969) have shown that stages II and III were completed within about 7,000–10,000 years (their ^{14}C dates). Stages IV and V were completed by an estimated 80,000–100,000 years. Later the growth of neomorphic spar yielded porphyrotopic and poikilotopic sparry calcite which is present in rocks with an age estimated on stratigraphic grounds as 130,000–170,000 years. These fabric and mineralogical changes were accompanied, as on Bermuda, by an increase in the lighter isotopes of carbon and oxygen in the sediments, $\delta^{13}C$ moving from $+3$ to -6 and $\delta^{18}O$ from -0.5 to -4. Again, as on Bermuda, there is a close positive correlation between the total content of aragonite and high-magnesian calcite and the content of the heavier isotopes of carbon and oxygen. Strontium was lost continuously from the Israeli carbonate sediments over a period of at least 300,000 years. There is also a good positive correlation between the content of low-magnesian calcite and that of the lighter carbon and oxygen isotopes (as on Bermuda). Of unusual interest is the evidence for replacement of quartz grains by calcite during stages II and III.

FABRIC AND MINERALOGICAL EVOLUTION OF CAINOZOIC LIMESTONES: GUAM, ENIWETOK AND LAU

Passive dissolution and precipitation

The diagenetic fabrics in the Eocene to Recent limestones of Eniwetok and Guam have been described in detail by SCHLANGER (1963, 1964) and those of Lau by

CRICKMAY (1945). The bore-hole records show cyclic changes from unaltered carbonate sediment passing gradually upward via a transitional state into lithified limestone and thence, through a sharp break, back into unaltered sediment (Fig.249, p.322). The lithified limestone is apparently at the stage V of Land (p.330): its sharp top coincides with a palaeontological discontinuity. The transitional limestones bear aragonite fossils in various stages of replacement by calcite, not only through the formation of visible dissolution cavities followed by filling with calcite cement, but also by calcitization *in situ*. This evidence is of fundamental importance because it indicates that the growth of neomorphic spar can be an early process acting at surface temperature and pressure. The fossils are mostly embedded in calcite cement and it is plain, from Schlanger's descriptions, that some of the cement arrived before the dissolution of aragonite had been completed, just as it did in Bermuda. One of the early signs of the dissolution of aragonite material is a striking change in its appearance: it alters to a loose powder of aragonite. SCHLANGER (1963, p.997) wrote of coral: "The boundary between the original aragonite and the replacing calcite closely resembles a stylolitic seam and is occupied by a thin film of powdery aragonite", and again (1963, p.994): "aragonite fossils in these [transition] zones commonly show a tendency to disintegrate into loose aragonite powder." If this disintegration follows simply upon the rotting and removal of the organic matrix (as proposed by SORBY, 1879), it is not obvious why calcite tests are not similarly affected.

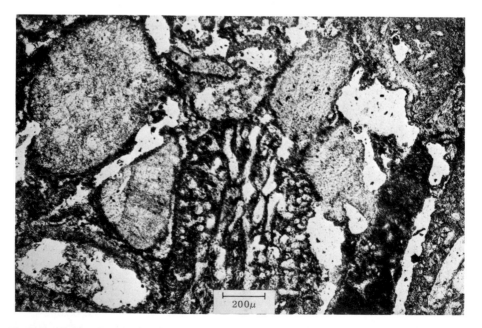

Fig.266. Lithified limestone in hole *F-1*, Eniwetok (Fig.249) showing secondary vugs and preservation of porosity in *Halimeda*. Slice.

The fully calcitic limestones of Eniwetok and Guam contain moulds of molluscs, *Halimeda* and coral (Fig.250, p.323), which are empty or partly or wholly filled with a calcite cement that may be fibrous or equant in habit. The rocks are also traversed by irregular channels and vugs (Fig.266), commonly lined with laminar fibrous calcite: these cavities truncate fossils, but it is not stated if calcite fossils or only aragonite fossils have been dissolved in this way.

Calcitization of aragonite skeletons and carbonate mud at Guam, Eniwetok and Lau

Data of Schlanger and Crickmay. SCHLANGER (1964, p.10) gave data on the process described often as replacement of aragonite by calcite *in situ*, an alteration of aragonite to calcite without the formation of visible cavities (i.e., the growth of neomorphic spar, p.481; also CRICKMAY, 1945). He was at pains to demonstrate that this is not a matter of dry, solid-state, polymorphic inversion (p.239), but is a wet polymorphic transformation requiring dissolution-precipitation. Time is not the only controlling factor (SCHLANGER, 1963, p.997): rather the replacement of aragonite by calcite in Eniwetok is related to the proximity of unconformities and to a history of localized subaerial exposure (p.322). Schlanger pointed out that the loss of Sr^{2+} from the corals also indicates a wet rather than a dry process (p.329). Sr^{2+} was also lost from the Bermudan limestones (p.332).

This *in situ* process of wet polymorphic transformation proceeds, we must assume, with the aid of a solution film and preserves a ghost of the original wall structure (Fig.340–342, p.488): it is quite distinct from the process leading to the development of passive voids. It is further discussed on p.486 where it is treated as one aspect of the growth of neomorphic spar.

Five stages in the progress of wet calcitization in the molluscan shell wall were described by SCHLANGER (1964, p.10) (compare description by Cullis on p.353):

(1) Almost unaltered fibrous yellow-brown, laminar, aragonite shell fragments, sharply outlined against the matrix of carbonate mud.

(2) A patchwork of coarse well-defined, polygonal, calcite crystals within the shell wall. The new crystals, which are clearly visible only with crossed polars, end sharply at the boundary of the wall.

(3) Calcite crystals easily distinguished in ordinary light, but wall structure is still distinct.

(4) The calcite mosaic is clearer and partly obscures the original structure.

(5) The calcite crystals extend beyond the wall into the carbonate mud matrix, but wall structure can still be seen vaguely as yellow-brown bands and fibres. In only rare fragments is all structure completely effaced.

HUDSON (1962) has compared the Jurassic bivalve *Neomiodon*, in its unaltered aragonitic state, with its calcitized product. It is significant that the aragonitic shell is preserved in shale while the calcitized shell occurs in a biosparite; a reflection, perhaps, of pore-water mobility and the protective action of amino acids (p.275). The planes of inclusions in the calcitized shell, which Hudson regarded as the remains of intercrystalline conchiolin, are similar to the layered structure of the unaltered shell.

Pseudo-pleochroism in the secondary calcite, ω colourless (or nearly so) and ε brown, has been described by P. R. BROWN (1961) and by HUDSON (1962). Hudson attributed the apparent absorption to the presence of inclusions ($< 1\ \mu$ across) derived from the original conchiolin. He found a reversed scheme of pseudo-pleochroism in the original aragonite, where α has the minimum absorption. P. R. BROWN (1961) and TEBBUTT (1967) recorded an intermediate stage of neomorphic growth in which the new sparry crystals are aragonite (Fig.343, 344, p.489). Cullis found that some of his neomorphic spars were aragonite (p.354). Brown also noted that, in the Upper Jurassic (Purbeckian) limestones of Dorset, the calcitization of aragonite shells *in situ* seemed to have taken place preferentially in biosparites deposited in a brackish environment.

The calcitization process: more details and discussion. The progress of wet calcitization of microcrystalline aragonite (the growth of neomorphic spar, p.481) is affected by skeletal mineralogy and fabric, and by local conditions. LAND (1966) noted that calcitization of aragonite, in molluscs and corals, is commoner in the Pleistocene limestones of Florida than in those of Bermuda. P. R. Brown and Tebbutt found the first stage of stabilization to be the growth of a coarse aragonite spar, whereas, in Guam, corals are calcitized first to a finely crystalline calcite, and then to a coarse mosaic. This two-stage process illustrates an important point, namely that the wet stabilization of $CaCO_3$ sediment apparently derives its energy from two sources, the trend toward mineralogical equilibrium (aragonite being metastable in the P–T stability field of calcite), and the greater stability of large crystals over small ones (below). Put differently, the two-stage process includes not only wet polymorphic transformation, but also wet recrystallization (p.476) wherein large calcite crystals grow at the expense of smaller calcite crystals. The whole process is examined in some detail on p.481.

Two points made by Schlanger deserve special attention: the recrystallization of calcite fossils and the extension of the secondary calcite crystals (neomorphic spar) of the mollusc wall outward into the carbonate mud matrix. He stated that foraminiferids and calcitic algae in the Guam limestones, though generally well preserved, do show some replacement by a more coarsely crystalline calcite. Since these skeletons were originally high-magnesian calcite the process seems at first glance to be distinct from calcitization of aragonite. On the other hand, the more magnesian of the high-magnesian calcites are more soluble than aragonite under near-surface pressures (Fig.218, p.237) and it may be that a dissolution–precipitation type of recrystallization, with some metasomatism and the involvement of solution films, can take place. This process would differ from calcitization of aragonite in that, as the more soluble mineral would be a magnesian-calcite, there would be an exchange between Mg^{2+} and Ca^{2+} ions. In such circumstances, where magnesian-calcite crystals are totally dissolved and replaced by new crystals, the dissolution would presumably be congruent. There is the further possibility, that the walls of foraminiferids and coralline algae are so finely crystalline that they are subject to wet recrystallization. Further studies of this problem are needed.

Enhanced solubility of carbonate mud crystals. The extension of secondary calcite crystals beyond the margins of the skeletal wall, into the enclosing carbonate mud (Fig.345, 346, p.492), is presumably the result of a continuation of the calcitization process. There are good reasons why the mud crystals should be more soluble than the large secondary crystals, even supposing there were no mineralogical differences between them. Some of the mud crystals will be so small as to be supersoluble (p.254): all will have more edges and coigns per unit area of surface than the larger crystals and therefore a greater concentration of unneutralized bonds. In addition, since many of the mud crystals are the products of abrasion, their surface layers will contain a residual elastic strain and will have, therefore, enhanced solubility (p.257). Finally, given the high concentration of the more soluble minerals, high-magnesian calcite and aragonite, in the mud, there seems good cause for the large secondary crystals in the shell surface to continue to grow at the expense of the mud.

Susceptibility to calcitization. CRICKMAY (1945, p.241) arranged the skeletal materials in the limestones of Lau and Guam in an order of decreasing susceptibility to calcitization. His order is given here, modified by the addition of *Halimeda* as proposed by SCHLANGER (1964), thus: corals > *Halimeda* > molluscs > pelagic foraminiferids > beach foraminiferids > larger foraminiferids > echinoids > coralline algae. (Compare the orders by Bathurst and by Banner and Wood, p.490.) The order is controlled by several factors, as Crickmay pointed out, and he listed mineralogy, wall structure and the presence of $MgCO_3$ in the calcite. Thus the materials are not arranged simply in order of either mineralogy or $MgCO_3$ content. The aragonite skeletons are understandably the most susceptible. On the other hand, the pelagic thin-walled foraminiferids (*Globigerina* and *Globorotalia*), which are low-magnesian calcite, are followed by the less susceptible thicker-walled tests (*Amphistegina*, etc.) although these are constructed of the more soluble high-magnesian calcite. The coralline algae have the highest content of $MgCO_3$, and are probably the most soluble of all, yet they are the least susceptible. The key to this seemingly illogical order may rest with the three little known factors of wall structure, the nature of the organic conchiolin sheaths (p.14) around the crystals and, thirdly, the small size of the ordered lattice domains in some of the apparently single crystals, that look homogeneous but are really crystallographically heterogeneous (see discussion on Towe and Cifelli, p.40).

Calcitization of carbonate mud at Guam and Eniwetok. Under the term "mud" SCHLANGER (1963, p.994 on Eniwetok) included carbonate sediments of clay, silt and fine sand grades. Presumably the same definition applies to his discussion of the "recrystallization" of mud in the limestone of Guam (SCHLANGER, 1964, p.10). In thin section the mud has a dark, semi-opaque appearance. The main alteration has been the conversion of the mud to coarse mosaics of clear calcite, a process described by Cullis from Funafuti (p.353). In some places the presence of large, sand-sized fossils has controlled the site of initial calcitization. Thus the new crystals have grown outward in lattice continuity with crystals in the surface of the fossil (as discussed earlier, p.347), forming a halo of coarse, oriented, secondary calcite not to be confused with space-filling cement. In other places, the patches of secondary mosaic are unrelated to the larger fossil fragments and simply grade outward into dark mud (p.353).

Growth of neomorphic spar at Guam and Eniwetok

The disposition of the secondary calcite mosaics in the mud does not imply passive precipitation of calcite cement in voids formed by earlier dissolution, and SCHLANGER (1964, p.10) treated the mosaics as products of *in situ* replacement by coarse calcite. There seems no reason to doubt that the formation of the isolated patches of spar is one aspect of the process which also produces the sparry outgrowths of calcitized shell into the surrounding mud, and the sparry crystals of the shell walls themselves. Reasons for the relative instability of at least the clay grade crystals are given on p.349. The nature of this neomorphic process and its products in ancient limestones are examined in Chapter 12.

On a final note, it is important to realize that the neomorphic changes in the carbonate sediments are irregular and patchy in Eniwetok and Guam, as they are in Bermuda and Florida. On the grand scale there is a pattern related to time, but in detail the pattern is untidy, reflecting local variations in movement and composition of ground water and the inevitable slowness and inefficiency of diagenetic processes.

FABRIC AND MINERALOGICAL EVOLUTION OF RECENT LIMESTONES: FUNAFUTI

Of all the researches into the early stages of near-surface diagenesis, none rivals, in variety, in detail, or in the clarity of its illustrations, the description by CULLIS (1904) of the mineralogical changes observed by him in thin-sections of cores of Recent limestones (and dolomites) from Funafuti Atoll in the Ellice Islands, 3,000 km east of New Guinea. The value of Cullis's work with the petrological microscope was greatly enhanced by his application of Meigen's solution which stains aragonite and distinguishes it from calcite and dolomite, and Lemberg solution which stains calcite and separates it from dolomite (HOLMES, 1921, p.262, 265; FRIEDMAN, 1959). Cullis's theme embraced the distribution of primary skeletal aragonite and calcite, the later addition of aragonite and calcite cements, the dissolution and the calcitization of the aragonite, the addition of dolomite cement, and dolomitization.

Cementation at Funafuti

On the subject of cementation Cullis made some useful generalizations. The amount of cement increases with depth in the bore-hole and, although both aragonite and calcite occur as cement, calcite is the commoner and is present in greater quantities. Aragonite and dolomite never occur together: calcite is present throughout the cores, but aragonite exists only in the upper cores, and dolomite only in the lower cores below about 195 m. The distribution of aragonite cement, always acicular, is limited. The mineral has grown only on clean *aragonite* surfaces: a film of calcilutite on an aragonite skeleton ensures that the cement is calcite. For this reason aragonite cement is virtually confined to primary cavities within aragonite skeletons. The precipitation of calcite and aragonite cements is controlled not only by the mineralogy of the substrate, but also by its fabric (see also SKEATS, 1902). The crystals of aragonite cement "are invariably deposited in continuity with the fibres of aragonite organisms" (Fig.267) but the calcite cement, though generally in lattice continuity with the calcite skeletal substrate (e.g., on echinoid plates), is not so everywhere (Fig.268). Calcite cement occurs in two main forms, as fibres (Fig.269) and as relatively equant crystals (Fig.268, 270). The equant crystals, from Cullis's drawings, appear to occur in the two habits usual in cements, scalenohedral and rhombohedral (Fig.270, 268). Cullis noted that the fibrous calcite is concentrated at horizons where the $MgCO_3$ content of the rock is high (reaching 18 mole %). On this point, it is possibly significant that Usdowski's experiments (p.260) showed a positive correlation between the development of radial-fibrous habit in calcite and the Mg^{2+}/Ca^{2+} ratio in the precipitating solution. This fibrous calcite in Funafuti was never found in lattice continuity with the substrate. Cullis's suggestion that the fibrous habit characterizes the most rapid growth is consistent with the ideas brought together on p.259 concerning the stability of $CaCO_3$ polymorphs. It is probable that some of these cements were precipitated from sea water, particularly those that were later covered by early internal sediment, as in Fig.268. Fibrous aragonite is a typical submarine cement (see discussion in FRIEDMAN, 1968b). Unfortunately the Titian yellow stain used by WINLAND (1968) to reveal high-magnesian calcite was not available to Cullis, so that it is not clear which, if any, of his calcite cements are high-magnesian and, therefore, of marine origin.

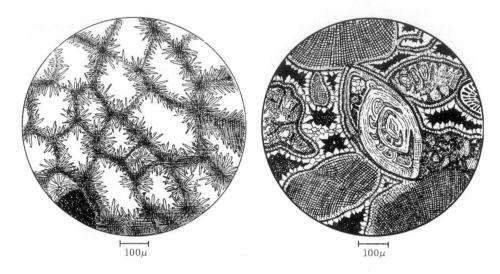

Fig.267. The aragonitic scleractinian, *Heliopora*, with fibres of aragonite cement in lattice continuity with the crystals in the trabeculae. Slice. Funafuti core, Ellice Island, Pacific. (From CULLIS, 1904.)

Fig.268. Biomicrite of foraminiferids and coralline algae. The late micrite matrix has buried an earlier precipitate of equant calcite cement. Slice. Funafuti core. (From CULLIS, 1904.)

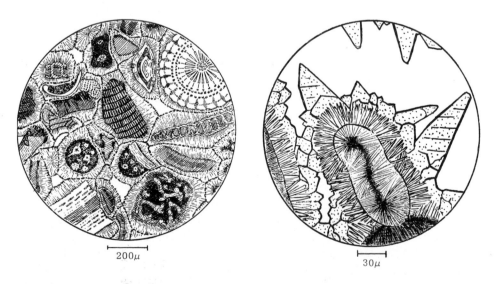

Fig.269. Porous calcarenite partly cemented by fibrous calcite. This coats even the unicrystalline echinoid spine which has no syntaxial rim. Slice. Funafuti core. (From CULLIS, 1904.)

Fig.270. Coral trabeculae coated with an early fibrous aragonite cement followed by an equant calcite cement. Slice. Funafuti core. (From CULLIS, 1904.)

Calcitization of calcilutite at Funafuti

The growth of neomorphic spar in the calcilutite (Cullis's "crystallization") is interesting. Cullis found that, in thin-sections of samples taken within a metre of the top of the boring, some of the cavities are filled with "finely divided calcareous detritus" which, in transmitted light, appears as "dark, more or less opaque areas" (CULLIS, 1904, p.398). From this description and from his drawings this material would appear to comprise carbonate material of clay and silt grade. However, in sections from deeper cores, some of this calcilutite has been replaced by a mosaic of crystals, in some places calcite, in others aragonite, which "closely resembles" the cement mosaic. The fully developed replacement mosaic is not everywhere readily distinguishable from cement, but can be identified generally by its relation to other fabrics, its content of "minute particles which have remained uncrystallized", and by included relics of organic structure. This growth of neomorphic spar, similar to that described by Schlanger from Guam (p.350), began simultaneously at numerous points and gradually changed the whole mass of calcilutite to a coarser, sparry mosaic (see also SKEATS, 1902). Nevertheless, though patches of secondary spar are common, complete replacement of the calcilutite is rare. Much calcilutite was found even in the deepest cores, "almost as fresh in appearance as that which has been but recently deposited" (CULLIS, 1904, p.399). The relation between the secondary spar in the calcilutite and the adjacent large skeletal grains recalls the fabrics described by Schlanger from Guam (p.347). Some of the coarse crystals which replaced the calcilutite are in lattice continuity with the crystals in the skeletons, as if crystals in skeletal surface acted as nuclei from which replacement of the calcilutite began. There is, however, a peculiarity in the Funafuti fabrics which is of particular interest: since the coarse crystals in the altered mud are either calcite or aragonite, it follows that, where they are optically continuous with crystals in the adjacent skeletons, this continuity is mineralogical: aragonite fibres in the substrate are extended as aragonite fibres into the mud (Fig.271), calcite crystals are similarly extended as calcite. In other words, where the growth of neomorphic spar in calcilutite was influenced by the crystal fabrics of the larger skeletal particles, this control was, not unnaturally, mineralogical. P. R. Brown and Tebbutt have recorded aragonite neomorphic spar as a replacement of aragonite skeleton (p.489).

Calcitization of primary and secondary aragonite at Funafuti

Closely associated with the growth of neomorphic spar in the calcilutite is the alteration of the skeletal aragonite to calcite neomorphic spar (Fig.341, 342, p.488): also the change to calcite of aragonite mosaics which are themselves replacements of calcilutite. The aragonite which replaced the calcilutite was altered to calcite *before* the aragonite of the skeletons was altered: for unknown reasons

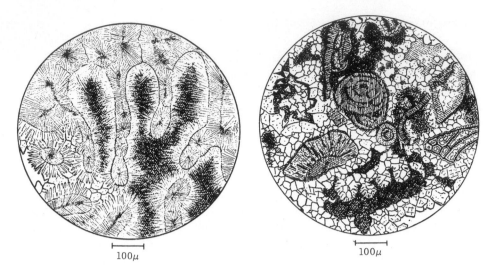

Fig.271. Aragonitic coral trabeculae with dark centres of calcification. In the spaces between the trabeculae aragonite fibres, syntaxial with those in the trabeculae, have partly replaced a micrite fill. Slice. Funafuti core. (From CULLIS, 1904.)

Fig.272. In a biomicrite the micrite has been partly replaced by neomorphic calcite spar. The intraparticle spar is cement. Slice. Funafuti core. (From CULLIS, 1904.)

the skeletal aragonite was the more stable in this situation. Commonly the calcite mosaic which has replaced the sclerodermites of the coral skeleton retains a pale brown colour which distinguishes it from the clear mosaic of the calcite cement.

The various stages in the calcitization of coral were described in unusual detail by CULLIS (1904, p.401). He wrote: "The first indication of loss of organic texture in the coral is the disappearance of the 'dark line', a series of comparatively dense and opaque centres from which the component fibres of the coral substance radiate [sclerodermites, p.25]. In a perfectly fresh coral these form a very conspicuous dark line, occupying the middle of the coral wall, but, as lower and lower cores are examined, this line becomes less and less distinct and presently disappears. The next stage is the gradual obliteration of the fibres themselves. Yet later, the substance of the coral is seen to be broken up by a number of sinuous lines into a granular mosaic; but, for a while, both the granular structure and something of the fibrous character may be discerned. Finally all traces of the fibres disappear, and the mass becomes an aggregate of irregularly rounded and mutually interfering grains of calcite." (Compare description by Schlanger on p.347.) The calcitized aragonite grains are distinguishable from the enclosing calcite mosaic because they are bounded by a narrow dark line. Cullis suggested that this line is composed of "finely divided detritus, which has not crystallized, or has done so only imperfectly". This description and his drawings recall alga-produced micrite envelopes (p.384).

Growth of neomorphic spar at Funafuti

In all these replacement fabrics there is no sign of associated development of passive cavities, no sign that dissolution of aragonite was not followed immediately by precipitation of calcite or aragonite. The replacement of calcilutite (Fig.272) and of skeletal material appears to be an *in situ* process, the same neomorphic process that Schlanger described from Guam (p.350). In Funafuti, we can assume that the same two driving forces are involved, the need for polymorphic equilibrium re quiring a change by dissolution-precipitation from aragonite to calcite, and wet recrystallization—the replacement of many tiny crystals by the enlargement of a few of their number. In Funafuti, recrystallization without wet polymorphic transformation has resulted in the replacement of fine aragonitic debris by coarser aragonite, or a fine calcite mosaic by a coarser calcite mosaic.

Dissolution of aragonite at Funafuti

Accompanying the *in situ* processes of wet polymorphic transformation and recrystallization in Funafuti, but distinct from it, there has been a considerable amount of dissolution of aragonite, with the formation of passive cavities. The early stage of this dissolution process was described by CULLIS (1904, p.403) as "the appearance, in the substance of the aragonite organisms, of numerous small and irregular holes, which impart a porous or spongy character to it". Between 70 m and 190 m the proportion of once aragonite skeletons, preserved now as calcite, is very small compared with the great volume of aragonite skeletons in the higher levels. The *total loss* by dissolution has clearly been great. It is worthy of note that Cullis does not include among the diagenetic processes active in Funafuti the filling of moulds of once aragonite skeletons by calcite cement. Instead he records that, between about 70 m and 190 m, the material is pure calcite, either originally calcite organisms, calcitized aragonite or calcite cement (little of this) and is largely unconsolidated: dissolution of aragonite has clearly not here been balanced by precipitation of calcite, as it has been in Bermuda during Land's stage IV (p.327).

Summary of diagenesis at Funafuti

Summarizing the changes which Cullis was able to trace downward in the Funafuti borings, the first sign of alteration, noticeable within about a metre of the surface, is the appearance of patches of coarse mosaic in the detrital calcilutite. This mosaic is the result, it would seem, of two processes acting together—wet polymorphic transformation of aragonite to calcite (calcitization) and recrystallization either of aragonite or calcite—both processes acting *in situ* via a system of solution films (p.498). Farther down, especially below 30 m, there are increasing signs of the calcitization of aragonite skeletons: aragonite cement becomes less common, is overlain by later calcite cement, and eventually disappears either as a result of dissolution or calcitization. Below 195 m the cores are a mixture of calcite and dolomite.

The relation between diagenetic stage and depth in Funafuti is most unlikely to be as simple as it appears, because this atoll, like those of Eniwetok, Guam and Bikini, has been subject to drastic changes of sea level and the diagenetic history must, therefore, be correspondingly complex.

EARLY VADOSE GEOPETAL SILT

Internal calcite silts in a Permian wackestone mound

The subject of subaerial, fresh water diagenesis has lately been illumined from an unexpected direction—by a study of the deposition of calcitic internal sediments of silt grade. These sediments, investigated by DUNHAM (1969a) in cores from a mound of carbonate wackestone ("reef") of Permian, Wolfcamp age, New Mexico, are not ordinary geopetal sediments of micrite, biomicrite or pelmicrite. The Permian internal sediments in question are strictly limited by the following characteristics:

(*1*) They are more or less equigranular calcite mosaics with median crystal diameters of 10–25 μ, extraordinarily well sorted, with virtually no sand-grade or clay-grade particles and no recognizable skeletal material. Dunham called them "crystal silts".

(*2*) They are geopetal, occupying always the lowest part of a one-time cavity.

(*3*) They occur in secondary voids as well as primary voids.

(*4*) They always overlie some early calcite cement and are normally covered by later calcite cement. Thus, they are clearly syndiagenetic, deposited after the onset of cementation but before its completion, and after the formation of secondary voids.

The crystal silts have interesting secondary characteristics. Whereas the normal micritic internal sediment usually has a horizontal upper surface (Dunham claimed) the crystal silts rarely have horizontal surfaces: instead, the surfaces dip in various directions at angles of up to 30 or 40°. In places they are cross-laminated. Cavities may be filled to the roof with silt. Dunham regarded these criteria as indicating deposition from *moving* water. The silts are peculiar in that they occur only as internal sediments (the internal K_2 of SANDER, 1936, 1951). They never form beds or appear as a primary component in the primary sediments (Sander's K_1). They are very uniform in texture throughout the Wolfcamp mound. The content of crystal silt in this mound decreases downward. There is nothing to indicate that the silt is neomorphic in origin (such as the criteria on p.513).

In a long, detailed and wide-ranging discussion, with many illustrations, Dunham concluded that the crystal silts must have been deposited in the vadose zone from meteoric water. Only in such near-surface porous rocks, for example, could water velocities be fast enough to carry silt. Other environments he examined but found unsatisfactory. In assessing the source of the silt Dunham was, perhaps, on less sure ground. He supposed that the silt-sized crystals were derived from the primary sediments of the mound (DUNHAM, 1969a, p.166). Formation by precipitation in suspension (p.515) was unlikely because the silt occurred only in interparticle pores and not in intraparticle pores such as the chambers of fusulinids. He supposed that the silt crystals were winnowed from the primary sediment by moving ground water, the sand-grade particles being mechanically trapped in the pore system and unable to move. The clay-grade particles he suggested succumbed "to chemical attack by vadose water", consequent upon their large surface area. The winnowing-succumbing process is difficult to test by inspection of rock fabrics, but there should be little difficulty in setting up an experimental model in the laboratory to see if it works. Crystal silts should be looked for in the Recent and Pleistocene limestones of such places as Bermuda, Florida and Barbados. These peculiar sediments, whatever their origin, have now been beautifully demonstrated and it is likely that they will be recognized in many other limestones.

SUMMARY OF SUBAERIAL, FRESH WATER DIAGENESIS

On the basis of the studies of the limestones of Bermuda, Eniwetok, Guam and Funafuti, all of which have been more or less lithified in a fresh water subaerial environment, it is now possible to synthesize a simple, tentative, scheme of subaerial diagenetic evolution. In all of these cases the original unconsolidated sediment was dominated by two mineralogical groups of skeleton-forming organisms, those constructed of aragonite, the molluscs, corals and *Halimeda*, and those built of high-magnesian calcite, most of the foraminiferids, coralline algae and echinoderms.

The first sign of change is a tendency for the grains to cohere and to form a friable though still highly porous mass. Under the microscope, low-magnesian calcite cement (identified by X-ray diffraction) can be seen in the various intragranular pores—the chambers of foraminiferids, the utricles of *Halimeda*, the pores of echinoderms—and in the intergranular pores around the points of contact of the grains. This early cementation is Land's stage II. There are no signs yet of dissolution and, therefore, the cement must be allochthonous.

The next critical change is the loss of Mg^{2+} from the high-magnesian calcites. As Mg^{2+} ions diffuse from the calcite lattice their places are taken, one must assume, by newly absorbed Ca^{2+} ions and, as part of this process, an equivalent number of CO_3^{2-} ions originally in the lattice are exchanged with CO_3^{2-} from the solution. This exchange of CO_3^{2-} along with the precipitation of the early allochthonous cement, is responsible for increasingly negative mean values of the isotope ratios $\delta^{13}C$ and $\delta^{18}O$ in the sediment. The original ratios in the submarine sediment are believed to have been around $\delta^{13}C = +2$ and $\delta^{18}O = -0.5$. A persistent negative shift follows the influx of CO_3^{2-} ions in rain and soil waters that are rich in atmospheric CO_2 and have high negative ratios around $\delta^{13}C = -25$ and $\delta^{18}O = -3.5$. When all the high-magnesian calcites have become stabilized as low-magnesian calcites, Land's stage III has been reached.

During stage IV very considerable mineralogical and fabric changes occur, as all the aragonite is dissolved and low-magnesian calcite is precipitated, both in the primary pores and in the secondary pores formed by dissolution of the aragonite grains. The critical factor controlling development during this stage is the extent to which potential moulds have already been constructed around the aragonite grains. If, when the loose sediment was lying on the sea floor, micrite envelopes (p.384) replaced the grains centripetally, these envelopes will now act as moulds: when the skeletal aragonite dissolves the envelope will remain as a stable residue. (Possible reasons for this stability are discussed on p.333). Other potential moulds will have been formed if, during stage II, the early cement of low-magnesian calcite was precipitated as an enveloping skin on the grains: again, this stable material will remain as a mould when the aragonite dissolves. Thus, by the time that the rock enters stage IV, there may be in existence a variety of moulds within which, at some later time, low-magnesian calcite cement may be precipitated to yield calcite casts of one-time aragonite grains.

It is obvious that some carbonate sediments will proceed to stage IV—which embraces the more or less concomitant dissolution of aragonite and precipitation of calcite—without having undergone the early cementation of stage II. Such an abbreviated course of evolution is outlined by Land for the Bermudan limestones that underlie certain red soils (p.330). It is likely to be followed, either where the supply of $CaCO_3$ needed for early, stage II, cementation is not available, or where the movement of ground water carries the dissolved $CaCO_3$ away into another formation. Long distance transport of dissolved $CaCO_3$ is possible because of the long residence time of Ca^{2+} and CO_3^{2-} in solution, which in turn results from the rapid rate of dissolution and the much slower rate of precipitation (p.448). Limestones lacking skins of early cement are consequently wanting in moulds for the dissolved aragonite grains, and many of these grains may disappear without trace. The Funafuti limestones, between 70–190 m, show loss of aragonite equivalent to Land's stage IV, but no balancing precipitation of calcite. The same lack of cement may lead to local collapse where aragonite grains dissolve in an uncemented grainstone.

During stage IV calcitization of aragonite and recrystallization (of aragonite and calcite) are also active. It is significant that they act together: each involves wet replacement of an unstable micron-sized mosaic by a coarser more stable mosaic, without the intervention of a passive cavity. The calcitization, in this case, is a wet polymorphic transformation of aragonite to calcite: the recrystallization is a growth of certain crystals at the direct expense of smaller ones of the same mineral species. Together these two processes comprise the process called growth of neomorphic spar.

The calcitization affects two groups of aragonite fabrics, the aragonite skeletons (molluscs, corals, *Halimeda*, etc.) and the aragonite cement, earlier precipitated from sea water. With the exception of the echinoids (but see p.50), all the skeletons are constructed of very tiny, quite distinct micron-sized crystals. These are gradually replaced, during wet polymorphic transformation, first by a mosaic of small calcite crystals and then, as some of these grow at the expense of others, by a coarser crystalline mosaic. Thus, even in the calcitization of an aragonite skeleton or aragonite cement, a certain amount of recrystallization is evident. This is particularly apparent where aragonite spar has replaced an aragonite skeleton. Growth of neomorphic spar is encouraged where extra large crystals are already in existence, either as primary detrital grains or as products of wet polymorphic transformation or recrystallization. Once these larger crystals exist, then the growth of neomorphic syntaxial rims can proceed. This is nowhere more obvious than where a halo of syntaxial calcite develops in the calcilutite around the mosaic of a calcitized skeleton, extending from the outer surface of the skeleton into the calcilutite. The crystals which have thus replaced the calcilutite are in lattice continuity with the crystals of the calcitized skeleton: the outer surfaces of these secondary crystals in the skeleton, occupying the site of a one-time aragonite fabric, must have acted as surfaces for heterogeneous nucleation. It seems certain that the growth of neomorphic spar is accompanied by a significant local reduction in porosity, which in turn requires either prolific development of vugs or a net influx of $CaCO_3$ from outside the system.

The final product of the processes acting during stage IV is a rigid structure, still porous (porosity in Bermuda is around 20%), but consisting only of low-magnesian calcite: this rock has attained Land's stage V. This low-magnesian calcite is partly original (lithoclasts, some foraminiferids), partly altered high-magnesian calcite, and partly cement fillings. Any further reduction of porosity, in the subaerial, fresh water environment, can take place only by the delivery of allochthonous carbonate, making yet further demands upon donor limestones (p.451).

Here it is worth turning aside for a moment to consider the lessons that can be learned from the researches described in this chapter. Research must be broadly based, keeping hold of the four main strands which are so closely interwoven in this complex subject:

(*1*) The rocks must be observed acutely and accurately: as the end products they have much to reveal.

(*2*) The nature, or probable nature, of the original sediments at the time of their deposition must be clearly assessed if subsequent changes are to be understood.

(*3*) Relating these two extremes are the chemical and physical processes of diagenesis.

(4) Lastly, and so often neglected, there are the composition and movement of the pore solutions: without these none of the changes discussed in this chapter could occur. We still have far to go. As MURRAY and PRAY (1965) wrote: "All too often, even a critical observation of the rocks has been overlooked; and present knowledge of the other factors [above] is commonly rudimentary compared to what is desirable for adequate interpretation."

Finally, a return to the cautionary note on which this chapter began. The data presented here are drawn mainly from studies of four tiny oceanic islands, and the resultant conclusions can only be applied with the greatest reserve to the vast spreads of limestones laid down in ancient epeiric seas. MEDAWAR (1967, p.154) has observed that "the activity that is characteristically scientific begins with an explanatory conjecture which at once becomes the subject of an energetic critical analysis". The synthesis attempted in this chapter is little more than an explanatory conjecture. The energetic critical analysis lies in the future.

FURTHER READING

Excellent introductions to the general field of diagenesis are given by MURRAY and PRAY (1965), in their short but apposite article at the beginning of the S.E.P.M. symposium, and by PURDY (1968) in his more detailed and lavishly illustrated treatment. The main thesis of this chapter rests on four important and very readable texts by FRIEDMAN (1964), GROSS (1964), LAND (1967) and LAND et al. (1967), all of which deal with the limestones of Bermuda. Recent papers of considerable interest on fresh water diagenesis of Pleistocene limestones are by MATTHEWS (1967, 1968) and GAVISH and FRIEDMAN (1969). SCHROEDER's (1969) paper is basic for work on dissolution. For petrographic detail CULLIS's (1904) report on the Funafuti cores is still pre-eminent and is supported by SCHLANGER's (1963) stratigraphic information for Eniwetok, his (1964) petrography for Guam and CRICKMAY's (1945) petrography for Lau. BATHURST (1964b) has described the role of micrite envelopes in Bermudan diagenesis. Other references are given in the text.

Chapter 9

DIAGENESIS ON THE SEA FLOOR

The earliest stages of diagenesis are to be seen clearly in carbonate particles lying on the sea floor, either at the sediment–water interface or just below it. Particles that are buried today may be re-exposed, as a result of erosion, tomorrow, or after months or years: burial is not a unique occurrence, and most particles spend thousands of years at or near the sediment–water interface, sometimes in motion, more often stationary, before they are covered for the last time. In this environment the carbonate sediment is subject not only to alternating erosion and deposition, but to prolonged contact with water that may be supersaturated for $CaCO_3$. It is also subject to the feeding activities of the benthos, and to a fluctuating level of CO_2 tension in the water caused by changes in the rates of evaporation, of respiration and photosynthesis of organisms and of bacterial decay. The more obvious effects on the carbonate particles are dissolution, cementation and micritization.

THE PORE WATER IN RECENT MARINE CARBONATE SEDIMENTS

The pore waters of most cores taken down through the carbonate sediments for about 1–2 m, in the Bahamas and Florida, show no evidence of progressive dissolution or precipitation of carbonate minerals (TAFT and HARBAUGH, 1964; BERNER, 1966a; TAFT, 1968). It had seemed that the mineral assemblage that is stable at the sediment–water interface is stable also in the subsurface where the particles are still bathed in sea water. Recently, however, FLEECE and GOODELL (1963) reported a significant downward reduction in the content of high-magnesian calcite in cores from Florida Bay. Reasons for this supposed stability of an inherently mestastable collection of solid phases are discussed on p.325.
    The chemical analysis of pore solutions is a highly sensitive method for detecting slight changes in dissolution or precipitation, changes which would not affect the sizes of particles or their surface topographies to a degree that would be visually detectable with a microscope. Dissolution of aragonite or high-magnesian calcite would release Sr^{2+} and Mg^{2+} into the pore water, but, in fact, no changes in Sr^{2+}/Cl^- or Mg^{2+}/Cl^- were found, except in certain areas where the results of Taft and Harbaugh are in contradiction with those of Berner. Whereas Taft and Harbaugh found increases in Sr^{2+}/Cl^- and Mg^{2+}/Cl^-, in southern Florida and off

the west coast of Andros Island (Fig.127, p.97), Berner found no such changes, even in a core taken near that of Taft and Harbaugh in Florida Bay. This contradiction has not been resolved. P_{CO_2} was high in all the cores taken, presumably as a result of bacterial decay.

It is interesting that cores, taken by Berner in the *brackish* water environment of the Everglades, showed an increase of Mg^{2+}/Cl^-. Berner attributed this change to loss of Mg^{2+} from high-magnesian calcite undergoing dissolution in water already low in Mg^{2+}.

DISSOLUTION ON THE RECENT SEA FLOOR

Dissolution in shallow tropical seas

In shallow tropical seas the water is supersaturated for $CaCO_3$ and, below low-water, carbonate particles are not dissolved: an exception must obviously be made for the colloidal debris of particle breakage and abrasion, some of which must be supersoluble (p.254).

Along the limestone coasts of tropical seas it is usual to find a low cliff, its top commonly standing from 1–3 m above high-water. This cliff has a pronounced overhang and has clearly been undercut laterally for distances which in places exceed 6 m. At first sight the cliff looks as if it were a result of dissolution in sea water and this appearance has, in the past, led to some uncertainty as to the ability of tropical sea waters to dissolve limestone. The undercut surface cannot be the result of mechanical abrasion because it occurs in quiet lagoons as well as on coasts exposed to strong wave action. REVELLE and EMERY (1957) have described the limestone exposed between tides, on Bikini and nearby atolls, and their work may be germane to an understanding of the origin of the cliff overhang. They note that the Bikini limestone surfaces consist of many small depressions separated by knife-edge ridges, more suggestive of dissolution than of abrasion. They have monitored the chemical changes of the sea water trapped in intertidal rock pools and their data for a pool on Guam show a marked fall in pH and oxygen during the night, which can only have resulted from cessation of photosynthesis among the plants in the pool and the accumulation of CO_2 in the water. Such a change would encourage dissolution of the floor of the limestone pool. That this happens is shown by the nightly increase in alkalinity (p.234) despite the almost constant level of the chlorinity.

It must also be apparent to those who have wandered, curious, among intertidal limestone pools that the rock is continuously under attack by endolithic algae and lichens. The overhang itself may well be, in part, a consequence of similar attack, aided by boring sponges (*Cliona*), bivalves and chitons.

Dissolution in deep sea water

This matter is dealt with in Chapter 6 (p.267) and is only referred to here for the sake of completeness.

CEMENTATION ON THE RECENT SEA FLOOR

The better known extensive deposits of Recent carbonate sediments—in the Bahamas, Florida, Cuba, Persian Gulf and elsewhere—are unconsolidated. Until lately this lack of lithification was regarded as the normal situation, but it is now clear that in appropriate circumstances Recent carbonate sediments are being cemented to form submarine hardgrounds, at shallow water depths of only a few metres or tens of metres. Records of deep water submarine cementation of Recent sediments, at hundreds or thousands of metres under the sea, are dealt with on p.375. The truth is that precipitation of cement is widespread in Recent, shallow water sediments but that the process does not everywhere lead to consolidation. Cement may be precipitated within grains, as in the cellular voids of coralline algae, or on the surfaces of grains that are too mobile for the slow development of adhesive cement contacts between grains to be effective. Only where grains lie quietly in contact for long periods can the process of intergranular cementation yield a rigid framework of grains. A peculiarity of Recent marine cements is their restriction to aragonite and high-magnesian calcite. The question of aragonite precipitation in sea water is examined on p.243. In so far as the precipitation of calcite is regarded as being inhibited by the very process that leads to the precipitation of aragonite, there is a paradox which has not been entirely resolved. GLOVER and SIPPEL (1967), when precipitating $CaCO_3$ from sea water by adding excess HCO_3^-, usually produced a mixture of aragonite and high-magnesian calcite.

Subtidal cementation in unconsolidated lime sediments

The oöid is the most obvious case of the precipitation of cement on a grain surface. The origins of this structure are discussed in Chapter 7 and the conclusion is reached that the oölitic coat is a cement of aragonite needles which fails to bind the oöids together because of their high mobility. It is, therefore, particularly interesting that BALL (1967) should have found that, in oölites which have been immobile "for some time", as in the troughs between the crests of megaripples described on p.121, some oöids are cemented lightly together by a fringe of aragonite needles with a preferred orientation of long axes normal to the oöid surface. JINDRICH (1969) found clusters of skeletal grains cemented by aragonite in a tidal channel in Florida Keys.

During a visit to the Cat Cay oölite, Bahamas, in 1961, I found numerous platey fragments of mollusc shell, a few centimetres or more across, lying on the oölite while exposed at low tide. Cemented to the under surface of each fragment were a few oöids. Presumably the process of cementation benefited in some way from the shelter given by the shell (see also TAYLOR and ILLING, 1969, p.79).

Hardened faecal pellets are products of cementation within a loosely bound mixture of carbonate mud and adhesive organic materials. Presumably micro-organisms have a role here, though it is clear that subaerial exposure and evapora-tion play some part. In the Abu Dhabi lagoon (p.199), for example, the proportion of hardened pellets increases with frequency of (intertidal) exposure. ILLING (1954, p.25) on the other hand found all stages of pellet hardening on the Bahamian sea floor in deposits which were all subtidal and not exposed to evaporation. This process of pellet intracementation is not understood and has not apparently been studied.

Aragonite and high-magnesian calcite cements, both with acicular crystals, are common inside the chambers of foraminiferids in the Bahamas, also in the utricles of *Halimeda* and in the cells of coralline algae (mineralogy by electron microprobe, GLOVER and PRAY, 1971). In each case the mineralogy of the cement matches that of the host (see also PURDY, 1964b, 1968). CLOUD (1962a) related the distribution of this type of cementation on the Great Bahama Bank west of Andros Island to that of hardened pellets, both being more frequent toward the Florida Straits, where the value for $a\text{Ca}^{2+} \cdot a\text{CO}_3^{2-}$ in the water is higher (p.290). He also remarked that the pore waters of the more westerly sediments had higher pH and Eh than those further east.

Both aragonite and high-magnesian calcite fill the deserted bore-holes left by dead endolithic algae (p.383) in grains of lime sand. Access of sediment to these tiny tubes (6 μ in diameter) seems unlikely and the fill is probably a precipitate. It is certainly lithified.

TAYLOR and ILLING (1969, p.79) noted the general tendency for precipitation of cement to take place in protected hollows and holes in the lime sands of the coastal regions of the Qatar Peninsula, Persian Gulf, especially in the cavities of corals and gastropods. Steinkernen, formed by the filling of gastropod chambers with cement or cemented sediment with subsequent loss of the shell wall, are com-mon in the intertidal and lagoonal sediments.

MACINTYRE et al. (1968) indicated that, at a depth of about 20 m off Barbados, lime sand is being cemented by high-magnesian calcite precipitated by a burrowing bivalve, *Gastrochaena*. The animal apparently lines its burrow with cemented faecal pellets. The cement is two-layered, having an inner layer of micritic calcite sur-mounted by a layer of spar (crystal diameters about 15 μ) with crystals elongate normal to the surface.

It is plain that in pellets, in body chambers and in algal bore holes there exists an environment distinct from, and only partly dependent on, that of the enclosing sea water. Two of its characteristics are: (*1*) the existence of surfaces free from abrasion, and (2) a concentration of organic tissue undergoing bacterial decay (OPPENHEIMER, 1960, 1961; PURDY, 1968, p.195). Studies of the pore solutions in these restricted environments should not be too difficult and might give useful results. Similar environments occur in reef pores (below) and in the pores of some immobile sediments (pp.373–375).

An origin for grapestone has already been advanced tentatively on p.316. Here again cementation has proceeded in a sharply localized environment, probably influenced by biological activity and the restriction of grain motion. Doubtless the familiar "necked" oöids are formed in the same way as grapestone.

The reasons for the absence of a bonding interparticle cement in many of the great spreads of carbonate mud and sand in subtidal, shallow waters, despite the supersaturation of the water for $CaCO_3$ in tropical seas, are not entirely obvious. Precipitation of a thin film of oölitic aragonite on particles is probably widespread (p.309) and, if this coat is to be regarded as an interparticle cement which has failed to bind the particles because of their agitation, then agitation of particles must be counted as one of the factors that hinder interparticle adhesion.

Subtidal cementation in reef porosity

Subtidal cementation in restricted marine environments (as in *Halimeda* utricles or the cavities of gastropods) is exemplified especially by the evidence of submarine cementation in the reefs of Bermuda (GINSBURG et al., 1968) and Jamaica (LAND and GOREAU, 1970). Off Bermuda, small patch reefs, built mainly of coralline algae and *Millepora*, rise to intertidal level. Internal fabric was revealed in Bermuda and Jamaica by breaking into the reef below low-water with a charge of explosive (Fig.273). From about 1 m below the top of the reef, porosity is filled partly with geopetal carbonate sand and mud, consisting of reef debris and some coccolith ooze, and partly with radial-fibrous acicular aragonite cement and both acicular and equant micritic high-magnesian calcite cement (19 mole % $MgCO_3$). The internal sediments are cemented and are hard rock (Fig.274). In places, layers of cemented geopetal sediment alternate with layers of pure cement. In the Jamaican reefs, off the north coast, laminated pelmicritic crusts of high-magnesian calcite line channels and cavities down to a depth of at least 70 m: internal sediments in reef cavities are cemented (Fig.275). In both Bermudan and Jamaican reefs, holes made by boring bivalves and by the sponge *Cliona* are more or less filled with cemented geopetal sediment overlain by cement.

The submarine, Recent, origin of the cement is indicated by: (1) its occurrence in geopetal sediments derived by fragmentation of the reef frame, (2) the absence of discontinuities or other signs of subaerial emergence, (3) the occurrence of cemented sediments in bores made by bivalves, and by *Cliona*, and (4) of cemented internal sediments bored by *Cliona* and *Lithophaga*, (5) alternation, in some cavities, of internal sediment with layers of pure cement, and (6) the ^{14}C age for

Fig.273. Fresh section excavated through a Jamaican reef at a water depth of 26 m. The top of the picture is about the old reef–water interface. Cementation is just detectable at *A* but at *B* the rock is massively cemented. See Fig.274. (Courtesy of L. Land.)

Fig.274. Close-up of a piece of rock from the section in Fig.273. The coral *C*, just left of the hand lens, has a ^{14}C age of 130 ± 140 years. The white flakes *H* are fronds of *Halimeda* embedded in the pelmicrite of high-magnesian calcite. The geopetal sediments *S* in cavities show that the specimen is upside down. (Courtesy of L. Land.)

the reef frame 3.5 m below the top at Bermuda: this is $3,190 \pm 210$ years, so that the cements must be younger than that. The Jamaican coral in Fig.274 has a ^{14}C age of 130 ± 140 years. Jamaican high-magnesian calcite gave $\delta^{13}C = +3$ and $\delta^{18}O = -0.5$, both values typical of precipitation in sea water, not meteoric water (Fig.261, p.339).

The nature of the chemical environment which controls the precipitation of high-magnesian calcite from sea water is not understood. It is clear, however, that the water in the reef is continually replaced by new supplies of sea water that is supersaturated for $CaCO_3$, from which cement could be precipitated. This renewal is made possible by the pumping action of waves and tides.

Beach rock

Along many coasts there are deposits of Recent calcarenite cemented with either aragonite or calcite. Inasmuch as some of these "beach rocks" have derived their cement from sea water, their examination throws useful light on the process of cementation in the sea. The rocks are intertidal or supratidal and KUENEN (1933, p.87) recalled that there are two main kinds, "beach sandstone" (beach rock in the strict sense) and "cay sandstone". Whereas the beach sandstone (beach rock) is normally intertidal and is demonstrably a beach sand cemented as a result of a

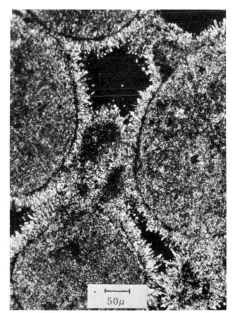

Fig. 275. Rhombohedral micrite of high-magnesian calcite (top right) encrusting a peloid of the same mineral (bottom left). Scanning electron mic. Jamaican reef. (Courtesy of L. Land.)

Fig.276. Beach rock with fibrous aragonite cement. Slice. Crossed polars. Recent. Berry Islands, Bahamas. (Courtesy of H. Buchanan.)

reaction involving sea water, the cay sandstone, occurring as it commonly does above high-water, owes its cementation to the effects of meteoric, fresh water. The fact that both types of calcarenite occur in close proximity along coasts, and that some cay rock is at present intertidal, has tended to obscure the essential distinction between two very different rock-forming processes: so much so that some writers have treated the two as one. (There is a useful discussion in McKee, 1958, p.251.) The problem of genesis has an added complexity since the coasts have been subjected to the Quaternary oscillations of sea level, so that a given calcarenite may, at one period, have been washed by sea water, while at another exposed to meteoric water.

Beach rock is here taken to be a beach sand that was cemented while it was still in the intertidal zone. Although beach rocks are best known from the tropics and subtropics, where the cements are generally aragonite and high-magnesian calcites, they are also known along the shores of the Black Sea and the Caspian Sea (ZENKOVITCH, 1967), the Netherlands (VAN STRAATEN, 1957) and many parts of Britain, especially the Cornish coast (B. B. CLARK, 1968).

There seems to be no doubt that two environments and two processes are concerned (ILLING, 1954, p.70). On or in the beach, aragonite or calcite is precipitated as cement: above high-water, beyond the reach of the sea, sand is cemented by low-magnesian calcite from meteoric water. Intermediate modes of formation are obviously possible, depending on the invasion of the sand by sea water or meteoric water at different times of the year and different states of the tide. Thus clear-cut distinctions are not everywhere discernable. There has been scarcely any investigation of the chemical reactions leading to the formation of beach rock: useful papers on petrography are few (e.g., GINSBURG, 1953; NESTEROFF, 1955a; FRIEDMAN, 1964, p.803; STODDART and CANN, 1965; GAVISH and FRIEDMAN, 1969; TAYLOR and ILLING, 1969). Thus any conclusions reached at the present time depend partly on analogy with other processes elsewhere and must necessarily be tentative. The lithification of cay sandstone falls into the general category of carbonate sediments lithified above sea level, as considered in Chapter 8. Only beach rock (beach sandstone) will be examined in detail here.

The typical beach rock consists of a layer of cemented calcarenite forming a hard surface on or in the beach, underlain by unconsolidated lime sand. The lamination and attitude of the rock are the same as those of the local unconsolidated beach sediment, to which the grains of the rock correspond also in composition and texture: it may be a calcarenite, a conglomerate or a breccia. The upper surface of the beach rock, if exposed, is typically indurated, pitted (as limestone pools, p.362), abraded and bored by penetrative algae and stained. The base of the rock, where it overlies unconsolidated sediment, is commonly sharp and plane. The degree of cementation varies throughout the rock.

Beach rock is distributed irregularly over the beach. On Bikini and nearby atolls it covers about 15% of the beach (EMERY et al., 1954, p.44). The disposition of beach rock is controlled by the local balance between the rates of three processes: sand accumulation, erosion and cementation. This is apparent from the descriptions of BRANNER (1904), DAVID and SWEET (1904), GARDINER (1930), KUENEN (1933), GINSBURG (1953), EMERY et al. (1954), ILLING (1954), NESTEROFF (1955a) and STODDART (1962a, b). Conclusive evidence that the precipitated cement (aragonite or high-magnesian calcite) is derived from the sea water comes from the observations of BRANNER (1904) and DALY (1924), who noted that beach rock can form where the original sediment contains little or no carbonate. That beach cementation is a contemporary process is clear from the inclusion in the rock of man-made articles such as beer bottles in the Bahamas and objects only seven to eight years old on Eniwetok (EMERY et al., 1954).

The resultant cement (Fig.276) has been described, from Florida, as a fringe of acicular aragonite crystals with lengths 0.005–0.4 mm (GINSBURG, 1953, p.88). Aragonite fringes, on calcarenite grains in reef islands of British Honduras (STODDART and CANN, 1965, p.244), were recorded as about 0.04 mm thick and composed of elongate platey crystals. These have a radial length up to 0.1 mm with the other two diameters up to 0.02 mm and 0.003 mm. ILLING (1954, p.70) described Bahamian aragonite cement crystals as radially oriented blades up to 10 μ long. The reasons why aragonite, and not calcite, should be precipitated are discussed on p.243. ILLING (1954, p.70) saw beach rock in the Bahamas having acicular aragonite cement with some granular calcite. In beach rocks along the Mediterranean coast of Israel, near Tel Aviv, the cements are aragonite and high-magnesian calcite, both minerals commonly occurring in the same rock (GAVISH and FRIEDMAN, 1969). TAYLOR and ILLING (1969) have discovered a variety of cements in beach rock along the shores of the Qatar Peninsula, Persian Gulf. These are: (1) clotted fabric of aragonite needles wherein randomly oriented needles are clumped together in discrete masses, (2) radial-fibrous aragonite, (3) radial-fibrous calcite, (4) crypto-crystalline aragonite, and (5) cryptocrystalline calcite. Some of the radial-fibrous cements are thicker on the upper and lower surfaces of the grains in the manner of micro-stalagmites and micro-stalactites. Does this imply growth as a result of evaporation above the water table? Calcite cements, on aragonite cements from which needles are missing, suggest replacement of aragonite by calcite through dissolution and precipitation. High-magnesian calcite cements in beach rock have also been described by FRIEDMAN (1968a) from the Gulf of Aqaba and the Canary Islands. It would be interesting to know if the high salinities, developed by evaporation, are in any way the cause of a loosening of the bonded water dipoles on the Mg^{2+} ion (p.244), thus allowing precipitation of calcite. DE GROOT (1969) has suggested that a *low* level of supersaturation permits the precipitation of high-magnesian calcite (p.374).

Recognition of fossil carbonate beach rocks is likely to be difficult, but it is worth bearing in mind that they should differ in several ways from carbonate sediments that are cemented in other environments. First, they have, or had, an aragonite or high-magnesian calcite cement. If the aragonite has been dissolved and replaced by calcite, then its one-time existence may be impossible to discover. If, however, it has been calcitized (presumably by wet transformation) then it is possible that some tell-tale relic fabric remains, though it must be admitted that such fabrics have not yet been recognized. Evidence of compaction should, of course, be absent. Complications may be expected where the primary cementation was incomplete or where primary fabrics have been modified by diagenesis in the fresh water, subaerial environment. It goes without saying that the calcarenites should normally consist of well sorted, abraded grains.

The chemical process leading to the formation of beach rock has been tacitly assumed, by many authors, to be a simple matter of an increase in the level of supersaturation for $CaCO_3$, as a result of evaporation and loss of CO_2 during subaerial exposure of the beach at low tide (as DANA, 1851, p.368). Experimental or field tests have scarcely been attempted. SCHMALZ's, (in press) work is a rare exception. Nowadays, however, thinking has become more sophisticated, as for example in the approach to the formation of submarine hardgrounds (p.373) and grapestone (p.316). We know that aragonite and high-magnesian calcite are widely precipitated from shallow subtropical and tropical sea water (p.364). Thus, rather than ask why grains are cemented on (or in) beaches, we should ask why they are not cemented everywhere. The answer is probably the same as that given for cementation of submarine hardgrounds in the Persian Gulf (p.374). Sand grains near the beach surface are too mobile to be cemented. Farther down, the pore water is relatively stagnant, nearly in equilibrium with the grains and isolated from fresh supplies of sea water. In between these two zones is one in which the grains are mechanically stable, but are bathed in supersaturated water frequently replenished from above by the pumping action of waves and tides. Here the grains will be cemented to make a hard layer. TAYLOR and ILLING (1969) found that, in beach rocks in the Qatar region, the content of cement decreases downward. The familiar exposed beach rock is probably wasting, by abrasion, by corrosion of the type described by Revelle and Emery (p.362), as a result of boring by chitons and bivalves, and micritization following attack by boring algae (p.383). The buried beach rock, like the hard layers around Qatar under the lime sand, have irregular upper surfaces, unstained, unbored and less indurated than the exposed rock (SCHMALZ, in press). Induration appears to be aided by the algal micritization process: this is inactive in the dark situation of buried sands. Where the chemistry of the pore water in a beach has been monitored over a period, it is apparent that other factors in addition to those mentioned play an important role (SCHMALZ, in press). Mixing of supersaturated meteoric, land-derived water with supersaturated sea water could lead to increased supersaturation so as to stimulate precipitation, as with mixing of ground waters (p.446). Metabolism of micro-organisms may be critically important (NESTEROFF, 1955b, 1956b; RANSON, 1955b).

Submarine hardgrounds (hard layers) in the Persian Gulf

The recently published descriptions of layers of cemented calcarenite covered by only a few metres or tens of metres of water, in the Persian Gulf, have probably removed for ever the doubts felt by geologists about the importance of submarine cementation in the formation of limestones in general. While the occurrence of deep oceanic crusts (p.375) could perhaps be regarded as having uncertain significance for the study of most ancient limestone successions, which are mainly of shallow water origin, the discovery of an area of about 70,000 km² in the Gulf from 1 to 60 m deep, wherein lime sands are being cemented today, means that submarine diagenesis must now be regarded as no less important than meteoric diagenesis in the lithification of limestones. Layers of calcarenite, 5–10 cm thick, have been found in otherwise unconsolidated lime sand, below tidal lagoons and below the subtidal sea floor, in and around the Qatar Peninsula (Fig.185, 186 on p.179, 182 resp.). Dredged samples indicate the probably widespread occurrence of similar layers west and south of the 60 m bathymetric contour in the Gulf (SHINN, 1969). The main research has been carried out by TAYLOR and ILLING (1969) of the firm of V. C. Illing and Partners, Cheam, England, and by SHINN (1969) and DE GROOT (1969) of Koninklijke/Shell Exploratie en Produktie Laboratorium, Rijswijk, The Netherlands.

The hard layers. The hard layers lie parallel to the sea floor and they are composed of lime sand no different from the few centimetres of uncemented lime sand that separates or overlies them. Broken surfaces are coloured light tan to amber. The content of cement is greatest at the upper surface and decreases downward. This upper surface—where it is or has been uncovered by loose sand—is smooth, greyish-black, abraded, commonly micritized by algal boring and precipitation (p.383), bored by molluscs, and encrusted by oysters and serpulids. The under surface is irregular and the cemented sediment peters out downward, the lowest cemented grains being concentrated around burrows. Where the upper surface has always been covered by sediment it is as irregular as the under surface. The cement, whether intergranular or intragranular, is either aragonite or high-magnesian calcite. The aragonite is either radial-fibrous, with long axes commonly 5–30 μ, or cryptocrystalline. The calcite cement may be either radial-fibrous parallel to c, though shorter and stumpier than the aragonite, or equigranular with diameters mainly 1–7 μ and a tendency for the c axis to lie normal to the surfaces of the grains.

Growth of hard layers. Careful field observations have shown that the cementation proceeds not only under water but, in places, some way down under the sediment surface, say 5–50 cm. The diachronous layers under the lagoons, when traced seaward, become more and more vuggy, as cement is more patchily developed, and eventually pass into separate lumps (TAYLOR and ILLING, 1969). Growth appears to have taken place by coalescence of lumps, as in the Chalk hardgrounds (p.406). The lumps are distinguished from the surrounding uncemented sand by their content of soft, white, silt-sized aggregates of cryptocrystalline aragonite or calcite. The lumps also bear radial-fibrous aragonite crystals, up to 5 μ long, or equigranular calcite crystals 1–2 μ across. Lumps exhumed by storm waves or burrowing animals have been micritized by blue-green algae and their under surfaces have been bored (as in Hallam's study, p.394, of exhumed Jurassic hard layers). Followed landward under the lagoon floor, the lumps increase in size, abundance and hardness, finally coalescing into layers which have upper surfaces as rough and irregular as their lower surfaces. The transition from lumps to layer may be complete in a distance of 10 m or perhaps as much as 1 km. There is no sign that the lumps are exotic lithoclasts: the horizontality and thinness of the layers distinguish them clearly from beach rock (TAYLOR and ILLING, 1969). To produce so many layers by subaerial or intertidal diagenesis would require impossible fluctuations of sea level, moreover the sediments in question do not show the typical structures of intertidal sands (e.g., laminae and fenestral porosity). Also, the skeletal biota is characteristically subtidal and two values for $\delta^{18}O$ give calculated temperatures appropriate to a submarine origin (SHINN, 1969). The Recent age of the cement is shown by pottery cemented into one layer, beneath which there is a sand containing burrowing molluscs. Radiocarbon dating of one of these molluscs gave 1,040 ± 180 years B.P. Thus cementation of the overlying hard layers must have begun less than a thousand years ago. The time for hardening is remarkably short when compared with the many thousands of years required for Ordovician (p.399) and Chalk (p.406) hardgrounds.

Crystal growth tectonics of the hard layers. Diving around the Qatar Peninsula, photographing and digging pits with a portable air hammer, Shinn found tens of square kilometres of floor "laced by polygonal fracture systems" in the hard layers. Polygons are from 10–40 m across and are separated by plant-encrusted anticlines which are commonly severely fragmented and overthrust. Smaller millimetre-sized and micron-sized fractures, filled with internal sediment or cement, occur throughout the hard layers, particularly near the thrusts and anticlines. Clearly the fracturing was caused by expansion of the hard layers which, in turn, was probably a result of pressure exerted by the growing cement crystals.

Internal fabrics in hard layers. The internal structure of the hard beds is complex. Rinds of cement, either aragonite or high-magnesian calcite, and layers of geopetal mud succeed each other in no particular order. Aragonite cement is dominant in some regions, calcite in others. Geopetal sediment of carbonate mud (skeletal detritus) has in many places been cemented and bored, and the bores have been filled with mud which is now also cemented and bored (as in the Bermudan reefs, p.365). Internal geopetal sediment is highly characteristic of the hard layers. Calcitization of aragonite cement, pellets and skeletons by high-magnesian calcite is apparent from the cross-cutting distribution of equigranular mosaics of calcite crystals, mostly 2–10 μ in diameter but in places attaining 100 μ. Relic patches of aragonite are distinctly in evidence. Boundaries between aragonite and the replacing calcite are sharp: apparently vague boundaries under the low power microscope are seen under higher power to be systems of interfacial embayments, with many islands of relic aragonite in the calcite and of new calcite crystals in the aragonite. The replacement process, though largely neomorphic in that it involves both polymorphic transformation of aragonite to calcite and recrystallization, is also partly metasomatic because there is an exchange of Mg^{2+} for Ca^{2+}. Some aragonitic molluscan shells have been recrystallized to vertical fibres of aragonite (SHINN, 1969) as noted by Tebbutt (p.348). It is particularly interesting that the geopetal internal muds, after first being cemented by high-magnesian calcite, are further altered to an equigranular calcite mosaic of uniformly sized crystals 5–10 μ, closely similar in appearance to that described from ancient limestones (p.513) and occurring as vadose silt (p.356). Indeed, it seems that ancient submarine hard layers, so rich in geopetal muds and pseudo-silts of calcite, may well be difficult to distinguish from limestone with geopetal silts formed in the meteoric diagenetic environment.

The hard layers as hardgrounds. The hard layer is clearly a Recent form of hardground, showing the appropriate characteristics (p.395)—a relatively level upper surface which is bored, encrusted and discoloured, a downward and lateral passage into patchily cemented sediment and finally into separate lumps or concretions. There is, however, no doubt that cementation acts not only at the sediment–water interface but also some tens of centimetres below it. Hard layers exposed on the sea floor are bored, abraded and altered by endolithic algae: these are in part retrogressive processes. Growth by coalescence of tiny concretions and along the walls of burrows produces the rough and irregular surfaces at the bases of all the hard layers and also on the tops of those which are covered by uncemented sediment.

Growth conditions for hard layers. The likely conditions for growth of the hard layers were given careful consideration by DE GROOT (1969), SHINN (1969) and TAYLOR and ILLING (1969). Their conclusions may be summarized as follows.

To begin with it is apparent that in warm, shallow sea water, supersaturated for $CaCO_3$, cement is normally precipitated—contrary to popular belief. But this cement does not lead to grain cohesion. It is concentrated in sheltered microenvironments such as the chambers of gastropods and foraminiferids and in sponge bores (p.363). In addition, many grains are coated with oölitic films a few microns thick (p.309): oölitic coats have been described elsewhere as frustrated cement prevented from causing cohesion by the vigour of grain movement. The degree of grain-to-grain cohesion attained obviously depends on both the level of agitation and the rate of precipitation. Where precipitation is fast enough and agitation slight enough, grapestone is known to form (p.316). It is also plain that precipitation is assisted by the presence of substantial areas of substrate of the same material as the precipitate, on which epitaxial growth can continue (p.428 and 435). In the Persian Gulf the three necessary conditions for lithification are provided by: (*1*) supersaturation of the sea water for both high-magnesian calcite and aragonite (DE GROOT, 1965), (*2*) freedom from bottom traction, and (*3*) the presence of suitable substrates. (It was noted by Taylor and Illing that the carbonate cements do not grow on quartz grains.) The sedimentary horizon at which cohesive inter-granular precipitation happens must be a thin one. Above it, grain agitation is too severe, below it stagnant water has achieved too great a degree of equilibrium with the grains so that precipitation is too slow. Slow sedimentation helps hardening: too rapid an accumulation will soon isolate the site of initial hardening from the overlying supply of supersaturated sea water. Once cohesive precipitation has started, permeability is reduced and the circulation of fresh supplies of supersaturated water is reduced: consequently the content of cement in a layer decreases downward. The hardened beds are thicker in the coarser sediments such as lime sands, because their permeability is greater and water circulation freer. Thin crusts are typical of lime muds, such as the abyssal coccolith oozes (p.375) in which permeability is slight.

It is worthwhile bearing in mind a tentative suggestion by R. M. LLOYD (in press) that the high positive values for $\delta^{13}C$ in the cements of the hardgrounds (mixed aragonite and calcite, $\delta^{13}C = +4.0$ to $+4.5$) are suggestive of algal precipitation (Fig.234, p.283).

The mineralogical nature of the cement—aragonite or high-magnesian calcite—is probably dependent on the level of saturation of the sea water for $CaCO_3$. High-magnesian calcites are mostly more soluble than aragonite (Fig.218, p.237) and should precipitate first from solutions that are supersaturated for both high-magnesian calcites and aragonite (but see Winland, p.258). Yet the kinetic effect of adsorption of hydrated Mg^{2+} on calcite favours the growth of aragonite (p.243). DE GROOT (1969) suggested that at very low supersaturation the growth rate of aragonite is so slow that it does not differ significantly from that of high-magnesian calcite (p.260). Indeed, during calcitization of aragonite, the calcite (possibly low in Mg^{2+}) is clearly the less soluble polymorph—a fact made obvious by accompanying dissolution of aragonite adjacent to the replacing calcite (TAYLOR and ILLING, 1969).

Submarine hardgrounds on the New Providence platform

On the New Providence platform, Bahamas, between Tongue of the Ocean and

Exuma Sound, TAFT et al. (1968) have found blocks and continuous sheets of cemented grapestone. The cement is acicular aragonite: the micritic aragonite cement so typical of grapestone was not recorded. The lithified sands occur in shallow water, 3–5 m deep, either exposed or under a cover of loose grapestone a few centimetres to about a metre thick (see also TAFT, 1967, 1968).

Cementation in deeper waters

The current enthusiasm for research on shallow water carbonates, in the Bahamas, the Persian Gulf, Bermuda and elsewhere, along with the spectacular revelations from the cores of Pacific atolls, has tended to divert interest from the less accessible, deeper water sediments, and to obscure the importance of the few but significant data on these deposits which have been gathered over the past seventy-five years. Recent work on cemented carbonate sediments, dredged from depths of 200–3,500 m, and studied by FRIEDMAN (1964), GEVIRTZ and FRIEDMAN (1966), MILLIMAN (1966), FISCHER and GARRISON (1967) and BARTLETT and GREGGS (1969), has redirected attention to the actuality of submarine cementation, at a time when the results of field studies of ancient limestones are forcing the geologist to re-examine the possibility that widespread submarine cementation took place at various times in the past (p.392). The review by Fischer and Garrison is a particularly helpful survey of knowledge in the realm of deeper water cementation.

All the known cemented sediments on the deeper sea floor today are pelagic coccolith oozes, with *Globigerina* as the dominant foraminiferid, and with some pteropods. The chief data, largely from Fischer and Garrison, are given in Table XVI.

Crusts of high-magnesian calcite. Two unusually detailed descriptions of cemented oozes have been given, by FISCHER and GARRISON (1967), and GARRISON and FISCHER (1969, p.29), for samples from the Mediterranean (2,055 m) and from a station off Barbados (280–440 m). The descriptions are given in shorter version below. The Mediterranean material, collected by the Austrian ship *Pola*, is one of a number of dredged samples of crusts with thicknesses of 1–8 cm. Under the crusts the oozes are unconsolidated. The degree of lithification varies from the scarcely detectable to a brittle limestone which rings when hammered, and in the sample examined by Fischer and Garrison the induration decreased downward. The upper surface of the crust varied from smooth to scoriaceous and was either black or brownish owing to the stain of iron and manganese oxides. The crusts were bored and, attached to the surface, there were bivalves and other sessile invertebrates. Lithification commonly extends downward along organic burrows into the underlying ooze. The general situation resembles that of the Ordovician and Chalk hardgrounds described by Lindström (p.397) and Bromley (p.399).

The sample collected off Barbados by R. J. Hurley, of the University of

TABLE XVI. RECORDS OF CEMENTED, SUBMARINE, CARBONATE OOZES

Locality	Depth (m)	Age	Dominant grains[1]	Cement	References
East Indies	229	—	Globigerina	—	Murray and Lee (1909)
Pacific	2,284	—		—	
Pacific	—	—		—	
East Mediterranean and Red Sea	327–3,310 (analysed sample from 2,055)	Miocene (Med.)	Globigerina, Pteropoda	high-mag. calcite (Med.)	Natterer (1892–1898), Fuchs (1894), De Windt and Berwerth (1904)
Eniwetok Atoll	800–1,000	Pliocene	Globigerina, Coccolithaceae	calcite	Bramlette et al. (1959)
Atlantis seamount, Mid-Atlantic Ridge	300	Pleistocene	Globigerina, Mollusca	high-mag. calcite	Friedman (1964, p.806)
Red Sea	517–1,700	Pleistocene?	Foraminiferida, Pteropoda	aragonite	Gevirtz and Friedman (1966, Blake Plateau p.149)
Blake Plateau, north of Bahamas	1,000	undated	Globigerina, Pteropoda	high-mag. calcite	
Atlantic seamounts	600–900	Plio–Pleistocene and Miocene	Globigerina, Mollusca	high-mag., low-mag. calcites	Milliman (1966)
Mid-Atlantic Ridge	—	—	sheared?	low-mag. calcite	Mecarini et al. (1968)
Barbados	280–440	Late Tertiary	Globigerina, Gastropoda	high-mag. calcite	Fischer and Garrison (1967)
Pacific seamounts	—	Tertiary and Cretaceous	Globigerina	—	Hamilton (1953, 1956), Hamilton and Rex (1959)
Mid-Atlantic Ridge	3,000	Miocene and Pliocene	Globogerinacea	CaCO₃	Cifelli et al. (1966)
North Atlantic	—	Quaternary	Globigerina	—	Bartlett and Greggs (1969)
Red Sea	89–2,704	Quaternary	complex	aragonite, high-mag. calcite	Milliman et al. (1969)

[1] In matrix of coccolith ooze unless otherwise indicated

Miami, was a lithified ooze in which were partly embedded "cannon-ball" nodules and "pancakes" of manganese oxides. Again, there is the association of crust with manganese oxides.

Both Mediterranean and Barbados samples (Fig.277) contained tests of low-magnesian calcitic foraminiferids, such as *Globigerina* or *Globorotalia*. In addition there were, in the Mediterranean samples, benthonic foraminiferids, and, in the Barbados samples, pteropods and bivalves (aragonite), echinoderms and melobesiid algae (both high-magnesian calcite). The matrix of each sample was a micrite, with crystal diameters of 0.5–3 μ (Mediterranean) and 0.5–5 μ (Barbados), measured on electron photomicrographs. X-ray diffraction patterns revealed the presence of low-magnesian calcite (from the *Globigerina* and *Globorotalia*), some aragonite (in the Barbados sample, presumably molluscan), and high-magnesian calcite. Some of the aragonite shells have undergone partial dissolution. In each sample euhedral rhombs of dolomite are visible. The Mediterranean crust contains some quartz silt, and in the Barbados rock there are particles resembling glass shards.

The precipitation of high-magnesian calcite in the lithification process of these two limestones is manifestly of critical importance. The other minerals of

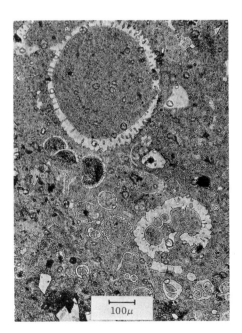
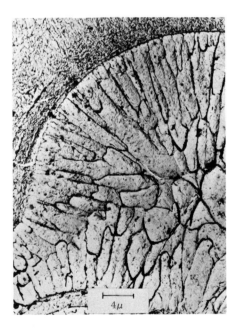

Fig.277. Crust of high-magnesian calcite micrite. Globigerinid tests in different stages of alteration varying from entire to grey and speckled or mere ghosts. Volcanic glass shards. Peel. Off Barbados. (From FISCHER and GARRISON, 1967.)

Fig.278. Crust of high-magnesian calcite micrite. Calcite cement in globigerinid chamber shows preferred orientation of longest crystal axes and increase of size away from test surface. Replica. Electron mic. Off Barbados. (From FISCHER and GARRISON, 1967.)

quantitative significance, the low-magnesian calcite and aragonite, can be accounted for in terms of the primary skeletal content. The preponderance of high-magnesian calcite is equated by Fischer and Garrison with the micrite, and probably the sparry fillings of tests (Fig.278), which together constitute the major bulk of these fine-grained limestones. A similar dominance of high-magnesian calcite was recorded by Friedman from the top of the Atlantis seamount and by Milliman from the lithified oozes of other Atlantic seamounts (Table XVI). Other factors which bear on an understanding of the lithification are the occurrence of iron and manganese oxides and the borings, all of which testify to the idea of a virtual cessation of sedimentation, as propounded by Lindström (p.399) and Bromley (p.406) in connection with the Ordovician and Chalk hardgrounds. Nor must the singularity of the depth of water be forgotten. Though the shallowest of the cemented oozes are not much deeper than some current estimates of the likely level of exposure of the sea floor during the Pleistocene glaciation, most of the crusts have certainly lain always under hundreds of metres of sea water.

Significant fabrics related to the diagenesis of the Mediterranean and Barbados limestones are exemplified in the description of the Barbados material by Fischer and Garrison. The body chambers of the tests are filled either with micrite or with calcite cement (both presumed to be high-magnesian calcite). Both the aragonitic test walls, and those of low-magnesian calcite, have commonly been partly or wholly replaced by the micrite. From a comparison of the X-ray data with the volumes of cement and micrite in the rocks, Fischer and Garrison concluded that the lithification process has involved a partial replacement of the original mixture of skeletal aragonite and skeletal low-magnesian calcite by a mixture of two fabrics, both composed of high-magnesian calcite. One of these is a space-filling cement (Fig.278), the other a micrite (Fig.277). This micrite appears to have replaced not only test walls but also the dominantly coccolithic mud, composed of low-magnesian calcite.

Whatever the process of lithification may have been, it is plain that it differed markedly from the subaerial, fresh water process described in Chapter 8. It seems to have consisted in part of a process of concentration of Mg^{2+} in the ooze, since the original material contains insufficient skeletal high-magnesian calcite to have yielded, by mechanical breakdown, anything but a dominantly low-magnesian calcitic ooze. If the calcite cement in the body chambers is really high-magnesian, as seems likely, then the solution from which it was precipitated must have had a suitably high content of free Mg^{2+}, in contrast to the virtual absence of Mg^{2+} in the rain water responsible for the diagenesis of emergent carbonate sediments, as described in Chapter 8. In such an environment the replacement of skeletal carbonate, and of the primary carbonate mud, by high-magnesian calcite must have proceeded by dissolution and precipitation in an aqueous solution with a high Mg^{2+} activity. The source of the dolomite crystals is not known, but if these are indigenous and not wind-blown detritus, then their growth again implies a solution rich in Mg^{2+}. On the other hand it is known that, in normal sea water, the high energy of hydration of the Mg^{2+} is such as to prohibit the growth of calcite and to allow only aragonite to precipitate (p.243). Though the deeper Mediterranean sample (2,055 m) may have come from an environment where, owing to great hydrostatic pressure, aragonite is more soluble than high-magnesian calcite (p.268), this is unlikely to be true for the Barbados sample which was no deeper than 440 m. Finally, the superficial nature of the lithification of deep sea oozes in general, points to a reaction involving sea water, as in the shallow-water, calcarenitic hardgrounds of the Persian Gulf (p.373). This reaction is also connected, perhaps, in some way with the precipitation of iron and manganese oxides. Biological intervention is not excluded. Further discussion of cementation processes is to be found on p.452.

Recently, a most interesting paper has appeared by BARTLETT and GREGGS (1969) which gives valuable insight into lithified layers found in cores drilled into the floor of the North Atlantic. These cores of oceanic sediment contain alternating unconsolidated and lithified layers, each couplet having a combined thickness of 3–6 cm. The unconsolidated layers are made of skeletal grains in a clay-grade calcitic matrix (coccolith ooze?). The grains include corals, gastropods, bivalves, pteropods and some foraminiferids: the larger grains are broken, the smaller intact. The lithified layers are *Globigerina*–pteropod oozes (hardened coccolith ooze?) with a little mollusc debris. The unconsolidated layers have a cold-water fauna, the lithified layers a warm-water fauna. Of particular significance is the contact between a lithified layer and the overlying unconsolidated layer. This junction is commonly "very undulating and stylolitic". In the lithified layer ferromagnesian oxides are concentrated near its upper surface. Age determinations, using ^{14}C and foraminiferids, indicate that the interface between lithified and overlying unconsolidated layer represents a time interval—at least in some instances—of about 30,000 years. The junctions between the unconsolidated layers and overlying lithified layers are gradational. It would appear, therefore, that the lithified layers are miniature hardgrounds (p.394).

Cementation with aragonite. Cementation of deep water carbonate sediments by aragonite has been described by GEVIRTZ and FRIEDMAN (1966) in their account of cores taken from the floor of the Red Sea at depths from 384 m to 1,737 m. The *dominant* sediment, representing normal conditions, is a pelagic calcilutite rich in tests of planktonic foraminiferids and pteropods. These lie in a lutite matrix, with a variable content of detrital quartz, silicates and organic matter, giving an alternation of light and dark laminae. The carbonates in these lutites are mainly both high-magnesian and low-magnesian calcites accompanied by less than 20% of aragonite (Fig.279), depending on the composition of the skeletal debris. This sediment is not cemented. The content of aragonite decreases with depth as pteropod tests are replaced by internal casts of calcite (FRIEDMAN, 1965c). MILLIMAN et al.

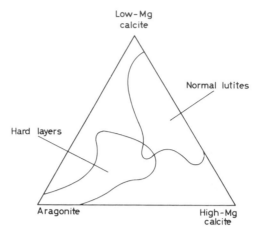

Fig.279. Mineralogical composition of the normal pelagic calcilutites and the hard layers in cores from the floor of the Red Sea. (After GEVIRTZ and FRIEDMAN, 1966.)

(1969) described the matrix as clusters of crystals of high-magnesian calcite (about 12.5 mole % $MgCO_3$), each crystal being a rhomb of diameter 0.1–0.2 μ.

At intervals in this normal succession appear hard layers (Fig.279) with thicknesses from less than 1 mm to 5 mm. They are amber-coloured, highly porous and have one surface smooth, the other irregular, commonly pitted. Regrettably, absence of information on way-up made it impossible for the authors to say which of these two surfaces was the upper. The pores in the hard layers are filled with radiating fibres of aragonite which have grown outward either from "pellet-like circular structures" (GEVIRTZ and FRIEDMAN, 1966, p.145), or commonly syntaxially, from the walls of foraminiferids and pteropods. Some dark-light banding in the hard layers is produced by variation in the amounts of clay and limonite.

In discussing the origin of the hard, aragonite-cemented layers, Gevirtz and Friedman related the discovery of hot saline waters in the axial trough of the Red Sea, by MANHEIM et al. (1965) and SWALLOW and CREASE (1965), to the precipitation of aragonite from hot hypersaline waters in the Dead Sea (NEEV, 1963, 1964; FRIEDMAN, 1965a; NEEV and EMERY, 1967). They concluded that the alternation, in the Red Sea, of unconsolidated lutites with hard layers, represents fluctuations in temperature and salinity at the bottom of the sea. The implication is that the increased solubility of the aragonite caused by the great pressure (p.271) is more than offset by the rise in temperature (up to 44°C) and the vastly enhanced salinity (> 270‰).

Gevirtz and Friedman did not record the presence of calcilutite in the hard layers, but their photographs of radial-fibrous aragonite show that the bunches of fibres pass into calcilutite at their younger ends. This can mean either that the lutite arrived after the fibres had grown or that the fibres have partly replaced the lutite. As lutite is a component in the lutite-supported, wackestone fabric of the normal Red Sea sediments, it seems more probable that the aragonite is a replacement.

MILLIMAN et al. (1969) have added to this picture of submarine cementation in hot hypersaline brines. During a cruise in *Chain* in the Red Sea, cores were taken from which flakes of aragonite-lithified sediment were extracted. These flakes were interpretated as fragments of lithified layers broken by the entrance of the corer. The lithified layers were located only at certain depths in the cores, in a horizon lying between radiocarbon age limits of 11,000 and 20,000 years B.P. This age coincides with a time when the sea level was at least 80 m lower than at present owing to the Würm glaciation. Extra high salinities at that time must be assumed since, even now, the Red Sea has temperatures greater than 21 °C and salinities more than 40‰: these persist throughout the well-mixed water column. These conditions depend on a high rate of evaporation, low inflow of fresh water and restricted exchange with the Indian Ocean caused by a sill at only 125 m. Milliman et al. suggested that more than half the deep-sea carbonates in the Red Sea may well be inorganic precipitates.

Stellate clusters of aragonite needles, up to 3 mm long, forming veins in manganese nodules, have been described by MCFARLIN (1967) from the Blake Plateau (north of Grand Bahama) at a depth of 600–800 m. Chemical and isotopic data indicate inorganic precipitation in an environment which was not significantly different from the present.

BORING ALGAE AND MICRITIZATION ON THE RECENT SEA FLOOR

The most widespread habitat of the non-calcareous, boring chlorophyte and cyanophyte algae is the limestone substrate. In intertidal and supratidal limestones, in places as diverse as Ireland, the Bahamas and the Persian Gulf, these plants live at a depth of a fraction of a millimetre below the rock surface. Their importance in the present discussion lies in their widespread occurrence in carbonate grains (Fig.280–283) and the accompanying alteration of these grains to micrite. NADSON (1927a, b) has reported these bored carbonate grains from many parts of the world, FRÉMY (1945) has shown their ubiquity in the Mediterranean, GINSBURG (1957) and SWINCHATT (1965) have described them from Florida, BATHURST (1966) and MONTY (1967) have recorded them from the Bahamas (Fig.285), also BATHURST (1966) from the Gulf of Batabano. It is plain that these endolithic algae cause widespread and wholesale destruction of all types of grain, skeletal, oölitic and peloidal. Grains are riddled with holes and are, as a result, particularly liable to abrasion. The consequences of this are seen in Bimini Lagoon, Bahamas, where the most densely bored grains are also the most rounded (BATHURST, 1967a, p.105).

Identification of the algae, probably coccoid cyanophytes, in the Bahamas, is difficult and tedious and the taxa described so far (p.297 and NEWELL et al., 1960)

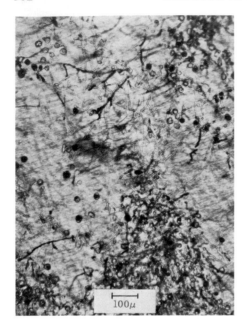
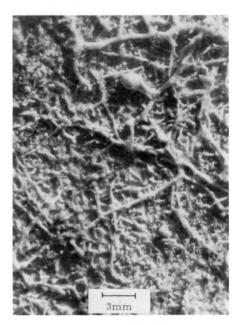

Fig.280. The green endolithic alga, *Gomontia polyrhiza*, in a Recent oyster shell. Filaments, about 4–5 µ diameter, and globose sporangia. Slice. Sussex shore. (Courtesy of R. G. Bromley.)

Fig. 281. Casts of algal bores in the scaphopod *Dentalium*, the aragonite of the shell having been dissolved. The casts are phosphatized chalk. Top centre is a crypt that housed a sporangium. Upper Cretaceous. (Courtesy of R. G. Bromley.)

may be only a small proportion of those engaged. It is also possible that some of the borers are fungi. Most of the bores described by Bathurst (1966) have transverse diameters of about 6 µ; larger bores up to 15 µ, though ubiquitous, are rare in Bimini Lagoon. Fungal bores are narrower, say 1–2 µ. Thus the possibility that some of the bores are those of arthropods, gastropods or sponges is ruled out by the tiny cross section. Holes bored by clionid sponges have diameters from 0.5 mm to more than 1 cm.

BROMLEY (1965) has summarized those features which have some value in distinguishing algal bores from fungal bores. While no one feature is diagnostic on its own, together a group of features add weight to the interpretation of bores:

"1. Fungal borings are on the whole finer than those of algae.

"2. The diameter of fungal borings is generally more or less constant while that of algae varies considerably.

"3. The mode of branching is characteristically 'false ramification' in algae, the thicker main borings giving the appearance of having been occupied by a bundle of threads which, separating individually, simulate branching. As a result of true ramification in fungi there is normally no sensible reduction in diameter from stem to branch. True ramification is also found in algae, however.

"4. The angle at which branches leave the major axis is very much more

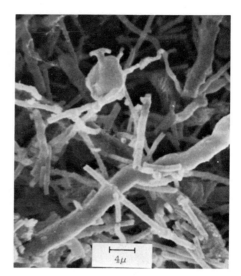

Fig.282. Large micrite-filled bores in a mollusc shell. The shell has been replaced by calcite cement. Slice. Upper Cretaceous. Monte Camposauro, Campagna, Italy. (Slice lent by B. D'Argenio.)

Fig.283. Casts of bores (mainly algal) in the grains of a biomicrite. The bores, lined with limonite, have been exposed by etching. Scanning electron mic. Jurassic. Somerset. (Courtesy of M. Gatrall.)

constant in fungi, and is often between 60° and 90°. Dichotomy is also common. Algae frequently branch very irregularly.

"5. Likewise, the articles [cells] in fungi tend to be more or less straight or gently curved, while those of algae are sometimes very irregular. Overcrowding probably induces irregularity in some species of both types, but some fungi appear to remain invariably straight."

The borers have a further effect on the grains, one that is more far reaching than mechanical destruction. The emptied bores, vacated by the borer presumably after death, are apt to be filled with micritic carbonate. By repeated boring, followed by vacation of the bore and the filling of it with micrite, carbonate grains are gradually and centripetally replaced by micrite (Fig.284, 285A). BATHURST (1966) has referred to this process as **micritization**. Its role in the preservation of casts of molluscan debris during subaerial, fresh water diagenesis is described on p.333. PURDY (1963a, p.348) recognized the general occurrence of this replacement process, but called it "recrystallization", a description which is no longer appropriate now that the details of the process have been worked out. MONTY (1967, p.79, 83, 89) recognized that algal boring caused the grains to change to cryptocrystalline carbonate, but did not explain the process. TAYLOR and ILLING (1969, p.79) recorded that the Recent calcarenites around the Qatar Peninsula in the Persian Gulf are bored by algae to a depth of about 200 μ. The bored rock is changed to a structureless, creamy to white, cryptocrystalline mass. Various workers have found

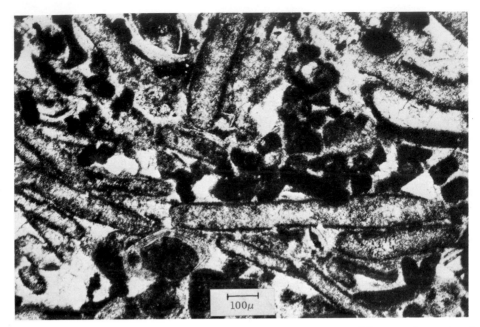

Fig.284. Micrite envelopes in a biosparite are filled with calcite cement. This cement has replaced, via a cavity stage, the dissolved core of aragonitic molluscan shell. Slice. Purbeckian. Dorset.

evidence, in material from the Persian Gulf, that the part of the grain *adjacent* to the bore can be micritized, possibly as a consequence of a bacteriological decay reaction (p.388).

Micritization in Bimini Lagoon

Centripetal replacement. The process of micritization was worked out using thin sections and peels of grains from the lagoon at Bimini, Bahamas (BATHURST, 1966). Though it was clear that all the types of grain in the lagoon have been micritized, the study was concentrated on the commoner grains, the skeletons of Soritidae, of *Halimeda* and of a variety of molluscs—grains, be it noted, of both calcite and aragonite. Many of these grains are now constructed of an outer rind or envelope of micrite enclosing a residual core of unaltered skeletal carbonate (Fig.285A). These incompletely micritized grains show most clearly the steps of the replacement process.

It is important first to demonstrate that the micrite envelope is a centripetal replacement and not a centrifugal accretion. Nowhere does it lie on an unaltered smooth surface of a skeleton, as something attached to it or added to it. Everywhere the contact between the envelope and the skeletal core is irregular, transecting the fabric of the skeleton. Where the trace of the outer surface of the skeleton would be

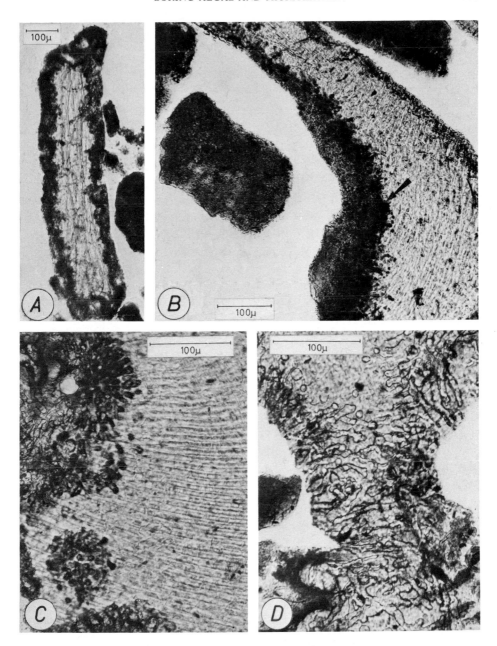

Fig.285. Bored and micritized grains of molluscan shell. Slices. Recent. Bimini Lagoon, Bahamas.
A. Shell fragment with micrite envelope. B. Shell fragment with dark layers of endolithic algae
(black spots by arrow) separating micrite envelope from unaltered crossed-lamellar structure.
A single intensely micritized grain to the left. C. Clusters of endolithic algae in crossed-lamellar
structure. D. Empty bores of endolithic algae. (From BATHURST, 1966.)

easily recognized if intact, as in unbroken miliolids or *Halimeda* segments, it is found to be incomplete (Fig.285A–D). Though the micrite generally lies around the skeletal core it may, in thin section, owing to the cut effect, be distributed irregularly anywhere in the skeleton.

Petrography of the envelope. In appearance the envelope is grey in transmitted light owing to its high density of Becke lines, and shows first order yellow and red polarization colours. The unaltered Soritidae and *Halimeda* are golden brown, other foraminiferids and molluscs are colourless. In thin section the envelope is seen as a patchy micrite. Patches are definable in terms of differences of polarization colour and positions of extinction: they tend toward equidimensional shape, having diameters mainly from 2.5 to 8 μ with a mode somewhere between 4–6 μ. Within each patch the micrite usually shows undulose extinction as of radiating needles: as the stage of the microscope is rotated the dark shadow swings across the patch. The detail is so fine that, though it can be seen by racking the objective up and down, it was found to be impossible to photograph it successfully in thin section. R. M. LLOYD (in press) has detected bundles of aragonite needles, using the scanning electron microscope.

In acetate peels the detail is sharper (Fig.289). Here the pattern of Becke lines reveals a scattering of more or less complete circles and ovals with shorter diameters close to 6 μ (Fig.289). The dimensions of the component micrite crystals, so far as they can be discerned, are mainly from 0.5 to 5 μ, but the larger "crystals" could well be aggregates of smaller crystals. The crystals are distinctly coarser than most of the calcite and aragonite crystals of the unaltered skeletons which have diameters mostly less than 1 μ.

The tubes. Looking next at the inner surface of the micrite envelope, the interface between micritized and unaltered skeleton, it is frequently possible to see numerous tubes embedded in the skeleton (also DE MEIJER, 1969, p.235). They run at various angles; some are empty (Fig.285D), some are filled with micrite. Others are filled with an opaque material that, in reflected light, appears colourless with a porcellaneous texture and rather paler than the surrounding carbonate of the skeleton. The diameters of the tubes are from 2.5 to 15 μ, though most are about 5–7 μ, not very different from the patches in the envelopes. The tubes, often short and stubby, may occur in clusters (Fig.285C).

The opaque tubes are important for two reasons. They are similar in appearance to those figured in photographs by NEWELL et al. (1960, plate 1–3) from the oölite of Browns Cay, which have been shown to include green and blue-green algae (p.297). They are also indistinguishable from the algal colonies visible in my own thin sections of Browns Cay oöids. KENDALL and SKIPWITH (1969a) were able to recognize a variety of algae inside carbonate grains, especially peneroplids, by treating the grains with dilute HF and thus transforming the carbonate to transparent CaF_2. From similar tubes in ancient limestones De Meijer has extracted filaments and coccoid colonies (p.91).

Micrite envelopes and tubes (whether empty, micrite-filled or opaque) are intimately and closely associated in space. Generally a skeletal grain is either unaltered or bears both micrite envelopes and tubes. Empty tubes and opaque tubes are common within the micrite envelope.

Mineralogy of the micrite. The mineralogy of the micrite of the envelopes is naturally of special interest and it was determined in the following way. In some samples of lime sand from the floor of Bimini Lagoon the micritization has gone almost to completion, so far as this stage can be recognized in thin section. In the sievefraction 150–90 μ, in particular, virtually all the grains are made of micrite yet do not possess the form or fabric of faecal pellets. The micrite is similar in appearance to that in the envelopes. Relic fabrics in the coarser sieve fractions suggest that the fully micritized grains are mainly whole or fragmental miliolids with a lesser amount of *Halimeda*. There seems little doubt, therefore, that a large part of the lime sand was initially high-magnesian calcite. Nevertheless, when grains from the 150–90 μ sieve fraction were analysed by X-ray diffraction, they showed 92–97% aragonite, the remainder being high-magnesian calcite ($MgCO_3$ 19 23 mole %). X-ray analysis by ILLING (1954, p.46) of Bahamian peloids showed aragonite with about 5% calcite, a result confirmed by KENDALL and SKIPWITH (1969a).

A pretty investigation by WINLAND (1968) showed that the content of high-magnesian calcite in micritized grains from Florida Bay increases directly with the extent of micritization. Clearly the micrite which fills the bores, in this case, is high-magnesian calcite. Yet Winland did not find high-magnesian calcite in micritized grains from Bimini Lagoon (where Bathurst concluded that the product is aragonite), though he did find it elsewhere toward Andros Island (H. D. Winland, personal communication, 1969). Selective dolomitization of Miocene micrite envelopes has been taken to imply that they were once high-magnesian calcite (BUCHBINDER and FRIEDMAN, 1970). E. D. Glover (personal communication, 1969) has found aragonite and high-magnesian calcite in the chambers of foraminiferids and in micrite envelopes in the Bahamas and the Florida reef tract. It is apparent, therefore, that the mineralogy of the micrite that fills the bores varies from place to place, for reasons that are at present not clear.

The existence of a mineralogical control over the micritic filling of algal bores is evident from Purdy's X-ray diffraction data: he found that, *in general*, aragonite skeletons were replaced by micritic aragonite and high-magnesian calcite skeletons by high-magnesian calcite (PURDY, 1968, p.188).

Summarizing the data, we can say that the micrite envelope, viewed in thin section, consists of patches that tend to be equidimensional with a modal diameter about 4–6 μ. In the peels the irregular outlines of whole or partial circles and ovals have diameters of 5–7 μ. Some of the smaller diameters must be the result of glancing sections or interference between patches or tubes.

Genesis of the envelope. Interpretation of the data, bearing in mind the intimate association, even intermingling, of envelopes and tubes, and the close similarity in diameter of patches and tubes, leads to the conclusion that they have a shared origin. The only reasonable order of events is: (*1*) boring and colonization by an alga, (*2*) death of alga and vacation of tube, (*3*) emplacement of micritic aragonite or high-magnesian calcite in tube by process unknown, to make, in fact, a micrite rod (Fig.287). It follows that a dense array of micrite-filled tubes will yield a micrite envelope, a mass of interfering micrite rods (Fig.287). Inasmuch as these were separately recognizable, the younger ones would have the more perfectly circular or oval cross sections, and the residue of older rods would show only as parts of their original forms. The process is illustrated diagrammatically in Fig.286,287. An earlier suggestion by D. J. Shearman (personal communication, 1964) that algae may, by their metabolism, be responsible in part for the precipitation of the new micrite is supported by R. M. LLOYD's (in press) finding that, among carbonate grains of the Florida–Bahamas platform, the oöids and other cryptocrystalline grains have values of $\delta^{13}C$ of $+3.2$ to $+5.3$, distinctly more positive than in faunal debris but similar to those for calcareous algae (Fig.261, 234, p.339, 283 resp.). Both D. J. Shearman (personal communication, 1964) and KENDALL and SKIPWITH (1969a) have noted a development of micritic alteration *adjacent* to the actual bore, as if algal metabolism were partly a cause of micritization (see also Winland, p.259).

KENDALL and SKIPWITH (1969a) compared the organic residue remaining after gentle digestion of unaltered tests of peneroplids in very dilute HCl with the organic residue left after digestion of micritized peneroplids. The first residue was a thin, diaphanous, soft and elastic membrane which retained the initial form of the test. This was presumably the organic matrix of the wall of the test. The residue of

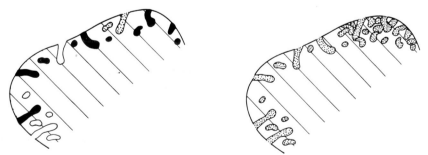

Fig.286. Diagram of a molluscan grain containing live endolithic algae (black), some empty bores vacated after death of algae and some bores subsequently filled with micrite (stippled).

Fig.287. More advanced stage of Fig.286. Bores all filled with micrite. In upper part of shell, further boring has taken place, the algae have died and the bores have been filled with micrite, yielding a more developed micrite envelope.

altered tests was much more translucent and closely resembled the mucilage already known to be a product of blue-green algae.

Although emphasis has been placed here on algal bores, it is commonly difficult to distinguish these from fungal bores (p.382). Recent work by GATRALL and GOLUBIĆ (1970) and GOLUBIĆ et al. (in press) has shown the importance of endolithic fungi. Indeed, micritization may be caused by a variety of processes: PURDY (1968) has stressed the possible influence of organic decomposition.

Micritization and the production of peloids

Micritized skeletal debris forms a considerable part of the Recent lime sand in Bimini Lagoon (Fig.285, 288), and the major part in the grapestone of the Cockroach Cay area, Berry Islands, Bahamas. Skeletal grains in the early stages of micritization are readily identifiable, but grains in a more advanced state of replacement need to be scanned carefully for traces of residual skeleton. The skeletal fabric is generally more finely crystalline than the secondary micrite, commonly showing low first order grey polarization colour and a golden brown colour in ordinary light (e.g., miliolids, *Halimeda*). It seems likely that a major proportion of the non-faecal peloids (p.86) in the Bahamian sediments are micritized skeletal debris. ILLING (1954, p.27) stressed the volumetric importance of "grains of aragonite matrix" (peloids) and suggested that they are accretionary aggregates of carbonate silt and clay-grade particles, and that the particles were cemented together while on the sea floor in a manner similar to the hardening of faecal pellets. It has yet to be shown, though, how the tiny particles could become aggregated in the first place to give sand sized accretions in the absence of an organic pelleting mechanism. More recently TAYLOR and ILLING (1969, p.73) have recognized the process of algal micritization in coastal areas of the Qatar Peninsula, Persian Gulf, where skeletal grains are altered to aragonitic micrite on a large scale.

Bored skeletal grains, weakened as they are mechanically, are apt to be highly rounded, and the micritized product yields a collection of rounded peloids similar in appearance to Illing's "grains of aragonite matrix" and to the "bahamite" (pelsparite) of BEALES (1958). It is not improbable that some of the bahamites are, for example, micritized skeletal sands.

The use of algal bores as bathymetric indicators has been mooted on various occasions. SWINCHATT (1969) has remarked on the change of abundance of algal-bored grains in carbonate sediments with depth of water. A reduction of abundance with depth is to be expected, because algae photosynthesize. Swinchatt worked in the Great Barrier Reef and, on the basis of his samples, he concluded that a great abundance of algal-bored grains in a sediment indicates almost certainly that deposition took place at less than 40 m, probably at less than 15–18 m. He referred to the Bahamas–Florida platform on which bored grains are abundant but depths

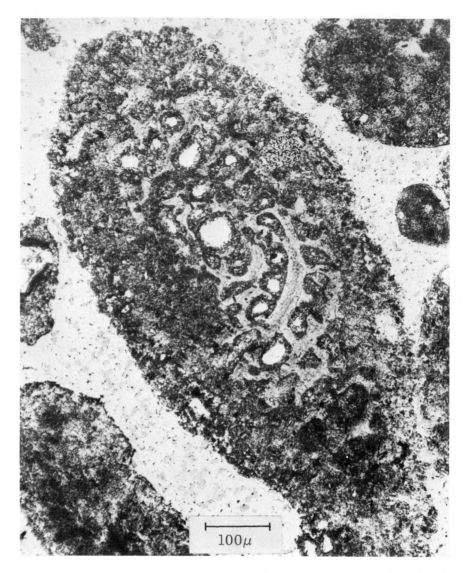

Fig.288. Soritid nearly replaced centripetally by micrite, only a few scattered relics of skeleton remaining. Some chambers only partially filled with micritic cement. Slice. Recent. Bimini Lagoon, Bahamas. (From BATHURST, 1966.)

do not normally exceed about 12 m. This bathymetric criterion can only be used if the sediment in question has not been reworked or carried to greater depths after an earlier stage of accumulation. Distinction from fungal bores is vital.

The role of this process of micritization in the development of the lamellae of unoriented aragonite in oöids is discussed on p.298.

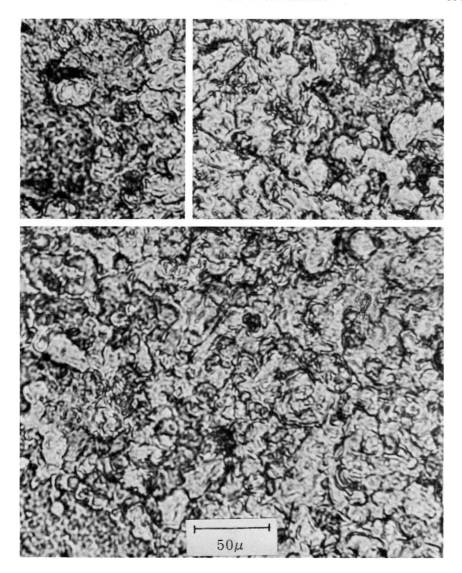

Fig.289. Pattern of more or less completely interfering circles and ovals in the fabric of micrite in a micrite envelope. Peel. Recent. Bimini Lagoon, Bahamas. (From BATHURST, 1966.)

Emplacement of micritic aragonite

The process whereby micritic aragonite or high-magnesian calcite is emplaced in the tubes (algal bores) is not known, but it is of interest to note that, in the laboratory, bacteria have been persuaded to cause precipitation of aragonite mud (DREW, 1914; GREENFIELD, 1963). In the tubes the algae, alive or dead, should provide an appropriate medium for bacterial growth. It has already been suggested that cavities in grains in sea water enclose a special physico-chemical environment which promotes the precipitation of micritic aragonite or high-magnesian calcite (p.364, 365, 374).

DIAGENESIS ON ANCIENT SEA FLOORS

Whereas the observation of modern carbonate sediments enables us to assert, without doubt, that a particular micritized grain or a cemented crust attained its present diagenetic state as a consequence of processes acting at the sediment–water interface on the sea floor, such certitude is seldom possible in the study of ancient limestones. This is because in ancient limestones three sets of diagenetic products are combined, those of submarine origin and those of subaerial and deep crustal origin (PURDY, 1968). A sure distinction between the three can only rarely be made. For example, calcite cements in turbidites are presumably not subaerial. Despite these obvious difficulties it is possible to draw some worthwhile conclusions with regard to submarine micritization, pellet cementation and the evolution of hard grounds.

Submarine cementation

Oöids and faecal pellets. On the subject of cementation in the sea—the growth of carbonate crystals on various surfaces—we can at least be certain that oöids past and present share the same origin, as inorganic precipitates. Aspects of this matter are examined in Chapter 7. With regard to faecal pellets there is little to be said here except that, as the fossil faecal pellets must have been rigid bodies (because their shapes were maintained after burial, contrary to terrigenous mud faecal pellets, e.g., H. B. MOORE, 1931, 1939), they must have been hardened by sea floor cementation in a manner similar to that affecting their Recent counterparts.

Mississippian lime mud: isotopic evidence. The application of isotopic data to the study of the history of the CO_3^{2-} ion groups in limestones is a method of great promise (e.g., Fig.261, p.339; MURATA et al. (1969). Its usefulness in the study of lithification of lime muds has lately been demonstrated by CHOQUETTE (1968) in a work of outstanding significance. From his analyses of $\delta^{18}O$ (p.280) and $\delta^{13}C$ in the Mississippian St. Genevieve Formation, in the Illinois basin, it seems that the lithification of the micrite (now low-magnesian calcite) must have taken place while the sediment was bathed in sea water—in other words, by submarine cementation. The limestones in question, from three bore holes, are variously composed of three components, micrite, microspar and dolomite. They show a transition from nearly pure micrite to rocks with increasing amounts of dolomite and interstitial microspar. Choquette cited evidence that indicates that the isotopic composition of the dolomite has remained practically unchanged since the mineral was formed in a marine or somewhat hypersaline pore water.

The interesting relation that concerns us here is the positive correlation between the degree of neomorphic alteration of the micrite (defined as the ratio microspar/micrite measured by point count) and the increase of the lighter oxygen and carbon isotopes. By extrapolation, the pure micrite can be shown to have $\delta^{18}O$ values between $+1$ and -1, little different from those in Recent marine carbonate sediments (Fig.261). The values of $\delta^{18}O$ for the microspar are roughly between -10 and -15, indicative of a process involving fresh water. The $\delta^{13}C$ for the pure micrite is about $+3$, in the range of values for Recent sediments (Fig.261). Choquette assumed that a major part of the volume of the micrite must consist of cement, and this means that, in view of the isotopic data, the CO_3^{2-} ion groups in the cement must have been derived from sea water. The very large proportion of cement in the micrite is shown by the lack of compaction features (p.439), and the high porosities of Recent lime muds, up to 70% or more (p.415). The source of the new CO_3^{2-} was presumably a combination of dissolved ions in the sea water and, possibly, ions liberated by dissolution of aragonite needle mud. Why calcite should have been precipitated in sea water is not clear (p.243). There is always the possibility that it was initially high-magnesian calcite (p.363), but this raises the question of how the Mg^{2+} was lost without the intervention of fresh water and the introduction of more of the lighter oxygen and carbon isotopes (p.335).

The St. Genevieve micrites have characteristics which suggest that they were deposited in shallow water no more than a few metres deep (CHOQUETTE, 1968). If further studies confirm Choquette's conclusions, then a major reapraisal will be necessary of the current assumption that the early cementation of other similarly shallow-water carbonate sediments has normally proceeded during times of emergence and exposure to fresh water.

Jurassic nodules: structural and petrographic evidence. Nodules regarded as lithoclasts of cemented carbonate ooze—a rubble of bits of submarine crust (p.375)—have been described from the Jurassic Adnet Beds in the Northern Calcareous Alps by GARRISON and FISCHER (1969). The nodules are embedded in a coccolith-bearing micrite. They have sharp boundaries, which truncate the contained fossils, and are enriched in iron and/or manganese minerals. From an examination of sedimentary structures, mineralogy, biota, and of ammonites and crinoid columnals partly dissolved prior to burial, of sedimentation rates and evidence of subsidence, Garrison and Fischer believed that the Adnet sea was bathyal (maximum depth 2,000 m) tending toward abyssal. The cementation of the nodules, or their crusty precursors, must, therefore, have been a submarine process.

In the Lower Jurassic of Dorset there is a discontinuous bed known as the Coinstone which shows interesting features of hardground formation (HALLAM, 1969). This is a horizon of limestone nodules in a marl. The flattish, ellipsoidal nodules pass laterally into marl: where the marl is more calcareous the Coinstone is more continuous. The nodules must have grown as concretions in the marl, yet their upper and lateral surfaces are bored and the bores are filled with marl. It is clear, therefore, that after the concretions had formed they were exhumed and exposed on the sea floor. Furthermore, the underface of at least one nodule has been bored and encrusted with serpulid tubes, indicating its temporary exposure on the sea floor. The Coinstone is a microspar with crystal diameters 5–25 μ: clay mineral content is about 15%. The various characteristics point to early cementation and the formation of a hardground which was, for a period, exposed on the sea floor. It is interesting that the crystal size in the nodules fits with the trend of the calcitization noted in the Recent hardgrounds in the Persian Gulf (p.373).

Palaeozoic bioherms. In Mississippian bioherms of North America, Great Britain and western Europe, the core facies (PRAY, 1958) have abundant, cloudy calcite cement which must have been precipitated under water, during growth of the bioherm. This is indicated by clasts of cemented core facies in flank breccias and conglomerates, by sediment–cement relations in the core itself, and by fabrics associated with limestone clastic dykes (PRAY, 1965).

Micritization

Micrite envelopes similar to those described from Recent carbonate sediments (p.384) have been described from Jurassic and Carboniferous limestones (Fig.258, p.334; BATHURST, 1964b). The similarity lies in the position of the envelope around the outside of the grain, its thickness, its micritic crystal size, and the inward extension of the micrite in the form of micritic rods which are presumably micrite-filled tubes (Fig.281, 282). As in Recent examples, the envelope cuts across the fabric of the skeletal carbonate: it is a replacement fabric and not an accretionary layer. The ancient envelopes differ from the Recent ones in that, instead of being composed dominantly of micritic aragonite or high-magnesian calcite, they now consist of a mosaic of equant crystals of low-magnesian micritic calcite. Possible reasons for this change are discussed on p.333.

The problem of hardgrounds

Despite the undoubted subaerial, fresh-water lithification of certain Pleistocene and Recent carbonate sediments, we are as yet in no position to dogmatize about the environment of lithification of ancient limestones in general. There is overwhelming evidence to show that large and extensive volumes of limestone can never have been exposed to the fresh-water, subaerial environment. The identification of the environment of lithification is at the moment highly controversial, and near the centre of this controversy lies the problem of the origin of hardgrounds. These are beds of limestone which show unmistakable evidence of having existed as hardened sea floors, as rock surfaces, at the sediment–sea water interface.

A bed of limestone is regarded as a **hardground** if its upper surface has been bored, corroded or eroded (by abrasion), if encrusting or other sessile organisms are attached to the surface, or if pebbles derived from the bed occur in the overlying sediment. These signs of the one-time existence of a hard floor are accompanied by other features which, though not strictly diagnostic, are characteristic: among these are crusts of, or impregnations by, glauconite, calcium phosphate, iron and manganese salts. Commonly the upper surface of the hardground coincides with a palaeontological non-sequence. All these qualities show that the hardground was lithified before deposition of the overlying sediment.

Two main problems face the student of hardgrounds. The first is the matter of their identification. The second is the recognition of the diagenetic environment in which lithification took place—either submarine or subaerial.

At Leckhampton Hill, near Cheltenham, in Gloucestershire, there is an oöbiosparite, called the Notgrove Freestone, of the Middle Inferior Oölite, which is overlain disconformably by a biosparite, the Upper Trigonia Grit of Upper Inferior Oölite age, deposited following the Upper Bajocian marine transgression.

Fig.290. Surface of an oöbiosparite, encrusted with oysters and bored. Major discontinuity on Notgrove Freestone, Middle Inferior Oölite, Jurassic. Leckhampton Hill, Gloucestershire. Width of picture 25 cm.

The surface of the Notgrove Freestone is encrusted with oysters and extensively bored (Fig.290). Between the Freestone and the Grit three zones are missing, representing a period of about three million years (ARKELL, 1956; HARLAND et al., 1964, p.205). The environment of lithification is uncertain.

ROSE (1970) has found that micrite-filled bores in a discontinuity surface at the base of the Doctor Burt Bed (base of the Edwards Formation, Cretaceous, central Texas) have themselves been bored and filled. Multiple truncations of earlier bores by later bores, showing sharp edges where a later bore cuts through the micritic filling of an earlier bore, indicate that the filling in the earlier bore was cemented before penetration of the later bore. Thus cementation proceeded while boring molluscs were active—a sure sign that cementation was submarine.

Bored surfaces in Middle Jurassic limestones in northeastern France have structures, textures, and age relationships with the overlying clays, which indicate that cementation took place on the sea floor and intertidally. The position of these hardgrounds at the tops of regressive sequences is interpreted by PURSER (1969) as a result of slow carbonate sedimentation accompanied by cementation of the sea floor before the onset of argillaceous, colder, or deeper-water sedimentation.

Hardgrounds in the Lower Jurassic Adnet Beds in the Northern Calcareous Alps were recognized by the following criteria (GARRISON and FISCHER, 1969): (a) unevenness of bedding planes associated with ferruginous crusts and truncated fossils, (b) angularity of fragments in what was identified as a solution rubble of limestone crusts, and (c) evidence of brittle fracture of beds during subaqueous slumping.

JAANUSSON (1961) has reviewed the main works on hardgrounds (discontinuity surfaces, *Diskontinuitätsflächen* of Heim, which existed as hard submarine surfaces). Jaanusson in 1961 emphasized the then apparent absence of submarine lithification in modern seas, though more recent researches have shown that submarine lithification has been more active than was realized at that time. There are limestones lying now below sea level which, by virtue of their stratigraphic relations, can never have emerged above sea level at any time in their histories. It is plain that some limestones cannot have experienced the subaerial, fresh water, environment as this is portrayed in Chapter 8 (though this does not exclude the possibility that the essential ingredient for successful lithification—a pore water low in Mg^{2+}— may have been available to these limestones as a result of some more deep-seated crustal process of brine-exchange).

On the subject of submarine lithification there is much still to be learned, but in the meantime it is rewarding to examine in detail two examples of ancient hardgrounds for which the evidence for submarine cementation is unusually strong. These are the Ordovician slump-folded limestones of Sweden described by M. LINDSTRÖM (1963) and the Chalk hardgrounds of northwestern Europe analysed by BROMLEY (1965, 1967b).

SOME ORDOVICIAN HARDGROUNDS IN SWEDEN

M. LINDSTRÖM (1963) has presented a strong case for the submarine cementation of some Ordovician calcilutites which are widely distributed in western and southern Sweden particularly in Öland and Västergötland. These lithified sediments once formed a hard sea floor and they qualify, therefore, for the description "hardground".

The rocks described by Lindström are a succession of fossiliferous calcilutites, with bed thicknesses of 2–20 cm, and with "intercalations" of marl or carbonate-poor shale. The succession is near horizontal and a little over 6 m thick. The large fossils, of early Arenigian age, are dominantly trilobites with some brachiopods and nautiloids, while the microfauna is mostly conodonts. Discontinuity surfaces are common, with characters that point to periods when sedimentation was almost at a standstill, as shown by borings by organisms and possibly pitting through inorganic dissolution, and by glauconitic replacement of the upper parts of the calcilutite beds. Lateral variation in lithology is slight, though beds may wedge out or pass into marls within about 10 m. A reasonable interpretation of the environment of deposition would seem to be that this was a body of well oxygenated sea water, with current velocities below the threshold for mud erosion. During certain periods sedimentation was so slow that, on a geological time scale, it can be regarded as having ceased. At these times the upper few centimetres of the calcilutite were lithified, presumably in the main by $CaCO_3$, since the rock contains 80–90% of calcite. It was also mineralized in places to give a dark green crust rich in glauconite. After lithification it was bored by animals and, probably, corroded.

The question which Lindström has tried so interestingly to answer is: did lithification take place subaerially, as a result of temporary emergence, or under the sea? He has brought to the solution of this question some detailed and acute observations concerning the history of some remarkable, contemporaneous folding, in which the discontinuity surfaces were involved.

The folds have wavelengths generally of 0.5 m or less, with an amplitude that varies slightly about 15 cm (Fig.291). They are mostly gentle though some are sharp-crested and overturned. Each fold is normally restricted to one bed, being covered and underlain by flat strata. The folds usually occur singly, passing laterally into undeformed beds. The marl has commonly acted, it would seem, as a lubricant for the movement of small Jura-type folds over flat beds, rather as a zone of *décollement* it fills the cores of some folds. The fold axes have a preferred azimuthal orientation trending roughly parallel to the regional strike. The folds have a tendency to overturn and it is conceivable that they are the result of subaqueous sliding downslope.

The special interest of these folds (Fig.291) resides in three factors: (*1*) the upper surface of the folded calcilutitic layer is a hardground, (*2*) the folded layer is mechanically fragile, commonly only 2–4 cm thick, and (*3*) despite prolonged

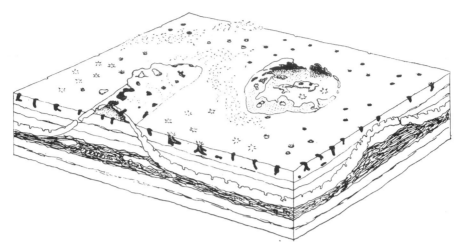

Fig.291. Reconstruction of an Ordovician sea floor, which is made of cemented calcilutite. The floor is bored and, when buried, will be a discontinuity surface. The corroded crests of the folds of another bored discontinuity surface rise above the general floor level. Thickness shown is about 30 cm. (From LINDSTRÖM, 1963.)

exposure at the sediment–water interface the folds show scarcely any signs of abrasion. Therefore, Lindström argued, the formation of the hardground cannot have been the result of emersion in the turbulent littoral zone, but must have proceeded in water deep enough to have a tranquil current regime in which these frail structures could survive for thousands of years. It is plain, from an examination of the faunal stages, that there were folds protruding above the general sediment surface throughout most of the Billingen stage of the Arenigian. We must now examine the folds in a little more detail.

Though by no means all the flat beds show signs of discontinuity, such as boring or corrosion, all the folds do. They are bored, pitted, have glauconitic coats and are thin compared with the flat beds. This thinness is especially obvious toward the crests where the folds were exposed to the sea water for the longest time before burial. Where a fold has been pierced or its crest broken, later sediment has filtered down into the domed space under the fold to give younger sedimentary laminae (Fig.291). Some later discontinuity surfaces, of post-folding age, are continuous through the fold, occurring both over its horizontal flanks and in the space underneath the dome. The folds comprise, therefore, a collection of shallow domes, their roofs thin, perforate, and having in places jagged broken edges. On the crests of the folds the beds are thinner and more corroded than elsewhere. Structures of this kind, urges Lindström, would not survive the abrasion associated with emersion above sea level. The quiet evolution of the hardgrounds becomes particularly apparent when one notes that, where a fold meets a younger hardground, there are no obvious signs of truncation or abrasion of any kind (Fig.291). The limbs of the fold continue unchanged through the younger discontinuity surface. The tranquil state of the water overlying the folded hardgrounds is even more strikingly demonstrated when it is realized how long these structures were exposed on the sea floor. Lindström makes a calculation of the rate of sedimentation, based on KULP's (1961) geological time scale and the distribution of faunal zones. The figure he arrives at is 1 mm in 1,000 years. It follows that the top of a high fold, say 20 cm amplitude, may have stood, frail and unprotected on the sea floor, for 200,000 years or more. (Bromley makes similar estimates for exposure times of Chalk hardgrounds, p.406; in the Recent Persian Gulf, p.372, exposure time is much shorter.)

From Lindström's evidence we must conclude, therefore, that before the calcilutites were hardened, they were folded; lengthy exposure at the sediment-water interface then resulted in hardening, impregnation by glauconite and phosphate, boring and corrosion. As sediment continued to accumulate, the flat surfaces were first buried, the flanks of the folds were next covered, and the crests remained exposed for the longest time. As the crest is more pitted, bored, glauconitized and corroded than the other parts of the hardground, it must be assumed that the processes of hardground formation were active throughout the period which began with the act of folding and ended with the burial of the crest. This conclusion is in accord with the widely held view that hardgrounds are formed during prolonged exposure of the floor to sea water when sedimentation is exceptionally slow—in this case 1 μ per year.

The reason for the early cementation and mineralization of the calcilutite is not indicated by the evidence. It is possible that the negligible rate of sedimentation permitted the development of a hardground as in the Persian Gulf (p.373). At the same time the growth of an organic (algal?) mucilage or mat might have maintained the reducing environment needed for the growth of glauconite (M. LINDSTRÖM, 1963, p.268).

THE CHALK HARDGROUNDS OF NORTHWEST EUROPE

Illuminating studies of hardgrounds in the Chalk of northwestern Europe have been made by VOIGT (1959) and BROMLEY (1965, 1967b, 1968). Bromley's unusually thorough work forms a convenient basis for a discussion of the interpretation of hardgrounds, especially the Turonian Chalk Rock of England.

The Chalk Rock (Fig.292) and other hardgrounds in the Chalk are beds of hard, jointed, well cemented chalk (coccolith micrite) in a succession that is otherwise composed of porous, lightly cemented, soft chalk. Both hard and soft chalks (Fig.293–295) contain, in addition to the coccolithic matrix, scattered foraminiferids, echinoids, brachiopods, sponges, calcitized molluscs, etc. The upper surface of the hardground shows signs of abrasion and corrosion, the rock is plentifully coated with, and replaced by, glauconite and calcium phosphate. It is bored, and

Fig.292. Three hardgrounds in the Chalk Rock combine to give a single prominent lithological unit. Turonian. Charnage Down, Wiltshire. (Courtesy of R. G. Bromley.)

is encrusted with bivalves (*Dimyodon*, *Spondylus*), bryozoans, serpulid worms and, less commonly, with foraminiferids, and, even more rarely, with small oysters and sponges. Deposition of the overlying soft chalk must, therefore, have been delayed long enough for the ooze of the hardground to be cemented, mineralized, corroded, encrusted, bored and eroded.

The nature of the hardgrounds poses three main questions: (*1*) why were these horizons cemented more densely than the rest of the chalk, (*2*) what accompanying changes, if any, were there in bathymetry and current regime, and (*3*) why was there a hiatus in sedimentation? These problems are intimately related and, in the discussion that follows, they are inevitably intertwined.

Porosity of the Chalk hardgrounds

The porosity of the hardened chalk ranges from 2.7 to 14%, and is considerably less than that of the soft chalk which has a range of 37–42% (BROMLEY, 1965). Since the chalk is a purely calcitic deposit (except for some argillaceous and siliceous horizons) it is apparent that, in the hardgrounds, from 60–90% or even more of the original porosity has been replaced by calcite cement. The differences in the porosities and crystal fabrics of the soft and hard chalks are seen in stereoscan photomicrographs by HANCOCK and KENNEDY (1967) and in Fig.293 and 294: the difficulty of identifying cement is obvious. In the field, while both chalks can stand as cliffs, brilliantly white, the soft chalk can be ground under the heel whereas the hard chalk can only be broken with a hammer.

The original porosity of the ooze will have exceeded the present porosity of the soft chalk by whatever quantity of cement this now contains: it has at least enough cement to make it rigid. Bearing in mind the existence of intragranular porosity as well as intergranular, a figure of 50% porosity for the initial ooze seems a modest estimate (see discussion in DUNHAM, 1962). This figure would apply to the contact packed grains after they had been compacted—below the upper levels of soupy mud, as suggested by GINSBURG (1957). It seems reasonable, therefore, to estimate the volume of cement in the hardgrounds as 35-45%.

The scant cementation of the soft chalk

A possible reason for the scant cementation of the Cretaceous soft chalk in Britain, outside Northern Ireland and Yorkshire, is that the original calcilutite was so poor in aragonite that there was an inadequate source of metastable carbonate which might, by dissolution and precipitation, give rise to a cement. After all, the chalk is composed of a stable low-magnesian calcite micrite of coccolith debris with a sand fraction of calcite skeletal grains: it is lacking in the remains of aragonite skeletons, only rare examples of which occur as moulds or calcite casts. This hypothesis appears to rest on two assumptions of uncertain validity. If the soft chalk is porous because the original micrite lacked aragonite, then it might be argued that the most densely cemented chalk of the hardgrounds must initially have had a higher content of aragonite: this has not been demonstrated. The second assumption is that the chalk was, actually, cemented by dissolution of metastable carbonate in fresh water and subsequent precipitation of calcite, after the fashion of the emergent Pleistocene limestones discussed in Chapter 8. Yet palaeontological work on the bathymetry of the Chalk sea has consistently opposed the idea of widespread temporary subaerial exposure of specific horizons.

The Chalk of Northern Ireland is uniformly hard with a porosity of only 5-6.5% (Fig.293, 295). Interesting light has been cast upon the causes of this peculiar hardness by WOLFE (1968). He distinguished three lithologies, the *Inoceramus* chalk, the White chalk and the Bioturbation chalk. Although evidence of bioturbation is particularly striking in the last, all three chalks have been much burrowed. The Bioturbation chalk shows scant compaction compared with the other chalks. Characteristically it bears many calcite cement casts of once aragonitic and siliceous skeletons. This lithology occurs everywhere for about 16 cm below discontinuity surfaces, the depth to which burrowers could be expected to penetrate. This chalk seems to have been cemented early during the eogenetic stage. By contrast, the *Inoceramus* and White chalks show considerable signs of compaction and flow and must have been cemented only after deep burial during the mesogenetic stage. All three chalks contain stylolites on all scales, visible in the field, in the

Fig.293. Hard chalk, with low porosity and lacking obvious skeletal structures. Fracture surface. Scanning electron mic. Campanian. Co. Antrim, Northern Ireland. (Courtesy of J. M. Hancock and W. J. Kennedy.)

thin section and with the electron microscope (Fig.321, p.463). Presumably the *Inoceramus* and White chalks had to await the pressure-solution stage before they became cemented. The overall hardness of the Irish chalk could well be a consequence of the widespread effects of pressure-solution and the consequent release of $CaCO_3$ in solution. The early cementation of the Bioturbation chalk may have been a result of enhanced exchange between pore water and open sea caused by burrowing—assuming that the cement was derived from sea water (p.374).

Two matters, the relative availability of aragonite and variation in depth, must now be lightly touched upon.

Fig.295. Hard chalk, with low porosity and scarce organic remains (coccolith middle top). Polished surface. Black patches are pores. Replica. Electron mic. Campanian. Co. Antrim, Northern Ireland. (Courtesy of M. Wolfe.)

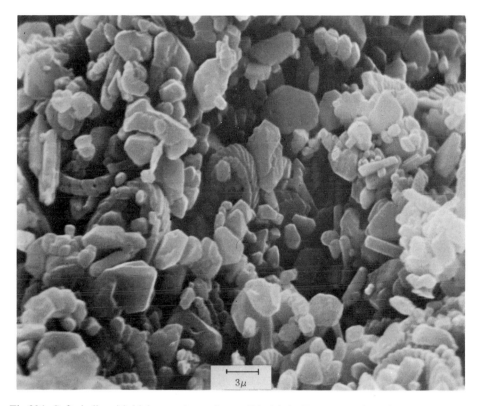

Fig.294. Soft chalk, with high porosity and coccolith debris. Fracture surface. Scanning electron mic. Santonian. Surrey. (Courtesy of J. M. Hancock and W. J. Kennedy.)

The role of skeletal aragonite

Even if sufficient aragonite was available in the chalk as a source of cement, it is not certain that it supplied the $CaCO_3$ for cementation of the hardground. Bromley's data show that the cementation of the chalk occurred before dissolution of much of the aragonitic macrofauna. Thus empty moulds or cement-filled casts of molluscan shells occur in the hardened chalk. They are absent from the soft chalk wherein the moulds could not be mechanically supported. Other casts in the hardgrounds are made of pyrite or calcium phosphate.

An approach to the possible role of aragonite as a source of cement was made by HANCOCK (1963). He stated that, if aragonite had been the source of the $CaCO_3$ in the cement, then this aragonite must have been a clay-grade, sediment: only such a fine sediment could have had the great surface area and resultant high solubility. Aragonite sediments of such fine particle size do not occur today below about 60 m: deeper aragonitic sediments are coarser. Therefore, concluded Hancock, since the Chalk sea was deeper than 60 m, it is inconceivable that cementation took place by dissolution and reprecipitation of aragonitic debris. This hypothesis rests on the assumption that aragonite particles of sand grade are not a potential source of dissolved carbonate, though this view is contradicted by the evidence of dissolution of aragonite shells in fresh water, as described in detail in Chapter 8. Nevertheless, Hancock seemed to feel that, for dissolution and precipitation to have gone on under sea water without the vigorous intervention of rain-derived fresh water, a source material of unusually high solubility must have existed, and he suggested an original ultra-fine calcite powder now recrystallized as cement. Why this was available to cement the Irish and Yorkshire hard chalks, for example, but not the southern English chalk is not clear. Nor does it seem likely that a sediment ever existed finer than the nannoplanktonic coccoliths, which are well preserved in the Chalk.

HUDSON (1967a) has made some interesting calculations that indicate the possibility that, on a Chalk sea floor between, say, 150–280 m, aragonite would have been soluble but not calcite. Such a critical situation would have allowed selective dissolution of aragonite particles, a process helped by vigorous bioturbation at the sediment–water interface. Indeed, the critical matters of water depth and aragonite solubility may have been the main factors controlling the distribution of hard and soft chalks. Where aragonite survived, there the Chalk was later cemented. There is a useful review of these problems in KENNEDY (1969).

Chalk hardgrounds emergent?

By comparison with the emergent Pleistocene limestones of Chapter 8 the Chalk hardgrounds differ in two important respects. Relative to the soft chalk above and below them, they show an increase in the preservation of aragonitic shells as casts or moulds, and a higher content of glauconite and phosphorite. In the emergent Pleistocene limestones aragonite skeletons are notably destroyed by dissolution, and mineralized hardgrounds are not reported. When these peculiarities of Chalk hardgrounds are added to the palaeontological evidence (REID, 1962a, b) and to the known depths at which glauconite and calcium phosphate occur in modern seas (CHARLES, 1953; BROMLEY, 1967a), the reasonable conclusion, following BROMLEY's (1965) work on the Chalk Rock, is that during the formation of the Chalk hardgrounds (in particular the horizon known in greatest detail, the Chalk Rock) the sea was never shallower than 50 m. At other times, during deposition of those coccolith oozes which have persisted as soft chalk, the depth of the floor may have been as much as 200 m. If the Chalk floor was not exposed above sea level, except perhaps locally where algal stromatolites grew in Northern Ireland, then the cementation of the chalk in the hardgrounds cannot have depended on dissolution of aragonite in magnesium-free fresh water and its reprecipitation as calcite cement.

Rate of deposition of the Chalk

The overall rate of deposition of the Chalk has been estimated as 1.5 cm/1,000 years by BLACK (1953), who rested his calculation on the thickness of the Chalk and its duration as known from radioactive dating. This is a minimum rate and makes no allowance for post-depositional compaction or the likelihood that the hardgrounds and bedding-planes indicate periods of slower or practically zero sedimentation. Estimates of the rate at which sediments actually accumulated at any one time vary: 50 cm/1,000 years (NESTLER, 1965, p.119) and 100 cm/1,000 years (MÜLLER, 1953, p.33).

Having an eye both to compaction and to hiati in sedimentation, it would be safer to assume a faster actual rate of accumulation than Black's estimate of the mean, and a figure of 50 cm/1,000 years will be adopted here. This would represent 500 μ/year or about 180 coccoliths a year piled one upon the other.

Chalk hardgrounds and non-deposition

Armed with an estimate of the rate at which most of the sediment was delivered, it is possible, first of all, to consider the oft made claim that the surface of a hardground represents a pause in sedimentation, a period of negative sedimentation or non-deposition. How else, it is argued, could a population of encrusting oysters, bryozoans and attached bivalves, with boring worms, clionid sponges and thallophytes, become established? This argument can now be countered by pointing out that no organism is going to be unduly handicapped by a depositional rate of 500 μ/year, the diameter of a single grain of sand. Most organisms have the means of ridding themselves of this small amount of sediment, either by locomotion or movement of cilia, etc. Nevertheless, it does seem that a severe restriction on the rate of sedimentation at certain times is implied by the need to permit cementation of the hardground. Each hardground has a sharp, clear-cut upper surface, and we are bound to conclude that below this surface the chalk was subjected to a thorough cementing process, *before the succeeding ooze was deposited*. One can either suppose that cementation at first kept pace with sedimentation and suddenly stopped— which makes nonsense—or that sedimentation stopped long enough for the bare surface of the hardground to become well cemented before the rain of coccoliths began again. It seems we are compelled to accept a stop–go type of deposition, where "stop" is defined, either as a rate so slow as to leave no trace, or as actual erosion. Signs of erosion are, in fact, typically abundant on the Chalk hardgrounds (BROMLEY, 1965). This conclusion fits, of course, with the generally accepted idea that hardgrounds mark temporary periods of shallowing. Bromley has approached the question of the length of the pause by noting that the *Holaster planus* zone is normally 18 m thick, but where it includes a hardground (the Chalk Rock), it is only 4.7 m thick. Thus some 14 m are missing and the implication is that the Chalk Rock hardground represents a depositional hiatus of "several hundreds of thousands of years" (BROMLEY, 1965). This figure is of the same order as Lindström's estimate of 200,000 years or more for the exposure of the crests of the folds in the Ordovician calcilutite prior to burial (p.399).

Cementation of Chalk hardgrounds

The field data. It is necessary now to attend to BROMLEY's (1965) description of the course of hardground cementation. He drew his conclusions from an examination of the changes to be seen as the Chalk succession is traced upward in the field, from soft chalk into a hardground. The first sign of change is a lumpiness in the soft chalk, irregular in its appearance and not necessarily related to bedding. Next, the lumps show "intense hardening" though the soft chalk matrix remains unaltered. At this stage the lumps are still clearly separated from each other and from the matrix, and have a rounded shape. Higher still, the hardened lumps are larger: they are no longer separated but are joined to form "a continuous network interlocking with the network of soft chalk between" (BROMLEY, 1965). Finally, at the top, no soft chalk remains and the completely hardened rock is jointed. The development resembles that more recently inferred from an examination of hardgrounds in the Persian Gulf (p.371).

The process of hardening was, it would seem, one of growth and coalescence of concretions. This upward succession of changes does not necessarily represent a time sequence of events in the conventional stratigraphic sense. Rather the history of concretionary growth is a separate sequence of events distinct from, and later than, the depositional history. *If* concretionary growth began at the sediment-water interface (and here we beg a big question), then the lower discrete lumps are the youngest structures. It is equally possible that they are relics of an earlier, incomplete development of a hardground.

One piece of information given by Bromley is of particular interest in connection with the process of hardening. He noted that the vertical graduation just described appears also as a horizontal transition where a hardground passes laterally, through a lumpy lithology, into soft chalk (compare Persian Gulf, p.371). It is clear, therefore, not only that the environment in which hardgrounds developed embraced a limited area of sea floor, but that conditions that favoured hardening faded laterally, across the sea floor. Bromley also showed that the lump stage was early, hardening beginning near the surface of the sediment, because, locally, some of the lumps were exhumed as the unconsolidated chalk was winnowed from between them—as in HALLAM's (1969) study of Jurassic hardgrounds (p.393). Their exposure to the open sea water, as poorly sorted unabraded pebbles, is clear from their superficial impregnation with glauconite or phosphate. This evidence supports the contention that the lumps at the base of the hardground do represent an earlier, uncompleted development.

The pattern of evolution. We must picture the process of hardening, therefore, as having been restricted regionally (FABRICIUS, 1968), possibly to topographic swells (p.409). Away from these areas the effectiveness of the process gradually faded. Bromley's evidence shows that the process began by the growth of concretions near enough to the sediment surface for them to be subsequently exhumed on occasion. In the lower parts of a hardground it is apparent that cementation advanced no further than the lump stage. As time ran on the process at higher levels was enabled to approach more nearly to completion. Either the chemical reactions were going at a faster rate, or the near-surface ooze remained near-surface for a longer time, because the rate of sedimentation was slower. At the top of a hardground, adjacent to a surface that was exposed to the open water for some hundreds of thousands of years, hardening proceeded to completion. This conclusion joins with Bromley's evidence for the near-surface origin of the lumps, to show that the cementation process was related to the existence of a sediment–water interface. It seems likely, then, that the cementation of the hardgrounds was a function of one, perhaps two, changes in the environment of deposition. In the first place the rate of deposition diminished so that the later surfaces of the ooze were exposed for longer periods to the appropriate chemical environment. Secondly, the chemical environment itself may have changed adjacent to the surface of the ooze, perhaps as a result of the shallowing of the water and the development of a new biochemical regime. It is not clear whether the chemical situation at the sediment–water interface changed as the hardgrounds evolved, so allowing precipitation of glauconite and phosphate (p.399) as crusts on the hardground or as superficial replacements of it. This change could have been biochemical. On the other hand, the hardening and mineralization may simply reflect unusually prolonged reactions in an unchanging chemical environment—reactions that were able to produce appreciable effects only during long periods of exceedingly slow sedimentation. As soon as the rate of sedimentation increased, the pore solutions that were once in communication with open sea water would have been isolated and the supply of reactants would have ceased. It is conceivable that hardground formation resulted from the completion of the process which, in our modern seas, leads to the development of grapestone and flakes (p.317). Are these the modern equivalents, despite differences in form, of the early lumps in the Chalk ooze? We are left enquiring what kind of process, working at the sediment–water interface, could have led to the growth of the concretions and their eventual coalescence into a hardground.

The rounded shape of the concretions, and their sharp separation from the soft matrix, indicate that they grew by centrifugal accretion of cement in the ooze. What is not clear is the source of this cement. The discreteness of the lumps implies centripetal migration of Ca^{2+} and HCO_3^- ions toward numerous centres and this in turn means that the available carbonate was distributed throughout the sediment. The source is unlikely to have been mineralogical: it has already been suggested that the sediments which became hardgrounds probably had no monopoly of aragonite (p.401) and that the dissolution of the aragonitic macrofauna was later than the hardening (p.404).

It can hardly be a coincidence that the hardening process was associated with a period of reduced sedimentation. Even so it is difficult to see how the chemistry of the pore water in the chalk ooze can have been affected by a reduced rate of accumulation. After all, a change in sedimentation rate from 500 μ/year to practically zero is hardly dramatic. What does seem highly probable is that the chemical situation and the change in sedimentation were both the result of some other transformation. This other transformation is generally thought to have been a shallowing of the Chalk sea, which led to a drastically reduced rate of sedimentation and a vast lengthening of the time during which the near-surface ooze was in effective contact with sea water (as in the Persian Gulf, p.374). The evidence for this shallowing will next be examined.

Shoaling and Chalk hardgrounds

There is no definite palaeontological evidence for a shallowing of the Chalk sea floor during the formation of the hardgrounds (general background in JEFFERIES, 1962): most of the faunal and floral changes reflect only a change from soft to hard substrate. This lack of evidence is not surprising since any change in depth would have taken place within the limits of 50 m and 250 m (BROMLEY, 1965), in a region of the sea unremarkable for extremes either of food supply or temperature. Bromley did observe, however, that throughout the Chalk Rock the shells have been bored by penetrative algae—yet he cannot recall ever having seen these borings in shells in the soft chalk. (This depth criterion is discussed on p.389.) He recorded also that, though skeletons are bored by Thallophyta, hardground and pebbles are not. Is this, he asked, because the soft sea floor was near to the maximum depth at which endolithic algae could photosynthesize, so that their colonization was restricted to the more translucent materials?

Another thing suggestive of shallowing is the pause in sedimentation. If a region of the sea floor had been raised so as to make a localized topographic swell, an important result is likely to have followed. The tidal water, as it flowed across the swell from one deeper region to another, would have been constricted and therefore would have moved with greater velocity and turbulence over the shallower floor. This situation could account not only for the absence of coccolith ooze, but for the erosion of the hardground, though this was obviously made easier by weakening of the rock by borers such as Thallophyta, clionid sponges and molluscs (BROMLEY, 1965). Borers have a similar effect on Recent reefs (p.108; STORR, 1964).

The Chalk hardground established

Finally, we must turn briefly to the vivid descriptions given by BROMLEY (1965) of the established hardground, its biota, its corrosion and erosion, and its mineralization—though scant justice can be done here to the detailed and lucid portrayal of complex relationships given in the original work (e.g., Fig.296).

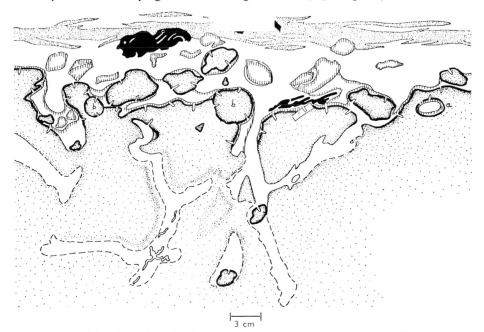

3 cm

Fig.296. Top of the Chalk Rock hardground, drawn from a weathered surface. Under an upper layer of clay (finely stippled) with flints (black), the soft chalk (white) extends in burrows into the hard chalk (coarsely stippled according to density of iron stain) and around pebbles of hard chalk. Glauconitized surfaces are shown as continuous thick black lines. Phosphatized surfaces are shown as vertical lines. Glauconite-lined bores are stylized. (Courtesy of R. G. Bromley.)

Erosion. Erosion of the hardground is indicated by the truncation of sponges which are planed level with the surface of the hardground, by the occurrence of pebbles, and by the convolute form of some surfaces (Fig.296). The surface of the hardground was, in places, convolute, though elsewhere plane, and the complex system of hollows and bosses was more susceptible to erosion than the plane surfaces. It is typically accompanied by an overlying accumulation of pebbles. Other pebbles are the result of erosion of the concretions which announced the oncoming of hardground conditions. These are neither well-sorted, nor abraded, but have rounded shapes. Their temporary exposure on the sea floor is shown by their glauconitic coats.

Bores and burrows. Boring of skeletons has occurred generally, but the hardened chalk was only lightly attacked. Bores are mainly straight and parallel sided, whereas burrows have rough walls, are curved and are deflected around pebbles and at buried surfaces of earlier hardgrounds. Bromley gives useful details of the system of canals, locally inflated into crypts, made by the boring sponge *Cliona* (including geopetal accumulations of spicules on the crypt floors). Undoubted bivalve borings he has not seen in the Chalk Rock, but there are fine bores that suggest the action of polychaete worms. BROMLEY (1967b) has published a very full description of burrows which were probably made by thalassinidean Crustacea. The burrows are visible as a network of passages, now filled, which were clearly formed in soft sediment (see also KENNEDY, 1967). Particularly common are the readily identifiable bores of a little crustacean, the order Acrothoracica of the Cirripedia. The animal was selective then as now, preferring, for example, the tests of echinoids to the shells of ammonites and brachiopods. Penetrative thallophytes were exceedingly active.

With his study of these thallophytes Bromley included a summary of the known features by which algal and fungal bores may be distinguished and these are quoted on p.382. These bores are in places so superficial that the roof has been eroded and only a groove remains. Their diameters rarely fall outside the range 5–80 μ. Bromley reminded us that the algae, like the fungi, put down "exploratory threads": these are nourished from the surface and are not green. Consequently the presence of bores in material beyond the reach of light is not necessarily a contra-indication of algal origin. The thallophytes were selective, attacking terebratulaceans and encrusting oysters, but tending to ignore rhynconellaceans and echinoids. It is clear that great quantities of skeletal material were destroyed by borers. Bromley inferred, moreover, that this ubiquitous process produced enormous quantities of clay grade or colloidal debris (see also Neumann, p.108) and that this could have had a significant effect on the level of supersaturation of $CaCO_3$ in the overlying sea water. One must add to this the effects of widespread algal photosynthesis, with abstraction of CO_2 from the water.

Glauconite. Glauconite has been recorded by Bromley in two forms, as an encrusting layer on hardground and pebbles (Fig.296), so presumed to be primary, and as a post-hardening superficial but incomplete replacement of secondary origin. It is variable in colour, either yellowish-green, dark green, almost black, or blue-green. The primary form occurs partly as a cavity filling, of pre-hardening age, in chambers of foraminiferids: commonly the test wall has been subsequently dissolved. The primary skin also encrusts the surfaces of cavities left by the dissolution of aragonite shells from the already hardened chalk, and the insides of bores, and is thus of post-hardening age. Where it occupies the axial canals of sponge spicules its formation pre-dated the dissolution of the spicule walls. It is obvious, therefore, that formation of glauconite went on over a long period, beginning before the chalk was cemented and continuing after this process was complete. Here, surely, if only it could be interpreted, is a critical clue to the chemical environment which led to the growth of concretions and eventual hardening of the ooze. Glauconite, though requiring reducing conditions for its development, is found in well oxygenated environments where there is a rich supply of organic matter. Thus it is found on open sea floors today, but inside the tests of foraminiferids where the microenvironment had at one time a negative redox potential. The establishment of an encrusting skin of glauconite over a hardground, even where the surface is convolute, suggests that the immediate chemical environment was also spread uniformly over the surface, plane or convolute: this in turn points, perhaps, to something living but adhesive. Gelatinous films known as pelogleas are deposited on subaqueous surfaces today by green or blue-green algae. The existence of some kind of organic film is worth considering, it would seem, for the hardground surface.

Phosphorite. Pink to brown phosphorite also occurs as a skin and as a replacement (Fig.296). It is detected readily with the aid of ultraviolet light (BROMLEY, 1965). Strangely enough, the primary skin only appears on glauconite, on phosphate-replaced chalk or coarsely crystalline calcite or on (one-time) aragonite, but never directly on pure chalk matrix. Phosphorite followed glauconite, except in coprolites which were phosphatized before hardening of the chalk.

Late diagenesis. Evidence of post-hardening dissolution and precipitation is plentiful. Bromley mentions cavities after aragonite shells, and after both calcitic and opaline sponge spicules.

The complex environment. Altogether, the very complete data given by Bromley set the stage for the entry of the petrographer-chemist who must now address himself to the problems of the source, transport and precipitation of $CaCO_3$ as cement. The environment was, however, far from simple and, as an example of the actual complexity it is fitting to close this section with a quotation from Bromley's thesis.

Bromley wrote: "In order to demonstrate the great variety of preservation which may occur in a single fossil a single example will be described, a large nautiloid shell from Hitch Wood, Hertfordshire. Its history can be inferred from the preservation of the various parts of the shell.

"1. After the death of the animal, the shell remained empty on the sea floor for a time, as in places the aragonite has been replaced by a very thin layer of phosphate at its surface both inside and out. The shell was encrusted by Polyzoa and annelids and bored by clionid sponges, algae and annelids.

"2. The shell was then buried in the sediment of the hardground and the body chamber and first 7 or 8 camerae filled with ooze, while the inner whorls remained empty.

"3. While the shell was still intact it was partially re-exposed, on the upper side as it rested, by removal of some sediment. The body chamber partly emptied of sediment and the shell of its upper side broke away. The first few camerae remained full of sediment which was phosphatised through the shell. Sediment filling the borings in the shell was also phosphatised.

"4. A part of the body chamber shell which was uppermost and probably exposed to the sea water on both sides was thoroughly glauconitised. This part of the shell was then probably buried, as the glauconitisation would have weakened it and it is not broken.

"5. The aragonite shell was then dissolved in the region of the camerae, exposing the phosphatised internal chalk to further phosphatisation finally forming a skin of phosphate (over which encrusting organisms grew simultaneously and later). With the removal of aragonite from this part, the empty inner whorls of the shell became exposed to the sea, and all trace of their walls and septa vanished, while phosphate skins were formed on the involute parts of the casts of the outer camerae.

"6. Renewed sedimentation then filled the inner whorls and top of the body chamber with white chalk in continuity with the external matrix. This was hardened.

"7. The aragonite shell of the body chamber and under side of the camerate portion of the shell was dissolved to leave an empty cavity.

"8. Finally, cavity sedimentation geopetally filled the bottom of the shell dissolution space."

FURTHER READING

Background is provided by ILLING (1954), PURDY (1964a, b, 1968). On the subject of beach rock the main works are by GINSBURG (1953), KAYE (1959), STODDART and CANN (1965), GAVISH and FRIEDMAN (1969) and TAYLOR and ILLING (1969).

Shallow water hardgrounds in the Persian Gulf are dealt with by SHINN (1969) and TAYLOR and ILLING (1969). There is a good introduction to deep water marine cementation by FISCHER and GARRISON (1967) and discussions of the cementation problem in CIFELLI et al. (1966) and THOMPSON et al. (1968). On boring algae the background is contained in NADSON (1927a, b) and FRÉMY (1945). Micrite envelopes are dealt with in BATHURST (1964b, 1966) with relevant matter in SHEARMAN and SKIPWITH (1965), MONTY (1967), WINLAND (1968) and KENDALL and SKIPWITH (1969a). On ancient hardgrounds there is the useful survey by JAANUSSON (1961), with two fundamental papers by M. LINDSTRÖM (1963) and CHOQUETTE (1968), works on Chalk hardgrounds (amongst many) by VOIGT (1959), HOLLMANN (1964) and BROMLEY (1967a, b, 1968), on Jurassic hardgrounds by PURSER (1969) and on synsedimentary hardgrounds in Triassic reefs by ZANKL (1969). Other references are given in the text.

Additional references not given in the preceding chapter

On the influence of organic matter in the formation of concretions and on estimated rates of growth of concretions, BERNER (1968a, b). On the destruction of grains by algae in Ordovician limestones, KLEMENT and TOOMEY (1967). On a connate origin for hot saline bottom waters, Red Sea, NEUMANN and CHAVE (1965). On the amount of $CaCO_3$ needed to effect cementation in a coastal eolianite, YAALON (1967). A discussion on the origin of an acicular aragonite cement in a quartzose calcarenite on the outer shelf, Delaware Bay, by ALLEN et al. (1969). Enrichment of the light isotope ^{12}C in the cement of this Recent sediment leads to the suggestion that, at a time of lowered sea level, methane escaped from submerged marshes and mixed with the ground water responsible for the precipitation of the acicular aragonite. On cementation on Bonaire, LUCIA (1968). On diagenesis in calcareous algae, MOBERLY (1970).

Chapter 10

CEMENTATION

INTRODUCTION

It is possible to make certain generalizations about the processes and products of cementation, with confidence that they apply in some measure to all three diagenetic environments: (*1*) the subaerial, fresh water eogenetic, (*2*) the submarine eogenetic, and (*3*) the deep crustal mesogenetic. The chemical and physical reactions in all three environments are bound to have much in common because, in each, the main process is one of precipitation, from solution, of an encrusting mass of crystals. In this chapter an attempt is made to discover, in a general way, how a loose carbonate sediment, which tickles us between our toes, can change into a hard rock suitable as a building stone for such august edifices as the Houses of Parliament. To say, simply, that the particles of the sediment were cemented together is to ignore the problems raised. Under what conditions does a cement grow? How are the Ca^{2+} and HCO_3^- ions transported to the site of growth? Whence comes the vast quantity of $CaCO_3$ precipitated? For the required amount of $CaCO_3$ is vast, if we bear in mind the likely primary porosities and the lack of compaction. Porosities of Recent carbonate muds and sands are normally 40–70% (GINSBURG, 1964; BATHURST, 1966; PRAY and CHOQUETTE, 1966) but those of most ancient limestones are less than 5%; most limestones show little or no sign of compaction by adjustment of packing or grain fracture (PRAY, 1960, 1969). We are forced, therefore, to conclude that the high primary porosity has been reduced to its present low value of about 3% almost entirely by the introduction of cement. This means that, as we gaze spellbound at the towering carbonate peaks in the southern Apennines or the Canadian Rockies, we must realize that at least half the volume of these mighty mountains is $CaCO_3$ cement which has been transported in solution to the place where we now find it. The questions of source and of transport of such enormous quantities of $CaCO_3$ have proved extremely difficult to answer, and the pages that follow will do little more than clarify some of the issues. These are topics of which our knowledge is as yet superficial and rudimentary: most of the work of understanding is still to come.

The precipitation of cement cannot be studied in isolation, because it is only one of the processes in the complex of reactions which accompany the stabilization (lithification) of carbonate sediments. It is probably true that stabilization is achieved largely, perhaps entirely, as a result of various reactions involving dissolution and precipitation. The dissolution of aragonite, accompanied by the formation of visible porosity, followed by reprecipitation as calcite cement, is examined in Chapter 8. A closely similar process, but without a visible pore stage, acting in pores measured in Ångstrom units, is dealt with in Chapter 12 on the development of neomorphic fabrics. Cements in general are considered here. Yet this separation into two processes—cementation and neomorphism—is artificial. The two causal processes are but end members of a spectrum of processes: the sizes of the pores in which reaction proceeds vary from the relatively enormous, such as the body chamber of a gastropod, to the smallest space in which dissolution and precipitation can take place, the intercrystalline boundary in a crystal mosaic. The dissolution-precipitation which accompanies cementation differs only in degree from that which accompanies the growth of neomorphic spar. In the fresh water eogenetic environment the full range of processes commonly acts on the same carbonate sediment at the same time. This is probably true of the submarine eogenetic and the mesogenetic environments also, though information on these is at present too scarce for us to draw conclusions. In reading the following pages it is necessary to remember always that cementation is only one among several major processes that contribute simultaneously to lithification.

RECOGNITION OF CEMENT IN THIN SECTION

The philosophy

Not all the sparry calcite in limestones is cement: some of it is neomorphic in origin and is examined in Chapter 12. This twofold origin compels us to place special emphasis on the *fabrics* of the sparry calcite, that we may distinguish with certainty a space-filling cement from a neomorphozed micrite or a calcitized aragonitic mollusc. The word "cement" is taken to include *all* passively precipitated, space-filling carbonate crystals which grow attached to a free surface: there is no other term in general use. It is equivalent to FOLK's (1965, p.26) "precipitated calcite" and his code "P". I have abandoned the terms "drusy" and "granular mosaic" (BATHURST, 1958) in favour of a rationalized single term. In doing so, I conform to the more enlightened current usage and, at the same time, emphasize the essential similarity of interparticle and intraparticle space-filling crystals. The process and the product are the same whether the cement unites two foraminiferids or fills their chambers. The sixteen fabric criteria assembled here have been taken from the publications of four authors, though it will be obvious that some of the criteria have been discovered independently by others.

   It is probably fair to say that the selection of certain fabrics as characteristic of cement has had a twofold basis. First of all, fabrics must be found which are critical, upon the presence or absence of which the decision concerning the origin of the sparry calcite (or aragonite or quartz) uniquely rests. Choice of these fabrics necessarily depends jointly on a knowledge of the fabrics that exist in limestones and on an understanding of the cementation process and its probable consequences. For example, cement spar is space-filling, thus it will be found in pores which were once open and not in those filled with mud. Changes of physico-chemical conditions with time will cause changes in the crystal fabric and yield the familiar generations of cement. Relics of neomorphism will be absent because the spar in question is not neomorphic in origin. These and other self evident assumptions are the basis of the criteria *1 11*.

Once a spar has been identified beyond any doubt as a cement on the basis of these *a priori* intuitive criteria, then it is possible to see whether it possesses other fabrics which, though possibly not yet justifiable on *a priori* grounds, are nevertheless present only in cements and can therefore be used as critical diagnostic criteria. Fabrics *1–11* (below) seem to depend largely on the intuitive argument, while fabrics *12–16* have been found generally to characterize cements already identified on the earlier criteria. Unhappily, very few of the criteria are entirely unequivocal, as will be seen in the discussion, but safety is in numbers and a satisfactory decision can generally be reached if several criteria are combined.

Fabric criteria for cement

(*1*) The spar is interstitial (interparticle) with well-sorted and abraded particles, which are in depositional contact with each other (Fig.297). Micrite, from which the spar might have evolved by aggrading neomorphism, is, therefore, unlikely to have been present in the original sediment.

(*2*) There are two or more generations of spar (p.432), a distribution unlikely to arise by neomorphism of fine grained carbonate (Fig.307, 308, 313).

(*3*) There are no relic structures such as are seen in neomorphic spar (Fig.312).

(*4*) Particles composed of micrite (e.g., pcloids) are not altered to spar.

(*5*) Micrite coats on particles are not altered to neomorphic spar.

(*6*) Mechanically deposited micrite is present but unaltered (Fig.298).

(*7*) Contacts between spar and particles are sharp (Fig.297–302).

(*8*) The margin of the sparry mosaic coincides with surfaces that were once free, such as the surfaces of skeletal particles or of oöids (Fig.297) or moulds of aragonitic shell fragments (Fig.298).

(*9*) The spar lines a cavity which it fills incompletely.

(*10*) The sparry mosaic occupies the upper part of a cavity whose lower part is occupied by a more or less flat-topped internal (geopetal) sediment (Fig.300).

(*11*) The mass of sparry mosaic has the form to be expected of a pore filling (Fig.301) or of an encrustation such as tufa or stalactite.

(*12*) The intercrystalline boundaries in the mosaic are made up of plane interfaces (Fig.301, 302, 313).

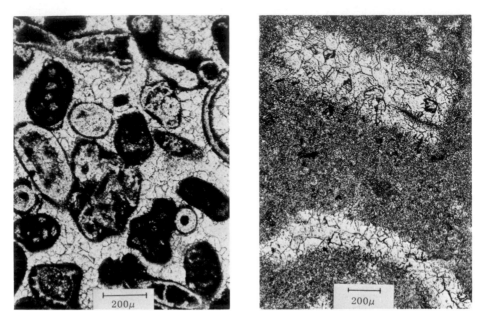

Fig.297. Biosparite with skeletal grains of foraminiferids, bryozoans, brachiopods. Foraminiferid chambers more or less filled with micrite—possibly a submarine cement. Slice. Carboniferous Limestone. Denbighshire.

Fig.298. Micrite containing calcite cement casts of molluscan shell fragments. Peel. Mississippian. New Mexico, U.S.A.

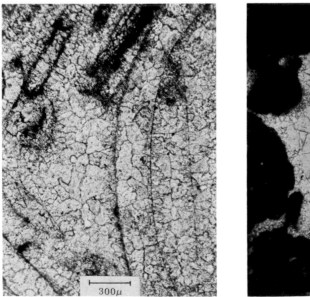

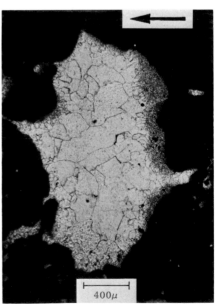

Fig.299   Fig.300

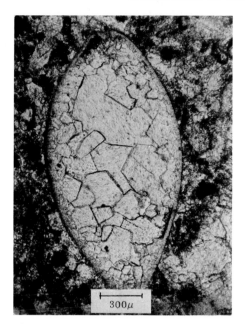

300μ

30μ

Fig.301. Ostracod filled with calcite cement having plane intercrystalline boundaries and increase of crystal size away from the surface of the test. Slice. Carboniferous Limestone. Denbighshire.

Fig.302. Calcite cement showing plane intercrystalline boundaries, increase of crystal size away from wall (bottom right) and enfacial junctions (arrows). Peel. Mississippian. New Mexico, U.S.A. (From BATHURST, 1964b.)

(*13*) The size of the crystals increases away from the initial substrate of the sparry mosaic (Fig.300–302, 313).

(*14*) The crystals of the mosaic have a preferred orientation of optic axes normal to the initial substrate of the mosaic.

(*15*) The crystals of the sparry mosaic have a preferred shape orientation with longest axes normal to the initial substrate of the mosaic (Fig.313). Extreme examples are radial-fibrous as in beach rock (Fig.276, p.367).

(*16*) The mosaics are characterized by a high percentage of enfacial junctions (Fig.302, 304) among the triple junctions: percentages so far recorded range from 30–73% (Table XVII; see also Fenninger, p.425). The recorded range for neomorphic spar is 2–5%.

(*17*) Fabrics characteristic of neomorphic spar are absent (Chapter 12).

Fig.299. Cement casts of mollusc shells embedded in cement in a biosparite. The thin lines separating the casts from the interparticle cement are possibly micrite envelopes that derived their mechanical strength from their high content of organic matter (p.333). Slice. Purbeckian. Dorset.

Fig.300. Cavity in micrite (fenestral porosity) filled with geopetal calcite siltstone (microspar) overlain by calcite cement. Arrow points to top. Spots are opaque pyrite. Slice. Carboniferous Limestone. Denbighshire.

TABLE XVII

TRIPLE JUNCTIONS IN SPARRY CALCITE MOSAICS
(After BATHURST, 1964b)

Type of mosaic	Age and location	Number of triple junctions with all angles < 180°	Number of triple junctions with one angle = 180° (enfacial)	% of enfacial junctions
Cement	Purbeckian, Dorset, England	65	55	46
	Callovian, Isle of Wight, England	124	83	40
	Bajocian, Lincolnshire, England	100	57	36
	Bajocian, Lincolnshire, England	101	51	33
	Bajocian, Lincolnshire, England	74	57	44
	Lower Carboniferous, Anglesey, Wales	41	50	55
	Lower Carboniferous, Yorkshire, England	80	50	38
	Mississippian, New Mexico, U.S.A.	67	59	47
	Locality unknown	103	44	30
	Purbeckian, Dorset, England	103	77	43
	Bajocian, Lincolnshire, England	113	70	38
	Mississippian, New Mexico, U.S.A.	36	50	58
Molluscan shell neomorphosed in situ	Wealden, Sussex, England	79	3	3.7
	Purbeckian, Dorset, England	171	1	4.5
	Purbeckian, Dorset, England	174	6	3.3
	Purbeckian, Dorset, England	177	3	1.7
Neomorphosed micrite	Pennsylvanian, New Mexico, U.S.A.	174	6	3.3
	Lower Carboniferous, Yorkshire, England	174	6	3.3
	Lower Carboniferous, Yorkshire, England	175	5	2.8

Discussion of fabric criteria

Some general points. In their paper on the Scurry Reef, BERGENBACK and
TERRIERE (1953) used the criteria numbered *1–5* as indications that the sparry
calcite was a cement. A few years later NELSON (1959) also stressed the characteristic
presence of two generations of cement and the improbability that mud could have
been deposited with abraded grains. He added two new criteria, *6* and *7*. Together
these seven criteria assume: (*a*) that it can be reliably demonstrated that micrite
was never present in the sediment, (*b*) that where micrite is present its unaltered
state shows that no neomorphism has taken place, (*c*) that the contact between a
neomorphozed micrite matrix and calcarenite grains would not be sharp, and (*d*)

that two generations of spar, a finely crystalline and a later more coarsely crystalline, would not be produced in neomorphic spar. These criteria are applicable to many limestones, but instances do occur where micrite is associated with well-sorted or abraded calcarenite grains, where peloids are unaltered though the micrite matrix has been changed to neomorphic spar, or where neomorphic spar is in sharp contact with grains. Only the presence of two generations of spar (p.432) seems a generally reliable standard, though the distinction of two generations is not always easy.

In 1958 I attempted to establish criteria that would distinguish cement from neomorphic spar. Like Nelson I stressed the importance of a sharp boundary between sparry mosaic and calcarenite grain, but added eight further criteria (8–15). The criteria numbered 12–15 are to be expected if the sparry crystals had been precipitated from a solution and had grown rooted on a free surface. Later, in 1964, I recorded the high frequency of enfacial junctions in cement compared with neomorphic spar: this criterion and the two generations of cement are probably the most reliable criteria available for distinguishing between the two types of spar.

The scale of a fabric is a matter of importance in its recognition. This is particularly so in an examination of intercrystalline boundaries and these will serve as an example. The boundaries between centimetre-sized crystals in a mosaic are readily distinguishable with a light microscope in thin section, but those between decimicron-sized or micron-sized crystals are difficult or impossible to analyze. Peels are useful here in that they reveal finer detail than a thin section. For the smaller crystals the electron microscope is a necessary tool (FISCHER et al., 1967). It is essential to bear in mind that details apparently absent with low magnification may be discovered if the magnification is increased. Stereoscan photographs by FENNINGER (1968) have been useful in this way.

The compromise boundary. SCHMIDEGG (1928) had shown by geometrical construction that where adjacent crystals meet, interrupting the freedom of each other's growth, they subsequently continue to grow but maintain contact along a plane interface which BUCKLEY (1951) was later to call a compromise boundary (Fig.303). Where two crystal faces, belonging to two adjacent crystals, are in contact they meet along a line. As each face grows this line moves in a direction dependent on the growth rates of the two faces and the angle between them. If the relations between angle and growth rates remain constant, the path of the line will lie in a single plane. When one of the faces ceases to exist (as in Fig.303), the compromise boundary will change direction and subsequent growth will give a plane with a new orientation. Plane interfaces are not always easily recognized in a thin section where they are oblique to the optic axes of the microscope, but they are readily discerned with the help of a Universal stage: GLOVER (1964) has shown how misleading the apparent morphology of interfaces can be unless checked in this way. Testing with Universal stage is essential where there is any doubt.

Fig.303. Diagram to illustrate the development of plane compromise boundaries between three crystals, each growing by the same increments. Note that the middle crystal is eventually buried by the other two.

The competitive growth fabric. The same competitive growth of adjacent crystals that yields compromise boundaries also gives rise to a reduction in the number of crystals in the mosaic, and a concomitant increase in their size, away from the initially free surface (criterion *13*) on which passive precipitation began. The more favourably oriented crystals obstruct the growth of less favoured crystals and then grow over them (Fig.303). Another result of the competition is that, as the more favourably oriented crystals continue to grow, the fabric of the mosaic is increasingly dominated by the favoured orientation. In this way evolve preferred orientations of crystallographic axes and shape axes (criteria *14* and *15*), a process of *Keimauswahl* (CORRENS, 1949a, p.141).

Criteria *12–15* are not infallible. It is important to realise that crystals compete for space, not only where they grow as a space-filling cement, but also where they replace micron-sized fabrics during the growth of neomorphic spar. Thus we must expect to see plane intercrystalline boundaries in neomorphic spar, also an increase of crystal size in the direction of growth and preferred lattice and shape orientations. All these fabrics have been seen in neomorphic spar. Nevertheless, these four fabrics are rarely well developed in neomorphic spar. My own limited experience of neomorphic spar that has been identified by other criteria suggests that plane intercrystalline boundaries are uncommon, whereas in cement they are practically universal. The curvature of intercrystalline boundaries in cement produced by etching of polished surfaces (FENNINGER, 1968: stereoscan photographs) could be an artifact caused by overdeep etching and rounding of adjacent crystals. Preferred shape and optic orientations, with an increase of crystal size, are to be found in the stellate form of neomorphic spar (p.503), but this fabric can hardly be confused with a space-filling mosaic, moreover, in the stellate mosaic the intercrystalline boundaries are not plane. The absence of plane intercrystalline boundaries in radiaxial mosaic (p.426) may be an indication of a secondary origin. As a reliable triple criterion for cement fabric, the combination of a high frequency of plane intercrystalline boundaries, two generations, with an increase of crystal size away from some clearly recognizable wall, does seem to be satisfactory, although SIPPEL snd GLOVER (1965) have noted the tendency for larger crystals to develop irregular surfaces and to adopt less geometric forms. This threefold relationship is apparent in the majority of cement mosaics which are identifiable on other grounds.

The previous discussion of competitive growth must now be extended to include the influence of the substrate on the relative growth rates of cement on different particles. L. C. Pray (personal communication, 1969) has coined the term **competitive cementation** to account for the observation that cements nucleated on different particles, and therefore on different substrates, occupy different volumes of the porosity. He has concluded, for example, that where a calcite substrate is available, as on an echinoid plate, growth will be quicker there than on, say, a substrate of aragonite, finely crystalline calcite or quartz (Fig.309). In a limestone containing echinoid grains and oöids (originally aragonite), most of the cement is normally syntaxial on the echinoids, because heterogeneous nucleation of calcite on calcite uses less work than nucleation on aragonite (see Lucia, Evamy and Shearman, p.429). Other substrate properties which influence preferential placement (Pray's term) of cement in a pore system are the crystal size and crystal orientation of the substrate, also the degree of lattice fit between the substrate and the cement. For example, calcite can more easily fit on dolomite than on quartz. An extreme case of a success story in competitive cementation is the poikilotopic fabric, where the various grains in an oöbiosparite are embedded in a small number of large cement crystals: each cement crystal encloses several grains and each is nucleated on that most favoured of substrates—an echinoderm grain.

The enfacial junction. A further criterion (*16*) for the identification of cement fabrics is based on the frequency of enfacial junctions among the triple junctions in a sparry mosaic (BATHURST, 1964b, p.362). A triple junction is the meeting place of three intercrystalline boundaries. Most junctions between interfaces are triple: they cannot be less in a mosaic and rarely comprise four or more interfaces. An **enfacial junction** is a triple junction where one of the three angles is 180° (Fig.304). Three crystals are shown in two dimensions in Fig.304, namely *a*, *b* and *c*, with plane intercrystalline boundaries *MO*, *NO* and *PO*. *MP* is a straight line and so this triple junction is an enfacial one. The term enfacial is derived from the french *en*,

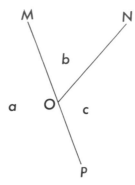

Fig. 304. Diagram of a triple junction between crystals *a*, *b* and *c*. As *MP* is a straight line, this triple junction is enfacial.

latin *in*, meaning "against", because the crystals *b* and *c* are believed to have grown against *MP* which is a face of crystal *a*. The argument for this runs as follows: although it is theoretically possible that *MP* is the result of a geometrical coincidence whereby two compromise boundaries, *MO* and *OP*, happened to evolve as a single, continuous, plane, the odds against this being a common occurrence are extremely high. Similarly it is unlikely that *MP* consists of a face *MO* of *b* and a compromise boundary *OP* between *c* and *a* (or *vice versa*). It is also improbable that *MO* and *PO* are faces of the crystals *b* and *c*, respectively, which were overgrown by *a*. Such coincidences are by their very nature exceedingly rare and cannot account for the observed high frequency of enfacial junctions in cements. There is, however, one event which does satisfy the geometrical conditions. The plane *MP* can be a face of crystal *a* which failed to grow while crystals *b* and *c* grew against it. Under these conditions the face *MP* of crystal *a* would have been inactive, a stationary barrier, during the development of the intercrystalline boundaries *MO* and *OP*. This event is the only one that does not depend on an unlikely geometrical coincidence. (If the face of *a* had continued to grow while the faces of *b* and *c* approached it, then *MÓ* and *OP* would be separate compromise boundaries, *a–b* and *a–c*, and the chances of their forming a continuous plane interface are absurdly small.)

Support for this interpretation of the origin of the enfacial junction came from ten measurements with a Universal stage (BATHURST, 1964b): (*1*) of the angle between the optic axis of crystal *a* and the pole of its supposed face *MP*, and (*2*) of the angle between this pole and the $(10\bar{1}1)$ cleavage in *a* (Table XVIII). Considering the level of accuracy possible with the Universal stage (TURNER, 1949, has the optic axis within 2–3°), the identification of *MP* as a crystal face was clear for nine out of the ten measurements, where the recorded angle is within 3° of the expected angle. The value for the tenth measurement was probably as close as can be expected, considering that it contains the errors of two angular measurements.

Enfacial junctions cannot be formed where three crystals grow together to

TABLE XVIII

CRYSTALLOGRAPHIC ORIENTATIONS OF THE FACE MP OF THE CRYSTAL a IN FIG.304*
(From BATHURST, 1964b)

Measurements with Universal stage		Nearest common face
$c \wedge$ pole of face	face $\wedge$ cleavage	
(a) 88 °		Prism: 90 °
(b) 59 °30′		(35$\bar{8}$4): 59 °55′
(c)	1 °42′	(10$\bar{1}$1)
(d)	1 °24′	(10$\bar{1}$1)
(e) 25 °		(01$\bar{1}$2): 26 °15′
(f) 55 °30′		(35$\bar{8}$4): 59 °55′
(g) 44 °		(10$\bar{1}$1): 44 °36′
(h)	3 °	(10$\bar{1}$1)
(i)	1 °54′	(10$\bar{1}$1)
(j) 47 °36′		(10$\bar{1}$1): 44 °36′

* Data from cement-filled cavities in Carboniferous Limestone, Great Orme, Denbighshire (means of five readings).

give compromise boundaries. It is apparent that where this happens a triple junction is formed with three angles none of which equals 180°.

It has been suggested to me that either crystal b or c may be a twin of a, the plane MO or OP being a twin plane. I have as yet made no measurements to test this idea. Though there is, theoretically, no reason why the growth of a should not include a twin plane, say between early a and a later b, the role of c and the interface ON in such a situation is not clear. Why did the twin not grow from all the surface a? How was it possible for c to interfere? If one tries to reconstruct a history beginning with the growth of b, the problems raised are equally difficult.

The importance of the enfacial junction is twofold. Its frequency among triple junctions in sparry mosaics is one of the least equivocal criteria for the distinction between cement and neomorphic spar (Table XVII). FENNINGER (1968) has applied this criterion to presumed calcite cement in the Tithonian limestones of Plassen, Austria. Frequencies of enfacial junctions among triple junctions ranged from 51.0–73.7%. The triple junction also sheds interesting light on the discontinuity of the cementation process (a matter discussed briefly on p.434). Further analysis of this fabric is needed beyond the small pilot study made by BATHURST (1964b).

The stability of calcite in geological time. One of the remarkable features of low-magnesian calcite cement is its long term stability (e.g., FRIEDMAN, 1968b, p.18). Judging by fabric studies with the light and electron microscopes, this type of cement retains its fabric, even its plane intercrystalline boundaries, from at least as far back as the Cambrian—so long as it does not enter the metamorphic

environment. In the low P–T and low shear conditions of the diagenetic environment, the fabrics of Palaeozoic cements remain indistinguishable from those of
Pleistocene cements in every detail visible to the microscopist. Indeed, slight changes
in calcite cement fabric have been used as sensitive criteria indicating the first incipient onset of metamorphism. The migration of crystal interfaces in calcite mosaics
during metamorphism was considered by VERNON (1968) who showed interesting
illustrations of curved interfaces. The Fe^{2+} in calcite seems also to be stable, despite
changes in redox potential of the pore water (EVAMY, 1969); so also the oxygen and
carbon isotopes (p.392).

    Radiaxial fibrous mosaic. In many limestones the first generation cement has
a kind of fabric which has been called **radiaxial** by BATHURST (1959b, p.511), who
described an example from the cavities in a Carboniferous knoll-reef. The characteristics of this fabric are so eccentric that they are worthy of special attention in
the hope that, when their origin is eventually understood, the fabric will be useful
as an indicator of diagenetic environment. The term radiaxial has been misquoted
by numerous writers since it was defined precisely in 1959: it does *not* refer to any
radial-fibrous fabric. It only refers to the peculiar combination of curved twins,
convergent optic axes and diverging subcrystals, *within* a cement crystal, here
described. This precise distinction is of the *utmost importance*, because radiaxial
mosaic probably has an origin altogether different from the simple radial-fibrous
crystals of aragonitic or calcitic cements.

    The crystals in radiaxial mosaic (as opposed to the subcrystals) are elongate
with length/breadth ratio (in thin section) as great as 7/1. They have a preferred
orientation of longest axes normal to the cavity wall (Fig.305). The crystal does not
extinguish uniformly between crossed polars but is generally composed of a number
of subcrystals having slightly different positions of extinction. Adjacent subcrystals
differ in extinction by 5–10° and the inter-subcrystalline boundaries are not plane.
Where subcrystals are not distinct the crystal shows undulose extinction. The
subcrystals are themselves elongate and, within each parent crystal, the subcrystals
diverge away from the cavity wall. Thus each crystal consists of a bundle of divergent subcrystals. Cutting across this bundle fabric there is a distinctly different
arrangement of optic axes in the crystal. The fast vibration directions *converge*
away from the wall: the twins, following this pattern, are curved with concave
surfaces facing away from the wall (Fig.306). The convergent pattern of fast vibration directions is easily recognized between crossed polars. As the stage is
rotated the extinction shadow swings across the crystal in a direction that is
opposite to that which is seen in the more familiar bundle of radiating cement fibres,
wherein the *c* axes are parallel to the lengths of the fibres and *diverge* from the wall.
This peculiarity distinguishes radiaxial crystal fabric from the more usual divergent
c axis fabric, with which it has been at times needlessly confused. The major crystals
which contain the radiaxial fabric may, indeed, be themselves radial-fibrous, but

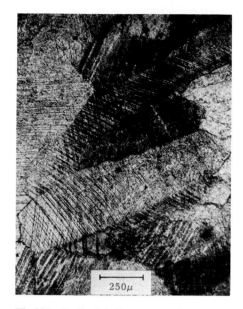

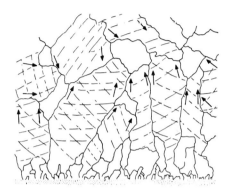

Fig.305. Radiaxial fibrous mosaic. Note curved twins. Wall to right. Slice. Carboniferous Limestone. Lancashire.

Fig.306. Drawing of radiaxial fibrous mosaic, showing fast vibration directions (arrows) converging away from the wall (stippled), also curved twins. Carboniferous Limestone. Lancashire. (From BATHURST, 1959b.)

we are not concerned with their orientation, but with the orientation of the arrangement of subcrystals or undulose extinction within them. It is useful to emphasize that radiaxial fabric refers to a fabric *within* a single large crystal. This fabric has two elements, divergent subcrystals and convergent optic axes.

Radiaxial fibrous mosaic is widely distributed in limestones as a space filler. It is normally followed by a second generation of cement with equant rhombohedral habit. The crystals of this second generation are commonly in lattice continuity with the radiaxial fibres, so that a fibre may pass into an equant crystal with an accompanying change from undulose to uniform extinction. This transition is in places marked by a zig-zag dust line, which presumably marks the final crystal faces of the radiaxial (first generation) crystals which were later buried by the syntaxial growth of the equant crystals of the second generation.

The accurate recognition of this peculiar fabric in limestones may be of special importance. It has been suggested, for example, that it is a recrystallized fibrous aragonite cement—but more evidence is needed.

THE PRECIPITATION OF CEMENT IN THE SUBAERIAL
FRESH WATER AND DEEP CRUSTAL ENVIRONMENTS

The growth of cement is controlled by four major factors: (*1*) the form of the substrate to which the crystals are attached, (*2*) the level of supersaturation in the solution, (*3*) the composition of the solution, and (*4*) the rate of movement of the solute ions past the growing crystal faces.

Influence of substrate

      Crystal size and mineralogy. In the pores of a sediment there is a variety of substrate upon which nucleation of cement can take place. The walls of pores vary in mineralogy, in the sizes and orientations of the crystals of which they are composed, and in the degree of cleanliness of the surfaces. Where the mineral in the substrate is the same as the cement, for example calcite, and is finely polycrystalline (such as the surface of a miliolid foraminiferid), and clean, then it must be supposed that the crystals in the surface of the miliolid act as substrates for syntaxial overgrowth (heterogeneous nucleation or epitaxy) of calcite. The growth of cement begins, therefore, at many points, and the result is a mosaic of crystals (Fig.297, 298, 300; BATHURST, 1958, p.15). The earliest cement crystals, at least, are presumably in lattice continuity with pre-existing crystals in the original free surface. As growth continues, competition results in the survival of the more favourably oriented crystals (Fig.301, 313), while others cease growing because they are obstructed or overgrown by the successful crystals. At the same time, plane compromise boundaries develop between crystals as they interfere with each other's growth. These matters are examined in more detail on p.421. The final result is a preferred orientation of *c* axes and longest shape axes normal to the original surface, and a reduction in crystal number and an increase in crystal size away from the surface.

      Where the substrate is not the same mineral as the cement, as when calcite grows on aragonite or quartz, or if the substrate is dirty (with a coat of micrite or organic mucilage), there are no nuclei in the free surface available for syntaxial overgrowth of calcite. In this case many calcite crystals nucleate on the surface as before but this happens slowly. The result is a mosaic similar to that on a polycrystalline calcite surface. The volume of this mosaic in the pore system is usually small, however, owing to its slow growth and the rapid expansion of other neighbouring cement crystals, which began growing earlier on more hospitable substrates.

      Where both polymorphs, aragonite and calcite, are available as substrates, then aragonite cement will grow on the aragonite substrate, and calcite on the calcite (p.351), provided that the composition of the pore solution and other factors at the time of growth allows this. GLOVER and PRAY (1970, in press, a, b) have found this rule applies to cements in the intraparticle porosity of Recent carbonate grains.

      Where large, millimetre-sized crystals exist in the substrate, the principle is the same, but the concentration of suitable nuclei in unit free surface is now greatly reduced compared with the micron-sized crystal mosaic in the miliolid. It is possible, however, to *see* that the cement crystals are syntaxial with host crystals in the substrate, whereas in considering the growth of cement mosaic on micron-sized crystals in the substrate we could only *suppose*, intuitively, that the cement crystals are syntaxial with their hosts. An extreme case is the sedimentary particle that is constructed of one crystal, e.g., a crinoid columnal or any piece of echino-dermal skeleton. Here the entire cement overgrowth can clearly be seen to have sprung from this one nucleus, and the calcite overgrowth is in lattice continuity with the original grain (Fig.309; Fig.76, p.52). This process of cementation by syntaxial (shared axes) overgrowth is the "cementation by enlargement" in HATCH et al. (1938, p.105) and the "secondary enlargement" of PETTIJOHN (1949, p.497). BATHURST (1958, p.21) called it "rim cementation", preferring the use of the word "cement" and the existence of convenient substantive and adjectival forms ("rim cement", "rim-cemented"). The use of the word "syntaxial" to indicate lattice continuity follows GOLDMAN (1952, p.7): the synonym "epitaxial" (for "epitaxy") is increasingly used.

Various skeletons besides echinoderms can be seen to have acted as sub-strates for syntaxial growth of cement: examples are the prismatic layers in gas-tropods, brachiopods and ostracods. It is well known that any surface constructed of a crystal lattice with strong preferred orientation can be the basis of a similarly oriented syntaxial overgrowth (FOLK, 1965, p.25).

If syntaxial growth of cement on substrate is so common, is it not sensible to conclude also that, even where there is no visible relation between cement and substrate, the cement is nevertheless syntaxial with the substrate? The important principle is that cement crystals will grow on pre-existing crystals of the same species in the substrate where these exist, even where the nucleus crystal in the substrate is too small to be detected. The question is further discussed on p.435.

Development of overgrowths on echinoderm grains. EVAMY and SHEARMAN (1965, 1969) have described in detail the evolution of rim cement on echinodermal grains, in particular on the crinoid columnal. This columnal is a porous struc-ture (Fig.75, p.51) with c axes parallel to the axis of the crinoid stem, perpendicular to the flat, mutually opposed surfaces of the columnals (Fig.307). The authors were able to plot the growth history of the syntaxial overgrowth because the composition of the solution from which the cement was precipitated had changed abruptly, from time to time, so that the amount of Fe^{2+} in the calcite had changed corres-pondingly. These changes are shown by alternate light and dark zones when the cement is treated with potassium ferricyanide, the iron-rich calcite being stained blue. As the interfaces between the zones are time planes, the history of the rim cementation can be described. Similar detail can be seen with the method of luminescence petrography (p.433).

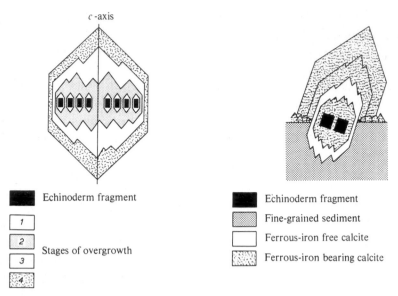

c -axis

Echinoderm fragment

1
2
Stages of overgrowth
3
4

Echinoderm fragment

Fine-grained sediment

Ferrous-iron free calcite

Ferrous-iron bearing calcite

Fig.307. Schematic representation of stages of syntaxial growth of calcite cement on an echinoderm fragment. (After EVAMY and SHEARMAN, 1965.)

Fig.308. As Fig.307, showing change in ferrous-iron content with age. After a period of iron free precipitation some lime mud was deposited in the remaining pores. The geopetal top of this mud was later buried by a further layer of iron rich cement. The mud supplied nuclei for many crystals of cement, but the syntaxial overgrowth on the echinoid continued to dominate the pore filling. (After EVAMY and SHEARMAN, 1965.)

The progress of syntaxial overgrowth is shown in Fig.307, 308. Growth begins on the echinodermal frame, the pores are soon filled and, as the volume of the syntaxial cement increases, so the number of pyramidal faces is reduced. Growth is also more rapid on the pyramidal faces and thus the cement rim is progressively elongated in the *c* direction. Evamy and Shearman gave, in addition, much useful detail about the development of compromise boundaries, the forms of which depend on the relative crystallographic orientations of the two original echinodermal nuclei and on the distance between them. This distance is significant because it is related to the degree of simplification of the zig-zag growth-front shown in Fig.307. The shorter the distance the more zig-zag the compromise boundary.

LUCIA (1962) noted that the volume of rim cement on echinodermal grains normally exceeds the volume of cement mosaic on the associated multicrystalline grains (Fig.309). Evamy and Shearman have found this also. In extreme cases the syntaxial rim is poikilotopic, many polycrystalline grains being enclosed within one crystal of rim cement. Commonly the rim fills the whole pore, extending either up to the surfaces of the polycrystalline grains or as far as the first generation cement fringes which encrust them. We must conclude, therefore, that the rate of growth of

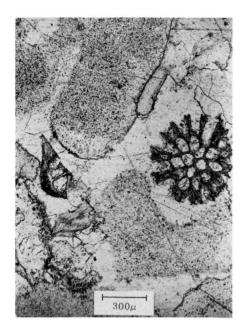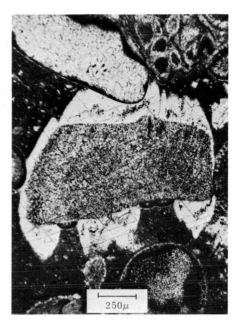

Fig.309. Biosparite showing the result of competitive cementation (p.423). The syntaxial over-growths on crinoid columnals have occupied most of the pre-existing porosity. On the bryozoan fragments there are only a few small cement crystals. See cleavage passing from crinoid to cement rim. Slice. Carboniferous Limestone. Denbighshire.

Fig.310. Syntaxial cement rim on an echinoderm fragment. The rim appears to be a space-filling cement, not a neomorphic replacement of the micrite. The growth of the cement has been prevented by an organic encrustation (algal?) in two places on the lower surface. Also the white line on the lowest grain is like a buried first generation cement. The fabric is packstone and the micrite probably arrived at a late stage in the cementation history. Slice. Bathonian. Ain, France. (Slice lent by D. J. Shearman.)

cement on large single crystals is greater than on a polycrystalline substrate of micron-sized crystals (Pray's competitive cementation, p.423). This is the same as saying, that the Ca^{2+} and HCO_3^- ions, given a choice between the two kinds of substrate, move more readily to the large single crystal. As cementation proceeds, the large crystal will grow even larger and so its favoured situation will be main-tained.

The growth of rim cement is prevented where the surface of the echinoderm is coated with micrite or where the superficial pores in the skeleton are filled with micrite (Fig.310, and 84 on p.55; LUCIA, 1962; EVAMY and SHEARMAN, 1965).

Discontinuities in growth of cement

Time intervals in cementation. In many limestones there is more than one generation of cement: that is to say, the laying down of the cement was demonstrably

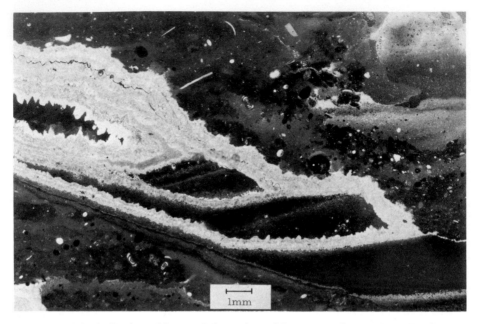

Fig. 311. Fabrics indicating a history of alternating calcite cementation on the walls of a cavity and geopetal sedimentation. Micrite. Upper Cretaceous. Monte Camposauro, Campania, Italy. (Slice lent by B. D'Argenio.)

discontinuous. Usually there are two generations. One generation is distinguishable from another, only if there is evidence that a *time interval* separated the growth of the first from that of the second. The simplest case is the interbedding of mechanically deposited sediment between one cement layer and the next (Fig.311). This may indicate only a small time interval, perhaps of minutes or days. Much longer intervals are indicated in limestones where shell fragments were broken during compaction and the first generation of tiny crystals appears only on the pre-fracture surfaces, whereas the fracture surfaces are covered directly by the larger crystals of the second generation (Fig.312). Thus the two periods of precipitation were separated by a period of compaction and fracture. The first generation crystals, besides being smaller, are commonly scalenohedral in habit (dog tooth spar), and the scalenohedral faces are preserved under the later overgrowth of second generation rhombohedra (Fig.313). Even where specific evidence of a time break is lacking, the sharp distinction in size and habit between the earlier and later crystals is strong presumptive evidence of a substantial time interval. With the help of stains (e.g., potassium ferricyanide) it is now usual to recognize a chemical distinction between the generations, the first being, commonly, unstained and the second ferrous iron-rich, stained blue and presumably precipitated in the reducing environment of the phreatic zone.

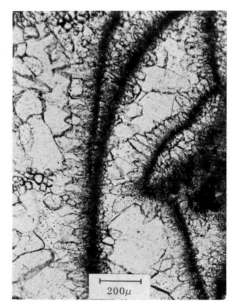

Fig.312. Molluscan grain fractured during compaction. The tiny cement crystals of the first generation appear only on the original grain surface (actually a micrite envelope). They are absent on the fracture surfaces which are covered only by the coarser crystals of the second generation. The primary shell structure of aragonite has been replaced by a neomorphic spar: its crystal mosaic is transected by lines that are relics of the original layered structure of the shell. Slice. Jurassic. Yorkshire.

Fig.313. Two generations of calcite spar on the dissepiments of a rugose coral. The earlier generation is scalenohedral in habit, the later rhombohedral with larger crystals. The faces of the scalenohedra are buried, but in the coarse spar it is not possible to say which of the plane intercrystalline boundaries are buried crystal faces and which are compromise boundaries. Slice. Carboniferous Limestone. Denbighshire.

Changes in composition of pore solution. In many limestones it is clear that relatively sudden changes took place in the compositions of the pore solutions. Staining of thin sections with potassium ferricyanide commonly produces bands of stained (shades of blue) and unstained calcite, deeper blue indicating higher concentration of iron in the calcite lattice (DICKSON, 1965, 1966; EVAMY and SHEARMAN, 1965; OLDERSHAW and SCOFFIN, 1967; DAVIES and TILL, 1968; COLLEY and DAVIES, 1969; EVAMY, 1969; NEAL, 1969).

The interfaces within crystals, between blue and colourless calcite, are normally plane and mark old crystal faces and they can be correlated from crystal to crystal. There is great scope, in tracing this chemical history of the pore water, for the application of luminescence petrography (SIPPEL and GLOVER, 1965; SIPPEL, 1968). With this method of examining limestones, structures can be seen which are invisible in transmitted or reflected light. Variations in the amount of Mn^{2+} and Fe^{2+} in the calcite lattice result in colour banding which reveals, often in fine detail, the history of the growing crystal and of the solution. As with stains, distinction between a sedimentary crystal and its syntaxial overgrowth is rendered particularly easy.

Closely allied to these studies of the chemical history of cement is the study of concretions. RAISWELL (1971) has shown how promising this approach can be. The progressive nature of concretionary growth preserves the history both of the pore water chemistry and the compaction of the sediment. Centrifugal changes in concentrations of elements and degree of compaction can be related: stages in different concretions can be correlated. The content of cement gives an estimate of the porosity at the time of precipitation. Where the content of cement decreases centrifugally it follows that compaction was proceeding during the precipitation of cement.

Other discontinuities. Other discontinuities occur in cement fabric but probably do not indicate either significant pauses in cementation or changes in composition of the solution. Enfacial junctions mark a hiatus in the growth of one crystal while it is overgrown by others, but this may be only a local affair involving simply the relation of one crystal to its neighbours and does not necessarily imply wholesale cessation of cementation. The alternation of cement with internal sediment most likely results from discontinuous sedimentation imposed on continuous cementation rather than the reverse. Fabric variations with time are unusually well shown in the travertines of Egypt (so-called "Egyptian alabasters") described by AKAAD and NAGGAR (1964, 1967).

Causes of discontinuities in cement fabric. The reasons for discontinuities in the growth of cement, whether time intervals, chemical changes in the solution, or other breaks, can only be guessed at. It seems likely that the earlier cement of relatively small crystals is commonly the product of dissolution of aragonite and precipitation of low-magnesian calcite in the fresh water, subaerial (eogenetic) environment (p.327), under low overburden. With increasing data on eogenetic submarine cements, it may soon be possible to recognize these also in ancient limestones. The eogenetic origin of early cement is supported by its pre-compaction age. If the porosity is not filled during the eogenetic phase, there may be a delay of thousands or millions of years before a new source of $CaCO_3$ becomes available in the mesogenetic environment. This new source may be the carbonate released during pressure-solution. HARMS and CHOQUETTE (1965) and DUNNINGTON (1967) have found a close positive relation between the frequency of stylolites in limestones and the amount of cement: as a stylolite is approached the porosity falls.

OLDERSHAW and SCOFFIN (1967) recognized two generations of calcite cement in certain Silurian and Carboniferous limestones. The first generation ($<$ 200 p.p.m. Fe^{2+} in the calcite) is not stained by potassium ferricyanide, the second generation is ferroan (200–500 p.p.m. Fe^{2+}) and yields a blue stain. The early, non-ferroan, calcite appears on primary grain surfaces, but not on compaction–fracture surfaces. The later, ferroan, calcite overlies the first generation and lies directly on fracture surfaces. Therefore an interval of time separated the two periods of precipitation and during this interval the sediment suffered compaction great enough to break grains. The later cement is separated from the early cement by what appear to be buried crystal faces of the first generation, since the triple junctions which separate them are enfacial (p.423), the earlier crystals being equivalent to a in Fig.304. In these same limestones there are cement casts of once aragonite shells and the authors supposed that dissolution of aragonite had released $CaCO_3$ for the early cement. Some of the casts are in micrite envelopes and, since few of the envelopes show any sign of collapse, Oldershaw and Scoffin concluded that aragonite dissolution and early cementation took place before compaction.

The two authors have given an important analysis of the origin of the iron-rich calcite cement in the Carboniferous limestones. The late, ferroan cement occurs uniformly in the argillaceous limestone, but is only found in the clay-free limestone where this is adjacent to an underlying shale. Higher levels of the clay-free limestone, far from the shale, show a second generation of cement low in iron. In the argillaceous limestone pressure-welded contacts between grains are more numerous than in the clay-free limestone. It is also apparent that the argillaceous limestone is relatively high in iron and that the clay under the clay-free limestone is similarly rich in iron. In the basal 5 m of the clay-free limestone, the iron content increases sharply toward the clay.

From this evidence Oldershaw and Scoffin inferred that the $CaCO_3$ of the second generation ferroan cement was derived by pressure–solution of the argillaceous, relatively iron-rich limestone and also from $CaCO_3$ released from the clay underlying the clay-free limestone.

These conclusions are important because they emphasize that cementation can have a long drawn out history, and that both eogenetic (subaerial or submarine) cementation and deep subsurface, mesogenetic cementation can each play volumetrically significant roles in the lithification of a single limestone.

Nucleation of cement crystals

The process of nucleation of cement crystals is obviously a matter of central importance. It has already been shown that the fabric of cement is influenced by the vectorial growth rates of adjacent crystals and by the initial host substrate. We must also ask to what extent the fabric is influenced, if at all, by nucleation of *new* crystals with unique orientations. In a cement-filled pore, are all the cement crystals syntaxial overgrowths on older crystals in the original substrate, in this case the pore wall? Or are some of them entirely new lattices, with unique orientations?

Theoretically, in a supersaturated solution, two processes of nucleation are supposed to operate. Where nucleation is **homogeneous** (SPRY, 1969, p.17), a piece of crystal lattice forms as a result of chance collision of solute ions. If the piece of lattice is larger than a critical size (say 6–9 molecules, NIELSEN, 1955) it will survive and continue to grow: if it is smaller it will be supersoluble (p.254). This ideal process of nucleation is probably never realized (NANCOLLAS and PURDIE, 1964, pp.11–12). It is doubtful if any solution is entirely free of minute stable particles, not necessarily the same material as the precipitate, but which could nevertheless act as hosts for growth of the precipitate. Furthermore, calculations indicate that homogeneous nucleation would require a level of supersaturation higher than is actually observed when nucleation takes place. WOLLAST (in press, a) has estimated that the probability of a collision-accreted nucleus exceeding the critical size is much too small to allow appreciable precipitation by this process.

Probably all nucleation is **heterogeneous** (epitaxial, syntaxial), the attachment of ions to an already existing host substrate as an overgrowth. The matter is usefully treated in terms of free surface energies.

The free energy of the interface between the nucleus and the containing phase (liquid or solid) is the most important single factor influencing the rate of nucleation. Although theory is still based on Kelvin's work on supposed homogeneous nucleation of water droplets, published in 1881, it is applicable also to heterogeneous nucleation. It is usual to treat the total energy E of a spherical nucleus of radius r as consisting of the algebraic sum of two parts (D. MCLEAN, 1965):

$$E = 4\pi r^2 \sigma - \tfrac{4}{3}\pi r^3 \Delta G \tag{20}$$

where σ is the free energy (regarded as isotropic) per unit area of interface, and $-\Delta G$ is the free energy liberated by each unit volume when it precipitates (or the difference between the chemical potential of an embryonic nucleus and that of a single molecule). The expression $4\pi r^2 \sigma$, dependent on r^2, increases at first more rapidly than $\tfrac{4}{3}\pi r^3 \Delta G$. Thus E rises from zero to a maximum value $+E$, before falling to negative values as the second term becomes greater (Fig.314). Differentiating the equation gives:

$$E_1 = \frac{16\pi\sigma^3}{3\Delta G^2} \tag{21}$$

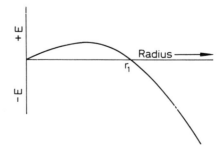

Fig.314. The energy E of the nucleus of a new phase related to its radius. (After MCLEAN, 1965.)

and this is the energy barrier which must be surmounted if the nucleus is to survive. Nuclei remaining smaller than the critical r_1 (Fig.314) will dissolve (see also eq.16, p.254).

Applying these ideas to the question of heterogeneous nucleation, it is clear that the value of σ, the free energy per unit area of interface, will be less for an interface between two structurally related solids (substrate and overgrowth) than for an interface between substrate and liquid, as in homogeneous nucleation. Therefore, in heterogeneous nucleation the energy barrier, E_1, is lower and so this type of nucleation takes place rather than homogeneous nucleation (see also WOLLAST's treatment).

WOLLAST (in press, a) has taken the matter further and has calculated, from a basis of interfacial energies, that microcracks and intercrystalline boundaries are favoured sites for nucleation and must exert a dominant influence on the processes of recrystallization and precipitation. He gave valuable theoretical support for the existence of a phenomenon the results of which are familiar to geologists from their studies of thin sections of rocks.

VAN DER MERWE (1949) developed a model for epitaxy in which a crystalline overgrowth can attach itself to a crystalline substrate of the same or a different lattice structure. The extreme case is, of course, the addition of isomorphous lattice, as calcite cement is added to crinoid columnals. Here the geometry and composition of the substrate lattice is perfectly matched by the overgrowth. Where substrate and overgrowth differ, both the surface ionic layers of the substrate and the rudimentary thin film of the overgrowth can be expected to adjust themselves to a limited extent, by elastic rearrangement. This question of fit is of special importance where carbonate crystals are nucleated on organic templates during calcification in shells and other biological systems. More recent developments of the theory of epitaxy have stressed the importance of pseudomorphism in thin, metastable, growth films that mimic the structure of the substrate (ANONYMOUS, 1969). An example is the growth of iron on copper, where the iron grows with a face-centred cubic structure like that of the underlying copper, although the structure of iron is normally body-centred cubic. At some thickness above the substrate the overgrowth reverts to its normal structure.

Although there is no doubt that nucleation of cement crystals is heterogeneous, it is far from clear how this process works. There would appear to be two extreme possibilities. In case I the cement crystal grows syntaxially on a host crystal in the original substrate. Where the crystal is large enough to be visible, it should be possible with the aid of serial sections to trace the cement crystal into its host. In case II the cement crystal grows syntaxially on a submicroscopic, *new* host crystal which has a *new* unique orientation. In some way this uniquely oriented host must have become attached to the surface of one of the cement crystals developed according to case I. We can only guess that perhaps the uniquely oriented host was a piece of lattice of colloidal size, suspended in the solution, which was adsorbed onto the surface of a cement crystal. Alternatively, it may be that irregularities in the surface of a cement crystal may act as suitable sites for the nucleation of uniquely oriented lattice. The importance of case II is that it permits the existence of a lattice discontinuity in the history of cement growth. This process is invoked in Chapter 7 to account for the appearance of new crystals during the growth of oöids (p.309). The true process of heterogeneous nucleation of cement crystals may well lie variously between the two extremes. An additional source of newly oriented crystals is the growth of twins, where the orientation of the new crystal differs from that of the substrate because, for example, of its rotation about a twin axis.

The appearance of new, uniquely oriented crystals during the dissolution-precipitation process in limestones is indicated by various data. Where radial-fibrous acicular crystals grow in bundles in which the needles diverge away from a point in the original substrate, we must assume that new crystals are nucleated from time to time, because otherwise the original acicular crystals would become fatter in order to fill the intercrystalline porosity produced as the needles diverged. It is a matter of observation that they do not become significantly fatter, that the number of crystals present at the outer surface of the bundle increases with the distance from the point at which growth began. Therefore new crystals must be "created" as growth proceeds.

In Chapter 6 we saw that the wet transformation of aragonite to calcite, described by Fyfe and Bischoff (p.240), proceeded as if there were a continual production of new nuclei accompanied by steady growth on existing nuclei. Pruna et al. found that the addition of new nuclei of ground calcite assisted the rate of transformation, the more finely ground the better (p.241). Ground $NaNO_3$, which is isomorphous with calcite, had a similar effect. Presumably, in certain circumstances, it is energetically more economical for precipitation to begin on a newly provided different substrate than on an old substrate. In the wet transformation of aragonite to calcite the reason is clear enough—calcite is the stable solid phase. In the growth of cement we have seen that like tends to grow preferentially on like, calcite on calcite, aragonite on aragonite. It may be that when the composition of the solution changes so that the composition of the precipitate changes, then this precipitate can no longer match adequately the lattice available in the old substrate. As an example, if early cement crystals were high-magnesian calcite, then the influx of a pore solution low in Mg^{2+} would lead to the precipitation of low-magnesian calcite which might not fit well onto the lattice of the high-magnesian calcite. Entropy might then demand that nucleation take place on a new host of low-magnesian calcite, colloidal pieces of which might well be present in the solution. The growth on new nuclei with unique orientations must be specially easy where the groundwater is dirty, full of a variety of potential hosts. Finally, the inhibiting effect on syntaxial overgrowth of a layer of impurity on an older crystal must also be born in mind.

Understanding of the process of nucleation in cements is clearly in an embryonic stage and, although, intuitively, it seems likely that cementation proceeds both by syntaxial overgrowth on substrate and by growth on newly supplied nuclei, there is much to be learned.

Influence of water chemistry on mineralogy and fabric of cement

Any attempt at the present time to relate cement fabrics and mineralogy to the chemistry of the pore water is likely to be rapidly outmoded by the accumulation of new data, some of it surprising. The controlling influences are certainly more varied than the simple relation between salinity and fabric postulated by FRIEDMAN (1968b). Indeed, salinity alone may be of little importance compared with the effect of the Mg^{2+} activity. The considerable influence of the substrate has already been examined. Organic compounds may exert great influence on intraparticle cementation (p.252; PURDY, 1968, p.195). Reducing conditions during precipitation permit the substitution of Fe^{2+} for Ca^{2+} in the calcite lattice (EVAMY, 1969).

THE SUPPLY AND DELIVERY OF CaCO₃ FOR CEMENTATION IN THE SUBAERIAL FRESH WATER AND DEEP CRUSTAL ENVIRONMENTS

The problem of the supply of CaCO₃

A clearer understanding of the origin and transport of the $CaCO_3$ which is precipitated as cement is an urgent requirement: not only does cementation play a major role in carbonate diagenesis, but it is also part of the history of several millions of cubic kilometres of limestone.

Significance of low compaction and high primary porosity. Two petrographic facts make the problem of the source of the $CaCO_3$ outstandingly difficult. The first is that limestones normally show scant evidence of compaction (Fig.315; PRAY, 1960; ZANKL, 1969). Exceptions, such as part of the Chalk of northern Europe (p.401) seem to be rare. The second fact is the known high porosity of unconsolidated Recent carbonate sediments (p.415), ranging commonly from 40 to 70%. We must therefore infer: (1) that cementation begins early, before compaction, and (2) that about half the volume of $CaCO_3$ in many limestones has been carried into the sediment from outside. Redistribution of carbonate by dissolution of aragonite and reprecipitation of calcite is, by itself, inadequate to fill the pores (p.327): it is accompanied only by an 8% increase in volume of $CaCO_3$. Thus initial porosities of 60% and 40% in pure aragonite sediment would be reduced only to 56.8% and 36.4% if all the dissolved aragonite were reprecipitated locally as calcite. The problem before us is twofold. How were such large quantities of

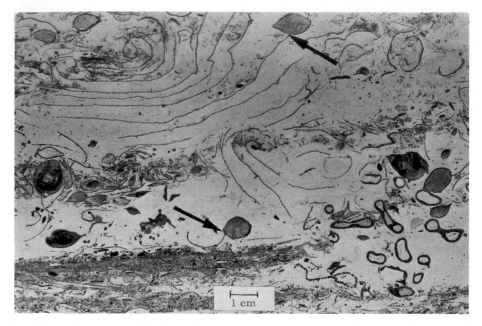

Fig.315. An illustration of the lack of compaction in biomicrites (wackestones). The long thin lines are the skeletons of fenestrate bryozoans, arranged spirally (especially top left). Neither shells nor bryozoans show the zig-zag walls, or sheared overlapping walls such as those found by WOLFE (1968) for fossils in compacted chalk which suffered up to 30% decrease in thickness. L. C. Pray has pointed out that the geometries of junctions such as those between bryozoan frond and brachiopod (at arrows) could not survive compaction. Slice. Mississippian. Illinois, U.S.A.
(Courtesy of L. C. Pray.)

cement delivered, and from what equally great repository of $CaCO_3$ were they derived?

   The vast scale of operations. It is possible to estimate the scale of transport by noting first that the content of $CaCO_3$ in aqueous solution in rock pores is unlikely to exceed 0.05 g/l for near-surface waters containing CO_2 at 25 °C (KRAUS-KOPF, 1967, p.65). Taking this as a starting point, we can say that, for a pore of unit volume to be filled with calcite cement (sp.gr. = 2.7), 54,000 unit volumes of solution will be needed—even if it be assumed that all the $CaCO_3$ in the solution passing through the pore is precipitated. Clearly, however, only a small fraction of the $CaCO_3$ is precipitated from the pore water in any one place, otherwise lateral transport of Ca^{2+} and HCO_3^- ions through the sediment would be impossible. PRAY (1966) put a probable figure of 10,000–50,000 pore volumes of solution to yield one fully cemented pore and DUNHAM (1969a, p.161) put 10,000—in the unlikely event that the process is 100% efficient and all $CaCO_3$ is precipitated. At 10% efficiency the figure is 100,000 volumes.
   An example will illustrate the scale of the process. Suppose that a solution of

CaCO$_3$, at 0.05 g/l enters at the top of a horizontal formation of unconsolidated carbonate grains with a porosity of 50%, and a thickness of 10 m, and then moves vertically down through it. For each cm^2 of upper surface of the formation across which the solution moves, there is an underlying column of pores with a volume of $1,000 \times 0.5 = 500$ cc. If these pores are to be filled with calcite, then the mass of calcite to be imported is $500 \times 2.7 = 1,350$ g. At a concentration of 0.05 g/l, the volume of solution required to pass into the sediment will be $1,350/0.05 = 27,000$ l for each cm^2 of the upper surface of the formation. If we assume a fast movement of pore water through the rock, say 300 cm/year, then the process of cementation will take 90,000 years. Real conditions are likely to be less extreme, with lower concentration, slower movements of water oblique to the horizontal, less efficient abstraction of CaCO$_3$, and reduction of flow rate following decrease of permeability, so that the complete process would take a great deal longer, perhaps hundreds of thousands of years. As the diameter of the pore channel approaches the thickness of the laminar boundary layer, fluid flow is replaced by transport of material by diffusion alone (HELING, 1968). It is interesting, therefore, to read Friedman's figure of 80,000–100,000 years for complete cementation of Pleistocene limestone along the Mediterranean coast of Israel.

Possible sources of CaCO₃. Sources for the great quantity of dissolved CaCO$_3$ needed to fill the porosity in a carbonate sediment are far from obvious. A few possibilities are examined here, but there is no denying that a source (or sources) of appropriate dimensions has yet to be recognized. Where the upper surface of a carbonate sediment is subaerially exposed, fresh water moving downward through the sediment will enable dissolution and precipitation to proceed with a net loss of material by dissolution in the upper part and a net gain by cementation in the lower. This means that if, for example, a bed 10 m thick with a porosity of 50% is to be fully cemented, then this must take place, theoretically, under an overburden at least as great as 10 m, even assuming that all the overlying 10 m of sediment is dissolved to yield CaCO$_3$ for cement. If we imagine a carbonate sediment gradually emerging above the sea, acted upon by rain, then, if the whole formation is to be cemented, there must be progressive dissolution of the upper layers (both particles and cement) in order to yield a supply of CaCO$_3$ for the layers below. A process of this type might yield karst topography on an enormous scale. It is of great interest that this has rarely been clearly recognized on any significant scale in the geological column. It is possible that the unconformities at Eniwetok and other atolls (p.323) represent such a development, but structural evidence is lacking.

The superficial cemented crusts that form in a year or so on loose carbonate sediment exposed to rain water (p.324) cannot depend upon a supply of CaCO$_3$ from the dissolution of overlying sediment. It must be supposed that bulk upward transport of CaCO$_3$ by fluid flow occurs as the result of evaporation and that the lost water is replaced through the pore system by capillary action. Pressure-solution is a source of dissolved $CaCO_3$ (p.470). The amount of $CaCO_3$ which has been released in some pressure-welded oösparites (Fig.318, p.462) can be calculated from measurements made in thin section, since the original volumes of the partly dissolved oöids can be readily estimated. Yet, unpublished data by various people show that this released $CaCO_3$ is far too little to account for the whole cementation of the oösparite. The amount of $CaCO_3$ released by stylolites may be considerable in some areas. In addition, it is necessary to emphasize that dissolution of aragonite and reprecipitation of the $CaCO_3$ as low-magnesian calcite yields only an 8% volume increase—an amount hopelessly insufficient to fill the original porosity of 40–70%.

Finally it must be admitted that none of the ways of supplying $CaCO_3$ considered here—the local dissolution of aragonite, influx of sea water, pressure-solution—are large enough to give the amount of cement that is known to be in limestones. The true process must be simple and ubiquitous, but at the moment it is entirely elusive.

The ideal pore system: calcite–CO_2–H_2O

The discussion up to now has assumed that, in some unspecified way, an under-saturated solution takes up $CaCO_3$ from one place and, after moving through a porous sediment, becomes supersaturated and loses some $CaCO_3$ as precipitated cement. It is now necessary to look beyond this simple anecdote to the true complexity of events, so far as these are known.

Phreatic pore waters, or more correctly, pore solutions, move with such extreme slowness that equilibrium is almost completely reached between the surfaces of the particles and the solution, because the rate of reaction at the particle surface is fast compared with flow transport (WEYL, 1958). In a monomineralic sediment the solution is believed to be virtually saturated for the minerals of which the adjacent particles are composed. Lateral changes in temperature or pressure will cause the solubility product to vary, so that, as the solution moves through the sediment, solid is dissolved in some places and precipitated in others. Nevertheless, the experiments which SIPPEL and GLOVER (1964, p.1417) made (p.445) led them to conclude that concentration gradients arising solely from flow transport of solutions are "relatively unimportant" in the dissolution alteration of carbonate rocks. It seems, from their results, and from the earlier work of WEYL (1958), that in a monomineralic carbonate sediment the system $CaCO_3$–CO_2–H_2O would be in most places stable, so that neither dissolution nor precipitation would take place and, as a result, the sediment would retain its primary mineralogy and porosity. Such extreme stability, as is well known, is the opposite of what we find, since carbonate sediments are remarkable for their susceptibility to diagenetic change through dissolution and precipitation. What is it, then, that makes a carbonate sediment such a vigorous chemical environment?

It is useful, first of all, to look rather more closely at the experiments and arguments of Weyl, and of Sippel and Glover. WEYL (1958) was concerned with the effect on a porous carbonate sediment of the movement through it of an aqueous solution containing CO_2 but no inhibiting ions such as Mg^{2+}. If the solution were undersaturated before entering the sediment, how far into the sediment would it move while continuing to dissolve particles? If the solution were supersaturated on entering, for what distance into the sediment would precipitation take place? In other words, how far would the solution have to travel through the pore system before it reached saturation and attained a steady state, such that neither the solution nor the particles suffered further change? The answer to this question is of high importance in any consideration of carbonate diagenesis, having special relevance to the concept of dissolution-precipitation during the subaerial exposure of limestones to fresh water, as outlined in Chapter 8. Intuitively we can answer that, if the flow velocity is slow, then equilibrium will be attained practically at once and the effect of the entering solution on the sediment will be negligible: but, as WEYL (1958, p.163) pointed out, the crucial need is to determine "how slow is slow?" for naturally occurring systems.

Dissolution: importance of ion transport. Weyl began by noting that the dissolution of $CaCO_3$ proceeds by four steps, thus:

(1) $CaCO_3 \rightarrow Ca^{2+} + CO_3^{2-}$

(2) $CO_2 \text{ (aqueous)} + H_2O + CO_3^{2-} \rightarrow 2HCO_3^-$

(3) $CO_2 \text{ (gas)} + H_2O_{\text{(liquid)}} \rightarrow CO_2 \text{ (aqueous)}$

(4) Movement of ions away from the solid, through the solution, by diffusion and fluid flow.

Steps *1–3* are the equations 4 and 5—broken into steps—on p.232 in Chapter 6.

He then set out to discover which of the four processes limits the rate of the whole dissolution reaction, extending earlier Russian work. GORTIKOV and PANTELEVA (1937) had measured the rate of dissolution of calcite crystals rotating in acids. Except for H_2CO_3, the rate of dissolution had increased linearly with the speed of rotation. A levelling off of the rate of solution in H_2CO_3 had been attributed by the authors to the limiting velocity of the reaction of CO_2 with water (Weyl's step *3*). SHTERNINA and FROLOVA (1952) had bubbled CO_2 into a vessel containing calcite crystals in distilled water continually stirred. They found that only after one month did the solution approach within 1% of saturation. Such a slow rate must, they thought, result from the slow change of CO_3^{2-} to HCO_3^- (Weyl's step *2*). In fact, as Weyl explained, two steps are involved here, his *2* and *3*.

In order to test his step *1*, Weyl directed a jet of CO_2-saturated water at a calcite crystal and measured with a microscope the size of the concavity which evolved around the place of impact, after various time intervals. The rate of dissolution of the calcite was related linearly to the rate of flow of the water jet. Clearly, therefore, the rate of dissolution in these experiments was not impeded by the rate of dissociation of calcite at the solid–liquid interface.

To test the possible influence of the rate of dissolution of fresh CO_2 gas, his step 3, Weyl bubbled CO_2 into a vessel of distilled water, both gas and water being at constant temperature. In two runs, CO_2 was bubbled first very slowly, at about 1 cc/min, and in the second run rapidly at 150 cc/min. In both experiments the relation of pH to time had the form expected for a dissolution rate limited by diffusion. In the slow experiment half-saturation was reached in about 210 min, but in the rapid experiment this time was reduced to 0.8 min. The difference between the two times was attributed by Weyl to the dependence of the slow experiment solely on diffusion, whereas, in the rapid experiment, dissolution was assisted by turbulence (fluid flow) and the greater area of gas–liquid interface. There was, therefore, no reason for supposing that the rate of dissolution of CO_2 was a limiting factor in the approach to saturation.

Weyl also repeated the experiment of Shternina and Frolova, but with more rapid stirring, so as to test the possible influence on calcite dissolution of the rate of change of CO_3^{2-} to HCO_3^{-}, as in his second step. CO_2 was bubbled vigorously through a vessel bearing distilled water and powdered calcite, rapidly stirred. Saturation reached 90% in 22 min in contrast to the 15 days found by Shternina and Frolova. The greater rate Weyl put down to the more rapid stirring and bubbling, as in the experiment with CO_2 and water, and to the finely powdered calcite which gave a much enhanced surface area compared with the 0.5 mm crystals of Shternina and Frolova. There seemed no reason to regard the rate of dissolution as being restricted by the rate of change of CO_3^{2-} to HCO_3^{-}. There is, however, good reason for supposing that the limiting factor is the rate of transport of ions, by diffusion and by fluid flow, away from the dissolving surface. The dominant importance of the rate of diffusion in controlling the rate of precipitation of $CaCO_3$ was also noted by HASSON et al. (1968) in their interesting study of the growth of $CaCO_3$ scale in the water pipes of industrial heat exchangers.

Uniform P–T: equilibration of pore solution. In the same paper WEYL (1958) went on to examine, theoretically and by experiment, the rate at which an undersaturated solution entering a calcite rock would reach saturation. Here the reaction of CO_2 with water (his step 3) can be neglected as the pore system is isolated from the atmospheric gas phase. The problem is one of diffusion of ions from a solid–liquid interface coupled with laminar flow in capillaries, in rock wherein the solubility of the mineral is uniform in space and does not vary with time.

In an elegant theoretical examination of straight circular and plane parallel capillaries, and in experiments with flow of water through a pack of calcite fragments, Weyl found that when water saturated with CO_2 entered a porous pack of calcite crystals, with the slow velocity expected in natural groundwater, it reached equilibrium practically instantaneously. As he put it, "we can assume that, under normal conditions below the water table, the water filling the pore spaces in the rock is always saturated with respect to the rock" (WEYL, 1958, p.175). This corroboration of truth long suspected has important implications which are examined in a later section (p.446). For the moment, it is necessary to emphasize that Weyl's conclusion applics to monomineralic rocks of low-magnesian calcite, to a Carboniferous limestone but not to a Recent lime sand. It also applies to the entry of a solution undersaturated for $CaCO_3$ and without any Mg^{2+}. As we shall see, different conclusions follow from the study of rocks containing a mixture of aragonite and calcite where the solution is supersaturated for the calcite (p.446).

Vertical P–T gradient: dissolution and precipitation during flow transport.
SIPPEL and GLOVER (1964) looked at the vertical flow of a solution through a rock in the same way as WEYL (1958), but added the complicating conditions of vertical geothermal and hydrostatic pressure gradients. They then set out to investigate the quantity of calcite dissolved or precipitated when a pore solution moves vertically along the thermal and pressure gradients, that is to say in a direction in which the solubility is increasing or decreasing.

The change in solubility with depth depends on both temperature and pressure and the authors carried out some experiments to determine these relations, using a modified Hassler cell that allowed aqueous solutions of CO_2 to move through a pack of ground calcite at various temperatures and pressures. The ranges of conditions were 30–80 °C and 3.5–179 bars, equivalent to an approximate crustal depth of 1,800 m, with two fixed concentrations of CO_2 at 0.3 and 0.1 atm (Fig.229, p.273).

There is a difficulty which is implicit in the argument so far, because the vertical fluid velocity and the geothermal gradient are not independent, and Sippel and Glover included an intriguing digression on the estimation of the effect of downward and upward flowing waters on the thermal gradient. They concluded that, although upward flowing waters cause a steepening of the gradient toward the top of a permeable formation, and downward flow causes a steepening in the lower part, nevertheless normal geothermal gradients can persist "over appreciable vertical distances" for flow rates of 30–300 cm/year. Such a range of flow rates probably includes most subsurface flows.

Finally, they estimated the changes in rock density which might be expected to result from a fast vertical flow, 300 cm/year, over a period of 10^6 years. Even with extreme conditions they found a change of only 13% in rock density (or an equal but opposite change of porosity). In more usual circumstances the change is unlikely to exceed about 4% in 10^6 years. It is, therefore, apparent that flow transport, as a device for bringing about cementation or dissolution in the pore system of a *monomineralic limestone*, is relatively unimportant. The significant fact about formation waters is that, owing to their slow movement, they are saturated as shown by Weyl. SIPPEL and GLOVER (1964, p.1416) remarked that "one can only be impressed by the difference between the great quantities of carbonate which can be removed locally when undersaturated solutions encounter carbonate rock and the resistance to alteration [dissolution or precipitation] which is shown by carbonate in association with saturated solutions". This conclusion is all the more remarkable if we reflect on the enormous quantities of cement which have been precipitated in limestones, cement for which we are bound to assume a largely extraformational source (p.439). Moreover, this cement was, in at least some of the Pleistocene and Tertiary limestones, implanted within only a few tens of thousands of years (about 80,000 years in the Pleistocene of Israel, p.345).

Mixing of saturated ground waters having different P_{CO_2}. A reaction which may well lead to substantial precipitation of cement is the mixing of two ground waters, each saturated for $CaCO_3$ but having different values of P_{CO_2} (BÖGLI, 1963; ARNTSON, 1964; implicit in DEFFEYES, 1965, p.419; THRAILKILL, 1968, p.32). The maximum undersaturation for $CaCO_3$ per unit volume of mixture will occur when about equal volumes of the two waters are mixed. RUNNELLS (1969) also noted that, depending on the shape of the solubility curve of a mineral as a function of added salts, mixing can cause either supersaturation or undersaturation.

WEYL (1964), in referring to physical models of the kind examined here, has commented that the work of the physicist "more often than not leads to predictions which are clearly contradicted by the geological evidence". Yet the attempt must be made, the discipline is essential, and with greater sophistication the verdicts of the experimentalist and the historian should converge. The experiments of Weyl, and Sippel and Glover, have not only stimulated interest and clarified our ideas in a field which has been too long neglected, but they have shown some of the important constraints within which future thinking must be conducted. We must conclude, it seems, that a saturated solution of $CaCO_3$, meandering sluggishly through the pore system of a monomineralic sediment, adjusting its concentration to vertical changes in pressure and geothermal gradient, is not capable of being the agent of large scale cementation such as would go to completion in no more than a few tens of thousands of years (e.g., Eniwetok, Guam, Funafuti, Israel and various hardgrounds). A model must instead be found which allows the operation of two regimes of ion transfer: (1) local dissolution and precipitation *within a formation* without recourse to an outside supply of $CaCO_3$, and (2) the transport of large quantities of $CaCO_3$ *between formations* and the precipitation of this material as a cement.

Pore systems with more than one solid phase: diffusion and fluid flow

Both the requirements of the preceding paragraph are satisfied by a sediment that contains grains of more than one solid carbonate phase, so that only one phase can be the least soluble and therefore the stable phase. Recent carbonate marine sediments consist mostly of aragonite, low-magnesian calcite and a number of different high-magnesian calcites. For the sake of simplicity we shall examine here an ideal system containing two solid phases, aragonite and low-magnesian calcite, but the theoretical treatment applies also to sediments with more than two solid phases. Two extreme systems will be studied in turn, the first having a stationary pore fluid, the second a moving pore fluid. We shall assume that the pore fluid in both cases contains CO_2 and thus HCO_3^-, but no Mg^{2+}, or other inhibiting ions, so that $CaCO_3$ can be dissolved and precipitated in a reasonably short geological time. We shall bear in mind that aragonite is more soluble than pure calcite. Finally, we shall assume constant pressure and temperature.

Stationary pore solution: transport by diffusion. In a rock composed of two $CaCO_3$ minerals of different solubilities the solution is not saturated. Instead it *tends toward* saturation for the more soluble polymorph, in this case aragonite. It is, therefore, supersaturated for the less soluble polymorph, calcite, and, as a result, new calcite is continually precipitated. Concurrently, a concentration gradient persists between growing calcite and dissolving aragonite: Ca^{2+} and HCO_3^- ions diffuse (GARRELS et al., 1949; GARRELS and DREYER, 1952) down gradient from aragonite to calcite. In this manner, without external intervention, a primary aragonite–calcite sediment can be changed, by a process of dissolution-precipitation (autolithification), to a rock composed of primary calcite and calcite cement. The natural system is, of course, more complex, but the principle is the same.

The rate of the combined dissolution–precipitation process, and the mean concentration of the concentration gradient, depend on the separate rates of dissolution and precipitation. If the rate of precipitation of calcite greatly exceeds the rate of dissolution of aragonite, the mean concentration will be nearer to the solubility product for calcite and the rate controlling process will be the dissolution of aragonite: and *vice versa*. It is important, therefore, to know whether these two rates are strongly discrepant and, if they are, which is the faster. It is also important to know to what extent, if at all, the rates of dissolution of the various solid phases depend on their solubilities. Although some such positive correlation is common among substances, there are many exceptions. SCHMALZ (1967), in an interesting study of the relevant kinetics, has described an experiment designed to reveal the velocities of dissolution and precipitation for a variety of carbonate solid phases (aragonite, reagent calcite and several high-magnesian calcites). The different carbonates were placed first separately, and then in pairs or in groups, in both distilled water and in sea water, under a controlled atmosphere of CO_2. The course of the reaction was followed by continuous recording of pH (eq.3, p.232) until a steady state was achieved. Schmalz's results for sea water are shown in Table XIX. Those for distilled water are similar. The alkalinity (p.234) of the steady state increases in the order of increasing solubility: calcite–aragonite–magnesian calcites (with decreasing Ca/Mg).

TABLE XIX

Solid phase	Symbol	Source	Steady state pH at P_{CO_2} of	
			$10^{+0.1}$ atm	$10^{-3.5}$ atm
Calcite	C	Reagent	5.68	8.30
Aragonite	A	Cymopolia	5.97	8.35
Magnesian calcites				
Ca/Mg = 39	M_1	Pinna	5.90	8.32
Ca/Mg = 5.5	M_2	Amphiroa	6.11	8.39
Mixtures				
C+A+M_1			5.98	8.34
C+A+M_1+M_2			6.10	8.38

Schmalz observed two important results: (*1*) the more soluble the phase the more rapidly it reached equilibrium, always dissolving at a rate greater than the precipitation rate for any less soluble phase, and (*2*) for a mixture of phases the steady state pH was matched closely with the most soluble phase in the mixture. The implication from these results is that, for carbonate mixtures, the rate of dissolution of the more soluble phase greatly exceeds the rate of precipitation of the less soluble phase. It is, therefore, the slow precipitation of the calcite cement which limits the velocity of the diagenetic process of dissolution-precipitation—a conclusion supported by the laboratory experiments of BISCHOFF (1969) on aragonite–calcite transformation.

The rate of precipitation of calcite will depend also on the availability of calcite nuclei in the sediment. MATTHEWS (1968) concluded that the slow initial nucleation of calcite controls the rate of diagenetic change. In the Pleistocene limestones of the subaerial terraces on Barbados, corals tend to remain aragonitic although the enclosing sediments have changed to low-magnesian calcite. This selective diagenesis Matthews attributed to precipitation of calcite on sedimentary calcite nuclei present in the sediment but not in the coral.

The progress of the diagenetic change from an aragonite–calcite sediment to a calcite rock will not be steady: it will be influenced by the quantity of aragonite remaining and by catalysis by dissolved ions in the water (p.242). A time will come when the amount of calcite in the rock is high and the residual aragonite is left only in a few, widely scattered places. Diffusion paths will thus be longer and the diagenetic exchange will be slower. If the calcite has been precipitated as a cement, the reduced pore size will allow narrower diffusion paths and so, again, the rate of exchange will be slowed (Fig.219, p.240).

The evolution of the rock fabric during the action of our simple system is of interest. If the process consists solely of calcitization of aragonite particles *in situ* and the accompanying precipitation of the 8% excess volume of $CaCO_3$ as calcite cement, then there seems to be no serious problem. Similarly, if aragonite is dissolved from within micrite envelopes and precipitated inside the same envelopes as a cement, with surplus calcite as an interparticle cement, the matter seems straightforward. Some collapse of micrite envelopes is to be expected and is, in fact, quite usual in some limestones (p.333). Suppose, however, that the aragonite is neither calcitized *in situ* nor enclosed in a micrite envelope. What happens to the rock fabric as the aragonite particles dissolve? There must be, one might suppose, a significant amount of collapse. In fact, further consideration suggests that, even in a sediment with abundant aragonite, collapse would not be significant. At least in the fresh water environment the first visible consequence of dissolution-precipitation is the appearance of low-magnesian calcite at interparticular contacts, so that a rigid framework is quickly developed and collapse is prevented. Later dissolution merely leads to the production of large secondary vugs (Fig.250, p.323; MATTHEWS, 1968).

Finally, there seems no doubt that, in this aragonite–calcite–CO_2–H_2O system, the primary porosity of 40–70% in the sediment cannot be filled with locally derived cement. Indeed, it is reasonable to conclude that the amount of *interparticle* cement which can be derived from the dissolution of aragonite and precipitation of calcite is approximately equal to the amount present in many limestones as a first generation calcite fringe (p.432). It seems certain, even on the basis of the preceding qualitative and very tentative discussion, that the final filling of the pores by second generation cement must depend largely on an outside source of $CaCO_3$.

The presence of high-magnesian calcite in real sediments does not invalidate our theoretical system, because the magnesian-calcite particles change metasomatically: they are not removed but are converted *in situ* to stable, low-magnesian calcite. Cementation is thereby encouraged by the production of additional substrates of low-magnesian calcite.

Moving pore solution: transport by diffusion and fluid flow. If we superimpose on the stationary system, discussed in the last section, a moving pore fluid, we find that the distance travelled by the ions, between the surface of the dissolving aragonite and the place of precipitation, is increased. This more realistic system is, again, similar to what seems to happen in the fresh water, eogenetic environment. Judgment of its applicability to submarine eogenetic or to mesogenetic conditions must await greater understanding of those conditions. The flow being laminar (discussion in THRAILKILL, 1968, p.25), the ions will move in two directions:

parallel to the flow because they are part of the fluid, and across the flow by diffusion. It would be possible for a pore solution, which is near to saturation for aragonite, to move out of the rock containing the aragonite into another rock. The relative slowness of the rate of precipitation of calcite (as found by Schmalz) would, presumably, allow solutions supersaturated for calcite to move long distances. The reluctance of calcite to precipitate may be enhanced by the tendency for successive atomic layers of overgrowth to be increasingly more soluble (p.252). Unfortunately Schmalz's measurements do not indicate how slow the rate of precipitation is or how it would compare with the rate of movement of groundwater. The geological indications are that the rate of precipitation must be slow compared with the fluid velocities: how else could large quantities of $CaCO_3$ be transported into a carbonate formation from outside, as we know has happened (p.439)? There is, therefore, no obvious difficulty in understanding how a pore solution, supersaturated for calcite, could move through a formation and allow cementation to take place at a distance from the initial source of solution saturated for aragonite.

Karst evolution and the supply of dissolved $CaCO_3$

In the interiors of a number of tropical islands, the calcite-cemented calcarenite has a highly irregular surface which indicates that the limestone has been extensively and deeply dissolved. DAVID and SWEET (1904) have described this low-lying corrosion surface from the interiors of islands in the Funafuti atoll. It seems that the Recent lime sands, presumably the usual mixture of aragonite, high-magnesian and low-magnesian calcites, have undergone two processes, one a cementation with calcite and the other large scale dissolution. A similar process has taken place in the oölitic facies of the Pleistocene Miami Limestone of Florida, where a karst topography is accompanied by calcite cementation, dissolution of the aragonite of the oöids, and the development not only of an oömouldic porosity but of vugs with diameters from a few millimetres to several centimetres (STANLEY, 1966; ROBINSON, 1967).

In all these rocks aragonite is present, and on the islands there must be skeletal high-magnesian calcite as well. DAVID and SWEET (1904, p.74) remark: "Obviously the calcium carbonate in solution must go somewhere." Presumably the water moves downward and outward into the ocean. There must be a downward movement of fresh water supersaturated for low-magnesian calcite from which, in time, calcite cement is precipitated. Here, again, we find limestones which, by dissolution of solid phases more soluble than low-magnesian calcite, act as sources of $CaCO_3$ for cementation in other limestones. This is just the kind of situation to be expected where an unconsolidated carbonate sediment emerges by uplift from the sea. As it rises, the fresh rain-water strikes its surface. Though this water, in its journey through the sediment pore-system, will become highly concentrated rapidly, according to Weyl (p.445), this very process involves an important amount of dissolution. Then, once the water is inside the sediment and moving slowly downward, the concentration gradients between particles of different solubilities will cause diffusion and further dissolution. Cementation will be slight, as the pore water is moving and the rate of precipitation is slow by comparison with the rate of dissolution, so there will be a net loss of $CaCO_3$. The surface of the rock will develop a karstic, corrosional topography, the inside of the rock, irregularly and lightly cemented, will evolve a secondary porosity. Perhaps somewhere down below, another carbonate sediment will benefit by the delivery of allochthonous $CaCO_3$, as its second generation of cement.

Yet truth must be stranger than fiction and the ideal system described, with several solid phases of $CaCO_3$ in moving CO_2-bearing water, though relatively complex, must be simpler than real systems. Particularly in the subaerial, meteoric, vadose environment, there is additional variability caused by the delivery of new water from sporadic rainfall, the release of CO_2 and organic acids by bacterial decay, and cooling brought about by changes in the weather: all these are factors tending to increase the dissolution of carbonates (discussion in DUNHAM, 1969a, p.163). It is known that the P_{CO_2} in the soil can be ten times greater that in the pores of the underlying calcarenite and the implied loss of CO_2 in the vadose zone must surely be accompanied by precipitation. Certainly there must be important differences between conditions in the phreatic zone (which have been the basis of the thinking in this chapter) and those in the overlying vadose zone, where the pores are only partly filled with water and there is an interface between the water and the pore atmosphere saturated with water vapour. A major problem is to find how CO_2 is lost from the vadose zone: it cannot escape upward, through the soil with its high P_{CO_2}.

Donor and receptor limestones

The cementation of a limestone with $CaCO_3$ derived from the dissolution of another limestone, as suggested in the previous section, raises the concept of donor and receptor limestones. From the donor limestone $CaCO_3$ is leached to yield cement for the receptor limestone. The rate of donation (loss) of $CaCO_3$ will depend on the proportion of particles of different solubilities and the rate of flow of groundwater. (LUCIA and MURRAY (1956) refer to selective leaching of micrite from a calcarenite.) It must be supposed that the dissolution of the donor limestone is either total, in which case it is lost for ever, or partial so that, at some later time, the donor can turn receptor and its pores can be filled with $CaCO_3$ leached from a new donor. Generally, one would guess, a receptor limestone, awaiting its second generation of cement, will have passed Land's stage IV (p.327), its only remaining carbonate will be low-magnesian calcite, and, therefore, this rock will never again yield a pore solution near-saturated for one of the more soluble carbonates and supersaturated for low-magnesian calcite cement. Monomineralic limestones cannot be donors.

This pretty picture of donor–receptor diagenetic development is, it must be admitted, largely hypothetical. Adequate supporting chemical and petrographic studies have not yet been made. Be that as it may, as a working hypothesis the idea seems useful; it appears to fit the known facts and forms a useful basis for further investigation. How frequently this type of donor–receptor process acted in the past is not clear. Possibly the unconformities in Eniwetok and Guam (p.322) represent such an evolution: from the cores alone it is difficult to tell. On the other hand, the stratigraphic record is not renowned for a high frequency of fossil karst surfaces. Indeed, the familiar well-bedded limestone sequence is, at first sight, quite unlike the product one would expect from the subaerial process here described.

SUBMARINE CEMENTATION

The question of submarine cementation has already been touched upon in Chapter 9, in connection with the new discoveries of cements in reefs and open sea floors, the growth of oöids, the hardening of faecal pellets, cement filling of the chambers of foraminiferids and other skeletons, also of algal bores, and the cementation of grapestone. In the same chapter the origins of hardgrounds in the Chalk and other carbonate successions were investigated, and some attention was paid to the recently accumulated data on cementation of *Globigerina*-coccolith oozes on the present deep sea floor. It is now pertinent to look again at these manifestations of submarine cementation in the light of the discussion in this chapter.

A striking peculiarity of all the submarine Recent and Tertiary cements is that they are not low-magnesian calcite: they are aragonite or high-magnesian calcites. The intraparticle cement in the faecal pellets is aragonite in the Bahamas, and the same mineral occurs in skeletal body chambers, in oöids and grapestone. Those crusts in the *Globigerina*-coccolith oozes that have been analyzed are composed of a micritic matrix of high-magnesian calcite which has partly replaced skeletons of aragonite and low-magnesian calcite. The aragonite cement in the oozes of the Red Sea, although possibly a special case in a specialized environment, conforms to the same general picture. The cements in the Bermudan and Jamaican reefs and the Persian Gulf hardgrounds are aragonite and high-magnesian calcite.

A second peculiarity of the submarine cements is the seemingly allochthonous source of the $CaCO_3$. The faecal pellets, when hard, appear smooth in outline and show no signs of dissolution that could have released $CaCO_3$. The cement filling of foraminiferal chambers and algal bores is obviously derived from a source outside the chambers and bores. Similarly, in grapestone, the cement is not accompanied by signs of dissolution of the grains. The signs of grain dissolution so characteristic of the meteoric, eogenetic environment are absent. The oozes cemented with high-magnesian calcite, though having a micritic matrix of uncertain origin, reveal chambers of *Globigerina* and other foraminiferids filled with what appears to be high-magnesian calcite. The unaltered ooze is dominated by the low-magnesian calcite of coccoliths with scattered tests of the Globigerinacea. This small quantity of aragonitic and high-magnesian calcitic skeletal debris seems hardly an adequate source, supposing it were dissolved, for such a great quantity of high-magnesian calcite cement. Again, in the Persian Gulf hardgrounds, the petrographic data indicate that the main cementation has not depended on a supply of locally dissolved carbonate.

All these factors point to the overlying sea water as the source of dissolved $CaCO_3$ for the cement. That cementation was submarine is not in doubt (Chapter 9). The absence of dissolution phenomena and the ready availability of fresh supplies of supersaturated sea water make the inference that the process involves precipi-tation from sea water inescapable—except for the matter of deep water cementation. It is not clear why precipitation of cement should be possible from water that is undersaturated for $CaCO_3$ (p.268). In concluding that sea water is the source of $CaCO_3$, we also presume the working of some kind of "pump" whereby a flow of sea water through the pores of the sediment is maintained. This flow may be con-trolled by such activities as tidal and wind-wave currents giving rise to bottom turbulence over a hydraulically rough bottom, aided by burrowing of benthonic organisms. From information on fabrics so far available, it is plain that the fabrics of submarine cements differ in many respects from most of the fabrics described earlier in this chapter (p.417). In particular, they are much more finely crystalline than the fabrics that have been identified as cements in ancient limestones. It may well be that we have been too narrow in outlook in our recognition of ancient cements and that many calcilutitic fabrics (micrites and microspars) are, in fact, submarine cements.

Perhaps the most plausible hypothesis of abyssal submarine cementation of crusts to have been offered so far is that by CIFELLI et al. (1966). They considered the effects on a carbonate ooze, near to equilibrium with its pore water, of an uplift of about 2,000 m involving a big reduction in pressure and a rise in tem-perature. Such a change could, they estimated, result in the precipitation, as cement, of as much as 200 mg of calcite per litre of sediment. Since a litre of sediment with a porosity of 50% would require, to fill its pores, about 13,500 g, then 200 mg is about $1 \times 10^{-3}\%$ of this. The process, on its present showing, would not yield anything like the kind of crust described by Fischer and Garrison (p.375).

The importance of submarine cementation in the past

The recent discovery of a vertical succession of hardgrounds in the floor of the Persian Gulf (p.371) compels us to re-examine the possibility that submarine cementation could have been responsible in the past for the lithification of great thicknesses of limestone. What influence did eogenetic marine cementation have on the formation of the Carboniferous and Jurassic limestones of Britain, the vast masses of Mesozoic and Tertiary carbonates in the Middle East and the Mississippian and Pennsylvanian of North America? Certainly a process of more or less continuous cementation of sediment, as it accumulates, has considerable attractions. The lack of compaction in most limestones is accounted for by the early development of a rigid framework, and the overlying sea water would have provided an inexhaustible supply of dissolved $CaCO_3$. Indeed, many geologists have assumed intuitively that the prime source of cement has always been sea water.

Cementation of a hardground need not have been complete. Once a framework had been constructed, further cementation leading to a reduced porosity of less than 5% (usual in limestones) may have been delayed until the limestone had been moved into a subaerial eogenetic or deep crustal, mesogenetic, environment. It will be argued that the cement fabrics described in this chapter, which are common in limestones, are known to characterize the fresh water, subaerial diagenetic environment (Chapter 8): this cannot be denied. On the other hand, can we be sure that all calcite cements in ancient limestones are primary? Could some of them, perhaps, have been precipitated from fresh water in *secondary* voids—exhumed primary voids—resulting from dissolution of primary aragonite cement? Certainly aragonite cements would have been dissolved in fresh water, although high-magnesian calcite cements might well have survived as space fillers, undergoing only metasomatic change to micritic low-magnesian calcite. We cannot yet say how much, if any, of the micrite matrix in packstones (grain-supported biomicrites) is Mg^{2+}-depleted submarine cement, similar in origin to that in the Persian Gulf, Bermuda and elsewhere. The fact that these problems remain unsolved shows how far we have yet to go in the understanding of limestone diagenesis. On uniformitarian grounds it is probable that there are many hardgrounds among ancient limestones, but we have been able to recognize only a few of these. It was clear, during the Bermuda Conference on Cementation in 1969, that it was not possible, at that time, to identify submarine cements, or fresh water cements, on the basis of their contained microfabrics.

EVIDENCE FOR NEAR-SURFACE VERSUS DEEP CRUSTAL CEMENTATION:
A PRAGMATIC APPROACH

Though the occurrence of cement in limestones of all ages is a commonplace, the determination of its place of origin, either eogenetic or mesogenetic, raises problems which have scarcely been examined. How can a distinction be drawn between a calcite cement precipitated within a few tens of metres of the surface and a cement formed several thousands of metres below the surface?

There seems no obvious reason why a near-surface calcite cement should differ in appearance from one of deep origin, purely as a result of different pressures and temperatures during growth. Nor is it yet entirely safe to assume that the first generation cement is a near-surface, fresh water product, as in Land's stage IV. My own experience is that it was normally precipitated prior to compactive fracture of grains, but this could happen in either the submarine or the subaerial eogenetic environment. On the other hand, a limestone having only a late stage, coarse sparry cement, without the fine crystals of an early cement, could perhaps be assumed to have lithified without emergence above sea level. There is need here for criteria by which a distinction could be made between two types of *second generation* cement, the eogenetic and the mesogenetic—in so far as a sharp distinction is possible where a gradation must exist. In a paper published in 1966, I attempted to produce some criteria, basing my argument on the probable effects of compaction while cementation was going on.

An experiment: data from micrite envelopes and grain contacts

In a small, tentative experiment, forty-four thin sections of biosparites and oösparites, of Carboniferous to Tertiary age, were examined, all containing micrite envelopes filled with calcite cement. In each section observations were made to discover: (*a*) if the micrite envelopes were entire or had been broken as a result of compaction while they were empty (p.333), and (*b*) if grains were in contact at points or showed signs of pressure-solution (solution-welding). The results are summarized in Table XX. The interesting thing about them is the remarkably sharp division into two types of fabric association. In one type, undamaged micrite envelopes accompany grains in puntal contact. In the other type, crushed envelopes are found with pressure-welded grains. Care must be exercised in drawing sweeping conclusions from such a small sample of limestones, but as a working hypothesis it could be supposed that many molluscan biosparites fall into two classes: those that underwent dissolution of aragonite and precipitation of calcite under a light

TABLE XX

SUMMARY OF DATA FROM 44 LIMESTONES ON RELATION BETWEEN OCCURRENCE OF WHOLE AND BROKEN
MICRITE ENVELOPES, SEPARATE AND PRESSURE-WELDED GRAINS
(From BATHURST, 1966)

	No grains pressure-welded		Some grains pressure-welded	
No envelopes broken	case 1	25	case 2	1
Some envelopes broken	case 3	1	case 4	17

overburden, and those wherein the process of dissolution-precipitation took place under an overburden great enough to break micrite envelopes and induce pressure-welding.

It is necessary to stress that there is no question of the skeletal aragonite inside the micrite envelopes having been calcitized *in situ*. In the rocks examined the calcite spar is typically cement, there are no relic, shell-wall fabrics (p.487) and, of course, where envelopes have collapsed their emptiness at some period is confirmed. This reconnaissance study was deliberately restricted to biosparites containing cement-filled micrite envelopes, because these indicate the action of the dissolution-precipitation process. It was essential to be sure that the envelopes really had been emptied so that they were susceptible to crushing should the overburden have been sufficient.

SUMMARY: THE STATE OF THINKING ON CEMENTATION

Owing to the high primary porosities of most carbonate sediments (40–70%) and their normal lack of compaction, the reduction of porosity to less than 5% by cementation requires early cementation, very large sources of $CaCO_3$ and a highly efficient means of transporting the $CaCO_3$ and precipitating it in the pores.

It is necessary, first of all, to be able to recognize cement and it is apparent that certain fabric criteria are useful. As a result of competitive growth the cement crystals have plane interfaces and increase in size away from the host substrate. Preferred orientations may be exhibited. These related criteria are not entirely unequivocal, since they do appear uncommonly in neomorphic spar, but taken together, with others listed, they generally lead to a firm identification. Among the fabric criteria, the presence of two crystal generations and the percentage of enfacial junctions among triple junctions seem to be unequivocal in distinguishing cement from neomorphic spar.

During the growth of cement, it is plain that the mineralogy and fabric of the host substrate influence not only the mineralogy and fabric of the cement, but the rate at which different types of cement will occlude the porosity. This is the theme of Pray's competitive cementation. The nucleation of cement crystals must involve both syntaxial overgrowth on the host substrate and the growth of new crystals with unique orientations. The relative importance of these two processes in various circumstances, and the mechanism of nucleation, are not well understood. Related to this problem is the recognition of a time break in the cementation of some limestones, separating the precipitation of first and second generation cements, formed before and after compactive fracture. Changes in the chemistry of the parent solutions can be traced in detail by staining cements for Fe^{2+} content, by luminescence petrography or by electron microprobe.

Concerning the supply of $CaCO_3$ to the pore system and its precipitation, in the phreatic zone or deeper, it seems that the critical requirements are a mixture of $CaCO_3$ solid phases of different solubilities, and fluid flow. Only in this way can there be an adequate transfer of $CaCO_3$ from donor to receptor limestone. Diffusion alone cannot perform the task of transportation. A purely monomineralic rock is too stable to permit dissolution-precipitation to proceed in its pore system on an appropriate scale. Such processes as local dissolution of aragonite and precipitation of calcite cement (autolithification) and pressure-solution, are helpless to fill the porosity of these carbonate sediments on the scale needed. With regard to cementation at the sea water–sediment interface, the mineralogy of the cements and the absence of dissolution phenomena, in addition to a number of factors discussed in Chapter 9, all point to precipitation from a continually renewed pore-filling of sea water. Finally, the results of a small pilot experiment with fabrics suggest that it is possible to distinguish between two groups of limestones, those that were cemented to rigid frameworks before compactive fracture of their grains, and those that were made rigid by cement after compactive fracture.

FURTHER READING

Basic works on the subject of cementation are those by BATHURST (1958, 1959b, 1964b), EVAMY and SHEARMAN (1965, 1969), FOLK (1965), PURSER (1969), SHINN (1969) and TAYLOR and ILLING (1969). Any consideration of the physical chemistry of dissolution and precipitation necessarily relies on the lucid studies by WEYL (1958, 1967), SIPPEL and GLOVER (1964) and SCHMALZ (1967). There are fine introductions to the subject by R. C. MURRAY (1960), WEYL (1964), ILLING et al. (1967) and LUCIA and MURRAY (1967). The possibility of a deep-seated origin for the $CaCO_3$ required for the second generation of calcite cement is examined by OLDERSHAW and SCOFFIN (1967). The book by FISCHER et al. (1967) contains many electron photomicrographs of great interest. THRAILKILL (1968) has provided a valuable background for consideration of subsurface water flow and chemistry in limestones. After this chapter was written, I read the stimulating work of MATTHEWS (1968) and was heartened to find that he had reached rather similar conclusions. Reading on submarine cementation is given in Chapter 9.

In September 1969 a group of 50 specialists attended a seminar on carbonate cementation in the Bermuda Islands. The thoughts expressed in the seminar are summarized in 60–70 short papers. This publication, more than any other, describes the status of cementation studies in 1969 (BRICKER, in press). Other useful references will be found at the ends of Chapters 6, 8 and 9. Finally, although it may seem strange to recommend a paper on holes for basic reading on cementation, the work by CHOQUETTE and PRAY (1970) on porosity in sedimentary carbonates is a deeply thoughtful and widely based introduction to thinking in this general area. Other references are given in the text.

Chapter 11

PRESSURE-SOLUTION

The role of pressure-solution in the post-depositional alteration of carbonate sediments has been remarkably little appreciated outside a small circle of oil geologists. Yet the growth of stylolites has caused changes in thickness comparable with the results of erosion: reductions of thickness in a vertical direction amounting to 20–35% are commonplace. In regions where laterally directed pressure has been active, the lateral shortening of beds by the formation of vertical stylolites has, in places, greatly exceeded the crustal shortening achieved by folding or other tectonic adjustment (DUNNINGTON, 1967). Stylolites act as permeability barriers which influence the distribution of aquifers and oil reservoirs, and the $CaCO_3$ released by stylolite growth is a major source of late diagenetic cement. Vertical thinning must have been accompanied by large scale migrations of pore fluids on the scale of whole tectonic basins (TRURNIT, 1968a).

THE GENERAL THEORY OF PRESSURE-SOLUTION

The theories of Thomson and Riecke

In 1862 Professor J. THOMSON of Queen's College, Belfast, presented a communication to the Royal Society: "On Crystallization and Liquefaction, as influenced by Stresses tending to change of form in Crystals." His simple, lucid approach merits attention, because differential dissolution of grains under load is a major process in diagenesis. He was concerned with the application of mechanical work to crystals in the form of stress, giving to them a potential energy. He began by considering ice immersed in ice-cold water, carefully distinguishing between pressure communicated to the ice through the water (hydrostatic pressure) and linear pressure in one or more directions applied to the ice directly. He formed the opinion that "any stresses whatever, tending to change the form of a piece of ice in ice-cold water . . ., must impart to the ice a tendency to melt away, and to give out its cold, which will tend to generate, from the surrounding water, an equivalent quantity of ice free from the applied stresses". He added the more general inference that "stresses tending to change the form of any crystals in the saturated solutions from which they have been crystallized must give them a tendency to dissolve away, and to generate, in substitution for themselves, other crystals free from the applied stresses". Experiments with halite in saturated brine confirmed these opinions. Thomson found that, by applying a load to such a mixture in a cylinder, using a suitably weighted piston and allowing excess brine to escape, he could produce "a hard mass like rock-salt". Thomson's final conclusion was: "If any substance, or any system of substances, be in a condition in which it is free to change its state . . ., and if mechanical work be applied to it (or put into it) as potential energy, in such a way as that the occurrence of the change of state will make it lose (or enable it to lose) (or be accompanied by its losing) that mechanical work from the condition of potential energy, without receiving other potential energy as an equivalent; *then the substance or system will pass into the changed state.*"

More than thirty years later RIECKE (1894, 1895) offered an expanded mathematical treatment based on similar ideas, showing how the melting point of a crystal would be depressed uniformly throughout the lattice, if a uniform linear stress were applied to it. Riecke's theory has more recently been expressed in terms of the effect of linear pressure on the solubility of a stressed lattice immersed in its saturated solution. Thus Correns and Steinborn (in CORRENS, 1949b) developed the equation:

$$RT \ln \frac{c}{c_\infty} = VP \tag{22}$$

where c and c_∞ are, respectively, the concentrations of the saturated solution, and of the supersaturated solution in equilibrium with the strained lattice (i.e., supersaturated with regard to the unstrained lattice). V is the molar volume of the crystal and P the linear pressure.

Grains under load and the Boussinesq equation

The relevance of the Thomson–Riecke relation becomes apparent when we consider a detrital grain in a sediment under load. The grain is not stressed uniformly throughout, but has stresses concentrated about the points—or surfaces—of contact with adjacent grains. The distribution of stress in a body acted on by a point load can be illustrated by the very simple case of a point vertical stress P in a material assumed elastic and isotropic (Fig.316). The general equation describing the distribution of stress in a uniform body of unlimited extent acted on by a vertical point load was derived by BOUSSINESQ (1876, 1885):

$$p_z = \left(\frac{3P}{2\pi z^2}\right) \cos^5 \theta \tag{23}$$

p_z being the stress (newtons) at a point in the body defined by the coordinate z and the angle θ; z is the vertical coordinate under the point of application of a load P (in kilograms) and θ is the angle between the coordinate z and the coordinate

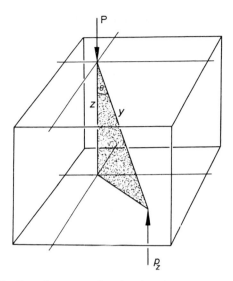

Fig.316. Illustration of the Boussinesq equation showing the relation of the applied vertical stress *P*, in an elastic, isotropic solid, to the stress *p* at a point in the body defined by the coordinate *z* and the angle *θ*. (After DALLAVALLE, 1943.)

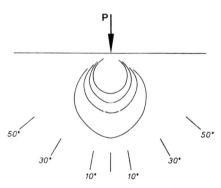

Fig.317. Two-dimensional distribution of stress in an elastic, isotropic material of unlimited extent, following the Boussinesq equation. Stress contours are at intervals of equal stress.

y joining the point in the body with the point at which stress is applied (Fig.316). From eq. 23 it can be seen that the stress p_z is not only inversely proportional to z^2, but is also proportional to $\cos^5\theta$ which decreases rapidly in value with increase of the angle *θ* (Fig.317).

GRAIN-TO-GRAIN PRESSURE-SOLUTION IN SEDIMENTS

Grain contacts, diffusion, precipitation

In natural conditions the variation of stress about a point load is much more complex than shown in Fig.317, because the body has finite size, is not perfectly elastic and is anisotropic. The applied stress, moreover, is likely to have an element of shear. Nevertheless, the principle of localization of maximum stress and of elastic strain about the contact between two grains is clear. The related increase of chemical potential, and thus of solubility, will be localized in a similar way. But localized strain in a crystal immersed in its saturated solution must lead to localized dissolution, and the process whereby grains undergo dissolution about their contacts is called **pressure-solution**.

The results of pressure-solution are most clearly seen in oölites, because the structure of the oöid is highly symmetrical (e.g., USDOWSKI, 1962; HENBEST, 1968): thus it is easy to judge in thin section where the original margin of the oöid lay and how much has been lost by pressure-solution (Fig.318, 319). During the action of pressure-solution, the point contacts between grains change to surface contacts, and it is the existence of these surface contacts that makes the resultant fabric relatively easy to recognize. The form of the surface contact varies according to the different solubilities of the material on either side of the surface. Less soluble grains

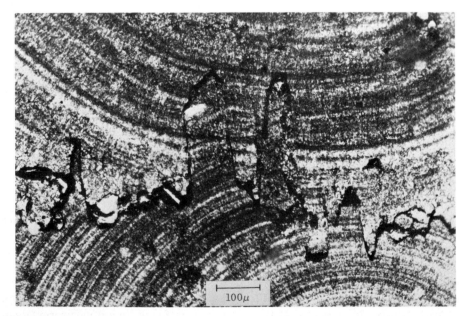

Fig.318. Pressure-solution contact between two oöids, showing concentration of dark or opaque impurities along the stylolitic contact. Amplitude of stylolite gives minimum vertical loss of thickness. Slice. Unteren Buntsandstein. Harz, Germany. (From USDOWSKI, 1962.)

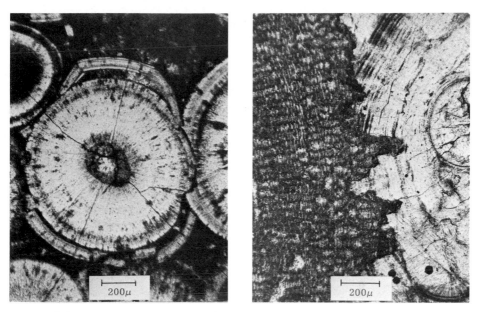

Fig.319. Oöids embedded in micrite showing outer layers exfoliated during compaction before micrite was lithified. Pressure-solution contact between oöids (right). Slice. Mississippian. Montana, U.S.A. (Slice lent by R. L. Folk.)

Fig. 320. Pressure-solution contact between two grains, a coralline alga (left) and a nummulitid. Slice. Miocene. Vitulano, Campania, Italy. (Slice lent by B. D'Argenio.)

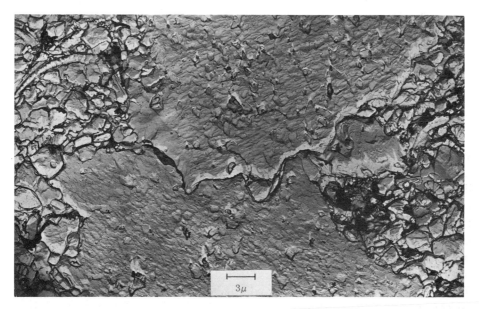

Fig.321. Pressure-solution contact between two calcite prisms of the bivalve *Inoceramus*. Replica. Electron mic. Co. Antrim, Northern Ireland. (From WOLFE, 1968.)

penetrate more soluble grains while suffering little change themselves (see section on stylolites, p.468). As the process continues the original centres of the grains approach each other (Fig.320, 321).

This process of pressure-solution can now be examined in more detail. It depends upon the decrease of strain in the grain surface away from the place of contact and on the related variation of the solubility product in the adjacent solution. It is convenient to consider first the ideal case of monomineralic iso-tropic grains. Owing to the sluggish movement of ground water the solution can be regarded as stationary. Where the grains are not under load, and are therefore unstrained, the water will be uniformly saturated. When a load is applied the situation changes. The solubility product constant, $[Ca^{2+}]\cdot[CO_3^{2-}]$, is now higher near to the grain-to-grain contact and, as a result, the ion activity product in the adjacent solution will likewise be higher there than elsewhere. A concentration gradient will exist and solute ions will diffuse away from the vicinity of the contact into the less concentrated solution occupying the adjacent voids. Yet, near the unstrained surfaces, the solubility product is lower, and thus the influx of diffusing ions will lead to local supersaturation and consequent precipitation of $CaCO_3$ on these unstrained surfaces. As THOMSON (1862) wrote, the crystals will tend "to dissolve away, and to generate, in substitution for themselves, other crystals free from the applied stresses".

In real conditions reprecipitation will take place on the surfaces on which heterogeneous nucleation is easiest wherever these may be. Its location will depend on the distribution among the grains of such factors as mineralogy and fabric anisotropy. The distance that solute ions will travel depends not only on the com-plex pattern of concentration gradients but on the movement of the solution itself. Diffusion is a slow process and bulk transport of solvent and solute together by fluid flow could be important. MAXWELL (1960) has demonstrated this with ex-periments on quartz cementation. The combined action of pressure-solution followed by precipitation is conveniently called **solution transfer** from the German *Lösungsumsatz*.

In an unconsolidated carbonate sediment under load, the distribution of strain at the surfaces of grains near their points of contact will be a function of grain orientation, size and shape, anisotropy of the crystal lattice and (in polycrystal-line grains) the fabric of the crystal mosaic. Despite this variation, the application of a dominantly vertical stress (overburden) causes preferential dissolution at those point contacts where the tangents to the contact surfaces lie near the horizon-tal. Preferential dissolution of upper and lower surfaces of grains may or may not be accompanied by preferential precipitation on their lateral surfaces. In either case the result is a preferred orientation of longer shape axes in the horizontal plane. There is no development of a preferred orientation of crystal lattices as the grains do not rotate and plastic deformation is not a part of the process. COOGAN (1969, 1970) has made a very interesting attempt to use the packing density in oölites, related to pressure-solution, as a rough time scale against which other diagenetic phenomena can be compared.

The effect of cement on pressure-solution

Grain-to-grain pressure-solution must be limited to conditions where the directional pressure transmitted from grain to grain is greater than the hydrostatic pore pressure of the solution. It cannot take place after the precipitation of the *second* generation of cement, the final pore filling, simply because the presence of this embracing material effectively prevents relative movement between grains. If pressure-solution acts after the delivery of the second generation of cement, then it does so by the formation of stylolites (p.468). On the other hand, grain-to-grain pressure-solution can act after precipitation of the *first* generation of cement because a thin fringe of crystals does not prevent grain-to-grain movement. In this event the cement fringe is itself involved and is locally dissolved. In Fig.322 and 323 the cement fringe was deposited on the oöids before pressure-solution began. As the load on the grains was increased the stress was applied not only to the oöids, but to their cement fringes. Owing to the existence of these fringes many oöids, though they approached one another during pressure-solution, never met because the cement fringe had not been entirely dissolved. Thus the oöids remain separated

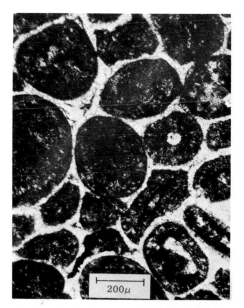
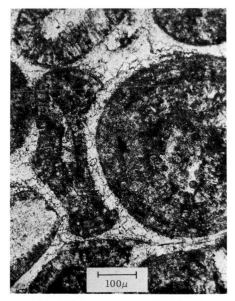

Fig.322. Oöids, fringed with early calcite cement, in pressure-solution contact. Slice. Carboniferous Limestone. Republic of Ireland.

Fig.323. As Fig.322. The right-hand oöid, with its cement fringe, has replaced a large part of the oöid to its left.

from each other in a situation wherein, at first sight, one would expect them to be in contact along surfaces. The somewhat polygonal outlines of the oöids is normal because, in many places, oöids have been penetrated so that their contacts are roughly plane. If two isotropic, isomineralic spheres are in contact under load, then, as pressure-solution progresses, the contacts become plane surfaces and simply enlarge, remaining plane throughout their development and the spheres are replaced by polyhedra. An unusually clear photomicrograph of this fabric is to be seen in CAYEUX (1935, fig.54).

The maintenance of pressure-solution

In order that pressure-solution shall continue to act it is necessary that stress be continually transmitted from grain to grain across the surface contact. At the same time a solution film must be maintained between the opposed surfaces so that dissolved ions may be transported. Attempts to solve these apparently contradictory requirements have led to the formulation of two hypotheses, one appealing to an undercutting action, the other to the persistence of a continuous solution film.

   The undercutting hypothesis. In 1958 I suggested that the two opposed grain surfaces that make up the pressure-solution contact must be topographically irregular. If this were so the surfaces would meet at a small number of points (Fig.324a), leaving free large areas of opposed surfaces, which would confine a system of connected pores filled with solution. This hypothesis seemed necessary in order to ensure that a layer of solution persisted between the opposed surfaces. If a solid–solid interface were to develop, then dissolution and transfer of ions throughout the interface would cease.
   Where two grains meet at a point it is obvious that removal of the solid adjacent to this point contact must have two results: the grains will approach each other and the point contact will change to a surface contact. Once this happens, the solution is excluded from the opposed surfaces and transfer of ions stops. However, the strained lattices adjacent to the contact surface still touch the solution around the edge of the surface. Thus dissolution continues at the edge of the surface contact, the edge migrates centripetally as the area of the contact surface is reduced, until a point contact is again formed. This is the undercutting hypothesis referred

a   *b*

Fig.324. Two dimensional relation between two grains (stippled) in pressure-solution contact in water according to: (*a*) the undercutting hypothesis, and (*b*) the solution film hypothesis.

to by WEYL (1959b). The difficulty here is the design of a satisfactory model on an atomic scale. Is there enough elasticity in crystals to allow the many micro-point contacts along a grain-to-grain interface (Fig.324a) to collapse at different times? Is the collapse of a micro-point contact the result of mechanical breaking, or is it the direct result of the supersolubility and dissolution of tiny pinnacles? Can these two processes be usefully differentiated?

The solution film hypothesis. WEYL (1959b) discarded the undercutting hypothesis, while admitting that it is difficult to rule it out, and substituted the solution film hypothesis. Stimulated by an extremely interesting discussion by HENNIKER (1949) on the properties of thin films of liquid, he proposed that a continuous film of solution separates the two grains during pressure-solution (Fig.324b). He referred to the experiments of BECKER and DAY (1916) and TABER (1916) in which they demonstrated that a crystal growing in a solution can lift a mass of inert material laid on top of it. The action of this **force of crystallization** (SPRY, 1969, p.149), Weyl argued, implies the maintenance of a film of solution between the crystal and its constraint. Moreover, this film must be able to support a shear stress and so cannot act mechanically as a liquid: at the same time diffusion of ions through it must be possible. The force of crystallization is responsible for the familiar action of the frost wedge, and probably is the cause of the lifting of the surface of the sabkha at Abu Dhabi (p.206) and the formation of thrust planes in hardgrounds (p.372). Weyl supposed that, if a solution film can be maintained between a growing crystal and its constraint, then a similar film must exist between a dissolving crystal and its constraint. It is tempting to retort that the experimental conditions of BECKER and DAY (1916) and TABER (1916), involving the growth of alum crystals from their saturated solutions at room temperature and pressure, are very different from the situation of a calcarenite at a depth of 2,000 m. In fact, the pressure acting to squeeze and break the film is the difference between the rock pressure transmitted by the grains and the smaller pore pressure of the solution: such a difference need not be large.

From the hypothesis of an intergranular surface film, WEYL (1959b) went on to derive a general equation for the transport of solute by diffusion and to examine models for pressure-solution and force of crystallization, giving by far the fullest mathematical treatments available, to which the reader is referred. We may note here that Weyl's proposed mechanism "indicates that crystallization will stop and pressure solution will take over if the average effective normal stress across the solution film exceeds the ratio of the supersaturation to the stress coefficient of solubility. Therefore, it is not possible to place a simple limit on this force in terms of a specific amount of overburden above which the force of crystallization cannot take place" (WEYL, 1959b, p.2023).

An application of Weyl's models to real conditions is awaited. For the moment, since it is not clear under what geological conditions the solution film can be broken, with a consequent change to a situation requiring the action of the undercutting process, the relative importance of the undercutting and solution film mechanisms for pressure-solution must remain an open question.

STYLOLITES

Form and growth

More than a century ago KLÖDEN (1828) described a fossil, *Stylolites sulcatus*, which we now know as a pressure-solution surface and which we call a stylolite (Fig.325). The surface is a complex interface between two rock masses, each mass having a number of roughly columnar or finger-like extensions which fit in between the opposed columnar extensions of the opposite mass. In this interdigitate structure the long axes of the columns are parallel to each other but perpendicular to the trend of the surface. Stylolites differ from grain-to-grain sutured contacts only in scale. Stylolites transect the whole rock rather than isolated grains: in biomicrites or biosparites, for example, stylolites are continuous surfaces cutting across both grains and micrite or grains and cement or across lamination (Fig.326). An interesting survey of early ideas on the formation of stylolites is given by MARSH (1867). Causal processes other than pressure-solution have been suggested as, for

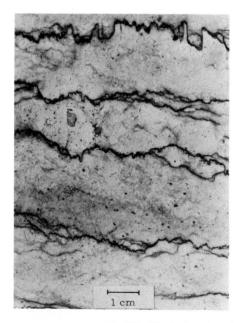 

Fig.325. Vertical section through stylolites in a limestone.

Fig.326. Multiple stylolites cutting laminae in a biosparite. Dark areas are insoluble residue. Slice.

example, by SHAUB (1939, 1949, 1953) who advocated a contraction-pressure hypothesis, and by PROKOPOVITCH (1952) who called for the action of dissolution in soft sediments. The logic of the argument in favour of pressure-solution has, however, gained wide acceptance, more especially after the careful analyses by WAGNER (1913), STOCKDALE (1922, 1926, 1936, 1943), DUNNINGTON (1954, 1967), W. W. BROWN (1959) and MANTEN (1966). One of the essential steps in under-standing the formative process is preparation of sections parallel to the axis of linear stress. Sections oblique to this axis give a highly misleading appearance and have doubtless been responsible for some of the claims favouring processes other than pressure-solution.

The initial surface at which dissolution begins is generally parallel to the bedding, though all orientations are known including surfaces perpendicular to the bedding (RIGBY, 1953). The controlling factor is the orientation of the axis of linear stress and this is generally vertical, being a simple consequence of overburden. Stylolites can in places be traced laterally into bedding planes. The zig-zag form of the surface is presumed to be a consequence of lateral variations, along the interface, of solubility differences of the rock across the interface. Where a more soluble part confronts a less soluble part across a solution film, then the more soluble part will dissolve and the less soluble part will move into the resultant space. The process is the same as grain-to-grain pressure solution, in that where the grains are equally soluble under stress they will both dissolve equally, but where they have different stress solubilities one will dissolve more rapidly than the other. The process is illustrated in Fig.327. Many pieces of skeletal carbonate are less

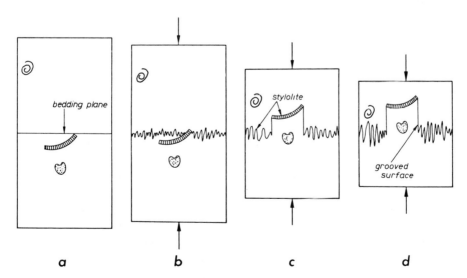

Fig.327. Diagram to illustrate the development of a stylolite from a bedding-plane, with pro-gressive approach of the gastropod to the brachiopod fragment and the crinoid columnal. The relatively insoluble brachiopod fragment is preserved at the advancing front of the lower limestone.

soluble under stress than a micritic matrix and so they often come to occupy the tips of the columns or fingers.

Theoretically the migration of ions away from the dissolving surfaces can proceed by two paths: either in the stylolite surface film in a fully cemented rock, or away from the surface, parallel to the linear stress axis, in a permeable rock. The stylolite surface is commonly a film of residual non-carbonate material rich in clay. DUNNINGTON (1967) has found that, where porosity and permeability are logged in boreholes through limestone sequences in the Middle East, the stylolites are situated, on the whole, at levels where porosity and permeability are low. HARMS and CHOQUETTE (1965) demonstrated the same relation using the sensitive GRAPE device for recording porosity in cores in the laboratory. GRAPE stands for Gamma-Ray Attenuation Porosity Evaluator (H. B. Evans, 1965). Away from stylolites there is a marked tendency for porosity and permeability to increase. This is an indication that dissolved $CaCO_3$ moved parallel to the stress axis and perpendicularly away from the general trend of the stylolite surface. Indeed, it may well be asked whether stylolites can evolve in a tight rock under non-metamorphic conditions: they are known in marbles. If the dissolved $CaCO_3$ had to escape by moving along the solution interface, where would it go? Stylolites are known to die out laterally. Moreover, the film of water between the opposed surfaces hardly seems a broad and generous highway for the transport of such large volumes of solute. It is important to note that, in the cores described by Dunnington, the vertical distance over which porosity and permeability have apparently been influenced by the growth of stylolites are measurable in centimetres. If diffusion alone cannot have accomplished a redistribution of $CaCO_3$ through such distances, then fluid flow through the rock must have been responsible, probably in a direction roughly parallel to the stylolite surfaces. Stylolites form impermeable barriers owing to their clay contents, and flow across them is inhibited. A paper of great interest in this respect is that by LERBEKMO and PLATT (1962), on a quartz sandstone constructed of alternate laminae of quartz and quartz-chert. The cherty laminae have undergone grain-to-grain pressure solution but the quartz laminae have been rim cemented. The authors concluded that dissolved silica moved from the pressure sutures in the cherty laminae to the open spaces of the pure quartz laminae, and was there precipitated as cement overgrowths. The distance travelled need only have been a matter of a few millimetres.

On the other hand, it is plain that stylolites must be post-cementation in origin, because they transect the intergranular micrite or cement in biomicrites or biosparites. It could be that stylolites begin in sediments which are only lightly-cemented and, by releasing $CaCO_3$, drive the cementation process to completion. Presumably the growth of a stylolite stops when the permeability in the adjacent sediment has fallen so low, as a result of cementation, that the transport of ions away from the solution film is practically inhibited.

The amplitude of a stylolite, the distance between finger tips on either side

of the structure, varies in general from less than a millimetre to several centimetres. The greatest amplitude I have seen, in the Carboniferous Limestone of
Yorkshire, was 1 m. The amplitude gives a *minimum* thickness of the material lost.
In the stylolites of large amplitude the sides of the columns, acting as piston-faults,
are grooved: the grooves are parallel to the axes of the columns.

Development of secondary clay seams (insoluble residues)

The growth of stylolites is obviously the cause of enormous losses of volume in a
carbonate succession and also a means of compaction. It is widely felt that the clay
seams in some limestones are not simply primary accumulations of detrital clay
but are, in part, stylolite surfaces along which clay (and other impurities) have been
concentrated as an insoluble residue. Clearly this is a difficult matter to determine:
it is not always easy to be sure about the origins of clay seams in limestones. In so
far as it is possible to identify purely stylolitic clay seams, the difference between the
ratio clay/$CaCO_3$ in the scam and in the rock yields an estimate of the amount of
$CaCO_3$ which has dissolved. It should be born in mind, however, that the dissolved
$CaCO_3$ was not necessarily lost to the system but may have been reprecipitated in
the pores on either side of the stylolite, and allowance must be made for this. It is
also apparent that the growth of stylolites not only results in a pronounced reduction of local porosity, but is a potent mechanism for driving water and oil out of
the rock (RAMSDEN, 1952; DUNNINGTON, 1954, p.46).

   BARRETT (1964) has described probable stylolitic clay seams in Oligocene
calcarenites of Waitomo-Te Anga, North Island, New Zealand. The seams are
roughly horizontal, from 2–8 cm apart, up to 5 mm thick, and are composed of
clay with quartz and glauconite. These non-carbonates are identical with non-
carbonates disseminated throughout the limestone. Some of the thinner seams are
markedly stylolitic in form: thicker seams are undulating. They all skirt around the
margins of the larger skeletal grains: some of these grains are truncated against the
clay seam: no grains cross the seam. Tension veins at right angles to the seams stop
at the thicker seams. The seams comprise 0.2–1% of the rock. By comparing the
clay content of the limestone with the thickness of seams Barrett estimated that,
even if all the seams were secondary, the released $CaCO_3$ could only account for
4–16% of the cement in the limestone. Barrett's records show that his limestones
fall naturally into two groups (Fig.328): (*1*) those with a low content of terrigenous
silt and clay but with many seams (crosses), and (*2*) those with a high content of
silt and clay but with no seams (dots). He concluded that the pure limestones must
have possessed initially high porosities and permeability and that the argillaceous
limestones had initially low porosities and permeabilities. These are interesting conclusions because it is commonly supposed that a high content of argillaceous material
predisposes limestones (or sandstones) to the development of stylolites. Barrett
emphasized the importance of another factor, permeability. In other words,

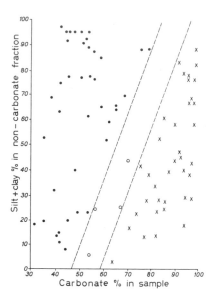

Fig.328. Percentage of silt and clay in the non-carbonate fractions of Oligocene calcarenites related to the percentage of carbonate. Terrigenous content high (dots), low (crosses). (After BARRETT, 1964.)

pressure-solution requires not only a water film but a route whereby solute ions can escape from the film into the adjacent rock.

A rather different situation was described by OLDERSHAW and SCOFFIN (1967, p.315) in the Silurian Wenlock Limestone of Shropshire. The limestones are interbedded with thin clay seams of undoubted primary occurrence: calculation from analyses also shows that, bearing in mind the clay content of the limestones and the thickness of the seams, these cannot be accounted for entirely on the basis of concentration of clay minerals by pressure-solution. Where the clay seams pass laterally into stylolites (which they commonly do) the growth of stylolites was obviously encouraged by the presence of a layer of clay-rich limestone. The authors also recorded that, where carbonate grains are coated with a primary clay film, the pressure-solution surface is highly sutured, but where the grains are clean, suturing is less pronounced. Quantitative estimation of the volume of $CaCO_3$ dissolved during stylolite growth is impossible because the clay content in the limestone is not homogeneous enough, but Oldershaw and Scoffin thought it probable, from visual estimation, that the supply of $CaCO_3$ released by pressure-solution was sufficient to cement completely the adjacent limestones.

Critical stress for pressure-solution

The question of the magnitude of directed stress which must be exceeded before pressure-solution can act has scarcely been examined. The problem is complicated by the influence of variables other than stress: it is obvious that the nature of the pore fluid is important (oil, gas or water), also the local porosity and permeability, surface areas of pore walls, rates of fluid flow, temperature and the concentration of clay in the rock and, potentially, in the stylolite. Dunnington records that, in most cases, overburdens of 600–900 m have been operative, although he has found stylolites in limestones which are *now* buried to only 300 m (DUNNINGTON, 1967, p.340, 344). In Guam, in the Alifan Limestone, there are microstylolites (14 were counted in a thickness of 10 mm) which show as thin red lines in reflected light owing to the presence of thin clay films. This limestone cannot have been buried under an overburden greater than about 90 m. SCHLANGER (1964, p.14) suggested that the special susceptibility of this biomicrite to pressure-solution may have been related to its relatively high clay content (compared with non-stylolitic limestones on neighbouring Eniwetok) and the strong lateral movement of ground water which was restricted in its movements by the underlying volcanic rocks. Pressure-welded grains in a Pleistocene raised beach at Easington, Durham, under only a few metres of boulder clay, probably suffered pressure-solution when covered by a thick overburden of ice (I. West, personal communication, 1970).

FURTHER READING

The obvious introduction to this field is the readable paper by DUNNINGTON (1967). In addition, the beautifully written paper by J. Thomson (1862) is delightful and good for essentials; then CORRENS (1949b) and the interesting theoretical work of WEYL (1959b). On stylolites, STOCKDALE's works (1922–1943) are highly relevant, as are papers by DUNNINGTON (1954), W. W. BROWN (1959), BARRETT (1964), RADWANSKI (1965), HENBEST (1968) and TRURNIT (1968a). PARK and SCHOT (1968a, b) have an interesting analysis of the relation of pressure-solution to the overall diagenetic history of limestones. Though they are works on sandstones, the papers by HEALD (1956) and A. THOMSON (1959) are helpful. The geometric relations of pressure-solution contacts in general are lucidly and thoroughly disentangled by TRURNIT (1967, 1968b, c). Other references are given in the text.

Chapter 12

NEOMORPHIC PROCESSES IN DIAGENESIS

> There is a mask of theory over the whole face of nature.
> *The Philosophy of the Inductive Sciences*
> W. WHEWELL (1840)

NEOMORPHIC PROCESSES

In 1965 Folk proposed the word "neomorphism" as a "comprehensive term of ignorance" to embrace "all transformations between one mineral and itself *or* a polymorph . . . whether the new crystals are larger or smaller or simply differ in shape from the previous ones. It does *not* include simple pore-space filling; older crystals must have gradually been consumed, and their place simultaneously occupied by new crystals of the same mineral or a polymorph." (FOLK, 1965, p.21.) The term, in fact, embraces two *in situ* processes: (*1*) polymorphic transformation, and (*2*) recrystallization. It is valuable because it can be used when it is certain that a carbonate fabric has been modified *in situ*, but it is not known whether polymorphic transformation or recrystallization or both were involved. The use of this term also allows the word "recrystallization" to retain its valuable restricted sense, referring to processes affecting a single, unchanging mineral species. It is necessary, at the same time, to realise how ample the conceptual portfolio of neomorphism is. It embodies, according to Folk, the processes listed below, though I have redescribed these in my own words. Only some of these processes are believed to have played a part in carbonate diagenesis. In particular, there is no evidence that dry polymorphic transformation has operated in the diagenetic, sub-metamorphic realm. The seemingly non-diagenetic processes are included in this chapter, because our understanding of diagenesis is still too rudimentary for us to discard them on decisive grounds, and their possible influence must be born in mind.

(*1*) Polymorphic transformations (Folk's "inversion").
  (*a*) Dry, *in situ*, transformation of aragonite to calcite (solid state inversion). Unknown in carbonate diagenesis.
  (*b*) Dry, *in situ*, transformation of calcite to aragonite (solid state inversion). Unknown in carbonate diagenesis.
  (*c*) Wet, *in situ*, transformation of aragonite to calcite. Common in carbonate diagenesis.
  (*a*) Wet, *in situ*, transformation of calcite to aragonite. Unknown in carbonate diagenesis.

(*2*) Recrystallization (mineralogy remains unchanged during the reaction: as used by Folk, but excluding the reaction high-magnesian → low-magnesian calcite).

(a) Primary recrystallization, dry, solid state (included in Folk's "degrading recrystallization"). Unknown in carbonate diagenesis: restricted to metamorphic environment.

(b) Grain growth, dry, solid state (included in Folk's "aggrading neomorphism"). Unknown in carbonate diagenesis: restricted to metamorphic environment.

(c) Wet recrystallization. Probably of importance in carbonate diagenesis only in the later stages of aggrading neomorphism (see below) where calcite crystals are growing at the expense of other calcite crystals.

(3) Aggrading neomorphism. The process whereby a mosaic of finely crystalline carbonate is replaced by a coarser (sparry) mosaic (FOLK, 1965). It is a complex process combining some of the *in situ* processes given above, namely polymorphic transformation and recrystallization. Despite Folk's definition which excludes passive dissolution and precipitation, these two processes are inevitably involved where a porous micrite is replaced by a neomorphic spar: the intention of the definition is not, however, seriously impaired by this slight complication. The products of these several processes cannot normally be distinguished and the inclusive term neomorphic spar is used here as another "comprehensive term of ignorance", to include all spar which has formed as an *in situ* replacement of a more finely crystalline crystal mosaic. It therefore includes FOLK's (1965) microspar and pseudospar. Though growth of neomorphic spar can, presumably, be dry or wet, the *assumption is made throughout this book that, in the diagenetic environment, it is always wet.*

These various processes must now be examined in more detail. The wet transformation which specially interests the student of carbonate diagenesis (as distinct from metamorphism) is that from aragonite to calcite, and it has already been dealt with (p.239, 347). The question of the dissolution of magnesian calcites has also been reviewed (p.335). There remain the matters of recrystallization and the growth of neomorphic spar.

RECRYSTALLIZATION

Recrystallization embraces any change in the fabric of a mineral or a monomineralic sediment. The mineral is the same after as before the reaction, according to the definition advocated by FOLK (1965, p.21) and SPRY (1969) and followed here. Three changes are possible, involving: (1) crystal volume, (2) crystal shape, and (3) crystal lattice orientation. This usage follows two of the most thoughtful of recent papers on the subject, by BANNER and WOOD (1964) and FOLK (1965). Dry, solid state processes are probably of little importance in limestone diagenesis, since the environment is inevitably a wet one. It is worthwhile drawing attention in this connection to an unfortunate use, as I believe, of the term "solid state" by FOLK

(1965). This term is used by physicists and chemists to indicate the crystalline state, a condition involving a regularity in the arrangement of the ions in a three-dimensional lattice. Liquids, and amorphous solids such as glass and pitch, are excluded from this category. It is, therefore, possibly misleading and confusing for geologists to use "solid state" as a synonym for "*in situ*". A wet reaction involving interstitial fluid (FOLK, 1965, p.24) cannot usefully be described as a solid state process without degrading the meaning of this term.

Primary recrystallization

This concept is taken from metallurgy, and was applied to limestone diagenesis by BATHURST (1958, p.24). Primary recrystallization is a dry process. It acts during the annealing of a deformed crystalline solid, that is to say in a crystal lattice which has been elastically strained while cold, either by bending or sintering. Nuclei of new unstrained crystals form in or near the interfaces between strained crystals or in glide planes. These new crystals grow until the old crystal mosaic has been replaced by a new, relatively strain-free mosaic with a nearly uniform crystal size. VOLL (1960, p.517) recorded patches of micritic calcite in crinoid columnals in a Dinantian

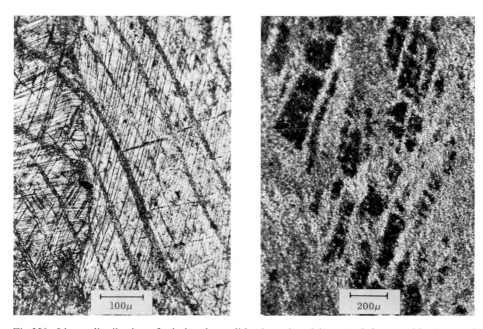

Fig.329. Linear distribution of micrite along glide planes in calcite crystals in a marble. Assumed primary recrystallization. Slice. Mt. Argastiria, Crete.

Fig.330. Linear distribution of micron-sized calcite crystals along glide planes in the calcite crystal of a crinoid columnal. Slice. Carboniferous Limestone. Co. Tipperary, Republic of Ireland. (From WARDLAW, 1962.)

limestone which he believed to be the product of primary recrystallization. In this connection it should be born in mind that echinodermal skeletons commonly contain some micrite, which has either filtered into the pores as a sediment, or has been precipitated in the pores as a fine-grained cement, or which evolved by algal micritization (p.383). It is probably only safe to conclude that the micrite in an echinodermal grain has grown by primary recrystallization if it is restricted to the proximity of intercrystalline boundaries or glide planes or strongly deformed parts of a crystal as in the Greek marble of Fig.329. Primary recrystallization is included in FOLK's term "degrading recrystallization" (1965, p.23), which he relates speci-fically to the process described by VOLL (1960).

The only description of primary recrystallization I know of (other than that by VOLL, 1960), which is adequately supported by evidence, is the careful analysis of a Dinantian limestone in the Armorican fold belt of southern Eire, by WARDLAW (1962). The fabrics were seen not only in strongly deformed limestones but also in weakly deformed limestones that are, in fact, commonly taken as undeformed. Up to 80% of the limestone seems to have been recrystallized to a mosaic of calcite, with crystal diameters mainly 5–15 μ. This mosaic has, for example, replaced crinoid columnals and the zooecia of bryozoans (Fig.330). The reason for con-cluding that the growth of the new micron-sized mosaic was by primary recrystal-lization is that the new mosaic can be seen aligned along shear fractures in large, deformed crystals of calcite and along the interfaces between twin lamellae. This is the distribution of finely crystalline mosaic which one would anticipate following primary recrystallization. Yet, as Wardlaw points out, caution is needed even in such an apparently straightforward case. Can the possibility of a phase change, of metasomatism, be excluded? Even if we assume that migration of ions along twin planes can be regarded as a dry, solid state, process, it is less easy to eliminate the possibility that cations other than Ca^{2+} played some part.

Wardlaw's paper is also the only unequivocal description known to me of a fabric produced by a process called, by ORME and BROWN (1963, p.61), **grain diminution.** These authors, following VOLL (1960, p.516), invented the new term (equivalent to primary recrystallization) to describe the *in situ* replacement of a deformed crystal by cryptocrystalline unstrained crystals of the same mineral. In their usage "grain" is synonymous with "crystal", as in the then current metal-lurgical usage: it does not refer to any detrital particle. Wardlaw's 5–15 μ mosaic has replaced crinoid columnals and, conventionally, these are regarded as single crystals. In addition, the fine mosaic is certainly an *in situ* replacement, because of its location along twin planes, etc. Therefore, it is claimed, a large crystal has been replaced by smaller ones, and recrystallization has caused a reduction in crystal size. Nevertheless, there is evidence (TOWE, 1967) that the single crinoidal crystal, so familiar to the microscopist, is, at least in part, composed of a multitude of tiny crystals having closely similar though not identical orientations (p.40). Thus, the whole concept of grain (meaning crystal) diminution must be approached with

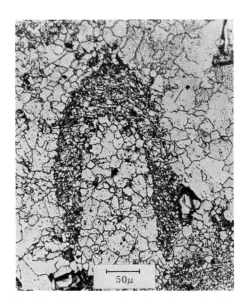 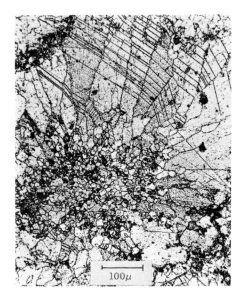

Fig.331. Oblique section through a crinoid columnal largely replaced by neomorphic sparry calcite. Peel. Carboniferous Limestone. Ingleton, Yorkshire.

Fig.332. Stellate mass of radial-fibrous spar, neomorphic alteration of a biomicrite. Core of equigranular microspar, lower left. Relic of unaltered micrite, top left. Peel. Carboniferous Limestone. Halkin Mountain, Denbighshire.

caution (e.g., in Fig.331). TOWE and CIFELLI (1967, p.744) have written a particularly apposite essay on the question of what is meant by a "crystal". All too frequently, a mineral grain that appears, to the naked eye or with a low-powered microscope, to be a single crystal, is discovered, under high-powered magnification or with electron microscopy or X-ray diffraction, to be, instead, an aggregate of crystallites with a highly preferred orientation. This polycrystalline ultrastructure is of the utmost importance to those concerned with mechanisms of crystal growth. Among carbonate petrologists, there has grown up in the last few years a special fascination for the term "grain diminution", which has been frequently misused to the detriment of a valuable concept. Once again, it is necessary to stress that the presence of micrite or microspar (p.513) within a skeletal crystal is, by itself, no evidence for recrystallization by grain diminution in the manner implied by the inventors of the term. Before the action of true crystal diminution, i.e., primary recrystallization, can be presumed to have taken place, it is necessary to *demonstrate* that the secondary, finely crystalline, calcite has not filtered in as sediment or been deposited during algal micritization—also that the pre-existing crystals were in truth larger than their supplanters. It is also particularly important to remember how very small the primary crystals in a skeleton are, especially in the coralline algal skeletons which are supposed to have undergone "grain diminution" (WOLF, 1965; COLLEY and DAVIES, 1969). In all the cases I have seen, the skeletal crystals are smaller than the crystals of the replacing micrite, however this may have evolved, except in echinoderms (Fig.331). The fact that foraminiferids described by Banner and Wood (p.490) have been replaced by micrite does not indicate a reduction in crystal size, as claimed by FOLK (1965, p.23)—nor does he now believe this (R. L. Folk, personal communication, 1969). FENNINGER (1968) has claimed that small equant crystals of calcite, in some Austrian Tithonian cements, are aligned, one on the other, in such a way as to indicate that they are the parts of once larger crystals. These larger crystals were elongated normally to the pore wall. He suggested that the small crystals are the products of "grain diminution" of the larger crystals. Yet his argument is based on stereoscan photographs of rather deeply etched polished surfaces. The traces of the interfaces are not sharp and no crystallographic data are given. The outlines of the supposed larger, parent crystals are far from obvious.

Grain growth

Like primary recrystallization, this, too, is a metallurgical term. It refers to the stage of annealing that follows the completion of primary recrystallization, when a selected few of the new unstrained crystals continue to grow at the expense of their neighbours. It is a dry, solid state process. The concept of grain growth is fully and admirably reviewed by FOLK (1965, pp.17–19). The term was introduced into carbonate petrology by me in 1958 to describe the familiar alteration of micritic calcite to sparry calcite (BATHURST, 1958, p.24). This was done after a great deal of hesitation provoked by the then uncertain role of water. I was, and I remain, impressed by the demonstrable truth that the replacement of micrite by sparry calcite has operated for the most part in a virtually tight limestone. After 1958, however, I became increasingly convinced, as my carbonate education was advanced by my wiser colleagues, that it is unreasonable to regard the diagenetic neomorphic (as distinct from metamorphic) fabrics in limestones as the products of dry processes. To begin with, there is too much water present. Also, the development of what appear to be crystal faces on the new secondary crystals, where they abut against unreplaced micrite, is not in keeping with a solid state process. "Such a texture is not characteristic of grain growth in metals, where the intercrystalline boundaries are curved. In fact, this anomaly points the main dilemma. Grain growth in the metallurgical sense is an anhydrous process and, although the evidence from limestones favors recrystallization in a lithified (consolidated) limestone, the role of water may be important." (BATHURST, 1964a, p.330.) In this book I have written of the growth of neomorphic spar when referring to the wet diagenetic process which I earlier described as "grain growth".

I do not mean that grain growth is unimportant in carbonate sediments, only that it belongs to the metamorphic phase of development with which this book is not concerned. The reader is referred to some observant, quantitative studies by KARCZ (1964) on grain growth in some Cambrian dolomites of the Isle of Skye, Scotland, and to the references in FOLK (1965).

Following my repudiation in 1964 of my hypothesis of grain growth in limestones, a full-scale attack on the hypothesis, devastating and necessary, was mounted by FOLK (1965), who thoroughly exorcized the term "grain growth" from the sphere of wet carbonate diagenesis. Nevertheless, we still need to explain the processes which bring about the replacement of micron-sized mosaics by sparry calcite. An explanation is attempted later in this chapter (below) in the section on aggrading neomorphism.

Wet recrystallization

Carbonate fabrics which are demonstrably the products of diagenetic *fabric* change, without accompanying change in mineralogy, seem to be rare. It is argued on p.501 that the later stages of the growth of neomorphic spar consist of the growth of calcite at the expense of calcite during boundary migration in a calcite mosaic. Primary calcite fabrics in skeletons are for the most part stable during diagenesis and most fabric changes have an important element of polymorphic transformation. It is also obvious that many diagenetic changes in carbonate sediments involve some exchange of Mg^{2+} (or other) ions and that it would be foolish to attribute these changes to single simple processes, with our present inadequate knowledge.

AGGRADING NEOMORPHISM: THE GROWTH OF NEOMORPHIC SPAR

Aggrading neomorphism (FOLK, 1965) is the process whereby finer crystal mosaics are replaced by coarser crystal mosaics of the same mineral or its polymorph, *in situ*, that is to say, without the intermediate formation of visible porosity (see discussion of "*in situ*" on p.494). It includes the familiar diagenetic alteration of micrite, or micron-sized skeletal fabrics, to sparry calcite (Fig.333, 335). This process was earlier called "grain growth", but this term, which refers to dry, solid state recrystallization (p.480), is regarded as inappropriate to describe a process which, in carbonate diagenesis, is obviously wet. The term "micrite enlargement" grew up among oil geologists in Tulsa, Oklahoma, during the 1960's and has been used and defined by FRIEDMAN (1964, 1966). It is not adopted here because of the implication that only micrite is replaced by sparry calcite. The intention in this chapter is to emphasize the wide applicability of aggrading neomorphism to the wet sparry replacement of all cryptocrystalline mosaics. The petrographic work of Cullis (p.353) and Schlanger (p.347) on Cainozoic carbonate sediments suggests that the growth of neomorphic spar begins in the *partly consolidated* sediment. It is apparent, therefore, that the process is not solely a matter of recrystallization, but involves also, in its earlier stages, the wet transformation of aragonite to

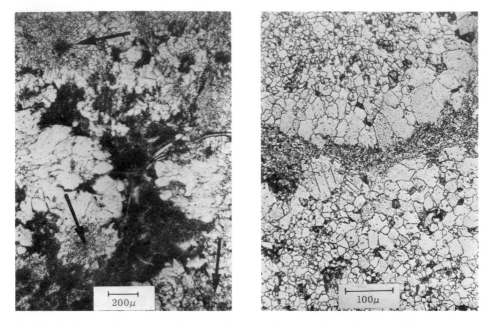

Fig.333. Relic of micrite embayed by several masses of neomorphic stellate spar. Microspar cores shown by arrows. Slice. Carboniferous Limestone. Ingleton, Yorkshire.

Fig.334. Relic of micrite embayed by neomorphic spar. Stellate radial-fibrous mass occupies upper half of figure with large core of microspar. Peel. Carboniferous Limestone. Ingleton, Yorkshire.

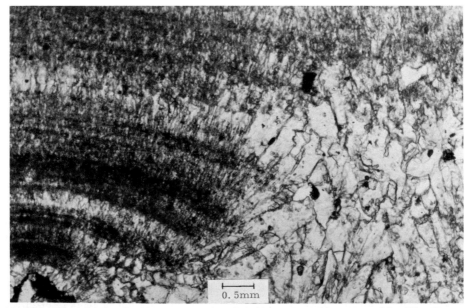

Fig.335. Calcite fabric of a mine oöid partly replaced by neomorphic sparry calcite. Slice. Recent. Styria, Austria. (Courtesy of M. Kirchmayer.)

calcite and some passive dissolution-precipitation (p.327). Thus it has a somewhat broader meaning than Folk's "aggrading neomorphism" as strictly defined but follows it in spirit.

The broad scope of aggrading neomorphism may appear regrettable, but it should be remembered that the term is used here to describe a process, or collection of processes, about which we know with certainty only the nature of the sparry product and, in some cases, the composition of the original material. Inferences concerning the processes which led to the sparry replacement can only be based on pathetically insubstantial grounds at the present time. Until these processes are better understood, aggrading neomorphism must remain something of a rag-bag.

The problem

The most striking visible diagenetic change that takes place in the evolution of a limestone is the increase in the content of sparry calcite. The addition of calcite spar in the form of cement has already been discussed. There remains the question of the growth of secondary spar by replacement of pre-existing micron-sized $CaCO_3$, *in situ*, that is to say without the formation of passive cavities (p.498). The micron-sized $CaCO_3$ may be an aragonite needle mud, the shell wall of a mollusc or foraminiferid, a micrite composed of a mixture of solid carbonate phases or any other micron-sized fabric. So far as is known, the secondary calcite spar has only directly replaced primary fabrics having micron-sized crystals, although this con-clusion follows mainly from work on limestones of Carboniferous age. The relation between primary crystal size and the growth of neomorphic spar in the Cainozoic limestones, which show the early stages of the process, is not known in detail. It is to an analysis of this replacement process that the following pages are directed.

If the products of this type of neomorphic alteration were easily recognized, then much of the growing literature on the subject would have been uncalled for. Instead, the process yields a sparry calcite that is always difficult to distinguish, sometimes impossible to distinguish, from the sparry calcite of cement. Over the years it has become all too plain that the sure identification of neomorphic spar requires a degree of care, the need for which has not always been appreciated. As BEALES (1965, p.56) wrote: "The same thin-section has been passed to different observers and one will see space-filling drusy calcite and others will see recrystal-lization and replacement." Attempts to support diagnosis with photomicrographs have fared little better, the pictures requiring for their comprehension, at times, a degree of antecedent belief amounting to missionary fervour. Even where identification of neomorphic spar is beyond doubt, there remains the further problem of working out the causal process or processes. Despite the widespread occurrence of neomorphic spar, its actual growth has not been observed either in the field or in the laboratory. We are faced with the familiar geological task of imagining hypothetical processes, when all we have to guide us are the products of their activities.

The fabrics of neomorphic spar

Although views on the causative processes in aggrading neomorphism have matured in recent years, my earlier study of the fabrics (BATHURST, 1958) retains its relevance and is given here with certain additions. The fabric description in this book is based primarily on that study (cf British Dinantian limestones) with additions from the works of Cullis and Schlanger on the cores from Funafuti, Eniwetok and Guam (p.347, 353), from studies by Banner and Wood (p.490) on the recrystallization of skeletal material in Miocene limestones of Papua, and from the fabrics described by Hudson (p.347) and P. R. Brown (p.348) from Jurassic limestones of Britain.

The philosophy behind the selection of diagnostic fabrics is the same as that behind the selection of fabrics for cement (p.416), and the reader is referred to that discussion.

Certain mosaics of micron-sized calcite crystals, once continuous, are now interrupted either by sparry calcite (decimicron–centimicron-sized) or by syntaxial rims on skeletal hosts, such as echinodermal grains or large brachiopodal crystals (Fig.345, 346). The original micron-sized materials include unconsolidated carbonate ooze, peloids, the walls of Foraminiferida, Mollusca, Zoantharia, calcareous Algae, oöids, and other materials of micritic crystal size.

Fabrics of the sparry calcite

The various known fabrics of neomorphic spar listed below are based mainly on studies of Carboniferous and other pre-Tertiary limestones.

(*1*) **Crystal diameters** (maximum apparent in thin section) range upward from about 4 μ. Diameters from 50 to 100 μ are common and greater diameters occur.

(*2*) The **contact** between unaltered, micron-sized material and secondary spar is generally abrupt (Fig.333, 334), but it can also be so gradual that the intermingling of fine and coarse crystals makes it impossible to draw a line of separation.

(*3*) The **crystal size** in the spar commonly varies irregularly and patchily from place to place (Fig.338). In this it differs from the regular vectorial change of crystal size in cements, and from the relatively uniform crystal size of well-sorted rim-cemented grains. Porphyrotopes are exceptional. It would be possible, following FOLK (1965, p.22), to speak of "porphyroid" neomorphic spar or equigranular neomorphic spar.

(*4*) **Radial-fibrous structure:** in some Dinantian limestones of North Wales a radial-fibrous arrangement of crystal shapes is apparent, as sub-spherical, stellate masses of spar that occur liberally scattered throughout certain pseudobreccias (Fig.332–334; p.503). These should not be confused with calcareous algae (MIŠÍK, 1968).

(*a*) The centre of each stellate mass consists of **microspar** (p.513), equigranular crystals of decimicron size (Fig.332).

(*b*) The central mass is surrounded by coarser, elongate crystals with their longer axes radially distributed.

(*c*) Many of the elongate crystals taper toward the centre of the stellate structure. The general appearance is of a poorly developed spherulite with a core of equigranular mosaic.

(5) The **intercrystalline boundaries** in Dinantian equigranular sparry calcites vary generally from curved to wavy (Fig.332, 336, 338; GERMANN, 1968, fig.12). The plane boundaries so typical of cement are uncommon. In the stellate structures, the intercrystalline boundaries are commonly plane.

(6) **Embayments:** some large crystals at the margins of the sparry masses embay the adjacent detrital micrite (Fig.333, 334). Many of these embayments are plane sided (saw-toothed). Relics of micron-sized mosaic commonly appear as wisps or threads in the spar. In places, the tests of Foraminiferida and other fossils have been so extensively transected by spar that only disconnected relics of the test remain. Nevertheless, these isolated parts retain their original orientation. In general, neomorphic spar transects older fabrics (Fig.331–336, 339), even having the form of veins (MIŠÍK, 1968).

(7) **Floating relics:** some micron-sized material (patches of micrite, skeletal walls, peloids, etc.) is entirely surrounded (in *three* dimensions) by spar (Fig.336).

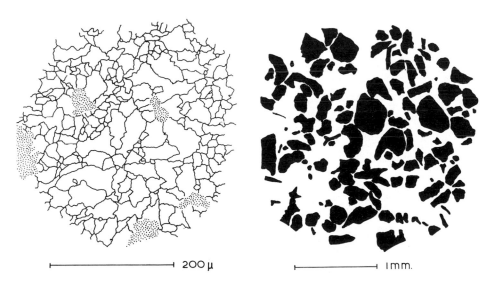

200 μ   1 mm.

Fig.336. Relic patches of micrite (stippled) in neomorphic spar. Drawing of slice traced on a microprojector. Carboniferous Limestone. Ingleton, Yorkshire. (From BATHURST, 1958.)

Fig.337. Silhouettes of skeletal sand grains in three dimensional contact. Prepared by dropping into a molten mounting medium. Drawing of slice on a microprojector. (From BATHURST, 1958.)

Great care is necessary in the recognition of these floating relics. In the past it has been too readily concluded that grains are embedded in neomorphic spar when more rigorous application of fabric criteria would now reveal this spar as cement.

The recognition of floating grains is particularly difficult because grain shapes are so variable. It is convenient in practice to compare the distribution of grains in two dimensions (thin section) with illustrations of sections through accumulations of contact-packed, irregularly shaped, grains such as in Fig.337 (also in DUNHAM, 1962). If the distribution in question is obviously looser than the packing in the illustrations, then it must be assumed that the sparry matrix has replaced an earlier mechanically deposited matrix. Of course borderline cases will be found where judgment is impossible, but happily these are few.

(8) **Calcitization of molluscan shells:** in the Cainozoic limestones of Eniwetok, Guam and Funafuti, skeletal structures (initially composed of micron-sized aragonite or high-magnesian calcite) are now transected by or wholly replaced by sparry calcite, as shown by SORBY (1879, p.68). Descriptions of this relationship from Funafuti, after Cullis, are given on p.353. Detailed descriptions by Schlanger of the stages of calcitization in molluscan walls are given on p.347. BATHURST (1964b, p.371) has described similar replacement fabrics in molluscan walls from Carboniferous and Jurassic limestones. The characteristics of these fabrics are given below.

Fig.338. Wall of foraminiferid embedded in neomorphic sparry calcite which has replaced a micrite. Note irregular distribution of crystal size and the scarcity of plane intercrystalline boundaries. Peel. Carboniferous Limestone. Ingleton, Yorkshire.

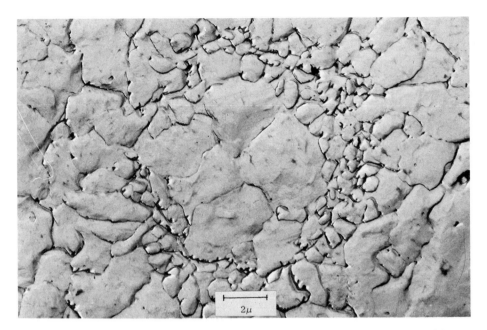

Fig.339. Coccolith (low-magnesian calcite) partly obscured by the growth of neomorphic spar. The partial preservation of the original fabric suggests that it was the primary crystals of the coccolith that either grew or dissolved as more favoured crystals replaced less favoured crystals. Replica. Fracture surface. Electron mic. Hallstätter Limestone, Triassic. Feuerkogel, Austria. (Courtesy of A. G. Fischer.)

(a) The wall of a molluscan shell is occupied partly or wholly by a mosaic of sparry calcite (Fig.340).

(b) The crystals of spar vary in maximum apparent diameter mainly from 10 to 300 μ where equant: elongate crystals have longest axes up to about 500 μ.

(c) The smallest crystals are commonly, but not invariably, concentrated along the margins of the shell wall.

(d) The intercrystalline boundaries are either wavy, gently curved (Fig.341, 342) or, uncommonly, plane (Fig.340).

(e) Triple junctions with one angle equal to 180° (enfacial junctions: p.423) are rare (Table XVII, p.420).

(f) In many spar-filled shell walls, a linear arrangement of inclusions (the remains of organic matrix; p.14) cuts across the calcite mosaic, continuing without deviation across the intercrystalline boundaries. Linear arrangements of intercrystalline boundaries also occur. These linear patterns resemble the layers in unaltered molluscan shell walls (Fig.341, 342). It is interesting that where sparry calcite has replaced *in situ* the radial-fibrous calcite of cave oöids (Fig.335) which lack organic matrix, there is no relic of the radial-fibrous fabric in the spar (Hahne et al., 1968).

(g) In some molluscan walls (e.g., in the Purbeckian in the south of

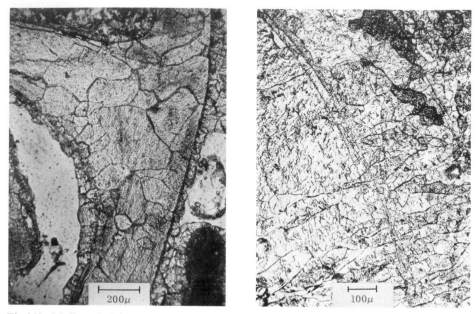

Fig.340. Mollusc shell fragment replaced by neomorphic sparry calcite. Traces of original layered structure transect the sparry mosaic. Fragment is coated with a layer of early calcite cement.

Fig.341. Shell of gastropod *Viviparus* replaced by neomorphic sparry calcite. Relics of layered structure transect sparry mosaic. Peel. Wealden. Sussex.

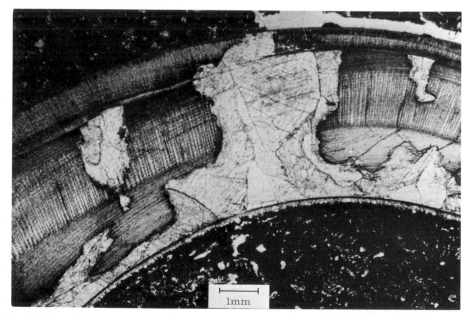

Fig.342. Aragonite shell of the gastropod *Strombus gigas* partly replaced by neomorphic sparry calcite. Note ghost of primary layers in the spar. Slice. Pleistocene. Ambergris Cay, British Honduras. (Courtesy of G. E. Tebbutt.)

England) the calcite spar is pale brown and pleochroic (P. R. Brown, p.348). Hudson has studied a similar mosaic in the Jurassic of the Isle of Skye, Hebrides, and has found that the colour and pleochroism seem to be related to inclusions of residual organic matter (p.348).

(*h*) Some spar crystals have undulose extinction. A crystal 150 μ long may have a range of angle of extinction of 10°. In places the undulose extinction is related to the shape of the shell wall.

(*i*) In some walls the crystals have a preferred elongation normal to the margins of the wall.

(*j*) Commonly, where altered shells have been broken during compaction, the separated parts, if reunited, would form a continuous crystal lattice. This means that the growth of the sparry calcite took place before compactive fracture.

(*9*) **Neomorphic aragonite in aragonite skeletons:** P. R. BROWN (1961), TEBBUTT (1967) and SHINN (1969, p.133) found that some aragonite skeletal structures have been replaced by sparry aragonite, respectively in Jurassic bivalves and in Pleistocene bivalves and corals. Tebbutt reported that the primary aragonite shell structure was distorted in the process (Fig.343, 344) and that ghosts of primary

Fig.343. Aragonite shell of *Lucina* (Bivalvia) partly replaced by fibrous sparry aragonite. Slice. Plane polarized light (see Fig.344). Pleistocene. Ambergris Cay, British Honduras. (Courtesy of G. E. Tebbutt.)

Fig.344. As Fig.343, but with crossed polars. Note apparent buckling of primary laminae adjacent to the neomorphic aragonite.

structure in the neomorphic spar had been displaced or rotated from their original positions. Moreover, where the coarse aragonite crystals reach the outer wall of the shell, they commonly extend beyond the original surface. There is at present no reason to believe that this process is an essential step toward calcitization.

(*10*) **Counts of triple junctions,** in calcitized molluscs of British Jurassic and Dinantian limestones, show that few of the triple junctions have an angle of 180°. The percentages of enfacial junctions (p.423), recorded by Bathurst for calcitized (spar-replaced) mollusc shells and sparry patches after micrite, range from 3 to 5% (7 samples) compared with 30 to 73% for calcite cement (p.420).

(*11*) **Overgrowths on cement crystals:** BATHURST (1959a, p.373) and BANNER and WOOD (1964, p.23) have shown that the replacement of micron-sized skeletal fabrics by calcite spar has taken place not only in biomicrites but in biosparites as well. In some limestones sparry *cement* crystals now extend syntaxially into what was once the micron-sized fabric of a peloidal allochem or skeletal wall (BATHURST, 1959a, p.373; P. R. BROWN, 1961; HUDSON, 1962). Thus a single crystal of spar may have a dual origin, the older part of it being cement, the younger part being a neomorphic replacement of some pre-existing micron-sized fabric.

(*12*) **Order of replacement:** it is apparent that certain fabrics were replaced more readily than others. In 1958 I put forward an order of increasing resistance to replacement in the Dinantian limestones (BATHURST, 1958, p.31) as follows: micrite, peloids and *Koninckopora* (a dasycladacean calcareous alga), Foraminiferida, oöids (see also Crickmay's order, p.349).

In a thorough and strikingly revealing study of neomorphism in Miocene limestones of Papua, BANNER and WOOD (1964) discovered a clear neomorphic sequence among taxa of microfossils: this was based on the study of 5,000 thin sections from 2,650 outcrop samples. The authors introduced their neomorphic sequence thus: "In these richly microfossiliferous rocks [limestones], it was observed that, in a particular sample, the wall structure of the alveolinids, for example, had been altered by recrystallization [sic], while, in the same thin-section, cycloclypeids, for example, had suffered no recognizable alteration; the reverse of this case was never observed to occur. Similarly, recrystallized alveolinids, miliolids and cycloclypeids occurred associated with unaltered elphidiids and, again, the opposite case was not found to occur. From the total of the observations, it was possible to summarise the results [Table XXI] . . . From this it can be seen that whenever miliolids were found to be recrystallized, any associated peneroplids, alveolinids, chlorophytan algae and scleractinian corals could be expected to be similarly recrystallized, but that the remainder of the microfossil assemblage might or might not be unchanged [stages *2–6* of above table]. By the time when the amphisteginids had suffered recrystallization, however, all the fossils, with the exception of the agglutinating foraminifera (e.g., textulariids and trochaminids) and some, at least, of the echinoderm debris, would be altered, some considerably so [stages *5–6* of the above table]. Therefore, the destruction of the microfossils by

TABLE XXI

STAGES IN NEOMORPHISM OF MICROFOSSILS*
(After BANNER and WOOD, 1964)

Primary skeletal fabric	Neomorphic stages = time					
	1	2	3	4	5	6
Chlorophyte algae	0	0	0	0	0	0
Scleractinians	0	0	0	0	0	0
Miliolids		0	0	0	0	0
Peneroplids		0	0	0	0	0
Alveolinids		0	0	0	0	0
Planorbulinids			0	0	0	0
Cycloclypeids			0	0	0	0
Bryozoans			?	0	0	0
Coralline algae			?	0	0	0
Rotaliids				0	0	0
Elphidiids					0	0
Amphisteginids					0	0
Echinoderms					?	0
Textulariids						0
Trochamminids						0

* 0 = skeletal fabric neomorphosed.

recrystallization of their skeletal calcite [sic] apparently takes place selectively, in an apparently constant order."

Though the neomorphism in the Papuan limestones always involved an increase in crystal size, the secondary mosaics vary from micrite (but coarser than the skeletal fabric) to spar. The order of susceptibility of skeletons to neomorphism given by Banner and Wood is somewhat similar to the order given by Crickmay and Schlanger (p.349). Banner and Wood include a detailed, carefully illustrated study of 4 thin sections, which, combined with their other photomicrographs, makes their paper probably the best illustrated account of neomorphic fabrics in existence.

Fabrics of the syntaxial calcite rim

These are examined under six headings.

(1) **Resemblance to cement rim:** the syntaxial neomorphic rim resembles, in its lattice continuity with its host and in its dimensions, the syntaxial cement rim (p.429), from which it may not always be readily distinguishable. The host may be

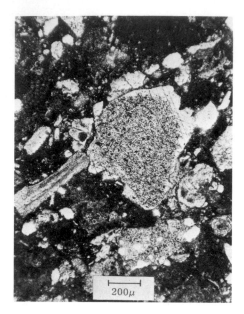

Fig.345. Echinoderm grain with syntaxial calcite rim. As the limestone is a wackestone the micrite cannot be a late arrival (as in Fig.310): thus the syntaxial rim must be a neomorphic replacement of the micrite. Slice. Miocene. Marina di Puolo, Sorrento Peninsula, Italy.

Fig.346. Part of calcitized molluscan grain, with syntaxial neomorphic (?) extensions of the calcite crystals into the enclosing micrite. Slice. Upper Cretaceous. Monte Camposauro, Campagna, Italy. (Slice lent by B. D'Argenio.)

an echinodermal grain (Fig.345), or a coarse crystal in the wall of a brachiopod, or a calcitized mollusc shell (Fig.346), etc.

(2) **The syntaxial rim**, or the host allochem where this is locally not rimmed, is in contact with a *pre-existing* matrix of micrite (unlike the cement rim), or with other rims or other allochems. Care must be taken to distinguish between a micrite matrix that was pre-rim and an internally deposited micrite which filtered into the pore system after the growth of a cement rim (as described by Evamy and Shearman, Fig.308).

(3) **The rim may transect** the fabric of a skeletal particle (Fig.347): more commonly it **embays** the surfaces of peloids.

(4) **Saw-toothed margin:** like the cement rims described by Shearman, this rim may have a saw-toothed outer margin, taking the form of small plane-sided spires with the points of the spires directed along the c axis (Fig.345).

(5) **Rims on calcitized shells:** in calcitized molluscan shell walls in the limestones of Eniwetok and Guam (p.347) and Funafuti (p.353) the secondary sparry calcite crystals in the shell wall may extend syntaxially into the adjacent micrite matrix. In Funafuti the mineralogy of the syntaxial extension, calcite or aragonite, matches that of the host.

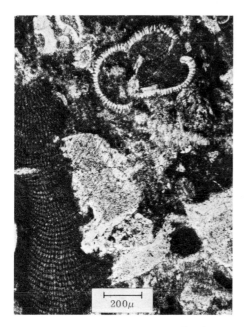 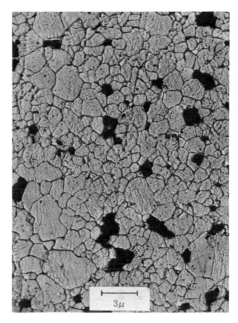

Fig.347. The clear syntaxial rim of a dusty echinoid grain appears to have replaced part of the cellular structure of a coralline alga. Since the limestone is poorly sorted, it seems likely that all the sediment was deposited at the same time and that the syntaxial rim is, therefore, a neomorphic replacement of micrite. Slice. Oligocene. Campania, Italy.

Fig.348. Micrite. Replica of polished surface. Electron mic. (black patches are artifacts). Viséan. Belgium. (Courtesy of C. Monty).

(6) **Cement syntaxial with neomorphic spar:** in calcitized molluscan shell walls in some British Jurassic limestones some of the sparry crystals in the wall are syntaxial with the adjacent sparry *cement* crystals (P. R. BROWN, 1961).

Interpretation of the fabrics of neomorphic spar

The neomorphic process, which was examined under the name of "grain growth" in some Dinantian limestones more than a decade ago (BATHURST, 1958), is treated in this book on a broader foundation. The replacement of micron-sized crystal fabrics of aragonite and calcite, *in situ*, by a sparry calcite having decimicron-sized and centimicron-sized crystals, has taken place in limestones of many ages, from Palaeozoic to Recent. These limestones have all been more or less changed from originally unconsolidated, carbonate sediments (assumed to have been a mixture of aragonite and a range of magnesian calcites) immersed in sea water, to a fabric of low-magnesian calcite with porosity commonly as low as 1–3%. They have not been recrystallized to any significant degree as a response to shear or high temperature: they are, conventionally, not metamorphic. The replacement has taken three courses: (*1*) patchy sparry replacement of detrital micrite, (*2*) sparry replacement of skeletal fabrics, such as the walls of molluscs or foraminiferids, (*3*) replacement of detrital micrite by the outward syntaxial growth of crystals in the walls of allochems. Two cases to illustrate course *3* are the syntaxial growth of a crinoid columnal, and the syntaxial growth of neomorphic spar crystals in molluscan shell walls.

Having made this general statement it is now necessary to examine the process of aggrading neomorphism in more detail, paying special attention to the following: *in situ* replacement, crystal enlargement, dissolution-precipitation, polymorphic transformation, migration of solution films (wet intercrystalline boundaries), the direction of crystal growth, and the relative instability of different micron-sized carbonate fabrics. In reading the pages that follow it is important to remember that a fundamental first step is the distinction of neomorphic spar from cement. We shall be seeking, not only to illumine the nature of the processes of neomorphism but, at the same time, to find criteria by which their fruits shall be known. Thus these pages can usefully be read in conjunction with the section on the recognition of cement in thin section (p.416). Throughout this examination, because the fabric data are derived for the most part from three distinct sources, it will be necessary to ensure that the separate aspects of the growth of neomorphic spar apply equally to the Eocene–Recent limestones of Eniwetok, Guam and Funafuti, to the Miocene limestones of Papua, and to the Jurassic and Dinantian limestones of Britain.

Evidence for *in situ* replacement. The possibility that the sparry calcite is a cement cavity filling or a rim-cemented deposit of detrital crystals is ruled out, in the British Jurassic and Dinantian limestones, where there is an irregular distribution of crystal size, a scarcity of plane intercrystalline boundaries and a low percentage of enfacial junctions among the triple junctions.

There are certain exceptions to this generalization which are covered by other evidence. In some spar-replaced mollusc shells there is a certain regularity in the increase of crystal size away from the shell surfaces, but as the sparry mosaic is commonly cut by the lines of inclusions or lines of crystal boundaries (layers in three dimensions), the question of it being either a cement replacement or a detrital filling of a passive cavity does not arise. The greater frequency of plane intercrystalline boundaries in some calcitized shell walls, than in the post-micrite spar, is also generally covered by this other evidence for *in situ* replacement. In the radial-fibrous structures, plane intercrystalline boundaries may be seen between the larger crystals but these structures, for other reasons, clearly have an *in situ* replacement origin. In common with the scattered patches of sparry mosaic, the radial-fibrous masses are large compared with the associated allochems, yet they do not have the smooth, eroded outlines of pebbles: they have instead highly irregular outlines. All these scattered patches of spar transect what seems to have been a once continuous fabric of carbonate detritus, and they contain within themselves irregularly shaped masses of this same detritus, as bits of biomicrite, biosparite (not so easy to recognize), etc. The enclosed masses appear to be surrounded by spar in three dimensions and they show various degrees of admixture with sparry crystals, as if partly altered. They have the appearance of relics. The margin of the spar, wherever it is in contact with micrite, may be saw-toothed, a feature which fits well with the growth of spar crystals at the expense of micrite.

Syntaxial rims, with or without saw-toothed margins, also occur as cement rims, for example on crinoids, so it is necessary to decide whether the enclosing micrite was deposited before or after the growth of the rim. My earlier contention (BATHURST, 1958), that the existence of delicate syntaxial spires on crinoid grains implied that overgrowth was post-depositional, was correct, but in concluding that the overgrowth was, therefore, a neomorphic replacement of the enclosing micrite, I ignored the possibility revealed by Evamy and Shearman (Fig.308) that the micrite might also be post-depositional and younger than the overgrowth. If the limestone is a mudstone or a wackestone, as defined by DUNHAM (1962, p.117), that is to say if the allochems (e.g., crinoidal grains) are mud-supported, then the sediment was deposited in a single act and the syntaxial rim must be an *in situ* replacement. It is assumed that the growth of the rim, with its delicate spires, was always post-depositional. If the limestone is Dunham's packstone, consisting of grain-supported allochems with a matrix of micrite, then growth of the syntaxial rim could have preceded the filtering of micrite into the intergranular porosity: thus the rim could be either a syntaxial cement or a syntaxial replacement of micrite. In this case other decisive evidence must be sought. If the filling of the pores between the allochems with micrite is incomplete, particularly if the micrite matrix is geopetally distributed, then the probability is high that the micrite filtered into the calcarenite after the allochems were deposited. The syntaxial rims are, thus, likely to be cement. LUCIA (1962, fig.4) and Evamy and Shearman (p.431) have noted that growth of cement is prevented where the potential host surface is covered by micrite. If the biomicrite is poorly sorted but if the size frequency distribution is not strongly bimodal, then the probability is high that the sediment was deposited in one act. In this case the syntaxial rim is likely to be an *in situ* replacement. There is, of course, no reason why the filtering of a micrite matrix into a grain-supported calcarenite should not, later on, be followed by replacement of the micrite *in situ* by a syntaxial rim nucleated on, for example, an echinodermal grain. Decisive evidence in support of a cement origin is given in Fig.308 (p.430) from the paper by Evamy and Shearman. Finally, if the syntaxial rim can be shown to transect a rigid object such as a skeletal grain, then the rim must be an *in situ* replacement (Fig.347).

It has been claimed that displacive crystal growth is a third possible way of producing syntaxial rims: this must be considered. FOLK (1962c, p.557) wrote that the calcite cement crystals inside ostracods, themselves embedded in shale, had

grown inward from the interior surfaces of the ostracod valves and had displaced the shale by pushing it in front of them. On the same page, Folk claimed
that in some pure limestones the growth of fibrous calcite has forced shells apart,
sometimes to the extent of doubling the volume of the limestone. As FOLK (1965,
p.24), wrote: "Sparry calcite may also create room for itself by shoving aside the
surrounding allochems or carbonate mud through the active force of crystal
growth." This process he called displacive precipitation. He added that the process
is relatively uncommon. Criteria for recognizing the products of the process in
limestones, *where carbonate would displace carbonate*, were not proffered. The
likelihood of such a process acting in rocks is open to considerable doubt (SPRY,
1969).

There seem to be three arguments against the acceptance of a displacive
origin for the syntaxial rims under discussion here.

(*1*) The growing rim needs a source of additional $CaCO_3$. Needless work
would be performed if the growing rim depended on dissolution of $CaCO_3$ at
some distance for its supply of ions, whereby it was enabled to push aside, without
dissolving, the adjacent $CaCO_3$ micrite. Surely the nearest, most soluble source of
$CaCO_3$, would be used, that is to say the adjacent micrite.

(*2*) The first effect of the exposure of a multiphase carbonate sediment to fresh
water is the precipitation of sufficient cement to make a rigid framework (p.326).

(*3*) Some syntaxial rims do transect skeletal structures and the interface
between fossil and rim is plane. This suggests *in situ* replacement of the fossil by the
syntaxial rim rather than filling of a region of broken, lost fossil by cement. Broken
surfaces are not normally plane: advancing crystal faces are.

For the limestones of Eniwetok, Guam and Funafuti, the available fabric
evidence for *in situ* replacement by sparry calcite is less detailed, and we have to
depend upon the judgments of Schlanger, Crickmay (p.347) and Cullis (p.353)
in their assessments of sparry replacement mosaics. They have described sparry
calcite mosaics which appear to have replaced finely crystalline sediment and molluscan shells in a manner similar to that discussed in connection with the British
limestones. In particular they show that replacement of micrite can proceed by the
syntaxial outgrowth of host crystals lodged in the surface of an allochem (p.347).
These host crystals may be primary or they may be secondary spar crystals. Cullis
made the interesting observation (p.353) that, in the Funafuti limestones, both
aragonite and calcite crystals act as hosts, so that the replacive syntaxial rim is not
simply calcite, as recorded elsewhere, but may be aragonite (see p.359 for discussion
on calcitized aragonite cement).

Workers on other limestones have given supporting evidence for replacement
in situ by calcite spar. In particular, the sequence of changes as *Austrotrillina
howchini* (Miliolidae) is replaced, layer by layer without visible displacement, has
been revealed by Banner and Wood and is strong evidence for *in situ* replacement.
In these Papuan limestones the fact that a single test of a foraminiferid can consist

in one place of unaltered fabric and, immediately adjacent to this, of calcite spar, suggests a process of replacement acting without the development of visible, passive cavities, though it does not prove it. In the Papuan examples, and in limestones generally, the occurrence of all stages of sparry replacement, from barely perceptible to complete, none of which are accompanied by visible cavities, is incompatible with replacement by dissolution and the precipitation of calcite cement: it can only be explained by *in situ* neomorphism. Other evidence includes the observations of P. R. Brown and Hudson on the *in situ* replacement of mollusc shells, as revealed by relics of laminar structure occurring as lines of inclusions transecting the fabric of the sparry calcite. A general remark may be appropriate here. Most of the authors referred to have noted that the secondary sparry mosaic is commonly rich in "inclusions" or is brown. The colour is soluble in organic solvents (e.g., acetone) and is probably a residue left from the primary skeletal structure. Nevertheless, the presence of dusty inclusions in a calcite mosaic cannot be relied upon as indicating *in situ* replacement, because many undoubted cement mosaics are rich in inclusions (water?), especially first generation cement.

Evidence for crystal enlargement. Wherever it has been possible to determine the crystal sizes in the original material, it has been clear that the replacing crystals are larger. Products of primary recrystallization (p.477) are, of course, excluded here. Details of crystal size in unaltered skeletal fabrics are given in Chapter 1. It is also apparent that the secondary, equigranular mosaic is less ordered geometrically than the intricately organized primary skeletal fabric. Even where the secondary mosaic is micrite (0.5–4 μ), this is coarser than the primary mosaic (as in the Papuan foraminiferids).

Evidence for dissolution-precipitation. Evidence that neomorphic spar grew by dissolution of certain crystals and precipitation from solution of others is, at present, necessarily indirect. The pores of carbonate sediments are mostly filled with water, except in the localized situation of terrestrial carbonate sediments above the water table or where they have oil or gas or other minerals. Given only a single alternative process, namely dry neomorphism in the solid state, it does not seem possible to conceive of crystal growth acting in the presence of a solution without the solution itself being involved. Moreover, the amount of work that must be done, or the energy threshold that must be exceeded, in order that growth shall begin, must on first principles, be lower for a wet than for a dry, solid state reaction. Field evidence in support of a wet reaction can be seen in the Eniwetok and Guam cores, wherein the products of neomorphism are associated with unconformities, and with a variety of dissolution cavities and cement fillings, and so probably with invasion of the sediment by fresh water. Schlanger related the accompanying loss of strontium in Guam to dissolution. Finally, it should be remembered that solid state crystal growth leading to development of a sparry calcite (by grain growth) would yield curved intercrystalline boundaries. The occurrence of plane boundaries in certain situations, as at the margins of sparry mosaics where they abut against micrite, and in some calcitized molluscan shells, can be regarded, therefore, as evidence favourable to a wet reaction.

Evidence for polymorphic transformation. Aggrading neomorphism has yielded calcite spar where the original material was micron-sized aragonite. This is plain from the data on Eniwetok and Guam (p.347) and Funafuti (p.353), and from the widespread sparry calcite replacement of mollusc shells in rocks of many ages. Therefore, the transformation of micron-sized aragonite to sparry calcite must be included within the ambit of aggrading neomorphism. That the transformation was wet is evident from previous discussions.

Argument for migration of solution films. If the processes of neomorphism are wet, then a system of solution-bearing cavities must have existed. Furthermore, since each crystal in a sparry mosaic has grown from a smaller crystal, all of the surface of each crystal must at some time have been bathed in an aqueous solution.

It is possible to argue that this cavity system must have lain between walls which were separated from each other by distances amounting only to a few molecular radii, for the following reasons. The crystals of, say, detrital aragonite needles or the crystals in shell walls (once sheathed in conchiolin which was later destroyed by oxidation), were originally either in contact only at a few points or separate from each other. The opposing crystal surfaces, as they grew, must have approached each other until they met. At this stage they must have formed either a compromise boundary (p.421) or a buried crystal face (p.422), as a result of the continued growth of one or both of the crystals. Now both of these types of intercrystalline boundary are regions of misfit between crystal lattices, regions where the adjacent lattices are elastically strained as a result of localized distortion. The elastic strain is accompanied by enhanced chemical energy. This type of intercrystalline boundary has been examined by metallurgists and even photographed with the field ion method (BRANDON et al., 1963). The results indicate that, in metals, the boundary is three to five atoms thick, and it must be presumed that in the calcite spar the order of magnitude of this disordered layer between crystals is about three to five unit cells thick.

It is also evident that, once all the crystals in a limestone make contact in the manner just described, any residual water can have resided only in one place—within this disordered layer that we call the intercrystalline boundary—and in this space only movement by diffusion is possible (see HELING, 1968).

The argument, therefore, is threefold. There must have been a system of pores: these pores were situated within the region of misfitting and distorted lattice that forms the intercrystalline boundary: these pores contained water. This intercrystalline water is referred to elsewhere in the book as a solution film (p.500). Within these film-like cavities (or wet intercrystalline boundaries) thermal agitation and other atomic vibrations will have ensured a continual exchange of ions between the two lattices. Where the exchange was unequal, there one crystal grew at the expense of another, by a process of wet boundary migration. This idea fits also with the known fact that such "microcracks" are kinetically favoured sites for nucleation (WOLLAST, 1971).

The misleading use of the term "coalescive neomorphism" (FOLK, 1965, p.22) to describe the process of neomorphism will doubtless by now be apparent. Coalescence (SPRY, 1969, p.156), as of water drops, implies a process in which two partners join together and share the product. In neomorphism one partner grows and destroys the other. Indeed, the process has been described in the metallurgical literature as cannibalism.

Reasons for the instability of micron-sized fabrics. The reasons why small crystals are less stable in solution than larger crystals have been discussed on p.349 in terms of supersolubility, greater density of edges and coigns, strain caused by abrasion, and the presence of the more soluble phases aragonite and high-magnesian calcites.

The processes of aggrading neomorphism

Standing upon the foregoing discussion it is possible to construct a working hypothesis to describe the processes of aggrading neomorphism, at least for the fresh water eogenetic environment for which we have most information. This hypothesis can only be a simple and crude affair, because our observational data are meagre: we know only that crystals of calcite spar appear in a partly lithified carbonate sediment, that they grow until the original fabric is entirely replaced by spar, and that some of them continue to grow at the expense of others until a centimicron-sized calcite mosaic has evolved. Our information is restricted for two reasons, partly because the Pleistocene and Recent limestones, in which the various early stages of neomorphism are to be found, have been examined only lightly from the neomorphic viewpoint, and partly because thin sections are more readily cut from well lithified rocks so that semi-consolidated sediments are apt to be ignored. The condition of the carbonate sediment at the onset of neomorphism is not known: at present there is no reason for supposing it to be other than multimineralogical, highly porous and wet. Finally, it is plain that neomorphism overlaps in time the lithification of micrite as described on p.504.

Aggrading neomorphism can conveniently be regarded as developing in three stages, though the evolution of these in any carbonate sediment is known to be irregular in both time and space. The three stages are here named: (*1*) nucleation, (*2*) framework, and (*3*) solution film. Their existence follows inescapably if it be assumed that the growth of secondary calcite spar begins in a lightly consolidated sediment and ends after the formation of a centimicron-sized spar with porosity down to 3 or 4%.

The nucleation stage. In the nucleation stage it must be supposed that certain relatively stable calcite crystals begin to grow at the expense of other crystals. The calcite crystals which could act as nuclei for neomorphism are either original detritus (fragments of calcite skeletons such as single prisms or whole echinoid plates), or secondary calcite spar formed by calcitization (earlier neomorphism) of, say, mollusc shells or, on a smaller scale, of aragonite needles. The reasons for presuming that growth occurs only on pre-existing nuclei (heterogeneous nucleation) are given on p.436. The relative stability of crystals must depend on mineralogy, $MgCO_3$ content, size, shape, strain caused by abrasion and the distribution of organic films (p.252), etc. The sediment, at the onset of nucleation, is assumed to be sufficiently cemented to be rigid, on the grounds that the very growth of calcite crystals implies a pore system within which precipitation is going on. The first visible result of the exposure of a carbonate sediment to fresh water is the appearance of a slight precipitate of low-magnesian calcite cement, causing the grains to cohere, as shown in Chapter 8 (p.326). This rapid development of a framework is the direct consequence of the unstable, multiphase composition of the sediment. The sediment is certainly highly porous and permeable. The source of carbonate for calcite growth is likely to be two-fold, locally dissolved crystals and exotic $CaCO_3$ brought in by fluid flow. Growth of the calcite crystals will take place by syntaxial precipitation partly in passive cavities (either primary pores or secondary, dissolution vugs) and partly in the solution films of intercrystalline boundaries— causing wet boundary migration (below). The micron-sized material replaced is both aragonitic and calcitic.

The framework stage. As the selected calcite crystals grow a time comes when a sufficient number of them have met (along compromise boundaries or buried crystal faces) to form a rigid framework of neomorphic calcite spar which is mechanically independent of the other residual components of the limestone. The passive porosity and the permeability are, at this time, greatly reduced and the role of the solution film, in the intercrystalline boundary, is correspondingly enhanced. This stage may not be so transitory as might be supposed, because by now the growth of the calcite crystals is much slower, since it depends increasingly on transport of ions by diffusion rather than by fluid flow.

The solution film stage. A stage is eventually reached when the rock is pure calcite spar, the only surviving porosity is the 1–2% in the solution films of the intercrystalline boundaries, and permeability is down to less than a millidarcy. The spar continues to coarsen as selected crystals grow at the expense of their neighbours by wet boundary migration, Ca^{2+} and CO_3^{2-} (or HCO_3^-) diffusing from one lattice to the other through the solution film. BAUSCH (1968) has shown how a clay mineral content of more than 2% in Jurassic limestones could inhibit this process. In this solution film stage of neomorphism the process is pure wet recrystallization, whereas in earlier stages other processes are involved, i.e. dissolution-precipitation (passive), wet transformation of aragonite to calcite, exchange of Mg^{2+} of high-magnesian calcites with Ca^{2+} in the pore solution.

It is essential to try to estimate the quantitative importance of wet boundary migration in the development of a centimicron-sized calcite mosaic. How much of the aggrading neomorphism is a result of dissolution-precipitation via a solution film? After all, it is this process which is normally implied in the use of the term *in situ* in connection with neomorphic fabrics. Progress will vary according to the initial density of nucleation. Where the calcite nuclei are far apart a porphyrotopic stage will precede the evolution of a sparry mosaic (FOLK's "porphyroid neomorphism", 1965, p.22). For a period calcite porphyrotopes will appear embedded in a markedly finer residue of micron-sized carbonate. If nucleation is relatively dense, the enlarging crystals will soon impinge on one another and porphyrotopic fabric will not develop (FOLK's "coalescive neomorphism", 1965, p.22). I earlier made an estimate of the amount of crystal growth that had taken place in the solution film stage (wet boundary migration) of Carboniferous limestones in Britain (BATHURST, 1958, p.29). Porphyrotopes are rare in the preserved intermediate stages and, for this reason, the distances between centres of the calcite nuclei cannot have exceeded a value of about 10 μ. Greater distances between initial centres of enlarged calcite crystals in a micrite matrix would have led to an obviously porphyrotopic fabric. From this figure I estimated that by the time that the crystal diameters had attained 15 μ (a generous estimate) the fabric must have reached the solution film stage. As the crystals in the fully developed calcite spar, as we see it now, have diameters from 50 to 100 μ, it is clear that, in terms of *volume* increase, from 97 to 99% or more of the growth of neomorphic spar had occurred during the solution film stage, as a result of wet boundary migration.

Experiments by GRÉGOIRE et al. (1969) on artificial neomorphism of nacre are not easy to assess, though the results are suggestive. Heating nacre of *Nautilus* to temperatures as high as 900 °C in vacuum or in air, they demonstrated a progressive change from small nacre tablets to large polyhedral crystals of calcite.

Fossilized stages. It is of the utmost importance that we should not confuse the fossilized intermediate states of neomorphism with the stages themselves. In diagenetically immature limestones (e.g., the Cainozoic limestones of Guam or Funafuti) the present fabric and mineralogy of the unaltered micrite may truly reflect the condition normally obtaining while neomorphism is active. However, in diagenetically mature limestones (e.g., Carboniferous of Europe, Pennsylvanian of the United States), which are low-magnesian calcite, the present micrite must be different from its mineralogically heterogeneous precursor which existed at the time when neomorphism was operating. In other words, when we see patches of secondary, sparry low-magnesian calcite associated with residual low-magnesian micritic calcite, we cannot logically presume that the spar is simply a product of recrystallization of the micrite. I think we are bound to assume, in view of the foregoing discussion, that the spar is the neomorphic product of a material that no longer exists, and that the original micrite (more finely crystalline, more porous, mineralogically heterogeneous) has since changed to low-magnesian calcite with porosity 1–2%. This is another way of saying that neomorphism probably goes on in some rocks during lithification.

The three end fabrics. It is curious that lithification of micrite and the growth of neomorphic spar yield, between them, a range of calcite fabrics which, once evolved, appear to resist strongly any further diagenetic change. These are arbitrarily classified as micrite (0.5–4 μ), microspar (5–50 μ) and pseudospar (50–100 μ), with the approximate crystal diameters given. A tentative explanation can be offered for this phenomenon. It does seem possible that the action of neomorphism is dependent on a supply of unstable crystals (aragonite or high-magnesian calcites or, simply, supersoluble). As soon as lithification is complete, when the limestone is practically pure low-magnesian calcite with porosity limited to solution films, then the supply of unstable crystals no longer exists. It could be that the development of neomorphic fabrics is tied to the development of lithification. The longer lithification is delayed, the more time there is for neomorphism to proceed. Delayed lithification is only possible where the nuclei of neomorphic spar are far apart, so that a groundmass of incompletely lithified micrite can persist between the encroaching neomorphic porphyrotopes for a prolonged period. A calcite pseudospar may represent the most delayed lithification and the most widely spaced nuclei, a microspar rather less delay and a micrite only slight delay and closely spaced nuclei. This hypothesis should not, however, be taken too seriously until it can be tested in the light of other data. (For an alternative by Folk see p.511.) We must set against it the rarity of porphyrotopes in limestones affected by neomorphism. On the other hand, in some Carboniferous biomicrites of North Wales, the radial-fibrous structures (p.484) do seem to represent a type of porphyrotopic neomorphism where large, radial-fibrous crystals have continued to grow at the expense of the surrounding micrite. The growth of these structures is further discussed on p.503.

Relative stability of fabrics. The selection of some crystals rather than others to be the nuclei of neomorphic spar is related, presumably, to the selective neomorphism of various types of material as described by Bathurst and by Banner and Wood (p.490). Bathurst's scale of susceptibility indicates, crudely, that stability increases with the degree of order of the crystal fabric. This, in turn, may influence the access of water to the intercrystalline boundaries. Crystals in shell walls may be protected by tightly packed envelopes of conchiolin. Banner and Wood noted that the first of their materials to suffer neomorphism were the tests of the Miliolina, with finely crystalline, porcellaneous, walls constructed of high-magnesian calcite. The more coarsely crystalline Rotaliina were affected later and, among these, the tests made of low-magnesian calcite were more stable than those of high-magnesian calcite.

The radial-fibrous masses. The growth of the stellate, or radial-fibrous, masses of spar (Fig.332) would appear to have progressed as follows—remembering that the fabric has a core of equigranular microspar, surrounded by elongate, sparry crystals with their long axes radially arranged. At first, a few of the crystals in the original micrite grew, by neomorphism, until a rather uniform mosaic of microspar had evolved. If we consider one of the new crystals of microspar, at the outer margin of the neomorphic sparry mosaic and abutting against unaltered micrite, then we can say of this crystal, as it continues to eat its neighbours: a time must come when the amount of energy required for the devouring of its microsparry neighbours (decimicron-sized crystals) will exceed by a critical amount the energy needed to devour smaller, micron-sized crystals. When this stage is reached the crystal will grow outward into the micrite more rapidly than it can grow laterally or inwardly at the expense of microspar. Thus it will become increasingly elongate and broad as it grows.

Fabric criteria for neomorphic spar

The distinction, in thin section, between the sparry mosaics of calcite formed by cementation and by aggrading neomorphism is at times extremely difficult. Too often I have encountered the enthusiastic student who, zealously *avant-garde* in his enlightened views on limestone fabrics, sees neomorphic spar everywhere. I earnestly recommend caution: specialists in this field at times disagree on the interpretation of sparry calcite. Moreover, it is all too easy to make a firm judgment when, as in so many geological situations, the past cannot return to disprove us. In many limestones the distinction between cement and neomorphic spar is straight-forward, because all the criteria point in one direction. In difficult cases it is important to employ as many criteria as possible. From the previous discussion on fabric interpretation it will be obvious that certain criteria, taken alone, can be misleading, for example, plane intercrystalline boundaries, vectorial change in crystal size, syntaxial rims. A neomorphic origin will be probable only if plane boundaries are rare, if crystal size increases centrifugally from a point, if syntaxial rims occur in a mudstone or wackestone—and so on. The likelihood of a neomorphic origin is rendered highly probable if enfacial junctions amount to less than 5% of all triple junctions.

LITHIFICATION OF MICRITE

The problem

The lithification of a carbonate mud, like the lithification of carbonate sediments in general, involves a change from a mixture of solid carbonate phases, bathed liberally in a pore solution, to a rock composed of low-magnesian calcite with a porosity of, perhaps, 2 or 3%. The problems are much the same whether we are dealing with a micrite or a calcarenite. The central difficulty is tantalizingly familiar: how to cement a carbonate ooze, while it is still largely uncompacted with initially a primary porosity of 50% or more. Slightness of compaction is indicated by the absence of crushing of delicate tests, the scarcity of drag or penetration effects among rigid allochems in a micritic matrix, and the general similarity between the intergranular matrix of micrite and the intragranular micrite occupying body chambers (PRAY, 1960). It is plain that, for many calcite-mudstones, lithification must have been contrapuntal, a weaving of two melodies, on the one hand the neomorphism of the original crystals, on the other the influx and precipitation of externally derived $CaCO_3$.

The initial composition of past oozes is not generally known, although studies with the electron microscope are revealing increasing detail in the less altered oozes, particularly coccolith oozes (HONJO and FISCHER, 1964; FISCHER, et al., 1967). It is likely that they ranged from aragonite needle muds like those on the Bahamas–Florida platform (p.277; Fig.231) and in the Persian Gulf (p.204), through mixed detrital carbonate oozes such as are known off British Honduras (p.212), to nearly pure calcitic oozes made of coccoliths and coccolith debris as in the Chalk (p.399; Fig.294). Particle size and shape would have varied with the origin (Chapter 2; discussion in FOLK, 1965, p.29). The well known tendency for organic matter to be concentrated in the finer sediments means that the silt and clay grade carbonates will have been rendered even more complex by the addition, not merely of organic matter, but of the bacteria, fungi and yeasts that accompany it. The loss of this non-carbonate material from muds by oxidation is not as efficient a process as in sand-grade, relatively well-circulated sediments, and the long term influence of residual organic products may, therefore, be more important in the neomorphism of carbonate oozes than of lime sands.

The many processes that, it must be assumed, proceed simultaneously during lithification are almost certainly wet ones. As the reduction of porosity continues, the water content must fall and it is sensible to expect that the reduction in the availability of solvent is responsible for a logarithmic slowing of the various processes. The relative importance of such processes is something we cannot yet judge. They include the wet transformation of aragonite to calcite (p.239), the dissolution of tiny supersoluble particles and prominences on grains (p.254), the transfer of Mg^{2+} from magnesian calcites (p.335), dissolution yielding passive voids or leading

Fig.349. Micrite. Replica of polished surface. Electron mic. Solnhofen Limestone, Upper Jurassic. Solnhofen, Germany. (Courtesy of W. Schwarzacher.)

Fig.350. Clotted structure *(structure grumeleuse)*. Are the clots of micrite mechanically deposited peloids in cement or micrite relics in neomorphic spar?

to compactive collapse, the influx of allochthonous $CaCO_3$ (p.439), pressure-solution (p.462), and the precipitation of cement (p.428)—in fact, all the paraphernalia of aggrading neomorphism.

The results of these processes are nevertheless clear (Fig.348, 349, 351). Many of the smallest particles (crystals), with shortest diameters as small as 0.1 μ, are lost, and the final calcite-mudstone has a new lower limit of crystal diameter probably about 0.5 μ and an upper limit of about 3–4 μ (BATHURST, 1959a, p.367; FOLK, 1965, p.29). There is commonly a mode around 1–2 μ (Fig.352). The restricted crystal size of the calcite-mudstones is combined with a tendency to equigranular texture, and surprisingly often, plane intercrystalline boundaries, features that have been amply demonstrated with the aid of the electron microscope. A list of studies based on electron photomicrographs is given at the end of this Chapter. Records of crystal size measurements are uncommon, a selection being given in Fig.352.

A note of caution is necessary here regarding the description of the crystal mosaics of micrites—and, indeed, of crystal mosaics, in general, which are: (*1*) monomineralic, (*2*) have no visible porosity, and (*3*) are equigranular. It is obvious in these circumstances that, any *plane* interface between two adjoining crystals must be either (*a*) a face of one of them, or (*b*) a compromise boundary.

Fig.351. Micrite (chinastone). Fracture surface. Scanning electron mic. Carboniferous Limestone. Clydach, Glamorgan. (Courtesy of W. J. Kennedy.)

Published descriptions of micrites (and commonly of dolomites) as idiomorphic or hypidiomorphic are based, it would seem, on the frequency of plane intercrystalline boundaries and not, as the terms imply, on the frequency of identified crystal faces. A simple arithmetic calculation will show that it is impossible theoretically for more than half the crystals in an equigranular non-porous mosaic to be euhedral. In natural conditions, where the crystals vary somewhat in size, shape and orientation and are bounded by numerous plane surfaces, the likelihood is that the proportion of euhedral crystals will be much lower, probably less than 10%. Subhedral crystals would not be so rare. What we need to have is a term which says that crystal mosaics have plane intercrystalline boundaries. I suggest that a new term "planomural" (Latin: *planus* = flat, *murus* = wall) be used in addition to the existing term "polygonal". I define these as follows.

Planomural: refers to a fabric of crystals and, with suitable prefix, indicates the proportion of plane intercrystalline boundaries among the total number of intercrystalline boundaries. Prefixes are taken from the Latin: *totus* = all, *multus* = many, *semi-* = half, *minor* = less, *nullus* = none.

Totiplanomural:	a fabric having all interfaces plane.
Multiplanomural:	a fabric having most interfaces plane.
Semiplanomural:	a fabric having half its interfaces plane.
Miniplanomural:	a fabric having a few of its interfaces plane.
Nulliplanomural:	a fabric having no plane interfaces.

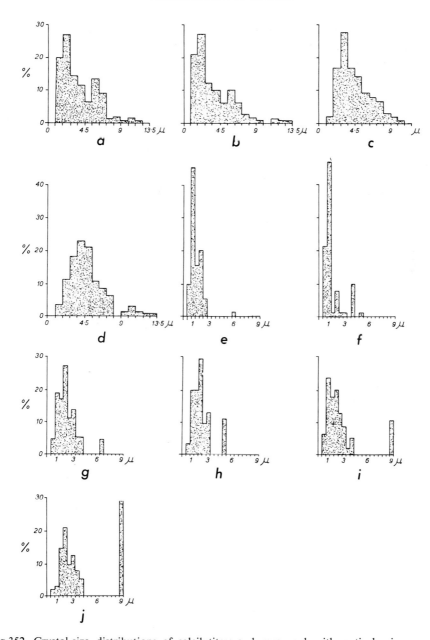

Fig.352. Crystal-size distributions of calcilutites: a–d measured with optical microscope on peels (from SCHWARZACHER, 1961); e–j measured on electron photomicrographs (after E. FLÜGEL, 1967b). a–d from Carboniferous Limestone, Northern Ireland. a. Reworked sediment in cavity. b. Bedded reef limestone. c. Massive limestone. d. Reef cover. e. Solnhofen Plattenkalk (Tithon). f. Calpionellen-Kalk (Tithon, Vocontischer Trog). g. Hallstätter Kalk (Nor, Salzkammergut). h. Neritischer Calpionellen-Kalk (Tithon, Vocontischer Trog). i. Girvanellen-Kalk (Perm, Karnische Alpen). j. Girvanellen-Kalk (Perm, Karnische Alpen).

For the description of a single crystal in the fabric, the existing word "polygonal" is convenient.

Polygonal: refers to a crystal enclosed by plane surfaces.

Degrees of polygonality can be indicated by adding the prefixes given above, thus, "multipolygonal" describes a crystal bounded mostly by plane surfaces.

The urgent reason for dwelling on this matter of the polygonality of equigranular, monomineralic, non-porous mosaics, is that the *exact* nature of the intercrystalline boundary is a question of profound importance in diagenetic studies. We can no longer afford the luxury of fabric descriptions which make statements about crystal interfaces for which there is no justification. It matters in fabric analysis whether an intercrystalline boundary is a crystal face or simply an unidentified plane interface.

Some processes

Implications of the form of intercrystalline boundaries. Observations that yield useful evidence of the processes of cementation and neomorphism of carbonate oozes are extremely scarce. Intercrystalline boundaries may be plane surfaces or curved. Plane intercrystalline boundaries may, as already shown, be crystal faces and/or compromise boundaries, but, despite claims that they are crystal faces, no evidence of identification has been produced. The planeness is certainly not only in the eye of the beholder, because in enlargements of electron photomicrographs, wherein crystals have diameters of 3–4 cm, a ruler can be laid along the boundary and the absence of detectable curvature discovered. Many apparently curved boundaries are probably plane surfaces lying obliquely to the axis of the microscope and outcropping on an irregular surface of the preparation. It is necessary to note that plane crystal interfaces can form either by passive growth at crystal–solution interfaces (i.e., rim cementation, p.429) or by growth *in situ* by neomorphism at the interface between a crystal face and a solution film (p.500).

There is here an awkward complication. It is well known that intercrystalline boundaries in metamorphosed rocks have simple forms which tend toward plane surfaces as equilibrium is approached. Illustrations in FISCHER et al. (1967) are instructive. It is necessary to learn to distinguish, therefore, between the fabrics of a micrite that has only experienced low temperature and pressure and of one that has undergone metamorphism. This should not be too difficult, but the appropriate research is awaited.

Other intercrystalline boundaries are amoeboid and thus it is probable that some crystal contacts are pressure-welded. SCHWARZACHER (1961, p.1500) has investigated the possibility that pressure-solution could have been responsible for the fabric of some silty micrites (pure calcite) in the Carboniferous Limestone of northwestern Ireland. He made acetate peels of etched, ground surfaces and in this

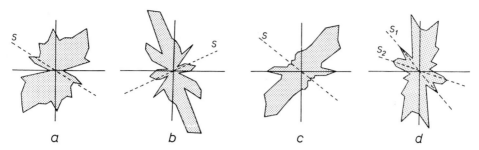

Fig.353. Orientation diagrams for longest axes of calcite crystals in Carboniferous calcilutites, Northern Ireland. a–c measured in vertical dip sections (*S* is trace of bedding-plane), d measured in vertical strike section. Number of crystals measured, respectively, 290, 100, 200, 100. (After SCHWARZACHER, 1961.)

way produced photomicrographs with exceptionally fine detail for light microscope preparations. He measured the longest axis within each crystal (in two dimensions) and found a strong maximum perpendicular to the bedding with a small submaximum in the bedding (Fig.353). This fabric is not what would be expected for pressure-solution but could, he suggested, have evolved in the presence of a pore solution moving vertically, along a hydrostatic pressure gradient. Pressure-solution might, he wrote, account for the submaximum. A difficulty in the way of interpreting any micrite fabric in terms of pressure-solution is the general paucity of compaction structures in these rocks. The early cementation which this implies seems to rule out the possibility of pressure-solution, at least as a major lithifying process. On the other hand pressure-welded contacts between micrite crystals are apparent in many illustrations by FISCHER et al. (1967), so the process is not unimportant.

Experiments of Hathaway and Robertson. The experiments of HATHAWAY and ROBERTSON (1961, p.301) and ROBERTSON (1965) are of uncertain relevance to the problem of the lithification of micrite—despite their inevitable fascination—because they imply a degree of compaction not generally found in nature (see also Grégoire et al., p.501). These authors subjected wet aragonite mud from the Bahamas (p.277) to various temperatures and pressures, in a cylinder from which surplus pore water could escape as the mud compacted. Temperatures ranged up to 400 °C (equivalent to a depth of about 20 km) and pressures to 3,450 bars (equivalent to an over-burden of about 10 km of average crust). Times ranged up to 63 days. Their series of electron photomicrographs (the mineralogy checked by X-ray diffraction) shows a change from aragonite to calcite accompanied by rounding of needles and the appearance of increasingly larger globular shaped masses of calcite (Fig.354). With their maximum pressure and time an equigranular mosaic of crystals was formed, with many plane intercrystalline boundaries. Photographs of the end product (fracture surface) are embarrassingly like those of natural micrites, such as those from the famous Upper Jurassic, Solnhofen Limestone of Bavaria, the Triassic Halstätter Limestone of Austria and the Carboniferous Limestone of

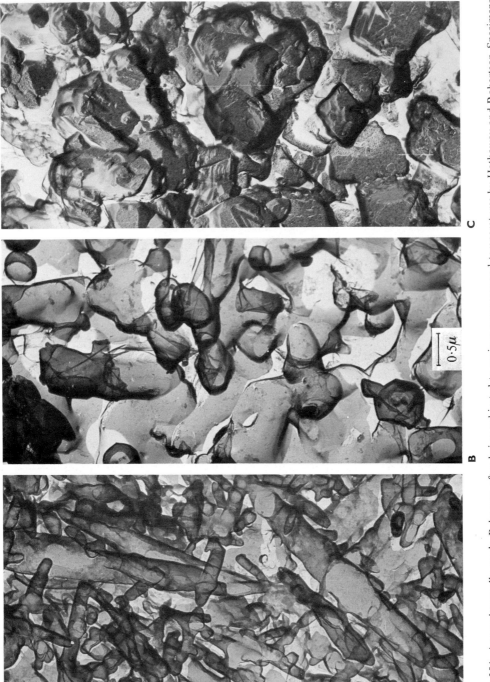

Fig.354. Aragonite needle muds, Bahamas, after being subjected to various pressures and temperatures by Hathaway and Robertson. Specimens now all calcite. A. 70 atm and 430°C for 5 h. B. 12.5 atm and 400°C for 2.6 days. C. 345 atm and 200°C for 63 days. (Courtesy of J. C. Hathaway.)

northern Ireland. Nevertheless, it is inconceivable that micrites in general were constructed with the heat and violence lavished upon the Bahamian muds by Hathaway and Robertson, but it is significant, as they point out, that an artificial calcite micrite can be produced in this way. Transformation of the aragonite was completed early in the process, so the remainder of the evolution consisted of wet recrystallization and cementation (the pore filling) of an accumulation of calcite crystals, giving a final specific gravity for the artificial rock of 1.9 (Solnhofen Limestone has sp. gr. of 2.6). The end product retains, therefore, a porosity of about 30% so that, to complete the lithification, a further 30% of $CaCO_3$ would have to be exotic in origin.

Significance of upper crystal size of calcite micrites. The widespread upper crystal diameter for the groundmass of lithified micrites at 3-4 μ, ignoring the included coarser skeletal debris, is intriguing. As I wrote earlier (BATHURST, 1959a, p.366), this "points to the existence of a universal threshold state at which fabric evolution stops and beyond which it can, but need not, continue". A possible reason for this, which would bear further investigation, is that a stage is reached in the combined neomorphism and cementation, when the porosity and permeability are so reduced that the transport of Ca^{2+} and CO_3^{2-} from one crystal face to another becomes slow even on a geological time scale. This stage would represent virtual stability. Some new driving force would be needed to induce further progress in neomorphism, such as elastic strain induced during deformation. Possibly a more plausible explanation has been made by FOLK (1965, p.36). He suggested that the crystal size of the micrite represents the *long* axes of the original crystals which "have mainly expanded in volume by fattening out rather than lengthening": substantial influx of allochthonous $CaCO_3$ would here be needed.

Structure grumeleuse

Before leaving the question of micrite lithification it is useful to glance for a moment at this clotted limestone (Fig.350), named and lucidly described by CAYEUX (1935, p.271) thus: ". . . elle montre de tout petits éléments calcaires, à pâte extrêmement fine, se détachant en gris sombre, de forme générale globuleuse ou irrégulière, dont les contours ne sont jamais franchement arrêtés, et sans différenciation d'aucune sorte. Ces matériaux, dont la microstructure est invariablement cryptocristalline, sont plongés dans une gangue de calcite incolore et grenue." Although the lucidity of Cayeux's succinct prose cannot be bettered, his description may be translated as follows: "**structure grumeleuse** appears as many little clots of an extremely finely crystalline calcite, standing out as dark grey in a matrix of colourless granular calcite. The clots are globular or irregular in shape, having outlines that are nowhere quite sharp and they lack internal differentiation of any kind." The fabric studied by Cayeux in the Carboniferous Limestone of France and, above all, of Belgium is known in limestones of many other places and ages. The two-component fabric, consisting of patches of micrite embedded in a matrix of microspar (p.513), occupies a central position in a spectrum of limestone fabrics. In one direction, from the *purely classificational* point of view, this clotted limestone passes into micrite, as the proportion of microspar falls and the boundaries between micrite patches and matrix become less obvious. In the other, it passes into *structure pseudoolithique* (CAYEUX, 1935), the familiar pelsparite of FOLK (1959). Cayeux noted particularly that the *grumeaux* (*grumeau* = clot or lump = peloid) are of silt grade, having diameters of about 50 μ. The passage from *structure grumeleuse* to micrite, again from a purely classificational aspect, is also marked by a merging of the micrite patches (Fig.350). CAYEUX's (1935, p.271) words cannot be improved: "Dans un stade de différenciation moins prononcé, les grumeaux, toujours séparés par de la calcite pure, contractent entre eux des adhérences multiples." ("In a less pronounced state of differentiation, the clots, though dispersed in clear calcite, touch one another in many places.") These various types of clotted calcilutites have been described and illustrated by BEALES (1958) from the Palaeozoic limestones of Alberta.

The existence of a range of petrographic types, from micrite through *structure grumeleuse*, or clotted limestone, to pelsparite, does not by itself mean that this variation represents the different stages in a continuous diagenetic evolution. Indeed the two leading writers on the question have expressed opposing views on the course of the diagenetic evolution. Cayeux, whose short but beautifully written section on this fabric I warmly recommend (CAYEUX, 1935, pp.271–272), believed that *structure grumeleuse* evolved by the growth of calcite crystals throughout the mass of an originally homogeneous micrite, and the gradual differentiation, thereby, of a more coarsely crystalline, continuous matrix separating discrete *residual* clots of microcrystalline (micritic) calcite. BEALES (1956, 1958, 1965), on the other hand, whose extensive researches into pellet limestones are an indispensable introduction to this field, thought that the processes operated the other way about, in that "many closely packed grains [peloids of this book] appear to have merged on recrystallization into a homogeneous microcrystalline rock differentiated with difficulty from calcilutite". The outstanding characteristic, and the most puzzling aspect, of this fabric, is the merged patches of micrite, with "des adhérences multiples". This obviously cannot be a primary fabric of mechanically deposited (p.546) peloids: it must be a secondary feature. Once this is granted, the field is open to speculation on the diagenetic evolution.

A factor that must have an important bearing on the development of any working hypothesis is the universality of clotted texture, the widespread occurrence of some degree of clotting in micrites of many ages and localities, despite differences of both depositional and diagenetic environments. Two additional factors also need to be kept in view. One is the ubiquitous formation of faecal pellets in modern carbonate sediments (p.85) allied to the common occurrence of heavily micritized skeletal debris (p.389). The other is the well-known nature of crystal growth fabrics. Where crystal growth starts at a number of points in a homogeneous crystal mosaic and spreads outward from these points, (FOLK's "porphyroid aggrading neomorphism", 1965, p.22) the resultant fabric is likely to contain radial elements (radial fibres or centrifugal increase in crystal size) and to yield, in ideal cases, a spherical growth front. An obvious example is the growth of spherulitic, radial-fibrous oöids or needle bundles (Fig.243, p.307). Having regard to these three factors, my own, purely intuitive, conclusion as to the origin of *structure grumeleuse* runs thus: the growth of sparry calcite, beginning at a number of points in a homogeneous carbonate mud (or mudstone), would produce a collection of relatively coarsely crystalline patches in a matrix of unaltered finely crystalline ooze. This is the reverse of what we find in *structure grumeleuse*. Furthermore, because of the ubiquitous occurrence of primary peloids, diagenetic processes are more likely to take the form of a reduction of the individuality of peloids rather than a fabrication of new ones. Such reduction may follow the merging of soft faecal pellets or the action of pressure-solution. *Structure grumeleuse* is common in algal stromatolites and it may well be that, in a peloidal calcarenite or calcisiltite trapped by algal filaments, photosynthesis can lead to the precipitation of additional aragonite micrite which, attaching itself to the existing peloids, will obscure their boundaries and cause merging (see grapestone, p.318). These are, nevertheless, only tentative conclusions and the elucidation of individual cases remains a matter of extraordinary difficulty, requiring patience and, maybe, the virtue of suspended judgment. With the help of the electron microscope some of the difficulties may be soluble by application of the fabric criteria for cements and neomorphic spar.

GROWTH OF MICROSPAR

Folk has described microspar as a limestone composed of equidimensional calcite crystals, with a relatively uniform size in the range 5 to 10 μ or less commonly to 50 μ, the commonest diameters being 5–6 μ (FOLK, 1959, p.32; 1962c, p.546, 561, 567; 1965, p.37). He has brought forward strong evidence to show that many microspars of Pennsylvanian, Cretaceous and Eocene age in the United States are the products of neomorphism of a pre-existing micrite. Recently I joined Dr. Folk in an examination of some of his thin sections and I do not doubt his conclusions with regard to these particular limestones. On the other hand, it does seem to me premature that we should regard all microspars as neomorphic products, as he would have us believe.

Folk's evidence for a neomorphic origin runs thus:

(*1*) Allochems float in three dimensions in the microspar so that it cannot be a cement.

(*2*) Though having uniform crystal size in any small area, microspar commonly passes, by gradual reduction of crystal size, into micrite.

(3) Microspar is commonly concentrated around the allochems in an otherwise micritic matrix.

(4) Some microspar adjacent to allochems has a radial-fibrous fabric.

(5) Some faecal pellets embedded in microspar have been replaced by identical microspar so that the only remaining evidence of their existence is an elliptical, brown organic stain.

Folk also brings as evidence the usual occurrence of clay-grade non-carbonates, generally clay minerals, as impurities in microspar. These tiny crystals, he rightly states, would not have been in hydraulic equilibrium with silt-sized carbonate crystals and thus, since the original sediment cannot have been an equilibrium mixture, the crystal size of the calcite must be a secondary feature. This testimony can be countered in part by arguing that clay crystals in sea water are normally clotted to give silt-sized pellets. A more serious consideration is that one of the special features of carbonate sediments is that they are (or were) *never* mixtures of grains in hydraulic equilibrium. Many carbonate sediments, being locally formed, were not dependent on water transport for their existence and their composition.

One of the most interesting characteristics of microsparites described by Folk is their usual association with clays which occur, not only as impurity, but as discrete shales. Indeed, some of the microspars described by Folk are nodules in shale. It may well be that the instability of argillaceous carbonate oozes, their ready susceptibility to neomorphism, is related to the presence of clay crystals with their tightly adsorbed water films. A sediment so persistently wet can hardly have the stability of a pure carbonate ooze. Microspar is also the main component of the calichified limestones on many hills in Texas (R. L. Folk, personal communication, 1969).

Folk's contention that microspars are secondary fabrics is further supported if one considers the range of possible source materials. The crystals which are released by the breakdown of skeletal carbonate to its ultimate components are practically all of micron-size. In the molluscan or the brachiopod shell, in the wall of the foraminiferid or bryozoan, in the calcareous alga or the trilobite, the component crystals are finer than microspar, and disintegration of these skeletons would tend to give a micritic sediment. Only the echinoderms, with some of the larger molluscan prisms, among skeletons, could be expected to yield calcarenites or calcisiltites composed of unit-crystal grains. It is worth mentioning that there is a peculiar exception to the generally equidimensional microspar: this is the fabric of elongate loaf-shaped crystals described by FOLK (1965, p.39), also known in the British Carboniferous Limestone: could these be skeletal in origin? An even more elongate habit is figured by FISCHER et al. (1967, fig.34).

An exception to the neomorphic microspar is surely the ubiquitous microspar that occurs as a geopetal, crystal mosaic deposit on the floors of fenestral cavities in bird's eye type micrites. In this category we include Dunham's syndiagenetic vadose silts (p.356). The mosaic of equant calcite crystals occupies the floor of a cavity, is covered by a space-filling calcite cement, and the filled cavity is enclosed by micrite. This micrite normally shows no signs of neomorphism and the junction between micrite and geopetal microspar is sharp (Fig.300, p.418). There is no reason to suppose, therefore, that this microspar has replaced a micrite. Moreover, owing to its crystal size it cannot be reworked micrite. It must have formed after the deposition of the micrite but before appreciable compaction (i.e., while cavities remained open). It is normally devoid of skeletal grains, and shows simple, commonly plane, intercrystalline boundaries. Yet, being geopetal, it was mechanically deposited. One possible mode of formation is chemical precipitation in suspension of discrete crystals, though it is not clear why, in cavities in micrites, the precipitation of a cloud of carbonate crystals should so frequently have occurred. Dunham has suggested (p.357) derivation from the primary sediment by combined winnowing and selective dissolution. It is not certain, of course, that the initial deposit was calcite. The mosaic may be a neomorphic alteration of a deposit of aragonite.

FURTHER READING

As a general background there can be no more profitable reading than FOLK's (1965) thoughtful and widely ranging review of recrystallization in ancient limestones. Additional material is contained in BATHURST (1958, 1959a), in the important papers by GARRELS and DREYER (1952), BANNER and WOOD (1964) and MATTAVELLI and TONNA (1967) and the stimulating review of fabrics by FISCHER et al. (1967). WEYL (1964, 1967) gives a clear and practical account of the basic matters of solution kinetics, fluid flow, diffusion and reaction rates. The question of the genesis of *structure grumeleuse*, with its far reaching implications, is dealt with by CAYEUX (1935, p.271) and BEALES (1956, 1958, 1965). Data on neomorphic spar in its early stages is to be found in the papers of CULLIS (1904), CRICKMAY (1945) and SCHLANGER (1964). The subject of crystal growth is covered in the reports of a number of conferences: ANONYMOUS (1958–1968), DOREMUS et al. (1958), PEISER (1967). Other references are given in the text.

Studies of calcilutites with the electron microscope

Pioneering work on micrites is contained in papers by GRUNAU and STUDER (1956), SEELIGER (1956), GRUNAU (1959), HATHAWAY and ROBERTSON (1961), ALBISSIN (1963), GRÉGOIRE and MONTY (1963), HONJO and FISCHER (1964), SHOJI and FOLK (1964), FLÜGEL and FENNINGER (1966), R. D. HARVEY (1966), FISCHER et al. (1967), FLÜGEL (1967a, b), FLÜGEL and FRANZ (1967), GARRISON and BAILEY (1967), FARINACCI (1968), FLÜGEL et al. (1968) and GILLOTT (1969). The most comprehensive survey of the field is contained in the stimulating book by FISCHER et al. (1967).

Additional references not given in the preceding chapter

On fabric selective neomorphism, CHANDA (1967a, b). On experimental thermal alteration of *Nautilus* shell, GRÉGOIRE (1964, 1968). On diagenetic fabrics, WEST (1965), NOVELLI and MATTAVELLI (1967) and R. C. L. WILSON (1967).

Chapter 13

RECENT DOLOMITES

INTRODUCTION

In 1952 Chave and Spotts (p.236) demonstrated that the Mg^{2+} in the magnesium-bearing calcite skeletons of invertebrates is in solid solution in the calcite lattice. At about the same time Graf and Goldsmith (p.238) reported the synthesis of non-ideal, calcian dolomite (their protodolomite). Since that time it has become increasingly obvious that an understanding of the development of the magnesian calcites, of dolomitization, and of magnesium metasomatism generally, can only proceed from a study of the total magnesium cycle in carbonate sediments. The discoveries of Recent dolomites in Australian playa lakes and in the sabkhas of the Persian Gulf and the West Indies have further emphasized the close interrelation of the various physical and chemical reactions in which magnesium takes part, both in open water and in the subsurface environment. The history of research into the origin and occurrence of the magnesium-bearing sedimentary carbonates has recently been exhaustively surveyed by FRIEDMAN and SANDERS (1967), so that no useful purpose would be served by presenting another general study of this topic. Instead, an attempt is made here to examine briefly one aspect of the field, in which progress has lately been especially rapid—namely the distribution and precipitation of dolomite in Recent sediments. This chapter will draw particularly on the researches in the Coorong of South Australia carried out in the University of Adelaide, on the work of members of The Imperial College, London, and of the Shell group of oil companies, in the Persian Gulf and West Indies, and the results of recent laboratory syntheses by a number of individuals.

Details of mineralogy have already been given in Chapter 6 (p.238). The reluctance of Mg^{2+} to form solids from supersaturated solution has already been noted (p.251), a tendency related, at least in part, to the high level of hydration of the magnesium ion in solution: after beryllium it is the smallest of the alkali earths and has a correspondingly low transference number.

CALCIUM-MAGNESIUM CARBONATE SEDIMENTS
OF THE COORONG DISTRICT, SOUTH AUSTRALIA

In this region of playa lakes and an ephemeral lagoon (Fig.355), a varied group of carbonate sediments is accumulating in an equally varied range of depositional

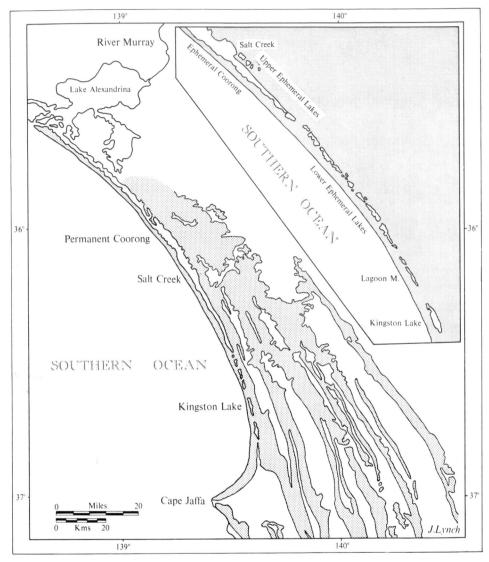

Fig.355. Map of the Coorong district, South Australia. Dunes shown stippled. Inset, enlarged map of part of district.

environments. Members of the Department of Geology in Adelaide University have brought together, over the last decade, a collection of data concerning the mineralogy of the magnesian carbonate sediments and their environments which has already led to an increased understanding of the processes of magnesian carbonate precipitation. So far eleven papers have appeared, variously by A. R. Alderman, H. C. W. Skinner, C. C. Von der Borch, B. J. Skinner and M. Rubin, and these are listed at the end of the chapter. As an open air laboratory on a grand scale the region rivals the much-studied Trucial coast. There is no doubt that six minerals are forming in and around the Coorong—aragonite, magnesian calcite, calcian dolomite, dolomite, magnesite and hydromagnesite—but their mode of formation is still far from clear. Researches so far carried out have raised about as many problems as they have solved. It is clear, however, that the chemical environment differs in important and interesting ways from that in the sabkha of the Persian Gulf.

The environment

South of Adelaide, between Lake Alexandrina and Cape Jaffa (Fig.355), an arcuate coast runs for 160 km, facing southwestward into the Southern Ocean. For most of its length this coast is the western shore of the Younghusband Peninsula, a stretch of Recent sand dunes forming a ridge about 1 km wide and about 120 km long. Behind this long but narrow barrier is the Coorong, a lagoon little wider than the Peninsula itself, which is connected with the sea at its northern end through Lake Alexandrina, where it is joined by the River Murray. Bounding the Coorong on its landward side is a further range of dunes and, beyond this, other nearly parallel ranges extend in a series of ridges up to a distance of 80 km from the coast. Between the ridges lie flat depressions from 0.4 to 8 km wide. The Coorong lagoon is only about 100 km long, but the valley between the two series of dunes extends southward for a further 50 km. Within this valley there are a number of playa lakes, the remains of what must once have been a longer, Early Quaternary, Coorong lagoon. There are other intermittent lakes and swamps between the dune ridges east of the Coorong. The southernmost 30 km of the Coorong, in the neighbourhood of Salt Creek, is itself intermittently flooded and evaporated to dryness in the summer. It is convenient to distinguish between this Ephemeral Coorong and the Permanent Coorong to the north.

It is the composition of the bottom sediments of the Ephemeral Coorong and neighbouring lakes which has aroused such great interest. When the flood waters have evaporated in the late summer, the floor is covered by a deposit of clay-grade carbonate, plastic when wet but hard and firm when dry. It is this that contains, in one place or another, the range of minerals given below. The magnesium and calcium carbonate crystals, seen with the electron microscope, have diameters from less than 1 μ to 20 μ. Other minerals are gypsum, halite, quartz, celestite, kaolinite, montmorillonite, illite, with organic matter and shells (both aragonite and calcite; SKINNER, 1963, p.452). Together these non-carbonate minerals never exceed 20% of the sediment. A high proportion of the halides in the lakes is believed to be derived from the sea as windblown salt, effectively collected by runoff (VON DER BORCH, 1965b).

The sand dunes, which are consolidated, have been cut in places by drains to enable water to escape seaward, so enabling some of the interdune areas to be used as pasture. The dune sand is a mixture of quartz and skeletal carbonate particles and is porous, so that rain water rapidly soaks in. The floors of the interdune areas are impermeable and on them the water accumulates to form the temporary swamps and lakes. These and the Ephemeral Coorong are gradually being filled, not only by the carbonates listed, but by windblown sediment and shells. The covering of halide salts and organic matter which overlies the carbonate muds, when they are first exposed, may be quickly blown away.

The organic life in the Coorong and the lakes is dominated by the subaquatic brackish to salt-water herb, *Ruppia maritima*, besides which there are copepods, small molluscs, sedges and small algae.

Rainfall varies from about 38 cm a year in the northern Coorong to 86 cm in the southern part of the district and is concentrated in the winter (April to September). The Ephemeral Coorong and the lakes become gradually desiccated in November, as the flood water is lost by evaporation, and are generally dry from December to March or April. Water temperatures in the lakes and lagoon are known to range between 10.5 and 28°C, but it is likely that higher temperatures occur for short periods in the shallower waters in the summer afternoons. The variation in pH of the water is remarkable, with a range from 8.2 (the value for open sea water) to 10.3. The wide range of the ratio Mg^{2+}/Ca^{2+} is also extraordinary, bearing in mind the sea water molar ratio of 5.4. In the various waters this ratio can be as low as 2.5 and as high as, perhaps, 20. Salinities are known from 10 to 274‰.

Minerals and chemical parameters

The pertinent details of location, mineralogy and chemical parameters are shown in Table XXII (ALDERMAN, 1965b). The water bodies in which the minerals are formed are given in order of age, that is to say the length of time since they were last connected with the open sea. The Ephemeral Coorong still has periodic connection with the sea through the Permanent Coorong. The Ephemeral Lakes, type III, on the other hand, are believed to have had the longest isolation from the sea.

In all, four factors have been found to increase together:
(*1*) Age of the water body (time since isolation from the sea).
(*2*) Mg^{2+}/Ca^{2+} in the water.
(*3*) Total Mg^{2+} in the sediment.
(*4*) (Less definitely) maximum pH in the water.

It is known that, during times of flood, the Mg^{2+} content of the water remains reasonably constant, but the Ca^{2+} content falls as $CaCO_3$ is precipitated, presumably either as skeletal material or as inorganic crystals. The final salts to form, mainly halides, are thus enriched in Mg^{2+} compared with the original sea water. It is believed that sufficient of these will escape deflation and remain on the floor, until the next annual flood, to enable the Mg^{2+}/Ca^{2+} ratio in the water for one

TABLE XXII

GEOGRAPHICAL DISTRIBUTION OF MINERALS AND CHEMICAL PARAMETERS IN THE COORONG AND NEIGHBOURING LAKES
(From ALDERMAN, 1965b)

Environment	Characteristic sediment	Max. pH of water	Approx. Mg/Ca in water at time of max. pH
Permanent Coorong	aragonite + mag. calcite	8.4	3–4
Ephemeral Coorong	aragonite + mag. calcite	8.6	5
Lower (northern) Ephemeral Lakes, type I	mag. calcite	9.1	6
Lower (southern) Ephemeral Lakes, type II	mag. calcite + protodolomite	9–10	7–8
Upper Ephemeral Lakes, type I	"ordered" dolomite	10	10
type II	"ordered" dolomite + magnesite	10.2	20–100
type III	aragonite + hydromagnesite	9.1	30

year to be transmitted to the flood waters of the succeeding year. The Mg^{2+}/Ca^{2+} ratio could thus be passed on as an inheritance year by year, increasing in value all the time, so that, the older the environment the higher its Mg^{2+}/Ca^{2+} ratio in the water. (This is not the only way of developing waters with high Mg^{2+}/Ca^{2+}. In the Persian Gulf (p.528) and in the West Indies (p.531) the ratio is raised by precipitation of gypsum.) Similarly, the total content of Mg^{2+} accumulating in the sediments increases with the age of the depositional environment. The higher values of pH have been found in regions of greater density of the aquatic vegetation, especially *Ruppia*. It is clear that the environments of the Upper Ephemeral Lakes, type III, do not fit into this evolutionary series. The precipitation of aragonite and hydromagnesite is proceeding in a special environment where the lake water has an unusually high ratio CO_3^{2-}/HCO_3^{-} caused by the introduction of ground water (ALDERMAN and VON DER BORCH, 1963). The mineralogy of the sediments also varies geographically. Aragonite-magnesian calcite muds are found in lagoons with permanent or intermittent connection with the sea, whereas the calcian dolomite-bearing and dolomite-bearing muds occur only in the isolated ephemeral lakes. Ratios of magnesian calcite to dolomite vary from 1/4 to 100/1. Dolomite was found everywhere accompanied by calcite: it was never recorded as the only carbonate phase. Not only is dolomite content higher in proportion to magnesian calcite in the older lakes, but, in cores through the carbonate muds, the dolomite increases downward, again at the expense of magnesian calcite (ALDERMAN and VON DER BORCH, 1960; SKINNER et al., 1963). A radiocarbon date of the upper surface of 50 cm of pure dolomite gave 300 ± 250 years (VON DER BORCH, 1965b). The rate of accumulation is estimated to be between 0.2 and 0.5 mm/year (SKINNER et al., 1963).

Skinner et al. regarded the dolomite as a protodolomite (p.238), but DEGENS and EPSTEIN (1964) identified it as well-ordered dolomite. X-ray diffraction analyses (SKINNER, 1963) show that the unit cell of the dolomite, which is calcian, is larger than that of stoichiometric dolomite, as would be expected for a lattice containing the larger Ca^{2+} ion in amounts up to 6% in excess of the stoichiometric amount. For a similar reason the unit cell of the magnesian calcite is on the small side for calcite, owing to the substitution of the smaller Mg^{2+} ion giving cation compositions from $Ca_{98}Mg_2$ to $Ca_{77}Mg_{23}$. The calcite of shells is quite distinct having a range from pure calcite to $Ca_{91}Mg_9$.

At certain times of the year the waters of the Ephemeral Coorong and the lakes are milky with suspended carbonate sediment. In view of the possibility that this is a primary precipitate, it is particularly interesting to read the analyses of suspended solids (November, December) and parent waters (November) made by SKINNER (1963). Some of her data are given in Table XXIII. It should be noted that, as in the floor sediments, dolomite never occurs without calcite in the suspension; also that the ratio magnesian calcite/dolomite varies from 1/1 to 100/1 whereas, in the floor sediments, the ratio is locally as low as 1/4. No aragonite was recorded in suspension in the Ephemeral Coorong, despite its presence in the floor sediment (Table XXII). Analyses of suspended carbonates from the Ephemeral Coorong and the Lower Ephemeral Lakes (in addition to those in Table XXIII) show a consistent north–south variation in the level of magnesium in the calcite and the proportion of dolomite in the suspension. The more southerly

TABLE XXIII

CHEMICAL PARAMETERS FOR WATERS CARRYING SUSPENSIONS OF MAGNESIAN CALCITE AND PROTO-
DOLOMITE DURING NOVEMBER IN THE COORONG AND NEIGHBOURING LAKES
(From SKINNER, 1963)

Environment	Suspension	pH of water	Mg/Ca water[1]	Salinity (%)
Ephemeral Coorong 54	mag. calcite	8.5	6.7	141
Kingston Lake K1	mag. calcite equals protodolomite	—	6.9	24
Kingston Lake 26		9.0	6.1	16
Kingston Lake 27		9.2	6.1	16

[1] Molar ratio calculated from Skinner's data.

(thus the older) the water body, the less magnesian is the calcite and the lower is the ratio calcite/dolomite. Also this ratio appears to be characteristic for particular lakes or particular parts of the Ephemeral Coorong. The disparity between Skinner's figures for Mg^{2+}/Ca^{2+} in the water (Table XXIII) and those given by Alderman (Table XXII) may, in part, reflect the different times of sampling. Skinner's samples were taken in November whereas maximum pH occurs in October. Water bodies may evaporate and precipitate at different rates and may receive different amounts of fresh runoff water, resulting in such major differences of salinity as seen in Table XXIII and, consequently, of Mg^{2+}/Ca^{2+}.

It is remarkable that the salinities of the waters containing the assumed precipitates (SKINNER, 1963) are not only highly varied but are commonly lower than that of normal sea water. There is here a marked contrast with conditions in the lagoons of the Persian Gulf where precipitation is closely related to the high level of concentration in hypersaline water.

The role of pH in the chemical evolution of the various waters and their sediments may be of critical significance. Values range from 8.4, slightly above that of sea water, to the very high maximum of about 10.2. The pH not only varies with time, but is restricted to a distinctive range for each water body. The pH rises from the onset of flooding in the winter (April to September) and almost invariably reaches a maximum about October. From then the pH declines, and it is doubtless important that this decline begins when the rising salinity has reached 60‰, because this is about the highest salinity that *Ruppia* can tolerate (ALDERMAN, 1965b). In view of the repeated observations that the milky suspensions of magnesian calcite and dolomite are found at a time when plant growth is most vigorous and are concentrated in areas of the denser stands of *Ruppia*, the possibility of a critical relation between pH, photosynthesis and precipitation must be considered.

Hypotheses

There can be no doubt that carbonate minerals are currently precipitated in the flooded Ephemeral Coorong and Ephemeral Lakes. The uncertainty lies in our understanding of the place of precipitation, its time and kinetics. It must always be extremely difficult to demonstrate the very act of precipitation in a natural environment. The appearance of a milky suspension, so readily observed and controlled in a test tube, may, on a lake floor, be the result of stirring by currents or wind-wave turbulence, photosynthetic bubbling from the leaves of *Ruppia* or the perambulations of ostracods (VON DER BORCH, 1965a). Indeed, such a slow rate of deposition as 0.2–0.5 mm/year could not yield a visible milkiness for days on end, and it seems clear that the maintenance of turbidity in such conditions requires continual disturbance of the mud floor. It is, therefore, not certain that the suspended sediments of November are freshly nucleated crystals. We are forced back upon circumstantial evidence—and this is rather scarce, though suggestive. The problem is complicated by the varied association of carbonate minerals, some of which are presumably primary though others are probably the products of later reactions between crystals and saline water.

It seems likely that salinity has not the same importance in the Coorong district that it has in the Persian Gulf, though to plead the low salinities at the time of the November milkiness may be to beg the question. On the other hand, the remarkably high pH of more than 10, associated with the most vigorous plant photosynthesis in October, does suggest that a reduction in the solubility product of $CaCO_3$ is brought about by abstraction of CO_2 from the water by plant metabolism, a factor apparently of scant importance in the Persian Gulf. The precipitation of aragonite with the magnesian calcite in the Ephemeral Coorong (Table XXII) is consistent with the known ionic balance of the water which is close to that of sea water (p.272). In the lakes, where the ratio Mg^{2+}/Ca^{2+} is higher, the inhibiting effect of the hydrated Mg^{2+} ions on the formation of magnesian calcite may not be so great. Deposition of magnesian calcite indicates one of three possibilities. Possibly primary aragonite is rapidly transformed by dissolution-precipitation to magnesian calcite with substitution of some Mg^{2+}, or the free energy of hydration of the Mg^{2+} is sufficiently reduced at the high values of Mg^{2+}/Ca^{2+}, and of salinity and temperature, to enable a magnesian calcite to grow, or the level of super-saturation for $CaCO_3$ is so low that calcite can grow (p.259). SKINNER (1963) concluded that the formation of the dolomite must be contemporary with sedimentation, because the dolomite is present in the surface sediments and, moreover, the ratio dolomite/calcite in these sediments in the Coorong increases regularly southward (i.e., with age of isolation). These factors, she emphasized, are inconsistent with delayed diagenetic dolomitization. Her assumption that the dolomite is a primary phase coprecipitating with magnesian calcite may, nevertheless, be questioned bearing in mind the possibility of dolomitization of aragonite or of magnesian carbonates such as may take place in Deep Spring Lake (p.539). In the Coorong, the process, though very slow, keeps pace with the equally slow sedimentation rate. VON DER BORCH (1965b) has put forward the suggestion that, as a lake dries up, the surrounding water table falls below the lake bed. In this condition the highly concentrated lake water must seep down through the lake floor sediments, dolomitizing as it goes. The possible existence of this seepage reflux mechanism (p.532) is supported by the high salinities of groundwaters collected from auger holes adjacent to the lakes during the dry season. A detailed study of the hydrology, akin to that carried out on Bonaire (p.532), would be of great interest.

A high rate of evaporation and photosynthesis in Lake Balaton, Hungary, are probably responsible for the high ratio Mg^{2+}/Ca^{2+} in the fresh water, and for the precipitation of high-magnesian calcite and a calcian dolomite (MÜLLER, 1970).

DOLOMITES OF THE PERSIAN GULF

The environment and mineralogy of the sabkha, bordering the Trucial coast embayment and the Qatar Peninsula, have been described in Chapter 4 where reference was made to dolomitic carbonate muds. Researches in the area have been pursued in considerable detail, yielding by far the most fundamental revelations concerning the growth of dolomite yet to appear. Radiocarbon dates show that all the sabkha sediments are less than 4,000 years old (M. Rubin in KINSMAN, 1966b). The succession of processes differs from that in the Coorong.

Chemical parameters

Dolomite grows under the sabkha surface and in the high algal flat. It replaces aragonite muds of lagoonal origin. The content of dolomite in the capillary zone above the water table increases landward until, near the landward edge of the sabkha, the sediment is nearly pure dolomite. Even at 1 m below the surface the sediment is nearly 50% dolomite. The pore water has two sources: some of it is flood water, but most of it is believed to have moved sideways through the sediment from the lagoon to replace water lost as a result of the intense evaporation from the sabkha surface. The concentration of ions in the pore water increases, both away from the lagoon, and vertically upward from the water table. The greatest lateral change takes place across the algal flat (p.199) such that, at the landward margin of the flat the chlorinity of the pore water is more than twice that at its seaward margin (Fig.356). Near the landward termination of the sabkha the chlorinity decreases slightly. The Mg^{2+}/Ca^{2+} molar ratio in the pore water also rises landward across the algal flats from about 4 or 5, reaching a peak just inside the sabkha close to its seaward margin, before falling gradually landward to about 3 (Fig.356). The peak values of this ratio are commonly between 12 and 20. The pH in the pore water falls landward, from about 7.2, at the seaward edge of the algal flat, to values in the sabkha which vary between about 6.8 and 6.0. The brine chemistry is summarized in Fig.356, 357. When changes away from the lagoon are plotted against *distance*,

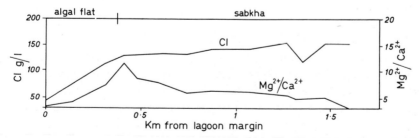

Fig.356. Variation of brine concentration (chloride in g/l) and the ratio Mg^{2+}/Ca^{2+} with distance from the outer edge of the algal flat. (After ILLING et al., 1965.)

the variations revealed are somewhat irregular because the chemical gradients in
the sabkha are modified by local circumstances: these include the extent and time
of the last flood, accumulation of rain water, position of the water table, and composition of the sediment, both textural and mineralogical. A more coherent chemical picture emerges when the parameters are plotted against *concentration*,
expressed in terms of standard sea water (Fig.357). The change in concentration
away from the lagoon is almost entirely in a landward direction and this parameter
is also directly related to the lateral change in brine chemistry. On the assumption
that the pore water has moved in from the lagoon, the concentration is also a record
of brine history.

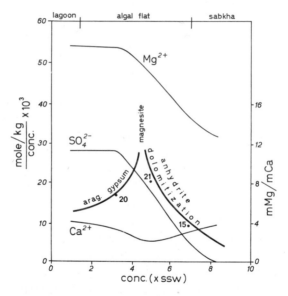

Fig.357. Relations between molar concentrations of ions, and the molar ratio Mg^{2+}/Ca^{2+}, and
the concentration of the brine in the Abu Dhabi sabkha. The ionic concentration is expressed in
terms of the total concentration in order to eliminate the changes caused by increases in concentration alone. (After KINSMAN, 1964b, 1966b.)

Dolomite

The distribution of minerals in the sediments is described in Chapter 4. The
dolomite has a range of composition between $Ca_{50}Mg_{50}$ and $Ca_{55}Mg_{45}$, and a
disordered lattice. KINSMAN (1964b) has tried to express the degree of disorder
quantitatively, using the ordering lattice reflection 01.5 and the non-ordering
reflection 11.3. He compared their intensities as shown in Fig.358, making the
assumption that the degree of ordering is proportional to the intensity of the ordering reflection. The two end points he took from standard data in GRAF (1961). On
this basis the ordering in the Trucial coast dolomites is in the range 40–95% and

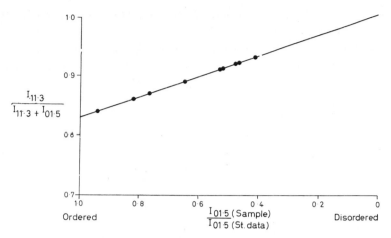

Fig.358. Disorder of dolomite expressed in terms of the intensities of the non-ordering and ordering lattice reflections 11.3 and 01.5. (After KINSMAN, 1964b.)

the mineral is a protodolomite as defined on p.238. This identification is supported by X-ray diffraction powder analyses showing absence of the superstructure reflections (221) and (111), see CURTIS et al. (1963).

Hypothesis

It is as well to begin by noting that the zone in which dolomite is forming, and in which sabkha diagenesis in general is believed to be active, extends little more than a metre below the surface. As the sabkha may be as much as 10 km wide, this active zone is, relatively, a very thin sheet and the distribution of chemical parameters is dominantly two dimensional. Above the fluctuating water table (in the vadose zone) the proportion of pores filled with brine decreases upward, through a region where some pores are filled with brine to one where the brine is restricted to a film on the detrital grains, especially around their contacts, and finally to dry sediment. As water is lost by evaporation in the upper sediment it is replaced by water withdrawn from the phreatic zone. Workers on the Persian Gulf sabkhas make the assumption that, in the absence of any other adequate source of water in this arid region, the sabkha brines are derived from the lagoon and move landward through the sediment below the water table. Studies based on brine samples from just below the water table have all yielded the same general results (KINSMAN, 1964b, 1966b; BUTLER, 1965, 1969; ILLING et al., 1965).

As the initial composition of the brine (as lagoonal water) is known, any diagenetic changes involving precipitation from the landward moving brine, or reaction between the brine and the solid phases, should be reflected in the brine chemistry (KINSMAN, 1964b, 1966b, p.313; BUTLER, 1965, 1969). This assumption is the cornerstone of the interpretation of the chemical data. Its logic is borne out by the constant ratio Na^+/Cl^- in all lagoon and brine samples and by the distribution of the chemical parameters.

In a traverse away from the lagoon the diagenetic (authigenic) minerals appear in the order aragonite–gypsum–magnesite–dolomite–anhydrite. It is plain from field and microscopical data that dolomite has formed at the expense of aragonite: as the aragonite content falls so the content of dolomite (and of anhydrite) increases. ILLING et al. (1965) found that the rhombs of dolomite have replaced selectively the aragonite mud, before the accompanying aragonite faecal pellets. Apparent replacement of the high-magnesian calcite of foraminiferids may, in fact, be only a replacement of the aragonite-filled bores (in micrite envelopes, p.386).

In Fig.357 the parameters for the water of the inner lagoon are shown on the far left. As the salinity of the inner lagoon is as high as 60‰ in Abu Dhabi, the concentration of 2 ($\times$ SSW) represents a position in the algal flat close to its seaward margin. The concentration of 7 marks the approximate junction between algal flat and sabkha.

In the algal flat Ca^{2+} decreases continuously landward because some aragonite is precipitated from the younger (seaward) brine and great quantities of gypsum are precipitated from the older brine. The onset of gypsum precipitation is shown by the sudden descent of the SO_4^{2-} line at a concentration of about 3.4. This is the concentration at which gypsum should start to form from sea water (POSNJAK, 1938, 1940). The continual loss of Ca^{2+} as the pore water moves landward causes a persistent rise in the ratio Mg^{2+}/Ca^{2+}, a factor of outstanding significance for the dolomitization process. This ratio in the lagoon is about 5.5, but at a concentration of about 4.5 $\times$ SSW the ratio attains a peak of 11, when dolomitization begins: this stage is marked by the downward bend of the Mg^{2+} line. The most active dolomitization proceeds between concentrations of 5 and 6 $\times$ SSW while the Mg^{2+}/Ca^{2+} ratio is still high, between 6 and 10 (KINSMAN, 1966b; BUTLER, 1969). In this connection some work by IRION and MÜLLER (1968) and MÜLLER and IRION (1969) is of interest: they recorded dolomitized lime biomicrites, in central Anatolia, in the sediments of a salt lake which now has water with the high Mg^{2+}/Ca^{2+} ratio of 150.

Kinsman (1964b) noted that two reactions can lead to dolomitization:

either

$$2CaCO_3 + Mg^{2+} \rightarrow CaMg(CO_3)_2 + Ca^{2+} \qquad (24)$$
$$\text{arag.} \qquad \text{brine} \qquad\qquad \text{dol.} \qquad\qquad \text{brine}$$

or

$$CaCO_3 + Mg^{2+} + CO_3^{2-} \rightarrow CaMg(CO_3)_2 \qquad (25)$$
$$\text{arag.} \qquad \text{brine} \quad \text{brine} \qquad \text{dol.}$$

Mass balance calculations indicate the dominant importance of the first

TABLE XXIV

PARTIAL ANALYSES OF TWO BRINES FROM THE ABU DHABI SABKHA
(From KINSMAN, 1964b)

	meq $\times$ 10³ conc.		
	sample 20	sample 21	loss
Ca^{2+}	17	12	5
Mg^{2+}	108	97	11
SO_4^{2-}	54	36	18

TABLE XXV

PARTIAL ANALYSES OF TWO BRINES FROM THE ABU DHABI SABKHA
(From KINSMAN, 1964b)

	meq $\times$ 10³ conc.		
	sample 21	sample 15	loss or gain
Ca^{2+}	12	18	+ 6
Mg^{2+}	97	66	−31
SO_4^{2-}	36	8	−28

reaction 24, accompanied by precipitation of $CaSO_4$ (KINSMAN, 1964b). It seems plain that reaction 24 can only be an abbreviated statement: Liebermann's thinking requires that all ions involved in the reaction be in solution. In Fig.357 and Table XXIV, in sample 20 there are 17 meq of Ca^{2+}, so that the maximum loss of SO_4^{2-} which could take place from the brine to form anhydrite (or gypsum) is also 17 meq. This is, indeed, nearly the amount by which SO_4^{2-} does fall to sample 21, but this simple reaction is modified by the loss of Mg^{2+} accompanying dolomitization. For the 11 meq of Mg^{2+} that replace Ca^{2+} sites in the aragonite lattice, 11 meq of Ca^{2+} must be released into solution. So the 18 meq of SO_4^{2-} which have gone to form anhydrite have used two sources of Ca^{2+}, 11 meq released by dolomitization and 5 meq pre-existing in the brine (the balance of 18 $\approx$ 16 is a good agreement considering the complexity of the chemical environment). Similarly, in Fig.357 and Table XXV, the 28 meq SO_4^{2-} which have been lost to the solution (between samples 21 and 15) to precipitate anhydrite have combined with 28 meq Ca^{2+} which were released when the 31 meq Mg^{2+} went to make dolomite. The surplus Ca^{2+} $(31-28 = 3meq)$ has accumulated in the brine so raising the level of Ca^{2+}. Again, the agreement $31-28 \approx 6$ is reasonable in the circumstances.

KINSMAN (1964b) estimated the quantity of brine which must pass through the aragonite mud to dolomitize it completely. The porosity being about 50% and the density of aragonite 2.9 g/cc, it follows that 1 cc of mud contains 1.45 g of solid or 0.58 g of Ca^{2+}. If the dolomite is $Mg_{50}Ca_{50}$, then 0.174 g of Mg^{2+} are needed. (Mg^{2+} in the original sediment is negligible.) This mass of Mg^{2+} is contained in 130 cc of normal sea water, but at a concentration between 5 and 6 × SSW this volume reduces to 20–25 cc. Yet the brine is rarely stripped of more than 40% of its Mg^{2+}, so this reduced volume must be increased again to 50–60 cc.

The implication which follows from the need for 50–60 cc of brine to dolomitize 1 cc of aragonite mud is of critical importance. A transport system is needed for the continual supply of brine. The only feasible system combines loss of water by evaporation from the sabkha surface and its replacement by sea water flowing through the mud from the lagoon.

The possibility of a return flow (seepage reflux), as the dense, concentrated brine flows seaward at a lower level, remains open and awaits further study. In the absence of a reflux only superficial dolomitization of the sediments can take place. Deeper dolomitization demands a seepage reflux circulation such as that postulated for Bonaire (p.532).

Not surprisingly, more thorough and sophisticated field work by Kinsman and others in the Abu Dhabi area is showing that the ideas portrayed in Fig.356 are overly simple. The supply of lagoon water to the sabkha is complicated by the penetration of wind-driven water from the lagoon across the sabkha along old channels: the influx of water from the continental side is greater than was supposed. Magnesite is being found in great quantities.

DOLOMITE OF BONAIRE ISLAND, SOUTHERN CARIBBEAN

The recent research into the origin of dolomite crusts on Bonaire is of outstanding interest, not only for the way it illumines the process of superficial dolomitization, but because of the well-documented argument about reflux dolomitization (DEFFEYES et al., 1964, 1965; LUCIA, 1968; HSÜ and SIEGENTHALER, 1969; R. C. MURRAY, 1969).

The environment

The Dutch island of Bonaire (Buen Ayre), lying off the Venezuelan mainland 200 km northwest of Caracas (12 °N), has an area of about 750 km^2. In the southern part of the island there is an extremely flat region of 30 km^2 where Recent carbonate sediments are accumulating on supratidal flats, in shallow hypersaline lakes, and as low dunes. The biggest lake, the Pekelmeer with an area of 2.5 km², occupies a position along the lee coast separated from the sea by a coral-rubble ridge. The water level of the lake is below sea level, so that sea water percolates through the rubble ridge into the lake. Sediments on the lake floor reflect, therefore, a twofold supply of water and carbonates, by storm floods and by seepage. They consist mainly of pelleted lime muds and silts, algal stromatolites and gypsum (details in LUCIA, 1968). Diagenesis of the Recent sediments has yielded two common products, one a hard, lacy limestone crust—the other a dolomite.

Textural and chemical parameters

Dolomite has been found in unconsolidated sediments and in crusts in the top few centimetres of the pelleted lime muds. The post-depositional origin of the dolomite is indicated by the dolomitization of gastropod shells and by the occurrence of dolomite preferentially in the mud matrix rather than in the pellets (as on Qatar, p.212). The porosity of both the crust and the unaltered mud is about 50%, but, since the crust is vuggy owing to leaching of pellets and shells, the porosity of its mud matrix must have been substantially reduced by pore-filling dolomite. It seems, therefore, that much of the dolomite is not a simple molar replacement of sedimentary grains, but has been precipitated as a cement through the combination of dissolved $CaCO_3$ with Mg^{2+} from the water, and the release of Ca^{2+} in exchange.

Concentrations of dissolved ions in the lake water were recorded as between 5 and 6 times normal sea water: in cores they vary between 2 and 6.8 × SSW. The ionic balance between Cl^-, Mg^{2+} and K^+ is maintined in all samples with little variation, but, as the concentration increases there is a substantial accompanying decrease in Ca^{2+} and a lesser decrease in SO_4^{2-}. Concomitantly there is a rise in the molar ratio Mg^{2+}/Ca^{2+} from 5 (normal sea water) to the very high figure of nearly 50.

Dolomite

The dolomite crystals have diameters of about 2 μ and the composition ranges from $Ca_{54}Mg_{46}$ to $Ca_{56}Mg_{44}$. Radiocarbon dates indicate that the dolomite is no more than 2,220 years old.

Hypothesis

DEFFEYES et al. (1965) saw that, if a $CaCO_3$ sediment is to be dolomitized by reaction 24 (p.528), the value of the ratio aMg^{2+}/aCa^{2+} must exceed that which is in equilibrium with both calcite and dolomite at the given temperature and pressure. Waters with larger ratios will dissolve $CaCO_3$ and precipitate dolomite until the ratio falls to the equilibrium value. The precipitation of a great part of the

Ca^{2+}, as gypsum and aragonite mud, brings about the required rise in Mg^{2+}/Ca^{2+} which remains high in the brines, presumably because of the much slower rate of dolomitization compared with evaporite precipitation. The relation of Mg^{2+}/Ca^{2+} to the chlorinity is, in fact, close to that expected to follow precipitation of gypsum and anhydrite.

Seepage reflux

In a detailed study of the hydrology DEFFEYES et al. (1965) investigated the relations between seepage of sea water through the rubble ridge into the lake, rainfall, runoff, evaporation and changes in salinity and of lake levels. Their calculated balance indicates that a counter flow of brine with Mg^{2+}/Ca^{2+} of about 30 should be carrying dense hypersaline lake water seaward through the carbonate sediments underlying the rubble ridge (Fig.359). The driving force for such a **seepage reflux** (a process put forward by ADAMS and RHODES, 1960) would be the greater density of the lake water over sea water. The authors were at pains to show that the rate of the operation is appropriate to the time available.

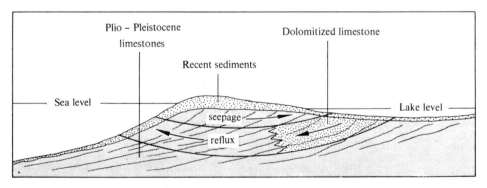

Fig.359. Diagrammatic illustration of a seepage-reflux cycle of the supposed Pekelmeer type. Sea water, entering the lake by seeping through a coral rubble ridge, is concentrated as a result of evaporation. Precipitation of gypsum in the lake raises the ratio Mg^{2+}/Ca^{2+} to at least 30. The now dense water sinks through the lake floor and, flowing seaward, dolomitizes the limestone through which it passes. Yet boring through the Pekelmeer sediments failed to confirm the process and experiments by Hsü indicate a reverse process of dolomitization by evaporative pumping.

In a most interesting postscript, the authors pointed out that, in northern Bonaire, beds of Plio-Pleistocene marine calcarenites, inclined toward the sea, have been dolomitized. Dolomite passes into limestone down dip, dolomite–limestone junctions transgress bedding planes, and the dolomite crystals are large, 40–80 μ across. This, they suggested, is the expected result of reflux of the Pekelmeer type. It implies the one-time existence of a lake similar to the Pekelmeer, though this lake floor and its sediments have since been eroded.

The seepage reflux hypothesis suffered something of a set-back when LUCIA

(1968) reported on the lithology of a bore hole through the floor of the Pekelmeer. No dolomite was found and the pore water was normal sea water at a depth where a hypersaline brine would have been expected. It also became known that a volcanic ash of low permeability underlies the upper 0.5–1 m of lake sediments and that the vertical flow of ground water is restricted to localized permeability channels where the ash is absent or broken (LUCIA, 1968; R. C. MURRAY, 1969). The upper ends of the permeability channels manifest themselves as springs around the lake and in the lake floor. Their distribution is apparently related to that of sink holes, in the underlying Pleistocene limestone, which descend well below sea level. Furthermore, the water that flows from the springs is sea water and this path is a more continuous and important one for the delivery of sea water into the lake than the flow through the coral-rubble ridge. Throughout most of the year the upward flow of ground water is the reverse of that required for the seepage reflux process. Nevertheless, when the water pressure differential across the volcanic ash is calculated, from variations in the density of the lake brine and the changes in sea and lake level throughout the year, it can be seen that a reversed, downward reflux may well operate for a short period, say two or three weeks, during the summer. Murray pointed out that there must be some regular escape of saline water from the lake or the evaporation would have led to the precipitation of the more soluble salts such as halite.

Opposed altogether to the seepage reflux process, even in its diminished form, Hsü and SIEGENTHALER (1969) postulated a reverse process of dolomitization which they called **evaporative pumping**. They based their ideas on a series of preliminary experiments in the laboratory. A glass-walled tank was half filled with sand. Toward one end the surface of the sand was nearly flat (the land): toward the other it sloped downward at about 10° (the sea floor). The sea was filled with a solution of NaCl, the land was saturated with fresh water. Over the land a 250 W lamp acted as the sun and caused the fresh water to evaporate. The levels of sea water and fresh water were kept constant by regulators leading from reservoirs. Once a steady state was formed a continuous flow of sea water moved landward into the sand to replace the fresh water lost by evaporation. When the level of the fresh water regulator was raised, so that the height of the fresh water balanced the density head of the salt water, the movement of the salt water would cease *when the lamp was turned off*. The authors concluded that evaporation provides the energy for the landward movement of subsurface water in an arid coastal region underlain by sediments with saline pore water. They recognized that the result of this process is the precipitation of an evaporitic crust on the tidal flat: some Recent and ancient evaporites may have originated in this way. This system of dolomitization would work in a way rather similar to that under the sabkhas of the Persian Gulf (p.527).

Hsü and Siegenthaler emphasized that evaporative pumping is not a vadose process depending on upward movement of water in capillaries as a result of surface tension. The sand surface of the land remained saturated throughout.

This is not to say that dolomitization by seepage reflux can never take place. MÜLLER and TIETZ (1966) have described what seems to be a relatively simple system of evaporation, seepage and dolomitization, in supratidal marine calcarenites and eolianites in the Canary Islands, which appears to fit the requirements of a seepage reflux process. The supratidal calcarenites are, from time to time, washed by high waves and spray. Some of the sea water remains in pools on the surface of the calcarenite and evaporates, causing the precipitation of gypsum and halite. In the underlying calcarenite, high-magnesian calcite grains of coralline algae have been altered to cryptocrystalline dolomite, primary porosity has been occluded by dolomite cement and large dolomite crystals (up to 2 mm) have grown *in situ* possibly by replacement of earlier cryptocrystalline dolomite. Textural relations show that the dolomite cement is younger than an early sparry calcite cement which is similar in appearance to the early meteoric cements of Bermuda and elsewhere. The coralline algae may have been dolomitized either directly or after their transformation to low-magnesian calcite. The situation as a whole seems to call for a reflux process in which waters with a high ratio Mg^{2+}/Ca^{2+} (a result of the precipitation of gypsum) percolate down through the calcarenite, and out again lower down the beach. The calcarenite forms a terrace, 0.5–1.0 m thick on basalt, and the refluxing pore water can escape seaward at the base of the terrace scarp. The only force needed is gravity, no other form of pump being necessary.

DOLOMITES OF THE BAHAMAS

The crust

The description by SHINN and GINSBURG (1964) and SHINN et al. (1965a, b) of dolomitized aragonite muds on Andros Island adds a further special facet to the developing concept of Recent supratidal dolomitization. Andros Island is mainly an agglomeration of Pleistocene limestone cays, mangrove swamp, ponds, tidal channels and supratidal lime muds (SHINN et al., 1969). The cays are clothed rather sparsely in scrub with occasional pines. The mangroves, though they may attain heights of 3 m or more, grow in many places to only about a metre above sea level. Many of the ponds are sink holes in the underlying karst topography. The supratidal lime muds are of marine origin and have settled out of suspension from sea water during spring tides and storms, when sea level was abnormally high (PURDY and IMBRIE, 1964). Deposited on the limestone cays, they raise the land surface and extend it laterally. In this way cays are welded together to form larger land areas. PURDY and IMBRIE (1964, p.38) write: "It would appear that most of Andros Is. has been constructed in this manner."

Unlike the muds on the neighbouring sea floor, these aragonitic lime muds are laminated. The lamination reflects both variation in the coarseness of flood deposits and the growth of intertidal algal mats (p.217). Typically these supratidal laminated muds have a fenestral porosity (birdseye) resulting from desiccation shrinkage and the pressure of gas bubbles (SHINN, 1968a). As a biological habitat the environment is inhospitable, so that the laminae are not destroyed by burrowing.

These laminated muds have been banked up to form supratidal mounds on the summits of which palms grow (known throughout Florida and the Bahamas as "palm hammocks"): the tops of the mounds are 30–60 cm above mean high water. The surfaces of these low mounds are very extensive, many having a scarcely detectable slope. On their lower flanks there are widespread crusts, 2–3 cm thick, which bear from 20% to over 80% of dolomite. Below high-water level these crusts have been traced seaward, by probing, where they pass underneath more recently deposited muds.

The typical crust is dark grey to black, brittle, with a pitted surface. A maximum thickness of 5 cm was recorded from an unweathered, buried crust. Similar crusts described by SHINN (1964) from the Florida Keys attain 10 cm. The crusts are cracked into polygonal slabs up to 80 cm across. Sedimentary structures in the crust, though commonly not obvious, include cracks, chips, laminae, burrows and shells of *Batillaria*, Peneroplidae and the land gastropod *Cerion*. Many of the shells are partly leached and can be crushed between the fingers. The sediment is otherwise composed of faecal pellets, the debris of foraminiferids, molluscs and non-carbonate organic matter. Vugs are lined with an acicular carbonate cement, giving a characteristic fenestral fabric. Exposed crusts are "extremely porous".

Radiocarbon ages for the crust show a progressive increase outward, from the exposed crust on the hammock which has the same ^{14}C age as living gastropod shells, to the crust buried by submarine muds. This buried crust yielded an age of about 2,000 years.

Dolomite

The dolomite crystals are tiny, 1–2 μ. X-ray diffraction patterns show a non-ideal composition of $Ca_{55}Mg_{45}$, order reflections being present but weak.

Hypothesis

The Bahamas study is primarily a record of environmental conditions of dolomite formation. The distribution of the crust, exposed and under the submarine muds, and its radiocrabon ages, suggest continuous dolomitization of supratidal sediments as a rising sea transgressed centripetally onto the hammocks. Dolomitization takes place on the lower flanks of the supratidal mounds and, as more mud accumulates, the older part of the crust is progressively buried while continued dolomitization causes the crust to grow toward the crest of the mound. Chemical parameters are not available but, in a similar environment in the Florida Keys (SHINN, 1968b), the pore water squeezed from the dolomitic sediment gave a molar ratio Mg^{2+}/Ca^{2+} of more than 40, and salinities which were 5–6 times that of normal sea water. SHINN et al. (1965) suggested that water lost by evaporation at the exposed surface is replaced by subsurface flow of sea water (as in Abu Dhabi, p.527), accompanied, very probably, by precipitation of gypsum and consequent rise in Mg^{2+}/Ca^{2+}. The absence of gypsum in such a humid climate may be due to the washing of the surface by rain, also by spring tides and storms.

THE GROWTH OF DOLOMITE IN THE RECENT ENVIRONMENT

Search for a mechanism

The search for a mechanism which would account for the growth of dolomite, either as a direct precipitate, a penecontemporaneous alteration product or a late diagenetic product, has proceeded in two distinct though related directions. The classic approach has been the study of the bulk thermodynamic properties of entire crystals and their solutions, but this has been joined lately by a growing curiosity about the special properties of the crystal surfaces. The surface, after all, is the part of a crystal which is involved in reaction with the adjacent aqueous solution, and so its peculiar structure and composition are highly relevant (p.242).

On theoretical grounds based on thermodynamic properties of the whole crystal, Hsü (1967) wrote that the possibility remains open that dolomite is meta-stable and that nesquehonite, $MgCO_3.3H_2O$, or hydromagnesite, $MgCO_3.Mg(OH)_2$ $.3H_2O$, co-precipitates with a $CaCO_3$ phase at the temperature and pressure of the earth's surface. Or it may be that dolomite is stable but its growth is so slow that these metastable pairs tend to form from supersaturated solutions.

The various theoretical studies and experiments based on the properties of whole crystals (to the properties of which the surfaces make an insignificant con-tribution) do, it must be admitted, bear a resemblance to the results of the field studies of Recent dolomites. The skillful extrapolation from known data by LIEBER-MANN (1967) enabled him, moreover, to design experiments which led to the pro-duction of magnesian calcites and non-ideal dolomites from artificial sea water in conditions closely similar to a natural environment.

The results of work in the Persian Gulf, Bonaire, Deep Spring Lake, Califor-nia (p.539)—and possibly the Coorong—show that dolomite is an alteration product of solid $CaCO_3$, possibly precipitating as the $CaCO_3$ is dissolved. The work of Liebermann is relevant to this concept.

LIEBERMANN's (1967) eventual conclusion, that dolomite should precipitate directly from sea water, at salinities greater than 4–6 times that of normal sea water (after $CaSO_4$ has been precipitated), and pH 8–9 following CO_2 extraction, is in reasonable agreement with field data. High pH seems to play a critical role in the Coorong district although, oddly enough, the high level of salinity seems less relevant to the Australian dolomites than to those of the Persian Gulf and Bonaire. Liebermann's first postulate was that precipitation of a solid from aqueous solution implies that the compound must be in true ionic solution at the moment of precipitation. His second postulate was that if dolomite (or any double salt) is to precipitate, its solubility must be lower than that of its constituents. This is so for dolomite over a wide range of CO_2 tension, salinity and temperature. His third, and last, postulate was that the relative solubilities of the constituents ($CaCO_3$ and $MgCO_3$) must be nearly equal at the moment of co-precipitation. He showed, on theoretical grounds, that the solubilities of $CaCO_3$ and $MgCO_3$ in sea water should converge at a salinity of about 220 ‰ and a pH of 9.

The process whereby dolomite is precipitated was treated by Liebermann as a two-stage reaction. In stage *1* solid $CaCO_3$ dissolves in hypersaline water and this is followed, in stage *2* by co-precipitation of $CaCO_3$ and $MgCO_3$. He suggested that in the sea, in shallow water, stage *1* could take place during the night when the temperature and pH are relatively low and the CO_2 tension is high (respiration). Stage *2* is favoured by high daytime temperatures, high pH and low CO_2 tension (photosynthesis). In the laboratory, for his stage *1*, Liebermann bubbled CO_2 through flasks of artificial sea water to which fine $CaCO_3$ had been added, at 5–10°C. The pH was adjusted by the addition of Na_2CO_3 solution and a few drops of ammonia. In the more successful experiments the concentration was 3 and 6 times that of normal sea water. The flasks were then transferred, for stage *2*, to an oven at 43°C. The durations of the two stages were 12 and 60 h respectively. In the experiments which gave a non-ideal dolomite, the two-stage process was repeated about 15 times. In a number of the experiments magnesian calcites were produced and, in one of these, the calcite contained possibly as much as 30 mole % of $MgCO_3$. Quantities of magnesian carbonates were small, being described as "little", "traces", etc., except for the disordered dolomite which amounted to 5 wt.%. Aragonite was the dominant carbonate precipitate at the concentration of normal sea water, but at all higher concentrations calcite was dominant. The pH, adjusted to between 8.6 and 7.5 at the start of stage *1*, fell to between 6.1 and 5.3 at the end of the stage. After completion of stage *2* the pH was again high as 8.7.–7.9.

Liebermann found that the elimination of $CaSO_4$ from the water was an essential preliminary. In some of his artificial sea water preparations $CaSO_4$ was excluded and it was these that yielded the greater quantities of magnesian carbonates. $CaSO_4$ inhibits the precipitation of dolomite since the excess Ca^{2+} in solution reduces the solubility product of $CaCO_3$ far below that of $MgCO_3$. In the Persian Gulf it is significant that dolomitization begins only where gypsum has been massively precipitated. The dominance of calcite in the carbonate precipitate where the concentration exceeds $3 \times SSW$ is interesting in view of its failure to come down in natural sea water (p.242), as in the Abu Dhabi lagoon, where the inorganic precipitate is aragonite (p.204). The lagoon water does not reach a concentration of $2 \times$ SSW and it may be that, at the higher concentrations, the Mg^{2+} ion is less strongly hydrated and does not inhibit the growth of calcite.

Theoretical calculations by SASS (1965) on a straight thermodynamic basis, and supporting experiments, show that normal sea water is supersaturated for calcite above 16°C and for dolomite above 28°C: thus dolomite should be precipitated if temperature rises above about 22°C. Nevertheless, if the sea water is in contact with pure calcite, then it is decidedly undersaturated for dolomite. This degree of undersaturation is reduced if the solid $CaCO_3$ is more soluble than pure calcite, as aragonite or a magnesian calcite. Contact with magnesian calcite also raises the activity of Mg^{2+} in the water, so favouring dolomite stability. SCHLANG-ER (1957), MÜLLER and TIETZ (1966) and LAND and EPSTEIN (1970) have recorded the preferential dolomitization of grains of coralline algae (high-magnesian calcite) in a calcarenite (p.534). Thus, sea water in effective contact with carbonate sediments made of aragonite and magnesian calcites, though less undersaturated for dolomite than if the sediment were pure calcite, must have its temperature and salinity substantially raised before it becomes supersaturated for dolomite.

Some laboratory experiments by WEYL (1967) with dolomite are of interest. He studied the steady state behaviour of packs of pure calcite and dolomite grains in flowing sea water. The pH was modified by the addition of acid or base to cause dissolution or precipitation. Both materials appear to be in equilibrium over an extended range of carbonate concentration, but the dolomite range is markedly greater than the calcite range and includes it. Weyl concluded that, on a 10 year to 100 year scale, normal sea water neither dolomitizes nor dedolomitizes a carbonate sediment of sand grade. Dolomite remained in equilibrium in a molality of carbonate as high as 0.002 mole. Non-ideal dolomite has been precipitated by DONAHUE and DONAHUE (1968) at room temperature and pressure, though the other conditions differed from those of any natural environment. In the laboratory $CaCO_3$ and $MgCO_3$ were dissolved in water by adding excess powdered mineral under P_{CO_2} of 12 atm. The solution was then allowed to flow into a beaker, 50 ml at a time, under atmospheric pressure at 36–38°C. Solids were excluded by a filter. After each addition the solution was evaporated and a new addition made until enough residue remained for X-ray diffraction analysis. This revealed calcite, magnesite and a non-ideal dolomite with composition $Ca_{55}Mg_{45}$. Further details have yet to be published.

Significance of the surface layer of a crystal

In a particularly important experiment (E_2) WEYL (1967) demonstrated that, whereas a clean surface of calcite in sea water is more soluble than a clean surface of dolomite, the precipitated overgrowth on dolomite is more soluble than the overgrowth on calcite (p.252). Thus, once the surfaces of the two mineral overgrowths have grown into pseudoequilibrium, they can coexist. Dolomite cannot dissolve because the pure dolomite is less soluble than calcite, neither can dolomite grow since the precipitate would be more soluble than both the pure calcite and its overgrowth. Weyl pointed out that this accounts for the persistence of detrital dolomite for several thousand years in the mixed carbonate sediment in Florida Bay.

The story of Deep Spring Lake. Investigations of dolomite crystals with regard to a possible surface layer have been carried out by PETERSON and BIEN (1963) and PETERSON et al. (1963, 1966) on the dolomites which seem unquestionably to be precipitating in the intermontane playa, Deep Spring Lake, California (118°00' W 37°15' N). Radiocarbon dates imply a Recent origin for the tiny crystals, less than 1 μ in diameter, which have a mean growth rate of some hundreds of Ångstrom units in 1,000 years. The good euhedral form of the rhombs implies that they are primary precipitates and are not replacements of pre-existing carbonates. The lake is about 30 cm deep in the winter and spring, but, in the summer and autumn only a small brine body remains. Much sulphate and carbonate is precipitated at a pH varying between 9.5 and 10.0.

PETERSON et al. (1963) took samples of *supposedly* pure dolomite sediment and leached the crystals in acetic acid (buffered with ammonium acetate at pH 5), in a series of steps, stopping the leaching at intervals by adding ammonium hydroxide: samples were analysed by X-ray diffraction after each leaching. The results pointed to a range of compositions from magnesian calcite to dolomite, as indicated by a progressive shift in position of the (211) peak. It was possible with care to remove only the most Ca^{2+}-rich material and the residue showed a distinct beneficiation of more nearly ordered dolomite. The crystals *appeared*, therefore, to change in composition from the surface inward, the lattice becoming increasingly more ordered. It could be supposed that this ordering cannot have taken place in the outermost part of the crystal, say 1–2 unit cells thick, because the lattice is warped owing to the unsatisfied bonds. Nor can the necessary ordering take place over long distances, throughout the crystal lattice, because long-range solid state diffusion would take too long. The authors therefore argued that, if ordering cannot proceed at the surface or throughout the main bulk of the lattice, it must go on in a thin layer just below the surface where the lattice is still sufficiently defective to allow movement of cations. In later experiments, with a similar leaching technique, PETERSON et al. (1966) showed an apparent progressive change from a magnesian calcite with composition approximately $Ca_{80}Mg_{20}$ to a dolomite with composition about $Ca_{55}Mg_{45}$. This change seemed to be completed within the outer 100 Å or so of the crystal. The data for successive radiocarbon analyses of the leached material fitted best a steady growth rate of 0.09 μ in a thousand years. The authors concluded that the 100 Å layer was a transient structure which moved outward as the crystal grew, leaving behind it a non-ideal, calcian dolomite. Ordering within the layer was assumed to develop by solid state diffusion inwards of Mg^{2+} ions and a reverse drift of Ca^{2+}.

If the conclusions of Peterson and his co-workers are accepted it is particularly intriguing to read the paper by DEGENS and EPSTEIN (1964) on the oxygen and carbon isotope ratios in coexisting calcites and dolomites. They examined a range of materials from the Bahamas, Florida Bay, the Coorong, the Alpine Dolomites, the Cretaceous of the Western Interior (U.S.A.), the Precambrian, etc. If these various dolomites were precipitated from the same waters as the calcites, their $\delta^{18}O$ values (p.280) should be heavier by 6–10‰, but they are not significantly different. Therefore, as the dolomites cannot be primary precipitates, they must be alteration products. The only available materials to be altered were aragonite and calcite: the similarity of the $CaCO_3$ and dolomite $\delta^{18}O$ values implies that the $\delta^{18}O$ value in the dolomite must be inherited from the $CaCO_3$ precursor. It would appear also that, if the CO_3^{2-} framework of the original lattice maintained its $\delta^{18}O$ value, it must have remained undisturbed and did not dissolve. Thus, dolomitization must have acted in the solid state, by diffusion of cations only associated with lattice defects.

Turning again to Deep Spring Lake, and the proposed calcitic surface layer on the dolomite crystals, the process of oxygen isotope selection and inheritance could be interpreted in the following way (PETERSON et al., 1966). The oxygen isotopes are selected by the calcitic surface layer and, as millimicrodolomitization proceeds, exchange between crystal and solution involves only the cations. The CO_3^{2-} units, once absorbed, remain in the crystal. Their $\delta^{18}O$ values were determined by the calcite and this value remains fixed despite the solid state migrations of Mg^{2+} and Ca^{2+} ions.

The interesting hypothesis of Peterson and co-workers rests, however, on the assumption that the samples of sediment analyzed by successive leaching were not contaminated by other carbonate minerals. Aragonite, by virtue of its acicular habit, was excluded but the likelihood that the rhombs included some calcite (despite examination by electron microscope) has been emphasized by the work of CLAYTON et al. (1968). They proposed that the calcitic material detected by Peterson et al. in the early stages of leaching belongs, not to a surface layer on the dolomites, but to separate, previously unrecognized, crystals of calcite in the sediment. The greater solubility of these would cause their dissolved products to appear preferentially in the early leached carbonate. (see also BERNER, 1967, p.57.)

From X-ray diffraction analyses Clayton et al. found that magnesian calcite particles are present sporadically in the sediment. Samples showing undoubted calcite were subjected to sequential isotopic analyses of serially leached crystals. It was apparent in every size range that there was a fast-reacting carbonate with $\delta^{18}O$ about -6, corresponding to calcite. They examined the $\delta^{18}O$ values of the various components of the sediment by relying on the different dissolution rates of calcite and dolomite, and of larger and smaller crystals. This sequential isotopic analysis shows typically the existence of a small amount of isotopically lighter material in the first-reacting fractions. Since this amount decreases in the finer size fractions *it cannot be derived from a surface layer of uniform thickness.*

CLAYTON et al. (1968) found that the calcite-water isotopic fractionation in Deep Spring Lake is only 1.0297, distinctly smaller than that for the local dolomite-water fractionation of 1.0351. This implies that the calcite and dolomite were precipitated under different conditions. This conclusion is strengthened by differences of nearly 5% in the $\delta^{13}C$ values for calcite and dolomite in one sample. These various results led the authors to conclude that the data do not require the dolomite to be a solid state replacement of a calcite precursor.

Dissolution of samples finer than 0.5 μ, equivalent to a surface layer of 100 Å, gave a Ca^{2+} excess of less than 2% over that expected for non-ideal dolomite. Furthermore, the equilibrium ion activity products in solutions showed dissolution behaviour consistent with a solid phase of the composition of dolomite.

The validity of the Peterson model has also been questioned in an important paper by BERNER (1965b) in which he noted that isotopic analyses of modern sediments (in DEGENS and EPSTEIN, 1964) show that the oxygen isotopic composition of co-existing calcite and dolomite are not, in fact, very different and certainly do not show the predicted 6‰ difference.

Clayton et al. suggested that the dolomite could have grown by alteration of detrital calcite, this alteration involving dissolution-precipitation and isotopic exchange with the water. This, of course, is the kind of process put forward by Liebermann. On the other hand, the apparent perfection of the dolomite rhombs described by Peterson et al. does not seem to fit with a replacement origin. More information is needed about the structural relation of a replacement dolomite rhomb to its host calcite rhomb.

SUMMARY

Summarizing the work of the last few years in the laboratory and in Recent environments, we can conclude that the necessary condition for dolomite formation in the top metre or so of Recent sediments is a high value for the molar ratio Mg^{2+}/Ca^{2+}, possibly between 6 and 10. A high activity of Mg^{2+} relative to Ca^{2+} in the solution is apparently necessary to give the chemical potential needed to bring about the high degree of order in the dolomite lattice. Presumably greatly enhanced supersaturation for dolomite is also required. The role of $CaSO_4$ precipitation in the evolution of an increased ratio Mg^{2+}/Ca^{2+} is apparent in the Persian Gulf and on Bonaire and has been emphasized in the experimental work of Liebermann.

At the temperatures and pressures current at the earth's surface there is probably insufficient energy to drive the ordering beyond the stage of a non-ideal dolomite. At all events the growth of dolomite is likely to be a slow process in any natural environment (estimated at 0.1 μ/1,000 years in Deep Spring Lake). It is probably for this reason that deposits of Recent dolomite are rare. Current rates of sedimentation are generally high relative to the rate of dolomite formation so that dolomite has little chance to accumulate in appreciable quantities, either as a primary precipitate or as a secondary, early diagenetic product, before it is buried and removed from the appropriate chemical environment. Stagnant subsurface pore solutions will be in equilibrium with the grains of the enclosing sediment and the attainment of high levels of supersaturation will be impossible. This situation may be partly offset by the action of a seepage reflux process, as dolomitizing solutions move downward through the pores of the sediment. The slowness of the reaction (or reactions) yielding dolomite may, in places, permit other carbonates having greater solubilities but faster growth rates to precipitate in larger quantities. High-magnesian calcite is a case in point. In the laboratory Glover and Sippel have precipitated magnesian calcites at room temperature and pressure with compositions that range from $Ca_{80}Mg_{20}$ to $Ca_{39}Mg_{61}$, some having the cationic proportions of dolomite or protodolomite but random positioning of the cations.

The importance of salinity is not clear. The normal to low salinities recorded from the Coorong and neighbouring lakes do not necessarily relate to the growth of the dolomite, since the time at which the growth takes place is not known. In fact, the action of a seepage reflux process is not unlikely. Certainly, highly super-saline brines must exist in the last stages of drying of the sediment surfaces and this condition could be critical. Salinities of 5–6 that of normal sea water are assumed essential for the dolomitization in the Persian Gulf and on Bonaire. As for pH, Eh and P_{CO_2}, there seems at present no reason for ascribing a universally critical role to these factors.

The field data seem to favour the growth of dolomite by the wet replacement of some $CaCO_3$ precursor. An attempt by Peterson and others to demonstrate progressive ordering in a solid precursor by solid state diffusion of cations has run into difficulties regarding the proposed existence of a calcitic layer on the dolomite rhombs. It seems likely that Liebermann is right in insisting on the aqueous dissolution of all ions at the moment of precipitation and that dolomitization is always a wet process, involving exchange of oxygen and carbon isotopes between precursor, solution and precipitate. The closeness of the physical association of dolomite and $CaCO_3$ in the field is, nevertheless, not everywhere obvious. Though fabrics indicating *in situ* dolomitization of pre-existing carbonate particles have been seen in the sediments of the Persian Gulf and Bonaire, they are suspiciously absent from those of Deep Spring Lake.

It is nowadays usual to remark that Recent carbonate sediments, with their numerous solid phases, should not be in chemical equilibrium, but Weyl's work has shown how the greater solubility of overgrowths on crystals can lead to a definite equilibrium between dolomite and calcite. Both Weyl and Sass have given reasons for the long term inability of normal sea water to dolomitize sedimentary particles.

FURTHER READING

The Coorong is mainly covered by ALDERMAN and SKINNER (1957), SKINNER (1963), ALDERMAN (1965b) and VON DER BORCH (1965a, b). The Persian Gulf is dealt with by ILLING et al. (1965), KINSMAN (1966b) and BUTLER (1969). DEFFEYES et al. (1965) cover Bonaire and SHINN et al. (1965a, b) the Bahamas. The problems of Deep Spring Lake are examined in B. F. JONES (1961, 1965), PETERSON and BIEN (1963), PETERSON et al. (1963, 1966), DEGENS and EPSTEIN (1964), BERNER (1965b) and CLAYTON et al. (1968). Laboratory and theoretical matters are considered by WEYL (1960, 1967), WEBER (1964a, b), E. SASS (1965), LIEBERMANN (1967), MURRAY and LUCIA (1967) and ROSENBERG et al. (1967).

SHEARMAN's (1966) work on evaporites is highly relevant, also FAIRBRIDGE's general discussions (1955, 1967). Other references are given in the text.

Additional references not given in the preceding chapter

On the Coorong, ALDERMAN (1959, 1965a) and ALDERMAN and VON DER BORCH (1961). On Recent dolomites in Texas, AMSBURY (1962) and FRIEDMAN (1966b). On magnesium and oxygen isotopes in calcites and dolomites, respectively, DAUGHTRY et al. (1962) and FRIEDMAN and HALL (1963). For a discussion of the source of magnesium and the course of dolomitization in limestones underlying the Bahamas–Florida platform, GOODELL and GARMAN (1969). On possible intertidal dolomitization in the Funafuti reef, the first consideration of intertidal dolomitization, REULING (1934). On the same problem armed with isotopic data, BERNER (1965b). On preferred crystallographic orientation in undeformed dolomites (c normal to bedding), E. SASS (1969).

Papers on calcitization of dolomites

For the reader interested in the processes of calcitization of dolomite (dedolomitization) the following references will be found useful. For petrography, SHEARMAN et al. (1961) and EVAMY (1963, 1967), MATTAVELLI (1966), MATTAVELLI and TONNA (1967), MATTAVELLI and NOVELLI (1968) and BRAUN and FRIEDMAN (1970). For chemical considerations DE GROOT (1967); also FAUST (1949), LUCIA (1961), BAUSCH (1965) and A. KATZ (1968).

Appendix

RECENT DEVELOPMENTS IN CARBONATE SEDIMENTOLOGY

PETROGRAPHY OF CARBONATE GRAINS

There are papers on molluscan nacre by MUTVEI (1969, 1970) and ERBEN (1972a), on the Chamacea by KENNEDY et al. (1970), on the Bivalvia by TAYLOR and LAY-MAN (1972), TAYLOR (1973) and TAYLOR et al. (1973), and on structures in the Bivalvia revealed after ion-beam thinning by TOWE and THOMPSON (1972). Cephalopod structures have been studied by DRUSHITS and KHIAMI (1970), ERBEN and REID (1961), ERBEN (1972b), MUTVEI (1972a, b) and BIRKELUND and HANSEN (1974). Fossil nacreous conchiolin has been examined by VOSS-FOUCART and GRÉGOIRE (1971).

On brachiopod shell structure there are works by WILLIAMS and WRIGHT (1970), and MACKINNON and WILLIAMS (1974) and the fifth of her series on proteins by JOPE (1973). On scleractinian structure there are SPIRO (1971a), a general work by SORAUF (1972) and a critical examination by FENNINGER and FLAJS (1974). The relationship of the stromatoporoids to modern sclerosponges is examined by STEARN (1972). On the Foraminiferida TOWE (1971) has a work on lamellar wall structure. Calcareous algae are generously covered in the Short Course by GINS-BURG et al. (1971b), while crystal development in the calcareous algae is treated by BLACK (1972) and Recent algal nodules by BOSELLINI and GINSBURG (1971) and ALEXANDERSSON (1974). New works on the ultrastructure of bryozoans are by ARMSTRONG (1970) and SANDBERG (1971 and in press a); aspects of their mineralogy are reported by POLUZZI and SARTORI (1973). The stratigraphic occurrence of calcis-pheres given on p.69 should include appearance in the Cretaceous Chalk as *Oligostegina*. Some details of serpulid structures can be found in GÖTZ (1931) and REGEN-HARDT (1961) and there is a work on generic and environmental control of serpulid mineralogy by BORNHOLD and MILLIMAN (1973). Serpulid ultrastructure is currently being investigated with the scanning electron microscope by N. Svendsen, Univer-sitetets Institut for historisk Geologi og Palaeontologi, København.

Works of a more general nature are the very useful book on petrography by HOROWITZ and POTTER (1971), papers on biogenic particles by HAY et al. (1970), STIEGLITZ (1972, 1973) and a chapter in MILLIMAN (1974). TOWE (1972) has an essay on shell structure and the organic matrix concept.

On non-skeletal grains there is a chapter in MILLIMAN (1974), also a paper on ultrastructure in calcilutites by LOREAU (1972a) and on ooids by LOREAU (1972b). A more refined explanation of the anomalous birefringence of the cortex in Recent marine ooids (p.78) is given by LOREAU (1972c) and by Loreau in PURSER (1973,

For references see pp.615–632.

p.324). Loreau and Purser give a range of ooid structures in PURSER (1973, p.279). The similar morphologies of aragonite crystals in codiaceans and in carbonate muds of Florida Bay are noted by PERKINS et al. (1972).

RECENT CARBONATE ENVIRONMENTS

Interest in shallow water Recent environments has been strongly biased toward reefs. A clearer picture is emerging of the intimate relation between physiography, ecology and sedimentation and of the balance between organic accretion and a combination of biological and mechanical destruction (GARRETT et al., 1971; SCOF-FIN, 1972b; ZANKL and SCHROEDER, 1972; GINSBURG and SCHROEDER, 1973; SCHROEDER and ZANKL, 1974). The role of encrusting bryozoans is described by CUFFEY (1972). Supporting research into the behaviour of corals *in vitro* and dia-genetic development of their microstructures, both biological and inorganic, has been carried out by HUBBARD and POCOCK (1972) and HUBBARD (1974a, b, 1975), complemented by the works of SCHROEDER (1972a) and SCHERER (1974a) on the effects of endolithic algae on diagenesis in microcavities. Work on mechanisms of sediment rejection and shifting by distension, tentacular activity and on a variety of microenvironments in cavities in coral colonies has been done by HUBBARD (1973). Relations between living corals have been illumined by LANG's (1973) work on interspecific aggression. Depredation by the crown-of-thorns starfish is examin-ed by CHESHER (1969), TALBOT and TALBOT (1971) and GLYNN (1973).

Application of these new ideas on coral reefs to the study of ancient reef rock has been an interesting revelation in the works of KREBS (1969, 1971, 1972), ZANKL (1969) and SCOFFIN (1972a).

On the reef complex *sensu lato* there is a useful book on atolls (to which reference is overdue) by WIENS (1962), some essential logic by DUNHAM (1970), also papers on zonation, ecology and sedimentation by TAYLOR and LEWIS (1970), JACKSON et al. (1971), ROSEN (1971), TAYLOR (1971a, b), GLYNN et al. (1972), JORDAN (1973), OGDEN et al. (1973), TAYLOR et al. (1973), LANG (1974), PORTER (1974) and GARRETT and DUCKLOW (1975). Papers by THOMASSIN (1971, 1973) not only give an interesting picture of ecology in the Tuléar reef complex but provide a useful bibliography of the French literature. Reefs as sediment producers are ex-amined by CHAVE et al. (1972). The use of submersibles is reflected in studies of the deeper parts of fore-reef slopes by CLIFTON (1973), GINSBURG and JAMES (1973) and BURNE (1974). Six collections of papers on reefs are available in STODDART and YONGE (1971), JONES and ENDEAN (1973), in the *Papers in Honor of Dr. Thomas Goreau* (SMITH et al., 1973), and in CAMERON et al. (1974), LAPORTE (1974) and MONTY (1974). There is a superbly illustrated book on the Great Barrier Reef by BENNET (1971). An extensive, up-to-date bibliography and a helpful chapter are to be found in MILLIMAN (1974). Recent papers by PURDY (1974a, b), containing a detailed survey of numerous modern reefs, examine the possible influence on con-

For references see pp.615–632.

temporary reef morphologies and facies of their karstic basements. A thorough study of the Holocene shallow marine and tidal flat sediments of part of Andros Island, and their stratigraphy, has been carried out by GEBELEIN (in press). The siting of *Thalassia* carpets and mangrove stands on sediments accumulated in sink holes is described by DODD and SIEMERS (1971).

New information on the supply of carbonate mud is contained in LAND (1970), PATRIQUIN (1972) and FÜTTERER (1974), while accumulation of carbonate muds and mud banks is dealt with by HOWARD et al. (1970), SCHOLLE and KLING (1972), BASAN (1973) and READ (1974a). Modern and relic biogenic sediments on continental shelves are compared in major works by CAULET (1972a, b) and GINS-BURG and JAMES (1974), complemented by the new Memoir on Shark Bay (LOGAN et al., 1974a).

Salutary reminders that carbonate sediments accumulate in temperate as well as tropical latitudes are provided by the interesting surveys of northern European regions by HOMMERIL (1971), LEES and BULLER (1972) and LEES (1974), with work on diagenesis in these waters that are undersaturated for calcium carbonate by ALEXANDERSSON (1972a, 1975b).

The carbonate sabkha environment has continued to arouse interest and JAUZEIN and PERTHUISOT (1972) describe a Quaternary transgressive–regressive sequence. Various aspects of pore-water diagenesis and the precipitation of new minerals are dealt with by BUSH (1973), BUTLER (1973), DE GROOT (1973), HSÜ and SCHNEIDER (1973) and HAGEN and LOGAN (in press).

An important new development, in the field of ancient sabkha-like diagenesis, deserves mention here, namely the tepee structure. These structures, with complex and enigmatic histories, seem to be ancient analogues of Shinn's "polygonal fracture systems" (p.372). Key papers are by ASSERETO and KENDALL (1971), STURANI (1971), BURRI et al. (1973) and SMITH (1974).

The Persian Gulf has been particularly well served by the book edited by PURSER (1973). This includes results of the 1965 Meteor expedition with data on sedimentation rates, petrographic analyses of photomicrographs of Recent carbonate sediments, some fascinating information about sedimentation near topographic highs, and much useful material on patterns of evolving sedimentary environments.

RECENT CARBONATE ALGAL STROMATOLITES

The far ranging studies of MONTY (1971, 1972, 1973), with their welcome biological insight, make it clear that a fuller understanding of Recent cyanophyte carbonate stromatolites can only come from a deeper understanding of *all* such Recent structures, both marine and fresh water and the transitions between them, and from an appreciation of their Phanerozoic history. He makes the interesting observation that cementation of stromatolites by algal-induced precipitation of

For references see pp.615–632.

carbonate is today characteristic mainly of lacustrine and other fresh-water stromatolites, adding that the carbonate accumulations in marine stromatolites are
(contrary to some recently expressed views) dominantly a result of trapping and
binding of sediment (Black and Gebelein on p.224; HOFFMAN, 1973). (See also
GARRETT, 1970; AWRAMIK, 1971; FRIEDMAN et al., 1973; WALTER et al., 1973.) A
very extensive and complete survey of the literature and thought on carbonate
stromatolites of all ages is to be found in Monty's three papers and those of P.
HOFFMAN (1973) and H. J. HOFMANN (1969, 1973), with the important new work by
LOGAN et al. (1974b).

The microfabrics of Recent stromatolites have only lately received close
attention in the works of HUBBARD (1972b), GEBELEIN and HOFFMAN (1973) and
HOFFMAN et al. (in press). There has been a growing certainty that the alternation
of calcite–dolomite laminae in many ancient stromatolites reflects initial control by
algae of the distribution of magnesium. Calcite represents the sediment-rich layers
and dolomite the once algal-rich layers.

CHEMICAL CONSIDERATIONS

The subject is well served by two new textbooks: that by GARRELS and MACKENZIE
(1971) providing general background, while BERNER (1971) is specially useful for
carbonate diagenesis. Details of carbonate mineralogy are given by LIPPMANN
(1973).

A series of five related papers, on the dissolution kinetics of calcium carbonate, examines saturation state parameters in a new approach to the problems of
the position and origin of the lysocline. The five are by BERNER and WILDE (1972),
MORSE and BERNER (1972), BERNER and MORSE (1974), and MORSE (1974 and
in press). The main outcome of this work is the authors' conclusion regarding the
lysocline (BERGER, 1968), a zone about 1,000 m above the compensation depth
(p.267) where dissolution of calcite foraminiferids is first clearly active. (The saturation level is much nearer the surface: p.270). The authors conclude that dissolution
above the lysocline is largely prevented by the existence on the calcite surface of
adsorbed phosphate. (See p. 243 for this controlling effects as discussed by Brooks
et al.). The lysocline is therefore attributed to the existence, at that level, of a degree
of undersaturation that allows steps on the crystal surface to penetrate easily
between the adsorbed phosphate ions, so enabling dissolution to proceed. In some
work on the effect of the magnesium ion on the ion activity product of Ca^{2+} and
CO_3^{2-} in sea water, MÖLLER and PAREKH (1975) draw significant conclusions
about a seeming low level of supersaturation for calcium carbonate in sea water.
On the other hand, direct saturometer measurements (p.256) by BEN-YAAKOV et al.
(1974) suggest that the compensation depth really marks a change from saturated
to undersaturated water.

On marine chemistry in general we have the useful books by RILEY and

CHESTER (1971) and by BROECKER (1974) and the review by MILLERO (1974). On the effect of high pressures on carbonate equilibria there are papers by DUEDALL (1972) and MILLERO and BERNER (1972). On saturation levels see LI et al. (1969) and ED-MOND and GIESKES (1970); on the calcium carbonate ionpair see LAFON (1970), and on a kinetic model for sea water BROECKER (1971). PYTKOWICZ (1971) goes further into the question of the influence of adsorbed organic films on sand grains upon particle–water reactions (p.252).

The problems of compensation depth (p.267) and lysocline are examined in a series of papers: SMITH et al. (1969), HEATH and CULBERSON (1970), PYTKOWICZ (1970), BERGER (1971, 1973b), MCINTYRE and MCINTYRE (1971), LISITZIN (1972), MOORE et al. (1973), SCHNEIDERMANN (1973a), SLITER et al. (1975), with a major overall view, in time and space, by BERGER and WINTERER (1974). Dissolution of oceanic carbonate sediments in earlier times, Tertiary to Pleistocene, and ancient compensation depths, are treated by BERGER (1973a), HAY (1973), RAMSAY et al. (1973) and VAN ANDEL and MOORE (1974).

The uptake of magnesium in skeletal material is examined by MILLIMAN et al. (1971) and WEBER (1974). The relation between magnesium uptake and growth is reported by CADOT et al. (1972) and WEBER (1973a). Growth rate of crystals clearly has an influence on partitioning (p.260) of both Mg^{2+}/Ca^{2+} and Sr^{2+}/Ca^{2+}, as became obvious during discussion in a recent Penrose Conference. The factors controlling the ratio Mg^{2+}/Ca^{2+} in surface layers of calcite are examined by MÖLLER (1973) and MÖLLER and RAJAGOPALAN (in press). The mechanism of incongruent dissolution of high-magnesian calcite (p.337) and the significance, in skeletal calcite, of adjacent domains having different molar quantities of magnesium ion, is analysed by MACQUEEN et al. (1974) and PLUMMER and MACKENZIE (1974). In-congruent dissolution in dolomite is usefully discussed by T. M. L. WIGLEY (1973). Four papers that deal with the inhibiting effect of magnesium on the precipitation of carbonates (p.243) are by CHRIST and HOSTETLER (1970), KATZ (1973), MÖLLER (1973), PYTKOWICZ (1973), SALES and FYFE (1973), SASTRI and MÖLLER (1974) and MÖLLER and PAREKH (1975).

In a provocative sortie into this area FOLK (1973, and extended 1974) has argued a causal relation between the Mg^{2+}/Ca^{2+} ratio in waters and the mineralogy and crystal habits of the precipitated solid phases, relying on alleged restriction of growth by selective absorption of magnesium ions into the lattice of the prism faces. However, J. A. D. Dickson, University of Nottingham, has analysed 40 well-crystallized calcite crystals of various habits, using near-surface samples of faces of known Miller indices. Contents of Mg^{2+}, Mn^{2+}, Sr^{2+} and Fe^{2+} show no correlation with habit (personal communication). Interesting arguments supporting a low ratio Mg^{2+}/Ca^{2+} in ancient sea water, thus allowing precipitation of marine low-magnesian calcite, are put forward by BROECKER (1974) and SANDBERG (in press b).

The role of strontium in the aragonite–calcite transformation is discussed by KATZ et al. (1972) who propose that diagenesis in sea water can yield limestones

For references see pp.615–632.

with as little as 500 p.p.m. of strontium. The behaviour of strontium in carbonate mineral growth and diagenesis is examined by USDOWSKI (1973). The interpretation of strontium levels in limestone sequences in terms of sedimentation and diagenesis is elaborated by RENARD (1972), also by VEIZER et al. (1971) and VEIZER and DEMOVIČ (1974). The relation of strontium uptake in the aragonite of corals to growth rate is noted by WEBER (1973b). Correlations between aragonite and strontium contents in limestones of Barbuda, West Indies, have been recorded by P. WIGLEY (1973), who finds that the change from aragonite to calcite during meteoric diagenesis does not lead to the accumulation of strontium in the pore water, as earlier found in Barbados (p.329), probably owing to the existence of a more open system.

The effect of the rate of stirring on the proportions of calcium carbonate polymorphs precipitated has been shown by the work of H. W. Douglas and L. Antl, in the Department of Chemistry, University of Liverpool (personal communication), to be in general agreement with the results of Pobeguin (p.259). Crystals were examined with the scanning electron microscope.

There has been a continued expansion in the study of the influence of organic materials on the kinetics of carbonate dissolution and precipitation. Emphasis has been placed on the role of organic films in the chemical isolation of grains from the ambient water (p.252). For example, in SUESS (1970), a hydrophobic adsorbed layer could be formed by a fatty acid. Its carboxyl end would react with the carbonate surface, but the water-repellent hydrocarbon end would project outward. Carboxyl groups can substitute for carbonate ions in the crystal lattice, and this may explain the adherence of aspartic acid-rich protein to carbonate surfaces. JACKSON and BISCHOFF (1971) reported that amino acids with free hydroxyl groups are adsorbed, selectively, on carbonate compounds. Acidic amino acids inhibit wet transformation of aragonite to calcite at 66 °C, by the deposition of overgrowths such as glutamic or aspartic acids. On the other hand, neutral and basic amino acids tend to accelerate the transformation catalytically. SMALLWOOD (in press a) measured the adsorption of substances in aqueous solutions onto surfaces of calcite and aragonite using microelectrophoresis. Calcium and magnesium ions were adsorbed by calcite to similar extents, but aragonite adsorbed calcium ions more strongly than magnesium ions. Carbonate and sulphate ions seemed to be adsorbed more strongly by aragonite than by calcite. In the presence of albumin and sodium alginate the two polymorphs had similar surface charges. In another paper SMALLWOOD (in press b) reports experiments on the growth of seed crystals added to supersaturated solutions of calcium carbonate. As previously found by Nancollas, the growth of calcite seems to be surface controlled but that of aragonite seems to proceed by different mechanisms. Albumin retarded the growth of aragonite seeds but not those of calcite. MITTERER (1971) found that organic layers, such as protein-polysaccharide complexes, appear to act as substrate-templates controlling the heterogeneous precipitation of the carbonate (p.40). Similar controls were examined by MITTERER (1972b) and SUESS and FÜTTERER (1972) and are apparent in

For references see pp.615–632.

ALLEVA et al. (1971). A high charge density on a large protein molecule could not only balance many charges at the surface of a carbonate grain, but with its own surplus ionizable groups could complex also dissolved calcium ions. Thus a sufficient quantity of adjacent adsorbed protein chains might raise the local concentration of calcium so as to stimulate nucleation of a mineral phase (MITTERER, 1972a). This research is inevitably related to the work of HUBBARD (1972a) on the tissue-clad and tissue-free surfaces of cavities in corals.

Two papers dealing with surface areas of sediments and their chemical significance are by ANDERSON et al. (1973) and SUESS (1973). HAMZA and BROECKER (1974) note a fractionation of oxygen isotopes between the surface of calcite and the CO_2 in the adjacent aqueous solution. This fractionation is 5–6% greater than that between the bulk calcite and the adjacent CO_2. MÖLLER and PAPENDORFF (1971) note the expectation of some fractionation of $^{44}Ca/^{40}Ca$ in diffusion processes, but not during dissolution-precipitation. In fact they detect a 5% shift of $\delta^{44}Ca$. The precipitation of an extraordinary variety of forms of calcite and aragonite (rafts, tubes, spherulites, pisolites, colloidal, sparry, acicular, etc.) was carried out in the presence of decaying organic matter by McCUNN (1972).

GROWTHS OF OOIDS

The growth processes in Recent, open marine ooids remain a good deal of a mystery (BATHURST, 1974). The aragonite fabric consists of needles, well-sorted, blunt-ended, lacking sharp edges, with preferred tangential orientation (frontispiece), and is well known in SEM photographs (e.g. Fabricius and Klingele, and Loreau, p.79; MARGOLIS and REX, 1971). To my mind, this fabric does not lend itself readily to an interpretation based on mechanical modification, in turbulent water, of earlier radial-fibrous or random fabrics (LOREAU 1972b, 1973; SASS et al., 1972; LOREAU and PURSER, 1973). The correlation between water turbulence in diverse environments along the Trucial Coast and ooid fabric is undeniable, but the hypothetical alteration series, from radial-fibrous or random to tangential, seems at variance with the considerable diversity in crystal size and habit in the figured stages. Could it really yield, as an end product, the well-sorted accumulation of crystals illustrated in the frontispiece? Some biochemical control of precipitation is suggested, for example by the work of MITTERER (p.253; 1971, 1972a), who showed that proteins, found in Recent marine ooids, are like those associated with skeletal calcification; also by the experimental precipitation of SUESS and FÜTTERER (1972).

The anomalous birefringence (p.79) is elegantly explained by LOREAU (1972c, 1973) and in LOREAU and PURSER (1973). The unexpected discovery of high-magnesian calcite ooids on the Great Barrier Reef has been reported by P. J. DAVIES (in press).

Fresh light has been shed on the growth of ooids in Great Salt Lake by

For references see pp.615–632.

KAHLE (1974), SANDBERG (in press b) and M. Scherer (University of Tübingen, personal communication). These workers confirm the results of earlier researchers that the ooids are dominantly composed of aragonite. Sandberg claims that the ooids have always been radial-fibrous aragonite (a view supported by Scherer). Neomorphic alteration of a pre-existing fabric would have yielded a sparry mosaic, as in calcitized aragonite shell walls, but never a radial-fibrous fabric. On this basis he believes that Mesozoic and earlier ooids, showing radial-fibrous calcite structures, have always been low-magnesian calcite, precipitated in sea water with Mg^{2+}/Ca^{2+} *less* than one. The subsequent reversal of this ratio may reflect increasing abstraction of calcium from sea water by growth of calcitic plankton. Cretaceous ooids, now low-magnesian calcite, which broke while on the sea floor after the development of radial-fibrous fabric, have been described by SIMONE (1974) from the Southern Apennines, Italy.

FRESH WATER DIAGENESIS, CEMENTATION AND NEOMORPHIC PROCESSES

There has been a notable sharpening of the distinction between three subsurface diagenetic environments: the marine phreatic, the fresh water phreatic and the fresh water vadose. The most extensive progress in this area has been made by a group of workers with Matthews of Brown University, who have continued to investigate the chemical system involved in the diagenesis of a series of uplifted Pleistocene coral and backreef terraces on Barbados (p.329; STEINEN and MATTHEWS, 1973; MATTHEWS, 1974; STEINEN, 1974). Cored boreholes reveal the complex histories of marine carbonate sediments that have been exposed at various times to the meteoric phreatic, the meteoric vadose and the marine phreatic diagenetic environments, during the up and down movement of the fresh water lens. During these changes, the cationic composition of the pore water will have reflected, at all times, the mineralogy of the grains with which it was free to react. Thus, along with the gradual evolution of the mineralogy and the fabric of the rock, there will have been an evolution in the composition of the chemistry of the pore water. This evolution of the pore water will have been further complicated by the pronounced lateral recharge that is known to occur, bringing in pore waters whose composition has already been influenced by their passage through carbonate sediments some kilometres away in a horizontal direction. Indeed, mass balance calculations indicate that substantial carbonate must have been derived from rocks outside the immediate column containing the analysed cores.

From all this it can readily be appreciated that, as compositions of pore waters change with time, there will be a corresponding change in the partitioning (p.260) of Mg^{2+}/Ca^{2+} in the cements (not forgetting parallel changes in $\delta^{13}C$ and $\delta^{18}O$, p.339). These changes have been detected, using the electron microprobe, in the low-magnesian calcite cements (BENSON and MATTHEWS, 1971; BENSON, 1974).

For references see pp.615–632.

Some of the changes can be correlated with particular generations of cement, others take place across single generations or single crystals. BENSON et al. (1972) have traced similar changes in cements of ancient limestones going back to the Devonian. The studies of various workers tend to support the meaningfulness of such analyses of ancient cements, encouraging the assumption that, once precipitated, a low-magnesian calcite cement, if it undergoes any change by solid state diffusion, does so at rates that are insignificant on a geological time scale (ANDERSON, 1969; BENSON and MATTHEWS, 1971; KATZ et al., 1972; MÖLLER and SASTRI, 1974; SANDBERG, in press a).

The diversity of diagenetic histories, both of solid phases and solutions, is of course very great. The chemical system in the Pleistocene limestones of Barbados is relatively closed, compared with the more open system on Barbuda (P. WIGLEY, 1973), and thus mineralogical histories and contemporary pore-water data on Barbados are likely to be more closely related (see also THORSTENSON and MACKEN-ZIE, 1974). Diffusion in pore systems and adsorption reactions are discussed by MANHEIM (1970). In some subsurface situations the throughput of groundwater is so rapid that approach toward equilibrium between solid phases and solutions is negligible. In some instances, where the total dissolved calcium carbonate happens to be a non-linear function of some other variable (such as salinity) then the mixing of two water masses (even if separately equilibrated) must cause either precipitation or dissolution (p.446; SCHMALZ, 1971; MATTHEWS, 1974). This factor may be significant in a zone of mixing where the fresh water phreatic zone overlies the marine phreatic. The thickness of a fresh water lens can vary from a few metres on Barbados to more than a thousand metres in Florida.

These researches into the fresh water diagenesis of Quaternary limestones have done much to clarify our thinking, but we should beware lest our view of the ancient become coloured unduly by studies carried out in the commodious play-ground of the Caribbean islands instead of, for example, in the harsher setting of the vast carbonate basin of the Gulf of Carpentaria, the remote depths of the Ceara Abyssal Plain or other situations more relevant to the history of many extensive ancient carbonate accumulations.

Another step forward of immense importance has been made in the new realm of cement stratigraphy, in which the technique of cathodoluminescence petro-graphy introduced by SIPPEL and GLOVER (p.433) is applied to cement fabrics in the context of their regional stratigraphic setting. MEYERS (1974) has applied this technique (which reveals, amongst other things, the zonal variation in the content of Mn^{2+} in calcite crystals, SOMMER, 1972) to an examination of the calcite cements in the Mississippian Lake Valley Formation of New Mexico. The zones are readily distinguishable on the basis of variations in their substructure, brightness and con-tents of manganese and ferrous iron (LINDHOLM and FINKELMAN, 1972). Meyers was able to correlate them through 100 m of vertical section and over a horizontal dis-tance of 16 km. The various cement generations can be dated in part by such features as truncation at unconformities (also FREEMAN, 1969, 1971), their truncated appearance in clasts, their burial by silts which have filtered down from overlying beds, and so on. The latest cement, which is ferroan, is clearly post-Mississippian, even post-Pennsylvanian, in age and is the most widely distributed in the Lake Valley limestones. Meyers has a valuable discussion on possible sources of carbonate, both autochthonous and allochthonous. In an interesting section he relates the distribution of the various cement zones to the positions of ancient phreatic lenses (see also the study of MAGARITZ, 1974, on carbon and oxygen isotope ratios at various levels in Senonian chalks). Other useful papers on cements are by MÜLLER (1970), by TALBOT (1971) on ferroan and non-ferroan cement generations, TALBOT (1972) on the preservation of structures, including probable marine aragonite cements, FREEMAN (1971) and JACKA (1974) on vadose textures, and LINDHOLM (1974) on the relation between fabric, crystal habit and magnesium content and time sequences. MURAVYOV (1970) has thrown out some stimulating ideas on the control of carbonate precipitation by non-carbonate impurities. Di GIROLAMO's (1968) study of the high proportion of enfacial junctions in cement (p.423) is a welcome confirmation of this hypothesis. LATTMAN and SIMONBERG (1971) show how the variation in texture among sedimentary layers influences cement distribution. LEVANDOWSKI et al. (1973) give an interesting discussion on movement of formation waters and the distribution and timing of cementation. SCHROEDER (1973) shows a sequence, in Pleistocene reef rock of Bermuda, of marine aragonite cement, followed by vadose sparry calcite from which neomorphic calcite extends into the aragonite; in dealing with calcitized gastropod walls, he shows how the neomorphic spar can have plane intercrystalline boundaries (p.485). An important work by SCHERER (in press) reveals the intimate relation between, on the one hand, mineralogy, habit and trace element content of cements and replacement calcite, and, on the other, a range of diagenetic environments such as the marine spray zone, mangrove swamp and the vadose and phreatic fresh water. WEST (1973) has made a detailed study of diagenesis in a temperate climate: the carbonate skeletal sands of some British raised beaches have undergone cementation by dissolution-precipitation, neomorphic replacement, and large-scale redistribution yielding solution pipes: three distinct episodes of cementation are apparent. The detailed fabric and mineralogical changes during diagenesis of Pleistocene corals are reported by SPIRO (1971b), JAMES (1974), PITTMAN (1974) and SCHERER (in press). In a field wherein experimental work is all too rare, THORSTENSON et al. (1972) have simulated processes of vadose and phreatic diagenesis, and ADELSECK et al. (1973) have produced etching and overgrowths in nanno oozes at elevated temperatures and pressures. A stimulating discussion of the growth mechanism of chalcedonic spherulites, which has general application to the growth of radial-fibrous carbonate cements, is given by MALEEV (1972) with emphasis on the splitting and deflection of fibres. An occurrence of aragonite as a fresh water precipitate is described by KONISHI and SAKAI (1972).

For references see pp.615–632.

In the closely related field of neomorphic replacement, the work of KENDALL and TUCKER (1973) has brought a new order into our understanding of cements that have undulose extinction and are cloudy with inclusions. In a closely reasoned examination of radiaxial fibrous mosaic (p.426), (*not* synonymous with radial-fibrous mosaic) and the orientation of its subcrystals and of its inclusion patterns, they argue that it must be an *in situ*, neomorphic, replacement of a pre-existing radial-fibrous fabric. It is likely that in many cases this precursor was a marine cement, aragonite or high-magnesian calcite, which was metastable and subject to replacement by low-magnesian calcite when bathed in water with a low ratio Mg^{2+}/Ca^{2+} (as in SCHNEIDERMANN et al., 1972). The neomorphic calcitization of coral skeletons is described by JAMES (1974) and SCHERER (in press). Work in progress by R. L. Folk (University of Texas, personal communication) on a layer of laminated flowstone (speleothem) has led him to conclude that, where aragonite cement is calcitized, the external crystal forms remain but the crystals are replaced by equant calcite mosaic.

Intermediate stages of neomorphic replacement, during meteoric calcitization of aragonite skeletons in Pleistocene limestones, have been observed with the help of the scanning electron microscope by SCHNEIDERMANN et al. (1972), SANDBERG et al. (1973) and SANDBERG (in press a). They found relics of the original aragonite skeleton in the new calcite crystals. Somewhat divergent views on the nature of the original mineralogy of Palaeozoic corals and on their diagenesis are argued by JAMES (1974) and SANDBERG (in press a).

The incipient stages of metamorphic change in limestones have been studied by BROWN (1972) who has recognised a coarsening of crystal size accompanied by an increase in the crystallinity of associated illites. There is an important survey of neomorphic calcite veins and their fabrics by MIŠÍK (1971). The whole subject of neomorphic replacement is usefully discussed by WARDLAW and REINSON (1971).

The mechanism of the change from high-magnesian calcite to low-magnesian calcite during fresh water diagenesis has been further illumined. SANDBERG (in press a) has studied the fabric changes undergone by a cheilostome bryozoan, *Schizoporella floridana*, in the Pleistocene Miami Limestone of Florida. Using the scanning electron microscope, he examined the calcite ultrastructure before and after the loss of magnesium in the fresh water diagenetic environment (p.326). There is a change from 4 to 5 mole % $MgCO_3$ in the original skeleton to about 0.5 mole % after alteration. Although no fabric change was detectable with the light microscope (as so often noted, p.327), the electron microscope showed pronounced changes in crystal shape and orientation, though little change in size. Sandberg concluded that the change must, therefore, be destructive and must act by dissolution-precipitation, a process of *in situ* calcitization (as suggested on p.337) akin to the calcitization of aragonite. He goes on to argue that preservation of ultrastructure in the Rugosa and Tabulata implies that they were always calcite (see also FENNINGER

For references see pp.615–632.

and FLAJS, 1974). This view of the transformation is supported by the discovery by
BENSON (1974) of two distinct calcite peaks (no gradation) in diagenetically altered
grains of echinoids and coralline algae from the Pleistocene of Barbados. PLUMMER
and MACKENZIE (1974), predicting solubilities from rate data, conclude that, in a
high-magnesian calcite, there are distinct domains, each characterized by a different
mole percentage of $MgCO_3$. Dissolution should proceed, they believe, in a stepwise
fashion, the more soluble domains going first. This is an interesting idea which may
help to explain the passing on of crystallographic information from the original
grain to its replacement (p.50). It is possible, for example, that immediate precipitation took place by heterogeneous nucleation on adjacent surviving domains, before
they in turn were dissolved (but see MACQUEEN et al. 1974). The inadequacy of any
solid state diffusion-controlled process to leach the magnesium so rapidly from
grains (this is an early stage of meteoric diagenesis, p.326) was demonstrated by AN-
DERSON (1969) who estimated that solid state diffusion of carbon and oxygen would
not take place significantly in less than 10^8 years unless a temperature of 200 °C was
exceeded. MÖLLER and SASTRI (1974), using Ca – ^{45}Ca exchange at the surfaces of
calcite crystals, found that the ion exchange involved only a monolayer—a
conclusion supported by the results of BRÄTTER et al. (1972). A useful point made
by MÖLLER and PAPENDORFF (1971) is that a diffusion process would result in some
isotopic fractionation whereas a dissolution-precipitation process would not. These
conclusions seem born out by the discovery by GOMBERG and BARNATTI (1970) of a
change from high-magnesian calcite to low-magnesian calcite in turbidite grains on
the deep-sea floor with accompanying isotopic adjustment. Incongruent dissolution
is also discussed by T. M. L. WIGLEY (1973).

Finally, MATTHEWS (1974) has put forward an interesting proposition to explain the common *in situ* replacement of high-magnesian calcites by calcite and the
dissolution–void–precipitation sequence of replacement of much aragonite. He has
suggested that the presence of high-magnesian calcite results in enough supersaturation with respect to low-magnesian calcite to make precipitation rapid and local,
whereas with respect to aragonite the level of supersaturation is not sufficient to
promote immediate local precipitation. However, we have yet to fit this hypothesis
with the known common *in situ* replacement of aragonite skeletons, or the low
magnesium content (4–5 mole %) in the original skeleton of *Schizoporella floridana*
which gives it a solubility less than that of aragonite.

In the rapidly evolving field of the formation of ancient concretions (localized
cementation), the significant works of RAISWELL (1971 and in press), discussed
briefly on p.434, are now available in full, also valuable studies by DEEGAN (1972),
MASSAAD (1972), SASS and KOLODNY (1972), NOBLE and HOWELLS (1974) and
DICKSON and BARBER (in press).

Hardgrounds in ancient limestones, commonly associated with concretionary
growth, are examined by FÜRSICH (1972), KENNEDY and KLINGER (1972), SCHLOZ
(1972) and BROMLEY (in press); there are many useful references in HSÜ and JEN-

KYNS (1974). Useful details of the growth of modern synsedimentary concretions, their chemistry and development of isotopic composition of carbon and oxygen are given by BOGGS (1972) and WHELAN and ROBERTS (1973). KOSTECKA (1972) outlines the growth of Triassic aragonite concretions and their pseudomorphism by calcite. Rotation of flakes in the estimation of porosity is discussed by OERTEL and CURTIS (1972). Cosmic fallout concentrations are analysed by CASTELLARIN et al. (1974).

We are, nonetheless, left with a significant question mark. HUDSON (1975) draws our attention to the representative sample of 272 Phanerozoic marine limestones that were analysed for $\delta^{13}C$ by KEITH and WEBER (1964). These limestones have a mean value of $\delta^{13}C = 0.56\%_0$ (standard deviation 1.55), which is not significantly different from that of modern marine carbonate sediments (Fig.261, p.339), but quite different from the negative values caused by meteoric diagenesis. Moreover, the limestones contain at least 50% cement and TAN and HUDSON (1974) have already shown that some late ferroan calcite cements have $\delta^{13}C - 1$–$3\%_0$. How, then, has the positive isotopic composition been maintained during diagenesis? How relevant are the studies of Pleistocene limestones on those buccaneer islands of Bahamas, Barbados, Barbuda and Bermuda—by Bob Matthews, Benson and others of Brown University? Admittedly submarine cementation could account for a small part of the deviation of the isotopic ratio from expected negative meteoric values. There could have been a little meteoric stabilization. The plain truth, however, seems to be that, for most Phanerozoic marine limestones, the lithification process was dominantly one of simple redistribution of the primary carbonate. The available mechanisms for this, as Hudson points out, are compaction of carbonate-bearing clays and pressure-solution—both acting in the mesogenetic environment, far removed from meteoric influences.

DIAGENESIS ON THE SEA FLOOR

Research developments since 1971 are reviewed by BATHURST (1974). Most progress has come through the examination of sediments and coral-algal reefs with the scanning electron microscope. This method not only reveals much of the fine detail of crystal fabrics, but yields also indirect evidence of chemical changes in the ambient water, such as supersaturation and undersaturation for calcium carbonate, that become apparent through the recognition of encrusted and etched material (ALEXANDERSSON, 1969, 1972a, b and c with subsequent discussion, and in press; SPIRO and HANSEN, 1970; GINSBURG et al., 1971a; SPIRO, 1971b; SCHROEDER, 1972b, 1974; GINSBURG and SCHROEDER, 1973; FRIEDMAN et al., 1974). The scanning microscope makes available for study another world that lies entirely beyond the resolving power of the light microscope. Moreover, the scanning microscope can be combined with the electron microprobe. The precipitates of aragonite and high-magnesian calcite revealed by this new approach show a considerable variety of crystal size, habit, distribution and growth sequence: these

For references see pp.615–632.

clearly have great significance for our understanding of the chemical reactions involving the various solid phases and the ambient solutions.

Mineralogy of the substrate is still seen to be a major control. Replacement of aragonite by high-magnesian calcite seems widespread (valuable discussion in SCHERER, 1974b). Locally elevated pH, accompanying cementation, is suggested by the presence of corroded quartz particles trapped in a cement of high-magnesian calcite (FRIEDMAN et al., 1974). Cement crystals are for the most part of micron size, and may be mixed with equally fine-grained sediment (e.g. coccoliths) and, under the light microscope, may resemble micrite, especially after diagenesis. This is a point to note in the examination of ancient limestones where, for example, the micrite filling in the zooecia of bryozoans, the chambers of foraminiferids or the pores in the echinodermata, may well be cement. Such finely crystalline cements are common in the intragranular porosity of shallow water carbonate sands, where these occur in warm seas that are supersaturated for calcium carbonate. Intergranular cementation, on the other hand, giving hardgrounds, flakes or grapestone, is more restricted in time and space, though equally finely crystalline. Rates of intragranular cementation are discussed by ALEXANDERSSON (1972b) and MOBERLY (1973).

A useful paper on grapestone deals with stages of binding and cementation (WINLAND and MATTHEWS, 1974). Regrettably, it makes no reference to the profound influence of cementation in vacated algal bores (micritization, p.384) on the whole fabric of grapestone, both grains and cements. These Bahamian grapestones show the clearest and most intense micritization of this type that I have ever seen (p.316, 389).

The postulated process of micritization of carbonate grains (p.381) has been substantiated through the works of Loreau (p.90), LLOYD (1971), MARGOLIS and REX (1971) and ALEXANDERSSON (1972a) who have revealed aragonite and high-magnesian calcite cements in vacated algal bores. The distribution of algal and fungal bores has been studied by PERKINS and HALSEY (1971), ROONEY and PERKINS (1972), GOBULIĆ (1973) and EDWARDS and PERKINS (1974), whose data indicate a probable maximum depth of algal boring today of about 50 m, though fungi live down to at least 450 m. FRIEDMAN et al. (1971) emphasized the possible importance of borers that can live below the photic zone, not only fungi but bacteria and heterotrophic algae. Evidence bearing on the mode of algal boring is presented by ALEXANDERSSON (1975a).

Evidence strongly suggestive of biochemical control of intragranular cementation has come from ALEXANDERSSON (1974), who has found cements in nodules of coralline algae in the North Sea. These cements are aragonite and high-magnesian calcite and were precipitated despite the undersaturation of the water for calcium carbonate. In this connection the works of SCHROEDER (1972a) and SCHERER (1974a) are of interest on the role of endolithic algae in cementation, also the results of WINLAND and MATTHEWS (1974) on precipitation of high-magnesian calcite in

For references see pp.615–632.

mucilaginous algal sheaths on grapestone surfaces and the papers of HUBBARD and POCOCK (1972) and HUBBARD (1974a, b and in press).

Significant progress in the understanding of active cementation of beach rock has come from several studies. DAVIES and KINSEY (1973) monitored the water chemistry of a rock pool on the Great Barrier Reef and their results indicate that cementation is related neither to temperature changes nor to photosynthesis and may be dominantly nocturnal. SCHMALZ (1971) investigated the mixing of meteoric ground water with sea water in beach sands of Eniwetok. He noted that, where the amount of dissolved calcium carbonate is a non-linear function of salinity, mixing of two supersaturated waters should yield extreme supersaturations for calcium carbonate. This would account for the rapid cementation of beach rock. MOORE (1973) also followed the mixing of sea water with meteoric water in beach rock on Grand Caymen Island and found a close relationship between water chemistry and cement chemistry in terms of Sr^{2+}/Ca^{2+}, Mg^{2+}/Ca^{2+}, $\delta^{18}O$ and $\delta^{13}C$. PURSER and LOREAU (1973) have given a detailed description of aragonite encrustations in the intertidal and supratidal zones off the Trucial Coast. Their evidence indicates inorganic precipitation in the marine vadose zone.

Concretions and hardgrounds are referred to on p. 556.

PRESSURE-SOLUTION

Important progress has been made by MOSSOP (1972) in his study of compaction within and around a Devonian bioherm. He found a close agreement between estimates of the amount of compaction made from an analysis of the irregular trend surfaces of horizons that were originally horizontal and the direct measurement of the amplitudes of stylolites. Interesting advances in theoretical understanding of pressure-solution have come through NEUGEBAUER's (1973, 1974) analysis of the situation in the Chalk of northern Europe. Theoretically, burial to 1,500 m or more should lead to substantial loss of porosity by pressure-welding, yet in the Chalk this has not happened (e.g. LOREAU, 1972a). Neugebauer constructed a model, relating overburden, particle geometry, mineralogy, the ratio Mg^{2+}/Ca^{2+} in the pore solution and evidence of dissolution and precipitation. He compared this model with DSDP data and concluded that, in favourable circumstances, an overburden of at least 2,000–4,000 m is required for the complete lithification of the Chalk by pressure-solution (see also SCHOLZ, 1973). Neugebauer's finding, that any large scale release of calcium and carbonate ions to give cement can only accompany the advanced stages of pressure-welding, is supported by the theoretical model of MANUS and COOGAN (1974).

General surveys of the role of pressure-solution in carbonate rocks have been given by GLOVER (1968, 1969). The matter is touched on by BERNOULLI (1972) and PLESSMAN (1972) in terms of tectonic stress.

An elegant analysis of pressure-solution under tectonic stress in a Chalk

For references see pp.615–632.

anticline in Dorset, by MIMRAN (1975) shows a positive correlation between stratal dip, density of the rock and flattening of once spherical calcispheres (*Oligostegina*). Volume loss may attain 70%. An illuminating study of the deformation of sharply folded limestone beds, by the evolution of stylolites, slickolites and slickensides, has been presented by CARANNANTE and GUZZETTA (1972). Other relevant works on deformation by pressure-solution are contained in papers by DURNEY (1972) and BEACH (1974).

SCHOLLE (1971) has noted that the stylolites in the Upper Cretaceous of the Northern Apennines formed too late to have been able to supply the main volume of cement (see also WACKS and HEIN, 1974).

Fabric information from DSDP cores is given by SCHNEIDERMANN (1973b) and MATTER (1974). Other useful papers are on mechanical compaction of sands by RITTENHOUSE (1971b), on compaction, ion filtration and osmosis by MAGARA (1974) and on high pore fluid pressures by BREDEHOEFT and HANSHAW (1968) and BERRY (1973). Some provocative ideas are provided by DEELMAN (1974). An interesting discussion of the state of stress in flat-lying sedimentary limestones has been presented by FRIEDMAN and HEARD (1974) who observe that the abundance of twin lamellae in calcite crystals is a function of depth of burial and have confirmed this conclusion in laboratory creep tests.

RECENT DOLOMITES

There are wide ranging and interesting studies of the processes of dolomitization, in Pleistocene–Recent carbonate sediments, by FISCHBECK and MÜLLER (1971), HANSHAW et al. (1971), BEHRENS and LAND (1972), BOURROUILH (1972, 1973), FROGET (1972), KONISHI et al. (1972), MÜLLER et al. (1972), COOK (1973), LAND (1973a and b), MÜLLER and FISCHBECK (1973) and RICHTER (1974). They cover such aspects as the Mg^{2+}/Ca^{2+} ratios in the pore water, the growth of protodolomites by replacement and precipitation, anomalous strontium contents, the role of meteoric water, mass balance considerations, sources of magnesium and the general diagenetic context of dolomitization and the importance of the time factor. Ideas on dolomite nucleation at calcite surfaces and the role of magnesium ions are presented by MÖLLER and WERR (1972) and SASTRI and MÖLLER (1974). Fractionation of oxygen and carbon isotopes is examined by FRITZ and SMITH (1970) and MURATA et al. (1972), and incongruent dissolution by T. M. L. WIGLEY (1973). There is an important analysis by LOVERING (1969).

The role of mixtures of fresh water and sea water in affecting dolomitization of calcite is the basis of a new model for dolomitization put forward by BADIOZAMANI (1973). Epitaxial overgrowths of calcite on dolomite are discussed by ZENGER (1973) and zoned dolomites are treated by KATZ (1971) and the fabrics of dolomites and dedolomites by WOLFE (1970). A critical review of supratidal dolomitization is presented by ZENGER (1972).

For references see pp.615–632.

Two papers are of considerable general significance for the study of older dolomites. One is a report on complexing of magnesium in algal-rich layers in stromatolites and the subsequent growth of dolomite, to give alternating laminae of calcite and dolomite, by GEBELEIN and HOFFMAN (1973). The other is a paper on the role of brines with elevated ratios Mg^{2+}/Ca^{2+} in late joint-controlled dolomitization by FREEMAN (1972).

DIAGENESIS UNDER THE DEEP SEA FLOOR

The Initial Reports of the Deep Sea Drilling Project have already yielded impressive new data regarding the sedimentary, palaeontological and diagenetic records for some hundreds of meters of sediments under the ocean floors. General sediment-ological surveys are by BEALL and FISCHER (1969), BERGER and VON RAD (1972), DAVIES and LAUGHTON (1972), LANCELOT and EWING (1972), LANCELOT et al. (1972) and BEALL et al. (1973). Other reports consider the differential dissolution of carbonate nannofossils (coccoliths, discoasters, foraminiferids, etc.), both pre-depositional and post-depositional, rates of deposition and fertility, the role of CO_2 released by decay and bacterial sulphate reduction, cementation and neo-morphic replacement and the growth of dolomite (GOMBERG and BONATTI, 1970; SCHNEIDERMANN, 1970; WISE and HSÜ, 1971; BERGER and VON RAD, 1972; WISE and KELTS, 1972; DAVIES and SUPKO, 1973; RAMSEY et al., 1973; COOK, 1974; GARRISON, 1974; SUPKO et al., 1974).

The detailed observations of MATTER (1974) with the scanning electron microscope have thrown light on etching of nannofossils and cement fabrics and also on the history of compaction, first by dewatering with increased packing and subsequently by dissolution-precipitation and pressure-welding. All this is auto-lithification (p.447) since no allochthonous source of calcium carbonate was available (see also KANEPS, 1973; SCHLANGER et al., 1973; WISE, 1973; SCHLANGER and DOUGLAS, 1974). Subsurface rates of dissolution appear to have exceeded rates of precipitation since the concentrations of calcium, magnesium and strontium in pore waters are greater than those in sea water. Calcium is enriched downward at the expense of magnesium, the ratio Mg^{2+}/Ca^{2+} remains mostly greater than unity but, locally, decreases downward (MANHEIM et al., 1970). The strontium concentra-tion is closely correlated positively with rate of deposition (MANHEIM and SAYLES, 1971; SAYLES and MANHEIM, 1975). Models of the factors governing the distribution of dissolved constituents during compaction have been constructed by BERNER (1975). Some extreme diagenetic changes are related to the proximity of the basalt basement and higher than usual heat flow. Causes of downhole variations in $\delta^{18}O$ and $\delta^{13}C$ are discussed by LLOYD and HSÜ (1972), ANDERSON and SCHNEIDERMANN (1973) and COPLEN and SCHLANGER (1973). There are four interesting papers on the alteration of deep sea sediments by cold and hydrothermal solutions (BOSTRÖM and HOROWITZ, 1972; THOMPSON, 1972; GARRISON et al., 1973; MILLIMAN and MÜLLER,

For references see pp.615–632.

1973). There are reports of cemented crusts by BARTLETT and GREGGS (1970a, b); precipitation of high-magnesian calcite at the sediment–water interface has been studied by SARTORI (1974). Three major works containing much valuable new material are FUNNEL and RIEDEL (1971), LISITZIN (1972) and HSÜ and JENKYNS (1974). Ancient calcareous oozes, exposed on land, and their diagenesis are examined by SCHOLLE (1971, 1974) and WACHS and HEIN (1974).

LAKE SEDIMENTS

In the last few years there have been important advances in our understanding of carbonate sedimentation and precipitation in a variety of lakes, particularly in Europe, Turkey, east Africa and Afghanistan. Most of this new work has been carried out in the Laboratorium für Sedimentforschung in the University of Heidelberg by G. Müller and co-workers. There is a summary by MÜLLER et al. (1972). Details of interest are hydrology, water chemistry, ratios of Mg^{2+}/Ca^{2+} and Sr^{2+}/Ca^{2+}, the sources and distributions of high-magnesian and low-magnesian calcites, monohydrocalcite, aragonite, protodolomite, etc. and the roles of plants and growth of oncolites, also subsequent diagenesis of sediments under the lake floors (MÜLLER, 1969a, b, c, 1970, 1971; MÜLLER and BLASCHKE, 1969; MÜLLER and STOFFERS, 1974; FÖRSTNER, 1973). There is a paper on oxygen and carbon isotopes in lacustrine carbonates in Texas by PARRY (1970). A major work on the Black Sea has been edited by DEGENS and ROSS (1974).

CALCRETE

Interest in calcrete (caliche, cancar, carbonate duricrusts, etc.) has grown rapidly in the last few years as geologists have appreciated the profound significance of these vadose carbonates. The whole subject is surveyed in a useful book on duricrusts in general by GOUDIE (1973) who offers one of the more helpful definitions, thus: calcrete is "A product of terrestrial processes within the zone of weathering in which" calcium carbonate has "accumulated in and/or replaced a pre-existing soil, rock, or weathered material, to give a substance which may ultimately develop into an indurated mass". Goudie's book regrettably has no pictures of calcrete but this lack is remedied in JAMES (1972), CASTELLARIN and SARTORI (1973), ESTEBAN (1974), WIEDER and YAALON (1974) and READ (1974b) who give also, with GOUDIE (in press), valuable petrographic details and describe textural evolution. General background of the weathering processes and their products is to be found in BREWER (1964).

Formation of calcrete by redistribution of the calcium carbonate in a parent limestone is helped by brecciation. This increases greatly the surface area available for attack by algae, fungi, bacteria and actinomycetes (KRUMBEIN, 1968). The newly formed fabrics are pisolites (of Thomas and Dunham, p.84), clots, laminated crusts

For references see pp.615–632.

(of Kornicker and of Multer and Hoffmeister, p.324), coated grains, networks of fine cracks and a variety of replacements. The new fabric is composed of low-magnesian calcite micrite or microspar. Recent calcretes, and some ancient ones, are reddish-brown. The whole may be complicated by reworking and dolomitization. Photographs with the light microscope and scanning electron microscope are in WIEDER and YAALON (1974). Photographs with the light microscope are also given in PIALLI (1971) and BOSELLINI and ROSSI (1974). There is an important basic paper on ancient calcretes by BERNOULLI and WAGNER (1971). An unusual calcrete-like deposit composed of aragonite and high-mangesian calcite on a supratidal sabkha is described by SCHOLLE and KINSMAN (1974). Other useful papers are by SWINEFORD et al. (1958), D'ARGENIO (1967, p.113), REEVES (1970), ESTEBAN (1972, 1974), GARDNER (1972) and GOUDIE (1972).

HYDROGEOLOGY AND KARST

Most problems facing the carbonate sedimentologist require for their solution some appreciation of the role of underground or pore water in the evolution of mineralogy and the morphology of sediments and rocks. Water as chemical reactant or physical transporter influences not only dissolution, precipitation and replacement, but maintenance or development of porosity and permeability on a scale ranging from fractions of a micron to caves in which a man may walk. The last few years have seen such a growth of valuable research in this varied field that it is now clearly desirable that some of the more significant works be brought together in one place. Obviously, however, there is no sharp boundary between this aspect of carbonate sedimentology and others already referred to on earlier pages.

The subject of karst is treated in a major book by SWEETING (1972), in a review by LEGRAND and STRINGFIELD (1973) and in a paper on Apennine karst by PASSERI (1972). The FAO (1972) have produced a multilingual glossary of terms. Palaeokarst is recorded in the works of D'ARGENIO (1967), PIRLET (1970), SEMINIUK (1971), CARANNANTE et al. (1974) and WALKDEN (1974). The sources of dissolved carbonate species in the water and their rates of uptake have been examined by PEARSON and HANSHAW (1970) and PITTY (1971). The question of just how many ground waters really are supersaturated for calcium carbonate has been raised by LANGMUIR (1971). The chemical variation and behaviour of ground waters has been studied by FRIEDMAN and GAVISH (1970) and PATERSON (1972) and the isotopic fractionation in precipitates by FANTIDIS and EHHALT (1970) and HENDY (1971). Vadose precipitation can lead to a variety of mineral deposits and replacements (THRAILKILL, 1971) and the influence of algae and mosses on the formation of travertines is substantial (GOLUBIĆ, 1969).

Chemical aspects of aquifers, such as elemental distribution, equilibria, process rates and radiocarbon ages, are covered in papers by HANSHAW et al. (1965), BACK and HANSHAW (1970, 1971), EDMUNDS (1973) and EDMUNDS et al.

For references see pp.615–632.

(1973). The relation between pore water composition and hydrostatic overpressure is analysed by DICKEY et al. (1972). Problems of flow and dispersion are dealt with by WILLIAMS (1970), PATERSON (1971), KLOTZ (1973) and PANDEY et al. (1974).

Some very interesting investigations of porosity by the use of pore casts and the scanning electron microscope have been carried out by PITTMAN and DUSCHATKO (1970) and WARDLAW (in press a). Other works on porosity and its evolution are by PITTMAN (1971), RITTENHOUSE (1971a), CHILINGAR et al. (1972), OLDERSHAW (1972) and WARDLAW (in press b).

In conclusion I would emphasis a pungent warning by HOFFMAN (1973): "Yesterday's heresy is today's dogma, and the bandwagon effect makes adherents to the new . . . paradigm easier to muster than proof."

For references see pp.615–632.

GLOSSARY OF SEDIMENTOLOGICAL TERMS
NOT DEFINED IN THE TEXT[1]

Allochem—Abbreviation of allochemical constituent (FOLK, 1959): "all materials that have formed by chemical or biochemical precipitation *within* the basin of deposition, but which are organized into descrete aggregated bodies and for the most part have suffered some transportation". Examples are oöids, skeletal particles and pellets.

Allodapic limestone—A deep water limestone believed to have been transported by a turbidity current (MEISCHNER, 1964).

Anhedral—Descriptive of the shape of mineral crystals on which crystal faces are absent. Modified from C.I.P.W. (CROSS et al., 1906) to conform with current usage by FRIEDMAN (1965d).

Biomicrite—A limestone composed of skeletal grains (allochems) in a matrix of micrite (FOLK, 1959).

Biosparite—A limestone composed of skeletal grains (allochems) in cement (FOLK, 1959; Fig.297, p.418).

Centimetre-sized—Refers to a fabric in which the crystal diameters are 10–100 mm (FRIEDMAN, 1965d).

Centimicron-sized—Refers to a fabric in which the crystal diameters are 100–1000 μ (FRIEDMAN, 1965d).

Decimicron-sized—Refers to a fabric in which the crystal diameters are 10–100 μ (FRIEDMAN, 1965d).

Eogenetic—The eogenetic stage of diagenesis applies to the time interval between final deposition and burial below the depth of significant influence by processes that either operate from the surface or depend for their effectiveness on proximity to the surface (CHOQUETTE and PRAY, 1970).

Euhedral—Descriptive of the shape of mineral crystals on which crystal faces are present. Modified from C.I.P.W. (CROSS et al., 1906) to conform with current usage by FRIEDMAN (1965d). Taken here to imply that a crystal is wholly bounded by crystal faces.

Fabric—The arrangement of crystals in a rock in terms of their sizes, shapes and crystallographic orientations. This use of the word follows KNOPF (1933) and FAIRBAIRN (1942) who thus translated the "*Gefüge*" of Sander (1936, 1951).

Fenestral—TEBBUTT et al. (1965) define "fenestra" as: "A primary or penecontemporaneous gap in rock framework, larger than grain-supported interstices. A fenestra may be an open space in the rock, or it may be completely or partially filled by secondarily introduced sediment or cement. The distinguishing characteristic of fenestrae is that the spaces have no apparent support in the framework of the primary grains forming the sediment." Examples in FISCHER, 1964 and SHINN, 1968a.

---

[1] Page references to definitions in the text are given in the Index in bold type.

Geopetal—SANDER (1951) defines as geopetal: "All the widely distributed spatial characters of a fabric that enable us to determine what was the relation of 'top' to 'bottom' at the time when the rock was formed." Probably the best known of these is sediment lying on the floor of a cavity which it partly fills (Fig.300, p.418).

Grainstone—A mud-free (micrite-free) rock in which the grains (allochems) are, necessarily, grain-supported (DUNHAM, 1962; Fig.297, p.418).

Grain-supported—Refers to the fabric of a rock in which the grains (allochems) are in contact with each other, even though they may have a mud (micrite) matrix (DUNHAM, 1962).

Matrix—Used here to describe the sedimentary, mechanically deposited material between grains (as distinct from precipitated cement; Fig.298, p.418).

Mechanical deposition—Descriptive of sedimentary particles which "have been developed in some way, entirely before the depositing process, for example as detritus, crystals, oölite grains" (SANDER, 1936, 1951). This definition implies no assumptions about the prior transport of the particles or their origins.

Mesogenetic—The mesogenetic stage of diagenesis applies to the time interval during which the rocks are buried below the major influence of the processes operating from or closely related to the surface (CHOQUETTE and PRAY, 1970).

Micrite—An abreviation of "microcrystalline ooze". One of the essentially normal precipitates, formed within the basin of deposition, and showing little or no evidence of significant transport; forms crystals 1–4 μ in diameter (FOLK, 1959, p.7–8). "In hand specimen, this is the dull and opaque ultra-fine-grained material that forms the bulk of 'lithographic' limestones and the matrix of chalk." (FOLK, 1962b, p.66; Fig.294, 298 on p.403, 418 resp.) The word "precipitate" here includes inorganic and biochemical precipitates (as the crystals in the shell wall of a gastropod later disaggregated to form ooze).

Micron-sized—Refers to a fabric in which crystal diameters are 0–10 μ (Friedman, 1965d).

Microspar—Neomorphic crystal mosaic with crystal diameters from 4 to 10 μ or even 50 μ. Used here for any mineral, not just for calcite as in FOLK's (1965, p.37) original definition.

Millimetre-sized—Refers to a fabric in which the crystal diameters are 1–10 mm (FRIEDMAN, 1965d).

Mud-supported—Refers to the fabric of a mudstone (micrite) in which the grains (allochems) are separated from each other by a micrite matrix (DUNHAM, 1962).

Oömicrite—A limestone composed of oöids (allochems) in a matrix of micrite (Folk, 1959).

Oösparite—A limestone composed of oöids (allochems) in cement (FOLK, 1959).

Packstone—A mudstone (micrite) in which the grains (allochems) are grain-supported (DUNHAM, 1962; see wackestone).

Passive dissolution—Term newly used here, akin to FOLK's (1965) passive precipitation, to refer to the dissolution of minerals in a rock to yield open spaces of the type in which passive precipitation could take place. To be distinguished from the kind of dissolution which accompanies the movement of a solution film during the development of neomorphic spar.

Passive precipitation—The precipitation of cement in open spaces (FOLK, 1965, p.24), as in between grains, in the body chambers of gastropods or in fenestrae. To be distinguished from the growth of neomorphic spar and the growth of crystals by displacement.

Pelmicrite—A limestone composed of peloids (allochems) in a matrix of micrite (FOLK, 1959). The word "peloid" is used here instead of "pellet" because Folk found that these grains in the pelsparite cannot always be identified as faecal pellets, but may have originated in other ways.

Peloid—An allochem formed of cryptocrystalline or microcrystalline material, irrespective of size or origin (MCKEE and GUTSCHICK, 1969). This useful term allows reference to grains composed of micrite or microspar without the need to imply any particular mode of formation.

Pelsparite—A limestone composed of peloids (allochems) in cement (FOLK, 1959). The word "peloid" is used here because Folk found that these grains in the pelsparite cannot always be identified as faecal pellets, but may have originated in other ways.

Poikilotopic—Descriptive of the fabric of a sedimentary rock in which the crystals are of more than one size and in which larger crystals enclose smaller crystals or detrital grains: the term is restricted to carbonate rocks which have undergone neomorphism or are precipitates (FRIEDMAN, 1965d).

Porphyrotopic—Descriptive of the fabric of a sedimentary rock in which the crystals are of more than one size and in which larger crystals (porphyrotopes) are enclosed in a groundmass of smaller crystals: the term is restricted to carbonate rocks which have undergone neomorphism or are precipitates (FRIEDMAN, 1965d).

Pseudospar—Neomorphic crystal mosaics with crystal diameters greater than those of microspar (i.e. > than about 10–50 μ). Used here for any mineral, not just for calcite as in FOLK's (1965, p.42) original definition (Fig.336, p.485).

Radial-fibrous—Descriptive of a fabric of crystal fibres arranged with their long axes radiating from a centre (opposite to tangential; Fig.242. p.306).

Skeletal—Descriptive of particles formed originally as biochemically precipitated skeletons, such as shells or coral trabeculae.

Spar—A mosaic of crystals larger than those in micrite, formed either as cement or as neomorphic spar: includes "microspar" and "pseudospar" (FOLK, 1959, 1962b, 1965).

Subhedral—Descriptive of the shape of mineral crystals on which crystal faces are partially developed. Taken here to imply that a crystal is partly bounded by crystal faces.

Syntaxial—Refers to the overgrowth on a crystal such that the original crystal and the overgrowth form a single larger crystal, sharing the same crystallographic axes. Synonymous with "optically continuous", "in lattice continuity" and "epitaxial". Applied, for example, to overgrowths of calcite on echinoid grains, whether passive cement or neomorphic in origin (BATHURST, 1958, following GOLDMAN, 1952).

Telogenetic—The telogenetic stage of diagenesis applies to the time interval during which long-buried rocks are located near the surface again, as a result of crustal movement and erosion, and are significantly influenced by processes associated with the formation of an unconformity (CHOQUETTE and PRAY, 1970).

Wackestone—A mudstone (micrite) in which the grains (allochems) are mud-supported (DUNHAM, 1962; see packstone).

REFERENCES

ABELSON, P. H., 1954. Amino acids in fossils. *Science*, 119: 576.

ADAMS, J. E. and RHODES, M. L., 1960. Dolomitization by seepage refluction. *Bull. Am. Assoc. Petrol. Geologists*, 44: 1912–1920.

AGASSIZ, A., 1894. A reconnoissance of the Bahamas and of the elevated reefs of Cuba in the steam yacht "Wild Duck", January to April, 1893. *Bull. Museum Comp. Zool. Harvard Coll.*, 26: 1–203.

AGASSIZ, A., 1896. The elevated reef of Florida. *Bull. Museum Comp. Zool. Harvard Coll.*, 28: 29–62.

AKAAD, M. K. and NAGGAR, M. H., 1964. Petrography of the Egyptian alabaster [calcite] of Wadi Al Assyuti. *Bull. Fac. Sci. Alexandria*, 6: 157–173.

AKAAD, M. K. and NAGGAR, M. H., 1967. Geology of the Wadi Sannur alabaster [calcite] and the general history of the Egyptian alabaster deposits. *Bull. Inst. Desert Egypte*, 13: 35–63.

AKIN, G. W. and LAGERWERFF, J. V., 1965a. Calcium carbonate equilibria in aqueous solutions open to the air. I. The solubility of calcite in relation to ionic strength. *Geochim. Cosmochim. Acta*, 29: 343–352.

AKIN, G. W. and LAGERWERFF, J. V., 1965b. Calcium carbonate equilibria in solutions open to the air. II. Enhanced solubility of $CaCO_3$ in the presence of Mg^{2+} and SO_4^{2-}. *Geochim. Cosmochim. Acta*, 29: 353–360.

ALBISSIN, M., 1963. Les traces de la déformation dans les roches calcaires. *Rev. Géograph. Phys. Géol. Dyn., Suppl.*, 5: 1–74.

ALDERMAN, A. R., 1959. Aspects of carbonate sedimentation. *J. Geol. Soc. Australia*, 6: 1–10.

ALDERMAN, A. R., 1965a. The problem of the origin of dolomite in sediments. *Indian Mineralogist*, 6: 14–20.

ALDERMAN, A. R., 1965b. Dolomitic sediments and their envi ronment in the South-East of South Australia. *Geochim. Cosmochim. Acta*, 29: 1355–1365.

ALDERMAN, A. R. and SKINNER H. C. W., 1957. Dolomite sedimentation in the South-East of South Australia. *Am. J. Sci.*, 255: 561–567.

ALDERMAN, A. R. and VON DER BORCH, C. C., 1960. Occurrence of hydromagnesite in sediments in South Australia. *Nature*, 188: 931.

ALDERMAN, A. R. and VON DER BORCH, C. C., 1961. Occurrence of magnesite–dolomite sediments in South Australia. *Nature*, 192: 861.

ALDERMAN, A. R. and VON DER BORCH, C. C., 1963. A dolomite reaction series. *Nature*, 198: 465–466.

ALLEN, J. R. L., 1968. *Current Ripples*. North-Holland, Amsterdam, 447 pp.

ALLEN, J. R. L., 1969. Some recent advances in the physics of sedimentation. *Proc. Geologists Assoc. Engl.*, 80: 1–42.

ALLEN, J. R. L., 1970, *Physical Processes of Sedimentation*. Allen and Unwin, London, 248 pp.

ALLEN, R. C., GAVISH, E., FRIEDMAN, G. M. and SANDERS, J. E., 1969. Aragonite-cemented sandstone from outer continental shelf off Delaware Bay: submarine lithification mechanism yields product resembling beachrock. *J. Sediment. Petrol.*, 39: 136–149.

ALLOITEAU, J., 1952. Madréporaires post-Paléozoïques: In: J. PIVETEAU (Editor), *Traité de Paléontologie, 1*. Masson, Paris, pp.539–684.

ALLOITEAU, J., 1957. *Contribution à la Systématique des Madréporaires Fossiles, I–II*. Centre Natl. Rech. Sci., Paris, 462 pp.

AMSBURY, D. L., 1962. Detrital dolomite in central Texas. *J. Sediment. Petrol.*, 32: 5–14.

ANDERSON, F. W., 1950. Some reef-building calcareous algae from the Carboniferous rocks of northern England and southern Scotland. *Proc. Yorkshire Geol. Soc.*, 28: 5–28.

ANDREWS, J. E., 1970. Structure and sedimentary development of the outer channel of the Great Bahama Canyon. *Geol. Soc. Am. Bull.*, 81: 217–226.

ANDREWS, J. E., SHEPARD, F. P. and HURLEY, R. J., 1970. Great Bahama Canyon. *Geol. Soc. Am. Bull.*, 81: 1061–1078.

ANONYMOUS, 1958–1968. *Growth of Crystals* (reports of a series of conferences in Russia—Consultants Bureau, New York), 1–63. Plenum, New York, N.Y.

ANONYMOUS, 1969. When is epitaxy possible? *Nature*, 223: 1206–1207.

ARKELL, W. J., 1956. *Jurassic Geology of the World*. Oliver and Boyd, Edinburgh, 806 pp.

ARMSTRONG, J., 1969. The cross-bladed fabrics of the shells of *Terrakea solida* Etheridge and Dun and *Streptorhynchus pelicanensis* Fletcher. *Palaeontology*, 12: 310–320.

ARNTSON, R. H., 1964. Effect of temperature and confining pressure on the solubility of calcite at constant CO_2 concentrations. *Geol. Soc. Am., Spec. Papers*, 76: 6–7 (abstract).

BAARS, D. L., 1963. Petrology of carbonate rocks. In: R. O. BASS and S. L. SHARPS (Editors), *Shelf Carbonates of the Paradox Basin: a Symposium—Four Corners. Geol. Soc., Field Conf., 4th*, pp.101–129.

BAAS-BECKING, L. G. M. and GALLIHER, E. W., 1931. Wall structure and mineralization in coralline algae. *J. Phys. Chem.*, 35: 467–479.

BAKER, G. and FROSTICK, A. C., 1951. Pisoliths, ooliths and calcareous growths in limestone caves at Port Campbell, Victoria, Australia. *J. Sediment. Petrol.*, 21: 85–104.

BALL, M. M., 1967. Carbonate sand bodies of Florida and the Bahamas. *J. Sediment. Petrol.*, 37: 556–591.

BALL, M. M., SHINN, E. A. and STOCKMAN, K. W., 1967. The geologic effects of hurricane Donna in south Florida. *J. Geol.*, 75: 583–597.

BANDY, O. L., 1964. Foraminiferal biofacies in sediments of Gulf of Batabano, Cuba, and their geologic significance. *Bull. Am. Assoc. Petrol. Geologists*, 48: 1666–1679.

BANERJEE, A., 1959. Petrography and facies of some Upper Viséan (Mississippian) limestones in North Wales. *J. Sediment. Petrol.*, 29: 377–390.

BANNER, F. T. and WOOD, G. V., 1964. Recrystallization in microfossiliferous limestones. *Geol. J.*, 4: 21–34.

BARRETT, P. J., 1964. Residual seams and cementation in Oligocene shell calcarenites, Te Kuiti Group. *J. Sediment. Petrol.*, 34: 524–531.

BARTLETT, G. A. and GREGGS, R. G., 1969. Carbonate sediments: oriented lithified samples from the North Atlantic. *Science*, 166: 740–741.

BATHURST, R. G. C., 1958. Diagenetic fabrics in some British Dinantian limestones. *Liverpool Manchester Geol. J.*, 2: 11–36.

BATHURST, R. G. C., 1959a. Diagenesis in Mississippian calcilutites and pseudobreccias. *J. Sediment. Petrol.*, 29: 365–376.

BATHURST, R. G. C., 1959b. The cavernous structure of some Mississippian *Stromatactis* reefs in Lancashire, England. *J. Geol.*, 67: 506–521.

BATHURST, R. G. C., 1964a. Diagenesis and paleoecology: a survey. In: J. IMBRIE and N. D. NEWELL (Editors), *Approaches to Paleoecology*. Wiley, New York, N.Y., pp.319–344.

BATHURST, R. G. C., 1964b. The replacement of aragonite by calcite in the molluscan shell wall. In: J. IMBRIE and N. D. NEWELL (Editors), *Approaches to Paleoecology*. Wiley, New York, N.Y., pp.357–376.

BATHURST, R. G. C., 1966. Boring algae, micrite envelopes and lithification of molluscan biosparites. *Geol. J.*, 5: 15–32.

BATHURST, R. G. C., 1967a. Oölitic films on low energy carbonate sand grains, Bimini lagoon, Bahamas. *Marine Geol.*, 5: 89–109.

BATHURST, R. G. C., 1967b. Depth indicators in sedimentary carbonates. *Marine Geol.*, 5: 447–471.

BATHURST, R. G. C., 1967c. Sub-tidal gelatinous mat, sand stabilizer and food, Great Bahama Bank. *J. Geol.*, 75: 736–738.

BATHURST, R. G. C., 1968. Precipitation of oöids and other aragonite fabrics in warm seas. In: G. MÜLLER and G. M. FRIEDMAN (Editors), *Recent Developments in Carbonate Sedimentology in Central Europe*. Springer, Berlin, pp.1–10.

BATHURST, R. G. C., 1969. Bimini Lagoon. In: H. G. MULTER (Editor), *Field Guide to some Carbonate Rock Environments: Florida Keys and Western Bahamas*. Fairleigh Dickinson Univ., Madison, N.J., pp.62–66.

BAUSCH, W. M., 1965. Dedolomitisierung und Recalcitisierung in fränkischen Malmkalken. *Neues Jahrb. Mineral., Monatsh.*, 3: 75–82.

BAUSCH, W. M., 1968. Clay content and calcite crystal size of limestones. *Sedimentology*, 10: 71–75.

BAVENDAMM, W., 1932. Die mikrobiologische Kalkfällung in der tropischen See. *Arch. Mikrobiol.*, 3: 205–276.

BAXTER, J. W., 1960. *Calcisphaera* from the Salem (Mississippian) Limestone in southwestern Illinois. *J. Paleontol.*, 34: 1153–1157.

BAYER, F. M., 1956. Octocorallia. In: R. C. MOORE (Editor), *Treatise on Invertebrate Paleontology, F. Coelenterata*. Geol. Soc. Am., Boulder, Colo., pp.166–231.

BÉ, A. W. H. and HEMLEBEN, C., 1970. Calcification in a living planktonic foraminifer. *Neues Jahrb. Geol. Paläontol., Abhandl.*, 134: 221–234.

BÉ, A. H. W., MCINTYRE, A. and BREGER, D. L., 1966. Shell microstructure of a planktonic foraminifer, *Globorotalia menardii* (D'Orbigny). *Eclogae Geol. Helv.*, 59: 885–896.

BEALES, F. W., 1956. Conditions of deposition of Palliser (Devonian) limestone of southwestern Alberta. *Bull. Am. Assoc. Petrol. Geologists*, 40: 848–870.

BEALES, F. W., 1958. Ancient sediments of Bahaman type. *Bull. Am. Assoc. Petrol. Geologists*, 42: 1845–1880.

BEALES, F. W., 1961. Modern sediment studies and ancient carbonate environments. *J. Alberta Soc. Petrol. Geologists*, 9: 319–330.

BEALES, F. W., 1963. Baldness of bedding surface. *Bull. Am. Assoc. Petrol. Geologists*, 47: 681–686.

BEALES, F. W., 1965. Diagenesis in pelletted limestones. In: L. C. PRAY and R. C. MURRAY (Editors), *Dolomitization and Limestone Diagenesis: a Symposium—Soc. Econ. Paleontologists Mineralogists, Spec. Publ.*, 13: 49–70.

BEALES, F. W., 1966. Field study of modern carbonate environments. *Bull. Can. Petrol. Geologists*, 14: 180–188.

BECKER, G. F. and DAY, A. L., 1916. Note on the linear force of growing crystals. *J. Geol.*, 24: 313–333.

BERGENBACK, R. E. and TERRIERE, R. T., 1953. Petrography and petrology of Scurry reef, Scurry County, Texas. *Bull. Am. Assoc. Petrol. Geologists*, 37: 1014–1029.

BERGER, W. H., 1967. Foraminiferal ooze. Solution at depths. *Science*, 156: 383–385.

BERGER, W. H., 1970. Planktonic foraminifera: selective solution and the lysocline. *Marine Geol.*, 8: 111–138.

BERNER, R. A., 1965a. Activity coefficients of bicarbonate, carbonate and calcium ions in sea water. *Geochim. Cosmochim. Acta.*, 29: 947–965.

BERNER, R. A., 1965b. Dolomitization of the mid-Pacific atolls. *Science*, 147: 1297–1299.

BERNER, R. A., 1966a. Chemical diagenesis of some modern carbonate sediments. *Am. J. Sci.*, 264: 1–36.

BERNER, R. A., 1966b. Diagenesis of carbonate sediments: interaction of magnesium in sea water with mineral grains. *Science*, 153: 188–191.

BERNER, R. A., 1967. Comparative dissolution characteristics of carbonate minerals in the presence and absence of aqueous magnesium ion. *Am. J. Sci.*, 265: 45–70.

BERNER, R. A., 1968a. Calcium carbonate concretions formed by the decomposition of organic matter. *Science*, 159: 195–197.

BERNER, R. A., 1968b. Rate of concretion growth. *Geochim. Cosmochim. Acta*, 32: 477–483.

BILLINGS, G. K. and RAGLAND, P. C., 1968. Geochemistry and mineralogy of the Recent reef and lagoonal sediments south of Belize (British Honduras). *Chem. Geol.*, 3: 135–153.

BIRKELUND, T., 1967. Submicroscopic shell structures in early growth-stages of Maastrichtian ammonites (Saghalinites and Scaphites). *Medd. Dansk. Geol. Foren.*, 17: 95–101.

BIRKELUND, T. and HANSEN, H. J., 1968. Early shell growth and structures of the septa and the siphuncular tube in some Maastrichtian ammonites. *Medd. Dansk Geol. Foren.*, 18: 71–78.

BISCHOFF, J. L., 1968a. Catalysis, inhibition, and the calcite–aragonite problem. II. The vaterite–aragonite transformation. *Am. J. Sci.*, 266: 80–90.

BISCHOFF, J. L., 1968b. Kinetics of calcite nucleation: magnesium ion inhibition and ionic strength catalysis. *J. Geophys. Res.*, 73: 3315–3321.

BISCHOFF, J. L., 1969. Temperature controls on aragonite–calcite transformation in aqueous solutions. *Am. Mineralogist*, 54: 149–155.

BISCHOFF, J. L. and FYFE, W. S., 1968. Catalysis, inhibition, and the calcite–aragonite problem. I. The aragonite–calcite transformation. *Am. J. Sci.*, 266: 65–79.

BLACK, M., 1933a. The algal sediments of Andros Island, Bahamas. *Phil. Trans. Roy. Soc. London, Ser. B.*, 222: 165–192.

BLACK, M., 1933b. The precipitation of calcium carbonate on the Great Bahama Bank. *Geol. Mag.*, 70: 455–466.

BLACK, M., 1953. The constitution of the Chalk. *Proc. Geol. Soc. London*, 1499: lxxxi–lxxxii.

BLACK, M., 1963. The fine structure of the mineral parts of Coccolithophoridae. *Proc. Linnean Soc. London*, 174: 41–46.

BLACK, M., 1965. Coccoliths. *Endeavour*, 24: 131–137.

BLACK, M. and BARNES, B., 1959. The structure of coccoliths from the English Chalk. *Geol. Mag.*, 96: 321–328.

BLACK, M. and BARNES, B., 1961. Coccoliths and discoasters from the floor of the south Atlantic Ocean. *J. Roy. Microscop. Soc.*, 80: 137–147.

BLACKMON, P. D. and TODD, R., 1959. Mineralogy of some Foraminifera as related to their classification and ecology. *J. Paleontol.*, 33: 1–15.

BOARDMAN, R. S. and CHEETHAM, A. H., 1969. Skeletal growth, intracolony variation, and evolution in Bryozoa: a review. *J. Paleontol.*, 43: 205–233.

BOETTCHER, A. L. and WYLLIE, P. J., 1968. The calcite–aragonite transition measured in the system CaO–CO$_2$–H$_2$O. *J. Geol.*, 76: 314–330.

BØGGILD, O. B., 1930. The shell structure of the mollusks. *Kgl. Danske Videnskab. Selskab, Mat. Fys. Medd.*, 9: 231–325.

BÖGLI, A., 1963. Beitrag zur Entstehung von Karsthöhlen. *Höhle*, 14: 63–68.

BOLIN, B., 1960. On the exchange of carbon dioxide between the atmosphere and the sea. *Tellus*, 12: 274–281.

BOSELLINI, A., 1965. Analisi petrografica della "Dolomia Principale" nel Gruppo di Sella (Regione Dolomitica). *Mem. Geopaleontol. Univ. Ferrara*, 1: 49–109.

BOUSSINESQ, J., 1876. Essai théorique sur l'équilibre des massifs pulvérulents, comparé à celui de massifs solides, et sur la poussée des terres sans cohésion. *Acad. Roy. Belg., Mém., Couronnes Savants Étrangers*, XL(4): 1–180.

BOUSSINESQ, J., 1885. *Applications des Potentials à l'Étude de l'Équilibre et du Mouvement des Solides Élastiques, etc.* Gauthier-Villars, Paris, 721 pp.

BRAMLETTE, M. N., 1961. Pelagic sediments. In: M. SEARS (Editor), *Oceanography*. Am. Assoc. Advan. Sci., Washington, D.C., Publ. 67, pp.345–390.

BRAMLETTE, M. N., FAUGHN, J. L. and HURLEY, R. J., 1959. Anomalous sediment deposition on the flank of Eniwetok Atoll. *Bull. Geol. Soc. Am.*, 70: 1549–1551.

BRANDON, D. G., WALD, M., SOUTHON, M. J. and RALPH, B., 1963. The application of field ion microscopy to the study of lattice defects. *J. Phys. Soc. Japan*, 18: 2–324.

BRANNER, J. C., 1904. The stone reefs of Brazil, their geological and geographical relations, with a chapter on the coral reefs. *Bull. Museum Comp. Zool. Harvard Coll.*, 44: 1–285.

BRAUN, M. and FRIEDMAN, G. M., 1970. Dedolomitization fabric in peels: a possible clue to unconformity surfaces. *J. Sediment. Petrol.*, 40: 417–419.

BRICKER, O. P. (Editor), 1971. *Carbonate Cements*. Johns Hopkins Press, Baltimore, Md., 376 pp.

BROECKER, W. S. and TAKAHASHI, T., 1966. Calcium carbonate precipitation on the Bahama Banks. *J. Geophys. Res.*, 71: 1575–1602.

BROMLEY, R. G., 1965. *Studies in the Lithology and Conditions of Sedimentation of the Chalk Rock and Comparable Horizons*. Thesis, Univ. London, London, 355 pp, unpublished.

BROMLEY, R. G., 1967a. Marine phosphorites as depth indicators. *Marine Geol.*, 7: 503–509.

BROMLEY, R. G., 1967b. Some observations on burrows of thalassinidean Crustacea in Chalk hardgrounds. *Quart. J. Geol. Soc. London*, 123: 157–177.

BROMLEY, R. G., 1968. Burrows and boring in hardgrounds. *Medd. Dansk Geol. Foren.*, 18: 247–250.

BROMLEY, R. G., 1970. Borings as trace fossils and *Entobia cretacea* Portlock, as an example. In: T. P. CRIMES and J. C. HARPER (Editors), *Trace Fossils–Geol. J.*, Spec. Issue, 4. Seel House Press, Liverpool, pp.49–90.

BROOKS, R., CLARK, L. M. and THURSTON, E. F., 1951. Calcium carbonate and its hydrates. *Phil. Trans. Roy. Soc. London, Ser. A*, 243: 145–167.

BROWN, P. R., 1961. *Petrology of the Lower and Middle Purbeck Beds of Dorset*. Thesis, Univ. Liverpool, Liverpool, 223 pp., unpublished.

BROWN, P. R., 1963a. Some algae from the Swan Hills reef. *Bull. Can. Petrol. Geologists.*, 11: 178–182.

BROWN, P. R., 1963b. Algal limestones and associated sediments in the basal Purbeck of Dorset. *Geol. Mag.*, 100: 565–573.

BROWN, P. R., 1964. Petrography and origin of some upper Jurassic beds from Dorset, England. *J. Sediment. Petrol.*, 34: 254–269.

BROWN, W. H., FYFE, W. S. and TURNER, F. J., 1962. Aragonite in California glaucophane schists, and the kinetics of the aragonite–calcite transformation. *J. Petrol.*, 3: 566–582.

BROWN, W. W., 1959. The origin of stylolites in the light of a petrofabric study. *J. Sediment. Petrol.*, 29: 254–259.

BRUNTON, C. H., 1969. Electron microscopic studies of growth margins of articulate brachiopods. *Z. Zellforsch.*, 100: 189–200.

BRYAN, W. H. and HILL, D., 1941. Spherulitic crystallization as a mechanism of skeletal growth in the hexacorals. *Proc. Roy. Soc. Queensland*, 52: 78–91.

BUCHBINDER, B. and FRIEDMAN, G. M., 1970. Selective dolomitization of micrite envelopes: a possible clue to original mineralogy. *J. Sediment. Petrol.*, 40: 514 517.

BUCHER, W. H., 1918. On oölites and spherulites. *J. Geol.*, 26: 593–609.

BUCKLEY, H. E., 1951. *Crystal Growth*. Wiley, New York, N.Y., 571 pp.

BURNS, J. H. and BREDIG, M. A., 1956. Transformation of calcite to aragonite by grinding. *J. Chem. Phys.*, 25: 1281.

BUSBY, R. F., 1962. Submarine geology of the Tongue of the Occan, Bahamas. *U.S. Naval Oceanog. Office, Tech. Rept.*, TR-108: 1–84.

BUTLER, G. P., 1965. *Early Diagenesis in the Recent Sediments of the Trucial Coast of the Persian Gulf*. Thesis, Imp. Coll. Sci. Technol., London, 163 pp., unpublished.

BUTLER, G. P., 1969. Modern evaporite deposition and geochemistry of coexisting brines, the sabkha, Trucial Coast, Arabian Gulf. *J. Sediment. Petrol.*, 39: 70–89.

BUTLER, G. P., KENDALL, C. G. ST. C., KINSMAN, D. J. J., SHEARMAN, D. J. and SKIPWITH, P. A. D'E., 1964. Recent anhydrite from the Trucial Coast of the Arabian Gulf. *Geol. Soc. London, Circ.*, 120: 3.

CAROZZI, A. V., 1960. *Microscopic Sedimentary Petrography*. Wiley, New York, N.Y., 485 pp.

CARRIKER, M. R., SMITH, E. H. and WILCE, R. T. (Editors), 1969. Penetration of calcium carbonate substrates by lower plants and invertebrates. *Am. Zool.*, 9: 629–1020.

CAYEUX, L., 1916. *Introduction à l'Étude Pétrographique des Roches Sédimentaires*. Imprimerie Nationale, Paris, 524 pp.

CAYEUX, L., 1933. Rôle des Trilobites dans la genèse des gisements de phosphate de chaux paléozoique. *Compt. Rend.*, 196: 1179–1182.

CAYEUX, L., 1935. *Les Roches Sédimentaires de France; Roches Carbonatées*. Masson, Paris, 463 pp.

CHANDA, S. K., 1967a. Selective neomorphism and fabric discontinuities in limestones. *J. Sediment. Petrol.*, 37: 688–690.

CHANDA, S. K., 1967b. Selective recrystallization in limestones. *Sedimentology*, 8: 73–76.

CHARLES, G., 1953. Sur l'origine des gisements de phosphates de chaux sédimentaires. *Congr. Géol. Intern. Compt. Rend., 19e, Algiers, 1952*: 163–184.

CHAUDRON, G., 1954. Contribution à l'étude des réactions dans l'état solide cinétique de la transformation aragonite-calcite. In: E. HEMLIN (Editor), *Proc. Intern. Symp. Reactivity Solids, Gothenburg, 1952*. Ingeniörsvetenskapsakademien och Chalmers Tekniska Högskola, Göteborg, pp.9–20.

CHAVE, K. E., 1952. A solid solution between calcite and dolomite. *J. Geol.*, 60: 190–192.

CHAVE, K. E., 1954a. Aspects of the biogeochemistry of magnesium. 1. Calcareous marine organisms. *J. Geol.*, 62: 266–283.

CHAVE, K. E., 1954b. Aspects of the biogeochemistry of magnesium. 2. Calcareous sediments and rocks. *J. Geol.*, 62: 587–599.

CHAVE, K. E., 1960. Carbonate skeletons to limestones: problems. *Trans. N.Y. Acad. Sci., Ser. II*, 23: 14–24.

CHAVE, K. E., 1964. Skeletal durability and preservation. In: J. IMBRIE and N. D. NEWELL (Editors), *Approaches to Paleoecology*. Wiley, New York, N.Y., pp.377–387.

CHAVE, K. E., 1965. Carbonates: association with organic matter in surface seawater. *Science*, 148: 1723–1724.

CHAVE, K. E., in press. Carbonate-organic interactions in sea water. In: *Organic Matter in Natural Waters*, Univ. Alaska Press.

CHAVE, K. E., 1970. Carbonate-organic interactions in sea water. In: D. W. HOOD (Editor), *Organic Matter in Natural Waters*. Univ. Alaska Press, College, Alaska, pp. 373–386.

CHAVE, K. E. and SCHMALZ, R. F., 1966. Carbonate–seawater reactions. *Geochim. Cosmochim. Acta*, 30: 1037–1048.

CHAVE, K. E. and SUESS, E., 1967. Suspended minerals in seawater. *Trans. N.Y. Acad. Sci., Ser. II*, 29: 991–1000.

CHAVE, K. E., DEFFEYES, K. S., WEYL, P. K., GARRELS, R. M. and THOMPSON, M. E., 1962. Observations on the solubility of skeletal carbonates in aqueous solutions. *Science*, 137: 33–34.

CHEETHAM, A. H., RUCKER, J. B. and CARVER, R. E., 1969. Wall structure and mineralogy of the cheilostome bryozoan *Metrarabdotus*. *J. Paleontol.*, 43: 129–135.

CHEN, C., 1964. Pteropod ooze from Bermuda Pedestal. *Science*, 144: 60–62.

CHENEY, E. S. and JENSEN, M. L., 1965. Stable carbon isotopic composition of biogenetic carbonates. *Geochim. Cosmochim. Acta*, 29: 1331–1346.

CHESTER, R. H., 1969. Destruction of Pacific corals by the sea star *Acanthaster planci*. *Science*, 165: 280–283.

CHILINGAR, G. V., BISSELL, H. J. and FAIRBRIDGE, R. W. (Editors), 1967a. *Carbonate Rocks: Origin, Occurrence and Classification*. Elsevier, Amsterdam, 471 pp.

CHILINGAR, G. V., BISSELL, H. J. and FAIRBRIDGE, R. W. (Editors), 1967b. *Carbonate Rocks: Physical and Chemical Aspects*. Elsevier, Amsterdam, 413 pp.

CHOQUETTE, P. W., 1968. Marine diagenesis of shallow marine lime-mud sediments: insights from δO^{18} and δC^{13} data. *Science*, 161: 1130–1132.

CHOQUETTE, P. W. and PRAY, L. C., 1970. Geological nomenclature and classification of porosity in sedimentary carbonates. *Bull. Am. Assoc. Petrol. Geologists*, 54: 207–250.

CIFELLI, R., BOWEN, V. T. and SIEVER, R., 1966. Cemented foraminiferal oozes from the mid-Atlantic Ridge. *Nature*, 209: 32-34.

CLARK, B. B., 1968. Geomorphological features of Mother Ivey's Bay near Padstow, with an account of the under-cliff bank and intertidal reef of cemented limesand at Little Cove. *Trans. Roy. Geol. Soc. Cornwall*, 20: 69–79.

CLARK, P. S., 1957. A note on calcite–aragonite equilibrium. *Am. Mineralogist*, 42: 564–566.

CLARKE, E. DE C. and TEICHERT, C., 1946. Algal structures in a Western Australia salt lake. *Am. J. Sci.*, 244: 271–276.

CLARKE, F. W. and WHEELER, W. C., 1922. The inorganic constituents of marine invertebrates. *U.S., Geol. Surv., Profess. Papers*, 124: 1–62.

CLAYTON, R. N., JONES, B. F. and BERNER, R. A., 1968. Isotope studies of dolomite formation under sedimentary conditions. *Geochim. Cosmochim. Acta*, 32: 415–432.

CLOUD, JR., P. E., 1955. Bahama Banks west of Andros Island. *Bull. Geol. Soc. Am.*, 66: 1542 (abstract).

CLOUD, JR., P. E., 1962a. Environment of calcium carbonate deposition west of Andros Island, Bahamas. *U.S., Geol. Surv., Profess. Papers*, 350: 1–138.

CLOUD, JR., P. E., 1962b. Behaviour of calcium carbonate in sea water. *Geochim. Cosmochim. Acta*, 26: 867–884.

CLOUD, JR., P. E., 1965. Carbonate precipitation and dissolution in the marine environment. In: J. R. RILEY and G. SKIRROW (Editors), *Chemical Oceanography*, 1. Academic Press, London, pp.127–158.

COLE, W. S., 1957. Larger Foraminifera from Eniwetok drill holes. *U.S., Geol. Surv., Profess. Papers*, 260-V: 742–784.

COLLEY, H. and DAVIES, P. J., 1969. Ferroan and non-ferroan calcite cements in Pleistocene–Recent carbonates from the New Hebrides. *J. Sediment. Petrol.*, 39: 554–558.

COOGAN, A. H., 1969a. Bahamian and Floridian biofacies. In: H. G. MULTER (Editor), *Field*

Guide to some Carbonate Rock Environments: Florida Keys and Western Bahamas. Fairleigh Dickinson Univ., Madison, N.J., 159 pp.

COOGAN, A. H., 1969b. Compaction effects in oolitic grainstone. *Bull. Am. Assoc. Petrol. Geologists*, 53: 713 (abstract).

COOGAN, A. H., 1970. Measurements of compaction in oolitic grainstone. *J. Sediment. Petrol.*, 40: 921–929.

CORRENS, C. W., 1949a. *Einführung in die Mineralogie.* Springer, Berlin, 414 pp.

CORRENS, C. W., 1949b. Growth and dissolution of crystals under linear pressure. *Discussions Faraday Soc.*, 5: 267–271.

COSTIN, J. M., 1965. Mixing and residence time on the Great Bahama Bank. *U.S. At. Energy Comm., Tech. Rept.*, CU-17-65: 1–18.

COX, L. R., 1960. General characteristics of Gastropoda. In: R. C. MOORE (Editor), *Treatise on Invertebrate Paleontology. I. Mollusca 1.* Geol. Soc. Am., Boulder, Colo., pp.85–169.

COX, L. R., NUTTALL, C. P. and TRUEMAN, E. R., 1969. General features of Bivalvia. Structure of the shell wall. In: R. C. MOORE (Editor), *Treatise on Invertebrate Paleontology. N. Mollusca 6.* Geol. Soc. Am., Boulder, Colo., pp.73–78.

CRAIG, G. Y., 1967. Size-frequency distributions of living and dead populations of pelecypods from Bimini, Bahamas, B.W.I. *J. Geol.*, 75: 34–45.

CRAIG, H., 1953. The geochemistry of the stable carbon isotopes. *Geochim. Cosmochim. Acta*, 3: 53–93.

CRICKMAY, G. W., 1945. Petrography of limestones. In: LADD, H. S. and HOFFMEISTER, J. E., 1945. Geology of Lau, Fiji. *Bernice P. Bishop Museum Bull.*, 181: 211–250.

CROSS, W., IDDINGS. J. P., PIRSSON, L. V. and WASHINGTON, H. S., 1906. The texture of igneous rocks. *J. Geol.*, 14: 692–707.

CULBERSON, C., KESTER, D. R. and PYTKOWICZ, R. M., 1967. High-pressure dissociation of carbonic and boric acids in seawater. *Science*, 157: 59–61.

CULKIN, F. and COX, R. A., 1966. Sodium, potassium, magnesium, calcium and strontium in sea water. *Deep-Sea Res.*, 13: 789–804.

CULLIS, C. G., 1904. The mineralogical changes observed in the cores of the Funafuti borings. In: T. G. BONNEY (Editor), *The Atoll of Funafuti.* Roy. Soc., London, pp.392–420.

CURL, R. L., 1962. The aragonite–calcite problem. *Bull. Natl. Speleol. Soc.*, 24: 57–73.

CURTIS, C. D. and KRINSLEY, D., 1965. The detection of minor diagenetic alteration in shell material. *Geochim. Cosmochim. Acta*, 29: 71–84.

CURTIS, R., EVANS, G., KINSMAN, D. J. J. and SHEARMAN, D. J., 1963. Association of dolomite and anhydrite in the Recent sediments of the Persian Gulf. *Nature*, 197: 679–680.

DACHILLE, F. and ROY, R., 1960. High-pressure phase transformations in laboratory mechanical mixers and mortars. *Nature*, 186: 34.

DAETWYLER, C. C. and KIDWELL, A. L., 1959. The Gulf of Batabano, a modern carbonate basin. *World Petrol. Congr., Proc., 5th, N.Y., 1959*, Sect. I: 1–21.

DALINGWATER, J. E., 1969. *Some Aspects of the Chemistry and Fine Structure of the Trilobite Cuticle.* Thesis, Univ. Manchester, Manchester, 95 pp., unpublished.

DALLAVALLE, J. M , 1943. *Micromeritics.* Pitman, New York, N.Y., 428 pp.

DALRYMPLE, D. W., 1966. Calcium carbonate deposition associated with blue-green algal mats, Baffin Bay, Texas. *Inst. Marine Sci. Publ.*, 10: 187–200.

DALY, R. A., 1917. Origin of the living coral reefs. *Scientia*, 22: 1–12.

DALY, R. A., 1924. The geology of American Samoa. *Papers Dept. Marine Biol., Carnegie Inst. Wash. Publ.*, 340: 95–143.

DANA, J. D., 1851. On coral reefs and islands. *Am. J. Sci., Ser. II*, 11: 357–372; 12: 25-51, 165–186, 329–338.

DANGEARD, L., 1936. Étude des calcaires par coloration et décalcification. Application à l'étude des calcaires oolithiques. *Bull. Soc. Géol. France*, (5)6: 237–245.

DANSGAARD, W., 1954. The O^{18}-abundance in fresh water. *Geochim. Cosmochim. Acta*, 6: 241–260.

DARWIN, C., 1962. Coral Islands. Reprint: introduction by D. R. Stoddart. *Atoll Res. Bull.*, 88: 1–20.

DAUGHTRY, A. C., PERRY, D. and WILLIAMS, M., 1962. Magnesium isotopic distribution in dolomite. *Geochim. Cosmochim. Acta*, 26: 857–866.

DAVID, T. W. E. and SWEET, G., 1904. The geology of Funafuti. In: T. G. BONNEY (Editor), *The Atoll of Funafuti*. Roy. Soc., London, pp.61–88.

DAVIDSON, S. C. and McKINSTRY, H. E., 1931. "Cave pearls", oölites, and isolated inclusions in veins. *Econ. Geol.*, 26: 289–294.

DAVIES, C. W. and JONES, A. L., 1955. The precipitation of silver chloride from aqueous solutions. 2. Kinetics of growth of seed crystals. *Trans. Faraday Soc.*, 51: 812–817.

DAVIES, D. K., 1968. Carbonate turbidites, Gulf of Mexico. *J. Sediment. Petrol.*, 38: 1100–1109.

DAVIES, G. R., 1967. *Recent and Pleistocene Carbonate Sedimentation, Eastern Shark Bay, Western Australia*. Thesis, Univ. Western Australia, Nedlands, W. Austr., 207 pp., unpublished.

DAVIES, G. R., 1970a. Carbonate bank sedimentation, eastern Shark Bay, Western Australia. *Am. Assoc. Petrol. Geologists, Mem.*, 13: 85–168.

DAVIES, G. R., 1970b. Algal-laminated sediments, Gladstone embayment, Shark Bay, Western Australia. *Am. Assoc. Petrol. Geologists, Mem.*, 13: 169–205.

DAVIES, P. J. and TILL, R., 1968. Stained dry cellulose peels of ancient and Recent impregnated carbonate sediments. *J. Sediment. Petrol.*, 38: 234–237.

DAVIS, B. L. and ADAMS, L. H., 1965. Kinetics of the calcite⇌aragonite transformation. *J. Geophys. Res.*, 70: 433–441.

DE CHARDIN, P. T., 1959. *The Phenomenon of Man*. Collins, London, 320 pp.

DEER, W. A., HOWIE, R. A. and ZUSSMAN, J., 1962. *Rock-Forming Minerals, 5*. Longmans, London, 371 pp.

DEFANT, A., 1961. *Physical Oceanography, 1*. Pergamon, Oxford, 729 pp.

DEFFEYES, K. S., 1965. Carbonate equilibria: a graphic and algebraic approach. *Limnol. Oceanog.*, 10: 412–426.

DEFFEYES, K. S., LUCIA, F. J. and WEYL, P. K., 1964. Dolomitization: observations on the island of Bonaire, Netherlands Antilles. *Science*, 143: 678–679.

DEFFEYES, K. S., LUCIA, F. J. and WEYL, P. K., 1965. Dolomitization of Recent and Plio–Pleistocene sediments by marine evaporite waters on Bonaire, Netherlands Antilles. In: L. C. PRAY and R. C. MURRAY (Editors), *Dolomitization and Limestone Diagenesis; a Symposium—Soc. Econ. Paleontologists Mineralogists, Spec. Publ.*, 13: 71–88.

DEFLANDRE-RIGAUD, M., 1957. A classification of fossil alcyonarian sclerites. *Micropaleontology*, 3: 357–366.

DEGENS, E. T., 1965. *Geochemistry of Sediments; a Brief Survey*. Prentice-Hall, Englewood Cliffs, N.J., 324 pp.

DEGENS, E. T. and EPSTEIN, S., 1962. Relationship between O^{18}/O^{16} ratios in coexisting carbonates, cherts, and diatomites. *Bull. Am. Assoc. Petrol. Geologists*, 46: 534–542.

DEGENS, E. T. and EPSTEIN, S., 1964. Oxygen and carbon isotope ratios in coexisting calcites and dolomites from recent and ancient sediments. *Geochim. Cosmochim. Acta*, 28: 23–44.

DE GROOT, K., 1965. Inorganic precipitation of calcium carbonate from sea-water. *Nature*, 207: 404–405.

DE GROOT, K., 1967. Experimental dedolomitization. *J. Sediment. Petrol.*, 37: 1216–1220.

DE GROOT, K., 1969. The chemistry of submarine cement formation at Dohat Hussain in the Persian Gulf. *Sedimentology*, 12: 63–68.

DE GROOT, K. and DUYVIS, E. M., 1966. Crystal form of precipitated calcium carbonate as influenced by adsorbed magnesium ions. *Nature*, 212: 183–184.

DE MEIJER, J. J., 1969. Fossil non-calcareous algae from insoluble residues of algal limestones. *Leidse Geol. Mededel.*, 44: 235–239.

DE WINDT, J. and BERWERTH, F., 1904. Untersuchungen von Grundproben der I., II., u. IV. Reise von S. M. Schiff "Pola" in den Jahren 1890, 1892, und 1893. *Jahrb. Oesterr. Wiss.*, 74: 285–294.

DICKSON, J. A. D., 1965. A modified staining technique for carbonates in thin section. *Nature*, 205: 587.

DICKSON, J. A. D., 1966. Carbonate identification and genesis as revealed by staining. *J. Sediment Petrol.*, 36: 491–505.

DISTECHE, A. and DISTECHE, S., 1967. The effect of pressure on the dissociation of carbonic acid from measurements with buffered glass electrode cells. *J. Electrochem. Soc.*, 114: 330–340.

DIXON, E. E. L. and VAUGHAN, A., 1911. The Carboniferous succession in Gower (Glamorganshire), with notes on its fauna and conditions of deposition. *Quart. J. Geol. Soc. London*, 67: 477–567.

DODD, J. R., 1963. Paleoecological implications of shell mineralogy in two pelecypod species. *J. Geol.*, 71: 1–11.

DODD, J. R., 1964. Environmentally controlled variation in the shell structure of a pelecypod species. *J. Paleontol.*, 38: 1065–1071.

DODD, J. R., 1965. Environmental control of strontium and magnesium in *Mytilus*. *Geochim. Cosmochim. Acta*, 29: 385–398.

DODD, J. R., 1966a. The influence of salinity on mollusk shell mineralogy: a discussion. *J. Geol.*, 74: 85–89.

DODD, J. R., 1966b. Processes of conversion of aragonite to calcite with examples from the Cretaceous of Texas. *J. Sediment. Petrol.*, 36: 733–741.

DODD, J. R., 1966c. Diagenetic stability of temperature-sensitive skeletal properties in *Mytilus* from the Pleistocene of California. *Geol. Soc. Am. Bull.*, 77: 1213–1224.

DODD, J. R., 1967. Magnesium and strontium in calcareous skeletons: a review. *J. Paleontol.*, 41: 1313–1329.

DONAHUE, J. D., 1962. The formation of cave pearls. *Missouri Speleology*, 4: 53–58.

DONAHUE, J. D., 1965. Laboratory growth of pisolite grains. *J. Sediment. Petrol.*, 35: 251–256.

DONAHUE, J. D., 1969. Genesis of oölite and pisolite grains: an energy index. *J. Sediment. Petrol.*, 39: 1399–1411.

DONAHUE, J. D. and DONAHUE, J. G., 1968. Dolomite synthesis at low temperature and pressure. *Geol. Soc. Am., Spec. Papers*, 101: 55 (abstract).

DOREMUS, R. H., ROBERTS, B. W. and TURNBULL, D., 1958. *Crystal Growth and Perfection of Crystals. Proc. Intern. Conf. Cooperstown, New York, August 27–29, 1958.* Wiley, New York, N.Y., 609 pp.

DREW, G. H., 1914. On the precipitation of calcium carbonate in the sea by marine bacteria, and on the action of denitrifying bacteria in tropical and temperate seas. *Papers Tortugas Lab., Carnegie Inst. Wash. Publ.*, 182: 7–45.

DUNHAM, R. J., 1962. Classification of carbonate rocks according to depositional texture. In: W. E. HAM (Editor), *Classification of Carbonate Rocks.* Am. Assoc. Petrol. Geologists, Tulsa, Okla., pp.108–121.

DUNHAM, R. J., 1969a. Early vadose silt in Townsend mound (reef), New Mexico. In: G. M. FRIEDMAN (Editor), *Depositional Environments in Carbonate Rocks: a Symposium—Soc. Econ. Paleontologists Mineralogists, Spec. Publ.*, 14: 139–181.

DUNHAM, R. J., 1969b. Vadose pisolite in the Capitan Reef (Permian), New Mexico and Texas. In: G. M. FRIEDMAN (Editor), *Depositional Environments in Carbonate Rocks: a Symposium—Soc. Econ. Paleontologists Mineralogists, Spec. Publ.*, 14: 182–191.

DUNHAM, R. J., 1971. Meniscus cement. In: O. P. BRICKER (Editor), *Carbonate Cements.* Johns Hopkins Press, Baltimore, Md., pp. 297–300.

DUNNINGTON, H. V., 1954. Stylolite development post-dates rock induration. *J. Sediment. Petrol.*, 24: 27–49.

DUNNINGTON, H. V., 1967. Aspects of diagenesis and shape change in stylolitic limestone reservoirs. *World Petrol. Congr., Proc., 7th., Mexico, 1967*, 2: 339–352.

EARDLEY, A. J., 1938. Sediments of Great Salt Lake Utah. *Bull. Am. Assoc. Petrol. Geologists*, 22: 1305–1411.

ELIAS, M. K. and CONDRA, G. E., 1957. *Fenestella* from the Permian of West Texas. *Geol. Soc. Am., Mem.*, 70: 1–158.

ELLIOTT, G. F., 1955. Fossil calcareous algae from the Middle East. *Micropaleontology*, 1. 125–131.

ELLIS, A. J., 1959. The solubility of calcite in carbon dioxide solutions. *Am. J. Sci.*, 257: 354–365.

EMERY, K. O., 1956. Sediments and water of the Persian Gulf. *Bull. Am. Assoc. Petrol. Geologists*, 40: 2354–2383.

EMERY, K. O., TRACEY, J. I. and LADD, H. S., 1954. Geology of Bikini and nearby atolls. *U.S.,* *Geol. Surv., Profess. Papers,* 260-A: 1–265.

EMILIANI, C., 1955. Mineralogical and chemical composition of the tests of certain pelagic foraminifera. *Micropaleontology,* 1: 377–380.

EMMONS, W. H., 1928. The state and density of solutions depositing metalliferous veins. *Trans.* *A.I.M.E.,* 76: 308–320.

EPSTEIN, S. and MAYEDA, T. R., 1953. Variation of O^{18} content of waters from natural sources. *Geochim. Cosmochim. Acta,* 4: 213–224.

EPSTEIN, S., BUCHSBAUM, R., LOWENSTAM, H. A. and UREY, H. C., 1951. Carbonate-water isotopic temperature scale. *Bull. Geol. Soc. Am.,* 62: 417–425.

EPSTEIN, S., BUCHSBAUM, R., LOWENSTAM, H. A. and UREY, H. C., 1953. Revised carbonate-water isotopic temperature scale. *Bull. Geol. Soc. Am.,* 64: 1315–1325.

ERBEN, H. K., FLAJS, G. and SIEHL, A., 1968. Über die Schalenstruktur von Monoplacophoren. *Abhandl. Akad. Wiss. Mainz, Math. Naturw. Kl.,* 1: 1–24.

ERDMANN, E., 1902. Stalagmit- och pisolitardade bildningar i Höganäs stenkolsgrufva, Skåne. *Geol. Fören. Stockholm Forh.,* 24: 501–507.

EVAMY, B. D., 1963. The application of a chemical staining technique to a study of dedolomitization. *Sedimentology,* 2: 164–170.

EVAMY, B. D., 1967. Dedolomitization and the development of rhombohedral pores in limestones. *J. Sediment. Petrol.,* 37: 1204–1215.

EVAMY, B. D., 1969. The precipitational environment and correlation of some calcite cements deduced from artificial staining. *J. Sediment. Petrol.,* 39: 787–793.

EVAMY, B. D. and SHEARMAN, D. J., 1965. The development of overgrowths from echinoderm fragments. *Sedimentology,* 5: 211–233.

EVAMY, B. D. and SHEARMAN, D. J., 1969. Early stages in development of overgrowths on echinoderm fragments in limestones. *Sedimentology,* 12: 317–322.

EVANS, G., 1966a. The Recent sedimentary facies of the Persian Gulf region. *Phil. Trans. Roy.* *Soc. London, Ser. A,* 259: 291–298.

EVANS, G., 1966b. Persian Gulf. In: R. W. FAIRBRIDGE (Editor), *The Encyclopedia of Oceanography, 1.* Reinhold, New York, N.Y., pp.689–695.

EVANS, G. and BUSH, P. R., 1969. Some sedimentological and oceanographic observations on a Persian Gulf lagoon. *U.N.E.S.C.O. Conference on Coast Lagoons, Mexico City, 1967:* 155–170.

EVANS, G., KENDALL, C. G. ST. C. and SKIPWITH, P. A. D'E., 1964a. Origin of the coastal flats, the sabkha, of the Trucial coast, Persian Gulf. *Nature,* 202: 759–761.

EVANS, G., KINSMAN, D. J. J. and SHEARMAN, D. J., 1964b. A reconnaissance survey of the environment of Recent carbonate sedimentation along the Trucial Coast, Persian Gulf. In: L. M. J. U. VAN STRAATEN (Editor), *Deltaic and Shallow Marine Deposits.* Elsevier, Amsterdam, pp.129–135.

EVANS, G., SCHMIDT, V., BUSH, P. and NELSON, H., 1969. Stratigraphy and geologic history of the sabhka, Abu Dhabi, Persian Gulf. *Sedimentology,* 12: 145–159.

EVANS, H. B., 1965. A device for continuous determination of material density and porosity. *Trans. Soc. Profess. Well Log Analysts, Ann. Logging Symp., 6th, Dallas, Texas, 1965,* B., pp.1–25.

FABRICIUS, F. H., 1964. Aktive Lage- und Ortsveränderung bei der Koloniekoralle *Manicena* *areolata* und ihre paläoökologische Bedeutung. *Senckenbergiana Lethaea,* 45: 299–323.

FABRICIUS, F. H., 1968. Calcareous sea bottoms of the Raetian and Lower Jurassic sea from the west part of the Northern Calcareous Alps. In: G. MÜLLER and G. M. FRIEDMAN (Editors), *Recent Developments in Carbonate Sedimentology in Central Europe.* Springer, Berlin, pp.240–249.

FABRICIUS, F. H. and KLINGELE, H., 1970. Ultrastrukturen von Ooiden und Oolithen: zur Genese und Diagenese quartärer Flachwasserkarbonate des Mittelmeeres. *Verh. Geol., B.–A.,* 4: 594–617.

FAIRBAIRN, H. W., 1942. *Structural Petrology of Deformed Rocks.* Addison-Wesley, Cambridge, Mass., 143 pp.

FAIRBRIDGE, R. W., 1950. Recent and Pleistocene coral reefs of Australia. *J. Geol.*, 58: 330–401.
FAIRBRIDGE, R. W., 1955. Warm marine carbonate environments and dolomitization. *Tulsa Geol. Soc. Dig.*, 23: 39–48.
FAIRBRIDGE, R. W., 1957. The dolomite question. In: R. J. LE BLANC and J. G. BREEDING (Editors), *Regional Aspects of Carbonate Deposition—Soc. Econ. Paleontologists Mineralogists, Spec. Publ.*, 5: 124–178.
FAIRBRIDGE, R. W., 1967. Coral reefs of the Australian region. In: J. N. JENNINGS and J. A. MABBUTT (Editors), *Landform Studies from Australia and New Guinea.* Cambridge Univ. Press, London, pp.386–417.
FAIRBRIDGE, R. W. and TEICHERT, C., 1948. The Low Isles of the Great Barrier: a new analysis. *Geograph. J.*, 111: 67–88.
FARINACCI, A., 1968. La tessitura della micrite nel calcare "Corniola" del Lias medio. *Atti Accad. Naz. Lincei, Rend. Classe Sci. Fis., Mat. Nat.*, 44: 68–73.
FAURE, G., CROCKET, J. H. and HURLEY, P. M., 1967. Some aspects of the geochemistry of strontium and calcium in the Hudson Bay and Great Lakes. *Geochim. Cosmochim. Acta*, 31: 451–461.
FAUST, G. T., 1949. Dedolomitization and its relation to a possible derivation of a magnesium-rich hydrothermal solution. *Am. Mineralogist*, 34: 789–823.
FENNINGER, A., 1968. Das Kalzitgefüge der sparitischen Kalke des Plassen (Tithonium, Nördliche Kalkalpen, Oberösterreich). *Sedimentology*, 10: 273–291.
FERAY, D. E., HEUER, E. and HEWATT, W. G., 1962. Biological, genetic, and utilitarian aspects of limestone classification. In: W. E. HAM (Editor), *Classification of Carbonate Rocks.* Am. Assoc. Petrol. Geologists, Tulsa, Okla., pp.20–32.
FIELD, R. M., 1928. The Great Bahama Bank. Studies in marine carbonate sediments. *Am. J. Sci.*, 16: 239–246.
FIELD, R. M. and collaborators, 1931. Geology of the Bahamas. *Bull. Geol. Soc. Am.*, 42: 759–784.
FIELD, R. M. and HESS, H. H., 1933. A bore hole in the Bahamas. *Trans. Am. Geophys. Union*, 14: 234–245.
FISCHER, A. G., 1964. The Lofer cyclothems of the Alpine Triassic. *Geol. Surv. Kansas, Bull.*, 169: 107–149.
FISCHER, A. G. and GARRISON, R. E., 1967. Carbonate lithification on the sea floor. *J. Geol.*, 75: 488–496.
FISCHER, A. G. and TEICHERT, C., 1969. Cameral deposits in cephalopod shells. *Univ. Kansas Paleontol. Contrib.*, 37: 1–30.
FISCHER, A. G., HONJO, S. and GARRISON, R. E., 1967. *Electron Micrographs of Limestones.* Princeton Univ. Press, Princeton, N.J., 141 pp.
FLEECE, J. B. and GOODELL, H. G., 1963. Carbonate geochemistry and sedimentology of the keys of Florida Bay, Florida. *Geol. Soc. Am., Spec. Papers*, 73: 6 (abstract).
FLÜGEL, H. W., 1967a. Die Lithogenese der Steinmühl-Kalke des Arracher Steinbruches (Jura, Österreich). *Sedimentology*, 9: 23–53.
FLÜGEL, E., 1967b. Elektronenmikroskopische Untersuchungen an mikritischen Kalken. *Geol. Rundschau*, 56: 341–358.
FLÜGEL, H. and FENNINGER, A., 1966. Die Lithogenese der Oberalmer Schichten und der imkritischen Plassen-Kalke (Tithonium, Nördliche Kalkalpen). *Neues Jahrb. Geol. Paläontol., Abhandl.*, 123: 249–280.
FLÜGEL, E. and FRANZ, H. E., 1967. Elektronenmikroskopischer Nachweis von Coccolithen im Solnhofener Plattenkalk (Ober-Jura). *Neues Jahrb. Geol. Paläontol., Abhandl.*, 127: 245–263.
FLÜGEL, E., FRANZ, H. E. and OTT, W. F., 1968. Review on electron microscope studies of limestones. In: G. MÜLLER and G. M. FRIEDMAN (Editors), *Recent Developments in Carbonate Sedimentology in Central Europe.* Springer, Berlin, pp.85–97.
FOLK, R. L., 1959. Practical petrographic classification of limestones. *Bull. Am. Assoc. Petrol. Geologists*, 43: 1–38.
FOLK, R. L., 1962a. Sorting in some carbonate beaches of Mexico. *Trans. N.Y. Acad. Sci., Ser. II*, 25: 222–244.

FOLK, R. L., 1962b. Spectral subdivision of limestone types. In: W. E. HAM (Editor), *Classification of Carbonate Rocks*. Am. Assoc. Petrol. Geologists, Tulsa, Okla., pp.62–84.

FOLK, R. L., 1962c. Petrography and origin of the Silurian Rochester and McKenzie Shales, Morgan County, West Virginia. *J. Sediment. Petrol.*, 32: 539–578.

FOLK, R. L., 1965. Some aspects of recrystallization in ancient limestones. In: L. C. PRAY and R. C. MURRAY (Editors), *Dolomitization and Limestone Diagenesis: a Symposium—Soc. Econ. Paleontologists Mineralogists, Spec. Publ.*, 13: 14–48.

FOLK, R. L., 1967. Sand cays of Alacran Reef, Yucatán, Mexico: Morphology. *J. Geol.*, 75: 412–437.

FOLK, R. L. and ROBLES, R., 1964. Carbonate sands of Isla Perez, Alacran reef complex, Yucatán. *J. Geol.*, 72: 255–292.

FOLK, R. L. and WARD, W. C., 1957. Brazos river bar: a study in the significance of grain size parameters. *J. Sediment. Petrol.*, 27: 3–26.

FORCE, L. M., 1969. Calcium carbonate size distribution on the west Florida shelf and experimental studies on the microarchitectural control of skeletal breakdown. *J. Sediment. Petrol.*, 39: 902–934.

FRANK, F. C., 1949. The influence of dislocations on crystal growth. *Discussions Faraday Soc.*, 5: 48–54.

FREEMAN, T., 1962. Quiet water oölites from Laguna Madre, Texas. *J. Sediment Petrol.*, 32: 475–483.

FRÉMY, P., 1945. Contribution à la physiologie des Thallophytes marins perforant et cariant les roches calcaires et les coquilles. *Ann. Inst. Océanog. (Paris)*, 22: 107–143.

FRIEDMAN, G. M., 1959. Identification of carbonate minerals by staining methods. *J. Sediment. Petrol.*, 29: 87–97.

FRIEDMAN, G. M., 1964. Early diagenesis and lithification in carbonate sediments. *J. Sediment. Petrol.*, 34: 777–813.

FRIEDMAN, G. M., 1965a. On the origin of aragonite in the Dead Sea. *Israel J. Earth Sci.*, 14: 79–85.

FRIEDMAN, G. M., 1965b. Terminology of *crystallization textures* and *fabrics* in sedimentary rocks. *J. Sediment Petrol.*, 35: 643–655.

FRIEDMAN, G. M., 1965c. Occurrence and stability relationships of aragonite, high-magnesian calcite, and low-magnesian calcite under deep-sea conditions. *Geol. Soc. Am. Bull.*, 76: 1191–1196.

FRIEDMAN, G. M., 1966. Occurrence and origin of Quaternary dolomite of Salt Flat, West Texas. *J. Sediment. Petrol.*, 36: 263–267.

FRIEDMAN, G. M., 1968a. Geology and geochemistry of reefs, carbonate sediments, and waters, Gulf of Aqaba (Elat), Red Sea. *J. Sediment Petrol.*, 38: 895–919.

FRIEDMAN, G. M., 1968b. The fabric of carbonate cement and matrix and its dependence on the salinity of water. In: G. MÜLLER and G. M. FRIEDMAN (Editors), *Recent Developments in Carbonate Sedimentology in Central Europe*. Springer, Berlin, pp.11–20.

FRIEDMAN, G. M. (Editor), 1969. *Depositional Environments in Carbonate Rocks—Soc. Econ. Paleontologists Mineralogists, Spec. Publ.*, 14: 1–209.

FRIEDMAN, G. M. and SANDERS, J. E., 1967. Origin and occurrence of dolostones. In: G. V. CHILINGAR, H. J. BISSELL and R. W. FAIRBRIDGE (Editors), *Carbonate Rocks. Origin, Occurrence and Classification*. Elsevier, Amsterdam, pp.267–348.

FRIEDMAN, I. and HALL, W. E., 1963. Fractionation of O^{18}/O^{16} between coexisting calcite and dolomite. *J. Geol.*, 71: 238–243.

FRIZZELL, D. L. and EXLINE, H., 1955. Monograph of fossil holothurian sclerites. *Missouri Univ., School Mines Met., Bull., Tech. Ser.*, 89: 1–204.

FRUTH, JR., L. S., ORME, G. R. and DONATH, F. A., 1966. Experimental compaction effects in carbonate sediments. *J. Sediment. Petrol.*, 36: 747–754.

FUCHS, T., 1894. Ueber einige von der oesterreichischen Tiefsee-Expedition S.M. Schiffes "Pola" in bedeutenden Tiefen gedredschten Cylindritesaehnliche Koerper. *Jahrb. Oesterreich. Wiss.*, 61: 11–22.

FÜCHTBAUER, H., 1970. Karbonatgesteine. In: H. FÜCHTBAUER and G. MÜLLER (Editors), *Sedimente und Sedimentgesteine*. Schweizerbart, Stuttgart, pp.275–417.

FÜCHTBAUER, H. and GOLDSCHMIDT, H., 1964. Aragonitische Lumachellen im bituminösen
    Wealden des Emslandes. *Beitr. Mineral. Petrog.*, 10: 184–197.
FUJIWARA, T., 1963. Palaeobiochemical studies on the organic substance remaining in various
    sorts of fossils. *Misc. Rept. Res. Inst. Nat. Resources Tokyo*, 58–59: 139–149.
FYFE, W. S. and BISCHOFF, J. L., 1965. The calcite–aragonite problem. In: L. C. PRAY and R. C
    MURRAY (Editors), *Dolomitization and Limestone Diagenesis: a Symposium—Soc. Econ.
    Paleontologists Mineralogists, Spec. Publ.*, 13: 3–13.

GALLOWAY, J. J., 1957. Structure and classification of the Stromatoporoidea. *Bull. Am. Paleontol.*,
    37: 341–470.
GARDINER, J. S., 1930. Studies in coral reefs. *Bull. Museum Comp. Zool. Harvard Coll.*,
    71: 1–16.
GARRELS, R. M. and CHRIST, C. L., 1965. *Solutions, Minerals, and Equilibria.* Harper and Row,
    New York, N.Y., 450 pp.
GARRELS, R. M. and DREYER, R. M., 1952. Mechanism of limestone replacement at low tempera-
    tures and pressures. *Bull. Geol. Soc. Am.*, 63: 325–379.
GARRELS, R. M. and THOMPSON, M. E., 1962. A chemical model for sea water at 25 °C and one
    atmosphere total pressure. *Am. J. Sci.*, 260: 57–66.
GARRELS, R. M., DREYER, R. M. and HOWLAND, A. L., 1949. Diffusion of ions through inter-
    granular spaces in water-saturated rocks. *Bull. Geol. Soc. Am.*, 60: 1809–1828.
GARRELS, R. M., THOMPSON, M. E. and SIEVER, R., 1960. Stability of some carbonates at 25 °C
    and one atmosphere total pressure. *Am. J. Sci.*, 258: 402–418.
GARRELS, R. M., THOMPSON, M. E. and SIEVER, R., 1961. Control of carbonate solubility by
    carbonate complexes. *Am. J. Sci.*, 259: 24–45.
GARRIDO, J. and BLANCO, J., 1947. Structure cristalline des piquants d'Oursin. *Compt. Rend.*,
    224: 485.
GARRISON, R. E. and BAILEY, E. H., 1967. Electron microscopy of limestones in the Franciscan
    Formation of California. *U.S., Geol. Surv., Profess. Papers*, 575-B: 94–100.
GARRISON, R. E. and FISCHER, A. G., 1969. Deep-water limestones and radiolarites of the Alpine
    Jurassic. In: G. M. FRIEDMAN (Editor), *Depositional Environments in Carbonate Rocks:
    a Symposium—Soc. Econ. Paleontologists Mineralogists, Spec. Publ.*, 14: 20–55.
GATRALL, M. and GOLUBIĆ, S., 1970. Comparative study on some Jurassic and Recent endolithic
    fungi using a scanning electron microscope. In: T. P. CRIMES and J. C. HARPER (Editors),
    Trace Fossils—Geol. J., Spec. Issue, 4. Seel House Press, Liverpool, pp.167–168.
GAVISH, E. and FRIEDMAN, G. M., 1969. Progressive diagenesis in Quaternary to Late Tertiary
    carbonate sediments: sequence and time scale. *J. Sediment. Petrol.*, 39: 980–1006.
GEBELEIN, C. D., 1969. Distribution, morphology, and accretion rate of Recent subtidal algal
    stromatolites, Bermuda. *J. Sediment. Petrol.*, 39: 49–69.
GEBELEIN, C. D. and HOFFMAN, P., 1968. Intertidal stromatolites and associated facies from Cape
    Sable, Florida. *Geol. Soc. Am., Spec. Papers*, 121: 109 (abstract).
GEBELEIN, C. D. and HOFFMAN, P., in press. Algal origin of dolomite laminations in stromatolitic
    limestones. *J. Sediment. Petrol.*
GERMANN, K., 1968. Diagenetic patterns in the Wettersteinkalk (Ladinian, Middle Trias),
    Northern Limestone Alps, Bavaria and Tyrol. *J. Sediment. Petrol.*, 38: 490–500.
GEVIRTZ, J. L. and FRIEDMAN, G. M., 1966. Deep-sea carbonate sediments of the Red Sea and
    their implications on marine lithification. *J. Sediment. Petrol.*, 36: 143–151.
GIBSON, T. G. and SCHLEE, J., 1967. Sediments and fossiliferous rocks from the eastern side
    of the Tongue of the Ocean, Bahamas. *Deep-Sea Res.*, 14: 691–702.
GILLOTT, J. E., 1969. Study of the fabric of fine-grained sediments with the scanning electron
    microscope. *J. Sediment. Petrol.*, 39: 90–105.
GINSBURG, R. N., 1953. Beach rock in south Florida. *J. Sediment. Petrol.*, 23: 85–92.
GINSBURG, R. N., 1956. Environmental relationships of grain size and constituent particles in
    some south Florida carbonate sediments. *Bull. Am. Assoc. Petrol. Geologists*, 40: 2384–2427.
GINSBURG, R. N., 1957. Early diagenesis and lithification of shallow-water carbonate sediments
    in south Florida. In: R. J. LeBLANC and J. G. BREEDING (Editors), *Regional Aspects of
    Carbonate Deposition—Soc. Econ. Paleontologists Mineralogists, Spec. Publ.*, 5: 80–99.

GINSBURG, R. N., 1964. South Florida carbonate sediments. *Guidebook for Field Trip No. 1 Geol. Soc. Am., Convention 1964*. Geol. Soc. Am., New York, N.Y., 72 pp.

GINSBURG, R. N. and GARRETT, P., 1969. Seminar on organism–sediment interrelationships. *Bermuda Biol. Sta. Res., Spec. Publ.*, 2: 1–151.

GINSBURG, R. N. and LOWENSTAM, H A., 1958. The influence of marine bottom communities on the depositional environment of sediments. *J. Geol.*, 66: 310–318.

GINSBURG, R. N., SHINN, E. A. and SCHROEDER, J. H., 1968. Submarine cementation and internal sedimentation within Bermuda reefs. *Geol. Soc.. Am., Spec. Papers*, 115: 78–79 (abstract).

GLOVER, E. D. and PRAY, L. C., 1971. High-magnesian calcite and aragonite cementation within modern subtidal carbonate sediment grains. In: O. P. BRICKER (Editor), *Carbonate Cements*. Johns Hopkins Press, Baltimore, Md., pp. 80–87.

GLOVER, E. D. and SIPPEL, R. F., 1967. Synthesis of magnesium calcites. *Geochim. Cosmochim. Acta*, 31: 603–613.

GLOVER, J. E., 1964. The universal stage in studies of diagenetic textures. *J. Sediment Petrol.*, 34: 851–854.

GOLDMAN, M. I., 1952. Deformation, metamorphism, and mineralization in gypsum-anhydrite cap rock, Sulphur Salt Dome, Louisiana. *Geol. Soc. Am., Mem.*, 50: 1–169.

GOLDSMITH, J. R., 1953. A "simplexity principle" and its relation to "ease" of crystallization. *J. Geol.*, 61: 439–451.

GOLDSMITH, J. R., 1959. Some aspects of the geochemistry of the carbonates. In: P. H. ABELSON (Editor), *Researches in Geochemistry*. Wiley, London, pp.336–358.

GOLDSMITH, J. R. and GRAF, D. L., 1958a. Structural and compositional variations in some natural dolomites. *J. Geol.*, 66: 678–693.

GOLDSMITH, J. R. and GRAF, D. L., 1958b. Relation between lattice constants and composition of the Ca-Mg carbonates. *Am. Mineralogist*, 43: 84–101.

GOLDSMITH, J. R., GRAF, D. L. and JOENSUU, O. I., 1955. The occurrence of magnesian calcites in nature. *Geochim. Cosmochim. Acta*, 7: 212–230.

GOLUBIĆ, S., 1969. Tradition and revision in the system of the Cyanophyta. *Verhandl. Intern. Ver. Limnol.*, 17: 752–756.

GOLUBIĆ, S., BRENT, G. and LECAMPION, T., 1970. Scanning electron microscopy of endolithic algae and fungi using a multipurpose casting embedding technique. *Lethaia*, 3: 203–209.

GOODELL, H. G. and GARMAN, R. K., 1969. Carbonate geochemistry of Superior Deep Test Well, Andros Island, Bahamas. *Bull. Am. Assoc. Petrol. Geologists*, 53: 513–536.

GOREAU, T. F., 1959. The ecology of Jamaican coral reefs. I. Species composition and zonation. *Ecology*, 40: 67–90.

GOREAU, T. F., 1961. Problems of growth and calcium deposition in reef corals. *Endeavour*, 20: 32–39.

GOREAU, T. F., 1963. Calcium carbonate deposition by coralline algae and corals in relation to their roles as reef-builders. *Ann. N.Y. Acad. Sci.*, 109: 127–167.

GOREAU, T. F., 1964. Mass expulsion of zooxanthellae from Jamaican reef communities after hurricane Flora. *Science*, 145: 383–386.

GOREAU, T. F. and GOREAU, N. I., 1959. The physiology of skeleton formation in corals. II. Calcium deposition by hermatypic corals under various conditions in the reef. *Biol. Bull.*, 117: 239–250.

GOREAU, T. F. and GOREAU, N. I., 1960a. The physiology of skeleton formation in corals. III. Calcification rate as a function of colony weight and total nitrogen content in the reef coral *Manicina areolata* (Linnaeus). *Biol. Bull.*, 118: 419–429.

GOREAU, T. F. and GOREAU, N. I., 1960b. The physiology of skeleton formations in corals. IV. On isotopic equilibrium exchanges of calcium between corallum and environment in living and dead reef-building corals. *Biol. Bull.*, 119: 416–427.

GOREAU, T. F. and HARTMAN, W. D., 1963. Boring sponges as controlling factors in the formation and maintenance of coral reefs. *Publ. Am. Assoc. Advan. Sci.*, 75: 25–54.

GOREAU, T. F. and YONGE, C. M., 1968. Coral community on muddy sand. *Nature*, 217: 421–423.

GORSLINE, D. S., 1963. Environments of carbonate deposition, Florida Bay and the Florida Straits. In: R. O. BASS (Editor), *Shelf Carbonates of the Paradox Basin: a Symposium—Four Corners Geol. Soc., Field Conf., 4th*, pp.130–143.

GORTIKOV, V. M. and PANTELEVA, L. I., 1937. Kinetics of solvation of calcium carbonate. *J. Gen. Chem. U.S.S.R. (Eng. Transl.)*, 7: 56–64.

GOTO, M., 1961. Some mineralo-chemical problems concerning calcite and aragonite, with special reference to the genesis of aragonite. *J. Fac. Sci., Hokkaido Univ., Ser. IV*, 10: 571–640.

GOTO, M., 1966. Persistance of μ-CaCO$_3$ in wet conditions. *J. Fac. Sci., Hokkaido Univ., Ser. IV*, 13: 287–292.

GRABAU, A. W., 1904. On the classification of sedimentary rocks. *Am. Geologist*, 33: 228–247.

GRADZIŃSKI, R. and RADOMSKI, A., 1967. Pisoliths from Cuban caves. *Rocznik Polsk. Towarz. Geol. (Ann. Soc. Géol. Pologne)*, 37: 243–265.

GRAF, D. L., 1960. Geochemistry of carbonate sediments and sedimentary carbonate rocks. I. Carbonate mineralogy–Carbonate sediments; II. Sedimentary carbonate rocks; III. Minor element distribution; IV-A. Isotopic composition–Chemical analyses; IV-B. Bibliography. *Illinois State Geol. Surv., Circ.*, 297: 1–39; 298: 1–43; 301: 1–71; 308: 1–42; 309: 1–55.

GRAF, D. L., 1961. Crystallographic tables for the rhombohedral carbonates. *Am. Mineralogist*, 46: 1283–1316.

GRAF, D. L. and GOLDSMITH, J. R., 1956. Some hydrothermal syntheses of dolomite and protodolomite. *J. Geol.*, 64: 173–186.

GRANDJEAN, J., GRÉGOIRE, C. and LUTTS, A., 1964. On the mineral components and the remnants of organic structures in shells of fossil molluscs. *Bull. Classe Sci., Acad. Roy. Belg.*, 50: 562–595.

GREEN, E. J., 1967. The stability of aragonite in seawater: thermodynamic influence of strontium. *Geochim. Cosmochim. Acta*, 31: 2445–2448.

GREENFIELD, L. J., 1963. Metabolism and concentration of calcium and magnesium and precipitation of calcium carbonate by a marine bacterium. *Ann. N.Y. Acad. Sci.*, 109: 23–45.

GREENWALD, I., 1941. The dissociation of calcium and magnesium carbonates and bicarbonates. *J. Biol. Chem.*, 141: 789–796.

GRÉGOIRE, C., 1957. Topography of the organic components in mother-of-pearl. *J. Biophys. Biochem. Cytol.*, 3: 797–808.

GRÉGOIRE, C., 1958a. Sur la structure, étudiée au microscope électronique, des constituants organiques du calcitostracum. *Arch. Intern. Physiol. Biochim.*, 66: 658–661.

GRÉGOIRE, C., 1958b. Essai de détection au microscope électronique des dentelles organiques dans les nacres fossiles (céphalopodes, gastéropodes et pélécypodes). *Arch. Intern. Physiol. Biochim.*, 66: 674–676.

GRÉGOIRE, C., 1959a. A study on the remains of organic components in fossil mother-of-pearl. *Inst. Roy. Sci. Nat. Belg., Bull.*, 35(13): 1–14.

GRÉGOIRE, C., 1959b. Conchiolin remnants in mother-of-pearl from fossil cephalopoda. *Nature*, 184: 1157–1158.

GRÉGOIRE, C., 1960. Further studies on structure of the organic components in mother-of-pearl, especially in pelecypods, 1. *Inst. Roy. Sci. Nat. Belg., Bull.*, 36(23): 1–22.

GRÉGOIRE, C., 1961a. Sur la structure submicroscopique de la conchioline associée aux prismes des coquilles de mollusques. *Inst. Roy. Sci. Nat. Belg., Bull.*, 37(3): 1–34.

GRÉGOIRE, C., 1961b. Structure of the conchiolin cases of the prisms in *Mytilus edulis* Linne. *J. Biophys. Biochem. Cytol.*, 9: 395–400.

GRÉGOIRE, C., 1962. On submicroscopic structure of the *Nautilus* shell. *Inst. Roy. Sci. Nat. Belg., Bull.*, 38(49): 1–71.

GRÉGOIRE, C., 1964. Thermal changes in the *Nautilus* shell. *Nature*, 203: 868–869.

GRÉGOIRE, C., 1966a. On organic remains in shells of Paleozoic and Mesozoic cephalopods (nautiloids and ammonoids). *Inst. Roy. Sci. Nat. Belg., Bull.*, 42(39): 1–36.

GRÉGOIRE, C., 1966b. Experimental diagenesis of the *Nautilus* shell. In: G. D. HOBSON and G. C. SPEERS (Editors), *Advances in Organic Geochemistry*. Pergamon, Oxford, pp.429–441.

GRÉGOIRE, C., 1967. Sur la structure des matrices organiques des coquilles de mollusques. *Biol. Rev. Cambridge Phil. Soc.*, 42: 653–688.

GRÉGOIRE, C., 1968. Experimental alteration of the *Nautilus* shell by factors involved in diagenesis and metamorphism. 1. Thermal changes in conchiolin matrix of mother-of-pearl. *Inst. Roy. Sci. Nat. Belg., Bull.*, 44(25): 1–69.

GRÉGOIRE, C. and MONTY, C., 1963. Observations au microscope électronique sur le calcaire à pâte fine entrant dans la constitution de structures stromatolithiques du Viséen moyen de la Belgique. *Ann. Soc. Géol. Belg., Bull.*, 85: 389–397.

GRÉGOIRE, C. and TEICHERT, C., 1965. Conchiolin membranes in shell and cameral deposits of Pennsylvanian cephalopods, Oklahoma. *Oklahoma Geol. Notes*, 25: 175–202.

GRÉGOIRE, C. and VOSS-FOUCART, M.-F., 1970. Proteins in shells of fossil cephalopods (nautiloids and ammonoids) and experimental simulation of their alterations. *Arch. Intern. Physiol. Biochim.*, 78: 191–203.

GRÉGOIRE, C., DUCHATEAU, G. and FLORKIN, M., 1955. La trame protidique des nacres des perles. *Ann. Inst. Océanog. (Paris)*, 31: 1–36.

GRÉGOIRE, C., GISBOURNE, C. M. and HARDY, A., 1969. Über experimentelle Diagenese der Nautilusschale. *Beitr. Elektronenmikroskop. Direktabb. Oberfl.*, 2: 223–238.

GROSS, M. G., 1964. Variations in the O^{18}/O^{16} and C^{13}/C^{12} ratios of diagenetically altered limestones in the Bermuda Islands. *J. Geol.*, 72: 170–194.

GROSS, M. G. and TRACEY, JR., J. I., 1966. Oxygen and carbon isotopic composition of limestones and dolomites, Bikini and Eniwetok Atolls. *Science*, 151: 1082–1084.

GRUNAU, H. R., 1959. *Mikrofazies und Schichtung Ausgewählter Jungmesozoischer, Radiolaritführender Sedimentserien der Zentral-Alpen*. Brill, Leiden, 179 pp.

GRUNAU, H. R. and STUDER, H., 1956. Elektronenmikroskopische Untersuchungen an Bianconekalken des Südtessins. *Experientia*, 12: 141–143.

GUILCHER, A., 1952. Morphologie sous-marine et récifs coralliens du Nord du banc Farsan (Mer Rouge). *Bull. Assoc. Géograph. Franç.*, 224–225: 52–63.

GUILCHER, A., 1956. Étude géomorphologique des récifs coralliens du nord-ouest de Madagascar. *Ann. Inst. Océanog. (Paris)*, 33: 1–136.

GUNN, R. K. and COOPER, B. S., 1964. Organic residues in Jurassic oolitic limestones. In: U. COLOMBO and G. D. HOBSON (Editors), *Advances in Organic Geochemistry*. Pergamon, Oxford, pp.145–148.

GUTSCHICK, R. C., 1954. Holothurian sclerites from the Middle Ordovician of northern Illinois. *J. Paleontol.*, 28: 827–829.

GUTSCHICK, R. C., 1959. Lower Mississippian holothurian sclerites from the Rockford Limestone of northern Indiana. *J. Paleontol.*, 33: 130–137.

HAGLUND, D. S., FRIEDMAN, G. M. and MILLER, D. S., 1969. The effect of fresh water on the redistribution of uranium in carbonate sediments. *J. Sediment. Petrol.*, 39: 1283–1296.

HAHNE, C., KIRCHMAYER, M. and OTTEMANN, J., 1968. "Höhlenperlen" (Cave Pearls), besonders aus Bergwerken des Ruhrgebietes. Modellfälle zum Studium diagenetischer Vorgänge an Einzelooiden. *Neues Jahrb. Geol. Paläontol., Abhandl.*, 130: 1–46.

HALL, J., 1883. Cryptozoön, N.G., cryptozoön proliferum, n. sp. *Ann. Rept. N.Y. State Museum Nat. Hist.*, 36: plate VI.

HALLAM, A., 1969. A pyritized limestone hardground in the Lower Jurassic of Dorset (England). *Sedimentology*, 12: 231–240.

HALLAM, A. and O'HARA, M. J., 1962. Aragonitic fossils in the Lower Carboniferous of Scotland. *Nature*, 195: 273–274.

HALLAM, A. and PRICE, N. B., 1966. Strontium contents of Recent and fossil aragonitic cephalopod shells. *Nature*, 212: 25–27.

HALLAM, A. and PRICE, N. B., 1968a. Further notes on the strontium contents of unaltered fossil cephalopod shells. *Geol. Mag.*, 105: 52–55.

HALLAM, A. and PRICE, N. B., 1968b. Environmental and biochemical control of strontium in shells of *Cardium edule. Geochim. Cosmochim. Acta*, 32: 319–328.

HAM, W. E. (Editor), 1962. *Classification of Carbonate Rocks—a Symposium*. Am. Assoc. Petrol. Geologists, Tulsa, Okla., 279 pp.

HAMILTON, E. L., 1953. Upper Cretaceous, Tertiary, and Recent planktonic Foraminifera from Mid-Pacific flat-topped seamounts. *J. Paleontol.*, 27: 204–237.

HAMILTON, E. L., 1956. Sunken islands of the mid-Pacific Mountains. *Geol. Soc. Am., Mem.*, 64: 1–97.

HAMILTON, E. L. and REX, R. W., 1959. Lower Eocene phosphatized *Globigerina* ooze from Sylvania Guyot, Marshall Islands. *U.S., Geol. Surv., Profess. Papers*, 260-W: 785–798.

HANCOCK, J. M., 1963. The hardness of the Irish Chalk. *Irish Naturalists' J.*, 14: 157–164.

HANCOCK, J. M. and KENNEDY, W. J., 1967. Photographs of hard and soft chalks taken with a scanning electron microscope. *Proc. Geol. Soc. London*, 1643: 249–252.

HARE, P. E., 1963. Amino acids in the proteins from aragonite and calcite in the shells of *Mytilus californianus*. *Science*, 139: 216–217.

HARLAND, W. B., SMITH, A. G. and WILCOCK, B., 1964. *The Phanerozoic Time-Scale*. Geol. Soc., London, 458 pp.

HARLAND, W. B. et al., 1967. *The Fossil Record*. Geol. Soc., London, 827 pp.

HARMS, J. C. and CHOQUETTE, P. W., 1965. Geologic evaluation of a gamma-ray porosity device. *Trans. Soc. Profess. Well Log Analysts, Ann. Logging Symp., 6th, Dallas, Texas, 1965, C*, pp.1–37.

HARPER, JR., C. W., and TOWE, K. M., 1967. Shell structure of the brachiopod *Pholidostrophia (Mesopholidostrophia) nitens* from Gotland. *J. Paleontol.*, 41: 1184–1187.

HARRISS, R. C. and PILKEY, O. H., 1966. Interstitial waters of some deep marine carbonate sediments. *Deep-Sea Res.*, 13: 967–969.

HARRIS, W. H. and MATTHEWS, R. K., 1968. Subaerial diagenesis of carbonate sediments: efficiency of the solution-precipitation process. *Science*, 160: 77–79.

HARTMAN, W. D. and GOREAU, T. F., 1966. *Ceratoporella*, a living sponge with stromatoporoid affinities. *Am. Zool.*, 6: 563–564.

HARVEY, H. W., 1955. *The Chemistry and Fertility of Sea Waters*. Cambridge Univ. Press, Cambridge, 224 pp.

HARVEY, R. D., 1966. Electron microscope study of microtexture and grain surfaces in limestones. *Illinois State Geol. Surv., Circ.*, 404: 1–18.

HASSON, D., AVRIEL, M., RESNICK, W., ROZENMAN, T. and WINDREICH, S., 1968. Mechanism of calcium carbonate scale deposition on heat-transfer surfaces. *Ind. Eng. Chem.*, 7: 59–65.

HATCH, F. H., RASTALL, R. H. and BLACK, M., 1938. *The Petrology of the Sedimentary Rocks*. Allen and Unwin, London, 383 pp.

HATHAWAY, J. C., 1967. Aragonite needles. *Geotimes*, 12(5) (photograph on cover).

HATHAWAY, J. C. and ROBERTSON, E. C., 1961. Microtexture of artificially consolidated aragonite mud. *U.S., Geol. Surv., Profess. Papers*, 424-C: 301–304.

HAWLEY, J. and PYTKOWICZ, R. M., 1969. Solubility of calcium carbonate in seawater at high pressures and 2 °C. *Geochim. Cosmochim. Acta.*, 33: 1557–1561.

HAY, W. W. and SANDBERG, P. A., 1967. The scanning electron microscope, a major breakthrough for micropaleontology. *Micropaleontology*, 13: 407–418.

HAY, W. W. and TOWE, K. M., 1962. Electronmicroscopic examination of some coccoliths from Donzacq (France). *Eclogae Geol. Helv.*, 55: 497–517.

HAY, W. W., TOWE, K. M. and WRIGHT, R. C., 1963. Ultramicrostructure of some selected foraminiferal tests. *Micropaleontology*, 9: 171–195.

HEALD, M. T., 1956. Cementation of Simpson and St. Peter sandstones in parts of Oklahoma, Arkansas, and Missouri. *J. Geol.*, 64: 16–30.

HELING, D., 1968. Microporosity of carbonate rocks. In: G. MÜLLER and G. M. FRIEDMAN (Editors), *Recent Developments in Carbonate Sedimentology in Central Europe*. Springer, Berlin, pp.98–105.

HENBEST, L. G., 1968. Diagenesis in oolitic limestones of Morrow (Early Pennsylvanian) age in northwestern Arkansas and adjacent Oklahoma. *U.S., Geol. Surv., Profess. Papers*, 594-H: 1–22.

HENNIKER, J. C., 1949. The depth of the surface zone of a liquid. *Rev. Mod. Phys.*, 21: 322–341.

HESS, F. L., 1929. Oölites or cave pearls in the Carlsbad caverns. *Proc. U.S. Natl. Museum*, 76: 1–5.

HILL, D., 1936. The British Silurian Rugose corals with acanthine septa. *Phil. Trans. Roy. Soc. London, Ser. B.*, 226: 189–217.

HILL, D., 1956a. Rugosa. In: R. C. MOORE (Editor), *Treatise on Invertebrate Paleontology. F. Coelenterata*. Geol. Soc. Am., Boulder, Colo., pp.233–324.

HILL, D., 1956b. Heterocorallia. In: R. C. MOORE (Editor), *Treatise on Invertebrate Paleontology.*
    F. Coelenterata. Geol. Soc. Am., Boulder, Colo., pp.324–327.
HILL, D. and STUMM, E. C., 1956. Tabulata. In: R. C. MOORE (Editor), *Treatise on Invertebrate*
    Paleontology. F. Coelenterata. Geol. Soc. Am., Boulder, Colo., pp.444–477.
HODGSON, W. A., 1966. Carbon and oxygen isotope ratios in diagenetic carbonates from marine
    sediments. *Geochim. Cosmochim. Acta*, 30: 1223–1233.
HOFFMEISTER, J. E. and MULTER, H. G., 1968. Geology and origin of the Florida Keys. *Bull.*
    Geol. Soc. Am., 79: 1487–1501.
HOFFMEISTER, J. E., JONES, J. I., MILLIMAN, J. D., MOORE, D. R. and MULTER, H. G., 1964. Living
    and fossil reef types of south Florida. *Guidebook for Field Trip No. 3 Geol. Soc. Am.,*
    Convention 1964. Geol. Soc. Am., New York, N.Y., 28 pp.
HOFFMEISTER, J. E., STOCKMAN, K. W. and MULTER, H. G., 1967. Miami Limestone of Florida
    and its Recent Bahamian counterpart. *Geol. Soc. Am. Bull.*, 78: 175–190.
HOLLMANN, R., 1964. Subsolutions-Fragmente. *Neues Jahrb. Geol. Paläontol., Abhandl.*, 119:
    22–82.
HOLMES, A., 1921. *Petrographic Methods and Calculations.* Murby, London, 515 pp.
HONJO, S. and BERGGREN, W. A., 1967. Scanning electron microscope studies of planktonic
    foraminifera. *Micropaleontology*, 13: 393–406.
HONJO, S. and FISCHER, A. G., 1964. Fossil coccoliths in limestone examined by electron micro-
    scopy. *Science*, 144: 837–839.
HOSKIN, C. M., 1963. Recent carbonate sedimentation on Alacran Reef, Yucatán, Mexico.
    Natl. Acad. Sci. Natl. Res. Council, Publ., 1089: i–xii, 1–160.
HOSKIN, C. M., 1966. Coral pinnacle sedimentation, Alacran Reef lagoon, Mexico. *J. Sediment.*
    Petrol., 36: 1058–1074.
HOSKIN, C. M., 1968. Magnesium and strontium in mud fraction of Recent carbonate sediment,
    Alacran Reef, Mexico. *Bull. Am. Assoc. Petrol. Geologists*, 52: 2170–2177.
HOSKINS, C. W., 1964. Molluscan biofacies in calcareous sediments, Gulf of Batabano, Cuba.
    Bull. Am. Assoc. Petrol. Geologists, 48: 1680–1704.
HOUBOLT, J. J. H. C., 1957. *Surface Sediments of the Persian Gulf near the Qatar Peninsula.* Thesis,
    Univ. Utrecht, Utrecht, 113 pp.
HSÜ, K. J., 1967. Chemistry of dolomite formation. In: G. V. CHILINGAR, H. J. BISSELL and
    R. W. FAIRBRIDGE (Editors), *Carbonate Rocks, Physical and Chemical Aspects.* Elsevier,
    Amsterdam, pp.169–191.
HSÜ, K. J. and SIEGENTHALER, C., 1969. Preliminary experiments on hydrodynamic movement
    induced by evaporation and their bearing on the dolomite problem. *Sedimentology*, 12:
    11–25.
HUDSON, J. D., 1962. Pseudo-pleochroic calcite in recrystallized shell-limestones. *Geol. Mag.*,
    99: 492–500.
HUDSON, J. D., 1967a. Speculations on the depth relations of calcium carbonate solution in
    Recent and ancient seas. *Marine Geol.*, 5: 473–480.
HUDSON, J. D., 1967b. The elemental composition of the organic fraction, and the water content,
    of some recent and fossil mollusc shells. *Geochim. Cosmochim. Acta*, 31: 2361–2378.
HUDSON, J. D., 1968. The microstructure and mineralogy of the shell of a Jurassic mytilid (Bi-
    valvia). *Palaeontology*, 11: 163–182.
HUDSON, J. D., 1970. Algal limestones with pseudomorphs after gypsum from the Middle Jurassic
    of Scotland. *Lethaia*, 3: 11–40.
HURLEY, R. J., SIEGLER, V. B. and FINK, JR., L. K., 1962. Bathymetry of the Straits of Florida
    and the Bahama Islands. 1. Northern Straits of Florida. *Bull. Marine Sci. Gulf Caribbean*,
    12: 313–321.

ILLING, L. V., 1954. Bahaman calcareous sands. *Bull. Am. Assoc. Petrol. Geologists*, 38: 1–95.
ILLING, L. V., 1963. Discussion of Recent anhydrite, gypsum, dolomite, and halite from the
    coastal flats of the Arabian shore of the Persian Gulf. *Proc. Geol. Soc. London*, 1607:
    64–65.
ILLING, L. V. and WELLS, A. J., 1964. Penecontemporary dolomite in the Persian Gulf. *Bull.*
    Am. Assoc. Petrol. Geologists, 48: 532–533 (abstract).

ILLING, L. V., WELLS, A. J. and TAYLOR, J. C. M., 1965. Penecontemporary dolomite in the Persian Gulf. In: L. C. PRAY and R. C. MURRAY (Editors), *Dolomitization and Limestone Diagenesis: a Symposium—Soc. Econ. Paleontologists Mineralogists, Spec. Publ.*, 13: 89–111.

ILLING, L. V., WOOD, G. V. and FULLER, J. G. C. M., 1967. Reservoir rocks and stratigraphic traps in non-reef carbonates. *World Petrol. Congr., Proc., 7th, Mexico, 1967*, 2: 487–499.

IMBRIE, J. and BUCHANAN, H., 1965. Sedimentary structures in modern carbonate sands of the Bahamas. *Soc. Econ. Palaeontologists Mineralogists, Spec. Publ.*, 12: 149–172.

IMBRIE, J. and PURDY, E. G., 1962. Classification of modern Bahaman carbonate sediments. In: W. E. HAM (Editor), *Classification of Carbonate Rocks*. Am. Assoc. Petrol. Geologists, Tulsa, Okla., pp.253–272.

INGERSON, E., 1962. Problems of the geochemistry of sedimentary carbonate rocks. *Geochim. Cosmochim. Acta*, 26: 815–847.

IRION, G. and MÜLLER, G., 1968. Huntite, dolomite, magnesite and polyhalite of Recent age from Tuz Gölü, Turkey. *Nature*, 220: 1309–1310.

JAANUSSON, V., 1961. Discontinuity surfaces in limestones. *Bull. Geol. Inst. Univ. Uppsala*, 40: 221–241.

JAMIESON, J. C., 1953. Phase equilibrium in the system calcite–aragonite. *J. Chem. Phys.* 21: 1385–1390.

JAMIESON, J. C., 1957. Introductory studies of high-pressure polymorphism to 24,000 bars by X-ray diffraction with some comments on calcite. II. *J. Geol.*, 65: 334–343.

JAMIESON, J. C. and GOLDSMITH, J. R., 1960. Some reactions produced in carbonates by grinding. *Am. Mineralogist*, 45: 818–827.

JANSEN, J. F. and KITANO, Y., 1963. The resistance of Recent marine carbonate sediments to solution. *J. Oceanog. Soc. Japan.*, 18: 208–219.

JEFFERIES, R. P. S., 1962. The palaeoecology of the *Actinocamax plenus* Subzone (Lowest Turonian) in the Anglo-Paris Basin. *Palaeontology*, 4: 609–647.

JENKYNS, H. C., 1972. Pelagic "oolites" from the Tethyan Jurassic. *J. Geol.*, 80: 21–23.

JINDRICH, V., 1969. Recent carbonate sedimentation by tidal channels in the lower Florida Keys. *J. Sediment. Petrol.*, 39: 531–553.

JOHNSON, J. H., 1946. Lime-secreting algae from the Pennsylvanian and Permian of Kansas. *Bull. Geol. Soc. Am.*, 57: 1087–1120.

JOHNSON, J. H., 1956. *Archaeolithophyllum*, a new genus of Paleozoic coralline algae: *J. Paleontol.*, 30: 53–55.

JOHNSON, J. H., 1961a. *Limestone-Building Algae and Algal Limestones*. Colo. School Mines, Boulder, Colo., 297 pp.

JOHNSON, J. H., 1961b. Fossil algae from Eniwetok, Funafuti and Kitā-Daito-Jima. *U.S., Geol. Surv., Profess. Papers*, 260-Z: 907–950.

JOHNSON, J. H., 1964. Fossil and Recent calcareous algae from Guam. *U.S., Geol. Surv., Profess. Papers*, 403-G: 1–40.

JOHNSTON, J. and WILLIAMSON, E. D., 1916. The role of inorganic agencies in the deposition of calcium carbonate. *J. Geol.*, 24: 729–750.

JONES, B. F., 1961. Zoning of saline minerals at Deep Spring Lake, California. *U.S., Geol. Surv., Profess. Papers*, 424-B: 199–209.

JONES, B. F., 1965. The hydrology and mineralogy of Deep Spring Lake, Inyo County, California. *U.S., Geol. Surv., Profess. Papers*, 502-A: 1–56.

JONES, J. A., 1963. Ecological studies of the southeastern Florida patch reefs, 1. Diurnal and seasonal changes in the environment. *Bull. Marine Sci. Gulf Caribbean*, 13: 282–307.

JOPE, M., 1967a. The protein of brachiopod shell. I. Amino acid composition and implied protein taxonomy. *Comp. Biochem. Physiol.*, 20: 593–600.

JOPE, M., 1967b. The protein of brachiopod shell. II. Shell protein from fossil articulates: amino acid composition. *Comp. Biochem. Physiol.*, 20: 601–605.

JOPE, M., 1969a. The protein of brachiopod shell. III. Comparison with structural protein of soft tissue. *Comp. Biochem. Physiol.*, 30: 209–224.

JOPE, M., 1969b. The protein of brachiopod shell. IV. Shell protein from fossil inarticulates: amino acid composition and disc electrophoresis of fossil articulate shell protein. *Comp. Biochem. Physiol.*, 30: 225–232.

JORDAN, G. F., 1952. Reef formation in the Gulf of Mexico off Apalachicola Bay, Florida. *Bull. Geol. Soc. Am.*, 63: 741–743.

KAHLE, C. F., 1965a. Strontium in oölitic limestones. *J. Sediment. Petrol.*, 35: 846–856.

KAHLE, C. F., 1965b. Aspects of diagenesis in oölitic limestones. *Geol. Soc. Am., Spec. Papers*, 82: 106 (abstract).

KALKOWSKY, E., 1908. Oolith und Stromatolith im norddeutschen Buntsandstein. *Z. Deut. Geol. Ges.*, 60: 68–125.

KANWISHER, J., 1963. On the exchange of gases between the atmosphere and the sea. *Deep-Sea Res.*, 10: 195–207.

KARCZ, I., 1964. Grain growth fabrics in the Cambrian dolomites of Skye. *Nature*, 204: 1080–1081.

KATO, M., 1963. Fine skeletal structures in the Rugosa. *J. Fac. Sci., Hokkaido Univ., Ser. IV*, 11: 571–630.

KATO, M., 1968. Note on the fine skeletal structures in Scleractinia and in Tabulata. *J. Fac. Sci., Hokkaido Univ., Ser. IV*, 14: 51–56.

KATZ, A., 1968. Calcian dolomites and dedolomitization. *Nature*, 217: 439–440.

KATZ, B., 1965. Circulation near the southern Berry Islands, Bahamas. *Tech. Rept. U.S. At. Energy Comm., Tech. Rept.*, CU-23-65, unpublished.

KAYE, C. A., 1959. Shoreline features and Quaternary shoreline changes, Puerto Rico. *U.S., Geol. Surv., Profess. Papers*, 317-B: 1–140.

KEITH, M. L. and PARKER, R. H., 1965. Local variation of ^{13}C and ^{18}O content of mollusk shells and their relatively minor temperature effect in marginal marine environments. *Marine Geol.*, 3: 115–129.

KEITH, M. L. and WEBER, J. N., 1964. Carbon and oxygen isotopic composition of selected limestones and fossils. *Geochim. Cosmochim. Acta*, 28: 1787–1816.

KEITH, M. L., ANDERSON, G. M. and EICHLER, R., 1964. Carbon and oxygen isotopic composition of mollusk shells from marine and fresh-water environments. *Geochim. Cosmochim. Acta*, 28: 1757–1786.

KELLER, W. D., 1937. "Cave pearls" in a mine near Columbia, Missouri. *J. Sediment. Petrol.*, 7: 108–109.

KELLERMAN, K. F. and SMITH, N. R., 1914. Bacterial precipitation of calcium carbonate. *J. Wash. Acad. Sci.*, 4: 400–402.

KENDALL, C. G. ST. C., 1966. *Recent Carbonate Sediments of the Western Khor al Bazam, Abu Dhabi, Trucial Coast.* Thesis, Imp. Coll. Sci. Technol., London, 273 pp., unpublished.

KENDALL, C. G. ST. C. and SKIPWITH, P. A. D'E., 1968a. Recent algal stromatolites of the Khor al Bazam, Abu Dhabi, the southwest Persian Gulf. *Geol. Soc. Am., Spec. Papers*, 101: 108 (abstract).

KENDALL, C. G. ST. C., and SKIPWITH, P. A. D'E., 1968b. Recent algal mats of a Persian Gulf lagoon. *J. Sediment. Petrol.*, 38: 1040–1058.

KENDALL, C. G. ST. C. and SKIPWITH, P. A. D'E., 1969a. Holocene shallow-water carbonate and evaporite sediments of Khor al Bazam, Abu Dhabi, southwest Persian Gulf. *Bull. Am. Assoc. Petrol. Geologists*, 53: 841–869.

KENDALL, C. G. ST. G. and SKIPWITH, P. A. D'E., 1969b. Geomorphology of a Recent shallow-water carbonate province: Khor al Bazam, Trucial Coast, southwest Persian Gulf. *Bull. Geol. Soc. Am.*, 80: 865–891.

KENDALL, C. G. ST. C., REES, G., SHEARMAN, D. J., SKIPWITH, P. A. D'E., TWYMAN, J. and KARIMI, M. Z., 1966. On the mechanical role of organic matter in the diagenesis of limestones. *Geologists' Assoc. Engl., Circ.*, 681: 1–2.

KENNAUGH, J. H., 1968. An examination of the cuticle of three species of Ricinulei (Arachnida). *J. Zool.*, 156: 393–404.

KENNEDY, W. J., 1967. Burrows and surface traces from the Lower Chalk of southern England. *Bull. Brit. Museum Geol.*, 15: 125–167.

KENNEDY, W. J., 1969. The correlation of the Lower Chalk of south-east England. *Proc. Geologists' Assoc. Engl.*, 80: 459–551.

KENNEDY, W. J. and HALL, A., 1967. The influences of organic matter on the preservation of aragonite in fossils. *Proc. Geol. Soc. London*, 1643: 253–255.

KENNEDY, W. J. and TAYLOR, J. D., 1968. Aragonite in rudists. *Proc. Geol. Soc. London*, 1645: 325–331.

KENNEDY, W. J., TAYLOR, J. D. and HALL, A., 1969. Environmental and biological controls on bivalve shell mineralogy. *Biol. Rev. Cambridge Phil. Soc.*, 44: 499–530.

KHVOROVA, I. V., 1946. On a new genus of Algae from the Middle Carboniferous deposits of the Moscow Basin. *Dokl. Akad. Nauk S.S.S.R.*, 53: 737–739.

KIELAN, Z., 1954. Les trilobites Mésodévoniens des Monts de Sainte-Croix. *Palaeontologia Polonica*, 6: 1–47.

KIMOTO, S. and HONJO, S., 1968. Scanning electron microscope as a tool in geology and biology. *J. Fac. Sci., Hokkaido Univ., Ser. IV*, 14: 57–59.

KINSMAN, D. J. J., 1964a. Reef coral tolerance of high temperatures and salinities. *Nature*, 202: 1280–1282.

KINSMAN, D. J. J., 1964b. *Recent Carbonate Sedimentation near Abu Dhabi, Trucial Coast, Persian Gulf*. Thesis, Imp. Coll. Sci. Technol., unpublished.

KINSMAN, D. J. J., 1964c. The Recent carbonate sediments near Halat el Bahrani, Trucial Coast, Persian Gulf. In: L. M. J. U. VAN STRAATEN (Editor), *Deltaic and Shallow Marine Deposits*. Elsevier, Amsterdam, pp.185–192.

KINSMAN, D. J. J., 1965. Dolomitization and evaporite development, including anhydrite, in lagoonal sediments, Persian Gulf. *Geol. Soc. Am., Spec. Papers*, 82: 108–109 (abstract).

KINSMAN, D. J. J., 1966a. Coprecipitation of Sr^{+2} with aragonite from sea water at 15–95 °C. *Geol. Soc. Am., Spec. Papers*, 87: 88 (abstract).

KINSMAN, D. J. J., 1966b. Gypsum and anhydrite of Recent age, Trucial Coast, Persian Gulf. In: J. L. RAU (Editor), *Second Symposium on Salt, 1*. Northern Ohio Geol. Soc., Cleveland, Ohio, pp.302–326.

KINSMAN, D. J. J., 1969. Interpretation of Sr^{2+} concentrations in carbonate minerals and rocks. *J. Sediment. Petrol.*, 39: 486–508.

KINSMAN, D. J. J. and HOLLAND, H. D., 1969. The co-precipitation of cations with $CaCO_3$. IV. The coprecipitation of Sr^{2+} with aragonite between 16° and 96 °C. *Geochim. Cosmochim. Acta*, 33: 1–17.

KIRCHMAYER, M., 1962. Zur Untersuchung rezenter Ooide. *Neues Jahrb. Geol. Palaeontol., Abhandl.*, 114: 245–272.

KIRCHMAYER, M., 1964. Höhlenperlen (cave pearls, perles des cavernes) Vorkommen, Definition sowie strukturelle Beziehung zu ähnlichen Sedimentsphäriten. *Oesterr. Akad. Wiss., Math. Naturw. Kl., Anz.*, 10: 223–229.

KIRTLEY, D. W. and TANNER, W. F., 1968. Sabellariid worms: builders of a major reef type. *J. Sediment. Petrol.*, 38: 73–78.

KITANO, Y., 1959. State of magnesium in calcium carbonate deposits in thermal springs. *J. Earth Sci., Nagoya Univ.*, 7: 65–79.

KITANO, Y., 1962a. The behavior of various inorganic ions in the separation of calcium carbonate from a bicarbonate solution. *Bull. Chem. Soc. Japan*, 35: 1973–1980.

KITANO, Y., 1962b. A study of the polymorphic formation of calcium carbonate in thermal springs with an emphasis on the effect of temperature. *Bull. Chem. Soc. Japan*, 35: 1980–1985.

KITANO, Y., 1963. Geochemistry of calcareous deposits found in hot springs. *J. Earth Sci., Nagoya Univ.*, 11: 68–100.

KITANO, Y., 1964. On factors influencing the polymorphic crystallization of calcium carbonate found in marine biological systems. In: Y. MIYAKE and T. KOYAMA (Editors), *Recent Researchers in the Fields of Hydrosphere, Atmosphere and Nuclear Geochemistry*. Maruzen, Tokyo, pp.305–319.

KITANO, Y. and FURUTSU, T., 1959. The state of a small amount of magnesium contained in calcareous shells. *Bull. Chem. Soc. Japan*, 33: 1–4.

KITANO, Y. and HOOD, D. W., 1962. Calcium carbonate crystal forms formed from sea water by inorganic processes. *J. Oceanog. Soc. Japan*, 18: 141–145.

KITANO, Y. and HOOD, D. W., 1965. The influence of organic material on the polymorphic crystallisation of calcium carbonate. *Geochim. Cosmochim. Acta*, 29: 29–41.

KITANO, Y. and KANAMORI, N., 1966. Synthesis of magnesian calcite at low temperatures and pressures. *Geochem. J.*, 1: 1–10.

KITANO, Y. and KAWASAKI, N., 1958. Behavior of strontium ion in the process of calcium carbonate separation from bicarbonate solution. *J. Earth Sci., Nagoya Univ.*, 6: 63–74.

KITANO, Y., PARK, K. and HOOD, D. W., 1962. Pure aragonite synthesis. *J. Geophys. Res.*, 67: 4873–4874.

KLEMENT, K. W., 1966. Studies on the ecological distribution of lime-secreting and sediment-trapping algae in reefs and associated environments. *Neues Jahrb. Geol. Palaeontol., Abhandl.*, 125: 363–381.

KLEMENT, K. W. and TOOMEY, D. F., 1967. Role of the blue-green alga *Girvanella* in skeletal grain destruction and lime-mud formation in the Lower Ordovician of west Texas. *J. Sediment. Petrol.*, 37: 1045–1051.

KLÖDEN, K. F., 1828. *Beiträge zur Mineralogischen und Geognostischen Kenntniss der Mark Brandenburg, I.* Dieterici, Berlin.

KNOPF, E. B., 1933. Petrotectonics. *Am. J. Sci.*, 25: 433–470.

KOBAYASHI, I., 1964. Microscopical observations on the shell structure of Bivalvia. I. *Barbatia obtusoides* (Nyst). *Sci. Rept. Tokyo Kyoiku Daigaku, Sect. C*, 8: 295–301.

KOBAYASHI, I., 1966. Submicroscopic observations on the shell structure of Bivalvia. II. *Dosinia (Phacosoma) japonica* Reeve. *Sci. Rept. Tokyo Kyoiku Daigaku, Sect. C*, 9: 189–210.

KOBAYASHI, I., 1968. The relation between the morphological structure types of shell tissues and the nature of the organic matrices in the bivalve molluscs. *Venus: Japan. J. Malacol.*, 27: 111–122.

KOBAYASHI, I., 1969. Internal microstructure of the shell of the bivalve molluscs. *Am. Zool.*, 9: 663–672.

KOBAYASHI, I. and KAMIYA, H., 1968. Submicroscopic observations on the shell structure of Bivalvia. III. Genus *Anadara. J. Geol. Soc. Japan*, 74: 351–362.

KONISHI, K., 1958. Some Devonian calcareous algae from Alberta, Canada. *Quart. Colo. School Mines*, 53: 85–109.

KONISHI, K. and WRAY, J., 1961. *Eugonophyllum*, a new Pennsylvanian and Permian algal genus. *J. Paleontol.*, 35: 659–666.

KORNICKER, L. S., 1958a. Bahamian limestone crusts. *Trans. Gulf Coast Assoc. Geol. Socs.*, 8: 167–170.

KORNICKER, L. S., 1958b. Ecology and taxonomy of Recent marine ostracodes in the Bimini area, Great Bahama Bank. *Texas, Univ., Inst. Marine Sci., Publ.*, 5: 194–300.

KORNICKER, L. S. and BOYD, D. W., 1962. Shallow-water geology and environments of Alacran reef complex, Campeche Bank, Mexico. *Bull. Am. Assoc. Petrol. Geologists*, 46: 640–673.

KORNICKER, L. S. and PURDY, E. G., 1957. A Bahamian faecal-pellet sediment. *J. Sediment. Petrol.*, 27: 126–128.

KRAUSKOPF, K. B., 1967. *Introduction to Geochemistry*. McGraw-Hill, New York, N.Y., 721 pp.

KRINSLEY, D., 1960a. Trace elements in the tests of planktonic foraminifera. *Micropaleontology*, 6: 297–300.

KRINSLEY, D., 1960b. Magnesium, strontium, and aragonite in the shells of certain littoral gastropods. *J. Paleontol.*, 34: 744–755.

KRINSLEY, D. and BIERI, R., 1959. Changes in the chemical composition of pteropod shells after deposition on the sea floor. *J. Paleontol.*, 33: 682–684.

KUENEN, PH. H., 1933. *The Snellius-Expedition, 5. Geological Results, 2: Geology of Coral Reefs*. Kemink, Utrecht, 125 pp.

KUENEN, PH. H., 1950. *Marine Geology*. Wiley, New York, N.Y., 568 pp.

KULP, J. L., 1961. Geologic time scale. *Science*, 133: 1105–1114.

KULP, J. L., TUREKIAN, K. and BOYD, D. W., 1952. Strontium content of limestones and fossils. *Bull. Geol. Soc. Am.*, 63: 701–716.

LABECKI, J. and RADWANSKI, A., 1967. Broken ooids in lagoonal Keuper deposits of the western margin of the Holy Cross Mts. *Bull. Acad. Polon. Sci., Sér. Sci. Géol. Géograph.*, 15: 93–99.

LADD, H. S., 1950. Recent reefs. *Bull. Am. Assoc. Petrol. Geologists*, 34: 203–214.

LADD, H. S., 1956. Coral reef problems in the open Pacific. *Proc. Pacific Sci. Congr. Pacific Sci. Assoc., 8th, Quezon City, 1953*, 2-A: 833–850.

LADD, H. S., 1958. Fossil land shells from western Pacific atolls. *J. Paleontol.*, 32: 183–198.

LADD, H. S., 1961. Reef building. *Science*, 134: 703–715.

LADD, H. S. and TRACEY, JR., J. I., 1957. Fossil land shells from deep drill holes on western Pacific atolls. *Deep-Sea Res.*, 4: 218–219.

LADD, H. S., INGERSON, E., TOWNSEND, R. C., RUSSELL, M. and STEPHENSON, H. K., 1953. Drilling on Eniwetok Atoll, Marshall Islands. *Bull. Am. Assoc. Petrol. Geologists*, 37: 2257–2280.

LADD, H. S., TRACEY, JR., J. I., WELLS, J. W. and EMERY, K. O., 1950. Organic growth and sedimentation on an atoll. *J. Geol.*, 58: 410–425.

LADD, H. S., TRACEY, JR., J. I. and GROSS, G., 1967. Drilling on Midway Atoll, Hawaii. *Science*, 156: 1088–1094.

LAGENHEIM, JR., R. L. and EPIS, R. C., 1957. Holothurian sclerites from the Mississippian Escabrosa limestone, Arizona. *Micropaleontology* 3: 165–170.

LALOU, C., 1957a. Studies on bacterial precipitation of carbonates in sea water. *J. Sediment. Petrol.*, 27: 190–195.

LALOU, C., 1957b. Étude expérimentale de la production de carbonates par les bactéries des vases de la baie de Villefranche-sur-Mer. *Ann. Inst. Océanog. (Paris)*, 33: 202–267.

LAND, L. S., 1966. *Diagenesis of Metastable Skeletal Carbonates*. Thesis, Marine Sci. Center, Lehigh Univ., Bethlehem, Pa., 141 pp.

LAND, L. S., 1967. Diagenesis of skeletal carbonates. *J. Sediment. Petrol.*, 37: 914–930.

LAND, L. S., 1970. Phreatic versus vadose meteoric diagenesis of limestones: evidence from a fossil water table. *Sedimentology*, 14: 175–185.

LAND, L. S. and EPSTEIN, S., 1970. Late Pleistocene diagenesis and dolomitization, north Jamaica. *Sedimentology*, 14(3/4): 187–200.

LAND, L. S. and GOREAU, T. F., 1970. Submarine lithification of Jamaican reefs. *J. Sediment. Petrol.*, 40: 457–462.

LAND, L. S., MACKENZIE, F. T. and GOULD, S. J., 1967. Pleistocene history of Bermuda. *Geol. Soc. Am. Bull.*, 78: 993–1006.

LANDER, J. J., 1949. Polymorphism and anion rotational disorder in the alkaline earth carbonates. *J. Chem. Phys.*, 17: 892–901.

LAPORTE, L. F., 1967. Carbonate deposition near mean sea-level and resultant facies mosaic: Manlius formation (Lower Devonian) of New York State. *Bull. Am. Assoc. Petrol. Geologists*, 51: 73–101.

LAPORTE, L. F., 1968. Recent carbonate environments and their paleoecologic implications. In: E. T. DRAKE (Editor), *Evolution and Environment*. Yale Univ. Press, New Haven, Conn., pp.229–258.

LAPORTE, L. F. and IMBRIE, J., 1964. Phases and facies in the interpretation of cyclic deposits. *Geol. Surv. Kansas, Bull.*, 169: 249–263.

LATIMER, W. M., 1952. *The Oxidation States of the Elements and their Potentials in Aqueous Solutions*. Prentice-Hall, New York, N.Y., 392 pp.

LAUBIER, L., 1965. Le coralligène des Albères. Monographie biocénotique. *Ann. Inst. Océanog. (Paris)*, 43: 137–316.

LE BLANC, R. J. and BREEDING, J. G. (Editors), 1957. *Regional Aspects of Carbonate Deposition—Soc. Econ. Paleontologists Mineralogists, Spec. Publ.*, 5: 1–178.

LECOMPTE, M., 1936. Contribution à la connaissance des "récifs" du Frasnien de l'Ardenne. *Mém. Inst. Géol. Univ. Louvain*, 10: 29–112.

LECOMPTE, M., 1937. Contribution à la connaissance des récifs du Devonien de l'Ardenne. Sur la présence de structures conservées dans des efflorescences cristallines du type "Stromatactis". *Bull. Muséum Roy. Hist. Nat. Belg.*, 13: 1–14.

LECOMPTE, M., 1952. Les stromatoporoides du Dévonien moyen et supérieur du basin de Dinant. *Inst. Roy. Soc. Nat. Belg., Mem.*, 117: 216–359.

LECOMPTE, M., 1956. Stromatoporoidea. In: R. C. MOORE (Editor), *Treatise on Invertebrate Paleontology. F. Coelenterata*. Geol. Soc. Am., Boulder, Colo., pp.107–144.

LEES, A., 1961. The Waulsortian "reefs" of Eire: a carbonate mudbank complex of Lower Carboniferous age. *J. Geol.*, 69: 101–109.

LEES, A., 1964. The structure and origin of the Waulsortian (Lower Carboniferous) "reefs" of west-central Eire. *Phil. Trans. Roy. Soc. London, Ser. B.*, 247: 483–531.

LENNARD-JONES, J. E. and DENT, B. M., 1928. The change in lattice spacing at a crystal boundary. *Proc. Roy. Soc. (London), Ser. A*, 121: 247–259.

LERBEKMO, J. F. and PLATT, R. L., 1962. Promotion of pressure-solution of silica in sandstones. *J. Sediment. Petrol.*, 32: 514–519.

LERMAN, A., 1965a. Paleoecological problems of Mg and Sr in biogenic calcites in light of recent thermodynamic data. *Geochim. Cosmochim. Acta*, 29: 977–1002.

LERMAN, A., 1965b. Strontium and magnesium in water and in *Crassostrea* calcite. *Science*, 150: 745–751.

LEUTWEIN, F. and WASKOWIAK, R., 1962. Geochemische Untersuchungen an rezenten marinen Molluskenschalen. *Neues Jahrb. Mineral., Abhandl.*, 99: 45–78.

LEWIS, M. S., 1968. The morphology of the fringing coral reefs along the east coast of Mahé, Seychelles. *J. Geol.*, 76: 140–153.

LEWIS, M. S., 1969. Sedimentary environments and unconsolidated sediments of the fringing coral reefs of Mahé, Seychelles. *Marine Geol.*, 7: 95–127.

LEWIS, M. S. and TAYLOR, J. D., 1966. Marine sediments and bottom communities of the Seychelles. *Phil. Trans. Roy. Soc. London, Ser. A*, 259: 279–290.

LIEBERMANN, O., 1967. Synthesis of dolomite. *Nature*, 213: 241–245.

LINCK, G., 1903. Die Bildung der Oolithe und Rogensteine. *Neues Jahrb. Mineral., Geol. Paläontol.*, 16: 495–513.

LINCK, G., 1909. Über die Bildung der Kalksteine. *Naturwiss. Wochschr.*, 8: 689–694.

LINDSTRÖM, G., 1901. Researches on the visual organs of the trilobites. *Kgl. Svenska Vetenskapsakad. Handl.*, 34: 1–89.

LINDSTRÖM, M., 1963. Sedimentary folds and the development of limestone in an Early Ordovician sea. *Sedimentology*, 2: 243–275.

LIPMAN, C. B., 1924. A critical and experimental study of Drew's bacterial hypothesis on $CaCO_3$ precipitation in the sea. *Papers Dept. Marine Sci., Carnegie Inst. Wash. Publ.*, 340: 179–191.

LIPMAN, C. B., 1929. Further studies on marine bacteria with special reference to the Drew hypothesis on $CaCO_3$ precipitation in the sea. *Papers Tortugas Lab., Carnegie Inst. Wash. Publ.*, 391: 231–248.

LIPPMANN, F., 1959. Darstellung und kristallographische Daten von $CaCO_3 . H_2O$. *Naturwissenschaften*, 46: 553–554.

LIPPMANN, F., 1960. Versuche zur Aufklärung der Bildungsbedingungen von Kalzit und Aragonit. *Fortschr. Mineral.*, 38: 156–161.

LIPPMANN, F., 1968a. Les structures cristallines des minéraux essentiels des roches carbonatées: clés aux problèmes de pétrogenèse. In: B. KUBLER and J.-P. SCHAER (Editors), *Texte de Trois Conférences de Mr. F. Lippmann*. Univ. Neuchatel, Neuchatel, pp. 1–6.

LIPPMANN, F., 1968b. L'influence du magnésium sur la formation de la calcite et de l'aragonite. In: B. KUBLER and J.-P. SCHAER (Editors), *Texte de Trois Conférences de Mr. F. Lippmann*. Univ. Neuchatel, Neuchatel, pp. 7–13.

LIPPMANN, F., 1968c. La synthèse des carbonates doubles de Ba–Mg et Pb–Mg à 20 degrés centigrades environ: modèle de dolomitisation. In: B. KUBLER and J.-P. SCHAER (Editors), *Texte de Trois Conférences de Mr. F. Lippmann*. Univ. Neuchatel, Neuchatel, pp. 14–21.

LLOYD, E. R., 1933. Coral reefs and atolls. *Bull. Am. Assoc. Petrol. Geologists*, 17: 85–87.

LLOYD, R. M., 1964. Variations in the oxygen and carbon isotope ratios of Florida Bay mollusks and their environmental significance. *J. Geol.*, 72: 84–111.

LLOYD, R. M., 1971. Some observations on Recent sediment alteration ("micritization") and the possible role of algae in submarine cementation. In: O. P. BRICKER (Editor), *Carbonate Cements*. Johns Hopkins Press, Baltimore, Md., pp. 72–79.

LOEBLICH, JR., A. R. and TAPPAN, H., 1964a. Foraminiferida. Morphology and biology. In: R. C. MOORE (Editor), *Treatise on Invertebrate Paleontology. C. Protista 2(1)*. Geol. Soc. Am., Boulder, Colo., pp.58–134.

LOEBLICH, JR., A. R. and TAPPAN, H., 1964b. Foraminiferida. Classification. In: R. C. MOORE (Editor), *Treatise on Invertebrate Paleontology. C. Protista 2(1)*. Geol. Soc. Am., Boulder, Colo , pp.140–163.

LOEBLICH, JR., A. R. and TAPPAN, H., 1964c. Foraminiferida. Systematic descriptions. In: R. C.

Moore (Editor), *Treatise on Invertebrate Paleontology. C. Protista 2 (1 and 2)*. Geol. Soc. Am., Boulder, Colo., pp. 164–782.

Logan, B., 1961. *Cryptozoon* and associate stromatolites from the Recent, Shark Bay, Western Australia. *J. Geol.*, 69: 517–533.

Logan, B. W., Rezak, R. and Ginsburg, R. N., 1964. Classification and environmental significance of algal stromatolites. *J. Geol.*, 72: 68–83.

Logan, B. W., Harding, J. L., Ahr, W. M., Williams, J. D. and Snead, R. G., 1969. Carbonate sediments and reefs, Yucatán shelf, Mexico. *Am. Assoc. Petrol. Geologists, Mem.*, 11: 1–198.

Loreau, J-P., 1969. Ultrastructures et diagenèses des oolithes marines anciennes (jurassiques). *Compt. Rend.*, 269: 819–822.

Loreau, J-P., 1970a. Ultrastructure de la phase carbonatée des oolithes marines actuelles. *Compt. Rend.*, 271: 816–819.

Loreau, J-P., 1970b. Contribution à l'étude des calcarénites hétérogènes par l'emploi simultané de la microscopie photonique et de la microscopie électronique à balayage. Problème particulier de la micritisation. *J. Microscopie*, 9: 727–734.

Lowenstam, H. A., 1950. Niagaran reefs of the Great Lakes area. *J. Geol.*, 58: 430–487.

Lowenstam, H. A., 1954. Factors affecting the aragonite/calcite ratios in carbonate-secreting marine organisms. *J. Geol.*, 62: 284–322.

Lowenstam, H. A., 1955. Aragonite needles secreted by algae and some sedimentary implications. *J. Sediment. Petrol.*, 25: 270–272.

Lowenstam, H. A., 1961. Mineralogy, O^{18}/O^{16} ratios, and strontium and magnesium contents of recent and fossil brachiopods and their bearing on history of the oceans. *J. Geol.*, 69: 241–260.

Lowenstam, H. A., 1963. Biologic problems relating to the composition and diagenesis of sediments. In: T. W. Donnelly (Editor), *The Earth Sciences—Problems and Progress in Current Research*. Univ. Chicago Press, Chicago, Ill., pp.137–195.

Lowenstam, H. A., 1964. Coexisting calcites and aragonites from skeletal carbonates of marine organisms and their strontium and magnesium contents. In: Y. Miyake and T. Koyama (Editors), *Recent Researches in the Fields of Hydrosphere, Atmosphere and Nuclear Geochemistry*. Maruzen, Tokyo, pp.373–404.

Lowenstam, H. A. and Epstein, S., 1957. On the origin of sedimentary aragonite needles of the Great Bahama Bank. *J. Geol.*, 65: 364–375.

Lucia, F. J., 1961. Dedolomitization in the Tansill (Permian) formation. *Geol. Soc. Am. Bull.*, 72: 1107–1109.

Lucia, F. J., 1962. Diagenesis of a crinoidal sediment. *J. Sediment. Petrol.*, 32: 848–865.

Lucia, F. J., 1968. Recent sediments and diagenesis of south Bonaire, Netherlands Antilles. *J. Sediment. Petrol.*, 38: 845–858.

Lucia, F. J. and Murray, R. C., 1967. Origin and distribution of porosity in crinoidal rock. *World Petrol. Congr., Proc., 7th Mexico, 1967*, 2: 409–423.

Lutts, A., Granjean, J. and Grégoire, C., 1960. X-ray diffraction patterns from the prisms of mollusk shells. *Arch. Intern. Physiol. Biochim.*, 68: 829–831.

Macdonald, G. J. F., 1956. Experimental determination of calcite–aragonite equilibrium relations at elevated temperatures. *Am. Mineralogist*, 41: 744–756.

Macintyre, I. G., Mountjoy, E. W. and D'Anglejan, B. F., 1968. An occurrence of submarine cementation of carbonate sediments off the west coast of Barbados, W. I. *J. Sediment. Petrol.*, 38: 660–663.

Mackin, J. H. and Coombs, H. A., 1945. An occurrence of "cave pearls" in a mine in Idaho. *J. Geol.*, 53: 58–65.

MacLintock, C., 1967. Shell structure of the Pattelloid and Bellerophontoid gastropods (Mollusca). *Peabody Museum Nat. Hist., Yale Univ., Bull.*, 22: 1–140.

Macneil, F. S., 1954. Organic reefs and banks and associated detrital sediments. *Am. J. Sci.*, 252: 385–401.

Maiklem, W. R., 1968. Some hydraulic properties of bioclastic carbonate grains. *Sedimentology*, 10: 101–109.

MAIKLEM, W. R., 1970. Carbonate sediments in the Capricorn Reef Complex, Great Barrier Reef, Australia. *J. Sediment. Petrol.*, 40: 55–80.

MAJEWSKE, O. P., 1969. *Recognition of Invertebrate Fossil Fragments in Rocks and Thin Sections.* Brill, Leiden, 101 pp.

MALONE, P. G. and DODD, J. R., 1967. Temperature and salinity effects on calcification rate in *Mytilus edulis* and its paleoecological implications. *Limnol. Oceanog.*, 12: 432–436.

MANHEIM, F. T., HATHAWAY, J. C., DEGENS, E. T., MCFARLIN, P. F. and JOKELLA, A., 1965. Geochemistry of Recent iron deposits in the Red Sea. *Geol. Soc. Am., Progr. Ann. Meeting*, 1965: p.100.

MANTEN, A. A., 1966. Note on the formation of stylolites. *Geol. Mijnbouw*, 45: 269–274.

MARSCHNER, H., 1969. Hydrocalcite ($CaCO_3 \cdot H_2O$) and nesquehonite ($MgCO_3.3H_2O$) in carbonate scales· *Science*, 165: 1119–1121.

MARSH, O. C., 1867. On the origin of the so-called lignilitites or epsomites. *Proc. Am. Assoc. Advan. Sci.*, 16: 135–143.

MARTIN, E. L. and GINSBURG, R. N., 1965. Radiocarbon ages of oolitic sands on Great Bahama Bank. *Proc. Intern. Conf. Radiocarbon Tritium Dating, 6th, Pullman, Wash., 1965*, pp.705–719.

MASLOV, P. B., 1956. Iskopaemye izvestkovye vodorosli SSSR. (Calcareous algae of the U.S.S.R.) *Tr. Inst. Geol. Nauk, Akad. Nauk S.S.S.R., Geol. Ser.*, 160: 1–301.

MATTAVELLI, L., 1966. Osservazioni petrografiche sulla sostituzione della dolomite con la calcite (dedolomitizazione) in alcune facies carbonate italiane. *Atti Soc. Ital. Sci. Nat. Museo Civico Storia Nat. Milano*, 105: 293–316.

MATTAVELLI, L. and NOVELLI, L., 1968. Petrografia e diagenesi della serie carbonato–argilloso–silicea di S. Fele. *Rend. Soc. Mineral. Ital.*, 24: 1–23.

MATTAVELLI, L. and TONNA, M., 1967. Osservazioni petrografiche su processi diagenetici in alcune facies carbonate mesozoiche italiane. *Rend. Soc. Mineral. Ital.*, 23: 245–273.

MATTER, A., 1967. Tidal flat deposits in the Ordovician of Western Maryland. *J. Sediment. Petrol.*, 37: 601–609.

MATTHEWS, R. K., 1966. Genesis of Recent lime mud in southern British Honduras. *J. Sediment. Petrol.*, 36: 428–454.

MATTHEWS, R. K., 1967. Diagenetic fabrics in biosparites from the Pleistocene of Barbados, West Indies. *J. Sediment. Petrol.*, 37: 1147–1153.

MATTHEWS, R. K., 1968. Carbonate diagenesis: equilibration of sedimentary mineralogy to the subaerial environment; coral cap of Barbados, West Indies. *J. Sediment. Petrol*, 38: 1110–1119.

MAXWELL, J. C., 1960. Experiments on compaction and cementation of sand. *Geol. Soc. Am., Mem.*, 79: 105–132.

MAXWELL, W. G. H. and SWINCHATT, J. P., 1970. Great Barrier Reef: variation in a terrigenous-carbonate province. *Geol. Soc. Am. Bull.*, 81: 691–724.

MAXWELL, W. G. H., DAY, R. W. and FLEMING, P. J. G., 1961. Carbonate sedimentation on the Heron Island reef, Great Barrier Reef. *J. Sediment. Petrol.*, 31: 215–230.

MAXWELL, W. G. H., JELL, J. S. and MCKELLAR, R. G., 1963. A preliminary note on the mechanical and organic factors influencing carbonate differentiation, Heron Island reef, Australia. *J. Sediment. Petrol.*, 33: 962–963.

MAXWELL, W. G. H., JELL, J. S. and MCKELLAR, R. G., 1964. Differentiation of carbonate sediments in the Heron Island reef. *J. Sediment. Petrol.*, 34: 294–308.

MAXWELL, W. G. J., 1968. *Atlas of the Great Barrier Reef.* Elsevier, Amsterdam, 258 pp.

MCCALLUM, J. S. and STOCKMAN, K. W., 1964. Water circulation. In: R. N. GINSBURG (Editor), *Guide Book for Field Trip No. 1. Geol. Soc. Am., Convention 1964.* Geol. Soc. Am., New York, N.Y., pp.11–13.

MCCAVE, I. N., 1969. Deposition of fine-grained sediment from tidal currents. *Nature*, 224: 1288–1289.

MCFARLIN, P. F., 1967. Aragonite vein fillings in marine manganese nodules. *J. Sediment. Petrol.*, 37: 68–72.

MCINTIRE, W. L., 1963. Trace element partition coefficients—a review of theory and applications to geology. *Geochim. Cosmochim. Acta*, 27: 1209–1264.

McIntyre, A. and Bé, A. W. H., 1967. Modern Coccolithophoridae of the Atlantic Ocean. 1. Placoliths and Cyrtoliths. *Deep-Sea Res.*, 14: 561–597.

McKee, E. D., 1958. Geology of Kapingamarangi Atoll, Caroline Islands. *Bull. Geol. Soc. Am.*, 69: 241–278.

McKee, E. D. and Gutschick, R. C., 1969. History of Redwall Limestone of northern Arizona. *Geol. Soc. Am., Mem.*, 114: 1–726.

McKee, E. D., Chronic, J. and Leopold, E. B., 1959. Sedimentary belts in lagoon of Kapingamarangi Atoll. *Bull. Am. Assoc. Petrol. Geologists*, 43: 501–562.

McLean, D., 1965. The science of metamorphism in metals. In: W. S. Pitcher and G. W. Flinn (Editors), *Controls of Metamorphism*. Oliver and Boyd, Edinburgh, pp.103–118.

McLean, R. F., 1967a. Measurements of beachrock erosion by some tropical marine gastropods. *Bull. Marine Sci. Gulf Carribbean*, 17: 551–561.

McLean, R. F., 1967b. Erosion of burrows in beachrock by the tropical sea urchin, *Echinometra lucunter*. *Can. J. Zool., Bull.*, 45: 586–588.

McLuhan, M. and Fiore, Q., 1967. *The Medium is the Massage*. Penguin Books, Harmondsworth, 159 pp.

Meadows, P. S. and Anderson, J. G., 1966. Micro-organisms attached to marine and freshwater sand grains. *Nature*, 212: 1059–1060.

Meadows, P. S. and Anderson, J. G., 1968. Micro-organisms attached to marine sand grains. *J. Marine Biol. Assoc. U.K.*, 48: 161–175.

Mecarini, G., Shimaoka, G. and Krause, D. C., 1968. Submarine lithification of Globigerina ooze. *Geol. Soc. Am., Spec. Papers*, 101: 269 (abstract).

Medawar, P. B., 1967. *The Art of the Soluble*. Methuen, London, 160 pp.

Meischner, K.-D., 1964. Allodapische Kalke, Turbidite in Riff-Nahen Sedimentations-Becken. In: A. H. Bouma and A. Brouwer (Editors), *Turbidites*. Elsevier, Amsterdam, pp.156–191 (Transl. from German).

Mellor, J. W., 1923. *A Comprehensive Treatise on Inorganic and Theoretical Chemistry*, 3. Longmans, Green, London, 927 pp.

Milliman, J. D., 1965. An annotated bibliography of recent papers on corals and coral reefs. *Atoll Res. Bull.*, 111: 1–55.

Milliman, J. D., 1966. Submarine lithification of carbonate sediments. *Science*, 153: 994–997.

Milliman, J. D., 1967a. Carbonate sedimentation on Hogsty Reef, a Bahamian atoll. *J. Sediment. Petrol.*, 37: 658–676.

Milliman, J. D., 1967b. The geomorphology and history of Hogsty Reef, a Bahamian atoll. *Bull. Marine Sci. Gulf Caribbean*, 17: 519–543.

Milliman, J. D., Ross, D. A. and Ku, T-L., 1969. Precipitation and lithification of deep-sea carbonates in the Red Sea. *J. Sediment. Petrol.*, 39: 724–736.

Mišík, M., 1968. Some aspects of diagenetic recrystallization in limestones. *Intern. Geol. Congr., 23rd, Prague, 1968, Rept. Session*, 8: 129–136.

Mitterer, R. M., 1968. Amino acid composition of organic matrix in calcareous oolites. *Science*, 162: 1498–1499.

Mitterer, R. M., 1969. The origin of calcareous oolites. *Geol. Soc. Am., Progr. S.E. Sect.*, 4: 54–55 (abstract).

Moberly, R., 1968. Composition of magnesian calcites of algae and pelecypods by electron microprobe analysis. *Sedimentology*, 11: 61–82.

Moberly, R., 1970. Microprobe study of diagenesis in calcareous algae. *Sedimentology*, 14: 113–123.

Molinier, R. and Picard, J., 1952. Recherches sus les herbiers de phanérogames marines du littoral Méditerranéen français. *Ann. Inst. Océanog. (Paris)*, 27: 157–234

Monaghan, P. H. and Lytle, M. A., 1956. The origin of calcareous oolites. *J. Sediment. Petrol.*, 26: 111–118.

Monty, C., 1965. Recent algal stromatolites in the windward lagoon, Andros Island, Bahamas. *Ann. Soc. Géol. Belg., Bull.*, 88: 269–276.

Monty, C., 1967. Distribution and structure of Recent stromatolitic algal mats, eastern Andros Island, Bahamas. *Ann. Soc. Géol. Belg., Bull.*, 90: 55–100.

MOOK, W. G. and VOGEL, J. C., 1968. Isotopic equilibrium between shells and their environment. *Science*, 159: 874–875.

MOORE, D. R., 1963. Distribution of the sea grass, *Thalassia*, in the United States. *Bull. Marine Sci. Gulf Caribbean*, 13: 329–342.

MOORE, D. R. and BULLIS, JR., H. R., 1960. A deep-water coral reef in the Gulf of Mexico. *Bull. Marine Sci. Gulf Caribbean*, 10: 125–128.

MOORE, G. W., ROBERSON, C. E. and NYGREN, H. D., 1962. Electrode determination of the carbon dioxide content of sea water and deep-sea sediment. *U.S., Geol. Surv., Profess. Papers*, 450-B: 83–86.

MOORE, H. B., 1931. The muds of the Clyde Sea area, III. Chemical and physical conditions; rate and nature of sedimentation, and fauna. *J. Marine Biol. Assoc. U.K.*, 17: 325–358.

MOORE, H. B., 1939. Faecal pellets in relation to marine deposits. In: P. D. TRASK (Editor), *Recent Marine Sediments*. Murby, London, pp.516–524.

MOORE, H. B., 1964. *Marine Ecology*. Wiley, New York, N.Y., 493 pp.

MOORE, W. E., 1957. Ecology of Recent foraminifera in northern Florida Keys. *Bull. Am. Assoc. Petrol. Geologists*, 41: 727–741.

MÜLLER, A. H., 1953. Bemerkungen zur Stratigraphie und Stratonomie der obersenonen Schreibkreide von Rügen. *Geologie (Berlin)*, 2: 25–34.

MÜLLER, G., 1970. High-magnesian calcite and protodolomite in Lake Balaton (Hungary) sediments. *Nature*, 226: 749–750.

MÜLLER, G. and FRIEDMAN, G. M. (Editors), 1968. *Recent Developments in Carbonate Sedimentology in Central Europe*. Springer, Berlin, 255 pp.

MÜLLER, G. and IRION, G., 1969. Subaerial cementation and subsequent dolomitization of lacustrine carbonate muds and sands from Paleo-Tuz Gölü ("Salt Lake"), Turkey. *Sedimentology*, 12: 193–204.

MÜLLER, G. and MÜLLER, J., 1967. Mineralogisch-sedimentpetrographische und chemische Untersuchungen an einen Bank-Sediment (Cross-Bank) der Florida Bay, U.S.A. *Neues Jahrb. Mineral., Abhandl.*, 106: 257–286.

MÜLLER, G. and TIETZ, G., 1966. Recent dolomitization of Quaternary biocalcarenites from Fuerteventura (Canary Islands). *Contrib. Mineral. Petrol.*, 13: 89–96.

MULTER, H. G., 1969. *Field Guide to some Carbonate Rock Environments: Florida Keys and Western Bahamas*. Fairleigh Dickinson Univ., Madison, N.J., 159 pp.

MULTER, H. G. and HOFFMEISTER, J. E., 1968. Subaerial laminated crusts of the Florida Keys. *Geol. Soc. Am. Bull.*, 79: 183–192.

MULTER, H. G. and MILLIMAN, J. D., 1967. Geologic aspects of sabellarian reefs, southeast Florida. *Bull. Marine Sci. Gulf Caribbean*, 17: 257–267.

MUNK, W. H. and SARGENT, M. C., 1948. Adjustment of Bikini Atoll to ocean waves. *Trans. Am. Geophys. Union*, 29: 855–860.

MUNK, W. H. and SARGENT, M. C., 1954. Adjustment of Bikini Atoll to ocean waves. *U.S., Geol. Surv., Profess. Papers*, 260-C: 275–280.

MURATA, K. J., FRIEDMAN, I. and MADSEN, B. M., 1969. Isotopic composition of diagenetic carbonates in marine Miocene formations of California and Oregon. *U.S., Geol. Surv., Profess. Papers*, 614-B: 1–24.

MURRAY, J. and HJORT, J., 1912. *The Depths of the Ocean*. Macmillan, London, 821 pp.

MURRAY, J. and LEE, G. V., 1909. The depth and marine deposits of the Pacific. *Mem. Museum Comp. Zool. Harvard Coll.*, 38: 1–169.

MURRAY, J. W., 1954. The deposition of calcite and aragonite in caves. *J. Geol.*, 62: 481–492.

MURRAY, J. W., 1965a. The Foraminiferida of the Persian Gulf. 1. *Rosalina adhaerens* sp. nov. 1965. *Ann. Mag. Nat. Hist., Ser. 13*, 8: 77–79.

MURRAY, J. W., 1965b. The Foraminiferida of the Persian Gulf. 2. The Abu Dhabi region. *Palaeogeog., Palaeoclimatol., Palaeoecol.*, 1: 307–332.

MURRAY, J. W., 1966a. The Foraminiferida of the Persian Gulf. 3. The Halat al Bahrani region. *Palaeogeog., Palaeoclimatol., Palaeoecol.*, 2: 59–68.

MURRAY, J. W., 1966b. The Foraminiferida of the Persian Gulf. 4. Khor al Bazam. *Palaeogeog., Palaeoclimatol., Palaeoecol.*, 2: 153–169.

MURRAY, J. W., 1966c. The Foraminiferida of the Persian Gulf. 5. The shelf off the Trucial coast. *Palaeogeog., Palaeoclimatol., Palaeoecol.*, 2: 267–278.

MURRAY, J. W., 1970a. The Foraminiferida of the Persian Gulf. 6. Living forms in the Abu Dhabi region. *J. Nat. Hist.*, 4: 55–67.

MURRAY, J. W., 1970b. The Foraminifera of the hypersaline Abu Dhabi Lagoon, Persian Gulf. *Lethaia*, 3: 51–68.

MURRAY, R. C., 1960. Origin of porosity in carbonate rocks. *J. Sediment. Petrol.*, 30: 59–84.

MURRAY, R. C., 1969. Hydrology of south Bonaire, N.A.—a rock selective dolomitization model. *J. Sediment. Petrol.*, 39: 1007–1013.

MURRAY, R. C. and LUCIA, F. J., 1967. Cause and control of dolomite distribution by rock selec tivity. *Geol. Soc. Am. Bull.*, 78: 21–35.

MURRAY, R. C. and PRAY, L. C., 1965. Dolomitization and limestone diagenesis—an introduction. In: L. C. PRAY and R. C. MURRAY (Editors), *Dolomitization and Limestone Diagenesis*: *a Symposium—Soc. Econ. Paleontologists Mineralogists, Spec. Publ.*, 13: 1–2.

MUTVEI, H., 1964. On the shells of *Nautilus* and *Spirula* with notes on the shell secretion in non-cephalopod molluscs. *Arkiv Zool.*, 16: 221–278.

MUTVEI, H., 1967. On the microscopic shell structure in some Jurassic ammonoids. *Neues Jahrb. Geol. Paläontol., Abhandl.*, 129: 157–166.

NADSON, G., 1927a. Les algues perforantes de la mer Noire. *Compt. Rend.*, 184: 896–898.

NADSON, G., 1927b. Les algues perforantes, leur distribution et leur rôle dans la nature. *Compt. Rend.*, 184: 1015–1017.

NANCOLLAS, G. H. and PURDIE, N., 1964. The kinetics of crystal growth. *Quart. Rev. (London)*, 18: 1–20.

NATTERER, K., 1892, 1893, 1894. Chemische Untersuchungen im oestlichen Mittelmeer, i. Reise S.M. Schiffes "Pola" im Jahre 1890. *Jahrb. Oesterr. Wiss.*, 59: 83–116: 60: 49–76; 61: 23–64.

NATTERER, K., 1898. Chemische Untersuchungen; Bericht d. Commision f. oceanographische Forschungen, Expedition S.M. Schiffes "Pola" in das Rothe Meer, noerdliche Haelfte (Okt. 1895–Mai 1896). *Jahrb. Oesterr. Wiss.*, 65: 445–570.

NEAL, W. J., 1969. Diagenesis and dolomitization of a limestone (Pennsylvanian of Missouri) as revealed by staining. *J. Sediment. Petrol.*, 39: 1040–1045.

NEEV, D., 1963. Recent precipitation of calcium salts in the Dead Sea. *Bull. Res. Council Israel, Sect. G*, 11: 153–154.

NEEV, D., 1964. *Geological Processes in the Dead Sea*. Thesis, Hebrew Univ., Jerusalem, 407 pp., unpublished.

NEEV, D. and EMERY, K. O., 1967. The Dead Sea. Depositional processes and environments of evaporites. *Israel Geol. Surv. Bull.*, 41: 1–147.

NELSON, H. F., 1959. Deposition and alteration of the Edwards Limestone, central Texas. In: J. T. LONSDALE (Editor), *Symposium on Edwards Limestone in Central Texas—Texas, Univ., Bur. Econ. Geol., Publ.*, 5905: 21–95.

NESTEROFF, W. D., 1955a. Les récifs coralliens du banc Farsan nord (Mer Rouge). *Ann. Inst. Océanog. (Paris)*, 30: 8–53.

NESTEROFF, W. D., 1955b. De l'origine des dépôts calcaires. *Compt. Rend.*, 240: 220–222.

NESTEROFF, W. D., 1956a. De l'origine des oolithes. *Compt. Rend.*, 242: 1047–1049.

NESTEROFF, W. D., 1956b. La substratum organique dans les dépôts calcaires, sa signification. *Bull. Soc. Géol. France, Ser. 6.*, 6: 381–390.

NESTLER, H., 1965. Die Rekonstruktion des Lebensraumes der Rügener Schreibkreide-Fauna (Unter-Maastricht) mit Hilfe der Paläoökologie und Paläobiologie. *Geologie*, 14: 1–147.

NEUMANN, A. C., 1965. Processes of Recent carbonate sedimentation in Harrington Sound, Bermuda. *Bull. Marine Sci. Gulf Caribbean*, 15: 987–1035.

NEUMANN, A. C., 1966. Observations on coastal erosion in Bermuda and measurements of the boring rate of the sponge, *Cliona lampa. Limnol. Oceanog.*, 11: 92–108.

NEUMANN, A. C. and CHAVE, K. E., 1965. Connate origin proposed for hot salty bottom water from a Red Sea basin. *Nature*, 206: 1346–1347.

NEUMANN, A. C. and LAND, L. S., 1969. Algal production and lime mud deposition in the Bight of Abaco: a budget. *Geol. Soc. Am., Spec. Papers*, 121: 219 (abstract).

NEUMANN, A. C., GEBELEIN, C. D. and SCOFFIN, T. P., 1969. The composition, structure, and erodability of subtidal mats, Abaco, Bahamas. *Bull. Am. Assoc. Petrol. Geologists*, 53: 734 (abstract).

NEUMANN, A. C., GEBELEIN, C. D. and SCOFFIN, T. P., 1970. The composition, structure and erodability of subtidal mats, Abaco, Bahamas. *J. Sediment. Petrol.*, 40: 274–297.

NEWELL, N. D., 1951. Organic reefs and submarine dunes of oölite sand around Tongue of the Ocean, Bahamas. *Bull. Geol. Soc. Am.*, 62: 1466 (abstract).

NEWELL, N. D., 1955. Bahamian platforms. *Geol. Soc. Am., Spec. Papers*, 62: 303–315.

NEWELL, N. D., 1965. Classification of the Bivalvia. *Am. Museum Novitates*, 2206: 1–25.

NEWELL, N. D. and IMBRIE, J., 1955. Biogeological reconnaissance in the Bimini area, Great Bahama Bank. *Trans. N.Y. Acad. Sci., Ser. II*, 18: 3–14.

NEWELL, N. D. and RIGBY, J. K., 1957. Geological studies on the Great Bahama Bank. In: R. J. LE BLANC and J. G. BREEDING (Editors), *Regional Aspects of Carbonate Deposition— Soc. Econ. Paleontologists Mineralogists, Spec. Publ.*, 5: 15–72.

NEWELL, N. D., RIGBY, J. K., WHITEMAN, A. J. and BRADLEY, J. S., 1951. Shoal-water geology and environments, eastern Andros island, Bahamas. *Bull. Am. Museum Nat. Hist.*, 97(1): 1–29.

NEWELL, N. D., IMBRIE, J., PURDY, E. G. and THURBER, D. L., 1959. Organism communities and bottom facies, Great Bahama Bank. *Bull. Am. Museum Nat. Hist.*, 117(4): 177–228.

NEWELL, N. D., PURDY, E. G. and IMBRIE, J., 1960. Bahamian oolitic sand. *J. Geol.*, 68: 481–497.

NICHOLSON, H. A., 1886–1892. *A Monograph of the British Stromatoporoids*. Palaeontol. Soc., London, 234 pp.

NIELSEN, J., 1955. The kinetics of electrolyte precipitation. *J. Colloid Sci.*, 10: 576–586.

NIGGLI, P., 1952. *Gesteine und Minerallagerstätten*. 2. Birkhäuser. Basel, 557 pp.

NISSEN, H., 1963. Röntgengefügeanalyse am Kalzit von Echinodermenskeletten. *Neues Jahrb. Geol. Paläontol. Abhandl.*, 117: 230–234.

NOËL, D., 1965. *Sur les Coccolithes du Jurassique Européen et d'Afrique du Nord*. Centre Natl. Recherche Sci., Paris, 209 pp.

NORRIS, R. M., 1953. Buried oyster reefs in some Texas bays. *J. Paleontol.*, 27: 569–576.

NORTH, W. J., 1954. Size distribution, erosive activities and gross metabolic efficiency of the marine intertidal snails, *Littorina planaxis* and *L. scutulata*. *Biol. Bull.*, 106: 185–197.

NOVELLI, L. and MATTAVELLI, L., 1967. Fenomeni diagenetici in livelli di arenarie della formazione "Collesano" (Sicilia). *Rend. Soc. Mineral. Ital.*, 23: 333–350.

NUGENT, JR., L. E., 1946. Coral reefs in the Gilbert, Marshall, and Caroline Islands. *Bull. Geol. Soc. Am.*, 57: 735–779.

OBERLING, J. J., 1955. Shell structure of west American Pelecypoda. *J. Wash. Acad. Sci.*, 45: 128–130.

OBERLING, J. J., 1964. Observations on some structural features of the pelecypod shell. *Mitt. Naturforsch. Ges. Bern*, 20: 1–60.

ODUM, H. T., 1957. Biochemical deposition of strontium. *Inst. Marine Sci. Publ.*, 4: 38–114.

ODUM, H. T. and ODUM, E. P., 1955. Trophic structure and productivity of a windward coral reef community on Eniwetok Atoll. *Ecol. Monographs*, 25: 291–320.

OLDERSHAW, A. E. and SCOFFIN, T. P., 1967. The source of ferroan and non-ferroan calcite cements in the Halkin and Wenlock Limestones. *Geol. J.*, 5: 309–320.

OMORI, M., KOBAYASHI, I. and SHIBATA, M., 1962. Preliminary report on the shell structure of *Glycimeris vestita* (Dunker). *Sci. Rept. Tokyo Kyoiku Daigaku*, 8: 196–202.

OPPENHEIMER, C. H., 1960. Bacterial activity in sediments of shallow marine bays. *Geochim. Cosmochim. Acta*, 19: 244–260.

OPPENHEIMER, C. H., 1961. Note on the formation of spherical aragonitic bodies in the presence of bacteria from the Bahama Bank. *Geochim. Cosmochim. Acta*, 23: 295–296.

ORME, G. R. and BROWN, W. W. M., 1963. Diagenetic fabrics in the Avonian limestones of Derbyshire and North Wales. *Proc. Yorkshire Geol. Soc.*, 34: 51–66.

ORR, A. P., 1933. Physical and chemical conditions in the sea in the neighbourhood of the Great Barrier Reef. *Sci. Rept. Great Barrier Reef Expedition, Brit. Museum Nat. Hist.*, 2: 37–86.

ORR, A. P. and MOORHOUSE, F. W., 1933. Variations in some physical and chemical conditions on and near Low Isles Reef. *Sci. Rept. Great Barrier Reef Expedition, Brit. Museum Nat. Hist.*, 2: 87–98.

OSTWALD, W., 1897. Studien über die Bildung und Umwandlung fester Körper, 1. *Z. Physik. Chem. (Leipzig)*, 22: 289–330.

OTTEMANN, J. and KIRCHMAYER, M., 1967. Über Höhlenperlen und die Mikroanalyse von Ooiden mit der Elektronensonde. *Naturwissenschaften*, 14: 360–365.

OWEN, B. B. and BRINKLEY, S. R., 1941. Calculation of the effect of pressure upon ionic equilibria in pure water and in salt solutions. *Chem. Rev.*, 29: 461–474.

PARK, P. K., 1966. Deep-sea pH. *Science*, 154: 1540–1542.

PARK, P. K., 1968. Seawater hydrogen-ion concentration: vertical distribution. *Science*, 162: 357–358.

PARK, W. C. and SCHOT, E. H., 1968a. Stylolitization in carbonate rocks. In: G. MÜLLER and G. M. FRIEDMAN (Editors). *Recent Developments in Carbonate Sedimentology in Central Europe*. Springer, Berlin, pp.66–74.

PARK, W. C. and SCHOT, E. H., 1968b. Stylolites: their nature and origin. *J. Sediment. Petrol.*, 38: 175–191.

PARKINSON, D., 1957. Lower Carboniferous reefs of northern England. *Bull. Am. Assoc. Petrol. Geologists*, 41: 511–537.

PEISER, H. S., 1967. *Crystal Growth. Proceedings of an International Conference on Crystal Growth, Boston, 20–24 June 1966*. Pergamon, Oxford, 856 pp.

PETERSON, M. N. A., 1966. Calcite: rates of dissolution in a vertical profile in the central Pacific. *Science*, 154: 1542–1544.

PETERSON, M. N. A. and BIEN, G. S., 1963. Radiocarbon age determinations of Recent dolomite from Deep Spring Lake, California. *Trans. Am. Geophys. Union*, 44: 108 (abstract).

PETERSON, M. N. A., BIEN, G. S. and BERNER, R. A., 1963. Radiocarbon studies of Recent dolomite from Deep Spring Lake, California. *J. Geophys. Res.*, 68: 6493–6505.

PETERSON, M. N. A., VON DER BORCH, C. C. and BIEN, G. S., 1966. Growth of dolomite crystals. *Am. J. Sci.*, 264: 257–272.

PETTIJOHN, F. J., 1949. *Sedimentary Rocks*, 1 ed. Harper and Row, New York, N.Y., 526 pp.

PIA, J., 1926. *Pflanzen als Gesteinsbildner*. Borntraeger, Berlin, 355 pp.

PIA, J., 1933. *Die Rezenten Kalksteine*. Akademische Verlag, Leipzig, 420 pp.

PILKEY, O. H., 1964. The size distribution and mineralogy of the carbonate fraction of United States south Atlantic shelf and upper slope sediments. *Marine Geol.*, 2: 121–136.

PILKEY, O. H., 1966. Mineralogy of Tongue of the Ocean sediments. *J. Marine Res. (Sears Found. Marine Res.)*, 24: 276–285.

PILKEY, O. H. and GOODELL, H. G., 1963. Trace elements in Recent mollusk shells. *Limnol. Oceanog.*, 8: 137–148.

PILKEY, O. H. and GOODELL, H. G., 1964. Comparison of the composition of fossil and Recent mollusk shells. *Geol. Soc. Am. Bull.*, 75: 217–228.

PILKEY, O. H. and HOWER, J., 1960. The effect of environment on the concentration of skeletal magnesium and strontium in *Dendraster*. *J. Geol.*, 68: 203–216.

PILKEY, O. H. and NOBLE, D., 1966. Carbonate and clay mineralogy of the Persian Gulf. *Deep-Sea Res.*, 13: 1–16.

PINGITORE, N. E., 1970. Diagenesis and porosity modification in *Acropora palmata*, Pleistocene of Barbados, West Indies. *J. Sediment. Petrol.*, 40: 712–721.

POBEGUIN, T., 1954. Contribution à l'étude des carbonates de calcium, précipitation du calcaire par les végétaux, comparaison avec le monde animal. *Ann. Sci. Nat. Botan. Biol. Végétale*, 15: 29–109.

POSNJAK, E., 1938. The system $CaSO_4$–H_2O. *Am. J. Sci.*, 35A: 247–272.

POSNJAK, E., 1940. Deposition of calcium sulfate from sea water. *Am. J. Sci.* 238: 559–568.

PRAY, L. C., 1958. Fenestrate bryozoan core facies, Mississippian bioherms, southwestern United States. *J. Sediment. Petrol.*, 28: 261–273.

PRAY, L. C., 1960. Compaction in calcilutites. *Bull. Geol. Soc. Am.*, 71: 1946 (abstract).

PRAY, L. C., 1965. Limestone clastic dikes in Mississippian bioherms, New Mexico. *Geol. Soc. Am., Spec. Papers*, 82: 154–155 (abstract).

PRAY, L. C., 1966. Informal comments on calcium carbonate cementation. *Soc. Econ. Paleontologists Mineralogists, Tech. Session on Lithification and Diagenesis, St. Louis Meetings, April 1966*, not published.

PRAY, L. C., 1968. Hurricane Betsy (1965) and nearshore carbonate sediments of the Florida Keys. *Geol. Soc. Am., Spec. Papers*, 101: 168–169 (abstract).

PRAY, L. C., 1969. Micrite and carbonate cement: genetic factors in Mississippian bioherms. *J. Paleontol.*, 43: 895 (abstract).

PRAY, L. C. and CHOQUETTE, P. W., 1966. Genesis of carbonate reservoir facies. *Bull. Am. Assoc. Petrol. Geologists*, 50: 632 (abstract).

PRAY, L. C. and MURRAY, R. C., 1965. *Dolomitization and Limestone Diagenesis: a Symposium—Soc. Econ. Paleontologists Mineralogists, Spec. Publ.*, 13: 1–180.

PRAY, L. C. and WRAY, J. L., 1963. Porous algal facies (Pennsylvanian), Honaker Trail, San Juan Canyon, Utah. In: R. O. BASS and S. L. SHARPS (Editors), *Shelf Carbonates of the Paradox Basin: a Symposium—Four Corners Geol. Soc., Field Conf., 4th*, pp.204–234.

PROKOPOVITCH, N., 1952. The origin of stylolites. *J. Sediment. Petrol.*, 22: 212–220.

PRUNA, M., FAIVRE, R. and CHAUDRON, G., 1948. Étude cinétique par dilatométrie isotherme de la transformation de l'aragonite en calcite. *Compt. Rend.*, 227: 390–391.

PUFFER, E. L. and EMERSON, K., 1953. The molluscan community of the oyster-reef biotope on the central Texas coast. *J. Paleontol.*, 27: 537–544.

PUGH, W. E., 1950. *Bibliography of Organic Reefs, Bioherms, and Biostromes*. Seismograph Service Corp., Tulsa, Okla., 139 pp.

PURDY, E. G., 1961. Bahamian oölite shoals. In: J. A. PETERSON and J. C. OSMOND (Editors), *Geometry of Sandstone Bodies*. Am. Assoc. Petrol. Geologists, Tulsa, Okla., pp.53–62.

PURDY, E. G., 1963a. Recent calcium carbonate facies of the Great Bahama Bank. 1. Petrography and reaction groups. *J. Geol.*, 71: 334–355.

PURDY, E. G., 1963b. Recent calcium carbonate facies of the Great Bahama Bank. 2. Sedimentary facies. *J. Geol.*, 71: 472–497.

PURDY, E. G., 1964a. Sediments as substrates. In: J. IMBRIE and N. D. NEWELL (Editors), *Approaches to Paleoecology*. Wiley, New York, N.Y., pp.238–271.

PURDY, E. G., 1964b. Diagenesis of Recent marine carbonate sediments. *Bull. Am. Assoc. Petrol. Geologists*, 48: 542–543 (abstract).

PURDY, E. G., 1968. Carbonate diagenesis: an environmental survey. *Geol. Romana*, 7: 183–228.

PURDY, E. G. and IMBRIE, J., 1964. Carbonate sediments, Great Bahama Bank. *Guidebook for Field Trip No. 2. Geol. Soc. Am., Convention 1964*. Geol. Soc. Am., New York, N.Y., 66 pp.

PURSER, B. H., 1969. Syn-sedimentary marine lithification of Middle Jurassic limestones in the Paris Basin. *Sedimentology*, 12: 205–230.

PYTKOWICZ, R. M., 1963. Calcium carbonate and the *in situ* pH. *Deep-Sea Res.*, 10: 633–638.

PYTKOWICZ, R. M., 1965a. Rates of inorganic calcium carbonate nucleation. *J. Geol.*, 73: 196–199.

PYTKOWICZ, R. M., 1965b. Calcium carbonate saturation in the ocean. *Limnol. Oceanog.*, 10: 220–225.

PYTKOWICZ, R. M., 1967. Carbonate cycle and the buffer mechanism of recent oceans. *Geochim. Cosmochim. Acta*, 31: 63–73.

PYTKOWICZ, R. M., 1968. The carbon dioxide–carbonate system at high pressures in the oceans. *Ann. Rev. Oceanog. Marine Biol.*, 6: 83–135.

PYTKOWICZ, R. M., 1970. On the carbonate compensation depth in the Pacific. *Geochim. Cosmochim. Acta*, 34: 836–839.

PYTKOWICZ, R. M. and CONNERS, D. N., 1964. High pressure solubility of calcium carbonate in sea water. *Science*, 144: 840–841.

PYTKOWICZ, R. M. and FOWLER, G. A., 1967. Solubility of foraminifera in seawater at high pressures. *Geochem. J.*, 1: 169–182.

PYTKOWICZ, R. M., DISTECHE, A. and DISTECHE, S., 1967. Calcium carbonate solubility in seawater at *in situ* pressures. *Earth Planetary Sci. Letters*, 2: 430–432.

RADWANSKI, A., 1965. Procesy wciskowe wosadach klastycznych i oolitowych. (Pitting processes in clastic and oolitic sediments.) *Rocznik Polsk. Towarz. Geol. (Ann. Soc. Géol.
Pologne)*, 35: 179–210.

RAISWELL, R., 1971. Cementation in some Cambrian concretions, South Wales. In: O. P.
BRICKER (Editor), *Carbonate Cements*. Johns Hopkins Press, Baltimore, Md., pp. 196–197.

RAMSDEN, R. M., 1952. Stylolites and oil migration. *Bull. Am. Assoc. Petrol. Geologists*, 36:
2185–2186.

RANSON, G., 1955a. Observations sur la consolidation des sédiments calcaires dans les régions
tropicales; consolidation récente de spicules d'Alcyonaires. *Compt. Rend.*, 240:
329–331.

RANSON, G., 1955b. La consolidation des sédiments calcaires dans les régions tropicales. *Compt.
Rend.*, 240: 640–642.

RANSON, G., 1955c. Observations sur les principaux agents de la dissolution du calcaire sous-
marin dans la zone côtière des îles coralliennes de l'archipel des Tuamotu. *Compt. Rend.*,
240: 806–808.

RANSON, G., 1955d. Observations sur l'agent essentiel de la dissolution du calcaire dans les
régions exondées des îles coralliennes de l'archipel des Tuamotu. Conclusions sur le
processus de la dissolution du calcaire. *Compt. Rend.*, 240: 1007–1009.

RANSON, G., 1958. Coraux et récifs coralliens (Bibliographie). *Bull. Inst. Océanog.*, 1121:
1–80.

RAUP, D. M., 1959. Crystallography of echinoid calcite. *J. Geol.*, 67: 661–674.

RAUP, D. M., 1960. Ontogenetic variation in the crystallography of echinoid calcite. *J. Paleontol.*,
34: 1041–1050.

RAUP, D. M., 1962a. The phylogeny of calcite crystallography in echinoids. *J. Paleontol.*, 36:
793–810.

RAUP, D. M., 1962b. Crystallographic data in echinoderm classification. *Systematic Zool.*, 11:
97–108.

RAUP, D. M., 1965. Crystal orientations in the echinoid apical system. *J. Paleontol.*, 39: 934–951.

RAUP, D. M., 1966a. Crystallographic data for echinoid coronal plates. *J. Paleontol.*, 40: 555–568.

RAUP, D. M., 1966b. The endoskeleton [of the echinoderm]. In: R. A. BOOLOOTIAN (Editor),
Physiology of Echinodermata. Interscience, New York, N.Y., pp.379–395.

RAUP, D. M. and SWAN, E. F., 1967. Crystal orientation in the apical plates of aberrant echinoids.
Biol. Bull., 133: 618–629.

RAW, F., 1952. A note on Ross' "Ontogenies of Three Garden City (Early Ordovician) trilobites".
J. Paleontol., 26: 854–857.

REID, R. E. H., 1962a. Sponges and the Chalk Rock. *Geol. Mag.*, 99: 273–278.

REID, R. E. H., 1962b. Relationships of fauna and substratum in the palaeoecology of the Chalk
and the Chalk Rock. *Nature*, 194: 276–277.

REULING, H-T., 1934. Der Sitz der Dolomitisierung. Versuch einer neuen Auswertung der
Bohr-Ergebnisse von Funafuti. *Abhandl. Senckenberg. Naturforsch. Ges.*, 428: 1–44.

REVELLE, R., 1934. Physico-chemical factors affecting the solubility of calcium carbonate in sea
water. *J. Sediment. Petrol.*, 4: 103–110.

REVELLE, R. and EMERY, K. O., 1957. Chemical erosion of beach rock and exposed reef rock.
U.S., Geol. Surv., Profess. Papers, 260-T: 699–709.

REVELLE, R. and FAIRBRIDGE, R., 1957. Carbonates and carbon dioxide. *Geol. Soc. Am., Mem.*,
67(1): 239–295.

REVELLE, R. and FLEMING, R. H., 1934. The solubility product constant of calcium carbonate
in sea-water. *Proc. Pacific Sci. Congr. Pacific Sci. Assoc.*, 5th, Victoria, Vancouver, 1933,
3: 2089–2092.

RICHTER, R., 1933. Crustacea (Paläontologie). In: *Handwörterbuch der Naturwissenschaften*, 2.
Fischer, Jena, pp.840–864.

RIECKE, E., 1894. Zur Lehre von der Quellung. *Nachr. Akad. Wiss. Goettingen, Math. Physik.
Kl. IIa*, 1894: 1–29.

RIECKE, E., 1895. Ueber das Gleichgewicht zwischen einem festen homogen deformierten Koerper
und einer flüssigen Phase. *Ann. Physik*, 54: 731–738.

RIGBY, J. K., 1953. Some transverse stylolites. *J. Sediment. Petrol.*, 23: 265–271.

RIGBY, J. K. and McINTIRE, W. G., 1966. The Isla de Lobos and associated reefs, Veracruz, Mexico. *Brigham Young Univ. Res. Studies, Geol. Ser.*, 7: 3–46.

RILEY, J. P. and SKIRROW, G., 1965. *Chemical Oceanography, 1.* Academic Press, London, 712 pp.

ROBERTSON, E. C., 1965. Experimental consolidation of carbonate mud. In: L. C. PRAY and R. C. MURRAY (Editors), *Dolomitization and Limestone Diagenesis: a Symposium—Soc. Econ. Paleontologists Mineralogists, Spec. Publ.*, 13: 170 (abstract).

ROBINSON, R. B., 1967. Diagenesis and porosity development in Recent and Pleistocene oolites from southern Florida and the Bahamas. *J. Sediment. Petrol.*, 37: 355–364.

ROME, D. R., 1936. Note sur la microstructure de l'appareil tégumentaire de *Phacops* (ph.) *accipitrinus maretiolensis* R. + E. RICHTER. *Inst. Roy. Sci. Nat. Belg., Bull.*, 12(31): 1–7.

ROSE, P. R., 1970. Stratigraphic interpretation of submarine versus subaerial discontinuity surfaces: an example from the Cretaceous of Texas. *Geol. Soc. Am. Bull.*, 81: 2787–2797.

ROSENBERG, P. E., BURT, D. M. and HOLLAND, H. D., 1967. Calcite–dolomite–magnesite stability relations in solutions: the effect of ion strength. *Geochim. Cosmochim. Acta*, 31: 391–396.

ROSS, C. A. and OANA, S., 1961. Late Pennsylvanian and Early Permian limestone petrology and carbon isotope distribution, Glass Mountains, Texas. *J. Sediment. Petrol.*, 31: 231–244.

ROSS, J. P., 1960. Larger cryptostome bryozoa of the Ordovician and Silurian, Anticosti Island, Canada, I. *J. Paleontol.*, 34: 1057–1076.

ROSS, J. P., 1963. Lower Permian bryozoa from Western Australia. *Palaeontology*, 6: 70–82.

ROTHPLETZ, A., 1892. On the formation of oolite. *Am. Geologist*, 10: 279–282 (Transl. from *Botan. Zentralblatt*, 35(1892): 265–268).

RUBEY, W. W., 1933. Settling velocities of gravel, sand, and silt particles. *Am. J. Sci.*, 25: 325–338.

RUBINSON, M. and CLAYTON, R. N., 1969. Carbon-13 fractionation between aragonite and calcite. *Geochim. Cosmochim. Acta*, 33: 997–1002.

RUCKER, J. B., 1968. Carbonate mineralogy of sediments of Exuma Sound, Bahamas. *J. Sediment. Petrol.*, 38: 68–72.

RUCKER, J. B. and CARVER, R. E., 1969. A survey of the carbonate mineralogy of cheilostome Bryozoa. *J. Paleontol.*, 43: 791–799.

RUNNELLS, D. D., 1969. Diagenesis, chemical sediments, and the mixing of natural waters. *J. Sediment. Petrol.*, 39: 1188–1201.

RUPP, A., 1967. Origin, structure, and environmental significance of Recent and fossil calcispheres. *Geol. Soc. Am., Spec. Papers*, 101: 186 (abstract).

RUSNAK, G. A., 1960. Some observations of recent oolites. *J. Sediment. Petrol.*, 30: 471–480.

SANDER, B., 1936. Beiträge zur Kenntniss der Anlagerungsgefüge (rhythmische Kalke und Dolomite aus der Trias). *Mineral. Petrog. Mitt.*, 48: 27–209.

SANDER, B., 1951. *Contributions to the Study of Depositional Fabrics: Rhythmically Deposited Triassic Limestones and Dolomites.* Am. Assoc. Petrol. Geologists, Tulsa, Okla. 207 pp.

SARGENT, M. C. and AUSTIN, T. S., 1954. Biologic economy of coral reefs. Bikini and nearby atolls, Marshall Islands. *U.S., Geol. Surv., Profess. Papers*, 260-E: 293–300.

SASS, D. B., 1967. Electron microscopy, punctae, and the brachiopod genus *Syringothyris* Winchell, 1863. *J. Paleontol.*, 41: 1242–1246.

SASS, E., 1965. Dolomite–calcite relationships in sea water: theoretical considerations and preliminary experimental results. *J. Sediment. Petrol.*, 35: 339–347.

SASS, E., 1969. Microphotometric determination of preferred orientation in undeformed dolomites. *Science*, 165: 802–803.

SAYLOR, C. H., 1928. Calcite and aragonite. *J. Phys. Chem.*, 32: 1441–1460.

SCHENK, P. E., 1967. The Macumber Formation of the Maritime Provinces, Canada—a Mississippian analogue to the Recent strand-line carbonates of the Persian Gulf. *J. Sediment. Petrol.*, 37: 365–376.

SCHINDEWOLF, O. H., 1942a. Zur Kenntnis der Heterophylliden, einer eigentümlichen palaeozoischen Korallengruppe. *Palaeontol. Z.*, 22: 213–316.

SCHINDEWOLF, O. H., 1942b. Zur Kenntnis der Polycoelien und Plerophyllen. *Abhandl. Reichsamts Bodenforsch.* (N.F.), 204: 324 pp.

SCHLANGER, S. O., 1957. Dolomite growth in coralline algae. *J. Sediment. Petrol.*, 27: 181–186.

SCHLANGER, S. O., 1963. Subsurface geology of Eniwetok Atoll. *U.S., Geol. Surv., Profess. Papers,* 260 BB: 991–1066.

SCHLANGER, S. O., 1964. Petrology of the limestones of Guam. *U.S., Geol. Surv., Profess. Papers,* 403-D: 1–52.

SCHMALZ, R. F. 1963. Role of surface energy in carbonate precipitation. *Geol. Soc. Am., Spec. Papers,* 76: 144–145 (abstract).

SCHMALZ, R. F., 1965. Brucite in carbonate secreted by the red alga *Goniolithon* sp. *Science,* 149: 993–996.

SCHMALZ, R. F., 1967. Kinetics and diagenesis of carbonate sediments. *J. Sediment. Petrol.,* 37: 60–67.

SCHMALZ, R. F., 1971. Formation of beachrock at Eniwetok Atoll. In: O. P. BRICKER (Editor), *Carbonate Cements.* Johns Hopkins Press, Baltimore, Md., pp. 17–24.

SCHMALZ, R. F. and CHAVE, K. E., 1963. Calcium carbonate: factors affecting saturation in ocean waters off Bermuda. *Science,* 139: 1206–1207.

SCHMIDEGG, O., 1928. Über geregelte Wachstumsgefüge. *Jahrb. Geol. Bundesanstalt (Austria),* 78: 1–52.

SCHMIDT, V., 1965. Facies, diagenesis, and related reservoir properties in the Gigas Beds (Upper Jurassic), northwestern Germany. In: L. C. PRAY and R. C. MURRAY (Editors), *Dolomitization and Limestone Diagenesis: a Symposium—Soc. Econ. Paleontologists Mineralogists, Spec. Publ.,* 13: 124–168.

SCHMIDT, W. J., 1921. Bau und Bildung der Perlmuttermasse. *Verhandl. Deut. Zool. Ges. (Leipzig),* 26: 59–60.

SCHMIDT, W. J., 1924a. Bau und Bildung der Perlmuttermasse. *Zool. Jahrb., Abt. Anat. Ontog. Tiere,* 45: 1–148.

SCHMIDT, W. J., 1924b. *Die Bausteine des Tierkörpers in polarisiertem Lichte.* Cohen, Bonn, 528 pp.

SCHMIDT, W. J., 1929. Bestimmung der Lage der optischen Achse in Biokristallen. In: E. ABDERHALDEN (Editor), *Handbuch der biologischen Arbeitsmethoden.* Urban and Schwarzenberg, Berlin, lief. 289, abt. 5,2: 1357–1400.

SCHNEIDER, E. D. and HEEZEN, B. C., 1966. Sediments of the Caicos outer ridge, the Bahamas. *Geol. Soc. Am. Bull.,* 77: 1381–1398.

SCHOLL, D. W., 1963. Sedimentation in modern coastal swamps, southwestern Florida. *Bull. Am. Assoc. Petrol. Geologists,* 47: 1581–1603.

SCHOLL, D. W., 1964. Recent sedimentary record in mangrove swamps and rise in sea level over the southwestern coast of Florida, 1. *Marine Geol.,* 1: 344–366.

SCHOLL, D. W., 1966. Florida Bay: a modern site of limestone formation. In: R. W. FAIRBRIDGE (Editor), *The Encyclopedia of Oceanography.* Reinhold, New York, N.Y., pp.282–288.

SCHOPF, T. J. M. and MANHEIM, F. T., 1967. Chemical composition of Ectoprocta (Bryozoa). *J. Paleontol.,* 41: 1197–1225.

SCHOUPPÉ, A. and STACUL, P., 1955. Die Genera *Verbeekiella* PENECKE, *Timorphyllum* GERTH, *Wannerophyllum* N. Gen., *Lophophyllidium* GRABAU aus dem Perm von Timor. *Palaeontographica, Supplementbände,* IV: 197–359.

SCHROEDER, J. H., 1969. Experimental dissolution of calcium, magnesium, and strontium from Recent biogenic carbonates: a model of diagenesis. *J. Sediment. Petrol.,* 39: 1057–1073.

SCHROEDER, J. H. and SIEGEL, F. R., 1969. Experimental dissolution of calcium, magnesium, and strontium from Holocene biogenic carbonates: a model of diagenesis. *Bull. Am. Assoc. Petrol. Geologists,* 53: 741 (abstract).

SCHROEDER, J. H., DWORNIK, E. J. and PAPIKE, J. J., 1969. Primary protodolomite in echinoid skeletons. *Geol. Soc. Am. Bull.,* 80: 1613–1616.

SCHWARZACHER, W., 1961. Petrology and structure of some Lower Carboniferous reefs in northwestern Ireland. *Bull. Am. Assoc. Petrol. Geologists,* 45: 1481–1503.

SCOFFIN, T. P., 1968. An underwater flume. *J. Sediment. Petrol.,* 38: 244–246.

SCOFFIN, T. P., 1970. The trapping and binding of subtidal carbonate sediments by marine vegetation in Bimini Lagoon, Bahamas. *J. Sediment. Petrol.,* 40: 249–273.

SEELIGER, R., 1956. Über mikroskopische Darstellung dichter Gesteine mit Hilfe von Oberflächenabdrücken. *Geol. Rundschau,* 45: 332–336.

SEIBOLD, E., 1962a. Untersuchungen zur Kalkfällung und Kalklösung am Westrand der Great Bahama Bank. *Sedimentology*, 1: 50–74.

SEIBOLD, E., 1962b. Das Korallenriff als geologisches Problem. *Naturw. Rundschau*, 15: 357–363.

SHARP, W. E. and KENNEDY, G. C., 1965. The system CaO–CO_2–H_2O in the two-phase region calcite + aqueous solution. *J. Geol.*, 73: 391–403.

SHAUB, B. M., 1939. The origin of stylolites. *J. Sediment. Petrol.*, 9: 47–61.

SHAUB, B. M., 1949. Do stylolites develop before or after the hardening of the enclosing rock. *J. Sediment. Petrol.*, 19: 26–36.

SHAUB, B. M., 1953. Stylolites and oil migration. *J. Sediment. Petrol.*, 23: 260–264.

SHAW, A. B., 1964. *Time in Stratigraphy*. McGraw-Hill, New York, N.Y., 365 pp.

SHEARMAN, D. J., 1963. Recent anhydrite, gypsum, dolomite and halite from the coastal flats of the Arabian shore of the Persian Gulf. *Proc. Geol. Soc. London*, 1607: 63–65.

SHEARMAN, D. J., 1966. Origin of marine evaporites by diagenesis. *Trans. Inst. Mining Met. (B)*, 75: 208–215.

SHEARMAN, D. J. and SHIRMOHAMMADI, N. H., 1969. Distribution of strontium in dedolomites from the French Jura. *Nature*, 223: 606–608.

SHEARMAN, D. J. and SKIPWITH, P. A. D'E., 1965. Organic matter in Recent and ancient limestones and its role in their diagenesis. *Nature*, 208: 1310–1311.

SHEARMAN, D. J., KHOURI, J. and TAHA, S., 1961. On the replacement of dolomite by calcite in some Mesozoic limestones from the French Jura. *Proc. Geologists' Assoc. (Engl.)*, 72: 1–12.

SHEARMAN, D. J., TWYMAN, J. and KARIMI, M. Z., 1970. The genesis and diagenesis of oolites. *Proc. Geologists' Assoc. (Engl.)*, 81: 561–575.

SHINN, E. A., 1963. Spur and groove formation on the Florida reef tract. *J. Sediment. Petrol.*, 33: 291–303.

SHINN, E. A., 1964. Recent dolomite, Sugarloaf Key. In: *Guidebook for Field Trip No. 1. Geol. Soc. Am., Convention 1964*. Geol. Soc. Am., New York, N.Y., pp.62–67.

SHINN, E. A., 1968a. Practical significance of birdseye structures in carbonate rocks. *J. Sediment. Petrol.*, 38: 215–223.

SHINN, E. A., 1968b. Selective dolomitization of recent sedimentary structures. *J. Sediment. Petrol.*, 38: 612–616.

SHINN, E. A., 1968c. Burrowing in recent lime sediments of Florida and the Bahamas. *J. Paleontol.*, 42: 879–894.

SHINN, E. A., 1969. Submarine lithification of Holocene carbonate sediments in the Persian Gulf. *Sedimentology*, 12: 109–144.

SHINN, E. A. and GINSBURG, R. N., 1964. Formation of Recent dolomite in Florida and the Bahamas. *Bull. Am. Assoc. Petrol. Geologists*, 48: 547 (abstract).

SHINN, E. A., GINSBURG, R. N. and LLOYD, R. M., 1965a. Recent supratidal dolomitization in Florida and the Bahamas. *Geol. Soc. Am., Spec. Papers*, 82: 183–184 (abstract).

SHINN, E. A., GINSBURG, R. N. and LLOYD, R. M., 1965b. Recent supratidal dolomite from Andros Island, Bahamas. In: L. C. PRAY and R. C. MURRAY (Editors), *Dolomitization and Limestone Diagenesis: a Symposium—Soc. Econ. Paleontologists Mineralogists, Spec. Publ.*, 13: 112–123.

SHINN, E. A., LLOYD, R. M. and GINSBURG, R. N., 1969. Anatomy of a modern carbonate tidal-flat, Andros Island, Bahamas. *J. Sediment. Petrol.*, 39: 1202–1228.

SHIRMOHAMMADI, N. H. and SHEARMAN, D. J., 1966. On the distribution of strontium in some dolomitized and dedolomitized limestones from the French Jura. *Mineral. Mag.*, 35: lxxii (abstract).

SHOJI, R. and FOLK, R. L., 1964. Surface morphology of some limestone types as revealed by electron microscope. *J. Sediment. Petrol.*, 34: 144–155.

SHTERNINA, E. B. and FROLOVA, E. V., 1952. The solubility of calcite in the presence of CO_2 and NaCl. *Izv. Sektora Fiz. Khim. Analiza, Inst. Obshch. Neorgan. Khim., Akad. Nauk S.S.S.R.*, 21: 271–287 (Transl. from Russian).

SIEGEL, F. R., 1960. The effect of strontium on the aragonite–calcite ratios of Pleistocene corals. *J. Sediment. Petrol.*, 30: 297–304.

SIEGEL, F. R., 1961. Variations of Sr/Ca ratios and Mg contents in Recent carbonate sediments of the northern Florida Keys area. *J. Sediment. Petrol.*, 31: 336–342.

SIEGEL, F. R., 1965. Aspects of calcium carbonate deposition in Great Onyx Cave, Kentucky. *Sedimentology*, 4: 285–299.

SIEGEL, F. R. and REAMS, M. W., 1966. Temperature effect on precipitation of calcium carbonate from calcium bicarbonate solutions and its application to cavern environments. *Sedimentology*, 7: 241–248.

SIEVER, R., BECK, K. C. and BERNER, R. A., 1965. Composition of interstitial waters of modern sediments. *J. Geol.*, 73: 39–73.

SIMKISS, K., 1964. Variations in the crystalline form of calcium carbonate precipitated from artificial sea water. *Nature*, 201: 492–493.

SIPPEL, R. F., 1968. Sandstone petrology, evidence from luminescence petrography. *J. Sediment. Petrol.*, 38: 530–554.

SIPPEL, R. F. and GLOVER, E. D., 1964. The solution alteration of carbonate rocks, the effects of temperature and pressure. *Geochim. Cosmochim. Acta*, 28: 1401–1417.

SIPPEL, R. F. and GLOVER, E. D., 1965. Structures in carbonate rocks made visible by luminescence petrography. *Science*, 150: 1283–1287.

SKEATS, E. W., 1902. The chemical composition of limestones from upraised coral islands, with notes on their microscopical structures. *Bull. Museum Comp. Zool. Harvard Coll.*, 42: 53–126.

SKINNER, H. C. W., 1963. Precipitation of calcian dolomites and magnesian calcites in the southeast of South Australia. *Am. J. Sci.*, 261: 449–472.

SKINNER, H. C. W., SKINNER, B. J. and RUBIN, M., 1963. Age and accumulation rate of dolomite-bearing carbonate sediments in South Australia. *Science*, 139: 335–336.

SKIPWITH, P. A. D'E., 1966. *Recent Carbonate Sediments of the Eastern Khor al Bazam, Abu Dhabi, Trucial Coast*. Thesis, Imp. Coll. Sci. Technol., London, 407 pp., unpublished.

SKIRROW, G., 1965. The dissolved gases—carbon dioxide. In: J. P. RILEY and G. SKIRROW (Editors), *Chemical Oceanography, 1*. Academic Press, London, pp.227–322.

SKOUGSTAD, M. W. and HORR, C. A., 1963. Occurrence and distribution of strontium in natural water. *U.S., Geol. Surv., Water Supply Paper*, 1496-D: 55–97.

SMITH, C. L., 1940. The Great Bahama Bank. 1. General hydrographic and chemical factors. 2. Calcium carbonate precipitation. *J. Marine Res. (Sears Found. Marine Res.)*, 3: 1–31; 147–189.

SMITH, C. L., 1941. The solubility of calcium carbonate in tropical sea water. *J. Marine Biol. Assoc. U.K.*, 25: 235–242.

SMITH, J. B., TATSUMOTO, M. and HOOD, D. W., 1960. Carbamino carboxylic acids in photosynthesis. *Limnol. Oceanog.*, 5: 425–431.

SMITH, S. V., DYGAS, J. A. and CHAVE, K. E., 1968. Distribution of calcium carbonate in pelagic sediments. *Marine Geol.*, 6: 391–400.

SOKAL, R. R. and MICHENER, C. D., 1958. A statistical method for evaluating systematic relationships. *Univ. Kansas Sci. Bull.*, 38: 1409–1438.

SORBY, H. C., 1879. The structure and origin of limestones. *Proc. Geol. Soc. London*, 35: 56–95.

SPOTTS, J. H., 1952. *X-ray Studies and Differential Thermal Analyses of some Coastal Limestones and Associated Carbonates of Western Australia*. Thesis, Univ. Western Australia, Perth, W. Austr., 23 pp.

SPRY, A., 1969. *Metamorphic Textures*. Pergamon, Oxford, 350 pp.

SQUIRES, D. F., 1958. Stony corals from the vicinity of Bimini, Bahamas, British West Indies. *Bull. Am. Museum Nat. Hist.*, 115: 217–262.

SQUIRES, D. F., 1964. Fossil coral thickets in Wairarapa, New Zealand. *J. Paleontol.*, 38: 904–915.

SQUIRES, D. F., 1965. Deep-water coral structure on the Campbell Plateau, New Zealand. *Deep-Sea Res.*, 12: 785–788.

STANLEY, S. M., 1966. Paleoecology and diagenesis of Key Largo Limestone, Florida. *Bull. Am. Assoc. Petrol. Geologists*, 50: 1927–1947.

STANTON, R. O., 1963. Upper Devonian calcispheres from Redwater and South Sturgeon Lake reefs, Alberta, Canada. *Bull. Can. Petrol. Geologists*, 11: 410–418.

STEARN, C., 1966. The microstructure of stromatoporoids. *Palaeontology*, 9: 74–124.

STEHLI, F. G., 1956. Shell mineralogy in Paleozoic invertebrates. *Science*, 123: 1031–1032.

STEHLI, F. G. and HOWER, J., 1961. Mineralogy and early diagenesis of carbonate sediments. *J. Sediment. Petrol.*, 31: 358–371.

STENZEL, H. B., 1962. Aragonite in the resilium of oysters. *Science*, 136: 1121–1122.

STENZEL, H. B., 1963. Aragonite and calcite as constituents of adult oyster shells. *Science*, 142: 232–233.

STENZEL, H. B., 1964. Living *Nautilus*. In: R. C. MOORE (Editor), *Treatise on Invertebrate Paleontology. K. Mollusca 3*. Geol. Soc. Am., Boulder, Colo., pp.59–93.

STERNBERG, T. E., FISCHER, A. G. and HOLLAND, H. D., 1959. Strontium content of calcites from the Steinplatte Reef complex, Austria. *Bull. Geol. Soc. Am.*, 70: 1681 (abstract).

STETSON, T. R., SQUIRES, D. F. and PRATT, R. M., 1962. Coral banks occurring in deep water on the Blake Plateau. *Am. Museum Novitates*, 2114: 1–39.

ST. JEAN, JR., J., 1964. Maculate tissue in Stromatoporoidea. *Geo'. Soc. Am., Spec. Papers*, 76: 143 (abstract).

ST. JEAN, JR., J., 1967. Maculate tissue in Stromatoporoidea. *Micropaleontology*, 13: 419–444.

STOCKDALE, P. B., 1922. Stylolites: their nature and origin. *Indiana, Univ. Studies*, 9: 1–97.

STOCKDALE, P. B., 1926. The stratigraphic significance of solution in rocks. *J. Geol.*, 34: 399–414.

STOCKDALE, P. B., 1936. Rare stylolites. *Am. J. Sci.*, 32: 129–133.

STOCKDALE, P. B., 1943. Stylolites: primary or secondary? *J. Sediment. Petrol.*, 13: 3–12.

STOCKMAN, K. W., GINSBURG, R. N. and SHINN, E. A., 1967. The production of lime mud by algae in south Florida. *J. Sediment. Petrol.*, 37: 633–648.

STODDART, D. R., 1962a. Physiographic studies on the British Honduras reefs and cays. *Geograph. J.*, 128: 161–171.

STODDART, D. R., 1962b. Three Caribbean atolls: Turneffe Islands, Lighthouse Reef and Glover's Reef, British Honduras. *Atoll. Res. Bull.*, 87: 1–151.

STODDART, D. R., 1962c. Catastrophic storm effects on the British Honduras reefs and cays. *Nature*, 196: 512–515.

STODDART, D. R., 1964. Carbonate sediments of Half Moon Cay, British Honduras. *Atoll. Res. Bull.*, 104: 1–16.

STODDART, D. R., 1965a. Re-survey of hurricane effects on the British Honduras reefs and cays. *Nature*, 207: 589–592.

STODDART, D. R., 1965b. The shape of atolls. *Marine Geol.*, 3(5): 369–383.

STODDART, D. R., 1969. Ecology and morphology of Recent coral reefs. *Biol. Rev. Cambridge Phil. Soc.*, 44: 433–498.

STODDART, D. R. and CANN, J. R., 1965. Nature and origin of beach rock. *J. Sediment. Petrol.*, 35: 243–247.

STOLKOWSKI, J., 1951. Essai sur le déterminisme des formes minéralogiques du calcaire chez les êtres vivants (calcaires coquilliers). *Ann. Inst. Océanog. (Paris)*, 26: 2–113.

STØRMER, L., 1930. Scandinavian Trinucleidae, with special reference to Norwegian species and varieties. *Norg. Videnskaps Akad. Oslo, Mat. Naturw. Kl.*, 4: 1–111.

STORR, J. F., 1964. Geology and oceanography of the coral-reef tract, Abaco Island, Bahamas. *Geol. Soc. Am., Spec. Papers*, 79: 1–98.

STROMEYER, F., 1813. De arragonite eiusque differentia a spatho calcareo rhomboidali chemica. *Comment. Soc. Reg. Sci. Göttingensis*, 2: 1–36.

SUESS, E., 1968. *Calcium Carbonate Interaction with Organic Compounds*. Thesis, Marine Sci. Center, Lehigh Univ., Bethlehem, Pa., 153 pp., unpublished.

SUESS, E., 1970. Interaction of organic compounds with calcium carbonate. I. Association phenomena and geochemical implications. *Geochim. Cosmochim. Acta*, 34: 157–168.

SUGDEN, W., 1963a. Some aspects of sedimentation in the Persian Gulf. *J. Sediment. Petrol.*, 33: 355–364.

SUGDEN, W., 1963b. The hydrology of the Persian Gulf and its significance in respect to evaporite deposition. *Am. J. Sci.*, 261: 741–755.

SUGIURA, Y., IBERT, E. R. and HOOD, D. W., 1963. Mass transfer of carbon dioxide across sea surfaces. *J. Marine Res., (Sears Found. Marine Res.)*, 21: 11–24.

SUMMERSON, C. H. and CAMPBELL, L. J., 1958. Holothurian sclerites from the Kendrick Shale of eastern Kentucky. *J. Paleontol.*, 32: 961–969.

SUNAGAWA, I., 1953. Variation of crystal habit of calcite with special reference to the relation between crystal habit and crystallization stage. *Rept. Geol. Surv. Japan*, 155: 1–66.

SVERDRUP, H. U., JOHNSON, M. W. and FLEMING, R. H., 1942. *The Oceans, their Physics, Chemistry and General Biology*. Prentice-Hall, New York, N.Y., 1087 pp.

SWALLOW, J. C. and CREASE, J., 1965. Hot salty water at the bottom of the Red Sea. *Nature*, 205: 165–166.

SWINCHATT, J. P., 1965. Significance of constituent composition, texture, and skeletal breakdown in some Recent carbonate sediments. *J. Sediment. Petrol.*, 35: 71–90.

SWINCHATT, J. P., 1969. Algal boring: a possible depth indicator in carbonate rocks and sediments. *Geol. Soc. Am. Bull.*, 80: 1391–1396.

SWITZER, G. and BOUCOT, A. J., 1955. The mineral composition of some microfossils. *J. Paleontol.*, 29: 525–533.

TABB, D. C., DUBROW, D. L. and MANNING, R. B., 1962. The ecology of northern Florida Bay and adjacent estuaries. *Florida Geol. Surv., Tech. Ser.*, 39: 1–81.

TABER, S., 1916. The growth of crystals under external pressure. *Am. J. Sci.*, 4: 532–556.

TAFT, W. H., 1967. Modern carbonate sediments. In: G. V. CHILINGAR, H. J. BISSELL and R. W. FAIRBRIDGE (Editors), *Carbonate Rocks, Origin, Occurrence and Classification*. Elsevier, Amsterdam, pp.29–50.

TAFT, W. H., 1968. Yellow Bank, Bahamas: a study of modern marine carbonate lithification. *Bull. Am. Assoc. Petrol. Geologists*, 52: 551 (abstract).

TAFT, W. H. and HARBAUGH, J. W., 1964. Modern carbonate sediments of southern Florida, Bahamas, and Espíritu Santo Island, Baja California: a comparison of their mineralogy and chemistry. *Stanford Univ. Publ., Univ. Ser., Geol. Sci.*, 8(2): 1–133.

TAFT, W. H., ARRINGTON, F., HAIMOVITZ, A., MACDONALD, C. and WOOLHEATER, C., 1968. Lithification of modern carbonate sediments at Yellow Bank, Bahamas. *Bull. Marine Sci. Gulf Caribbean*, 18: 762–828.

TARUTANI, T., CLAYTON, R. N. and MAYEDA, T. K., 1969. The effect of polymorphism and magnesium substitution on oxygen isotope fractionation between calcium carbonate and water. *Geochim. Cosmochim. Acta*, 33: 987–996.

TATSUMOTO, M., WILLIAMS, W T., PRESCOTT, J. M. and HOOD, D. W., 1961. Amino acids in samples of surface sea water. *J. Marine Res. (Sears Found. Marine Res.)*, 19: 89–95.

TAVENER-SMITH, R., 1969. Skeletal structure and growth in the Fenestellidae (Bryozoa). *Palaeontology*, 12: 281–309.

TAYLOR, J. D., 1968. Coral reef and associated invertebrate communities (mainly molluscan) around Mahé, Seychelles. *Phil. Trans. Roy. Soc. London, Ser. B.*, 254: 129–206.

TAYLOR, J. D., KENNEDY, W. J. and HALL, A., 1969. The shell structure and mineralogy of the Bivalvia. Introduction Nuculacea–Trigonacea. *Bull. Brit. Museum Zool., Suppl.*, 3: 1–125.

TAYLOR, J. M. C. and ILLING, L. V., 1969. Holocene intertidal calcium carbonate cementation, Qatar, Persian Gulf. *Sedimentology*, 12: 69–107.

TAYLOR, J. M. C. and ILLING, L. V., 1971. Alteration of Recent aragonite to magnesian calcite cement, Qatar, Persian Gulf. In: O. P. BRICKER (Editor), *Carbonate Cements*. Johns Hopkins Press, Baltimore, Md., pp. 36–39.

TEBBUTT, G. E., 1967. *Diagenesis of Pleistocene Limestone on Ambergris Cay, British Honduras*. Thesis, Rice Univ., Houston, Texas, 133 pp., unpublished.

TEBBUTT, G. E., 1969. Diagenesis of Pleistocene limestone on Ambergris Cay, British Honduras. *Bull. Am. Assoc. Petrol. Geologists*, 53: 745 (abstract).

TEBBUTT, G. E., CONLEY, C. D. and BOYD, D. W., 1965. Lithogenesis of a distinctive carbonate rock fabric. In: R. B. PARKER (Editor), *Contributions to Geology*. Univ. Wyoming, Laramie, Wyo., pp.1–13.

TEICHERT, C., 1958. Cold- and deep-water coral banks. *Bull. Am. Assoc. Petrol. Geologists*, 42: 1064–1082.

TEICHERT, C., 1964. Morphology of hard parts. In: R. C. MOORE (Editor), *Treatise on Invertebrate Paleontology. K. Mollusca 3*. Geol. Soc. Am., Boulder, Colo., pp.13–59.

THOMAS, C., 1965. Origin of pisolites. *Bull. Am. Assoc. Petrol. Geologists*, 49: 360 (abstract).

THOMAS, L. P., MOORE, D. R. and WORK, R. C., 1961. Effects of hurricane Donna on the turtle grass beds of Biscayne Bay, Florida. *Bull. Marine Sci. Gulf Caribbean*, 11: 191–197.

THOMPSON, G. and BOWEN, V. T., 1969. Analyses of coccolith ooze from the deep tropical Atlantic. *J. Marine Res. (Sears Found. Marine Res.)*, 27: 32–38.

THOMPSON, G., BOWEN, V. T., MELSON, W. G. and CIFELLI, R., 1968. Lithified carbonates from the deep-sea of the equatorial Atlantic. *J. Sediment. Petrol.*, 38: 1305–1312.

THOMPSON, T. G. and CHOW, T. J., 1955. The strontium–calcium atom ratio in carbonate-secreting marine organisms. *Deep-Sea Res., Suppl.*, 3: 20–39.

THOMSON, A., 1959. Pressure solution and porosity. In: H. A. IRELAND (Editor), *Silica in Sediments*. Soc. Econ. Paleontologists Mineralogists, Tulsa, Okla., pp.92–110.

THOMSON, J., 1862. On crystallization and liquefaction, as influenced by stresses tending to change of form in crystals. *Phil. Mag.*, 24: 395–401.

THORP, E. M., 1936. Calcareous shallow-water marine deposits of Florida and the Bahamas. *Papers Tortugas Lab., Carnegie Inst. Wash. Publ.*, 452: 37–143.

THORSTENSON, D. C., 1969. Pore-water chemistry of carbonate sediments from Harrington Sound, Bermuda. *Bull. Am. Assoc. Petrol. Geologists*, 53: 746 (abstract).

THRAILKILL, J., 1968. Chemical and hydrologic factors in the excavation of limestone caves. *Geol. Soc. Am. Bull.*, 79: 19–45.

TOWE, K. M., 1967. Echinoderm calcite: single crystal or polycrystalline aggregate. *Science*, 157: 1048–1050.

TOWE, K. M. and CIFELLI, R., 1967. Wall ultrastructure in the calcareous foraminifera: crystallographic aspects and a model for calcification. *J. Paleontol.*, 41: 742–762.

TOWE, K. M. and HAMILTON, G. H., 1968a. Ultrastructure and inferred calcification of the mature and developing nacre in bivalve mollusks. *Calcified Tissue Res.*, 1: 306–318.

TOWE, K. M. and HAMILTON, G. H., 1968b. Ultramicrotome-induced deformation artifacts in densely calcified material. *J. Ultrastruct. Res.*, 22: 274–281.

TOWE, K. M. and HARPER, JR., C. W., 1966. Pholidostrophiid brachiopods: origin of the nacreous luster. *Science*, 154: 153–155.

TOWE, K. M. and MALONE, P. G., 1970. Precipitation of metastable carbonate phases from sea-water. *Nature*, 226: 348–349.

TRACEY, JR., J. I., LADD, H. S. and HOFFMEISTER, J. E., 1948. Reefs of Bikini, Marshall Islands. *Bull. Geol. Soc. Am.*, 59: 861–887.

TRAGANZA, E. D., 1967. Dynamics of the carbon dioxide system on the Great Bahama Bank. *Bull. Marine Sci. Gulf Caribbean*, 17: 348–366.

TRICHET, J., 1968. Étude de la composition de la fraction organique des oolites. Comparaison avec celle des membranes des bactéries et des cyanophycées. *Compt. Rend.*, 267: 1492–1494.

TRURNIT, P., 1967. Morphologie und Entstehung diagenetischer Druck-Lösungserscheinungen. *Geol. Mitt.*, 7: 173–204.

TRURNIT, P., 1968a. Druck-Lösungsstadien innerhalb der Entwicklung einer Geosynklinale. *Neues Jahrb. Geol. Paläontol., Monatsh.*, 6: 376–384.

TRURNIT, P., 1968b. Analysis of pressure solution contacts and classification of pressure-solution phenomena. In: G. MÜLLER and G. M. FRIEDMAN (Editors), *Recent Developments in Carbonate Sedimentology in Central Europe*. Springer, Berlin, pp.75–84

TRURNIT, P., 1968c. Pressure-solution phenomena in detrital rocks. *Sediment. Geol.*, 2: 89–114.

TSUJII, T., 1960. Studies on the mechanism of shell- and pearl-formation in Mollusca. *J. Fac. Fisheries, Prefect. Univ. Mie*, 5: 1–70.

TSUJII, T., SHARP, D. G. and WILBUR, K. M., 1958. The submicroscopic structure of the shell of the oyster *Crassostrea virginica*. *J. Biophys. Biochem. Cytol.*, 4: 275–279.

TUREKIAN, K. K., 1957. Salinity variations in sea water in the vicinity of Bimini, Bahamas, British West Indies. *Am. Museum Novitates*, 1822: 1–12.

TUREKIAN, K. K. and ARMSTRONG, R. L., 1960. Magnesium, strontium, and barium concentrations and calcite–aragonite ratios of some recent molluscan shells. *J. Marine Res. (Sears Found. Marine Res.)*, 18: 133–151.

TUREKIAN, K. K. and ARMSTRONG, R. L., 1961. Chemical and mineralogical composition of fossil molluscan shells from the Fox Hills Formation, South Dakota. *Geol. Soc. Am. Bull.*, 72: 1817–1828.

TUREKIAN, K. K. and KULP, J. L., 1956. The geochemistry of strontium. *Geochim. Cosmochim. Acta*, 10: 245–296.

TURMEL, R. and SWANSON, R., 1964. Rodriguez Bank. In: *Guidebook for Field Trip No. 1. Geol. Soc. Am., Convention 1964*. Geol. Soc. Am., New York, N.Y., pp.26–33.

TURNER, F. J., 1949. Preferred orientation of calcite in Yule marble. *Am. J. Sci.*, 247: 593–621.

ULRICH, E. O., 1890. Paleozoic bryozoa. *Illinois State Geol. Surv., Circ.*, 8: 283–688.

UMBGROVE, J. H. F., 1947. Coral reefs of the East Indies. *Bull. Geol. Soc. Am.*, 58: 729–777.

UREY, H. C., LOWENSTAM, H. A., EPSTEIN, S. and MCKINNEY, C. R., 1951. Measurements of paleotemperatures and temperatures of the Upper Cretaceous of England, Denmark, and the southeastern United States. *Bull. Geol. Soc. Am.*, 62: 399–416.

USDOWSKI, H.-E., 1962. Die Entstehung der kalkoolithischen Fazies des norddeutschen Unteren Buntsandsteins. *Beitr. Mineral. Petrog.*, 8: 141–179.

USDOWSKI, H.-E., 1963. Der Rogenstein des norddeutschen Unteren Buntsandsteins, ein Kalk-oölith des marinen Faziesbereichs. *Fortschr. Geol. Rheinland Westfalen*, 10: 337–342.

VAN DER MERWE, J. H., 1949. Misfitting monolayers and oriented overgrowth. *Discussions Faraday Soc.*, 5: 201–214.

VAN STRAATEN, L. M. J. U., 1957. Recent sandstones on the coasts of the Netherlands and of the Rhône delta. *Geol. Mijnbouw*, 19: 196–213.

VAN STRAATEN, L. M. J. U. and KUENEN, PH. H., 1958. Tidal action as a cause of clay accumulation. *J. Sediment. Petrol.*, 28: 406–413.

VATAN, A., 1947. Remarques sur la silicification. *Compt. Rend. Soc. Géol. France*, 5: 99–101.

VAUGHAN, T. W., 1910. A contribution to the geologic history of the Floridian plateau. *Papers Tortugas Lab., Carnegie Inst. Wash. Publ.*, 133: 99–185.

VAUGHAN, T. W., 1914a. Building of the Marquesas and Tortugas Atolls and a sketch of the geologic history of the Florida reef tract. *Papers Tortugas Lab., Carnegie Inst. Wash. Publ.*, 182: 55–67.

VAUGHAN, T. W., 1914b. Preliminary remarks on the geology of the Bahamas, with special reference to the origin of the Bahaman and Floridian oolites. *Papers Tortugas Lab., Carnegie Inst. Wash. Publ.*, 182: 47–54.

VAUGHAN, T. W., 1919. Corals and the formation of coral reefs. *Smithsonian Inst., Ann. Rept.*, 1917: 189–276.

VAUGHAN, T. W. and WELLS, J. W., 1943. Revision of the suborders families, and genera of the Scleractinia. *Geol. Soc. Am., Spec. Papers*, 44: 1–363.

VERNON, R. H., 1968. Microstructure of high-grade metamorphic rocks at Broken Hill, Australia. *J. Petrol.*, 9: 1–22.

VINOGRADOV, A. P., 1953. *The Elementary Chemical Composition of Marine Organisms*. Sears Found. Mar. Res., New Haven, Conn., 647 pp. (Transl. from Russian).

VOIGT, E., 1959. Die ökologische Bedeutung der Hartgründe ("Hardgrounds") in der oberen Kreide. *Palaeontol. Z.*, 33: 129–147.

VOLL, G., 1960. New work on petrofabrics. *Liverpool Manchester Geol. J.*, 2: 503–567.

VON ARX, W. S., 1948. The circulation systems of Bikini and Rongelap lagoons. *Trans. Am. Geophys. Union*, 29: 861–870.

VON DER BORCH, C. C., 1965a. The distribution and preliminary geochemistry of modern carbonate sediments of the Coorong area, South Australia. *Geochim. Cosmochim. Acta*, 29: 781–799.

VON DER BORCH, C. C., 1965b. Source of ions for Coorong dolomite formation. *Am. J. Sci.*, 263: 684–688.

VON ZITTEL, K. A., 1887. *Traité de Paléontologie, II (1). Paléozoologie*. Oldenbourg, Munich, 897 pp.

WADA, K., 1958. The crystalline structure on the nacre of pearl oyster shell. *Bull. Japan. Soc. Sci. Fisheries*, 24: 422–427.

WADA, K., 1959. On the arrangement of aragonite crystals in the inner layer of the nacre. *Bull. Japan. Soc. Sci. Fisheries*, 25: 342–345.

WADA, K., 1961. Crystal growth of molluscan shells. *Bull. Natl. Pearl Res. Lab.*, 7: 703–828.

WAGNER, G., 1913. Stylolithen und Drucksuturen. *Geol. Palaeontol. Abhandl.* (N.F.), 11 (2): 101–128.

WAINWRIGHT. S. A., 1963. Skeletal organization in the coral, *Pocillopora damicornis. Quart. J. Microscop. Sci.*, 104: 169–183.

WAINWRIGHT, S. A., 1964. Studies of the mineral phase of coral skeleton. *Exptl. Cell Res.*, 34: 213–230.

WALCOTT, C. D., 1906. Algonkian formations of northwestern Montana. *Bull. Geol. Soc. Am.*, 17: 1–28.

WALCOTT, C. D., 1914. Precambrian Algonkian algal flora. *Smithsonian Inst. Misc. Collections*, 64: 77–156.

WALTHER, J., 1888. Die Korallenriffe der Sinaihalbinsel. *Abhandl. Sächs. Akad. Wiss. Leipzig, Math. Naturw. Kl.,* 14: 437–506.

WALTHER. J., 1891. Die Denudation in der Wüste und ihre Bedeutung. *Abhandl. Sächs. Akad. Wiss. Leipzig, Math. Naturw. Kl.*, 16: 345–569.

WANG, H. C., 1950. A revision of the Zoantharia Rugosa in the light of their minute skeletal structures. *Phil. Trans. Roy. Soc. London, Ser. B.*, 234: 175–246.

WANGERSKY, P. J. and JOENSUU, O., 1964. Strontium, magnesium, and manganese in fossil foraminiferal carbonates. *J. Geol.*, 72: 477–483.

WARDLAW, N. C., 1962. Aspects of diagenesis in some Irish Carboniferous limestones. *J. Sediment. Petrol.*, 32: 776–780.

WATABE, N., 1965. Studies on shell formation. XI. Crystal–matrix relationships in the inner layers of mollusk shells. *J. Ultrastruct. Res.*, 12: 351–370.

WATABE, N. and WADA, K., 1956. On the shell structure of Japanese pearl oyster, *Pinctada martensii* Dunker. I. Prismatic layer I. *Rept. Fac. Fisheries, Prefect. Univ. Mie*, 2: 227–231.

WEBER, J. N., 1964a. Carbon isotope ratios in dolostones: some implications concerning the genesis of secondary and "primary" dolostones. *Geochim. Cosmochim. Acta*, 28: 1257–1265.

WEBER, J. N., 1964b. Trace element composition of dolostones and dolomites and its bearing on the dolomite problem. *Geochim. Cosmochim. Acta*, 28: 1817–1868.

WEBER, J. N., 1967. Factors affecting the carbon and oxygen isotopic composition of marine carbonate sediments. I. Bermuda. *Am. J. Sci.*, 265: 586–608.

WEBER, J. N., 1968a. Fractionation of the stable isotopes of carbon and oxygen in calcareous marine invertebrates—the Asteroidea, Ophiuroidea and Crinoidea. *Geochim. Cosmochim. Acta*, 32: 33–70.

WEBER, J. N., 1968b. Quantitative mineralogical analysis of carbonate sediments: comparison of X-ray diffraction and electron probe microanalyser methods. *J. Sediment. Petrol.*, 38: 232–234.

WEBER, J. N., 1969. The incorporation of magnesium into the skeletal calcites of echinoderms. *Am. J. Sci.*, 267: 537–566.

WEBER, J. N. and KAUFMAN, J. W., 1965. Brucite in the calcareous alga *Goniolithon. Science*, 149: 996–997.

WEBER, J. N. and RAUP, D. M., 1966a. Fractionation of the stable isotopes of carbon and oxygen in marine calcareous organisms—the Echinoidea. I. Variation of C^{13} and O^{18} content within individuals. *Geochim. Cosmochim. Acta*, 30: 681–703.

WEBER, J. N. and RAUP, D. M., 1966b. Fractionation of the stable isotopes of carbon and oxygen in marine calcareous organism—the Echinoidea. II. Environmental and genetic factors. *Geochim. Cosmochim. Acta*, 30: 705–736.

WEBER, J. N. and RAUP, D. M., 1968. Comparison of C^{13}/C^{12} and O^{18}/O^{16} in the skeletal calcite of Recent and fossil echinoids. *J. Paleontol.*, 42: 37–50.

WEBER, J. N. and SCHMALZ, R. F., 1968. Factors affecting the carbon and oxygen isotopic composition of marine carbonate sediments. III. Eniwetok Atoll. *J. Sediment. Petrol.*, 38: 1270–1279.

WEBER, J. N. and WOODHEAD, P. M. J., 1969. Factors affecting the carbon and oxygen isotopic composition of marine carbonate sediments. II. Heron Island, Great Barrier Reef, Australia. *Geochim. Cosmochim. Acta*, 33: 19–38.

WELLS, A. J., 1962 Recent dolomite in the Persian Gulf. *Nature*, 194: 274–275.

WELLS, A. J. and ILLING, L. V., 1964. Present-day precipitation of calcium carbonate in the Persian Gulf. In: L. M. J. U. VAN STRAATEN (Editor), *Deltaic and Shallow Marine Deposits.* Elsevier, Amsterdam, pp.429–435.

WELLS, J. W., 1956. Scleractinia. In: R. C. MOORE (Editor), *Treatise on Invertebrate Paleontology. F. Coelenterata.* Geol. Soc. Am., Boulder, Colo., pp.328–444.

WELLS, J. W., 1957a. Coral reefs. In: *Treatise Marine Ecology Paleontology, 1. Ecology—Geol. Soc. Am., Mem.,* 67: 609–631.

WELLS, J. W., 1957b. Corals. *Geol. Soc. Am., Mem.,* 67(1): 1087–1104.

WENTWORTH, C. K., 1922. A scale of grade and class terms for clastic sediments. *J. Geol.,* 30: 377–392.

WEST, I. M., 1965. Macrocell structure and enterolithic veins in British Purbeck gypsum and anhydrite. *Proc. Yorkshire Geol. Soc.,* 35: 47–57.

WEST, I. M., BRANDON, A. and SMITH, M., 1968. A tidal flat evaporitic facies in the Viséan of Ireland. *J. Sediment. Petrol.,* 38: 1079–1093.

WETHERED, E., 1890. On the occurrence of the genus *Girvanella* in oolitic rocks, and remarks on oolitic structure. *Quart. J. Geol. Soc. London,* 46: 270–281.

WETHERED, E. B., 1895. The formation of oolite. *Quart. J. Geol. Soc. London,* 51: 196–206.

WEYL, P. K., 1958. The solution kinetics of calcite. *J. Geol.,* 66: 163–176.

WEYL, P. K., 1959a. The change in solubility of calcium carbonate with temperature and carbon dioxide content. *Geochim. Cosmochim. Acta,* 17: 214–225.

WEYL, P. K., 1959b. Pressure solution and the force of crystallization—a phenomenological theory. *J. Geophys. Res.,* 64: 2001-2025.

WEYL, P. K., 1960. Porosity through dolomitization: conservation-of-mass requirements. *J. Sediment. Petrol.,* 30: 85–90.

WEYL, P. K., 1961. The carbonate saturometer. *J. Geol.,* 69: 32–44.

WEYL, P. K., 1963. In book review: "Environment of Calcium Carbonate Deposition West of Andros Island Bahamas" by P. E. Cloud (1962). *Limnol. Oceanog.,* 8: 494.

WEYL, P. K., 1964. The solution alteration of carbonate sediments and skeletons. In: J. IMBRIE and N. NEWELL (Editors), *Approaches to Paleoecology.* Wiley, New York, N.Y., pp.345–356.

WEYL, P. K., 1967. The solution behaviour of carbonate materials in sea water. *Studies Tropical Oceanog., Univ. Miami,* 5: 178–228.

WHEWELL, W., 1840. *The Philosophy of the Inductive Sciences, 1–2.* Parker, London, 586 and 523 pp.

WHITEHEAD, A. N., 1942. *Adventures of Ideas.* Pelican Books, Harmondsworth, 349 pp.

WIENS, H. J., 1962. *Atoll Environment and Ecology.* Yale Univ., New Haven, Conn., 532 pp.

WILBUR, K. M., 1964. Shell formation and regeneration. In: K. M. WILBUR and C. M. YONGE (Editors), *Physiology of the Mollusca, I.* Academic Press, New York, N.Y., pp.243–282.

WILBUR, K. M. and SIMKISS, K., 1968. Calcified shells. *Comprehensive Biochem.,* 26A: 229–295.

WILBUR, K. M. and YONGE, C. M., 1964. *Physiology of the Mollusca, I.* Academic Press, New York, N.Y., 473 pp.

WILBUR, K. M. and YONGE, C. M., 1966. *Physiology of the Mollusca, II.* Academic Press, New York, N.Y., 645 pp.

WILLIAMS, A., 1956. The calcareous shell of the Brachiopoda and its importance to their classification. *Biol. Rev. Cambridge Phil. Soc.,* 31: 243–287.

WILLIAMS, A., 1965. Shell structure and ornamentation. In: R. C. MOORE (Editor), *Treatise on Invertebrate Paleontology. H. Brachiopoda 1.* Geol. Soc. Am., Boulder, Colo., pp.65–85.

WILLIAMS, A., 1966. Growth and structure of the shell of living articulate brachiopods. *Nature,* 211: 1146–1148.

WILLIAMS, A., 1968a. Evolution of the shell structure of articulate brachiopods. *Palaeontol. Assoc. London, Spec. Papers,* 2: 1–55.

WILLIAMS, A., 1968b. A history of skeletal secretion among articulate brachiopods. *Lethaia,* 1: 268–287.

WILLIAMSON, W. C., 1880. On the organization of the fossil plants of the Coal-Measures. X. Including an examination of the supposed radiolarians of the Carboniferous rocks. *Phil. Trans. Roy. Soc. London,* 171: 493–539.

WILSON, J. B., 1963. The relation of shell beds to living molluscan faunas. *Trans. Dumfriesshire Galloway Nat. Hist. Antiquarian Soc.*, 40: 98–101.

WILSON, J. B., 1967. Palaeoecological studies on shell-beds and associated sediments in the Solway Firth. *Scot. J. Geol.*, 3: 329–371.

WILSON, R. C. L., 1967. Diagenetic carbonate fabric variations in Jurassic limestones of southern England. *Proc. Geologists' Assoc. Engl.*, 78: 535–554.

WINCHELL, A. N., 1946. *Elements of Optical Mineralogy*. Wiley, New York, N.Y., 459 pp.

WINLAND, H. D., 1968. The role of high Mg calcite in the preservation of micrite envelopes and textural features of aragonite sediments. *J. Sediment. Petrol.*, 38: 1320–1325.

WINLAND, H. D., 1969. Stability of calcium carbonate polymorphs in warm, shallow seawater. *J. Sediment. Petrol.*, 39: 1579–1587.

WISE, JR., S. W., 1969a. Patterns of calcification in nacreous layer of pelecypods and gastropods. *Bull. Am. Assoc. Petrol. Geologists*, 53: 751 (abstract).

WISE, JR., S. W., 1969b. Study of molluscan shell ultrastructures. In: O. JOHARI (Editor), *Scanning Electron Microscopy 1969*. IIT Res. Inst., Chicago, pp.205–216.

WISE, JR., S. W., 1970. Microarchitecture and deposition of gastropod nacre. *Science*, 167: 1486–1488.

WISE, JR., S. W., in press. Microarchitecture and mode of formation of nacre (mother-of-pearl) in pelecypods, gastropods, and cephalopods. *Eclogae Geol. Helv.*

WISE, JR., S. W. and HAY, W. W., 1968a. Scanning electron microscopy of molluscan shell ultra-structures. I. Techniques for polished and etched sections. *Trans. Am. Microscop. Soc.*, 87: 411–418.

WISE, JR., S. W. and HAY, W. W., 1968b. Scanning electron microscopy of molluscan shell ultrastructures. II. Observations of growth surfaces. *Trans. Am. Microscop. Soc.*, 87: 419–430.

WOLF, K. H., 1965. "Grain-diminution" of algal colonies to micrite. *J. Sediment. Petrol.*, 35: 420–427.

WOLFE, M. J., 1968. Lithification of a carbonate mud: Senonian Chalk in Northern Ireland. *Sediment. Geol.*, 2: 263–290.

WOLLAST, R. 1971. Kinetic aspects of the nucleation and growth of calcite from aqueous solutions. In: O. P. BRICKER (Editor), *Carbonate Cements*. Johns Hopkins Press, Baltimore, Md., pp. 264–273.

WOLLAST, R., DEBOUVERIE, D. and DUVIGNEAUD, P. H., 1971. Influence of Sr and Mg on the stability of calcite and aragonite. In: O. P. BRICKER (Editor), *Carbonate Cements*. Johns Hopkins Press, Baltimore, Md., pp. 274–277.

WOOD, A., 1941a. "Algal dust" and the finer-grained varieties of Carboniferous Limestone. *Geol. Mag.*, 78: 192–200.

WOOD, A., 1941b. The Lower Carboniferous calcareous algae *Mitcheldeania* Wethered and *Garwoodia*, gen. nov. *Proc. Geologists' Assoc. Engl.*, 52: 216–226.

WOOD, A., 1942. The algal nature of the genus *Koninckopora* Lee; its occurrence in Canada and western Europe. *Quart. J. Geol. Soc. London*, 98: 205–221.

WOOD, A., 1949. The structure of the wall of the test in the Foraminifera. *Quart. J. Geol. Soc. London*, 104: 229–252.

WOOD, A., 1957. The type-species of the genus *Girvanella* (calcareous algae). *Palaeontology*, 1: 22–28.

WOOD, A., 1964. A new dasycladacean alga, *Nanopora*, from the Lower Carboniferous of England and Kazakhstan. *Palaeontology*, 7: 181–185.

WRAY, J. L., 1964. *Archaeolithophyllum*, an abundant calcareous alga in limestones of the Lansing Group (Pennsylvanian), southeastern Kansas. *Geol. Surv. Kansas, Bull.*, 170: 1–13.

WRAY, J. L., 1968. Late Paleozoic phylloid algal limestones in the United States. *Intern. Geol. Congr., 23rd, Prague, 1968, Proc.*, 8: 113–119.

WRAY, J. L. and DANIELS, F., 1957. Precipitation of calcite and aragonite. *J. Am. Chem. Soc.*, 79: 2031–2034.

WYMAN, JR., J., SCHOLANDER, P. F., EDWARDS, G. A. and IRVING, L., 1952. On the stability of gas bubbles in sea water. *J. Marine Res.*, 11: 47–62.

YAALON, D. H., 1967. Factors affecting the lithification of eolianite and interpretation of its environmental significance in the coastal plain of Israel. *J. Sediment. Petrol.*, 37: 1189–1199.

YONGE, C. M., 1930. *A Year on the Great Barrier Reef.* Putnam, London–New York, 246 pp.

YONGE, C. M., 1951. The form of coral reefs. *Endeavour*, 10: 136–144.

YONGE, C. M., 1968. Living corals. *Phil. Trans. Roy. Soc. London, Ser. B.*, 169: 329–344.

ZANKL, H., 1969. Structural and textural evidence of early lithification in fine-grained carbonate rocks. *Sedimentology*, 12: 241–256.

ZELLER, E. J. and WRAY, J. L., 1956. Factors influencing precipitation of calcium carbonate. *Bull. Am. Assoc. Petrol. Geologists*, 40: 140–152.

ZEN, E-AN., 1957. Partial molar volumes of some salts in aqueous solutions. *Geochim. Cosmochim. Acta*, 12: 103–122.

ZENKOVITCH, V. P., 1967. *Processes of Coastal Development.* Oliver and Boyd, Edinburgh, 738 pp.

REFERENCES IN THE APPENDIX

ADELSECK, C. G., GEEHAN, G. W. and ROTH, P. H., 1973. Experimental evidence for the selective dissolution and overgrowth of calcareous nannofossils during diagenesis. *Geol. Soc. Am., Bull.*, 84: 2755–2762.

ALEXANDERSSON, T., 1969. Recent littoral and sublittoral high-Mg calcite lithification in the Mediterranean. *J. Sediment. Petrol.*, 39: 47–61.

ALEXANDERSSON, T., 1972a. Micritization of carbonate particles: processes of precipitation and dissolution in modern shallow marine sediments. *Bull. Geol. Insts. Univ. Uppsala, N.S.*, 3: 201–236.

ALEXANDERSSON, T., 1972b. Intragranular growth of marine *aragonite* and Mg-*calcite*: evidence of precipitation from supersaturated seawater. *J. Sediment. Petrol.*, 42: 441–460.

ALEXANDERSSON, T., 1972c. Mediterranean beachrock cementation: marine precipitation of Mg-calcite. In: D. J. STANLEY (Editor), *The Mediterranean Sea: A Natural Sedimentation Laboratory*. Dowden, Hutchinson and Ross, Stroudsburg, Pa., pp. 203–223.

ALEXANDERSSON, T., 1974. Carbonate cementation in coralline algal nodules in the Skagerak, North Sea: biochemical precipitation in undersaturated waters. *J. Sediment. Petrol.*, 44: 7–26.

ALEXANDERSSON, T., 1975a. Marks of unknown carbonate-decomposing organelles in cyanophyte borings. *Nature*, 254: 212, 237–238.

ALEXANDERSSON, T., 1975b. Etch patterns on calcareous grains: petrographic evidence of marine dissolution of carbonate minerals. (Accepted for *Science*.)

ALLEVA, J. J., ALLEVA, F. R. and FRY, B. E., 1971. Calcium carbonate concretions: cyclic occurrence in the hamster vagina. *Science*, 174: 600–603.

ANDERSON, T. F., 1969. Self-diffusion of carbon and oxygen in calcite by isotope exchange with carbon dioxide. *J. Geophys. Res.*, 74: 3918–3932.

ANDERSON, T. F. and SCHNEIDERMANN, N., 1973. Stable isotope relationships in pelagic limestones from the central Caribbean: Leg 15, Deep Sea Drilling Project. In: N. T. EDGAR et al. (Editors), *Initial Reports of the Deep Sea Drilling Project, Vol. XV*. U. S. Government Printing Office, Washington, D.C., pp. 795–803.

ANDERSON, T. F., BENDER, M. L. and BROECKER, W. S., 1973. Surface areas of biogenic carbonates and their relation to fossil ultrastructure and diagenesis. *J. Sediment. Petrol.*, 43: 471–477.

ARMSTRONG, J., 1970. Zoarial microstructures of two Permian species of the bryozoan genus *Stenopora*. *Palaeontology*, 13: 581–587.

ASSERETO, R. L. and KENDALL, C. G. St. C., 1971. Megapolygons in Landinian limestones of Triassic of southern Alps: evidence of deformation by penecontemporaneous desiccation and cementation. *J. Sediment. Petrol.*, 41: 715–723.

AWRAMIK, S. M., 1971. Precambrian columnar stromatolite diversity: reflection of Metazoan appearance. *Science*, 174: 825–827.

BACK, W. and HANSHAW, B. B., 1970. Comparison of chemical hydrogeology of the carbonate peninsulas of Florida and Yucatan. *J. Hydrol.*, 10: 330–368.

BACK, W. and HANSHAW, B. B., 1971. Rates of physical and chemical processes in a carbonate aquifer. In: *Advances in Chemistry*, 106: 77–93.

BADIOZAMANI, K., 1973. The Dorag dolomitization model—application to the Middle Ordovician of Wisconsin. *J. Sediment. Petrol.*, 43: 965–984.

BALCON, J., 1973. Sédimentation et diagenèse des carbonates. *Bull. Centre Rech. Pau-SNPA*, 7: 97–289.

BARTLETT, G. A. and GREGGS, R. G., 1970a. The Mid-Atlantic Ridge near 45°00′ north. VIII. Carbonate lithification on oceanic ridges and seamounts. *Can. J. Earth Sci.*, 7: 257–267.

BARTLETT, G. A. and GREGGS, R. G., 1970b. A reinterpretation of stylolitic solution surfaces deduced from carbonate cores from San Pablo Seamount and the Mid-Atlantic Ridge. *Can. J. Earth Sci.*, 7: 274–279.

BASAN, P. B., 1973. Aspects of sedimentation and development of a carbonate bank in the Barracuda Keys, south Florida. *J. Sediment. Petrol.*, 43: 42–53.

BATHURST, R. G. C., 1973. Problèmes généraux posés par la diagenèse des sédiments carbonatés. *Bull. Centre Rech. Pau-SNPA*, 7: 99–110.

BATHURST, R. G. C., 1974. Marine diagenesis of shallow water calcium carbonate sediments. In: F. A. DONATH, F. G. STEHLI and G. W. WETHERILL (Editors), *Annual Review of Earth and Planetary Sciences, Vol. 2*. Annual Reviews Inc., Palo Alto, Calif., pp. 257–274.

BEACH, A., 1974. A geochemical investigation of pressure solution and the formation of veins in a deformed greywacke. *Contrib. Mineral. Petrol.*, 46: 61–68.

BEALL, A. O. and FISCHER, A. G., 1969. Sedimentology. In: M. EWING et al. (Editors), *Initial Reports of the Deep Sea Drilling Project, Vol. I*. U.S. Government Printing Office, Washington, D.C., pp. 521–593.

BEALL, A. O. et al., 1973. Sedimentology. In: J. L. WORZEL et al. (Editors), *Initial Reports of the Deep Sea Drilling Project, Vol. X*. U.S. Government Printing Office, Washington, D.C., pp. 699–716.

BEHRENS, E. W. and LAND, L. S., 1972. Subtidal Holocene dolomite, Baffin Bay, Texas. *J. Sediment. Petrol.*, 42: 155–161.

BENNET, I., 1971. *The Great Barrier Reef*. Lansdowne, Melbourne, 183 pp.

BENSON, L. V., 1974. Transformation of a polyphase sedimentary assemblage into a single phase rock: a chemical approach. *J. Sediment. Petrol.*, 44: 123–135.

BENSON, L. V. and MATTHEWS, R. K., 1971. Electron microprobe studies of magnesium distribution in carbonate cements and recrystallized skeletal grainstones from the Pleistocene of Barbados, West Indies. *J. Sediment. Petrol.*, 41: 1018–1025.

BENSON, L. V., ACHAUER, C. W. and MATTHEWS, R. K., 1972. Electron microprobe analyses of magnesium and iron distribution in carbonate cements and recrystallized sediment grains from ancient carbonate rocks. *J. Sediment. Petrol.*, 42: 803–811.

BEN-YAAKOV, S., RUTH, E. and KAPLAN, I. R., 1974. Carbonate compensation depth: relation to carbonate solubility in ocean waters. *Science*, 184: 982–984.

BERGER, W. H., 1968. Planktonic foraminifera: selective solution and paleoclimatic interpretation. *Deep-Sea Res.*, 15: 31–43.

BERGER, W. H., 1971. Sedimentation of planktonic foraminifera. *Marine Geol.*, 11: 325–358.

BERGER, W. H., 1973a. Deep-sea carbonates: Pleistocene dissolution cycles. *J. Foram. Res.*, 3: 187–195.

BERGER, W. H., 1973b. Deep-sea carbonates: evidence for a coccolith lysocline. *Deep-Sea Res.*, 20: 917–921.

BERGER, W. H. and PARKER, F. L., 1970. Diversity of planktonic foraminifera in deep-sea sediments. *Science*, 168: 1345–1347.

BERGER, W. H. and VON RAD, U., 1972. Cretaceous and Cenozoic sediments from the Atlantic Ocean. In: D. E. HAYES et al. (Editors), *Initial Reports of the Deep Sea Drilling Project, Vol. XIV*. U.S. Government Printing Office, Washington, D.C., pp. 787–954.

BERGER, W. H. and WINTERER, E. L., 1974. Plate stratigraphy and the fluctuating carbonate line. In: K. J. HSÜ and H. C. JENKYNS (Editors), *Pelagic Sediments: on Land and under the Sea— Int. Assoc. Sedimentol., Spec. Publ.*, 1: 11–48.

BERNER, R. A., 1971. *Principles of Chemical Sedimentology*. McGraw-Hill, New York, N.Y., 240 pp.

BERNER, R. A., 1975. Diagenetic models of dissolved species in the interstitial waters of compacting sediments. *Am. J. Sci.*, 275: 88–96.

BERNER, R. A. and MORSE, J. W., 1974. Dissolution kinetics of calcium carbonate in sea water IV. Theory of calcite dissolution. *Am. J. Sci.*, 274: 108–134.

BERNER, R. A. and WILDE, P., 1972. Dissolution kinetics of calcium carbonate in sea water I. Saturation state parameters for kinetic calculations. *Am. J. Sci.*, 272: 826–839.

BERNOULLI, D., 1972. North Atlantic and Mediterranean Mesozoic facies: a comparison. In: C. D. HOLLISTER et al. (Editors), *Initial Reports of the Deep Sea Drilling Project, Vol. XI*. U.S. Government Printing Office, Washington, D.C., pp. 801–871.

BERNOULLI, D. and WAGNER, C. W., 1971. Subaerial diagenesis and fossil caliche deposits in the Calcare Massiccio Formation (Lower Jurassic, Central Apennines, Italy). *Neues Jahrb. Geol. Paläontol. Abh.*, 138: 135–149.

BERRY, F. A. F., 1973. High fluid potentials in California Ranges and their tectonic significance. *Am. Assoc. Petrol. Geol., Bull.*, 57: 1219–1249.

BIRKELUND, T. and HANSEN, H. J., 1974. Shell ultrastructures of some Maastrichtian Ammonoidea and Coleoidea and their taxonomic implications. *K. Dan. Vidensk. Selsk. Biol. Skr.*, 20:1–34.

BLACK, M., 1972. Crystal development in Discoasteraceae and Braarudosphaeraceae (planktonic algae). *Palaeontology*, 15: 476–489.

BLATT, H., MIDDLETON, G. and MURRAY, R., 1972. *Origin of Sedimentary Rocks*. Prentice-Hall, Englewood Cliffs, N.J., 634 pp.

BOGGS, S., 1972. Petrography and geochemistry of rhombic, calcite pseudomorphs from mid-Tertiary mudstones of the Pacific Northwest, U.S.A. *Sedimentology*, 19: 219–235.

BORNHOLD, B. D. and MILLIMAN, J. D., 1973. Generic and environmental control of carbonate mineralogy in serpulid (polychaete) tubes. *J. Geol.*, 81: 363–373.

BOSELLINI, A. and GINSBURG, R. N., 1971. Form and internal structure of Recent algal nodules (rhodolites) from Bermuda. *J. Geol.*, 79: 669–682.

BOSELLINI, A. and ROSSI, D., 1974. Triassic carbonate buildups of the Dolomites, northern Italy. In: L. F. LAPORTE (Editor), *Reefs in Space and Time: Selected Examples from the Recent and Ancient—Soc. Econ. Paleontol. Mineral., Spec. Publ.*, 18: 209–233.

BOSTRÖM, K. and HOROWITZ, A., 1972. Origin of pH variations and inorganic carbonates in pelagic sediments. *Geol. Fören. Stockh. Förh.*, 94: 515–535.

BOURROUILH, F., 1972. Diagenèse récifale: calcitisation et dolomitisation: leur répartition horizontale dans un atoll soulevé, île Lifou. Territoire de la Nouvelle Calédonie. *Cah. ORSTOM, Sér. Géol.*, 4: 121–148.

BOURROUILH, F., 1973. Les dolomies et leurs genèses. *Bull. Centre Rech. Pau-SNPA*, 7: 111–135.

BRAITHWAITE, C. J. R., TAYLOR, J. D. and KENNEDY, W. J., 1973. The evolution of an atoll: the depositional and erosional history of Aldabra. *Phil. Trans. Roy. Soc. London, Ser. B*, 266: 307–340.

BRÄTTER, P., MÖLLER, P. and RÖSICK, U., 1972. On the equilibrium of coexisting sedimentary carbonates. *Earth Planet. Sci. Lett.*, 14: 50–54.

BREDEHOEFT, J. D. and HANSHAW, B. B., 1968. On the maintenance of anomalous fluid pressures: 1. Thick sedimentary sequences. *Geol. Soc. Am., Bull.*, 79: 1097–1106.

BREWER, R., 1964. *Fabric and Mineral Analysis of Soils*. Wiley, New York, N.Y., 470 pp.

BROECKER, W. S., 1971. A kinetic model for the chemical composition of sea water. *Quaternary Res.*, 1: 188–207.

BROECKER, W. S., 1974. *Chemical Oceanography*. Harcourt, Brace, Jovanovich, New York, N.Y., 214 pp.

BROMLEY, R. G., in press. Trace fossils at omission surfaces. In: R. W. FREY (Editor), *Study of Trace Fossils*. Springer, New York, N. Y.

BROWN, P. R., 1972. Incipient metamorphic fabrics in some mud-supported carbonate rocks. *J. Sediment. Petrol.*, 42: 841–847.

BURNE, R. V., 1974. The deposition of reef-derived sediment upon a bathyal slope: the deep off-reef environment, north of Discovery Bay, Jamaica. *Marine Geol.*, 16: 1–19.

BURRI, P., DU DRESNAY, R. and WAGNER, C. W., 1973. Tepee structures and associated diagenetic features in intertidal carbonate sands (Lower Jurassic, Morocco). *Sediment. Geol.*, 9: 221–228.

BUSH, P., 1973. Some aspects of the diagenetic history of the sabkha in Abu Dhabi, Persian Gulf. In: B. H. PURSER (Editor), *The Persian Gulf: Holocene Carbonate Sedimentation and Diagenesis in a Shallow Epicontinental Sea*. Springer, Berlin, pp. 395–408.

BUTLER, G. P., 1973. Strontium geochemistry of modern and ancient calcium sulphate minerals. In: B. H. PURSER (Editor), *The Persian Gulf: Holocene Carbonate Sedimentation and Diagenesis in a Shallow Epicontinental Sea*. Springer, Berlin, pp. 423–452.

CADOT, H. M., VAN SCHMUS, W. R. and KAESLER, R. L., 1972. Magnesium in calcite of marine Ostracoda. *Geol. Soc. Am., Bull.*, 83: 3519–3521.

CAMERON, A. M. et al. (Editors), 1974. *Proceedings of the Second International Symposium on Coral Reefs*. Great Barrier Reef Committee, Brisbane (2 vols.), vi+630 and 753 pp.

CARANNANTE, G. and GUZZETTA, G., 1972. Stiloliti e sliccoliti come meccanismo di deformazione delle masse rocciose. *Boll. Soc. Nat. Napoli*, 81: 157–170.

CARANNANTE, G., FERRERI, V. and SIMONE, L., 1974. Le cavità paleocarsiche cretaciche di Dragoni (Campania). *Boll. Soc. Nat. Napoli*, 83: 1–11.

CASTELLARIN, A. and SARTORI, R., 1973. I ciclotemi carbonatici infraliassici di S. Massenza (Trento). *G. Geol.*, 39: 221-241.

CASTELLARIN, A., DEL MONTE, M. and FRASCARI, F., 1974. Cosmic fallout in the "hard grounds" of the Venetian regions. *Ann. Mus. Geol. Bologna*, (2a) 39: 333–345.

CAULET, J. P., 1972a. Les sédiments organogènes du précontinent algérien. *Mém. Mus. Hist. Nat.*, *Sér. C.*, 25: 1–289.

CAULET, J. P., 1972b. Recent biogenic calcareous sedimentation on the Algerian continental shelf. In: D. J. STANLEY (Editor), *The Mediterranean Sea: A Natural Sedimentation Laboratory*. Dowden, Hutchinson and Ross, Stroudsburg, Pa., pp. 261–277.

CHAVE, K. E., SMITH, S. V. and ROY, K. J., 1972. Carbonate production by coral reefs. *Marine Geol.*, 12: 123–140.

CHESHER, R. H., 1969. Destruction of Pacific corals by the sea star *Acanthaster planci*. *Science*, 165: 280–283.

CHILLINGAR, G. V., MANNON, R. W. and RIEKE, H. H. (Editors), 1972. *Oil and Gas Production from Carbonate Rocks*. Elsevier, New York, N. Y., 408 pp.

CHRIST, C. L. and HOSTETLER, P. B., 1970. Studies in the system $MgO–SiO_2–CO_2–H_2O$ (II): the activity-product constant of magnesite. *Am. J. Sci.*, 268: 439–453.

CLIFTON, H. E., 1973. Role of reef fauna in sediment transport and distribution—studies from Tektite I and II. *Helgoländer Wiss. Meeresunters.*, 24: 91–101.

COOK, P. J., 1973. Supratidal environment and geochemistry of some Recent dolomite concretions, Broad Sound, Queensland, Australia. *J. Sediment. Petrol.*, 43: 998–1011.

COOK, P. J., 1974. Geochemistry and diagnesis of interstitial fluids and associated oozes, Deep Sea Drilling Project, Leg 27, Site 262, Timor Trough. In: J. J. VEEVERS et al. (Editors), *Initial Reports of the Deep Sea Drilling Project, Vol. XXVII*. U.S. Government Printing Office, Washington, D.C., pp. 463–480.

COPLEN, T. B. and SCHLANGER, S. O., 1973. Oxygen and carbon isotope studies of carbonate sediments from Site 167, Magellan Rise, Leg 17. In: E. L. WINTERER et al. (Editors), *Initial Reports of the Deep Sea Drilling Project, Vol. XVII*. U.S. Government Printing Office, Washington, D.C., pp. 505–509.

CUFFEY, R. J., 1972. The roles of bryozoans in modern coral reefs. *Geol. Rundsch.*, 61: 542–550.

D'ARGENIO, B., 1967. Geologia del gruppo Taburno-Camposauro (Appennino Campano). *Soc. Naz. Sci. Fis. Mat.*, *(3)*, 6: 1–218.

DAVIES, P. J., in press. High-Mg calcite ooids from the Great Barrier Reef.

DAVIES, P. J. and KINSEY, D. W., 1973. Organic and inorganic factors in Recent beach rock formation, Heron Island, Great Barrier Reef. *J. Sediment. Petrol.*, 43: 59–81.

DAVIES, T. A. and LAUGHTON, A. S., 1972. Sedimentary processes in the North Atlantic. In: T. A. DAVIES et al. (Editors), *Initial Reports of the Deep Sea Drilling Project, Vol. XII*. U.S. Government Printing Office, Washington, D.C., pp. 905–934.

DAVIES, T. A. and SUPKO, P. R., 1973. Oceanic sediments and their diagenesis: some examples from deep-sea drilling. *J. Sediment. Petrol.*, 43: 381–390.

DEEGAN, C. E., 1972. The mode of origin of some late diagenetic sandstone concretions from the Scottish Carboniferous. *Scott. J. Geol.*, 7: 357–365.

DEELMAN, J. C., 1974. Granulo-mechanical aspects of lithification. *Neues Jahrb. Geol. Paläontol. Abh.*, 147: 237–268.

DEGENS, E. T. and ROSS, D. A., 1974. The Black Sea—geology, chemistry, and biology. *Am Assoc. Petrol. Geol., Mem.*, 20.

DE GROOT, K., 1973. Geochemistry of tidal flat brines at Umm Said, SE Qatar, Persian Gulf. In: B. H. PURSER (Editor), *The Persian Gulf: Holocene Carbonate Sedimentation and Diagenesis in a Shallow Epicontinental Sea*. Springer, Berlin, pp. 377–394.

DICKEY, P. A., COLLINS, A. G. and FAJARDO, M. I., 1972. Chemical composition of deep formation waters in southwestern Louisiana. *Am. Assoc. Petrol. Geol., Bull.*, 56: 1530–1533.

DICKSON, J. A. D. and BARBER, C., in press. Petrography, chemistry and origin of early diagenetic concretions in the Lower Carboniferous of the Isle of Man. *Sedimentology*.

DI GIROLAMO, P., 1968. Contributo allo studio dei mosaici calcitici riempienti le cavità della diagenesi precoce di alcuni rocce carbonatiche cretaciche dell' Appennino meridionale. *Boll. Soc. Nat. Napoli*, 77: 5–25.

Dodd, J. R. and Siemers, C. T., 1971. Effect of late Pleistocene karst topography on Holocene sedimentation and biota, Lower Florida Keys. *Geol. Soc. Am., Bull.*, 82: 211–218.

Drushits, V. V. and Khiami, N., 1970. Structure of the septa, protoconch walls and initial whorls in early Cretaceous ammonites. *Paleontol. J.*, 1: 26–38.

Duedall, I. W., 1972. The partial molal volume of calcium carbonate in sea water. *Geochim. Cosmochim. Acta*, 36: 729–734.

Dunham, R. J., 1970. Stratigraphic reefs versus ecologic reefs. *Am. Assoc. Petrol. Geol., Bull.*, 54: 1931–1932.

Durney, D. W., 1972. Solution-transfer, an important geological deformation mechanism. *Nature*, 235: 315–317.

Edmond, J. M. and Gieskes, J. M. T. M., 1970. On the calculation of the degree of saturation of sea water with respect to calcium carbonate under *in situ* conditions. *Geochim. Cosmochim. Acta*, 34: 1261–1291.

Edmunds, W. M., 1973. Trace element variations across an oxidation-reduction barrier in a limestone aquifer. In: E. Ingerson (Editor), *Proc. Symp. Hydrogeochem. Biogeochem. Vol 1. Hydrogeochemistry*. Clarke, Washington, D.C., pp. 500–526.

Edmunds, W. M., Lovelock, P. E. R. and Gray, D. A., 1973. Interstitial water chemistry and aquifer properties in the Upper and Middle Chalk of Berkshire, England. *J. Hydrol.*, 19: 21–31.

Edwards, B. D. and Perkins, R. D., 1974. Distribution of microborings within continental margin sediments of the southeastern United States. *J. Sediment. Petrol.*, 44: 1122–1135.

Erben, H. K., 1972a. Über die Bildung und das Wachstum von Perlmutt. *Biomineralisation*, 4: 15–46.

Erben, H. K., 1972b. Die Mikro- und Ultrastruktur abgedeckter Hohlelemente und die Conellen des Ammoniten-Gehäuses. *Paläontol. Z.*, 46:1–2, 6–19.

Erben, H. K. and Reid, R. E. H., 1971. Ultrastructure of shell, origin of conellae and siphuncular membranes in an ammonite. *Biomineralisation*, 3: 22–31.

Esteban, M., 1972. Presencia de caliche fósil en la base Eoceno de los Catalánides, provincias de Tarragona y Barcelona. *Acta Geol. Hisp.*, 7: 164–168.

Esteban, M., 1974. Caliche textures and *Microcodium. Suppl. Boll. Soc. Geol. Ital.*, 92: 105–125.

Fantidis, J. and Ehhalt, D. H., 1970. Variations of the carbon and oxygen isotopic composition in stalagmites and stalactites: evidence of non-equilibrium isotopic fractionation. *Earth Planet. Sci. Lett.*, 10: 136–144.

FAO, 1972. *Glossary and Multilingual Equivalents of Karst Terms*. UNESCO, Paris, 72 pp.

Fenninger, A. and Flajs, G., 1974. Zur Mikrostruktur rezenter und fossiler Hydrozoa. *Biomineralisation*, 7: 69–99.

Fischbeck, R. and Müller, G., 1971. Monohydrocalcite, hydromagnesite, nesquehonite, dolomite, aragonite and calcite in speleothems of the Fränkische-Schweiz, Western Germany. *Contrib. Mineral. Petrol.*, 33: 87–92.

Folk, R. L., 1973. Carbonate petrography in the post-Sorbian age. In: R. N. Ginsburg (Editor), *Evolving Concepts in Sedimentology*. Johns Hopkins, Baltimore, Md., pp. 118–158.

Folk, R. L., 1974. The natural history of crystalline calcium carbonate: effect of magnesium content and salinity. *J. Sediment. Petrol.*, 44: 40–53.

Förstner, U., 1973. Petrographische und geochemische Untersuchungen an afghanischen Endseen. *Neues Jahrb. Mineral. Abh.*, 118: 268–312.

Freeman, T., 1969. Cement-composition discontinuity in the Cambrian of Texas and its stratigraphic implications. *Geol. Soc. Am., Bull.*, 80: 2095–2096.

Freeman, T., 1971. Morphology and composition of an Ordovician vadose cement. *Nat. Phys. Sci.*, 233: 133–134.

Freeman, T., 1972. Sedimentology and dolomitization of Muschelkalk carbonates (Triassic), Iberian Range, Spain. *Am. Assoc. Petrol. Geol., Bull.*, 56: 434–453.

Friedman, G. M. and Gavish, E., 1970. Chemical changes in interstitial waters from sediments of lagoonal, deltaic, river, estuarine, and salt water marsh and cove environments. *J. Sediment. Petrol.*, 40: 930–953.

FRIEDMAN, G. M., GEBELEIN, C. D. and SANDERS, J. E., 1971. Micritic envelopes of carbonate grains are not exclusively of photosynthetic algal origin. *Sedimentology*, 16: 89–96.

FRIEDMAN, G. M., AMIEL, A. J., BRAUN, M. and MILLER, D. S., 1973. Generation of carbonate particles and laminites in algal mats—example from sea-marginal hypersaline pool, Gulf of Aqaba, Red Sea. *Am. Assoc. Petrol. Geol., Bull.*, 57: 541–557.

FRIEDMAN, G. M., AMIEL, A. J. and SCHNEIDERMANN, N., 1974. Submarine cementation in reefs: example from the Red Sea. *J. Sediment. Petrol.*, 44: 816–825.

FRIEDMAN, M. and HEARD, H. C., 1974. Principal stress ratios in Cretaceous limestones from Texas Gulf Coast. *Am. Assoc. Petrol. Geol., Bull.*, 58: 71–78.

FRITZ, P. and SMITH, D. G. W., 1970. The isotopic composition of secondary dolomites. *Geochim. Cosmochim. Acta*, 34: 1161–1173.

FROGET, C., 1972. Exemples de diagenèse sous-marine dans les sédiments Pliocènes et Pléistocènes: dolomitisation, ferruginisation (Méditerranée nord-occidentale, sud de Marseille). *Sedimentology*, 19: 59–83.

FÜCHTBAUER, H., 1975. *Sediments and Sedimentary Rocks 1*. Schweizerbart, Stuttgart, 464 pp.

FUNNEL, B. M. and RIEDEL, W. R., 1971. *The Micropalaeontology of Oceans*. Cambridge Univ. Press, Cambridge, 828 pp.

FÜRSICH, F., 1971. Hartgründe und Kondensation im Dogger von Calvados. *Neues Jahrb. Geol. Paläontol. Abh.*, 138: 313–342.

FÜTTERER, D. K., 1974. Significance of the boring sponge *Cliona* for the origin of the fine grained material of carbonate sediments. *J. Sediment. Petrol.*, 44: 79–84.

GARDNER, L. R., 1972. Origin of the Mormon Mesa caliche, Clark County, Nevada. *Geol. Soc. Am., Bull.*, 83: 143–156.

GARRELS, R. M. and MACKENZIE, F. T., 1971. *Evolution of Sedimentary Rocks*. Norton, New York, N.Y., xvi+397 pp.

GARRETT, P., 1970. Phanerozoic stromatolites: noncompetitive ecologic restriction by grazing and burrowing animals. *Science*, 169: 171–173.

GARRETT, P. and DUCKLOW, H., 1975. Coral diseases in Bermuda. *Nature*, 253: 349–350.

GARRETT, P., SMITH, D. L., WILSON, A. O. and PATRIQUIN, D., 1971. Physiography, ecology, and sediments of two Bermuda patch reefs. *J. Geol.*, 79: 647–668.

GARRISON, R. E., 1974. Sedimentation and diagenesis of pelagic sediments: observations from the deep sea floor and in mountain ranges. *Ann. Soc. Géol. Belg., Bull.*, 97: 163–168.

GARRISON, R. E., HEIN, J. R. and ANDERSON, T. F., 1973. Lithified carbonate sediment and zeolithic tuff in basalts, Mid-Atlantic Ridge. *Sedimentology*, 20: 399–410.

GEBELEIN, C. D., in press. Holocene sedimentation and stratigraphy, southwest Andros Island, Bahamas, *Int. Sediment. Congr., 9th, Nice, 1975.*

GEBELEIN, C. D. and HOFFMAN, P., 1973. Algal origin of dolomite laminations in stromatolitic limestone. *J. Sediment. Petrol.*, 43: 603–613.

GINSBURG, R. N., 1974. Introduction to comparative sedimentology of carbonates. *Am. Assoc. Petrol. Geol., Bull.*, 58: 781–786.

GINSBURG, R. N. and JAMES, N. P., 1973. British Honduras by submarine. *Geotimes*, 18(5): 23–24.

GINSBURG, R. N. and JAMES, N. P., 1974. Holocene carbonate sediments of continental shelves. In: C. BURKE and C. DRAKE (Editors), *Continental Margins*. Springer, Berlin, pp. 137–157.

GINSBURG, R. N. and SCHROEDER, J. H., 1973. Growth and submarine fossilization of algal cup reefs, Bermuda. *Sedimentology*, 20: 575–614.

GINSBURG, R. N., MARSZALEK, D. S. and SCHNEIDERMANN, N., 1971a. Ultrastructure of carbonate cements in a Holocene algal reef of Bermuda. *J. Sediment. Petrol.*, 41: 472–482.

GINSBURG, R. N., REZAK, R. and WRAY, J. L., 1971b. *Geology of Calcareous Algae (Notes for a Short Course). Sedimenta I*. Comparative Sedimentology Laboratory, Univ. Miami, i–iii, 1–61.

GLOVER, J. E., 1968. Significance of stylolites in dolomitic limestones. *Nature*, 217: 835–836.

GLOVER, J. E., 1969. Observations on stylolites in Western Australian rocks. *J. R. Soc. W. Aust.*, 52: 12–17.

GLYNN, P. W., 1973. *Acanthaster*: effect on coral reef growth in Panama. *Science*, 180: 504–506.

GLYNN, P. W., STEWART, R. H. and McCOSKER, J. E., 1972. Pacific coral reefs of Panamá: structure, distribution and predators. *Geol. Rundsch.*, 61: 483–519.

GOLUBIĆ, S., 1969. Cyclic and noncyclic mechanisms in the formation of travertine. *Verh. Int. Verein. Limnol.*, 17: 956–961.

GOLUBIĆ, S., 1973. The relationship between blue-green algae and carbonate deposits. In: N. G. CARR and B. A. WHITTON (Editors), *The Biology of Blue-Green Algae*. Blackwell, Oxford, pp. 434–472.

GOMBERG, D. N. and BONATTI, E., 1970. High-magnesium calcite: leaching of magnesium in the deep sea. *Science*, 168: 1451–1453.

GÖTZ, G., 1931. Bau und Biologie fossiler Serpuliden. *Neues Jahrb. Mineral. Geol. Paläontol.*, 66: 385–438.

GOUDIE, A., 1972. The chemistry of world calcrete deposists. *J. Geol.*, 80: 449–463.

GOUDIE, A., 1973. *Duricrusts in Tropical and Subtropical Landscapes*. Clarendon, Oxford, 174 pp.

GOUDIE, A., in press. Petrographic characteristics of calcretes (caliches): modern analogues of ancient cornstones.

HAGEN, G. M. and LOGAN, B. W., 1974. History of Hutchinson embayment tidal flat, Shark Bay, Western Australia. *Am. Assoc. Petrol Geol., Mem.*, 22: 283–315.

HAMZA, M. S. and BROEKER, W. S., 1974. Surface effect on the isotopic fractionation between CO_2 and some carbonate minerals. *Geochim. Cosmochim. Acta,* 38: 669–681.

HANSHAW, B. B., BACK, W. and RUBIN, M., 1965. Carbonate equilibria and radiocarbon distribution related to groundwater flow in the Floridan limestone aquifer, U.S.A. *Proc. Symp., Dubrovnik, Oct. 1965, Vol. II*. IASH/UNESCO, 1967.

HANSHAW, B. B., BACK, W. and DEIKE, R. G., 1971. A geochemical hypothesis for dolomitization by ground water. *Econ. Geol.*, 66: 710–724.

HAY, W. W., 1973. Significance of paleontological results of Deep-Sea Drilling Project Legs 1-9. *Am. Assoc Petrol. Geol., Bull.*, 57: 55–62.

HAY, W. W., WISE, S.W. and STIEGLITZ, R. D., 1970. Scanning electron microscope study of fine grain size biogenic particles. *Trans. Gulf-Coast Assoc. Geol. Socs.*, 20: 287–302.

HEATH, G. R. and CULBERSON, C., 1970. Calcite: degree of saturation, rate of dissolution, and the compensation depth in the deep oceans. *Geol. Soc. Am., Bull.*, 81: 3157–3160.

HENDY, C. H., 1971. The isotopic geochemistry of speleothems—I. The calculation of the effects of different modes of formation on the isotopic composition of speleothems and their applicability as palaeoclimatic indicators. *Geochim. Cosmochim. Acta*, 35: 801–824.

HOFFMAN, P., 1973. Recent and ancient algal stromatolites: seventy years of pedagogic cross-pollination. In: R. N. GINSBURG (Editor), *Evolving Concepts in Sedimentology*. Johns Hopkins, Baltimore, Md., pp. 178–191.

HOFFMAN, P., LOGAN, B. W. and GEBELEIN, C. D., in press. Algal mats, cryptalgal fabrics and structures, Hamelin Pool, Shark Bay. *Am. Assoc. Petrol. Geol., Mem.*

HOFMANN, H. J., 1969. Attributes of stromatolites. *Geol. Surv. Can., Pap.*, 69–39: 1–58.

HOFMANN, H. J., 1973. Stromatolites: characteristics and utility. *Earth-Sci. Rev.*, 9: 339–373.

HOMMERIL, P., 1971. Dynamique du transport des sédiments calcaires dans la partie nord du golfe normand-breton. *Bull. Soc. Géol. France, Sér. 7.*, 12: 31–41.

HOROWITZ, A. S. and POTTER, P. E., 1971. *Introductory Petrography of Fossils*. Springer, Heidelberg, 302 pp.

HOWARD, J. F., KISSLING, D. L. and LINEBACK, J. A., 1970. Sedimentary facies and distribution of biota in Coupon Bight, Lower Florida Keys. *Geol. Soc. Am., Bull.*, 81: 1929–1946.

HSÜ, K. J. and JENKYNS, H. C. (Editors), 1974. *Pelagic Sedimentation: on Land and under the Sea— Int. Assoc. Sedimentol., Spec. Publ.*, 1: 1–447.

HSÜ, K. J. and SCHNEIDER, J., 1973. Progress report on dolomitization—hydrology of Abu Dhabi sabkhas, Arabian Gulf. In: B. H. PURSER (Editor), *The Persian Gulf: Holocene Carbonate Sedimentation and Diagenesis in a Shallow Epicontinental Sea*. Springer, Berlin, pp. 409–422.

HUBBARD, J. A. E. B., 1972a. Cavity formation in living scleratinian reef corals and fossil analogues. *Geol. Rundsch.*, 61: 551–564.

HUBBARD, J. A. E. B., 1972b. Stromatolitic fabric: a petrographic model. *Int. Geol. Congr., 24th, Montreal, 1972, Rep. Sect.*, 7: 380–396.

HUBBARD, J. A. E. B., 1973. Sediment-shifting experiments: a guide to functional behaviour in colonial corals. In: R. S. BOARDMAN, A. H. CHEETHAM and W. A. OLIVER (Editors), *Animal Colonies*. Dowden, Hutchinson and Ross, Stroudsburg, Pa., pp. 31–42.

HUBBARD, J. A. E. B., 1974a. Barnes' technique amended for analyzing fabric and cavity development in coral reef communities. *J. Paleontol.*, 48: 769–777.

HUBBARD, J. A. E. B., 1974b. Coral colonies as micro-environmental indicators. *Ann. Soc. Géol. Belg.*, 97: 143–152.

HUBBARD, J. A. E. B., 1975. Life and the afterlife of reef corals: a timed study of incipient diagenesis. *Int. Sediment. Congr., 9th, Nice, 1975*, in press.

HUBBARD, J. A. E. B. and POCOCK, Y. P., 1972. Sediment rejection by recent scleractinian corals: a key to palaeo-environmental reconstruction. *Geol. Rundsch.*, 61: 598–626.

HUDSON, J. D., 1975. Carbon isotopes and limestone cements. *Geology*, 3: 19–22.

JACKA, A. D., 1974. Differential cementation of a Pleistocene carbonate fanglomerate, Guadalupe Mountains. *J. Sediment. Petrol.*, 44: 85–92.

JACKSON, J. B. C., GOREAU, T. F. and HARTMAN, W. D., 1971. Recent brachiopod-coralline sponge communities and their paleoecological significance. *Science*, 173: 623–625.

JACKSON, T. A. and BISCHOFF, J. L., 1971. The influence of amino acids on the kinetics of the recrystallization of aragonite to calcite. *J. Geol.*, 79: 493–497.

JAMES, N. P., 1972. Holocene and Pleistocene calcareous crust (caliche) profiles: criteria for subaerial exposure. *J. Sediment. Petrol.*, 42: 817–836.

JAMES, N. P., 1974. Diagenesis of scleractinian corals in the subaerial vadose environment. *J. Sediment. Petrol.*, 48: 785–799.

JAUZEIN, A. and PERTHUISOT, J.-P., 1972. Quelques réflexions sur la sédimentation marine à propos de la Sebkha el Melah (Zarzis-Tunisie). *C. R. Soc. Géol. France*, 2: 86–87.

JONES, O. A. and ENDEAN, R. (Editors), 1973. *Biology and Geology of Coral Reefs. Vol. 1 Geology 1*. Academic Press, New York, N.Y., 410 pp.

JOPE, M., 1973. The protein of brachiopod shell—V. N-terminal end groups. *Comp. Biochem. Physiol.*, 45B: 17–24.

JORDAN, C. F., 1973. Carbonate facies and sedimentation of patch reefs off Bermuda. *Am. Assoc. Petrol. Geol., Bull.*, 57: 42–54.

KAHLE, C. F., 1974. Ooids from Great Salt Lake, Utah, as an analogue for the genesis and diagenesis of ooids in marine limestones. *J. Sediment. Petrol.*, 44: 30–39.

KANEPS, A., 1973. Carbonate chronology for Pliocene deep-sea sediments. In: L. F. MUSICH and O. E. WESER (Editors), *Initial Reports of the Deep Sea Drilling Project, Vol. XVIII*. U.S. Government Printing Office, Washington, D.C., pp. 873–881.

KATZ, A., 1971. Zoned dolomite crystals. *J. Geol.*, 79: 38–51.

KATZ, A., 1973. The interaction of magnesium with calcite during crystal growth at 25–90°C and one atmosphere. *Geochim. Cosmochim. Acta*, 37: 1563–1586.

KATZ, A., SASS, E., STARINSKY, A. and HOLLAND, H. D., 1972. Strontium behaviour in the aragonite–calcite transformation: an experimental study at 40–98°C. *Geochim. Cosmochim. Acta*, 36: 481–496.

KEITH, M. L. and WEBER, J. N., 1964. Carbon and oxygen isotopic composition of selected limestones and fossils. *Geochim. Cosmochim. Acta*, 28: 1787–1816.

KENDALL, A. C. and TUCKER, M. E., 1973. Radiaxial fibrous calcite: a replacement after acicular carbonate. *Sedimentology*, 20: 365–389.

KENNEDY, W. J. and KLINGER, H. C., 1972. Hiatus concretions and hardground horizons in the Cretaceous of Zululand (South Africa). *Palaeontology*, 15: 539–549.

KENNEDY, W. J., MORRIS, N. J. and TAYLOR, J. D., 1970. The shell structure, mineralogy and relationships of the Chamacea (Bivalvia). *Palaeontology*, 13: 379–413.

KLOTZ, D., 1973. Untersuchungen zur Dispersion in porösen Medien. *Z. Dtsch. Geol. Ges.*, 124: 523–533.

KONISHI, K. and SAKAI, H., 1972. Fibrous aragonite of fresh-water origin in sealed Pliocene *Glycymeris yessoensis*. *Jap. J. Geol. Geogr.*, 42: 19–30.

KONISHI, K., KANESHIMA, K., NAKAGAWA, K. and SAKAI, H., 1972. Pleistocene dolomite and associated carbonates in south Okinawa, the Ryukyu Islands. *Geochem. J.*, 6: 17–36.

KOSTECKA, A., 1972. Calcite paramorphs in the aragonite concretions. *Ann. Soc. Géol. Pol.*, 42: 289–296.

KREBS, W., 1969. Early void-filling cementation in Devonian fore-reef limestones (Germany). *Sedimentology*, 12: 279–299.

KREBS, W., 1971. Devonian reef limestones in the eastern Rhenish Schiefergebirge. In: G. MÜLLER (Editor), *Sedimentology of Parts of Central Europe*. Kramer, Frankfurt, pp. 45–81.

KREBS, W., 1972. Facies and development of the Meggen reef (Devonian, West Germany). *Geol. Rundsch.*, 61: 647–671.

KRUMBEIN, W. E., 1968. Geomicrobiology and geochemistry of the "Nari-lime-crust" (Israel). In: G. MÜLLER and G. M. FRIEDMAN (Editors), *Recent Developments in Carbonate Sedimentology in Central Europe*. Springer, Berlin, pp. 138–147.

LAFON, G. M., 1970. Calcium complexing with carbonate ion in aqueous solutions at 25°C and 1 atmosphere. *Geochim. Cosmochim. Acta*, 34: 935–940.

LANCELOT, Y. and EWING, J. I., 1972. Correlation of natural gas zonation and carbonate diagenesis in Tertiary sediments from the north west Atlantic. In: C. D. HOLLISTER et al. (Editors), *Initial Reports of the Deep Sea Drilling Project, Vol. XI*. U.S. Government Printing Office, Washington, D.C., pp. 791–799.

LANCELOT, Y., HATHAWAY, J. C. and HOLLISTER, C. D., 1972. Lithology of sediments from the western North Atlantic. In: C. D. HOLLISTER et al. (Editors), *Initial Reports of the Deep Sea Drilling Project, Vol. XI*. U.S. Government Printing Office, Washington, D.C., pp. 901–949.

LAND, L. S., 1970. Carbonate mud: production by epibiont growth on *Thalassia testudinum*. *J. Sediment. Petrol.*, 40: 1361–1363.

LAND, L. S., 1973a. Contemporaneous dolomitization of Middle Pleistocene reefs by meteoric water, North Jamaica. *Bull. Marine Sci. Gulf Caribb.*, 23: 64–92.

LAND, L. S., 1973b. Holocene meteoric dolomitization of Pleistocene limestones, North Jamaica. *Sedimentology*, 20: 411–424.

LANG, J. C., 1973. Interspecific aggression by scleractinian corals. 2. Why the race is not only to the swift. *Bull. Marine Sci. Gulf Caribb.*, 23: 260–279.

LANG, J. C., 1974. Biological zonation at the base of a reef. *Am. Sci.*, 62: 272–281.

LANGMUIR, D., 1971. The geochemistry of some carbonate ground waters in central Pennsylvania. *Geochim. Cosmochim. Acta*, 35: 1023–1045.

LAPORTE, L. F. (Editor), 1974. *Reefs in Space and Time: Selected Examples from the Recent and Ancient—Soc. Econ. Paleontol. Mineral., Spec. Publ.*, 18: 1–256.

LATTMAN, L. II. and SIMONBERG, E. M., 1971. Case-hardening of carbonate alluvium and colluvium, Spring Mountains, Nevada. *J. Sediment. Petrol.*, 41: 274–281.

LEES, A., 1974. Contrasts between warm- and cold-water shelf carbonates: significance in the interpretation of ancient limestones. *Ann. Soc. Géol. Belg.*, 97: 159–161.

LEES, A. and BULLER, A. T., 1972. Modern temperate-water and warm-water shelf carbonate sediments contrasted. *Marine Geol.*, 13: 67–73.

LEGRAND, H. E. and STRINGFIELD, V. T., 1973. Karst hydrology—a review. *J. Hydrol.*, 20: 97–120.

LEVANDOWSKI, D. W., KALEY, M. E., SILVERMAN, S. R. and SMALLEY, R. G., 1973. Cementation in Lyons Sandstone and its role in oil accumulation, Denver basin, Colorado. *Am. Assoc. Petrol. Geol., Bull.*, 57: 2217–2244.

LI, Y. H., TAKAHASHI, T. and BROECKER, W. S., 1969. Degree of saturation of $CaCO_3$ in the oceans. *J. Geophys. Res.*, 74: 5507–5525.

LINDHOLM, R. C., 1974. Fabric and chemistry of pore filling calcite in septarian veins: models for limestone cementation. *J. Sediment. Petrol.*, 44: 428–440.

LINDHOLM, R. C. and FINKLEMAN, R. B., 1972. Calcite staining: semiquantitative determination of ferrous iron. *J. Sediment. Petrol.*, 42: 239–242.

LIPPMANN, F., 1973. *Sedimentary Carbonate Minerals*. Springer, Berlin, 228 pp.

LISITZIN, A. P., 1972. *Sedimentation in the World Ocean—Soc. Econ. Paleontol. Mineral., Spec. Publ.*, 17: i–xiii, 1–218.

LLOYD, R. M., 1971. Some observations on Recent sediment alteration ("micritization") and the possible role of algae in submarine cementation. In: O. P. BRICKER (Editor), *Carbonate Cements*. Johns Hopkins, Baltimore, Md., pp. 72–79.

LLOYD, R. M. and HSÜ, K. J., 1972. Stable-isotope investigations of sediments from the DSDP III cruise to South Atlantic. *Sedimentology*, 19: 45–58.

LOGAN, B. W., HOFFMAN, P. and GEBELEIN, C. D., 1974a. Evolution and diagenesis of Quaternary carbonate sequences, Shark Bay, Western Australia. *Am. Assoc. Petrol. Geol., Mem.*, 22: 1–358.

LOGAN, B. W., HOFFMAN, P. and GEBELEIN, C. D., 1974b. Algal mats, cryptalgal fabrics and structures, Hamelin Pool, Shark Bay, Western Australia. *Am. Assoc. Petrol. Geol., Mem.*, 22: 140–194.

Loreau, J.-P., 1972a. Pétrographie de calcaires fins au microscope électronique à balayage: introduction à une classification des "micrites". C. R., 274: 810–813.

Loreau, J.-P., 1972b. Relation entre structure, ultrastructure et milieu des oolithes de la Trucial Coast (Golfe persique). Mécanismes de l'oolithisation. Ann. Soc. Géol. Belg., 95: 393–398.

Loreau, J.-P., 1972c. Une explication de la biréfrigence anormalement faible de l'aragonite du cortex des oolithes marines. Bull. Soc. Fr. Minéral. Cristallogr., 95: 595–602.

Loreau, J.-P., 1973. Nouvelles observations sur la genèse et la signification des oolithes. Sci. Terre, 18: 213–244.

Loreau, J.-P. and Purser, B. H., 1973. Distribution and ultrastructure of Holocene ooids in the Persian Gulf. In: B. H. Purser (Editor), The Persian Gulf: Holocene Carbonate Sedimentation and Diagenesis in a Shallow Epicontinental Sea. Springer, Berlin, pp. 279–328.

Lovering, T. S., 1969. The origin of hydrothermal and low temperature dolomite. Econ. Geol., 64: 743–754.

Mackinnon, D. I. and Williams, A., 1974. Shell structure of terebratulid brachiopods. Palaeontology, 17: 179–202.

Macqueen, R. W., Ghent, E. D. and Davies, G. R., 1974. Magnesium distribution in living and fossil specimens of the echinoid Peronella lesueuri Agassiz, Shark Bay, Western Australia. J. Sediment. Petrol., 44: 60–69.

Magara, K., 1974. Compaction, ion filtration, and osmosis in shale and their significance in primary migration. Am. Assoc. Petrol. Geol., Bull., 58: 283–290.

Magaritz, M., 1974. Lithification of chalky limestone: a case study in Senonian rocks from Israel. J. Sediment. Petrol., 44: 947–954.

Maleev, M. N., 1972. Diagnostic features of spherulites formed by splitting of a single-crystal nucleus. Growth mechanism of chalcedony. Mineral. Petrog. Mitt., 18: 1–16.

Manheim, F. T., 1970. The diffusion of ions in unconsolidated sediments. Earth Planet. Sci. Lett., 9: 307–309.

Manheim, F. T. and Sayles, F. L., 1971. Interstitial water studies on small core samples. Deep Sea Drilling Project, Leg 8. In: J. I. Tracey et al. (Editors), Initial Reports of the Deep Sea Drilling Project, Vol. VIII. U.S. Government Printing Office, Washington, D.C., pp. 857–869.

Manheim, F. T., Chan, D. M., Kerr, D. and Sunda, W., 1970. Interstitial water studies on small core samples, Deep Sea Drilling Project, Leg 3. In: A. E. Maxwell et al. (Editors), Initial Reports of the Deep Sea Drilling Project, Vol. III. U.S. Government Printing Office, Washington, D.C., pp. 663–666.

Manus, R. W. and Coogan, A. H., 1974. Bulk volume reduction and pressure-solution derived cement. J. Sediment. Petrol., 44: 466–471.

Margolis, S. and Rex, R. W., 1971. Endolithic algae and micrite envelope formation in Bahamian oölites as revealed by scanning electron microscopy. Geol. Soc. Am., Bull., 82: 843–852.

Massaad, M., 1972. Les concrétions de "l'Aalénien". Schweiz. Mineral. Petrog. Mitt., 53: 405–459.

Matter, A., 1974. Burial diagnesis of pelitic and carbonate deep-sea sediments from the Arabian Sea. In: R. B. Whitmarsh et al. (Editors), Initial Reports of the Deep Sea Drilling Project, Volume XXIII. U.S. Government Printing Office, Washington, D.C., pp. 421–469.

Matthews, R. K., 1974. A process approach to diagenesis of reefs and reef associated limestones. In: L. F. Laporte (Editor), Reefs in Time and Space—Soc. Econ. Paleontol. Mineral., Spec. Publ., 18: 234–256.

McCunn, H. J., 1972. Calcite and aragonite phenomena precipitated by organic decay in high lime concentrate brines. J. Sediment. Petrol., 42: 150–154.

McIntyre, A. and McIntyre, R., 1971. Coccolith concentrations and differential solution in oceanic sediments. In: B. M. Funnell and W. R. Riedel (Editors), The Micropaleontology of Oceans. Cambridge Univ. Press, Cambridge, pp. 253–261.

Meyers, W. J., 1974. Carbonate cement stratigraphy of the Lake Valley Formation (Mississippian) Sacramento Mountains, New Mexico. J. Sediment. Petrol., 44: 837–861.

Millero, F. J., 1974. The physical chemistry of seawater. In: F. A. Donath, F. G. Stehli and G. W. Wetherill (Editors), Annual Review of Earth and Planetary Sciences, Vol. 2. Annual Reviews, Inc., Palo Alto, Calif., pp. 101–150.

MILLERO, F. J. and BERNER, R. A., 1972. Effects of pressure on carbonate equilibria in sea water. *Geochim. Cosmochim. Acta*, 36: 92–98.

MILLIMAN, J. D., 1974. *Marine Carbonates*. Springer, Berlin, 375 pp.

MILLIMAN, J. D. and MÜLLER, J., 1973. Precipitation and lithification of magnesian calcite in the deep-sea sediments of the eastern Mediterranean Sea. *Sedimentology*, 20: 29–45.

MILLIMAN, J. D., GASTNER, M. and MÜLLER, J., 1971. Utilization of magnesium in coralline algae. *Geol. Soc. Am., Bull.*, 82: 573–580.

MIMRAN, Y., 1975. Fabric deformation induced in Cretaceous chalks by tectonic stresses. *Tectonophysics*, 26: 309–316.

MIŠÍK, M., 1971. Observations concerning calcite veinlets in carbonate rocks. *J. Sediment. Petrol.*, 41: 450–460.

MITTERER, R. M., 1971. Influence of natural organic matter on $CaCO_3$ precipitation. In: O. P. BRICKER (Editor), *Carbonate Cements*. Johns Hopkins, Baltimore, Md., pp. 252–258.

MITTERER, R. M., 1972a. Biogeochemistry of aragonite mud and oolites. *Geochim. Cosmochim. Acta*, 36: 1407–1422.

MITTERER, R. M., 1972b. Calcified proteins in the sedimentary environment, In: H. R. V. GAERTNER and H. WEHNER (Editors), *Advances in Organic Geochemistry, 1971*. Pergamon, Oxford, pp. 441–451.

MOBERLY, R., 1973. Rapid chamber-filling growth of marine aragonite and Mg-calcite. *J. Sediment. Petrol.*, 43: 634–635.

MÖLLER, P., 1973. Determination of the composition of surface layers of calcite in solutions containing Mg^{2+}. *J. Inorg. Nucl. Chem.*, 35: 395–401.

MÖLLER, P. and PAPENDORFF, H., 1971. Fractionation of calcium isotopes in carbonate precipitates. *Earth Planet. Sci. Lett.*, 11: 192–194.

MÖLLER, P. and PARFKH, P. P., 1975. Influence of magnesium on the ion activity product of calcium and carbonate dissolved in sea water: a new approach. *Marine Chem.*, 3.

MÖLLER, P. and RAJAGOPALAN, G., in press, Cationic distribution and structural changes of mixed Mg-Ca layers on calcite crystals. *J. Phys. Chem., N.F.*

MÖLLER, P. and SASTRI, C. S., 1974. Estimation of the number of surface layers of calcite involved in Ca–^{45}Ca isotopic exchange with solution. *Z. Phys. Chem. Frankf.*, 89: 80–87.

MÖLLER, P. and WERR, G., 1972. Influence of anions on Ca^{2+}–Mg^{2+} surface exchange process on calcite in artificial sea water. *Radiochim. Acta*, 18: 144–147.

MONTY, C. L. V., 1971. An autoecological approach of intertidal and deepwater stromatolites. *Ann. Soc. Géol. Belg.*, 94: 265–276.

MONTY, C. L. V., 1972. Recent algal stromatolitic deposits, Andros Island, Bahamas. Preliminary report. *Geol. Rundsch.*, 61: 742–783.

MONTY, C. V. L., 1973. Precambrian background and Phanerozoic history of stromatolitic communities, an overview. *Ann. Soc. Géol. Belg.*, 96: 585–624.

MONTY, C. L. V. (Editor), 1974. Aspects of reef and sedimentological studies. *Ann. Soc. Géol. Belg.*, 97: 139–183.

MOORE, C. H., 1973. Intertidal carbonate cementation, Grand Cayman, West Indies. *J. Sediment. Petrol.*, 43: 591–602.

MOORE, T. C., HEATH, G. R. and KOWSMANN, R. O., 1973. Biogenic sediments of the Panama Basin. *J. Geol.*, 83: 458–472.

MORSE, J. W., 1974. Dissolution kinetics of calcium carbonate in sea water. III: A new method for the study of carbonate reaction kinetics. *Am. J. Sci.*, 274: 97–107.

MORSE, J. W., in press. Dissolution kinetics of calcium carbonate in sea water. V: Effects of natural inhibitors and position of the lysocline. *Am. J. Sci.*, 274.

MORSE, J. W. and BERNER, R. A., 1972. Dissolution kinetics of calcium carbonate in sea water: II. A kinetic origin for the lysocline. *Am. J. Sci.*, 272: 840–851.

MOSSOP, G. D., 1972. Origin of the peripheral rim, Redwater reef, Alberta. *Bull. Can. Petrol. Geol.*, 20: 238–280.

MÜLLER, G., 1969a. High strontium contents and Sr/Ca-ratios in Lake Constance waters and carbonates and their sources in the drainage area of the Rhine river (Alpenrhein). *Mineral. Deposita*, 4: 75–84.

MÜLLER, G., 1969b. Diagenetic changes in interstitial waters of Holocene Lake Constance sediments. *Nature*, 224: 258–259.

MÜLLER, G., 1969c. Sedimentbildung im Plattensee/Ungarn. *Naturwissenschaften*, 56: 606–615.
MÜLLER, G., 1970a. Aragonite precipitation in a freshwater lake. *Nat. Phys. Sci.*, 229: 18.
MÜLLER, G., 1970b. Petrology of the Cliff Limestone (Holocene), North Bimini, Bahamas. *Neues Jahrb. Mineral. Monatsh.*, 11: 507–523.
MÜLLER, G., 1971. Sediments of Lake Constance. In: G. MÜLLER (Editor), *Sedimentology of Parts of Central Europe*. Kramer, Frankfurt, pp. 237–252.
MÜLLER, G. and BLASCHKE, R., 1969. Zur Entstehung des Tiefsee–Kalkschlammes im Schwarzen Meer. *Naturwissenschaften*, 56: 561–562.
MÜLLER, G. and FISCHBECK, R., 1973. Possible natural mechanism for protodolomite formation. *Nat. Phys. Sci.*, 242: 139–141.
MÜLLER, G. and STOFFERS, P., 1974. Mineralogy and petrology of Black Sea basin sediments. *Am. Assoc. Petrol. Geol., Mem.*, 20: 200–248.
MÜLLER, G., IRION, G. and FÖRSTNER, U., 1972. Formation and diagenesis of inorganic Ca-Mg-carbonates in the lacustrine environment. *Naturwissenschaften*, 59: 158–164.
MURATA, K. J., FRIEDMAN, I. and CREMER, M., 1972. Geochemistry of diagenetic dolomites in Miocene marine formations of California and Oregon. *U.S. Geol. Surv. Prof. Pap.*, 724–C: 1–12.
MURAVYOV, V. I., 1970. Formation of carbonate cement in clastic rocks. *Sedimentology*, 15: 139–145.
MUTVEI, H., 1969. On the micro- and ultra-structure of the conchiolin in the nacreous layer of some Recent and fossil molluscs. *Acta Univ. Stockh. Contrib. Geol.*, 20: 1–17.
MUTVEI, H., 1970. Ultrastructure of the mineral and organic components of molluscan nacreous layers. *Akad. Wiss. Lit. Mainz. Biomin. Repts.*, 2: 49–72.
MUTVEI, H., 1972a. Ultrastructural relationships between the prismatic and nacreous layers in *Nautilus* (Cephalopoda). *Akad. Wiss. Lit. Mainz. Biomin. Repts.*, 4: 81–86.
MUTVEI, H., 1972b. Ultrastructural studies on cephalopod shells. Pt. I: The septa and siphonal tube in *Nautilus*. *Bull. Geol. Inst. Univ. Uppsala*, 3: 237–261.

NEUGEBAUER, J., 1973. The diagenetic problem of chalk. The role of pressure solution and pore fluid. *Neues Jahrb. Paläontol. Abh.*, 143: 223–245.
NEUGEBAUER, J., 1974. Some aspects of cementation in chalk. In: K. J. HSÜ and H. C. JENKYNS (Editors), *Pelagic Sediments: on Land and under the Sea—Int. Assoc. Sedimentol., Spec. Publ.*, 1: 149–176.
NOBLE, J. P. A. and HOWELLS, K. D. M., 1974. Early marine lithification of the nodular limestones in the Silurian of New Brunswick. *Sedimentology*, 21: 597–609.

OERTEL, G. and CURTIS, C. D., 1972. Clay-ironstone concretion preserving fabrics due to progressive compaction. *Geol. Soc. Am., Bull.*, 83: 2597–2606.
OGDEN, J. C., BROWN, R. A. and SALESKY, N., 1973. Grazing by the echinoid *Diadema antillarum* Philippi: formation of halos around West Indian patch reefs. *Science*, 182: 715–717.
OLDERSHAW, A. E., 1972. Microporosity control in microcrystalline carbonate rocks, southern Ontario, Canada, *Int. Geol. Congr., 24th, Montreal, 1972, Rep. Sess.*, 6: 198–207.

PANDEY, G. N., TEK, M. R. and KATZ, D. L., 1974. Diffusion of fluids through porous media with implications in petroleum geology. *Am. Assoc. Petrol. Geol., Bull.*, 58: 291–303.
PARRY, W. T., REEVES, C. C. and LEACH, J. W., 1970. Oxygen and carbon isotopic composition of West Texas lake carbonates. *Geochim. Cosmochim. Acta*, 34: 825–830.
PASSERI, L., 1972. Ricerche sulla porosità delle rocce carbonatiche nella zona di M. Cucco (Appennino umbromarchigiano) in relazione alla genesi della canalizzazione interna. *Grotte D'Italia, Ser. 4*, 3: 5–38.
PATERSON, K., 1971. Some considerations concerning percolation waters in the Chalk of north Berkshire. *Trans. Cave Res. Group G.B.*, 13: 277–282.
PATERSON, K., 1972. Responses in the chemistry of spring waters in the Oxford region to some climatic variables. *Trans. Cave Res. Group G.B.*, 14: 132–140.
PATRIQUIN, D. G., 1972. Carbonate mud production by epibionts on *Thalassia*: an estimate based on leaf growth rate data. *J. Sediment. Petrol.*, 42: 687–689.

PEARSON, F. J. and HANSHAW, B. B., 1970. Sources of dissolved carbonate species in ground water and their effects on carbon-14 dating. In: *Isotope Hydrology, 1970.* Int. Atomic Energy Agency, Vienna, pp.271–286.

PERKINS, R. D. and HALSEY, S. D., 1971. Geologic significance of microboring fungi and algae in Carolina shelf sediments. *J. Sediment. Petrol.*, 41: 843–853.

PERKINS, R. D., MCKENZIE, M. D. and BLACKWELDER, P. L., 1972. Aragonite crystals within codiacean algae: distinctive morphology and sedimentary implications. *Science*, 175: 624–626.

PIALLI, G., 1971. Facies di piana cotidale nel Calcare Massiccio dell'Appennino umbro-marchigiano. *Boll. Soc. Geol. Italy*, 90. 481–507.

PIRLET, H., 1970. L'influence d'un karst sous-jacent sur la sédimentation calcaire et l'intérét de l'étude des paléokarsts. *Ann. Soc. Géol. Belg.*, 93: 247–254.

PITTMAN, E. D., 1971. Microporosity in carbonate rocks. *Am. Assoc. Petrol. Geol., Bull.*, 55: 1873–1881.

PITTMAN, E. D., 1974. Porosity and permeability changes during diagenesis of Pleistocene corals, Barbados, West Indies. *Geol. Soc. Am., Bull.*, 85: 1811–1820.

PITTMAN, E. D. and DUSCHATKO, R. W., 1970. Use of pore casts and SEM to study pore geometry. *J. Sediment. Petrol.*, 40: 1153–1157.

PITTY, A. F., 1971. Rate of uptake of calcium carbonate in underground karst water. *Geol. Mag.*, 108: 537–543.

PLESSMANN, W., 1972. Horizontal-Stylolithen im französisch-schweizerischen Tafel- und Fattenjura und ihre Einpassung in den regionalen Rahmen. *Geol. Rundsch.*, 61: 332–347.

PLUMMER, L. N. and MACKENZIE, F. T., 1974. Predicting mineral solubility from rate data: application to the dissolution of magnesian calcites. *Am. J. Sci.*, 274: 61–83.

POLUZZI, A. and SARTORI, R., 1973. Carbonate mineralogy of some Bryozoa from Talbot Shoal (Strait of Sicily, Mediterranean). *G. Geol., Ann. Mus. Geol. Bologna*, 39: 11–15.

PORTER, J. W., 1974. Community structure of coral reefs on opposite sides of the isthmus of Panama. *Science*, 186: 543–545.

PURDY, E. G., 1974a. Karst-determined facies patterns in British Honduras: Holocene carbonate sedimentation model. *Am. Assoc. Petrol. Geol., Bull.*, 58: 825–855.

PURDY, E. G., 1974b. Reef configurations: cause and effect. In: L. F. LAPORTE (Editor), *Reefs in Time and Space—Soc. Econ. Paleontol. Mineral., Spec. Publ.*, 18: 9–76.

PURSER, B. H., 1973. *The Persian Gulf—Holocene Carbonate Sedimentation and Diagenesis in a Shallow Epicontinental Sea.* Springer, Berlin, 471 pp.

PURSER, B. H. and LOREAU, J.-P., 1973. Aragonitic, supratidal encrustations on the Trucial Coast, Persian Gulf. In: B. H. PURSER (Editor), *The Persian Gulf—Holocene Carbonate Sedimentation and Diagenesis in a Shallow Epicontinental Sea.* Springer, Berlin, pp.343–376.

PYTKOWICZ, R. M., 1970. On the carbonate compensation depth in the Pacific Ocean. *Geochim. Cosmochim. Acta*, 34: 836–839.

PYTKOWICZ, R. M., 1971. Sand–seawater interactions in Bermuda beaches. *Geochim. Cosmochim. Acta*, 35: 509–515.

PYTKOWICZ, R. M. 1973. Calcium carbonate retention in supersaturated seawater. *Am. J. Sci.*, 273: 515–522.

RAISWELL, R., 1971. The growth of Cambrian and Liassic concretions. *Sedimentology*, 17: 147–171.

RAISWELL, R., in press. The formation of carbonate concretions by sulphate-reducing bacteria. *J. Sediment. Petrol.*

RAMSEY, T. S., SCHNEIDERMANN, N. and FINCH, J. W., 1973. Fluctuations in the past rates of carbonate solution in Site 149: a comparison with other oceanic basins and an interpretation of their significance. In: N. T. EDGAR et al. (Editors), *Initial Reports of the Deep Sea Drilling Project, Vol. XV.* U.S. Government Printing Office, Washington, D.C., pp.805–811.

READ, J. F., 1974a. Carbonate bank and wave-built platform sedimentation, Edel Province, Western Australia. *Am. Assoc. Petrol. Geol., Mem.*, 22: 1–60.

READ, J. F., 1974b. Calcrete deposits and Quaternary sediments, Edel Province, Western Australia. *Am. Assoc. Petrol. Geol., Mem.*, 22: 250–282.

REEVES, C. C., 1970. Origin, classification and geologic history of caliche on the southern High Plains, Texas and eastern New Mexico. *J. Geol.*, 78: 352–362.

REGENHARDT, H., 1961. Serpulidae (Polychaeta sedentaria) aus der Kreide Mitteleuropas, ihre ökologische, taxionomische und stratigraphische Bewertung. *Mitt. Geol. Staatsinst. Hamburg*, 30: 5–115.

RENARD, M., 1972. Interprétation des teneurs en strontium des carbonates du Lutétien supérieur, à Saint-Vaast-les-Mello (Oise). *Bull. Inf. Géol. Bass. Paris*, 34: 19–29.

RICHTER, D. K., 1974. Zur subaerischen Diagenese von Echinidenskeletten und das relative Alter pleistozäner Karbonatterrassen bei Korinth (Griechenland). *Neues Jahrb. Geol. Paläontol. Abh.*, 146: 51–77.

RILEY, J. P. and CHESTER, R., 1971. *Introduction to Marine Chemistry*. Academic Press, London, 465 pp.

RITTENHOUSE, G., 1971a. Pore-space reduction by solution and cementation. *Am. Assoc. Petrol. Geol., Bull.*, 55: 80–91.

RITTENHOUSE, G., 1971b. Mechanical compaction of sands containing different percentages of ductile grains: a theoretical approach. *Am. Assoc. Petrol. Geol., Bull.*, 55: 92–96.

ROONEY, W. S. and PERKINS, R. D., 1972. Distribution and geologic significance of microboring organisms within sediments of the Arlington Reef Complex, Australia. *Geol. Soc. Am., Bull.*, 83: 1139–1150.

ROSEN, B. R., 1971. Principle feature of reef coral ecology in shallow water environments of Mahé, Seychelles. In: D. R. STODDART and M. YONGE (Editors), *Regional Variation in Indian Ocean Coral Reefs*. Symp. Zool. Soc., Lond., 28, pp.163–183.

SANDBERG, P. A., 1971. Scanning electron microscopy of cheilostome bryozoan skeletons; techniques and preliminary observations. *Micropaleontology*, 17: 129–151.

SANDBERG, P. A., in press a. Bryozoan diagenesis: bearing on the nature of the original skeleton of rugose corals. (To be published in *J. Paleontol.*)

SANDBERG, P. A., in press b. New interpretations of Great Salt Lake ooids and of ancient non-skeletal carbonate mineralogy. (To be published in *Sedimentology.*)

SANDBERG, P. A., SCHNEIDERMANN, N. and WUNDER, S. J., 1973. Aragonitic ultrastructural relics in calcite-replaced Pleistocene skeletons. *Nat. Phys. Sci.*, 245: 133–134.

SARTORI, R., 1974. Modern deep-sea magnesian calcite in the central Tyrrhenian Sea. *J. Sediment. Petrol.*, 44: 1313–1322.

SASS, E. and KOLODNY, Y., 1972. Stable isotopes, chemistry and petrology of carbonate concretions (Mishash Formation, Israel). *Chem. Geol.*, 10: 261–286.

SASS, E., WEILER, Y. and KATZ, A., 1972. Recent sedimentation and oolite formation in the Ras Matarma Lagoon, Gulf of Suez. In: D. J. STANLEY (Editor), *The Mediterranean Sea: A Natural Sedimentation Laboratory*. Dowden, Hutchinson and Ross, Stroudsburg, Pa., pp.279–292.

SASTRI, C. S. and MÖLLER, P., 1974. Study of the influence of Mg^{2+} ions on Ca-^{45}Ca isotope exchange on the surface layers of calcite single crystals. *Chem. Phys. Lett.*, 26: 116–120.

SAYLES, F. L. and FYFE, W. S., 1973. The crystallization of magnesite from aqueous solution. *Geochim. Cosmochim. Acta*, 37: 87–99.

SAYLES, F. L. and MANHEIM, F. T., 1975. Interstitial solutions and diagenesis in deeply buried marine sediments: results from the Deep Sea Drilling Project. *Geochim. Cosmochim. Acta*, 39: 103–127.

SCHERER, M., 1974a. The influence of two endolithic micro-organisms on the diagenesis of recent coral skeletons. *Neues Jahrb. Geol. Paläontol. Monatsh.*, 9: 557–566.

SCHERER, M., 1974b. Submarine recrystallization of a coral skeleton in a Holocene Bahamian reef. *Geology*, 2: 499–500.

SCHERER, M., in press. Cementation and replacement of Pleistocene corals from the Bahamas and Florida: diagenetic influence of non-marine environments. *Neues Jahrb. Geol. Paläontol. Abh.*

SCHLANGER, S. O. and DOUGLAS, R. G., 1974. The pelagic ooze-chalk-limestone transition and its implications for marine stratigraphy. In: K. J. HSÜ and H. C. JENKYNS (Editors), *Pelagic Sediments: on Land and under the Sea—Int. Assoc. Sedimentol., Spec. Publ.*, 1: 117–148.

SCHLANGER, S. O., DOUGLAS, R. G., LANCELOT, Y., MOORE, T. C. and ROTH, P. H., 1973. Fossil preservation and diagenesis of pelagic carbonates from the Magellan Rise, central north

Pacific Ocean. In: P. H. ROTH and J. R. HERRING (Editors), *Initial Reports of the Deep Sea Drilling Project, Vol. XVII*. U.S. Government Printing Office, Washington, D.C., pp. 407–427.

SCHLOZ, W., 1972. Zur Bildungsgeschichte der Oolithenbank (Hettangium) in Baden-Württemberg. *Inst. Geol. Paläontol. Univ. Stuttgart, arb. N.F.*, 67: 101–212.

SCHMALZ, R. F., 1971. Formation of beach rock at Eniwetok Atoll. In: O. P. BRICKER (Editor), *Carbonate Cements*. Johns Hopkins, Baltimore, Md., pp.17–24.

SCHNEIDERMANN, N., 1970. Genesis of some Cretaceous carbonates in Israel. *Israel J. Eearth-Sci.*, 19: 97–115.

SCHNEIDERMANN, N., 1973a. Deposition of coccoliths in the compensation zone of the Atlantic Ocean. In: L. A. SMITH and J. HARDENBOL (Editors), *Proc. Symp. Calcareous Nannofossils —Soc. Econ. Paleontol. Mineral. (Gulf Coast Sect.)*, pp. 140–151.

SCHNEIDERMANN, N., 1973b. Pelagic limestones of the central Caribbean, Leg 15. In: N. T. EDGAR et al. (Editors), *Initial Reports of the Deep Sea Drilling Project, Vol. XV*. U.S. Government Printing Office, Washington, D.C., pp.773–793.

SCHNEIDERMANN, N., SANDBERG, P. A. and WUNDER, S. J., 1972. Recognition of early cementation of aragonite skeletal carbonates. *Nat. Phys. Sci.*, 240: 88–89.

SCHOLLE, P. A., 1971. Diagenesis of deep-water carbonate turbidites, Upper Cretaceous Monte Antola Flysch, Northern Apennines, Italy. *J. Sediment. Petrol.*, 41: 233–250.

SCHOLLE, P. A., 1974. Diagenesis of Upper Cretaceous chalks from England, Northern Ireland and the North Sea. In: K. J. HSÜ and H. C. JENKYNS (Editors), *Pelagic Sediments: on Land and under the Sea—Int. Assoc. Sedimentol., Spec. Publ.*, 1: 177–210.

SCHOLLE, P. A. and KINSMAN, D. J. J., 1974. Aragonitic and high-Mg calcite caliche from the Persian Gulf—a modern analog for the Permian of Texas and New Mexico. *J. Sediment. Petrol.*, 44: 904–916.

SCHOLLE, P. A. and KLING, S. A., 1972. Southern British Honduras: lagoonal coccolith ooze. *J. Sediment. Petrol.*, 42: 195–204.

SCHOLZ, R. W., 1973. Zur Sedimentologie und Kompaktion der Schreibkreide von Lägerdorf in SW-Holstein. *Neues Jahrb. Mineral. Abh.*, 118: 111–133.

SCHROEDER, J. H., 1972a. Calcified filaments of an endolithic alga in Recent Bermuda reefs. *Neues Jahrb. Geol. Paläontol. Monatsh.*, 1: 16–33.

SCHROEDER, J. H., 1972b. Fabrics and sequences of submarine carbonate cements in Holocene Bermuda cup reefs. *Geol. Rundsch.*, 61: 708–730.

SCHROEDER, J. H., 1973. Submarine and vadose cements in Pleistocene Bermuda reef rock. *Sediment. Geol.*, 10: 179–204.

SCHROEDER, J. H., 1974. Carbonate cements in Recent reefs of the Bermudas and Bahamas: keys to the past? *Ann. Soc. Géol. Belg.*, 97: 153–158.

SCHROEDER, J. H. and ZANKL, H., 1974. Dynamic reef formation: a sedimentological concept based on studies of Recent Bermuda and Bahama Reefs. In: A. M. CAMERON et al. (Editors), *Proc. 2nd Int. Symp. on Coral Reefs, Vol. 2*. Barrier Reef Committee, Brisbane, pp.413–428.

SCOFFIN, T. P., 1972a. Cavities in the reefs of the Wenlock Limestone (Mid-Silurian) of Shropshire, England. *Geol. Rundsch.*, 61: 565–578.

SCOFFIN, T. P., 1972b. Fossilization of Bermuda patch reefs. *Science*, 178: 1280–1282.

SEMINIUK, V., 1971. Subaerial leaching in the limestones of the Bowan Park Group (Ordovician) of central western New South Wales. *J. Sediment. Petrol.*, 41: 939–950.

SIMONE, L., 1974. Genesi e significato ambientale degli ooidi a struttura fibroso-raggiata di alcuni depositi mesozoici dell-area Appennino-dinarica e delle Bahamas meridionali. *Boll. Soc. Geol. Ital.*, 93: 513–545.

SLITER, W. V., BÉ, A. W. H. and BERGER, W. H. (Editors), 1975. *Dissolution of Deep-Sea Carbonates—Cushman Foundation for Foraminiferal Research, Spec. Publ.*, 13.

SMALLWOOD, P. V., in press a. Some aspects of the surface chemistry of calcite and aragonite. I. An electrokinetic study.

SMALLWOOD, P. V., in press b. Some aspects of the surface chemistry of calcite and aragonite. II. Crystal growth.

SMITH, D. B., 1974. Origin of tepees in Upper Permian shelf carbonates of Guadalupe Mountains, New Mexico. *Am. Assoc. Petrol. Geol., Bull.*, 58: 63–70.

SMITH, R. A., GRAHAM, E. H. and BAYER, F. M., 1973. Papers in Honor of Dr. Thomas F. Goreau. *Bull. Marine Sci. Gulf Caribb.*, 23: 1–464.

SMITH, S. V., DYGAS, J. A. and CHAVE, K. E., 1969. Distribution of calcium carbonate in pelagic sediments. *Marine Geol.*, 6: 391–400.

SOMMER, S. E., 1972. Cathodoluminescence of carbonates, 1. Characterization of cathodoluminescence from carbonate solid solutions. *Chem. Geol.*, 9: 257–273.

SORAUF, J. E., 1972. Skeletal microstructure and microarchitecture in Scleractinia (Coelenterata). *Palaeontology*, 15: 88–107.

SPIRO, B. F., 1971a. Ultrastructure and chemistry of the skeleton of *Tubipora musica* Linné. *Bull. Geol. Soc. Den.*, 20: 279–284.

SPIRO, B. F., 1971b. Diagenesis of some scleractinian corals from the Gulf of Elat, Israel. *Bull. Geol. Den.*, 21: 1–10.

SPIRO, B. F. and HANSEN, H. J., 1970. Note on early diagenesis of some scleractinian corals from the Gulf of Elat, Israel. *Bull. Geol. Soc. Den.*, 20: 72–78.

STEARN, C. W., 1972. The relationship of the stromatoporoids to the sclerosponges. *Lethaia*, 5: 369–388.

STEINEN, R. P., 1974. Phreatic and vadose diagenetic modification of Pleistocene limestone: petrographic observations from subsurface of Barbados, West Indies. *Am. Assoc. Petrol. Geol., Bull.*, 58: 1008–1024.

STEINEN, R. P. and MATTHEWS, R. K., 1973. Phreatic vs. vadose diagenesis: stratigraphy and mineralogy of a cored borehole on Barbados, W. I. *J. Sediment. Petrol.*, 43: 1012–1020.

STIEGLITZ, R. D., 1972. Scanning electron microscopy of the fine fraction of Recent carbonate sediments from Bimini, Bahamas. *J. Sediment. Petrol.*, 42: 211–226.

STIEGLITZ, R. D., 1973. Carbonate needles: additional organic sources. *Geol. Soc. Am., Bull.*, 84: 927–930.

STODDART, D. R. and YONGE, M. (Editors), 1971. *Regional Variation in Indian Ocean Coral Reefs.* Symp. Zool. Soc. Lond., 28: 584 pp.

STURANI, C., 1971. Ammonites and stratigraphy of the "Posidonia Alpina" beds of the Venetian Alps. *Mem. Ist. Geol. Mineral. Padova*, 28: 1–190.

SUESS, E., 1970. Interaction of organic compounds with calcium carbonate—I. Association phenomena and geochemical implications. *Geochim. Cosmochim. Acta*, 34: 157–168.

SUESS, E., 1973. Interaction of organic compounds with calcium carbonate—II. Organo-carbonate association in Recent sediments. *Geochim. Cosmochim. Acta*, 37: 2435–2447.

SUESS, E. and FÜTTERER, D., 1972. Aragonitic ooids: experimental precipitation from seawater in the presence of humic acid. *Sedimentology*, 19: 129–139.

SUPKO, P. R., STOFFERS, P. and COPLEN, T. B., 1974. Petrography and geochemistry of Red Sea dolomite. In: R. B. WHITMARSH et al. (Editors), *Initial Reports of the Deep Sea Drilling Project, Vol. XXIII.* U.S. Government Printing Office, Washington, D.C., pp.867–878.

SWEETING, M. M., 1972. *Karst Landforms.* Macmillan, London, 362 pp.

SWINEFORD, A., LEONARD, A. B. and FRYE, J. C., 1958. Petrology of the Pliocene pisolitic limestone in the Great Plains. *Bull. Geol. Surv. Kansas*, 130: 97–116.

TALBOT, F. H. and TALBOT, M. S., 1971. The crown-of-thorns starfish (Acanthaster) and the Great Barrier Reef. *Endeavour*, 30: 38–42.

TALBOT, M. R., 1971. Calcite cements in the Corallian Beds (Upper Oxfordian) of southern England. *J. Sediment. Petrol.*, 41: 261–273.

TALBOT, M. R., 1972. The preservation of scleractinian corals by calcite in the Corallian Beds (Oxfordian) of southern England. *Geol. Rundsch.*, 61: 731–742.

TAN, F. C. and HUDSON, J. D., 1974. Isotopic studies on the palaeoecology and diagenesis of the Great Estuarine Series (Jurassic) of Scotland. *Scott. J. Geol.*, 10: 91–128.

TAYLOR, J. D., 1971a. Intertidal zonation at Aldabra Atoll. *Philos. Trans. R. Soc. Lond., Ser. B.*, 260: 173–213.

TAYLOR, J. D., 1971b. Reef associated molluscan assemblages in the western Indian Ocean. In: D. R. STODDART and M. YONGE (Editors), *Regional Variation in Indian Ocean Coral Reefs.* Symp. Zool. Soc. Lond., 28, pp. 501–534.

TAYLOR, J. D., 1973. The structural evolution of the bivalve shell. *Palaeontology*, 16: 519–534.

TAYLOR, J. D. and LAYMAN, M., 1972. The mechanical properties of bivalve (Mollusca) shell structures. *Palaeontology*, 15: 73–87.

TAYLOR, J. D. and LEWIS, M. S., 1970. The flora, fauna and sediments of the marine grass beds of Mahé, Seychelles. *J. Nat. Hist.*, 4: 199–220.

TAYLOR, J. D., KENNEDY, W. J. and HALL, A., 1973. The shell structure and mineralogy of the Bivalvia II. Lucinacea-Clavagellacea conclusions. *Bull. Brit. Mus. (Nat. Hist.) Zool.*, 22: 253–294.

THOMASSIN, B. A., 1971. Les facies d'épifaune et d'épiflore des biotopes sédimentaires des formations coralliennes dans la région de Tuléar (sud-ouest de Madagascar). In: D. R. STODDART and M. YONGE (Editors), *Regional Variation in Indian Ocean Coral Reefs*. Symp. Zool. Soc. Lond., 28, pp.371–396.

THOMASSIN, B. A., 1973. Peuplements des sables fins sur les pentes internes des récifs coralliens de Tuléar (S.-W. de Madagascar). *Téthys Suppl.*, 5: 157–220.

THOMPSON, G., 1972. A geochemical study of some lithified carbonate sediments from the deep-sea. *Geochim. Cosmochim. Acta*, 36: 1237–1253.

THORSTENSON, D. C. and MACKENZIE, F. T., 1974. Time variability of pore water chemistry in recent carbonate sediments, Devil's Hole, Harrington Sound, Bermuda. *Geochim. Cosmochim. Acta*, 38: 1–19.

THORSTENSON, D. C., MACKENZIE, F. T. and RISTVET, B. L., 1972. Experimental vadose and phreatic cementation of skeletal carbonate sand. *J. Sediment. Petrol.*, 42: 162–167.

THRAILKILL, J., 1971. Carbonate deposition in Carlsbad Caves. *J. Geol.*, 79: 683–695.

TOWE, K. M., 1971. Lamellar wall construction in planktonic foraminifera. In: A. FARINACCI (Editor), *Proc. Planktonic Conference, 2nd, Rome, 1970*. Tecnoscienza, Rome, pp.1213 1224.

TOWE, K. M., 1972. Invertebrate shell structure and the organic matric concept. *Akad. Wiss. Lit. Mainz Biomin. Repts.*, 4: 1–14.

TOWE, K. M. and THOMPSON, G. R., 1972. The structure of some bivalve shell carbonates prepared by ion-beam thinning. A comparison study. *Calc. Tiss. Res.*, 10: 38–48.

USDOWSKI, E., 1973. Das geochemische Verhalten des Strontiums bei der Genese und Diagenese von Ca-Karbonat- und Ca-Sulfat- Mineralen. *Contrib. Mineral. Petrol.*, 38: 177–195.

VAN ANDEL, T. H. and MOORE, T. C., 1974. Cenozoic calcium carbonate distribution and calcite compensation depth in the central equatorial Pacific Ocean. *Geology*, 2: 87–92.

VEIZER, J. and DEMOVIČ, R., 1974. Strontium as a tool in facies analysis. *J. Sediment. Petrol.*, 44: 93–115.

VEIZER, J., DEMOVIČ, R. and TURAN, J., 1971. Possible use of strontium in sedimentary carbonate rocks as a paleoenvironmental indicator. *Sediment. Geol.*, 5: 5–22.

VOSS-FOUCART, M. F. and GRÉGROIRE, C., 1971. Biochemical composition and submicroscopic structure of matrices of nacreous conchiolin in fossil cephalopods (nautiloids and ammonoids). *Bull. Inst. R. Sci. Nat. Belg.*, 47: 1–42.

WACHS, D. and HEIN, J. R., 1974. Petrography and diagenesis of Franciscan limestones. *J. Sediment. Petrol.*, 44: 1217–1231.

WALKDEN, G. M., 1974. Palaeokarstic surfaces in Upper Visean (Carboniferous) limestones of the Derbyshire Block, England. *J. Sediment. Petrol.*, 44: 1232–1247.

WALTER, M. R., GOLUBIĆ, S. and PREISS, W. V., 1973. Recent stromatolites from hydromagnesite and aragonite depositing lakes near the Coorong Lagoon, South Australia. *J. Sediment. Petrol.*, 43: 1021–1030.

WARDLAW, N. C., in press a. Pore systems in carbonates as revealed by pore casts and capillary pressure data. *Am. Assoc. Petrol. Geol.*, Bull.

WARDLAW, N. C., in press b. Water above the transition zone in carbonate oil reservoirs. *Bull. Can. Petrol. Geol.*

WARDLAW, N. C. and REINSON, G. E., 1971. Carbonate and evaporite deposition and diagenesis, Middle Devonian Winnipegosis and Prairie Evaporite Formations of south-central Saskatchewan. *Am. Assoc. Petrol. Geol., Bull.*, 55: 1759–1786.

WEBER, J. N., 1973a. Temperature dependence of magnesium in echinoid and asteroid skeletal calcite: a reinterpretation of its significance. *J. Geol.*, 81: 543–556.

WEBER, J. N., 1973b. Incorporation of strontium into reef coral skeletal carbonate. *Geochim. Cosmochim. Acta*, 37: 2173–2190.

WEBER, J. N., 1974. Skeletal chemistry of scleractinian reef corals: uptake of magnesium from sea water. *Am. J. Sci.*, 274: 84–93.

WEST, I. M., 1973. Carbonate cementation of some Pleistocene temperate marine sediments. *Sedimentology*, 20: 229–249.

WHELAN, T and ROBERTS, H. H., 1973. Carbon isotope composition of diagenetic carbonate nodules from freshwater swamp sediments. *J. Sediment. Petrol.*, 43: 54–58.

WIEDER, M. and YAALON, D. H., 1974. Effect of matrix composition on carbonate nodule crystallization. *Geoderma*, 11: 95–121.

WIENS, H. J., 1962. *Atoll Environment and Ecology*. Yale Univ. Press, New Haven, Conn., 532 pp.

WIGLEY, P., 1973. The distribution of strontium in limestones on Barbuda, West Indies. *Sedimentology*, 20: 295–304.

WIGLEY, T. M. L., 1973. The incongruent solution of dolomite. *Geochim. Cosmochim. Acta*, 37: 1397–1402.

WILLIAMS, A. and WRIGHT, A. D., 1970. Shell structure of the Craniacea and other calcareous inarticulate Brachiopoda. *Palaeontol. Assoc. Lond., Spec. Pap.*, 7: 1–51.

WILLIAMS, R. E., 1970. Groundwater flow systems and accumulation of evaporite minerals. *Am. Assoc. Petrol. Geol., Bull.*, 54: 1290–1295.

WINLAND, D. and MATTHEWS, R. K., 1974. Origin and significance of grapestone, Bahama Islands. *J. Sediment. Petrol.*, 44: 921–927.

WISE, S. W., 1973. Calcareous nannofossils from cores recovered during Leg 18, Deep Sea Drilling Project: biostratigraphy and observations of diagenesis. In: L. F. MUSICH and O. E. WESER (Editors), *Initial Reports of the Deep Sea Drilling Project, Vol. XVIII*. U.S. Government Printing Office, Washington, D.C., pp. 569–615.

WISE, S. W. and HSÜ, K. J., 1971. Genesis and lithification of a deep sea chalk. *Ecl. Geol. Helv.*, 64: 273–278.

WISE, S. W. and KELTS, K. R., 1972. Inferred diagenetic history of a weakly silicified deep sea chalk. *Trans. Gulf Coast Assoc. Geol. Socs.*, 22: 177–203.

WOLFE, M. J., 1970. Dolomitization and dedolomitization in the Senonian chalk of Northern Ireland. *Geol. Mag.*, 107: 39–49.

ZANKL, H., 1969. Der Hohe Göll. Aufbau und Levensbild eines Dachsteinkalk-Riffes in der Obertrias der nördlichen Kalkalpen. *Abh. Senckenb. Naturforsch. Ges.*, 519: 1–123.

ZANKL, H. and SCHROEDER, J. H., 1972. Interaction of genetic processes in Holocene reefs off North Eleuthera Island, Bahamas. *Geol. Rundsch.*, 61: 520–541.

ZENGER, D. H., 1972. Significance of supratidal dolomitization in the geologic record. *Geol. Soc. Am., Bull.*, 83: 1–12.

ZENGER, D. H., 1973. Syntaxial calcite borders on dolomite crystals, Little Falls Formation (Upper Cambrian), New York. *J. Sediment. Petrol.*, 43: 118–124.

INDEX*

Abaco Is., 106, 142, 144
Abaco Sound, 124
Abrasion in laboratory, 80, 88
—, in sea, 88, 96, 102, 120, 134, 136, 184, 187, 191, 194, 195, 199, 213, 302, 309, 315, 318, 368, 370, 371, 417
—, strain caused by, 500
Abu Dhabi, 93, 179, 180, 181, 184, 189–202, 204–212, 216
— —, algal flats, 190, 191, 199–202, 203, 205, 206, 207, 208, 209, 216, 525, 528
— —, anhydrite, 205, 206, 207, 208, 211
— —, aragonite, 191, 194, 195, 196, 198, 199, 201, 204, 205, 206, 207, 208, 211, 212, 216, 263
— —, beach rock, 191
— —, brine chemistry, 206, 207, 208, 525–530
— —, chlorinity, 179
— —, commentary on, 209
— —, creeks and swamps, 190, 191, 199–202
— —, deltas, 180, 190, 191, 194, 195, 196, 199, 209
— —, dolomite, 178, 183, 191, 196, 205, 206, 207, 208, 211, 212, 251, 525–530, 536, 537, 538
— —, Foraminiferida, 199, 212, 215, 216
— —, gypsum, 178, 201, 204, 205, 206, 207, 208, 211, 212
— —, halite, 204, 205, 206, 207, 211
— —, inner island shores, 196
— —, inner shelf, 191
— —, lagoon, 190, 191, 193, 194, 195, 196, 198, 200, 201, 209
— —, magnesite, 206
— —, oöids, 78, 191, 194, 195
— —, oölite, 180, 191, 193, 194, 195, 196, 199
— —, reefs, 180, 190, 191, 193
— —, sabkha diagenesis, 206–208, 209, 251, 525–530
— —, sabkha sediments, 95, 181, 190 205–208
— —, salinity, 179, 193, 206
— —, seaward island beaches, 194, 196

— —, tidal channels, 190, 193, 194, 195, 196, 199, 209
— —, tides, 181, 191, 194, 195, 196, 198, 199, 201, 202, 204, 205, 206, 209
Acanthaster, **51**
Acaste, **74**
Acetabularia, 59, 69, **116**, 117, 130, 135 136, 170, 278
—, and calcisphere cysts, 70
Achistridae, structure, 57
Acmaea, **9**, 17
Acropora community, 103, 105
—, diagenesis, 329, 448
—, on sea floor, **108**, 109, **110**, 113, 149, 150, **151**, 153, 154, **156**, 191, 193
—, strontium in, 262
Acrothoracica burrows, 410
Actinostroma, 31, 34
Actinostromaria, 31
Adnet Beds, 393, 396
Adsorbed ion effect, calcite growth, 243–250
— — —, dolomite growth, 250–252
Agaricia, 109, **111**, 115, 157
Agglutinated structure, Foraminiferida, **39**, 41, 43, 45
Aggrading neomorphism, 346, 347, 350, 353, 359, 417, **476**, 481–503, 505, 513
Ahermatypic coral, 26
Ain, 81, 311, 431
Alacran reef, 106, 141, 143, 144
Alcyonacea, structure, 29
Alcyonaria, magnesium in, 236
—, on sea floor, 94, 102, 107, 108, **109**, 113, 115, 116, 117, 120, 129, 135, 139, **152**, 154, 157, 158, 163
—, structure, 28, 55, 76
Alexandrine, Lake, 518, 519
Algae, bathymetric indicator, 389
—, boring, 77, 80, 86, 88, 90, 102, 108, 113, 116, 117, 120, 123, 134, 139, 143, 191, 195, 205, 212, 213, 296 297, 298, 317, 318, 319, 328, 364, 368, 370, 371, 381–392, 412, 413

---

* Numbers in bold type refer either to illustrations or to definitions.

— —, point-count of constituents, 214
— —, reefs, 145, 149, 212, 213
— —, sediment mineralogy, 212
— —, skeletal calcilutites, 212–215, 276
— —, strontium in sediments, 214, 263
Broken Beds, 85
Brogniartella, **74**
Browns Cay, 99, 104, 105, 122
— — oölite, 77, 78, 80, 104, 121, 297, 301, 302, 305, 309, 313, 314, 316, 386
Brucite in *Goniolithon*, 335
Bryozoa, neomorphism, 491
—, on sea floor, 94, 115, 136, 154, 157, 163, 165, 170, 191, 195
—, recrystallization, 478
—, structure, 70–73, 75, 431, 440, 514
Buffering, 233
Buliminacea, structure, 42
Buliminella, 199
Buliminidae, structure, 42
Bumastus **74**
Buntsandstein, 462
Burgundia, **32**, 38
Burrows, *see Alpheus*
—, *see Callianassa*
—, *see* Crab

Caicos Ridge, 143
Calcancoridae, structure, 57
Calcarenite, VIII, 100
Calcareous algae, aragonite mud from, 162, 204, 279–284
— —, binding effect, 115, 126, 127, 134, 150, 163
— —, magnesium in, 236
— —, on Bahamas sea floor, 94, 102, 107, 108, 109, 111, 112, 113, 114, 115, 116, 117, 119, 120, 121, 126, 127, 128, 130, 132, 134, 135, 136, 137, 139, 140, 141, 142, 143, 144, 145
— —, on British Honduras sea floor, 213, 214, 215
— —, on Florida sea floor, 149, 150, 153, 154, 157, 158, 159, 161, 162, 163, 164
— —, on Gulf of Batabano floor, 170, 172, 174, 175, 176, 177
— —, on Persian Gulf floor, 183, 191, 193 195, 198, 209
— —, on Shark Bay floor, 165
— —, oxygen isotope ratios, 279–284
— —, strontium in, 162, 262
— —, structure, 58–69, 75, 76
— Alps, 396
Calcarinidae, structure, 42
Calcian dolomite, *see* Protodolomite
Calcification of aragonite, *see* Calcitization
— of dolomite, 534

Calcilutite, petrography, 87–90
—, studies with electron microscope, 515, 516
—, *see* Micrite
Calcispheres, 69, **70**, 75
Calcite cement, *see* Cement, calcite
— dissolution, 248, 250, 267–271
—, equilibrium with dolomite, 252, 538, 539
—, free energy of formation, 245, 258
—, high-magnesian, change to low-magnesian, 326, 327, 331, 335–338
—, —, dissolution, 248, 250, 336, 337, 338
—, —, in skeletons, 2
— in Brachiopoda, 20, 21, 22, 23, 24
— in Bryozoa, 72, 73
— in calcareous algae, 58, 59, 61, 67
— in calcispheres, 69
— in Echinodermata, 50, 51, 55, 56, 478
— in Foraminiferida, 41, 42, 44, 45, 48, 49
— in Heterocorallia, 28
— in Hydrozoa, 31
— in micrite envelopes, 90
— in Mollusca, 3, 7, 8, 9, 10, 11, 13, 16, 17, 18, 19, 20, 21, 22, 23
— in Octocorallia, 28, 29, 30, 31
— in oöids, 77, 81, 82, 83, 306, 311
— in peloids, 86
— in Rugosa, 26
— in sediments, 137, 184, 188, 194, 204, 206, 208, 211, 212, 277
— in Stromatoporoidea, 31, 33
— in Tabulata, 28
— in Trilobita, 73, 74
—, inhibition of precipitation, 243–250
—, magnesium in, 235
— mud, 87, 88
— mudstone, 86, 88, 89
—, pleochroism, *see* Pseudopleochroism
, precipitation, 163, 204, 217, 223, 225, 226, 237, 239–250, 252, 293, 370, 536, 542
— silt, 356, 357
—, solubility, 231, 237, 257, 258, 268, 273, 274, 293
—, solubility product, 231
—, stability, 234, 236, 241, 242, 243, 250, 251, 252, 253, 258–260
—, strontium in, 241, 242, 260–265
Calcitization of aragonite, 90, **239–242**, 246, 251, 329, 332, 337, 347–349, 350, 351, 354, 356, 359, 373, 374, 420, 438, 448, 449, 456, 475, 481, 486–489, 490, 492, 493, 494, 496, 498, 501, 504, 509, 511
— of dolomite, 543
Calcium, free energy of hydration, **245** 250
—, in lagoon water, 520, 521, 523
—, in pore water, 415, 447, 500, 501, 511, 525, 526, 528, 529, 530, 531, 532, 534, 536